Characteristic Infrared Absorption F[...] for Some Common Groups

Frequency range (cm^{-1})	Group	Class of compounds
Stretching vibrations		
3700–3200	O—H	alcohols, phenols
3500–3100	$\diagdown$N—H	1° and 2° amines, amides
3320–3000	≡C—H	terminal alkynes
3100–3000	=C—H	alkenes
	$\diagup$C—H	aromatics
3000–2800	—C—H	alkanes
3000–2500	O—H----O	carboxylic acids (H bonded)
2260–2240	C≡N	nitriles
2260–2100	C≡C	alkynes
1820–1600	$\diagdown$C=O	aldehydes, ketones, carboxylic acids, and derivatives
1680–1500	C=C	alkenes, aromatics
1560–1490; 1360–1320	—NO$_2$	nitro compounds
1200–1000	C—O	alcohols, ethers, carboxylic acids, esters
Bending vibrations		
1470–1430	—CH$_2$—; C—H	
1470–1375	CH$_3$	
960–900	C=C—H	
1600–1500	aromatic ring (often weak)	
900–700	Ar—H	

2nd edition

Elements of Organic Chemistry

Isaak Zimmerman

Bronx Community College of the City University of New York

Henry Zimmerman

New York City Technical College of the City University of New York

Macmillan Publishing Co., Inc.

NEW YORK

Collier Macmillan Publishers

LONDON

To our dear parents

Macmillan Publishing Co., Inc.
866 Third Avenue, New York, New York 10022

Collier Macmillan Canada, Inc.

Library of Congress Cataloging in Publication Data

Zimmerman, Isaak.
   Elements of organic chemistry.

   Previous ed.: Henry Zimmerman, Isaak Zimmerman, 1977.
   Includes index.
   1. Chemistry, Organic.   I. Zimmerman, Henry.
II. Title.
QD251.2.Z58   1982   547   81-5780
ISBN 0-02-479640-9   AACR2

Printing: 345678   Year: 4567890

ISBN 0-02-479640-9

Preface

This revision of *Elements of Organic Chemistry,* like its predecessor, is intended for a first course in organic chemistry. It is especially suited to the needs of students specializing in the life sciences, allied health fields, agricultural sciences, and other related curricula.

We have retained the basic philosophy of the first edition. Foremost, we wish this book to be a teaching text; one that students find readable, and can study and learn from with little dependence on the teacher. To this end, we have taken the same systematic, but *more selective,* approach to our material as in the earlier edition, keeping constantly in mind the student to whom this book is addressed. As in the first edition, the discussion of topics is organized around functional groups. In the selection of our material, we have stressed those aspects of organic chemistry pertinent to health, the environment, and biochemistry, areas of special interest to the career objectives of the students enrolled in the course. The practical uses of organic compounds as drugs, food additives, pesticides, plastics, and other products, as well as their occurrence in nature, are discussed throughout the text.

Nevertheless, a number of changes have been made in this second edition, with the hope to improve on the earlier one.

- The number of preparative methods and reactions presented has been purposely reduced. Some reactions, such as the Wurtz reaction for preparing alkanes, have been omitted because of their obsolescence or lack of practical applications. Others, such as the Gabriel synthesis of amines, have been deleted because of the time limitations inherent in a brief course and because we did not feel that they were important enough for nonchemistry majors.

iii

Some mechanisms, such as those involving carbocation rearrangements, have been omitted because we felt they were beyond the scope of a short organic chemistry course. We trust we have been judicious in our selection of reactions and reaction mechanisms presented.

- The number of problems within the body of each chapter has been increased considerably. Problems within a chapter serve to test the student's mastery of the subject section by section. Selective answers to these problems are provided at the end of the book as a further learning tool.

  The exercises at the end of each chapter have also been expanded. These exercises cover material for the whole chapter and sometimes previous chapters. To guide the student in solving the end-of-chapter exercises we have indexed the exercises to the appropriate sections. This indexing should also help the instructor in choosing assignments.

  Answers to all problems and exercises, with detailed explanations, are given in the *Solutions Manual for Elements of Organic Chemistry.*
- To reinforce the mastery of new concepts as they are presented we have increased the number of worked-out examples.
- A summary of methods of preparation and reaction has been provided in a special display box in almost every chapter. The summary allows the instructor the freedom to choose which topics he or she wishes to deal with. The displays also helps the student to focus on the main reactions found in each chapter.
- New pedagogical aids are a detailed summary and a list of key terms at the end of each chapter.
- In addition to reorganizing and updating a number of chapters, we have written two new chapters, "Organic Halogen Compounds" (Chapter 8) and "Spectral Methods of Structure Determination" (Chapter 17).

Some of the highlights are:

In Chapter 1, after briefly reviewing atomic structure and bonding, we present in detail, and in a deliberate manner, the various methods of writing organic chemical formulas following the rules of covalence. Often, treatment of this topic is abbreviated. Our experience tells us that students taking the course are often confused when they encounter organic chemical formulas for the first time. We hope, by our methodical approach, to help students overcome their natural apprehension when faced with the strange and seemingly complex "hieroglyphics" of organic chemistry. To emphasize the orderliness of organic chemistry, we end the chapter with the classification of compounds into families according to functional groups.

Chapters 2–5 deal with hydrocarbons. In Chapter 2, on the alkanes, we consciously deleted the methods of preparation, primarily because alkanes are obtained most readily from natural sources rather than from the laboratory and because we did not wish to overwhelm the student in his/her first encounter with a family of organic compounds. Instead, we concentrated on physical properties, structural isomerism, and practical applications. Before introducing the few reactions that alkanes undergo, we wrote a separate section on the types of bond-breaking and bond-making.

In Chapter 3, dealing with alkenes, we have introduced a brief discussion of the *E,Z* system of nomenclature to supplement the treatment of the classical *cis-trans* method for naming geometric isomers. The chemistry of alkynes is treated in Chapter 4, with special stress on the similarities of this class of compounds with the alkenes treated in the previous chapter. Dienes and the concept of resonance are also dealt with in Chapter 4. The resonance concept and how it helps us in understanding the chemistry of aromatic compounds is discussed again in Chapter 5.

To emphasize the three-dimensional nature of organic compounds we now present stereochemistry in Chapter 6. A new feature here is the methodical introduction of the R,S system of nomenclature in addition to the classical D,L system.

After having covered compounds containing only carbon and hydrogen, we reach Chapter 7, which deals with alcohols, phenols, and thiols. In the treatment of alcohols and phenols we consider them as organic derivatives of water. Special emphasis is placed on the chemistry common to both classes of compounds and what separates them from one another. By this time we are ready to deal with organic halogen compounds in Chapter 8. Chapter 8 correlates the material developed in the previous seven chapters with respect to the methods of preparing this class of compounds, the reactions they undergo, and their stereochemistry. A section on organometallic compounds provides the springboard for the application of Grignard reagents to the synthesis of compounds encountered earlier and to be found in the coming chapters.

Chapters 9–16 deal with ethers, aldehydes and ketones, carbohydrates, carboxylic acids and derivatives, lipids, amines, amino acids, peptides and proteins, and nucleic acids. The organization of these chapters is essentially the same as that of the earlier edition. However, we now treat aldehydes and ketones in one chapter (Chapter 10) and carboxylic acids and derivatives in one chapter (Chapter 12). In keeping with our philosophy of integrating the biochemical topics as soon as possible after discussion of the appropriate organic chemical families, carbohydrates are treated in Chapter 11 right after aldehydes and ketones; lipids in Chapter 13 immediately following carboxylic acids and derivatives; and amino acids, peptides, and proteins in Chapter 15 following treatment of amines.

Chapter 17 deals with uv–visible, ir, and nmr spectral methods of identification and structure determination. Bearing in mind that these topics are addressed to nonchemistry majors, we restricted ourselves to the key pieces of information that can be obtained from each technique. We deliberately omitted a discussion of mass spectrometry. Although this chapter is placed at the end of the book, the material can be taken up any time after the first few chapters, at the discretion of the instructor. To permit greater flexibility, we have written a number of additional exercises at the end of the chapter; these additional exercises are indexed to specific chapters in the text.

There is more material included in this book than is possible to cover in one semester. We feel this not to be a disadvantage. On the contrary, this allows a freedom of choice for subject matter on the part of the instructor.

Several of the changes in this edition were made in response to discussions, suggestions, and critical comments from colleagues, reviewers, and users of the previous edition. We thank them. In particular, we wish to express our gratitude to Professors Elmer E. Jones, Northeastern University; L. Salisbury, Kean College; M. Treblow, University of Pittsburgh; and G. Wilson, Western Kentucky University. Our thanks also go to those who reviewed our manuscript: Professors James O. Schreck, University of North Colorado; Wes Borden, University of Washington; Robert Coley, Montgomery College Maryland; Nathan Lerner, South Connecticut State College; and Daniel O'Brien, Texas A&M University.

Our colleagues at Bronx Community College, Professors J. Buckley,

D. Gracian, R. Miller, J. G. Riley, and H. Stein, and at New York City Technical College, O. Gaglione, Alla O. Romano, and S. Dreier, deserve our thanks for their encouragement and advice. A special thanks goes to Evelyn Shapiro whose words of advice were always a source of inspiration.

It has been a special pleasure to work with Elisabeth Belfer, our Production Supervisor, whose editorial thoroughness and excellent suggestions for improvement of the text we greatly appreciated. We are indebted to Greg Payne and Tom Vance, Editors, for their consistent interest and expert guidance in helping move the book through to its final form.

The nmr spectra in Chapter 17 were provided by Varian Associates, Inc., Palo Alto, California, and the ir spectra were provided by Sadtler Research Laboratories, Inc., Philadelphia. We thank them both for permission to use and adapt their materials.

We want especially to thank our wives and children for their patience and understanding during the long months of neglect while the manuscript was in preparation.

We shall be grateful for continued suggestions for improvement of the text in future editions.

ISAAK ZIMMERMAN
HENRY ZIMMERMAN

Contents

3 Unsaturated Hydrocarbons I: Alkenes
59

Unsaturated Hydrocarbons II: Dienes, Polyenes, *4* and Alkynes
94

Benzene and Aromatic Compounds *5*
120

Organic Halogen Compounds 8
212

Ethers and Epoxides 9
238

Aldehydes and Ketones 10
255

11 Carbohydrates
289

12 Carboxylic Acids and Their Derivatives
316

Lipids **13**
349

Amines and Other Nitrogen Compounds 14
372

15   Amino Acids, Peptides, and Proteins
401

16   Nucleosides, Nucleotides, and Nucleic Acids
436

Spectral Methods of Structure Determination *17*

462

Selected Answers to Problems

Index

1

Bonding, Structural Formulas, and Molecular Shapes

The subject of organic chemistry is unique in that it deals with vast numbers of substances, both natural and synthetic, that directly influence our welfare and standard of living. Organic chemistry is crucial to our economy as the source of countless manufactured products that are essential to our comfort and well-being. The clothes we wear; the petroleum products we use to run our machines; the paper, rubber, wood, plastics, paint, cosmetics, insecticides, and vitamins and drugs that we use every day—all are examples of organic compounds. The chemical substances that make up the organs of our bodies, the food we eat for nourishment, and the chemical reactions that take place inside our bodies are also organic in nature. Organic chemistry is a subject that is fundamental to medicine, biology, and other related disciplines such as nursing, dental hygiene, and medical laboratory technology. Because it is almost impossible to think of an aspect of our daily lives that is not somehow influenced by organic chemistry, the relevance in your study of this exciting and dynamic subject should be quite apparent.

Organic Chemistry: A Modern Definition 1.1

From observation of the chemical makeup of many organic compounds it was recognized that one constituent common to all was the element **carbon.** Today **organic chemistry** is defined as the *study of carbon/hydrogen-contain-*

ing compounds and their derivatives. Petroleum and coal are two vast natural reservoirs from which many organic materials are extracted. Both were formed, over long periods of time, from the decay of plants and animals.

1.2 The Uniqueness of Carbon

Although carbon ranks only twelfth in abundance among the elements and constitutes less than 0.1% of the earth's crust, oceans, and atmosphere, the number of its compounds far exceeds that of all known inorganic compounds. Inorganic compounds are compounds formed from elements other than carbon. There are only about 90,000 known inorganic compounds, whereas the number of known organic compounds is several million, and thousands of new ones are synthesized and described each year. Thus, it is not surprising that a special branch of chemistry is entirely devoted to the study of the compounds of carbon.

What is unique about the element carbon? Why does it form so many compounds? The answers to these questions lie in the structure of the carbon atom and the position of carbon in the periodic table. These factors enable it to form strong bonds with other carbon atoms and with other elements—most commonly hydrogen, oxygen, nitrogen, and the halogens. As a result, there exist numerous stable carbon-containing substances of various sizes and shapes (see Fig. 1.1).

Each organic compound has its own characteristic set of physical and chemical properties, which depend on the *structure* of the molecule. The structure of a molecule, in turn, depends on how the atoms composing it are bonded to each other. Because this relationship between properties and structure is fundamental to a good understanding of organic chemistry, it is appropriate that we review the topics of atomic structure and chemical bonding.

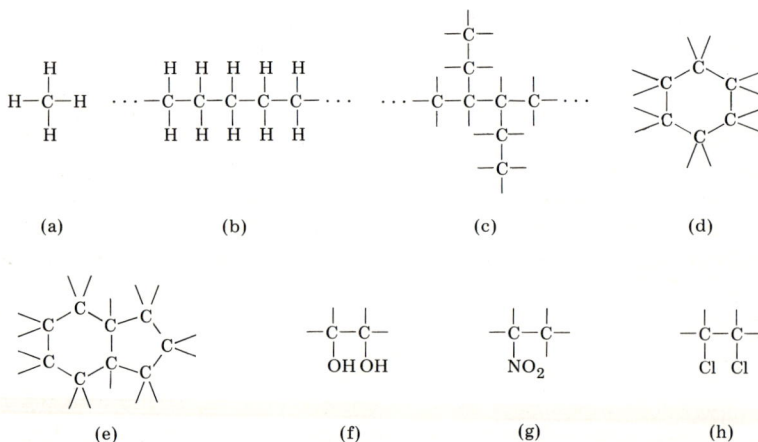

Figure 1.1 **(a)** A representation of methane, the simplest organic molecule. **(b)** The combination of many carbons with hydrogen in a straight-chain arrangement. **(c)** The combination of carbons in branched chains. **(d, e)** The combinations of carbons in rings of different shapes and sizes. **(f–h)** The combination of carbon with elements other than hydrogen.

Atomic Structure 1.3

A Electrons and Energy Levels

Atoms consist essentially of three fundamental particles: **neutrons, protons,** and **electrons.** Neutrons and protons are found in the nucleus; electrons are outside the nucleus. Neutrons are particles that have no charge, and protons are positively charged particles. The **atomic number** of an element indicates the number of protons. Since an atom is electrically neutral, this means that electrons are negatively charged and that the number of electrons must be equal to the number of protons.

Electrons are distributed around the nucleus in successive **shells,** or principal **energy levels,** of increasing radius. The electrons in levels close to the nucleus have lower energy than do electrons in levels farther from the nucleus.

The various energy levels are designated by capital letters or whole numbers (n). The first energy level, the one closest to the nucleus and therefore the one having the lowest energy, is called the K level and corresponds to $n = 1$. The second energy level, somewhat farther from the nucleus and with a somewhat higher energy content than the K level, is called the L level and corresponds to $n = 2$. The third energy level, the M level, corresponds to $n = 3$; the fourth energy level, the N level, corresponds to $n = 4$—and so on.

Each energy level has a given capacity for electrons. The K level may contain a maximum of 2 electrons and never more. The L energy level has a maximum capacity for 8 electrons, the M level for 18 electrons, and the N level for 32 electrons. (The maximum capacity of a shell is equal to $2n^2$ electrons, where n is the number of the energy level.)

B Arrangement of Electrons in Energy Levels

Having learned how many electrons are needed to fill a particular shell, let us consider the order in which electrons enter the various energy levels.

As expected, the first 2 electrons enter the K shell and the next 8, the L shell. After the L energy level is filled, electrons do *not* fill the third energy level (M level) to capacity (18 electrons) before the fourth energy level (N level) is started. In fact, there are *never more than 8 electrons in the outermost energy level of an atom.* The reason is that 8 electrons in the outermost shell give atoms their greatest stability (noble gas configuration). For example, the element potassium (atomic number 19) has its 19 electrons distributed about the nucleus as

Shell	K	L	M	N
Number of electrons	2	8	8	1
Rather than	2	8	9	0

and the electron distribution (electronic configuration) for calcium (atomic number 20) is

Shell	K	L	M	N
Number of electrons	2	8	8	2
Rather than	2	8	10	0

Table 1.1 Electron Distribution for the First Twenty Elements

Atomic number	Element	Energy level			
		K	*L*	*M*	*N*
1	H	1			
2	He	2			
3	Li	2	1		
4	Be	2	2		
5	B	2	3		
6	C	2	4		
7	N	2	5		
8	O	2	6		
9	F	2	7		
10	Ne	2	8		
11	Na	2	8	1	
12	Mg	2	8	2	
13	Al	2	8	3	
14	Si	2	8	4	
15	P	2	8	5	
16	S	2	8	6	
17	Cl	2	8	7	
18	Ar	2	8	8	
19	K	2	8	8	1
20	Ca	2	8	8	2

The distribution of electrons in the various energy levels for the first twenty elements is shown in Table 1.1.

C Valence Electrons: Electron-Dot Structures

Valence electrons are those electrons located in the outermost energy level, the valence shell. In general, the chemical properties of an element depend on its valence electrons. For this reason, atoms are often depicted by **electron-dot structures.** In such structures the symbol of the element represents the *core* of the atom (the nucleus and all electrons *except* the valence electrons), and the valence electrons are shown as dots, crosses, or small circles around the symbol. To place the electrons around the symbol in the correct manner, follow these simple rules.

1. Imagine that the element's symbol has four sides around it, each with room for two electrons.
2. Pair the first two valence electrons on one side of the symbol.
3. Place the third, fourth, and fifth valence electrons, one at a time, on the remaining three sides.
4. For elements that have more than five valence electrons, the three sides are filled up to a maximum of eight.

For example,

Elements with one valence electron	H·	Li·	etc.
Elements with two valence electrons	He:	Be:	etc.
Elements with three valence electrons	Ḃ:	Ȧl:	etc.
Elements with four valence electrons	·Ċ:	·Ṡi:	etc.
Elements with five valence electrons	·N̈:	·P̈:	etc.
Elements with six valence electrons	·Ö:	·S̈:	etc.
Elements with seven valence electrons	:F̈:	:C̈l:	etc.
Elements with eight valence electrons	:N̈e:	:Är:	etc.

Chemical Bonding 1.4

In 1916 G. N. Lewis proposed a theory of chemical bonding that accounted for many of the facts regarding the reactivity, or the lack of reactivity, of many elements. Lewis pointed out that the noble gases were particularly stable elements, and he ascribed their lack of reactivity to their having their valence shells filled with electrons: two in the case of helium and eight for the other noble gases. All other elements *do* enter into chemical reactions, and they do so because their valence shells are only partially filled. According to Lewis, in interacting with one another atoms can achieve a greater degree of stability by rearrangement of the valence electrons to acquire the outer-shell structure of the closest noble gas in the periodic table. This can be achieved in either of two ways: (1) through transfer of electrons between atoms (ionic bonding) or (2) through sharing of electrons between atoms (covalent bonding).

A   Ionic Bonding

Elements at opposite ends of the periodic table attain the noble gas configuration by *transferring* electrons to one another. In the electron-transfer process, elements at the left of the periodic table give up their valence electrons and become positively charged ions, or *cations,* and those at the right gain the electrons, thus becoming negatively charged ions, or *anions.* The electrostatic force of attraction between oppositely charged ions constitutes the **ionic bond.**

General equation

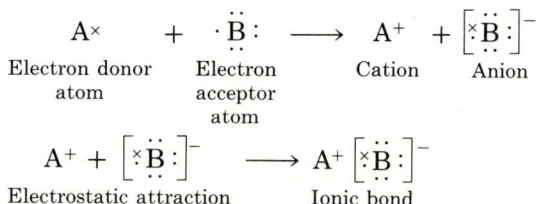

$$A^× \quad + \quad ·\ddot{B}: \quad \longrightarrow \quad A^+ \quad + \quad \left[×\ddot{B}: \right]^-$$

Electron donor atom   Electron acceptor atom   Cation   Anion

$$A^+ + \left[×\ddot{B}: \right]^- \quad \longrightarrow \quad A^+ \left[×\ddot{B}: \right]^-$$

Electrostatic attraction   Ionic bond

Specific example

$$Na\times + \cdot \overset{\cdots}{\underset{\cdots}{Cl}} : \longrightarrow \qquad Na^+ \qquad + \qquad \left[\overset{\cdots}{\underset{\cdots}{\overset{\times}{Cl}}} : \right]^-$$

$$\qquad\qquad\qquad\qquad\qquad \text{Sodium ion} \qquad\qquad \text{Chloride ion}$$
$$\qquad\qquad\qquad\qquad \text{(neon configuration)} \qquad \text{(argon configuration)}$$

$$Na^+ + \left[\overset{\cdots}{\underset{\cdots}{\overset{\times}{Cl}}} : \right]^- \longrightarrow \quad Na^+ \left[\overset{\cdots}{\underset{\cdots}{\overset{\times}{Cl}}} : \right]^-$$

$$\qquad\qquad\qquad\qquad \text{Sodium chloride}$$
$$\qquad\qquad\qquad\qquad \text{(an ionic compound)}$$

The vast majority of ionic compounds are inorganic substances. In the solid state they exist as high-melting-point crystals owing to the strong electrostatic forces that hold ions together. When the crystals are dissolved or melted, these interionic forces are overcome, and the result is a liquid in which individual ions move randomly about (Fig. 1.2). In contrast, when crystals of a covalent compound such as sucrose (table sugar) are dissolved, we find distributed throughout the solution individual molecules of sugar with definite size and shape.

Problem 1.1 Write the electron-dot formula showing the combination of magnesium and oxygen to form magnesium oxide, MgO. Clearly indicate the charges on the cation and on the anion.

B   Covalent Bonding

Elements that are close to each other in the periodic table attain the stable noble gas configuration, not through a transfer of electrons but by sharing valence electrons between them. The chemical bond formed when two atoms share

The NaCl crystal structure

In solution

In solution the crystal breaks up into individual ions

○ = Cl

● = Na

Figure 1.2 A sodium chloride crystal has a cubic shape in the solid state but no specific shape in solution.

one pair of electrons is called a **covalent bond.** Atoms in most substances are held together by bonds of this type.

The simplest example of a molecule with a covalent bond is hydrogen gas, H_2. In hydrogen gas each hydrogen atom fills the first energy level (which can hold no more than two electrons) by sharing the combined valence electrons.

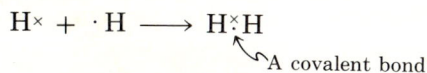

$$H\times + \cdot H \longrightarrow H\overset{\times}{\cdot}H$$

↖A covalent bond

Chlorine gas, Cl_2, is another molecule in which a covalent bond is formed between two like atoms.

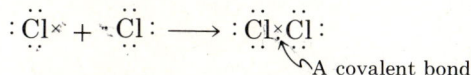

$$:\overset{\cdot\cdot}{Cl}\times + \overset{\cdot\cdot}{\cdot Cl}: \longrightarrow :\overset{\cdot\cdot}{Cl}\overset{\cdot\cdot}{\times}Cl:$$

↖A covalent bond

In Cl_2 each chlorine atom is surrounded by eight valence electrons, as is argon, its closest noble gas.

Because it is often tedious to write electron-dot diagrams of molecules, we shall introduce a simplified notation. A shared electron pair between two atoms, or single covalent bond, will be represented by a dash (—), and the electrons not involved in bonding (the nonbonding electrons) will be omitted unless needed to make a point. Thus, hydrogen gas is represented as H—H and chlorine gas as Cl—Cl.

In molecules that consist of two like atoms, such as elemental hydrogen and elemental chlorine, the bonding electrons are shared equally. This is because both atoms have the same **electronegativity.** The electronegativity is a measure of the attraction the nucleus of an atom has for electrons in its outer shell. Figure 1.3 shows the electronegativity values for the elements proposed by Linus Pauling. You see that from left to right, within a period, the electronegativity values increase, and from top to bottom, within a group, the electronegativity values decrease. Fluorine, with a value of 4.0, is the most electronegative element.

When two unlike atoms form a covalent bond, the bonding electrons are no longer shared equally. In the C—F bond, for example, the electron pair is shared unequally between the carbon and fluorine atoms. The greater electronegativity of fluorine (4.0) causes the electron pair to be closer to the fluorine atom than to the carbon atom. Such a bond, in which an electron pair is shared unequally, is

H 2.1																	He —
Li 1.0	Be 1.5											B 2.0	C 2.5	N 3.0	O 3.5	F 4.0	Ne —
Na 0.9	Mg 1.2											Al 1.5	Si 1.8	P 2.1	S 2.5	Cl 3.0	Ar —
K 0.8	Ca 1.0	Sc 1.3	Ti 1.5	V 1.6	Cr 1.6	Mn 1.5	Fe 1.8	Co 1.8	Ni 1.8	Cu 1.8	Zn 1.6	Ga 1.6	Ge 1.8	As 2.0	Se 2.4	Br 2.8	Kr —
Rb 0.8	Sr 1.0	Y 1.2	Zr 1.4	Nb 1.6	Mo 1.8	Tc 1.9	Ru 2.2	Rh 2.2	Pd 2.2	Ag 1.9	Cd 1.7	In 1.7	Sn 1.8	Sb 1.9	Te 2.1	I 2.5	Xe —
Cs 0.7	Ba 0.9	57–71 1.1–1.2	Hf 1.3	Ta 1.5	W 1.7	Re 1.9	Os 2.2	Ir 2.2	Pt 2.2	Au 2.4	Hg 1.9	Tl 1.8	Pb 1.8	Bi 1.9	Po 2.0	At 2.2	Rn —
Fr 0.7	Ra 0.9																

Figure 1.3 Electronegativity values of the elements (Pauling's scale).

called a **polar covalent bond.** In a polar covalent bond the more electronegative atom assumes a partial negative charge and the less electronegative atom assumes a partial positive charge. The polarity of a bond may be indicated by the symbol $+\!\!\longrightarrow$. The head of the arrow points in the direction of the more electronegative atom. The tail, marked with a plus sign, is located at the less electronegative atom. More frequently, the partial charges are denoted by the Greek letter symbols $\delta+$ and $\delta-$ (pronounced *delta plus* and *delta minus*).

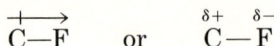

$$\overset{+\longrightarrow}{C-F} \quad \text{or} \quad \overset{\delta+ \quad \delta-}{C-F}$$

As we proceed through our study of organic chemistry, we will find that polar bonds exert special effects on the physical and chemical properties of organic molecules. In general, most reactions involve changes in polar covalent bonds (C—O, C—Cl, etc.) while nonpolar covalent bonds (C—C, C—H, etc.) remain unaltered.

Problem 1.2 Show the partial charges by placing the $\delta+$ and $\delta-$ symbols on the appropriate atoms in the following polar covalent bonds.
(a) H—Cl   (b) O—H   (c) C—Cl
(d) N—H   (e) C—O   (f) C—N

C   Coordinate Covalent Bonding

In the covalent bonding discussed so far each of the two atoms contributed one electron to the electron pair shared between them. There are molecules in which one atom supplies *both* electrons to another atom in the formation of a covalent bond. A covalent bond thus formed is called a **coordinate covalent bond.** For example, when ammonia, $:NH_3$, reacts with a proton, H^+, to form an ammonium ion, NH_4^+, the nitrogen atom in ammonia supplies both electrons to the new bond.

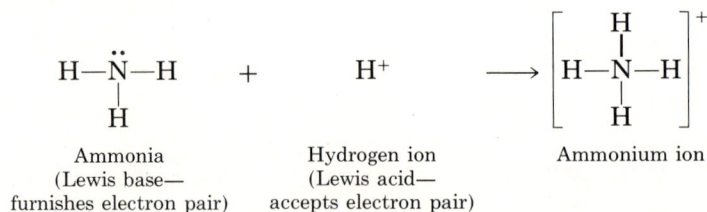

$$
\begin{array}{ccccc}
\overset{\displaystyle ..}{H-N-H} & + & H^+ & \longrightarrow & \left[\begin{array}{c} H \\ | \\ H-N-H \\ | \\ H \end{array}\right]^+ \\
\underset{\displaystyle H}{|} & & & &
\end{array}
$$

        Ammonia   Hydrogen ion   Ammonium ion
      (Lewis base—   (Lewis acid—
  furnishes electron pair)   accepts electron pair)

The species that furnishes the electron pair to form a coordinate covalent bond is called a **Lewis base.** The species that accepts the electron pair to complete its valence shell is called a **Lewis acid.** In subsequent chapters we shall have numerous occasions to refer to the concept of Lewis acids and Lewis bases to explain how chemical reactions occur.

Problem 1.3 Each of the following interactions involves the formation of a coordinate covalent bond. Indicate (1) the structure of the product formed and (2) which species acts as a Lewis acid and which acts as a Lewis base.

(a) $H-\overset{\cdot\cdot}{\underset{\cdot\cdot}{O}}-H + H^+ \longrightarrow$

(b) $H-\overset{\cdot\cdot}{\underset{\underset{H}{|}}{N}}-H + F-\overset{|}{\underset{F}{B}}-F \longrightarrow$

(c) $Cl-\overset{|}{\underset{Cl}{Al}}-Cl + \overset{\cdot\cdot}{\underset{\cdot\cdot}{:Cl:^-}} \longrightarrow$

How Many Bonds to an Atom? *1.5* Covalence Number

The number of covalent bonds an atom can form with other atoms is called its **covalence number.** The covalence numbers of atoms commonly found in organic compounds are listed in Table 1.2.

Table 1.2 Covalence Numbers for Typical Elements in Organic Compounds

Element	Number of valence electrons	Number of electrons in filled valence shell	Covalence number
H	1	2	1
C	4	8	4
N	5	8	3
O	6	8	2
F, Cl, Br, I (halogens)	7	8	1

Note that the covalence number for an element is equal to the number of electrons needed to fill its valence shell. For example, hydrogen, which has one valence electron, needs one more electron to fill its outermost shell. The covalence number for hydrogen is therefore one. Carbon, which needs four electrons to fill its valence shell, has a covalence number of four, and so on down the list.

Problem 1.4 Assuming that the method of assigning covalence numbers described is valid, write the covalence numbers for **(a)** S; **(b)** P; and **(c)** Ne.

Covalence Number and Structural Formula *1.6*

A molecular formula tells us what *kind* of atoms and how *many* of each kind of atom are present in a particular molecule. The molecular formula for ethanol (the drinkable alcohol), C_2H_6O, tells us that each molecule of ethanol contains two carbon atoms, six hydrogen atoms, and one oxygen atom. A **structural formula,** also called a **constitutional formula,** shows how the atoms in a particular molecule are connected or bonded together. The secret of writing correct structural formulas for organic molecules is to remember the covalence numbers of the component atoms. Carbon, with a covalence number of four, must *always have four bonds in any organic compound, and each of the other elements present must share the number of bonds indicated by its own covalence number* as listed in Table 1.2.

Following are examples of organic molecules in which each element satisfies its covalence number by sharing one electron pair—forming a **single bond**—with another atom connected to it. Carbon has four single bonds, hydrogen and chlorine one each, oxygen two, and nitrogen three.

Open-chain (acyclic) compounds

Ring (cyclic) compounds

A carbon atom may also share more than one pair of electrons with another carbon atom or with other elements, such as oxygen and nitrogen, to form *multiple bonds*. If two pairs of electrons are shared, a **double bond** is formed; if three pairs of electrons are shared, a **triple bond** is formed. A double bond is represented by two dashes (=) and a triple bond by three dashes (≡).

Organic molecules with one double bond

Organic molecules with two double bonds

$$H—C≡C—H \qquad H—C≡N \qquad H—\underset{\overset{|}{H}}{\overset{\overset{H}{|}}{C}}—C≡N$$

To summarize,

1. Structural formulas show how atoms are connected to one another in a molecule.
2. The structural formula of a compound is correct only if each element satisfies its covalence number.
3. The covalence number may be satisfied by forming single or multiple bonds.

Now consider a specific example.

Example 1.1 Given the skeletal structure

$$C—C=C—C—O$$

and assuming that only hydrogen atoms are missing, **(a)** draw the correct structural formula and **(b)** write the molecular formula of the completed structure.

Solution **(a)** (1) Starting from the left, count how many bonds each atom already has. The first carbon has only one bond (a single bond); the second carbon has three bonds (one single and one double bond); the third carbon also has three bonds (one double and one single bond); the fourth carbon has two bonds (two single bonds); and the oxygen atom has one bond.
(2) Determine how many extra bonds each atom needs to fulfill its covalence number. Since carbon has a covalence number of four and oxygen of two (Table 1.2), it means that the first carbon needs three more bonds; the second carbon needs one more bond, as does the third; the fourth carbon atom needs two extra bonds; and the oxygen needs one extra bond.
(3) Place the missing bonds.

$$—\overset{|}{\underset{|}{C}}—\overset{|}{C}=\overset{|}{C}—\overset{|}{\underset{|}{C}}—O—$$

(4) Put in the missing hydrogens to obtain the correct structural formula.

$$H—\underset{\overset{|}{H}}{\overset{\overset{H}{|}}{C}}—\overset{\overset{H}{|}}{C}=\overset{\overset{H}{|}}{C}—\underset{\overset{|}{H}}{\overset{\overset{H}{|}}{C}}—O—H$$

(b) Counting the number of atoms of each kind, the molecular formula of the structure in **(a)** is

$$C_4H_8O$$

Problem 1.5 Given the skeletal structures and assuming that only hydrogen atoms are missing, (1) draw the correct structural formula and (2) write the correct molecular formula for each of the following.

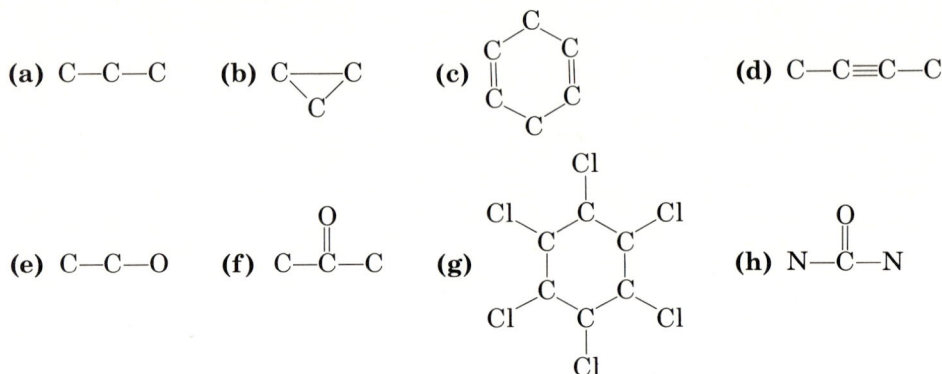

(a) C—C—C

(b) C⟍△⟋C with C below

(c) ring of carbons

(d) C—C≡C—C

(e) C—C—O

(f) C—C(=O)—C

(g) hexachloro ring structure

(h) N—C(=O)—N

1.7 Condensed Structural Formulas

Up to this point we have used expanded structural formulas, in which all bonds are shown, to represent organic molecules. Although very useful in visualizing structures, this method of representation is time-consuming and requires much space. One way to simplify the writing of organic structures is to include only the bonds of multivalent atoms and leave out the bonds of monovalent elements (hydrogen and halogens). The resulting structures are called *partially condensed formulas*. If we omit all bonds except carbon–carbon multiple bonds, we have *fully condensed formulas*. Examples are given in Table 1.3.

Cyclic compounds can also be represented by partially condensed and fully condensed structural formulas, as shown in Table 1.4. In the fully condensed formulas each corner represents a CH_2 for singly bonded carbon atoms and a CH when carbon is linked to another carbon by a double bond.

Problem 1.6 Write a partially and a fully condensed structural formula for each of these structures.

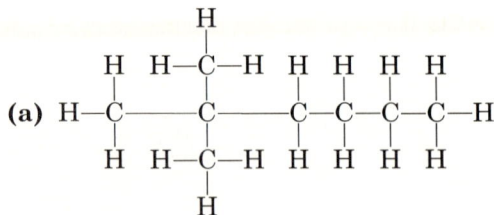

(a) expanded structural formula of a branched alkane

(b)

$$\text{H}-\underset{\underset{\displaystyle \text{H}}{|}}{\overset{\overset{\displaystyle \text{H}}{|}}{\text{C}}}-\text{C}\equiv\text{C}-\underset{\underset{\displaystyle \text{H}}{|}}{\overset{\overset{\displaystyle \text{H}}{|}}{\text{C}}}-\underset{\underset{\displaystyle \text{H}}{|}}{\overset{\overset{\displaystyle \text{H}}{|}}{\text{C}}}-\underset{\underset{\displaystyle \text{H}}{|}}{\overset{\overset{\displaystyle \text{H}}{|}}{\text{C}}}-\underset{\underset{\displaystyle \text{H}}{|}}{\overset{\overset{\displaystyle \text{H}}{|}}{\text{C}}}-\text{H}$$

(c)

$$\text{H}-\underset{\underset{\displaystyle \text{H}}{|}}{\overset{\overset{\displaystyle \text{H}}{|}}{\text{C}}}-\underset{\underset{\displaystyle \text{H}}{|}}{\overset{\overset{\displaystyle \text{H}}{|}}{\text{C}}}-\underset{\underset{\displaystyle \text{H}}{|}}{\overset{\overset{\displaystyle \text{H}}{|}}{\text{C}}}-\overset{\overset{\displaystyle \text{O}}{\|}}{\text{C}}-\underset{\underset{\displaystyle \text{H}}{|}}{\overset{\overset{\displaystyle \text{H}}{|}}{\text{C}}}-\underset{\underset{\displaystyle \text{H}}{|}}{\overset{\overset{\displaystyle \text{H}}{|}}{\text{C}}}-\text{H}$$

(d)

Table 1.3 Examples of Condensed Formulas

Expanded structural formula	Partially condensed formula	Fully condensed formula
H–C(H)(H)–C(H)(H)–O–H	CH_3-CH_2-OH	CH_3CH_2OH
H–C(H)(H)–C(H)(H)–C(H)(H)–C(H)(H)–H	$CH_3-CH_2-CH_2-CH_3$	$CH_3CH_2CH_2CH_3$ *or* $CH_3(CH_2)_2CH_3$
H–C(H)(H)–C(H)(CH_3)–C(H)(H)–H	$CH_3-CH-CH_3$ with CH_3 above	$(CH_3)_2CHCH_3$ *or* $(CH_3)_3CH$
H–C(H)(H)–C(H)(H)–C(=O)H	CH_3-CH_2-CH with $=O$ above	CH_3CH_2CHO
H–C(H)(H)–C(=O)–C(H)(H)–H	CH_3-C-CH_3 with $=O$ above	CH_3COCH_3 *or* $(CH_3)_2CO$
H–C(H)(H)–C(=O)–O–H	CH_3-C-OH with $=O$ above	CH_3COOH

Table 1.4   Some Condensed Cyclic Formulas

Expanded formula	Partially condensed formula	Fully condensed formula

Problem 1.7   Given the following condensed structural formulas, write the expanded structures.

(a) $(CH_3)_2CHC{\equiv}CCH_2OCH_3$

(b)

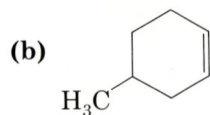

Shapes of Organic Molecules: *1.8*
Orbital Picture of Covalent Bonds

The Lewis model of the covalent bond and the concept of covalence number have been helpful in giving us a clearer picture of chemical bonding and structural formulas. But there is one aspect of organic chemistry on which the Lewis theory sheds no light at all—molecular geometry—and without also considering the shapes and sizes of organic molecules we cannot begin to discuss their chemistry at even the simplest level. The distinct geometry of organic compounds is a direct result of the covalent bonds involved. For this reason, it is necessary that we modify the Lewis theory of bonding and describe the electronic arrangement in atoms and molecules in terms of orbitals.

A   Atomic Orbitals

In Section 1.3 we considered the arrangement of electrons within various energy levels, but did not discuss the regions in space occupied by the electrons. Calculations based on spectroscopic studies of atoms have shown that electrons within each energy level are located in orbitals. An **atomic orbital** represents a specific region in space in which an electron is most likely to be found. Atomic orbitals are designated in the order in which they are filled by the letters s, p, d, and f. The first energy level (K shell) has only one orbital, the $1s$. The second energy level (L shell) has four orbitals, one $2s$ and three $2p$ orbitals. The third energy level (M shell) has nine orbitals: one $3s$, three $3p$, and five $3d$ orbitals. The atoms we will encounter in most organic compounds have only s and p orbitals. An s orbital is a spherically shaped electron cloud with the atom's nucleus at its center; a p orbital is a dumbbell-shaped electron cloud with the nucleus between the two lobes. Each p orbital is oriented along one of three perpendicular coordinate axes, that is, in the x, y, or z direction. The p orbitals are designated as $2p_x$, $2p_y$, and $2p_z$ if they are located in the L shell, and as $3p_x$, $3p_y$, and $3p_z$ if located in the M shell. The shapes of s and p orbitals are illustrated in Figure 1.4.

The energies of the electrons in orbitals increase in the order shown in Figure 1.5. Note that the three $2p$ orbitals are of equal energy, and so are the three $3p$ orbitals.

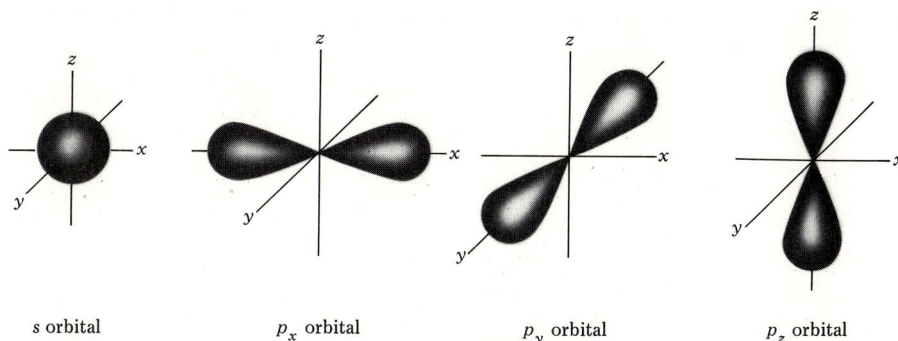

s orbital p_x orbital p_y orbital p_z orbital

Figure 1.4 Shapes of s and p orbitals.

Energy content
of orbital
increases

$$3p_x \text{ —} \quad 3p_y \text{ —} \quad 3p_z \text{ —}$$
$$3s \text{ —}$$
$$2p_x \text{ —} \quad 2p_y \text{ —} \quad 2p_z \text{ —}$$
$$2s \text{ —}$$
$$1s \text{ —}$$

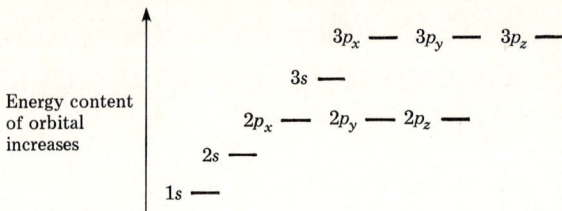

Figure 1.5 An energy level diagram of atomic orbitals showing the order in which they become filled with electrons.

When filling the atomic orbitals, keep in mind that

1. An atomic orbital can contain no more than two electrons (shown by ↑ and ↓).
2. Electrons fill orbitals of lower energy first (a $1s$ orbital before a $2s$ orbital, a $2s$ orbital before any of the three equivalent $2p$ orbitals, and so on).
3. No orbital is filled by two electrons until all the orbitals of equal energy have at least one electron.

The electronic configuration of carbon (atomic number 6) can be represented as shown in Figure 1.6. More simply, this distribution is shown by the notation

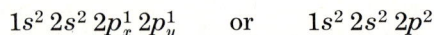

$$1s^2 \, 2s^2 \, 2p_x^1 \, 2p_y^1 \qquad \text{or} \qquad 1s^2 \, 2s^2 \, 2p^2$$

The superscripts indicate the numbers of electrons in the atomic orbitals. Table 1.5 shows the electronic configurations of the first ten elements in the periodic table.

Problem 1.8 Using s and p notation, write the electronic configurations for **(a)** Na (atomic number 11) and **(b)** Cl (atomic number 17).

B   Molecular Orbitals

A covalent bond consists of the overlap between two atomic orbitals to form a **molecular orbital.** A molecular orbital encompasses the nuclei of two atoms. Like atomic orbitals, a molecular orbital can accommodate no more than two electrons. For example, the molecular orbital of H_2 is formed when the two $1s$

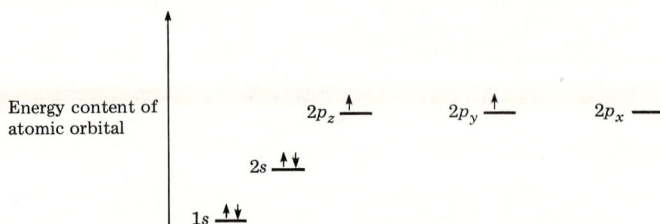

Energy content of
atomic orbital

$$2p_z \, \uparrow\text{—} \qquad 2p_y \, \uparrow\text{—} \qquad 2p_x \text{ —}$$
$$2s \, \uparrow\downarrow$$
$$1s \, \uparrow\downarrow$$

Figure 1.6 Energy level diagram for carbon.

Table 1.5 Electronic Configuration of the First- and Second-Row Elements in the Periodic Table

Atomic number	Symbol	Electronic configuration	1s	2s	$2p_x$	$2p_y$	$2p_z$
1	H	$1s^1$	↑				
2	He	$1s^2$	↑↓				
3	Li	$1s^2 2s^1$	↑↓	↑			
4	Be	$1s^2 2s^2$	↑↓	↑↓			
5	B	$1s^2 2s^2 2p_x^1$	↑↓	↑↓	↑		
6	C	$1s^2 2s^2 2p_x^1 2p_y^1$	↑↓	↑↓	↑	↑	
7	N	$1s^2 2s^2 2p_x^1 2p_y^1 2p_z^1$	↑↓	↑↓	↑	↑	↑
8	O	$1s^2 2s^2 2p_x^2 2p_y^1 2p_z^1$	↑↓	↑↓	↑↓	↑	↑
9	F	$1s^2 2s^2 2p_x^2 2p_y^2 2p_z^1$	↑↓	↑↓	↑↓	↑↓	↑
10	Ne	$1s^2 2s^2 2p_x^2 2p_y^2 2p_z^2$	↑↓	↓↑	↑↓	↑↓	↑↓

orbitals from each hydrogen atom overlap. This molecular orbital is cylindrically symmetrical about the axis that joins the two nuclei. Molecular orbitals that have this cylindrical or sausage shape are called **sigma (σ) orbitals,** and the bond between the two atoms is called a **sigma bond** (σ bond) (Fig. 1.7).

Sigma bonds can be formed not only from a combination of two *s* atomic orbitals, as in H_2, but also from the end-on overlap of two *p* atomic orbitals or from the overlap of an *s* atomic orbital with a *p* atomic orbital. (Another type of bond, called the pi (π) bond, which involves the side-side overlap between two *p* atomic orbitals, will be discussed in Chapter 3.)

Bond Energy and Bond Length *1.9*

Since atoms achieve a stable noble gas configuration when they combine to form molecules, we conclude that a molecule is more stable than the isolated constituent atoms. This stability is apparent in the release of energy during the

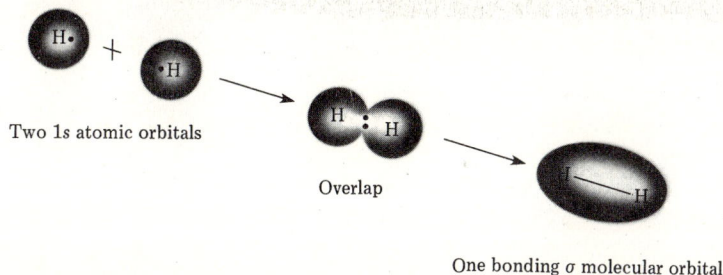

Two 1s atomic orbitals

Overlap

One bonding σ molecular orbital

Figure 1.7 Overlap of atomic orbitals between two hydrogen atoms to form a sigma molecular orbital (= sigma bond).

Table 1.6 Bond Dissociation Energies of Some Simple Molecules

Molecule	Bond dissociation energy (kcal/mole)
H—H	104
Cl—Cl	58
H—OH	111
H—CH$_3$	101
H$_3$C—CH$_3$	83
H$_2$C=CH$_2$	146
HC≡CH	200
H$_3$C—OH	89
H$_2$C=O	166

Table 1.7 Bond Length of Some Covalently Bonded Atoms

Bond	Bond length (Å)
H—H	0.72
C—H	1.09
O—H	0.96
C—C	1.54
C=C	1.34
C≡C	1.20
C—O	1.43
C=O	1.22

formation of the molecular bond. The amount of energy released when a bond is formed is called the **heat of formation** or the **bond energy.** Conversely, the same amount of energy would have to be supplied to break the bond. The amount of energy that must be absorbed to break a bond is called the **bond dissociation energy.** For a given pair of atoms, the greater the overlap of the atomic orbitals, the stronger the bond and the greater the amount of bond dissociation energy. In the case of H$_2$, the bond dissociation energy is 104 kcal/mole. Table 1.6 shows the dissociation energies of certain bonds in some simple molecules. Note that when two atoms are held together by a single bond, the bond dissociation energy is lower than when they are held together by more than one bond.

The distance between nuclei in the molecular structure is called the **bond length.** For a given pair of atoms, the bond length depends upon the extent of overlap of their atomic orbitals. For H$_2$, the bond length is 0.72 Ångstrom (1 Å = 10^{-8} cm). Table 1.7 shows the bond lengths between some covalently bonded atoms. Note that when atoms are held together by more than one bond, the bond lengths become shorter.

Problem 1.9 Predict which of the carbon–oxygen or carbon–nitrogen bonds in each pair of structures has (1) the greater bond dissociation energy and (2) the longer bond length.

(a) H$_3$C—O—CH$_3$ and O=C=O **(b)** CH$_3$CH$_2$—OH and CH$_3$$\overset{\overset{\displaystyle H}{|}}{C}$=O

(c) HC≡N and H$_2$C=NH

1.10 sp^3 Hybridization: The Tetrahedral Carbon

Now that we have examined atomic and molecular orbitals, we can address ourselves to our main topic of interest, namely, shapes of organic molecules. Let us consider first the simplest organic molecule, methane.

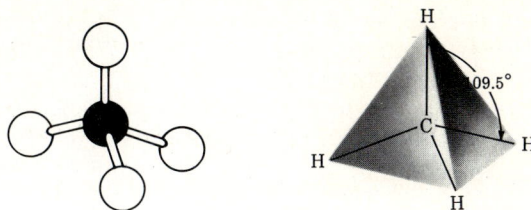

Figure 1.8 The tetrahedral structure of methane.

Based on experimental evidence, methane is known to consist of a carbon atom bonded covalently to four hydrogen atoms, thus having the molecular formula CH_4. Each of the four carbon–hydrogen bonds is identical: each has the same strength, 101 kcal/mole, and length, 1.09 Å. We also know that the four bonds are directed toward the corners of a regular **tetrahedron** with all H—C—H bond angles equal to 109.5° (Fig. 1.8). The tetrahedron is a pyramid-like structure with the carbon atom at the center and each of the four attached atoms or groups of atoms located at a corner.

The picture of the bonded carbon atom we have described is inconsistent with the one predicted from the electronic configuration of the isolated or ground-state carbon, which is

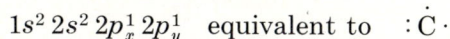

$$1s^2\, 2s^2\, 2p_x^1\, 2p_y^1 \quad \text{equivalent to} \quad :\overset{\displaystyle \cdot}{C}\cdot$$

As you can see, there are only two half-filled p orbitals in the ground-state carbon. We should therefore expect carbon to form not *four* but only *two* covalent bonds, as in $\overset{\displaystyle H}{\underset{|}{C}}$—H, with a bond angle of 90°.

To explain this discrepancy, Linus Pauling proposed that it is possible, by providing the required amount of energy to the ground-state carbon (Fig. 1.9a), to promote one electron from the $2s$ orbital to the empty $2p_z$ orbital. The resulting carbon is said to be in an excited or activated state. In this state the carbon atom has four unpaired electrons (Fig. 1.9b), which should account for the formation of four covalent bonds. However, the four bonds would not all be the same; three would be formed from $2p$ electrons, and the fourth from a $2s$ electron. But we know from experimentation that the four bonds are identical. Pauling therefore further proposed that the four excited-state orbitals mix together, or **hybridize,** to create four equivalent $2s2p^3$ ($2sp^3$ for short) hybrid orbitals,

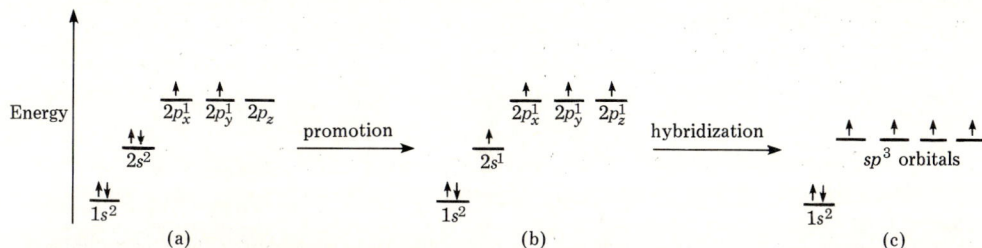

Figure 1.9 **(a)** The electronic ground state. **(b)** The activated state. **(c)** The sp^3-hybridized state of carbon.

each directed toward one corner of a tetrahedron. The merging of an s orbital and three p orbitals to give four sp^3 orbitals ($\frac{1}{4}s$ and $\frac{3}{4}p$ characteristic) is referred to as sp^3 **hybridization** (Fig. 1.9c).

Figure 1.10 shows in (a) the formation of a σ bond from the overlap of one sp^3 orbital with the s orbital of one hydrogen atom and in (b) the overlap of four sp^3 orbitals with the s orbitals of four hydrogen atoms to form the tetrahedral methane molecule.

The tetrahedral shape allows for the most effective overlap between the orbitals, thus forming strong bonds. Whenever carbon is singly bonded to other atoms, it utilizes sp^3-hybridized orbitals and assumes a tetrahedral shape. For example,

Cl	Cl	H H	H H H
H—C—Cl	Cl—C—Cl	H—C—C—H	H—C—C—C—H
Cl	Cl	H H	H H H
Chloroform	Carbon tetrachloride	Ethane	Propane

all represent molecules in which each carbon atom is sp^3 hybridized. Chloroform and carbon tetrachloride, like methane, consist of a single tetrahedron. Ethane is made up of two tetrahedra, and propane consists of three tetrahedra.

In molecules where the carbon atom is doubly or triply bonded, it utilizes a

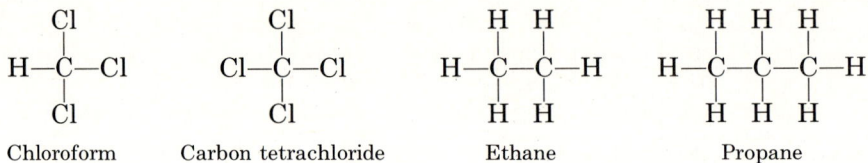

(a)

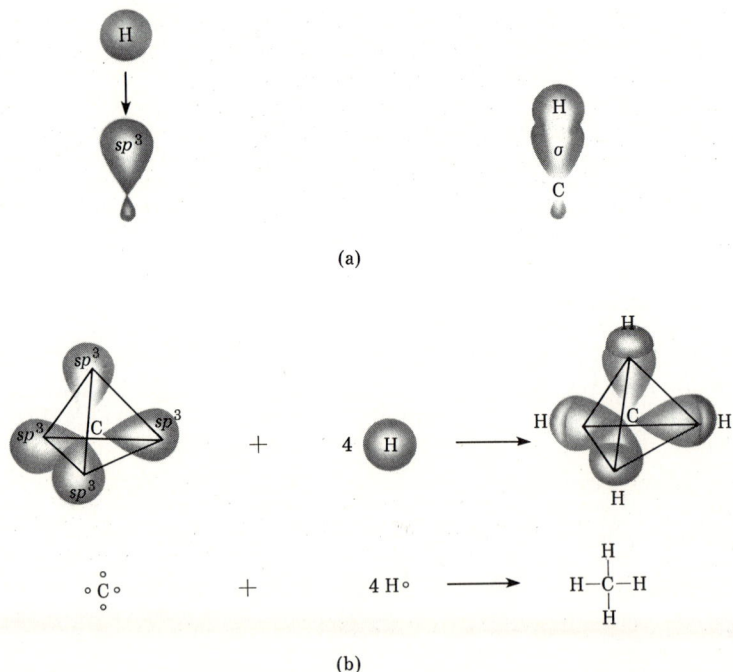

(b)

Figure 1.10 **(a)** Overlap of the s orbital of a hydrogen atom with an sp^3 orbital of carbon to form a σ bond. **(b)** Overlap of four sp^3 orbitals of carbon with the s orbitals of four hydrogen atoms to form the tetrahedral methane molecule.

different kind of hybrid orbital. For this reason, such molecules are no longer tetrahedral, and we shall discuss their geometry in Chapter 3.

No matter what the geometry, it is difficult to represent three-dimensional structures on a two-dimensional surface. Therefore, as a matter of convenience, we will continue to use planar structures unless three-dimensional structures are necessary to the discussion.

Functional Groups *1.11*

There are several million organic compounds. The study of their chemistry is made possible only because we are able to classify them into a limited number of families depending on the **functional groups** present. A functional group is a reactive portion of an organic molecule, an atom, or a group of atoms that confers on the whole molecule its characteristic properties. All compounds with the same functional group belong to one family. Members of a given organic family react in a similar and predictable manner. An example of a functional group, indicating an *alcohol,* is the hydroxyl group, OH, attached to a singly bonded carbon atom, as in $-\overset{|}{\underset{|}{C}}-OH$. Thus, the first three of the following structures represent specific alcohols of increasing carbon chain length. The fourth structure, R—OH, indicates the general formula for all alcohols; R stands for a carbon chain of any length attached to the functional group.

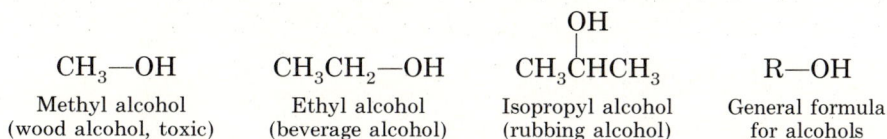

$$CH_3-OH \qquad CH_3CH_2-OH \qquad CH_3\overset{OH}{\underset{|}{C}}HCH_3 \qquad R-OH$$

Methyl alcohol   Ethyl alcohol   Isopropyl alcohol   General formula
(wood alcohol, toxic)   (beverage alcohol)   (rubbing alcohol)   for alcohols

Table 1.8 lists some of the functional groups and corresponding classes of compounds. In Chapter 2 we will discuss the chemistry of alkanes, the first class of compounds listed.

Problem 1.10 Identify the class of compound represented by each structure.

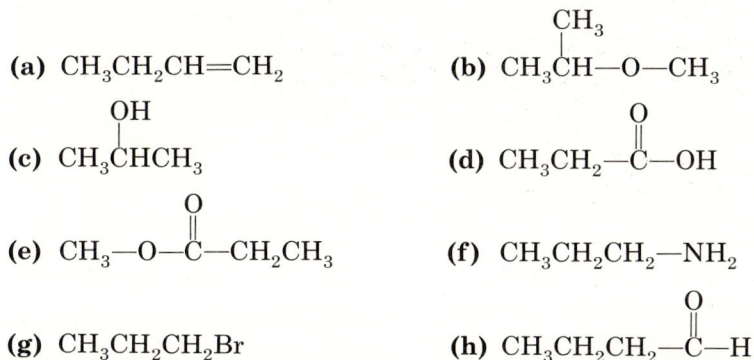

(a) $CH_3CH_2CH{=}CH_2$

(b) $CH_3\overset{CH_3}{\underset{|}{C}}H-O-CH_3$

(c) $CH_3\overset{OH}{\underset{|}{C}}HCH_3$

(d) $CH_3CH_2-\overset{O}{\overset{\|}{C}}-OH$

(e) $CH_3-O-\overset{O}{\overset{\|}{C}}-CH_2CH_3$

(f) $CH_3CH_2CH_2-NH_2$

(g) $CH_3CH_2CH_2Br$

(h) $CH_3CH_2CH_2-\overset{O}{\overset{\|}{C}}-H$

Table 1.8 Functional Groups and Classes of Organic Compounds

Class	General formula	Functional group	Specific examples
Alkane	RH	C—C (single bond)	H_3C—CH_3
Alkene	[a] R—CH=CH_2	C=C (double bond)	H_2C=CH_2
Alkyne	[a] R—C≡CH	C≡C (triple bond)	HC≡CH
Alkyl halide	RX	—X (X = F, Cl, Br, I)	H_3C—Cl
Alcohol	R—OH	—OH	H_3C—OH
Ether	R—O—R′	$-\overset{\displaystyle |}{\underset{\displaystyle |}{C}}-O-\overset{\displaystyle |}{\underset{\displaystyle |}{C}}-$	H_3C—O—CH_3
Aldehyde	[a] $R-\overset{\displaystyle O}{\overset{\|}{C}}H$	$-\overset{\displaystyle O}{\overset{\|}{C}}-H$	$H-\overset{\displaystyle O}{\overset{\|}{C}}-H,\ H_3C-\overset{\displaystyle O}{\overset{\|}{C}}-H$
Ketone	$R-\overset{\displaystyle O}{\overset{\|}{C}}-R'$	$-\overset{\displaystyle |}{\underset{\displaystyle |}{C}}-\overset{\displaystyle O}{\overset{\|}{C}}-\overset{\displaystyle |}{\underset{\displaystyle |}{C}}-$	$H_3C-\overset{\displaystyle O}{\overset{\|}{C}}-CH_3$
Carboxylic acid	[a] $R-\overset{\displaystyle O}{\overset{\|}{C}}-OH$	$-\overset{\displaystyle O}{\overset{\|}{C}}-OH$	$H-\overset{\displaystyle O}{\overset{\|}{C}}-OH,\ H_3C-\overset{\displaystyle O}{\overset{\|}{C}}-OH$
Ester	[a] $R-\overset{\displaystyle O}{\overset{\|}{C}}-OR$	$-\overset{\displaystyle O}{\overset{\|}{C}}-OR$	$H-\overset{\displaystyle O}{\overset{\|}{C}}-OCH_3,\ H_3C-\overset{\displaystyle O}{\overset{\|}{C}}-OCH_3$
Amine	R—NH_2	$-\overset{\displaystyle |}{\underset{\displaystyle |}{C}}-NH_2$	H_3C—NH_2

[a] In these classes of compounds R can also be H.

Problem 1.11 There are three compounds with the molecular formula C_3H_8O. Two are alcohols, and one is an ether. Draw their expanded structural formulas and identify the family to which each belongs.

Summary of Concepts and Reactions

Organic chemistry is the study of carbon/hydrogen-containing compounds and their derivatives.   [Sec. 1.1]

Carbon is unique among the elements for its ability to bond infinitely with itself to form compounds of various sizes and shapes as well as to bond with many other elements.   [Sec. 1.2]

Atoms consist essentially of three fundamental particles: neutrons, protons, and electrons. The atomic number of an element indicates the number of protons.
[Sec. 1.3A]

Electrons are distributed around the nucleus in shells or energy levels.   [Sec. 1.3A]

There are never more than 8 electrons in the outermost energy level of an atom.
[Sec. 1.3B]

The outer-shell electrons are called valence electrons, and they are depicted by dots around the symbol of the element (electron-dot structures).   [Sec. 1.3C]

Atoms can achieve a noble gas configuration by forming ionic bonds or covalent bonds.
[Sec. 1.4A, B]

Electronegativity is a measure of the attraction the nucleus of an atom has for its valence electrons.   [Sec. 1.4B]

A coordinate covalent bond is formed when one atom supplies 2 electrons to form a bond.   [Sec. 1.4C]

A Lewis base is a species that has 2 electrons available for bonding.   [Sec. 1.4C]

A Lewis acid is a species that can accept an electron pair to complete its valence shell.
[Sec. 1.4C]

Covalence number indicates the number of covalent bonds an atom can form with other atoms.   [Sec. 1.5]

A structural or constitutional formula shows how the atoms in a particular molecule are connected or bonded together.   [Sec. 1.6]

Expanded structural formulas can be simplified by condensing them.   [Sec. 1.7]

An atomic orbital represents a specific region in space where an electron is most likely to be found.   [Sec. 1.8A]

Overlap of two atomic orbitals forms a molecular orbital.   [Sec. 1.8B]

Molecular orbitals with a cylindrical or sausage shape are called sigma (σ) orbitals, and the covalent bond between the two atoms is called a σ bond.   [Sec. 1.8B]

Bond formation is an energy-releasing process. The amount of energy released when a bond is formed is called the heat of formation or the bond energy.   [Sec. 1.9]

The amount of energy that must be absorbed to break a bond is called the bond dissociation energy. (Bond dissociation energy is numerically equal to heat of formation.)
[Sec. 1.9]

When carbon is bonded to four other atoms, it makes use of sp^3 hybridization and will be at the center of a regular tetrahedron.   [Sec. 1.10]

A functional group is a reactive portion of an organic molecule that confers on the whole molecule its characteristic properties.   [Sec. 1.11]

Key Terms

carbon
organic chemistry
neutron
proton
electron
atomic number
shell
energy level
valence electron
electron-dot structure
ionic bond
covalent bond

electronegativity
polar covalent bond
coordinate covalent bond
Lewis base
Lewis acid
covalence number
structural formula
constitutional formula
single bond
double bond
triple bond
atomic orbital

molecular orbital
sigma (σ) orbital
sigma bond
heat of formation
bond energy
bond dissociation energy
bond length
tetrahedron
hybridize
sp^3 hybridization
functional group

Valence Electrons and Electron-Dot Structures [Sec. 1.3]

1.1 Write an electron-dot structure for each compound.

(a) H_2O (b) NH_3 (c) CH_4 (d) KBr

(e) $CaCl_2$ (f) H_2CO (g) CO_2 (h) HCN

Ionic, Covalent, and Polar Covalent Bonding [Sec. 1.4A, B]

1.2 List which compound(s) in Exercise 1.1 contain(s) (a) only ionic bonds, (b) only nonpolar covalent bonds, (c) only polar covalent bonds, and (d) both nonpolar and polar covalent bonds.

1.3 Show the partial charges by placing $\delta+$ and $\delta-$ symbols on the atoms involved in a polar covalent bond.

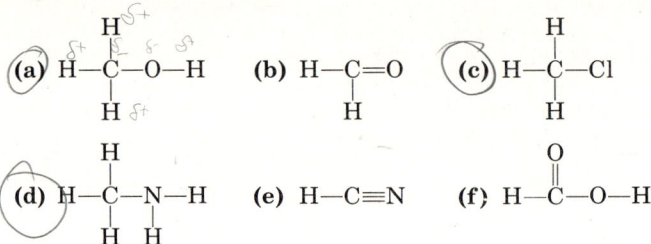

(a) $H-\overset{\underset{|}{H}}{\underset{\underset{H}{|}}{C}}-O-H$ (b) $H-\overset{|}{\underset{\underset{H}{|}}{C}}=O$ (c) $H-\overset{\underset{|}{H}}{\underset{\underset{H}{|}}{C}}-Cl$

(d) $H-\overset{\underset{|}{H}}{\underset{\underset{H}{|}}{C}}-\overset{\underset{|}{}}{\underset{\underset{H}{|}}{N}}-H$ (e) $H-C\equiv N$ (f) $H-\overset{\overset{O}{\|}}{C}-O-H$

1.4 Arrange the hydrogen halides (HI, HBr, HCl, HF) in order of polarity, from the most polar to the least polar.

Coordinate Covalent Bonding; Lewis Acid and Lewis Base [Sec. 1.4C]

1.5 For each of the following Lewis acid-base interactions (1) indicate the structure of the product formed and (2) identify the Lewis acid and the Lewis base.

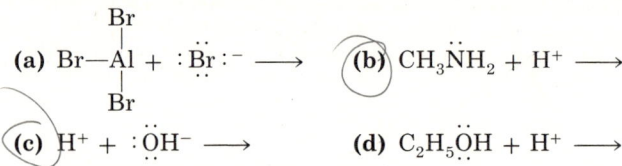

(a) $Br-\overset{\overset{Br}{|}}{\underset{\underset{Br}{|}}{Al}} + \,:\!\ddot{B}r\!:^- \longrightarrow$ (b) $CH_3\ddot{N}H_2 + H^+ \longrightarrow$

(c) $H^+ + \,:\!\ddot{O}H^- \longrightarrow$ (d) $C_2H_5\ddot{O}H + H^+ \longrightarrow$

Covalence Number and Structural Formula [Secs. 1.5, 1.6]

1.6 Given the skeletal structure, and assuming that only hydrogen atoms are missing, draw the correct structural formula for each of the following.

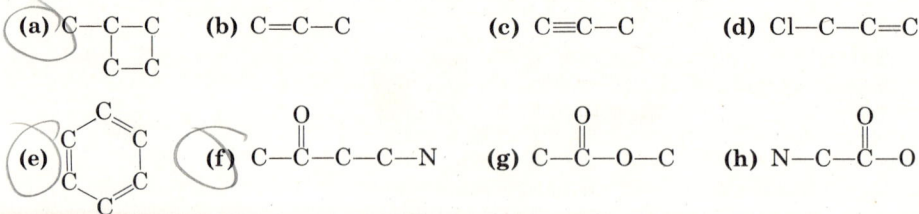

(a) $\overset{\underset{|}{C}}{\underset{\underset{C}{|}-C}{C}}-C-C$ (b) $C=C-C$ (c) $C\equiv C-C$ (d) $Cl-C-C=C$

(e) skeletal ring structure of C atoms   (f) $C-\overset{\overset{O}{\|}}{C}-C-C-N$ (g) $C-\overset{\overset{O}{\|}}{C}-O-C$ (h) $N-C-\overset{\overset{O}{\|}}{C}-O$

1.7 Check the following structures to see whether or not they represent possible compounds within the rules of covalence. State either "possible" or "impossible" for each.

(a) $CH_3CH_2CH_2CH_3$
$\overset{\underset{|}{}}{\underset{CH_3}{}}$

(b) $\underset{\underset{CH_2}{\diagdown\diagup}}{CH_3\!-\!CH}$

(c) $CH_3CH_2{-}O{-}CH_2{-}O{-}CH_3$

(d)
$$CH_3\underset{\underset{OH}{|}}{C}HCH_2\overset{\overset{O}{\|}}{C}{-}OH$$

(e)
$$CH_3CH_2\underset{\underset{Cl}{|}}{C}H_2Cl$$

(f) $CH_3CH_2{=}CH_2$

(g) $CH_3{-}NH_2{-}CH_3$

(h) $CH_3C{\equiv}CCH_2CH_3$

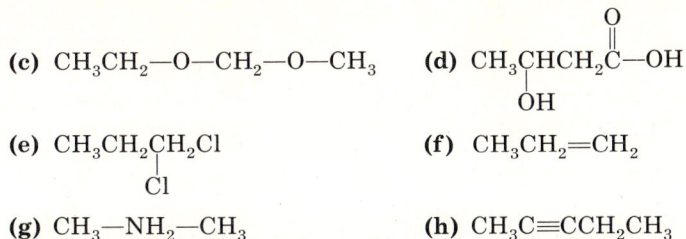

1.8 Draw a correct structural formula for each "impossible" representation in Exercise 1.7 by either adding or removing hydrogen atoms.

1.9 Draw an expanded structural formula corresponding to each molecular formula.
(a) $CHCl_3$ **(b)** CH_2F_2 **(c)** CH_4O **(d)** CH_5N **(e)** C_2H_2
(f) C_2H_4 **(g)** C_2H_6 **(h)** C_2H_5Cl **(i)** CH_3N **(j)** H_3NO

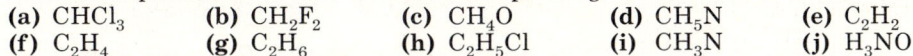

Condensed and Expanded Structural Formulas   [Sec. 1.7]

1.10 Convert each of the following expanded structural formulas into (1) a partially condensed and (2) a fully condensed formula.

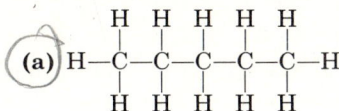

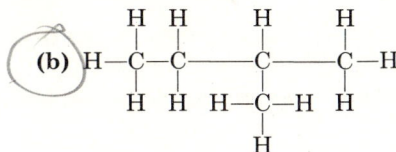

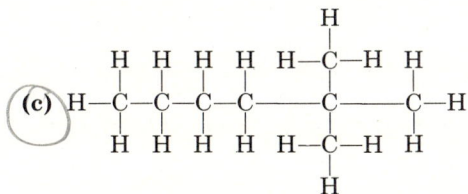

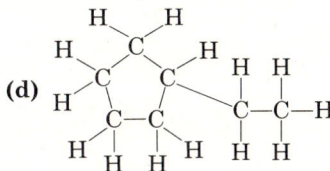

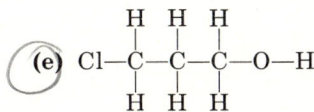

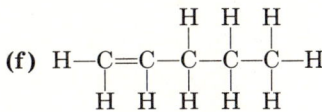

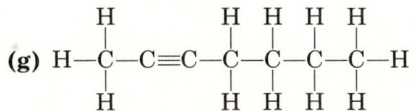

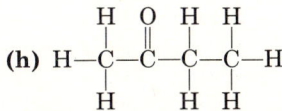

(a)

(b)

(c)

(d)

(e)

(f)

(g)

(h)

1.11 Draw the fully expanded structures corresponding to the partially condensed formulas (a–d) and fully condensed formulas (e–h).

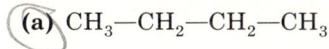

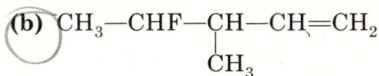

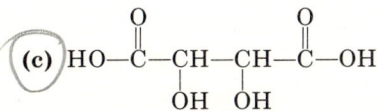

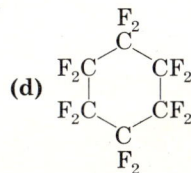

(a) $CH_3{-}CH_2{-}CH_2{-}CH_3$

(b) $CH_3{-}CHF{-}\underset{\underset{CH_3}{|}}{C}H{-}CH{=}CH_2$

(c) $HO{-}\overset{\overset{O}{\|}}{C}{-}\underset{\underset{OH}{|}}{C}H{-}\underset{\underset{OH}{|}}{C}H{-}\overset{\overset{O}{\|}}{C}{-}OH$

(d)

(e) $(CH_3)_2CH(CH_2)_4CH_3$

(f) $CH_3CH{=}CHCHO$

(g) $CH_3C{\equiv}CCH_2COOH$

(h)

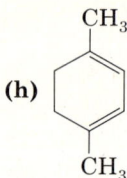

Bond Length, Bond Strength, and Bond Angle [Secs. 1.9, 1.10]

1.12 For each structure, predict which of the two bonds shown has (1) the greater bond dissociation energy and (2) the longer bond length.

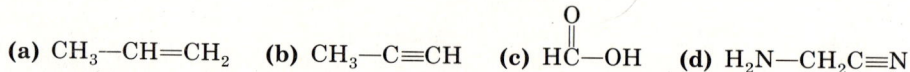

(a) $CH_3{-}CH{=}CH_2$ **(b)** $CH_3{-}C{\equiv}CH$ **(c)** $\overset{\overset{\textstyle O}{\|}}{HC}{-}OH$ **(d)** $H_2N{-}CH_2C{\equiv}N$

1.13 (a) What is the size of the $H{-}C{-}H$ bond angle in methane, CH_4?
   (b) What would you expect the size of the $F{-}C{-}F$ bond angle in carbon tetrafluoride, CF_4, to be?

Hybridization and Shape of Molecules [Sec. 1.10]

1.14 Indicate (1) the type of hybridized orbital utilized by carbon in each of the following structures and (2) the shape of each molecule.

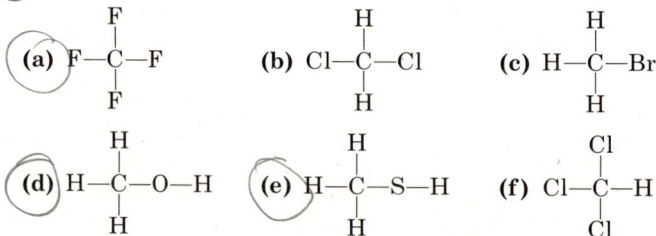

(a)
$$\overset{\textstyle F}{\underset{\textstyle F}{F{-}C{-}F}}$$
(b)
$$\overset{\textstyle H}{\underset{\textstyle H}{Cl{-}C{-}Cl}}$$
(c)
$$\overset{\textstyle H}{\underset{\textstyle H}{H{-}C{-}Br}}$$

(d)
$$\overset{\textstyle H}{\underset{\textstyle H}{H{-}C{-}O{-}H}}$$
(e)
$$\overset{\textstyle H}{\underset{\textstyle H}{H{-}C{-}S{-}H}}$$
(f)
$$\overset{\textstyle Cl}{\underset{\textstyle Cl}{Cl{-}C{-}H}}$$

Functional Groups and Classification of Compounds [Sec. 1.11]

1.15 Name the class to which each of the following compounds belongs.

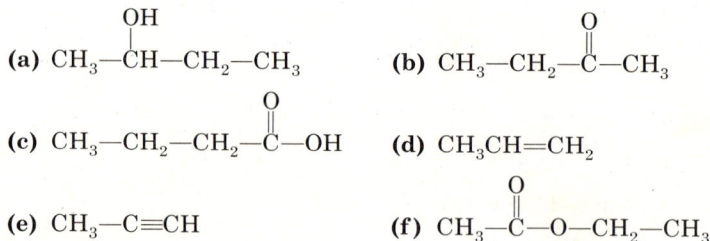

(a)
$$\overset{\overset{\textstyle OH}{|}}{CH_3{-}CH{-}CH_2{-}CH_3}$$
(b)
$$\overset{\overset{\textstyle O}{\|}}{CH_3{-}CH_2{-}C{-}CH_3}$$

(c)
$$\overset{\overset{\textstyle O}{\|}}{CH_3{-}CH_2{-}CH_2{-}C{-}OH}$$
(d) $CH_3CH{=}CH_2$

(e) $CH_3{-}C{\equiv}CH$
(f)
$$\overset{\overset{\textstyle O}{\|}}{CH_3{-}C{-}O{-}CH_2{-}CH_3}$$

1.16 Group together those compounds that you expect to behave chemically in a similar manner.

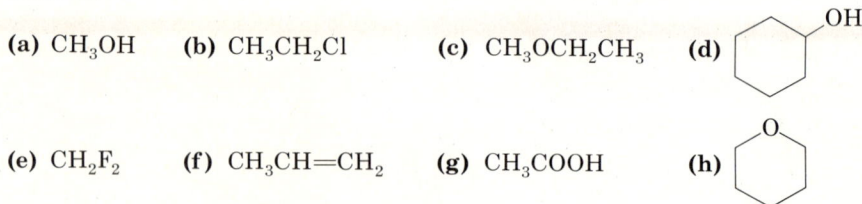

(a) CH_3OH **(b)** CH_3CH_2Cl **(c)** $CH_3OCH_2CH_3$ **(d)**

(e) CH_2F_2 **(f)** $CH_3CH{=}CH_2$ **(g)** CH_3COOH **(h)**

(i) C_2H_4 **(j)** $CHBr_3$ **(k)** $HOCH_2CH_2OH$ **(l)**

Covalence Number, Structural Formula, and Functional Group

[Secs. 1.5, 1.6, 1.11]

1.17 (a) One alcohol and one ether correspond to C_2H_6O. Draw their structures.

(b) One aldehyde and one ketone correspond to C_3H_6O. Draw their structures.

(c) One carboxylic acid and one ester correspond to $C_2H_4O_2$. Draw their structures.

2

Saturated Hydrocarbons: Alkanes

A large group of organic compounds, known as **hydrocarbons,** contain only the two elements carbon and hydrogen. Based on their structural features, the hydrocarbons are divided into two main classes, the *aliphatics* and *aromatics.*

Aliphatic hydrocarbons are subdivided into three families: *alkanes, alkenes,* and *alkynes.* Each family is characterized by a different functional group. **Alkanes,** the topic of this chapter, are characterized by the *carbon–carbon single bond.* Alkanes are also known as *saturated hydrocarbons* because each carbon is bonded to four other atoms, the maximum number of atoms to which any carbon can be attached.

Alkanes show a general lack of chemical reactivity, although they are extremely flammable. The high combustibility of alkanes is one reason for their importance; alkanes constitute the fuels we use in heating our homes and in running our machines.

2.1 The Three Simplest Alkanes: Methane, Ethane, Propane

The simplest member of the alkane family is *methane,* CH_4, a molecule introduced in Section 1.10. Methane has the shape of a tetrahedron with an sp^3-hybridized carbon at the center. All four C—H bonds in the molecule are equivalent. Figure 2.1 shows different representations of the methane molecule.

The second member of the alkane series, *ethane,* C_2H_6, is shown in Figure

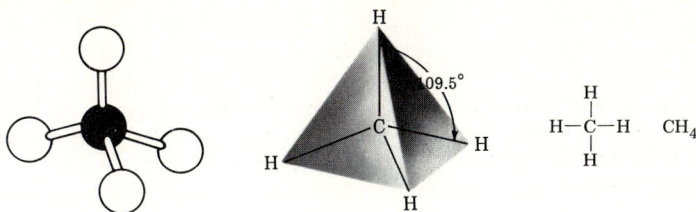

Figure 2.1   Tetrahedral and planar representations of methane.

2.2. Each carbon in ethane is sp^3 hybridized, and the bond joining the two carbons is called an sp^3–sp^3 molecular orbital. In fact, every carbon atom in any alkane makes use of sp^3-hybridized orbitals, since each is always linked via single bonds to four other atoms.

Note that ethane is larger than methane by a CH_2, or methylene group.

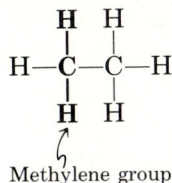

Methylene group

This is not to say that ethane is prepared in such a way from methane, but simply to illustrate the structural relation between the two compounds.

Next in the series, with three carbons, is *propane,* C_3H_8 (Fig. 2.3). Again, note that propane is larger than ethane by a methylene group. The bent representation of propane in Figure 2.3 indicates that carbon–carbon single bonds are free to rotate in any direction. Because of free rotation, an alkane can assume many different forms depending on the amount of twist or turn of a carbon with

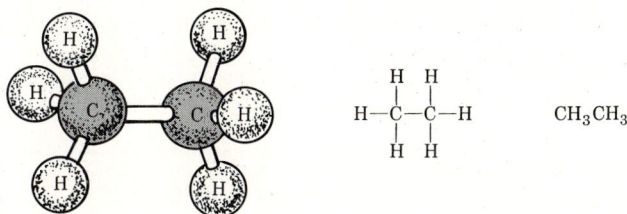

Figure 2.2   Representations of ethane.

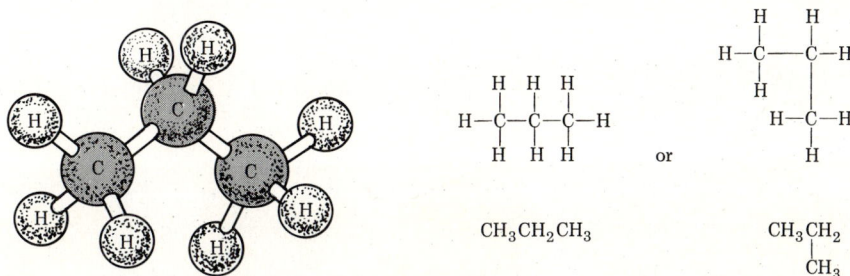

Figure 2.3   Representations of propane.

respect to another. Structures that differ from one another only because of the rotation of one or more carbon atoms are said to be in different **conformations.** Each structure is referred to as a **conformer.** At room temperature conformers cannot be separated because the change from one conformation to another requires so little expenditure of energy that it occurs with extreme rapidity.

2.2 Other Alkanes: Homologous Series C_nH_{2n+2}

If we keep adding CH_2 groups, one at a time, it is possible to build an infinite series of compounds, called a **homologous series.** In a homologous series each individual member differs from its next neighbor by a constant value, a CH_2 in this case. Each member of a homologous series is called a **homolog.** Methane, ethane, and propane are the first three homologs of the alkane family, whose general formula is C_nH_{2n+2}. In this general formula n is the number of carbons. The names and the molecular formulas of the first ten alkanes are shown in Table 2.1. It is important that you memorize these names because the names of many other organic compounds of various classes are derived from them.

Table 2.1 Names and Molecular
Formulas of the First Ten Alkanes

Name	Molecular formula
Methane	CH_4
Ethane	C_2H_6
Propane	C_3H_8
Butane	C_4H_{10}
Pentane	C_5H_{12}
Hexane	C_6H_{14}
Heptane	C_7H_{16}
Octane	C_8H_{18}
Nonane	C_9H_{20}
Decane	$C_{10}H_{22}$

Problem 2.1 Write the molecular formula of a 40-carbon alkane.

2.3 Structural Isomerism

Does a molecular formula fully identify a substance? Consider the following data.

Compound	Boiling point (°C)	Melting point (°C)	Molecular formula
I	0	−138	C_4H_{10}
II	−12	−145	C_4H_{10}

Obviously, compounds I and II are different substances. Yet they both have the same molecular formula, C_4H_{10}, indicating that they are both butanes. Different compounds with identical molecular formulas are called **isomers,** and the phenomenon is called **isomerism.**

Given a set of molecular models, it is possible to construct two structures with different carbon skeletons corresponding to C_4H_{10}.

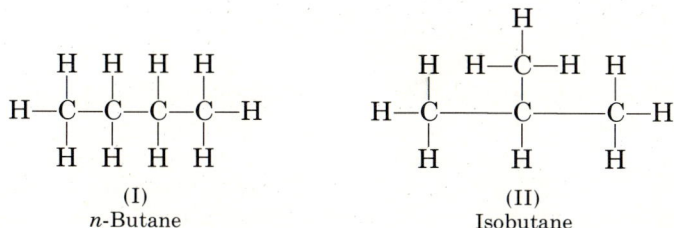

$$(I)$$
n-Butane

$$(II)$$
Isobutane

Compound I is called *normal* or *n*-butane. The prefix *n*- indicates that the carbon atoms are arranged in a continuous chain. Compound II, named isobutane, is a branched-chain hydrocarbon. Isomers such as *n*-butane and isobutane, which differ in the sequence of atoms bonded to each other, are more specifically called **structural** or **constitutional isomers.**

Alkanes beyond butane are all capable of structural isomerism. Thus, there are three isomeric pentanes, C_5H_{12}: *n*-pentane, isopentane, and neopentane.

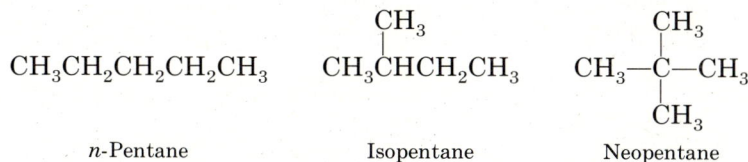

$$CH_3CH_2CH_2CH_2CH_3$$

n-Pentane

Isopentane

Neopentane

With higher alkanes the number of structural isomers increases rapidly (see Table 2.2).

Table 2.2   Number of Possible Structural Isomers of Alkanes

Name	Molecular formula	Number of isomers
Methane	CH_4	1
Ethane	C_2H_6	1
Propane	C_3H_8	1
Butanes	C_4H_{10}	2
Pentanes	C_5H_{12}	3
Hexanes	C_6H_{14}	5
Heptanes	C_7H_{16}	9
Octanes	C_8H_{18}	18
Nonanes	C_9H_{20}	35
Decanes	$C_{10}H_{22}$	75
Pentadecanes	$C_{15}H_{32}$	4,347
Eicosanes	$C_{20}H_{42}$	366,319

Structural isomerism is not confined to alkanes only. In fact, the phenomenon exists among all classes of organic compounds. For example,

		Molecular formula
CH$_3$CH$_2$—OH	CH$_3$—O—CH$_3$	C$_2$H$_6$O
Ethyl alcohol	Methyl ether	

$$CH_3CH_2-\overset{\overset{\displaystyle O}{\|}}{C}-H \qquad CH_3-\overset{\overset{\displaystyle O}{\|}}{C}-CH_3 \qquad C_3H_6O$$

Propionaldehyde   Acetone

In dealing with questions of structural isomerism, you must always keep in
mind that

1. Structures may be drawn in different conformations (Sec. 2.1). In such cases
   the various conformers represent the *same* molecule and not separate structural isomers. For example,

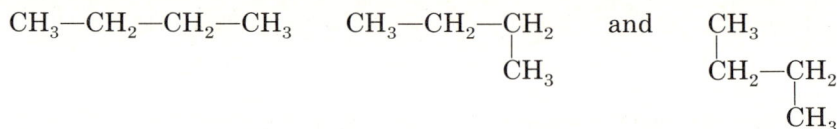

$$CH_3-CH_2-CH_2-CH_3 \qquad CH_3-CH_2-\underset{\underset{\displaystyle CH_3}{\|}}{CH_2} \qquad \text{and} \qquad \overset{\displaystyle CH_3}{\underset{\underset{\underset{\displaystyle CH_3}{\|}}{\displaystyle CH_2-CH_2}}{\|}}$$

   are the same molecule, *n*-butane.

2. Every atom in each structural isomer must have the correct number of bonds
   (Sec. 1.5).

Problem 2.2   Indicate whether the following pairs of structures are (1) the
same, (2) structural isomers, or (3) entirely unrelated.

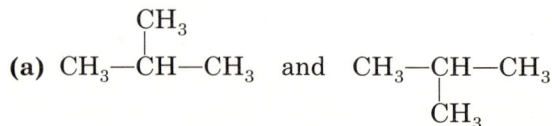

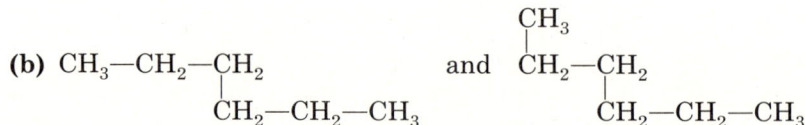

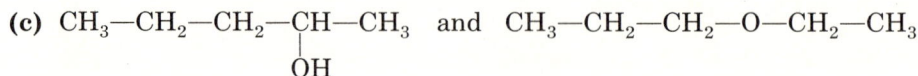

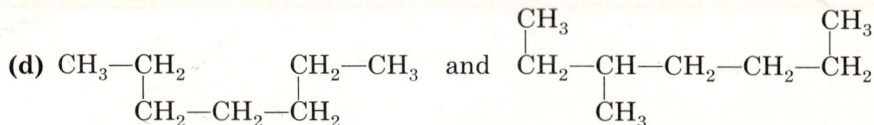

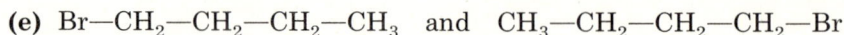

(a) CH$_3$—$\underset{\underset{\displaystyle CH_3}{\|}}{\overset{\overset{\displaystyle CH_3}{\|}}{CH}}$—CH$_3$ and   CH$_3$—$\underset{\underset{\displaystyle CH_3}{\|}}{CH}$—CH$_3$

(b) CH$_3$—CH$_2$—$\underset{\underset{\displaystyle CH_2-CH_2-CH_3}{\|}}{CH_2}$ and $\overset{\overset{\displaystyle CH_3}{\|}}{CH_2}$—$\underset{\underset{\displaystyle CH_2-CH_2-CH_3}{\|}}{CH_2}$

(c) CH$_3$—CH$_2$—CH$_2$—$\underset{\underset{\displaystyle OH}{\|}}{CH}$—CH$_3$ and   CH$_3$—CH$_2$—CH$_2$—O—CH$_2$—CH$_3$

(d) CH$_3$—$\underset{\underset{\displaystyle CH_2-CH_2-CH_2}{\|}}{CH_2}$ CH$_2$—CH$_3$ and $\overset{\overset{\displaystyle CH_3}{\|}}{CH_2}$—$\underset{\underset{\displaystyle CH_3}{\|}}{CH}$—CH$_2$—CH$_2$—$\overset{\overset{\displaystyle CH_3}{\|}}{CH_2}$

(e) Br—CH$_2$—CH$_2$—CH$_2$—CH$_3$ and   CH$_3$—CH$_2$—CH$_2$—CH$_2$—Br

(f) H$_2$C CH$_2$
        CH$_2$
H$_2$C—CH$_2$ and CH$_2$=CH—CH$_2$—CH$_2$—CH$_3$

(g) CH$_3$—CH$_2$—C≡C—CH$_2$—CH$_3$ and

CH$_2$
H$_2$C   CH
H$_2$C   CH
CH$_2$

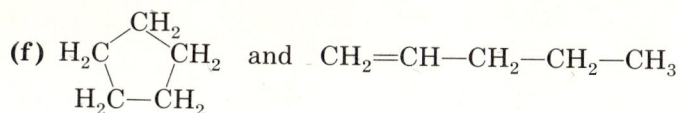

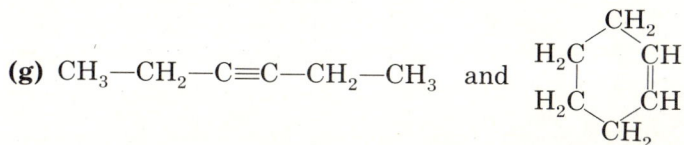

Problem 2.3 Write partially condensed structures for the five isomeric hexanes, C$_6$H$_{14}$.

Classes of Carbons and Hydrogens 2.4

The two isomeric butanes contain three different classes of carbons and hydrogens. A **primary (1°) carbon** is one that is bonded to only one other carbon. Carbons bonded to two other carbons are called **secondary (2°) carbons,** and those bonded to three other carbons are called **tertiary (3°) carbons.** Hydrogens are also referred to as 1°, 2°, or 3° according to the type of carbon they are bonded to. *n*-Butane has two 1° carbons with six 1° hydrogens and two 2° carbons and four 2° hydrogens. In isobutane there are three 1° carbons and nine 1° hydrogens but only one 3° carbon with one 3° hydrogen.

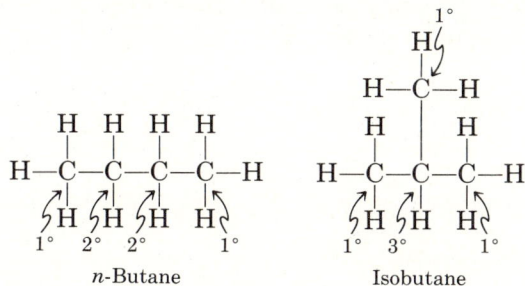

n-Butane   Isobutane

Problem 2.4 How many 1°, 2°, and 3° carbons and hydrogens, if any, are there in propane?

Problem 2.5 When a carbon is bonded to four other carbon atoms, it is called a **quaternary (4°) carbon.** Look at the structure of neopentane (page 31) and determine how many 1°, 2°, 3°, and 4° carbons and hydrogens, if any, there are in this molecule.

2.5 Alkyl Groups

Consider the complicated alkane

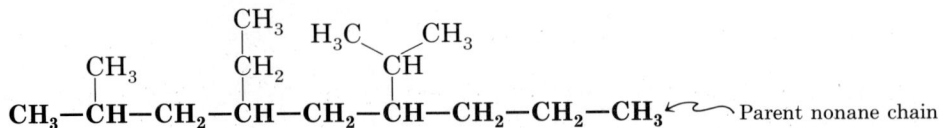

$$
\begin{array}{c}
\qquad\qquad CH_3 \qquad H_3C \qquad CH_3 \\
\qquad CH_3 \qquad\quad | \qquad\qquad \diagdown\ \diagup \\
\qquad\ | \qquad\qquad CH_2 \qquad\quad CH \\
\qquad\ | \qquad\qquad\ | \qquad\qquad\ | \\
CH_3-CH-CH_2-CH-CH_2-CH-CH_2-CH_2-CH_3
\end{array}
$$

Parent nonane chain

Its molecular formula is $C_{15}H_{32}$, and Table 2.2 indicates that there are 4347 possible isomers. Therefore 4347 names are needed to identify each one unambiguously. This seemingly impossible task can be accomplished by identifying and naming first the longest continuous carbon chain* and then the branches that are attached to it. In our example the longest continuous chain has nine carbons and is therefore a nonane. The branches, or alkyl substituents, have names of their own. An **alkyl group** is an alkane from which a hydrogen has been removed. The symbol **R** is used to represent an alkyl group, as in R—OH, the general formula for alcohols (Sec. 1.11).

Individual alkyl groups are named by replacing the suffix -*ane* of the parent alkane by -*yl*. Thus the alkyl group CH_3—, derived from methane, is called the *methyl* group. Similarly, *ethyl*, CH_3CH_2—, is the alkyl group derived from ethane.

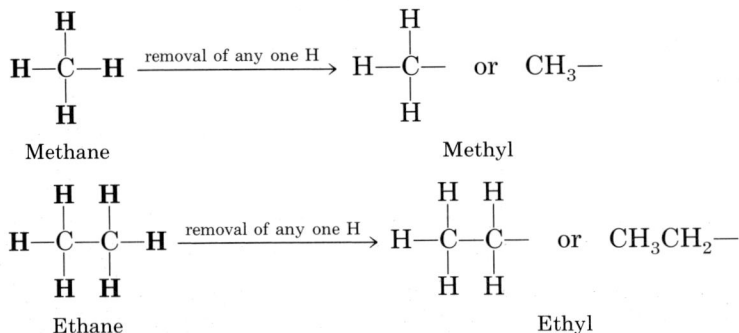

$$
\begin{array}{ccc}
\qquad H & & H \\
\quad\ | & & | \\
H-C-H & \xrightarrow{\text{removal of any one H}} & H-C- \quad \text{or} \quad CH_3- \\
\quad\ | & & | \\
\qquad H & & H
\end{array}
$$

Methane · Methyl

$$
\begin{array}{ccc}
\quad H\ \ H & & H\ \ H \\
\quad\ |\ \ \ | & & |\ \ \ | \\
H-C-C-H & \xrightarrow{\text{removal of any one H}} & H-C-C- \quad \text{or} \quad CH_3CH_2- \\
\quad\ |\ \ \ | & & |\ \ \ | \\
\quad H\ \ H & & H\ \ H
\end{array}
$$

Ethane · Ethyl

Propane has two kinds of hydrogens: the six in the two methyl groups and the two in the CH_2 group.

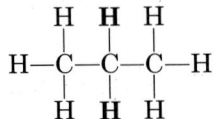

$$
\begin{array}{c}
H\ \ H\ \ H \\
|\ \ \ |\ \ \ | \\
H-C-C-C-H \\
|\ \ \ |\ \ \ | \\
H\ \ H\ \ H
\end{array}
$$

Removal of a hydrogen from either one of the two end carbons gives the *n*-propyl group, while removal of a hydrogen from the middle carbon yields the isopropyl group.

---

* The longest continuous chain may be determined by following, with one's finger, the sequence of carbon–carbon bonds that includes the maximum number of carbon atoms without ever doubling back over the same bond.

$$H-\overset{\overset{\displaystyle H}{|}}{\underset{\underset{\displaystyle H}{|}}{C}}-\overset{\overset{\displaystyle H}{|}}{\underset{\underset{\displaystyle H}{|}}{C}}-\overset{\overset{\displaystyle H}{|}}{\underset{\underset{\displaystyle H}{|}}{C}}-H \xrightarrow[\text{CH}_3\text{ hydrogen}]{\text{removal of any one}} H-\overset{\overset{\displaystyle H}{|}}{\underset{\underset{\displaystyle H}{|}}{C}}-\overset{\overset{\displaystyle H}{|}}{\underset{\underset{\displaystyle H}{|}}{C}}-\overset{\overset{\displaystyle H}{|}}{\underset{\underset{\displaystyle H}{|}}{C}}- \quad \text{or} \quad CH_3CH_2CH_2-$$

Propane   n-Propyl

$$H-\overset{\overset{\displaystyle H}{|}}{\underset{\underset{\displaystyle H}{|}}{C}}-\overset{\overset{\displaystyle H}{|}}{\underset{\underset{\displaystyle H}{|}}{C}}-\overset{\overset{\displaystyle H}{|}}{\underset{\underset{\displaystyle H}{|}}{C}}-H \xrightarrow[\text{CH}_2\text{ hydrogen}]{\text{removal of either}} H-\overset{\overset{\displaystyle H}{|}}{\underset{\underset{\displaystyle H}{|}}{C}}-\overset{\overset{\displaystyle H}{|}}{C}-\overset{\overset{\displaystyle H}{|}}{\underset{\underset{\displaystyle H}{|}}{C}}-H \quad \text{or} \quad CH_3CHCH_3$$

Propane   Isopropyl

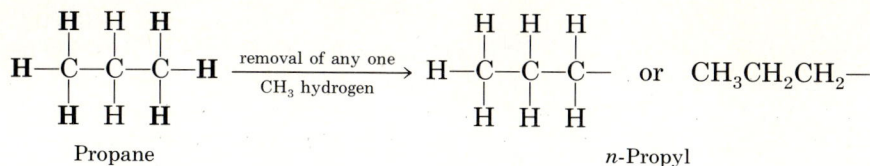

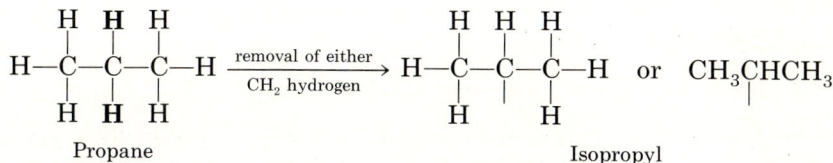

There are four alkyl groups that can be derived from the butanes, two from
n-butane and two from isobutane, because each isomer has two kinds of hydrogens.

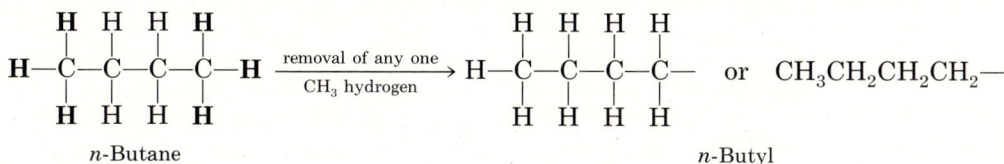

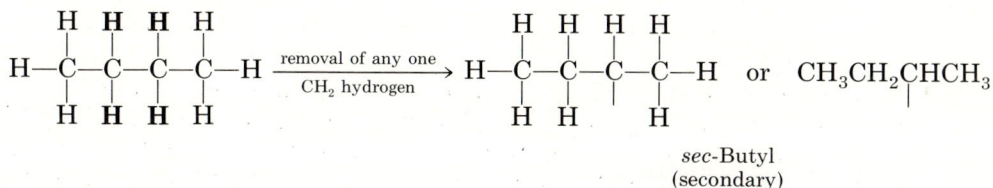

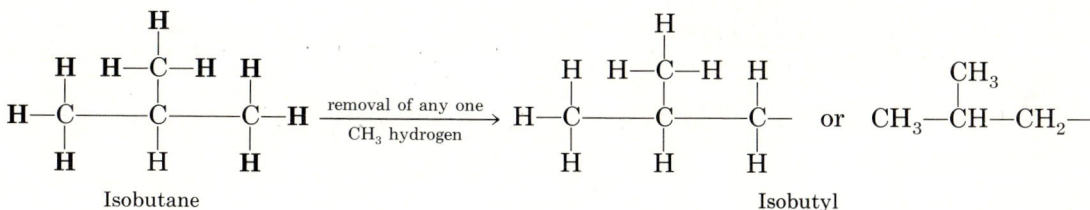

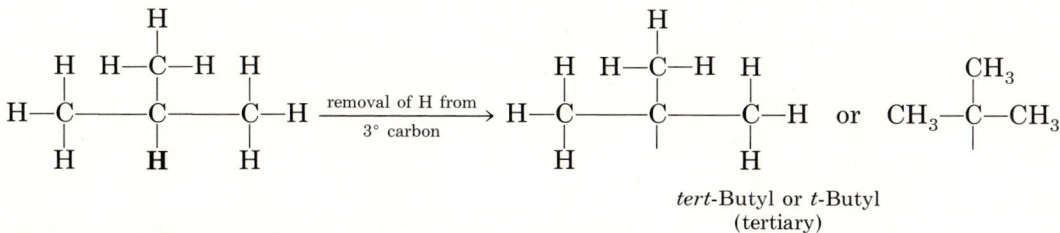

$$H-\overset{\overset{\displaystyle H}{|}}{\underset{\underset{\displaystyle H}{|}}{C}}-\overset{\overset{\displaystyle H}{|}}{\underset{\underset{\displaystyle H}{|}}{C}}-\overset{\overset{\displaystyle H}{|}}{\underset{\underset{\displaystyle H}{|}}{C}}-\overset{\overset{\displaystyle H}{|}}{\underset{\underset{\displaystyle H}{|}}{C}}-H \xrightarrow[\text{CH}_3\text{ hydrogen}]{\text{removal of any one}} H-\overset{}{\underset{}{C}}-\overset{}{\underset{}{C}}-\overset{}{\underset{}{C}}-\overset{}{\underset{}{C}}- \quad \text{or} \quad CH_3CH_2CH_2CH_2-$$

n-Butane   n-Butyl

$$H-\overset{\overset{\displaystyle H}{|}}{\underset{\underset{\displaystyle H}{|}}{C}}-\overset{\overset{\displaystyle H}{|}}{\underset{\underset{\displaystyle H}{|}}{C}}-\overset{\overset{\displaystyle H}{|}}{\underset{\underset{\displaystyle H}{|}}{C}}-\overset{\overset{\displaystyle H}{|}}{\underset{\underset{\displaystyle H}{|}}{C}}-H \xrightarrow[\text{CH}_2\text{ hydrogen}]{\text{removal of any one}} H-\overset{}{\underset{}{C}}-\overset{}{\underset{}{C}}-\overset{}{\underset{}{C}}-\overset{}{\underset{}{C}}-H \quad \text{or} \quad CH_3CH_2CHCH_3$$

sec-Butyl
(secondary)

$$H-\overset{}{\underset{\underset{\displaystyle H}{|}}{C}}-\overset{\overset{\displaystyle H-C-H}{|}}{\underset{\underset{\displaystyle H}{|}}{C}}-\overset{}{\underset{\underset{\displaystyle H}{|}}{C}}-H \xrightarrow[\text{CH}_3\text{ hydrogen}]{\text{removal of any one}} H-\overset{}{\underset{}{C}}-\overset{}{\underset{}{C}}-\overset{}{\underset{}{C}}- \quad \text{or} \quad CH_3-\overset{\overset{\displaystyle CH_3}{|}}{CH}-CH_2-$$

Isobutane   Isobutyl

$$H-\overset{}{\underset{\underset{\displaystyle H}{|}}{C}}-\overset{\overset{\displaystyle H-C-H}{|}}{\underset{\underset{\displaystyle H}{|}}{C}}-\overset{}{\underset{\underset{\displaystyle H}{|}}{C}}-H \xrightarrow[\text{3° carbon}]{\text{removal of H from}} H-\overset{}{\underset{}{C}}-\overset{}{\underset{}{C}}-\overset{}{\underset{}{C}}-H \quad \text{or} \quad CH_3-\overset{\overset{\displaystyle CH_3}{|}}{\underset{\underset{\displaystyle CH_3}{|}}{C}}-CH_3$$

tert-Butyl or t-Butyl
(tertiary)

Having identified the most common alkyl groups, let us go back to our
complicated alkane (page 34) and see if we can now give it an unambiguous
name. The alkyl branches on the parent nonane chain are methyl, ethyl, and
isopropyl; the compound could therefore be named methylethylisopropylno-
nane. Although this name does give us much information about the compound,
it is still ambiguous: we do not know the locations of the branches on the parent

chain. Thus, you can see that naming a complicated compound is difficult. We need a systematic set of rules. Fortunately, a set of rules, the IUPAC system of nomenclature, is available.

2.6 The IUPAC System of Nomenclature

As the number and complexity of organic compounds increased, it became obvious that the early names used to describe them were confusing and inadequate. This problem led eventually to the development of systematic nomenclature rules by the International Chemical Congress, which met in Geneva in 1892. These rules and their subsequent revisions are known today as the **IUPAC system of nomenclature** (International Union of Pure and Applied Chemistry). Although the IUPAC name of a compound makes much more sense, many of the unsystematic names that existed prior to the development of the IUPAC system are still in use today. As a result most organic compounds are known by two or more names: the older unsystematic names, which are referred to as **common** or **trivial,** and the IUPAC names. Isobutane and neopentane are examples of common names. Usually, for compounds of five carbons or less, the common nomenclature is employed. Larger compounds are identified by their IUPAC names.

The IUPAC rules that apply to alkanes also form the basis for the nomenclature of all other organic families.

1. Select as the parent structure the longest continuous chain, and consider the compound to have been derived from this structure by the replacement of hydrogens by various alkyl substituents.

In the following example the longest continuous chain contains six carbons; it is therefore a hexane. The substituent is an alkyl branch. Note again that the longest continuous chain is not necessarily straight. It may be bent in various directions because of the free rotation between single-bonded carbon atoms.

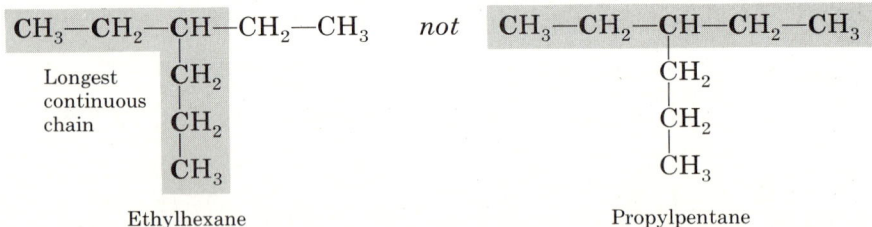

$$CH_3-CH_2-CH-CH_2-CH_3 \quad \textit{not} \quad CH_3-CH_2-CH-CH_2-CH_3$$

Longest continuous chain:
$$\begin{array}{c} CH_2 \\ | \\ CH_2 \\ | \\ CH_3 \end{array} \qquad\qquad \begin{array}{c} CH_2 \\ | \\ CH_2 \\ | \\ CH_3 \end{array}$$

Ethylhexane   Propylpentane

2. Number the carbons in the parent chain starting from whichever end will give the lowest number for the point of attachment of the substituent.

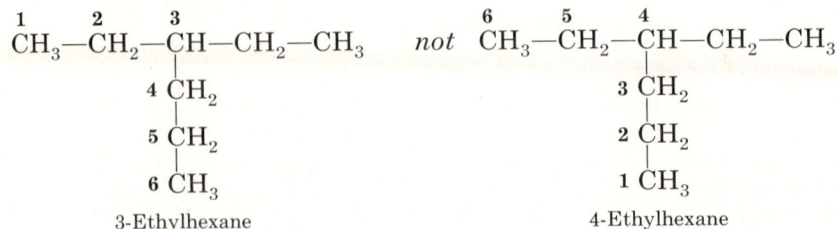

$$\overset{1}{C}H_3-\overset{2}{C}H_2-\overset{3}{C}H-CH_2-CH_3 \quad \textit{not} \quad \overset{6}{C}H_3-\overset{5}{C}H_2-\overset{4}{C}H-CH_2-CH_3$$

$$\begin{array}{c} 4\ CH_2 \\ | \\ 5\ CH_2 \\ | \\ 6\ CH_3 \end{array} \qquad\qquad \begin{array}{c} 3\ CH_2 \\ | \\ 2\ CH_2 \\ | \\ 1\ CH_3 \end{array}$$

3-Ethylhexane   4-Ethylhexane

To name the compound, first indicate the position of the substituent on the parent carbon chain by a number (in this example the position number is "3"). The number is followed by a hyphen and the combined name of the substituent (ethyl) and the parent carbon chain (hexane), giving 3-ethylhexane as the full name.

3. If the same alkyl substituent occurs more than once on the parent carbon chain, the prefixes *di-, tri-, tetra-, penta-,* and so on, are used to indicate two, three, four, five, and so on. The positions of these substitutents are indicated by appropriate numbers separated by commas. If the same substituent occurs twice on the same carbon, the number is repeated. For example,

$$
\begin{array}{ccccc}
& & & CH_3 & \\
5 & 4 & 3 & 2| & 1 \\
CH_3 & —CH—CH_2 & —C—CH_3 & & \text{Parent carbon chain (pentane)}\\
& | & | & & \\
& CH_3 & CH_3 & &
\end{array}
$$

2,2,4-Trimethylpentane

4. If different alkyl substituents are attached on the parent carbon chain, they are named either in order of increasing complexity or in alphabetical order. (It does not matter which method is chosen, as long as the compounds are named consistently.) The chain is numbered from the direction that gives the lowest possible number to one substituent group. For example, the following molecule is named 4-methyl-3,3-diethyl-5-*n*-propyloctane if the order of complexity is used, and 3,3-diethyl-4-methyl-5-*n*-propyloctane if the alphabetical order system is used.

$$
\begin{array}{cccccccc}
& & & & & CH_3 & & \\
& & & & CH_3 & CH_2 & & \\
8 & 7 & 6 & 5 & 4| & 3| & 2 & 1 \\
CH_3 —CH_2 —CH_2 —CH —CH —C —CH_2 —CH_3 & & & & & & & \text{Parent chain (octane)}\\
& & & | & & | & & \\
& & & CH_2 & & CH_2 & & \\
& & & | & & | & & \\
& & & CH_2 & & CH_3 & & \\
& & & | & & & & \\
& & & CH_3 & & & &
\end{array}
$$

5. If substituents other than alkyl groups are also present on the parent carbon chain, then all substituents are named alphabetically. The names for some of the more common nonalkyl substituents are

—F	fluoro	—NO$_2$	nitro	—CN	cyano
—Cl	chloro				
—Br	bromo	—NH$_2$	amino		
—I	iodo				

The following compound is therefore named 3-bromo-2-chloro-4-methyl-pentane.

$$
\begin{array}{ccccc}
& Br & & & \\
& | & & & \\
CH_3 —CH—CH—CH—CH_3 & & & & \\
& | & & | & \\
& CH_3 & & Cl &
\end{array}
$$

The following examples will help you to understand and apply the IUPAC rules of nomenclature.

Example 2.1 Write the structural formula of the compound whose IUPAC name is 2,5,5-trimethyl-4-ethylheptane.

Solution (1) Because the parent chain is always indicated by the last part of the compound's name, the chain in this case is a heptane. Write a seven-carbon chain without bothering yet about the hydrogen atoms.

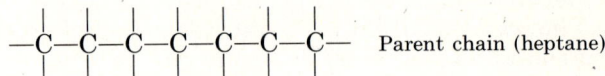

$$-\overset{|}{\underset{|}{C}}-\overset{|}{\underset{|}{C}}-\overset{|}{\underset{|}{C}}-\overset{|}{\underset{|}{C}}-\overset{|}{\underset{|}{C}}-\overset{|}{\underset{|}{C}}-\overset{|}{\underset{|}{C}}- \quad \text{Parent chain (heptane)}$$

(2) Number the carbons starting from either end.

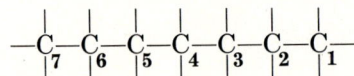

$$-\underset{|7}{C}-\underset{|6}{C}-\underset{|5}{C}-\underset{|4}{C}-\underset{|3}{C}-\underset{|2}{C}-\underset{|1}{C}-$$

(3) There are three methyl substituents, one on carbon 2 and two on carbon 5.

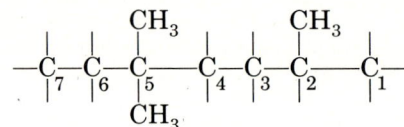

$$
\begin{array}{ccccccc}
 & & CH_3 & & & CH_3 & \\
 & & | & & & | & \\
-\underset{|7}{C}-\underset{|6}{C}-\underset{|5}{C}-\underset{|4}{C}-\underset{|3}{C}-\underset{|2}{C}-\underset{|1}{C}- \\
 & & | & & & & \\
 & & CH_3 & & & &
\end{array}
$$

(4) There is one ethyl substituent on carbon 4.

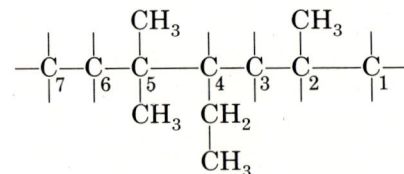

$$
\begin{array}{ccccccc}
 & & CH_3 & & & CH_3 & \\
 & & | & & & | & \\
-\underset{|7}{C}-\underset{|6}{C}-\underset{|5}{C}-\underset{|4}{C}-\underset{|3}{C}-\underset{|2}{C}-\underset{|1}{C}- \\
 & & | & | & & & \\
 & & CH_3 & CH_2 & & & \\
 & & & | & & & \\
 & & & CH_3 & & &
\end{array}
$$

(5) Now add the missing hydrogen atoms on the parent chain to get the correct structure of the compound.

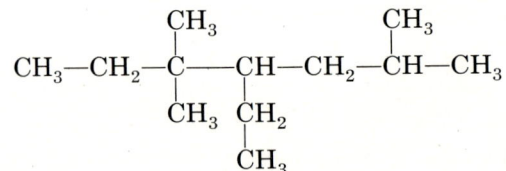

$$
\begin{array}{c}
CH_3CH_3 \\
| | \\
CH_3-CH_2-C-\!\!-\!\!-CH-CH_2-CH-CH_3 \\
| | \\
CH_3CH_2 \\
| \\
CH_3
\end{array}
$$

Example 2.2 The incorrect IUPAC name of a compound is 2,2-diethylbutane. Write the structural formula and correct name of the compound.

Solution (1) The parent carbon chain is called butane; it therefore has four carbons.

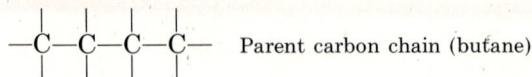

$$-\overset{|}{\underset{|}{C}}-\overset{|}{\underset{|}{C}}-\overset{|}{\underset{|}{C}}-\overset{|}{\underset{|}{C}}- \quad \text{Parent carbon chain (butane)}$$

(2) Number the chain.

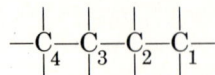

$$-\underset{|4}{C}-\underset{|3}{C}-\underset{|2}{C}-\underset{|1}{C}-$$

(3) There are two ethyl substituents on carbon 2.

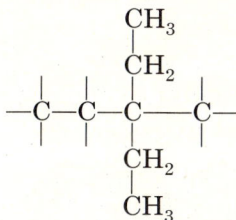

$$\begin{array}{c} \overset{\displaystyle CH_3}{|} \\ \overset{\displaystyle CH_2}{|} \\ -C-C-C-\!\!\!-C- \\ | \quad | \quad | \quad | \\ CH_2 \\ | \\ CH_3 \end{array}$$

(4) The longest continuous chain is not four carbons but five carbons. It is therefore a pentane.

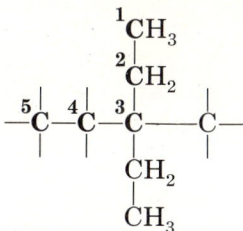

$$\begin{array}{c} \overset{1}{CH_3} \\ | \\ \overset{2}{CH_2} \\ | \\ \overset{5}{C}-\overset{4}{C}-\overset{3}{C}-\!\!\!-C- \\ | \quad | \quad | \quad | \\ CH_2 \\ | \\ CH_3 \end{array}$$

(5) When the parent chain is renumbered, we find a methyl substituent and an ethyl substituent on carbon 3. Named according to alphabetical order, the compound is 3-ethyl-3-methylpentane. Writing the missing hydrogens gives the correct structural formula.

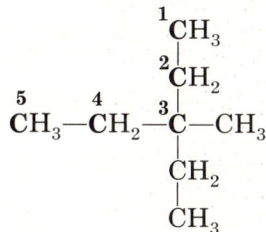

$$\begin{array}{c} \overset{1}{CH_3} \\ | \\ \overset{2}{CH_2} \\ | \\ \overset{5}{CH_3}-\overset{4}{CH_2}-\overset{3}{C}-CH_3 \\ | \\ CH_2 \\ | \\ CH_3 \end{array}$$

3-Ethyl-3-methylpentane

Problem 2.6 Give the IUPAC name for each of the following compounds. List the alkyl substituents in alphabetical order.

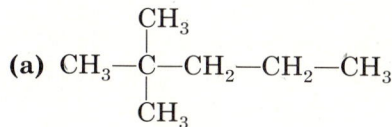

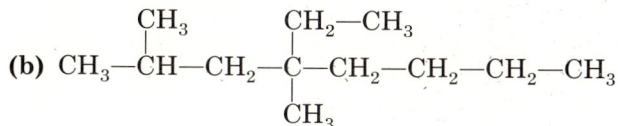

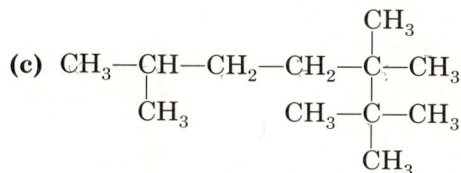

(a)
$$\begin{array}{c} CH_3 \\ | \\ CH_3-C-CH_2-CH_2-CH_3 \\ | \\ CH_3 \end{array}$$

(b)
$$\begin{array}{c} CH_3 \qquad\qquad CH_2-CH_3 \\ | \qquad\qquad\quad | \\ CH_3-CH-CH_2-C-CH_2-CH_2-CH_2-CH_3 \\ | \\ CH_3 \end{array}$$

(c)
$$\begin{array}{c} \qquad\qquad\qquad\qquad CH_3 \\ \qquad\qquad\qquad\qquad | \\ CH_3-CH-CH_2-CH_2-C-CH_3 \\ | \qquad\qquad\qquad | \\ CH_3 \qquad\quad CH_3-C-CH_3 \\ \qquad\qquad\qquad | \\ \qquad\qquad\qquad CH_3 \end{array}$$

$$\text{(d)} \quad CH_3-CH_2-\overset{\displaystyle |}{\underset{\displaystyle \underset{CH_3}{\overset{\displaystyle |}{CH}}\underset{CH_3}{}}{CH}}-CH_2-\overset{\displaystyle \overset{CH_3}{|}}{\underset{\displaystyle \underset{CH_3}{|}}{C}}-CH_2-CH_3$$

$$\text{(e)} \quad CH_3-\overset{\displaystyle \overset{CH_3}{|}}{CH}-CH_2-CH-CH_2-\overset{\displaystyle \overset{CH_3 \, CH_3}{\diagdown \diagup}}{\underset{CH}{}}-CH_2-CH_2-CH_3$$

Problem 2.7 Write the structural formula for each of the following compounds.

(a) 2-Methylpentane (isohexane)   (b) 2,2-Dichlorobutane
(c) 3,4-Dimethyl-5-ethyl-6-isopropylnonane   (d) 2-Bromo-3-methylpentane
(e) 2,4-Dimethyl-5-ethyl-4-*t*-butylheptane

Problem 2.8 These compounds are incorrectly named. Write the structure and correct name for each compound.

(a) 4,4-Dimethylpentane   (b) 2-Methyl-2-*sec*-butylpentane
(c) 2-Ethylpropane   (d) 2-*t*-Butylbutane
(e) 3-Methyl-5-isopropylhexane

2.7 Sources of Alkanes

The two principal sources of alkanes are petroleum and natural gas, both of which are products of the decay of animal, vegetable, and marine matter. Petroleum and natural gas constitute the chief sources of alkanes up to 40 carbons as well as of many aromatic, alicyclic (cyclic aliphatic hydrocarbons), and heterocyclic (cyclic molecules with more than one kind of atom in the ring) compounds.

A Petroleum Refining

Petroleum is a viscous oil liquid that varies in appearance from a dark yellow to a brown to a greenish black color. Its various components are separated and purified by a process called *refining*. This is usually done by distilling the petroleum into fractions of different boiling ranges and then treating the distilled petroleum in various ways to remove the undesirable components. The most volatile components (that is, those with the lowest boiling points) come out first. The less volatile components come out next, and the highest boiling components (those that boil at temperatures above 400°C) remain behind as residues. Gasoline boils at temperatures between 40 and 200°C. Kerosene (used as jet engine fuel) boils at temperatures ranging from 175 to 325°C. Lubricating oil, asphalt, and petroleum coke are the highest boiling residues that remain behind. Table 2.3 lists the useful components of refined petroleum.

Commercially, gasoline is by far the most important fraction of petroleum refining. It is not, however, a major fraction of the distillate. One industrial method used to increase the quantity of gasoline is a process known as cracking or *pyrolysis* (from the Greek *pyro,* fire; *lysis,* breaking) (see Fig. 2.4). In this process the higher-molecular-weight alkanes in petroleum, usually the kerosene and gas oil fractions, are heated in a chamber to very high temperatures (400–700°C), at which point they are broken into a mixture of smaller hydrocarbons, some of which have molecular sizes in the range of gasoline constituents.

The refined products of petroleum, known as *petrochemicals,* have great industrial importance. They are used as raw materials in the manufacture of many useful finished products, as Figure 2.5 indicates.

B   Octane Number

Gasoline performance in internal combustion engines has for many years been rated on the basis of the **octane number** scale. The higher the octane number assigned to a fuel, the better its performance and the lower the incidence of "knock" in the engine. *Knocking* is the familiar sharp "ping" that occurs in an automobile engine when driving up a steep grade or attempting to accelerate too rapidly. Knocking is caused by the premature ignition of the fuel–air mixture before completion of the compression stroke, resulting in reduced power and in engine wear.

In setting up the octane number scale, *n*-heptane, a poor fuel that causes severe knocking, was arbitrarily assigned an octane rating of zero. 2,2,4-Trimethylpentane (known as isooctane by the petroleum industry), an excellent fuel with no knocking tendency, was assigned an octane rating of 100. The octane ratings of other fuels are determined by comparing their knocking tendency with that of a synthetic blend of 2,2,4-trimethylpentane and *n*-heptane. The familiar "regular" gasoline with an octane rating of 90 has a knocking character-

Table 2.3   Some Components of Refined Petroleum

Fraction	Boiling range (°C)	Carbon content
Gas	Below 20	C_1–C_4
Petroleum ether	20–60	C_5–C_6
Naphtha	60–100	C_6–C_7
Gasoline	40–200	C_5–C_{10}
Kerosene	175–325	C_{11}–C_{18}
Gas oil	300–500	C_{15}–C_{40}
Lubricating oil, asphalt, petroleum coke, and paraffins	Above 400	C_{15}–C_{40}

Figure 2.4   A fluid catalytic cracking unit used in the production of gasoline. [Courtesy Mobil Oil Corp.]

istic equivalent to that of a mixture of 10% *n*-heptane and 90% 2,2,4-trimethyl-pentane.

When the system of rating a fuel was first established, no other known hydrocarbons gave better engine performance than 2,2,4-trimethylpentane. Since then, however, new hydrocarbons that are superior fuels have been discovered and given an octane rating higher than 100. Hydrocarbons that surpass 2,2,4-trimethylpentane in knock performance are rated not by their knocking tendency but by their *power output* as compared with the power developed by the standard octane. Thus, octane numbers over 100 are based differently than those under 100. Table 2.4 lists octane numbers of some hydrocarbons. Note that octane numbers decrease with increasing chain length and increase with increasing branching.

The octane number of a poor fuel can also be improved by blending it with small amounts of additives. Commercial Ethyl Fluid, which is added to gasolines to improve their octane ratings, consists of approximately 59% tetraethyllead,

Cumulated Value of Product

| $2 Billion | $8 Billion | $16 Billion | $100 Billion |

Primary Hydrocarbons

Basic Petrochemical Feedstocks
(produced in refineries and petrochemical plants)

Monomers:
"Building Blocks" for Plastic Resins and Synthetic Rubber
(produced in the organic chemical industry)

Polymers:
Plastic Resins and Synthetic Rubber
(produced in the synthetic resin and rubber industries)

Plastic and Rubber Products; Moldings, Extrusions, and Coatings
(produced and used in over 300 manufacturing industries)

Natural Gas, Refinery Gas, and Other Produced Gas

Methane
Ethane
Propane
Butane

Petroleum and Coal-Tar Oils

Naphthas and paraffin oils (saturated straight carbon chains)

Cyclic (aromatic) oils (various carbon-ring chemicals)

Separation, cracking, and reforming in refineries and petrochemical plants

Olefins:
Unsaturated Hydrocarbons

Ethylene
Propylene
Butenes
Butadienes
Isoprene

Cyclics:
Ring Hydrocarbons

Benzene
Toluene
Xylenes
Naphthalene

Vinyl chloride monomer

Butadiene monomer

Styrene monomer

Phenol monomer

Formaldehyde monomer

Polyvinyl chloride (PVC) resin (vinyl chloride polymer)

Synthetic rubber (butadiene-styrene copolymer)

Phenolic resin (phenol formaldehyde polymer)

Heat-Workable Plastics Products (Thermoplastic Products)
Phonograph records, plastic toys, vinyl flooring, artificial leather, auto upholstery, and others

Rubber Products
Auto tires and tubes, rubber gaskets, rubber hose and tubing, and others

Heat-Resistant Plastic Products (Thermoset Products)
Electric appliance parts, electric switch and control hardware, cookware handles and others

The six-carbon benzene ring

| O Oxygen atom | H Hydrogen atom | Cl Chlorine atom |

● The carbon (C) atom with zero to four hydrogen atoms attached: HCH HC— —C— —C≡

● The unsaturated carbon-to-carbon double (olefinic) bond, petrochemistry's coupling points.

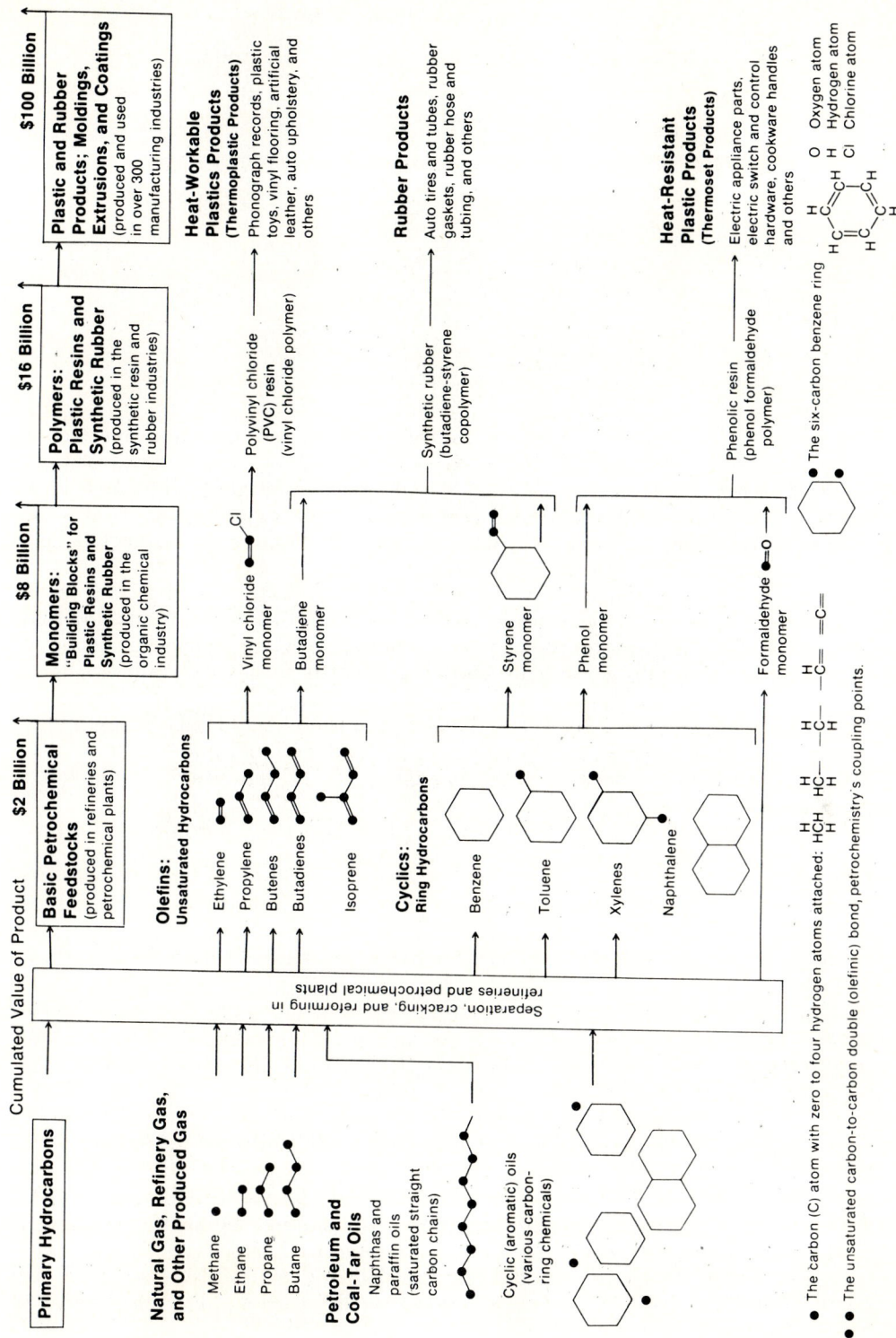

Figure 2.5 Manufactured products made from petrochemicals. [From The New York *Times,* Jan. 6, 1974. © 1974 by The New York Times Company. Reprinted by permission.]

43

Table 2.4 Octane Number of Some Hydrocarbons

Hydrocarbon	Octane number
n-Hexane	26
n-Heptane	0
n-Octane	-20
n-Nonane	-35
2-Methylpentane	73
2-Methylhexane	45
2-Methylheptane	24
2,2-Dimethylhexane	77
2,3-Dimethylbutane	93
2,2,4-Trimethylpentane	100
2,2,3-Trimethylpentane	116

13% ethylene bromide, 24% ethylene chloride, and 4% kerosene and dye. Tetra-ethyllead, $(C_2H_5)_4Pb$, is an efficient antiknock agent but has one disadvantage: its combustion product, lead oxide, is reduced to metallic lead that clogs the cylinder valves of an engine. Some lead, a dangerous pollutant, is also discharged into the air through the exhaust pipe. To remove the deposited lead, ethylene bromide and ethylene chloride are added with the tetraethyllead into the gasoline. Each additive reacts with the lead in the engine and transforms it into gaseous lead bromide and lead chloride at the combustion temperature. These gases are then discharged through the exhaust pipe. The net effect is to clean the engine but also to pollute the air even more. Although stringent federal emission standards were passed to combat pollution caused by the automobile, they have of late been relaxed because of the energy crisis. The emission control equipment put in automobiles since 1973 has increased gasoline consumption, and their effect in actually lessening the amount of pollution is questionable.

Other additives such as TCP (tricresyl phosphate) and boron hydrides have also enhanced the performance of many gasolines.

C Natural Gas

Natural gas consists of the low-molecular-weight alkanes from C_1 to C_8. It is composed primarily of methane (80%); the other constituents are ethane (13%), propane (3%), butane (1%), C_5 through C_8 alkanes (0.5%), and nitrogen (2.5%). Natural gas is a cleaner fuel than petroleum because it contains almost none of the sulfur compounds usually found in petroleum. The combustion of natural gas, unlike the combustion of many petroleum products, produces very little sulfur dioxide, SO_2, a troublesome air pollutant in many large cities.

The propane and butane components of natural gas can be removed by liquefaction and compressed into cylinders to be sold as bottled gas, which is used for fuel in many rural areas. In heavily populated urban centers, the natural gas is distributed by means of pipelines. Natural gas is also converted into many other important organic compounds, such as alcohols, aldehydes, ketones, carboxylic acids, and alkyl halides.

Methane is also called marsh gas because it rises as bubbles on the surface of marshes and peat bogs. Like propane, butane, and other alkanes, methane is highly flammable, and its spontaneous ignition in marshes is responsible for the ghostly looking blue flames called will-o'-the-wisps.

Physical Properties of Alkanes *2.8*

By physical properties we mean those properties that can be observed without the compound undergoing a chemical reaction.

A   Physical States and Solubilities

Alkanes occur at room temperature as gases, liquids, and solids. Alkanes from C_1 to C_4 are gases; most C_5 to C_{17} alkanes are liquids; and the C_{18} and larger alkanes are wax-like solids.

Alkanes are nonpolar compounds. Their solubility characteristic may be predicted by what is commonly known as the "like dissolves like" rule. What this rule means is that nonpolar compounds are soluble in other nonpolar solvents and that polar compounds are generally soluble in other polar solvents. Thus, alkanes are soluble in the nonpolar solvents carbon tetrachloride, CCl_4, and benzene, C_6H_6, but they are insoluble in polar solvents such as water.

B   Boiling Points

When a substance boils, it changes from a liquid to a gas, a process requiring energy. The boiling points of normal hydrocarbons increase with increasing molecular weight. One explanation for this trend is that as molecules become larger there are more forces of attraction between them, and more energy is needed to go from the liquid to the gaseous state. Except for the very small alkanes, the boiling point rises 20–30°C for each addition of a carbon atom to the chain (Fig. 2.6).

Among isomeric alkanes, the straight-chain compound has the highest boiling point. For the other isomers, the greater the number of branches, the lower the boiling point. For example,

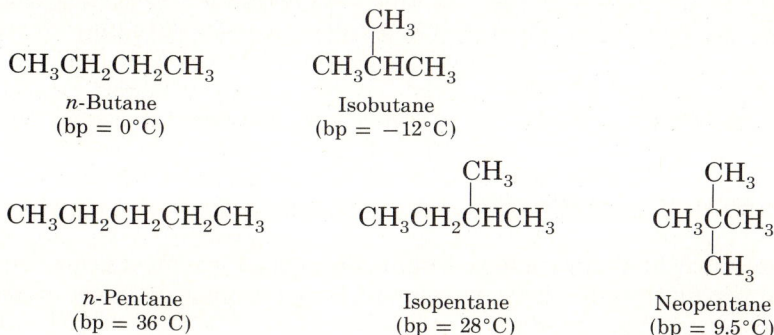

$$CH_3CH_2CH_2CH_3 \qquad\qquad CH_3\overset{\overset{\displaystyle CH_3}{|}}{C}HCH_3$$

n-Butane (bp = 0°C)   Isobutane (bp = −12°C)

$$CH_3CH_2CH_2CH_2CH_3 \qquad CH_3CH_2\overset{\overset{\displaystyle CH_3}{|}}{C}HCH_3 \qquad CH_3\overset{\overset{\displaystyle CH_3}{|}}{\underset{\underset{\displaystyle CH_3}{|}}{C}}CH_3$$

n-Pentane (bp = 36°C)   Isopentane (bp = 28°C)   Neopentane (bp = 9.5°C)

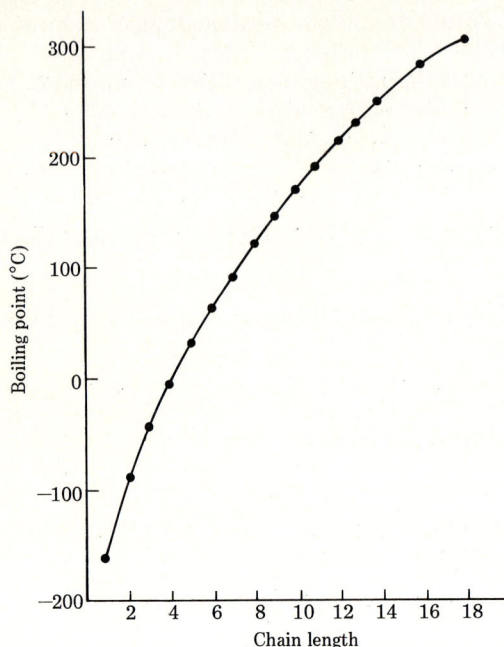

Figure 2.6 Boiling points of normal alkanes.

The decrease in boiling points with increased branching can be explained as follows: as branching increases, the molecule becomes more compact, decreasing its surface area. With a smaller surface area, the intermolecular forces of attraction are diminished, requiring less energy to go from the liquid state to the gaseous state.

Problem 2.9 Arrange the following compounds in order of increasing boiling points: *n*-hexane; 2,2-dimethylbutane; 2-methylpentane.

C   Melting Points

The melting points of alkanes also increase with increasing molecular weight. However, unlike the change in boiling point, there is no regularity in the change in melting point with the number of carbon atoms in a molecule. Also, differences in melting points between straight-chain and branched-chain compounds follow no regular pattern.

2.9 Preparation of Alkanes

A great number of alkanes can be obtained in pure form most economically by fractional distillation of crude petroleum. In some cases it is necessary to synthesize an alkane that cannot be obtained from natural sources, and laboratory methods of preparation are available. These consist of converting an-other class of compounds into alkanes. For this reason, we shall not discuss the preparations of alkanes until we encounter other classes of organic com-pounds.

Before proceeding to the reactions of alkanes, which involve the breaking and making of bonds, we shall first look into the various ways a bond can be broken and the notations used to indicate the movement of electrons.

Notations for Bond Breaking and Bond Making 2.10

A covalent bond can be broken in either of two ways, *homolytically* or *heterolytically*. Homolytic cleavage occurs when a bond splits evenly into two neutral fragments, each with one odd electron. A fishhook arrow $\frown$ indicates the movement of a single electron. Heterolytic cleavage occurs when a bond breaks unevenly, forming two oppositely charged fragments. A full-headed curved arrow $\frown$ denotes the movement of an electron pair. Both homolytic and heterolytic cleavages are illustrated with the hypothetical molecule $A—\overset{|}{\underset{|}{C}}—$.

Homolytic cleavage

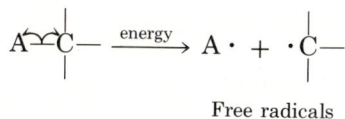

$$A\frown\overset{|}{\underset{|}{C}}— \xrightarrow{\text{energy}} A\cdot \;+\; \cdot\overset{|}{\underset{|}{C}}—$$

Free radicals

Heterolytic cleavage

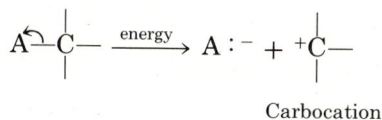

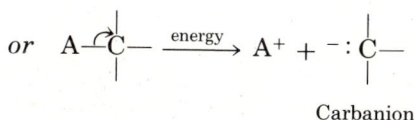

$$A\frown\overset{|}{\underset{|}{C}}— \xrightarrow{\text{energy}} A\colon^- \;+\; {}^+\overset{|}{\underset{|}{C}}—$$

Carbocation

$$or \quad A\frown\overset{|}{\underset{|}{C}}— \xrightarrow{\text{energy}} A^+ \;+\; {}^-\colon\overset{|}{\underset{|}{C}}—$$

Carbanion

Each species produced by homolytic cleavage has an odd electron and is called a **free radical.** In the case of heterolytic cleavage, depending on how cleavage has occurred, the carbon is either a positively charged ion called a **carbocation,** ${}^+\overset{|}{\underset{|}{C}}—$, or a negatively charged ion called a **carbanion,** ${}^-\overset{|}{\underset{|}{C}}—$. Free radicals, carbocations, and carbanions are reactive species that appear only as short-lived intermediates in organic reactions. We will encounter free radicals in halogenation, the first reaction of alkanes.

2.11 Reactions of Alkanes

Compared with most other classes of compounds, saturated hydrocarbons undergo very few reactions, and this is why they were originally called *paraffinic* hydrocarbons (from the Latin *parum,* little; *affinis,* affinity). We shall consider two important reactions that alkanes do undergo, namely, **halogenation** and **combustion.** In the former reaction only carbon–hydrogen bonds are broken; in the latter both carbon–hydrogen and carbon–carbon bonds are ruptured.

These reactions are presented here in a general form. They are discussed in greater detail in the next three sections.

1. Halogenation

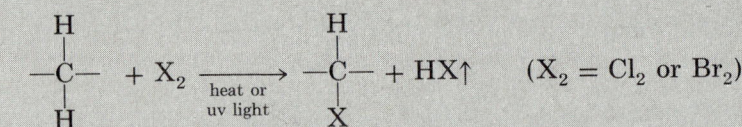

An alkane

[Secs. 2.12, 2.13]

2. Combustion

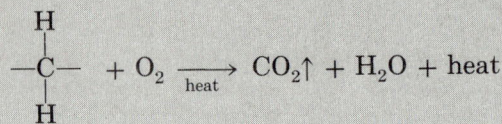

An alkane

[Sec. 2.14]

2.12 Halogenation

Halogenation is the typical **substitution reaction** of alkanes. It involves the replacement of hydrogen by halogen, usually chlorine or bromine, giving alkyl chlorides or alkyl bromides. Fluorine reacts explosively with alkanes and is thus an unsuitable reagent for the preparation of alkyl fluorides. Iodine is too unreactive and is not used in the halogenation of alkanes. Only chlorine and bromine react readily with alkanes under easily controllable conditions and are therefore used in the reaction.

Halogenation of alkanes takes place at high temperatures or under the influence of ultraviolet light. The reaction is illustrated by the general equation

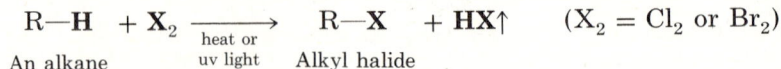

An alkane   Alkyl halide

Chlorination of an alkane usually gives a mixture of products because more than one hydrogen may be substituted by chlorine atoms. For example, the chlorination of methane yields a mixture of four different products.

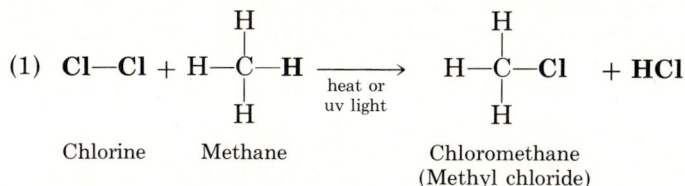

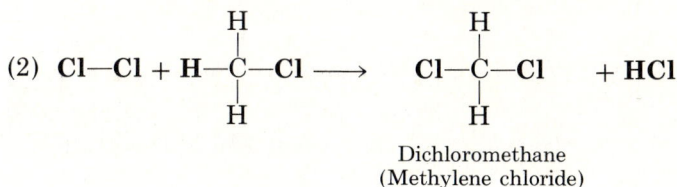

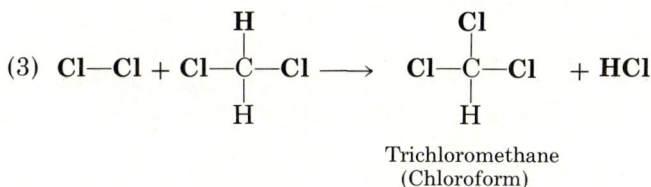

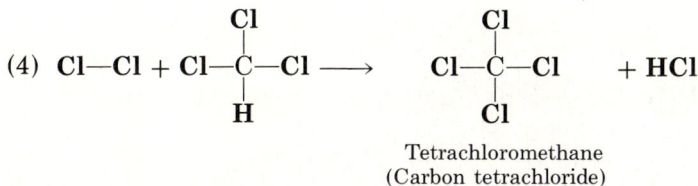

(1)   Cl—Cl + H—C—H $\xrightarrow[\text{uv light}]{\text{heat or}}$ H—C—Cl + HCl

   Chlorine   Methane   Chloromethane
                            (Methyl chloride)

(2)   Cl—Cl + H—C—Cl ⟶ Cl—C—Cl + HCl

                        Dichloromethane
                        (Methylene chloride)

(3)   Cl—Cl + Cl—C—Cl ⟶ Cl—C—Cl + HCl

                        Trichloromethane
                        (Chloroform)

(4)   Cl—Cl + Cl—C—Cl ⟶ Cl—C—Cl + HCl

                        Tetrachloromethane
                        (Carbon tetrachloride)

To form a *monohalogenated* product, where only one H atom is replaced by either Cl or Br, the reaction is conducted with a small amount of halogen and an excess of hydrocarbon. Under these conditions, the supply of halogen is exhausted as the first hydrogen of the hydrocarbon is replaced so there is practically no possibility for substitution of the remaining hydrogen atoms.

Both methane and ethane give only one monochlorinated product because in each compound all hydrogen atoms are equivalent. When propane is chlorinated, two monochlorinated products, 1-chloropropane and 2-chloropropane, are formed.

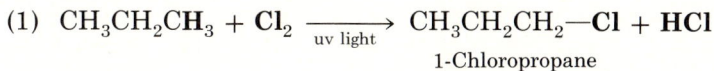

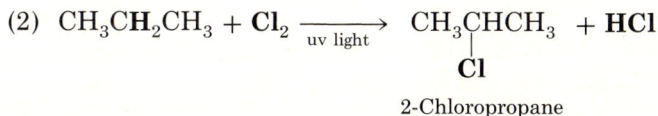

(1) $CH_3CH_2CH_3 + Cl_2$ $\xrightarrow{\text{uv light}}$ $CH_3CH_2CH_2$—Cl + HCl

                        1-Chloropropane

(2) $CH_3CH_2CH_3 + Cl_2$ $\xrightarrow{\text{uv light}}$ CH_3CHCH_3 + HCl
                                           |
                                           Cl

                        2-Chloropropane

Problem 2.10 Name and give the structures of all the products of the mono-chlorination of **(a)** butane and **(b)** pentane.

Problem 2.11 Name and give the structures of all the products of the mono-bromination of **(a)** isobutane and **(b)** 2,2-dimethylbutane.

2.13 Mechanism of Halogenation of Alkanes

A **reaction mechanism** is a detailed step-by-step model of exactly how a given reaction is believed to take place. Halogenation of alkanes proceeds by a **free-radical chain mechanism.** Recall (Sec. 2.10) that a free radical is any reactive species with an odd electron. The mechanism involves three distinct steps: (1) a chain-initiation step, (2) a chain-propagating step, and (3) a chain-terminating step. Using chlorine and methane as our reactants, we shall examine these steps and see how they contribute to the overall reaction.

1. Chain-initiation step. Homolytic cleavage of a chlorine molecule forms two highly reactive chlorine free radicals.

$$Cl\frown Cl \longrightarrow 2\,Cl\cdot$$

(Note that the weaker Cl—Cl bond, rather than the stronger H—CH$_3$ bond, is broken in the initial step. See Table 1.4.)

2. Chain-propagating step. The chlorine free radical attacks a hydrogen atom from methane, forming a methyl free radical and HCl.

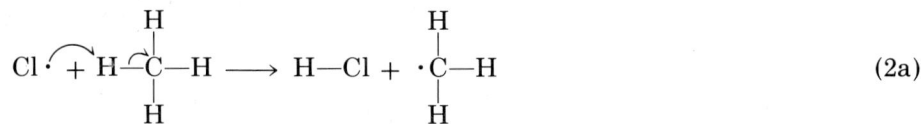

$$
Cl\cdot + H\!-\!\overset{\displaystyle H}{\underset{\displaystyle H}{\overset{|}{\underset{|}{C}}}}\!-\!H \longrightarrow H\!-\!Cl + \cdot\overset{\displaystyle H}{\underset{\displaystyle H}{\overset{|}{\underset{|}{C}}}}\!-\!H \qquad (2a)
$$

The reactive methyl free radical quickly reacts with another molecule of chlorine to give methyl chloride and another chlorine free radical.

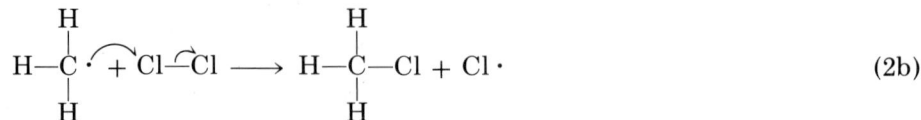

$$
H\!-\!\overset{\displaystyle H}{\underset{\displaystyle H}{\overset{|}{\underset{|}{C}}}}\!\cdot + Cl\frown Cl \longrightarrow H\!-\!\overset{\displaystyle H}{\underset{\displaystyle H}{\overset{|}{\underset{|}{C}}}}\!-\!Cl + Cl\cdot \qquad (2b)
$$

Back to (2a)

The chlorine free radical can go on to repeat reaction (2a). The reaction sequence (2a) to (2b) to (2a) to (2b) and so on is referred to as a chain reaction. It will continue indefinitely until one of the reactants is completely consumed or until two reactive intermediates react together.

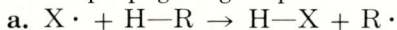

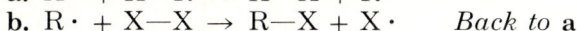

3. *Chain-terminating step.* The termination of the chain reaction sequence by reaction of two reactive intermediates may occur when two chlorine atoms react together to form a chlorine molecule, when two methyl free radicals react together to produce a new alkane, which is a dimer of the parent alkane, or when an alkyl free radical reacts with a chlorine atom to give an alkyl halide. The three reactions are

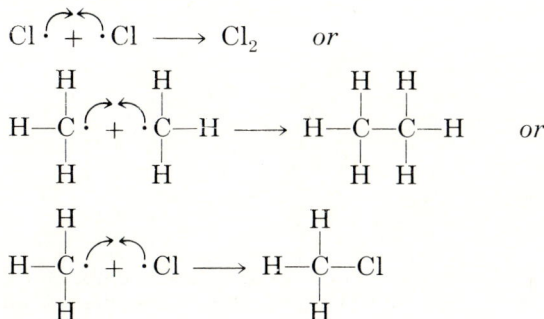

$$Cl \cdot + \cdot Cl \longrightarrow Cl_2 \quad or$$

$$H-\underset{\underset{H}{|}}{\overset{\overset{H}{|}}{C}} \cdot + \cdot \underset{\underset{H}{|}}{\overset{\overset{H}{|}}{C}}-H \longrightarrow H-\underset{\underset{H}{|}}{\overset{\overset{H}{|}}{C}}-\underset{\underset{H}{|}}{\overset{\overset{H}{|}}{C}}-H \quad or$$

$$H-\underset{\underset{H}{|}}{\overset{\overset{H}{|}}{C}} \cdot + \cdot Cl \longrightarrow H-\underset{\underset{H}{|}}{\overset{\overset{H}{|}}{C}}-Cl$$

The net effect of each of the possible occurrences is to remove from circulation one of the reactive species.

The overall general mechanism of free-radical halogenation is summarized here.

1. Chain-initiation step:

$$X_2 \xrightarrow[\text{uv light}]{} 2 X \cdot$$

2. Chain-propagating step:
   a. $X \cdot + H-R \rightarrow H-X + R \cdot$
   b. $R \cdot + X-X \rightarrow R-X + X \cdot$ *Back to* **a**
3. Chain-terminating step:
   a. $X \cdot + \cdot X \rightarrow X_2$ *or*
   b. $R \cdot + \cdot R \rightarrow R-R$ *or*
   c. $R \cdot + \cdot X \rightarrow R-X$

Problem 2.12 Monochlorination of ethane follows the same course as the chlorination of methane. Write the mechanism for the chlorination of ethane.

Combustion of Alkanes   *2.14*

When ignited in the presence of excess oxygen, alkanes are oxidized to carbon dioxide and water. A large quantity of heat (the heat of combustion) is liberated in the process. In fact, the heat is the most important product of this reaction because it is the source of power used to warm our homes and run our machines.

General equation

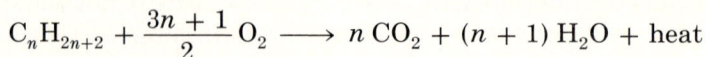

$$C_nH_{2n+2} + \frac{3n+1}{2}O_2 \longrightarrow n\,CO_2 + (n+1)\,H_2O + heat$$

Specific examples

Complete combustion of methane

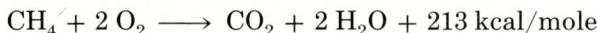

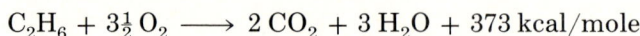

$$CH_4 + 2\,O_2 \longrightarrow CO_2 + 2\,H_2O + 213\;kcal/mole$$

Complete combustion of ethane

$$C_2H_6 + 3\tfrac{1}{2}\,O_2 \longrightarrow 2\,CO_2 + 3\,H_2O + 373\;kcal/mole$$

Complete combustion of propane

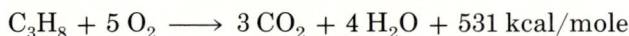

$$C_3H_8 + 5\,O_2 \longrightarrow 3\,CO_2 + 4\,H_2O + 531\;kcal/mole$$

Note that about 160 kcal of heat is liberated for each methylene group added to the hydrocarbon chain. The mechanism of this reaction is not well understood, but is believed to involve a complex series of free-radical reactions.

The incomplete combustion of alkanes liberates poisonous carbon monoxide (CO) or carbon in the form of soot; both are major contributors to air pollution. Incomplete combustion of fuels frequently occurs in automobile engines, where it results in carbon deposit in the pistons and in expulsion of toxic carbon monoxide in the exhaust fumes. This is why it is dangerous to drive an automobile with all the windows closed, or to warm up the car in a closed garage. The incomplete combustion of methane is illustrated in the following equations.

$$(1)\quad 2\,CH_4 + 3\,O_2 \longrightarrow 2\,CO + 4\,H_2O$$

$$(2)\quad CH_4 + O_2 \longrightarrow C\,(soot) + 2\,H_2O$$

Although soot is a nuisance when it forms in automobile engines, large quantities are used industrially, mainly in the manufacture of tires. This soot is produced by conducting the second reaction under controlled conditions.

Problem 2.13 Write the equation for the complete combustion of *n*-butane, and estimate the value of its heat of combustion.

2.15 Cycloalkanes: Nomenclature

Cycloalkanes are saturated hydrocarbons that exist in the form of a ring. Except for the three- and four-membered cyclic alkanes, which are very reactive, cyclic alkanes are as unreactive as their open-chain analogs. For this reason they are also called *cycloparaffins*. In the petroleum industry they are known as *naphthenes* because they are isolated from the naphtha fraction of petroleum.

Cycloalkanes are named by adding the prefix *cyclo-* to the name of the open-chain hydrocarbon that has the same number of carbon atoms as in the ring. For example, the three-carbon cycloalkane is called *cyclopropane,* and the four-carbon cycloalkane is called *cyclobutane.* Expanded and partially condensed structural formula representations of cyclic alkanes are given in Section 1.7.

When only one substituent is attached to the ring, we name the substituent first and then name the ring. For example,

CH$_3$ / CH$_2$CH$_3$

Methylcyclopropane   Ethylcyclobutane

If two or more substituents are attached to the ring, their positions are specified by numbers. The number 1 is assigned to one of the ring carbons bearing a substituent; then the rest of the ring is numbered in a way that will give the lowest number(s) for the position of the other substituent(s).

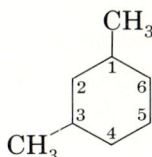

not

1,3-Dimethylcyclohexane
(correct)

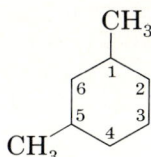

1,5-Dimethylcyclohexane
(incorrect)

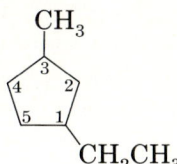

1-Ethyl-3-methylcyclopentane

Problem 2.14 Write the condensed structural formula for each of the incorrectly named compounds, and give the correct name for each.
(a) 1,3-Dichlorocyclopropane   **(b)** 1,4-Dimethylcyclobutane
(c) 3,5-Dibromocyclopentane   **(d)** 6,6-Dibromocyclohexane
(e) 1,5-Dimethylcyclohexane   **(f)** 2,2-Dichloro-5-methylcyclohexane

---

Geometric Isomerism in Cycloalkanes *2.16*

In open-chain alkanes very little energy is needed to twist around the carbon–carbon bond, and we say there is *free rotation* (Sec. 2.1). In contrast, in cycloalkanes the carbons are held together in a ring, and so much energy is needed for carbon–carbon bonds to rotate that the ring would break before complete rotation takes place. Therefore, we say that there is *no free rotation* in cycloalkanes. The lack of free rotation of single-bonded carbons in a ring produces a kind of isomerism called **geometric isomerism.** A cycloalkane with two substituents on different carbons of the ring can exist as a *cis* isomer or as a *trans* isomer. In the *cis* isomer the two substituents are on the same side of the ring. For example,

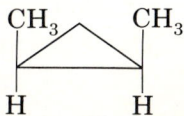

cis-1,2-Dimethylcyclopropane
(bp = 37°C)

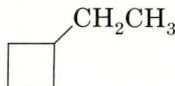

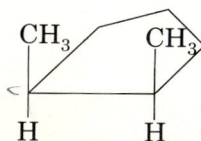

cis-1,2-Dimethylcyclopentane
(bp = 99°C; mp = −62°C)

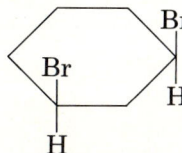

cis-1,3-Dibromocyclohexane
(mp = 112°C)

In the *trans* isomer the two substituents are on opposite sides of the ring.

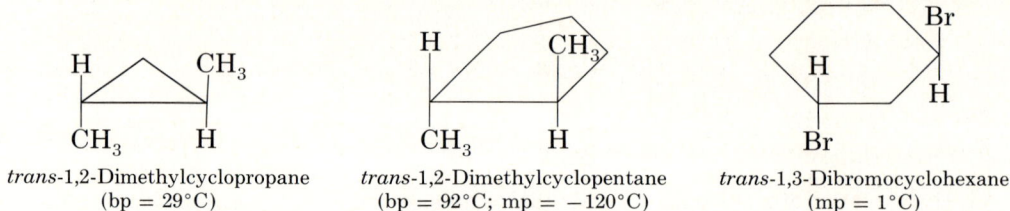

trans-1,2-Dimethylcyclopropane
(bp = 29°C)

trans-1,2-Dimethylcyclopentane
(bp = 92°C; mp = −120°C)

trans-1,3-Dibromocyclohexane
(mp = 1°C)

Note that *cis* and *trans* isomers have the same sequence of atoms bonded to each other and are therefore *not* structural isomers. *Cis-trans* isomers differ only in their **configurations,** that is, the orientation in space of the substituents around a carbon. For this reason geometric isomers are referred to as *stereoisomers.* (The subject of stereoisomerism is dealt with in greater detail in Chapter 6.) Nevertheless, *cis* and *trans* isomers represent distinct molecules with different physical properties.

Problem 2.15 Draw a condensed structural formula for each compound.
(a) *cis*-1-Chloro-2-methylcyclopentane
(b) *trans*-1,3-Difluorocyclohexane
(c) 1,1-Dimethylcyclohexane
(d) 1,1-Dibromo-3-methylcyclopentane (*Note:* There are no *cis* or *trans* in this case. Why?)

Problem 2.16 Which compounds can exist as geometric isomers?
(a) 1,2-Dibromocyclobutane   **(b)** 1,1-Dimethylcyclohexane
(c) 1,2-Dimethylcyclohexane   **(d)** 1,1-Dichloro-2-methylcyclohexane

Summary of Concepts and Reactions

Alkanes are saturated hydrocarbons consisting only of C—C and C—H single bonds.
They have the general formula C_nH_{2n+2}.   [Secs. 2.1, 2.2]
Alkanes form a homologous series.   [Sec. 2.2]
All carbon atoms in alkanes are sp^3 hybridized and have a tetrahedral shape.
                                                                            [Sec. 2.1]
There is free rotation about the C—C single bond in open-chain alkanes.   [Sec. 2.1]
Conformations are structures that differ by rotation about a C—C single bond. At room
temperature individual conformers cannot be separated.   [Sec. 2.1]
Different compounds with identical molecular formulas are called isomers, and the phe-
nomenon is called isomerism.   [Sec. 2.3]
Isomers that differ in the sequence of atoms bonded to each other are called structural
or constitutional isomers.   [Sec. 2.3]
There are four classes of carbons: primary (1°), secondary (2°), tertiary (3°), and quater-
nary (4°).   [Sec. 2.4]
An alkyl group (R) is an alkane from which a hydrogen has been removed.
                                                                            [Sec. 2.5]

Alkanes are known by two or more names, common names and IUPAC names.
[Sec. 2.6]

The two principal sources of alkanes are petroleum and natural gas. The components of petroleum are separated by a process called refining. [Sec. 2.7A]

Pyrolysis or cracking is an industrial method used to break large molecules into smaller and more useful molecules. [Sec. 2.7A]

The refined products of petroleum are called petrochemicals. [Sec. 2.7A]

Knocking is produced by a fuel with a low octane number rating. [Sec. 2.7B]

Methane is the major constituent of natural gas. [Sec. 2.7C]

The solubility of alkanes may be predicted by means of the "like dissolves like" rule. [Sec. 2.8A]

The physical properties of alkanes depend on chain length and degree of branching. [Sec. 2.8B]

A covalent bond can be broken during the course of an alkane reaction in either of two ways, homolytically or heterolytically. [Sec. 2.10]

Alkanes undergo two types of reactions, halogenation and combustion. [Sec. 2.11]

Halogenation proceeds via a free-radical chain mechanism: (a) $R—H + X \cdot \rightarrow R \cdot + HX$; (b) $R \cdot + X_2 \rightarrow RX + X \cdot$; back to (a). [Sec. 2.13]

Complete combustion of alkanes yields CO_2, H_2O, and heat. [Sec. 2.14]

Cycloalkanes are saturated hydrocarbons that exist in the form of a ring. [Sec. 2.15]

The lack of free rotation of singly bonded carbons in a ring gives rise to a kind of isomerism called geometric isomerism. [Sec. 2.16]

Geometric isomers in cycloalkanes can be formed by two substituents on the same side (cis) or opposite sides (trans) of the ring. [Sec. 2.16]

Key Terms

hydrocarbons	primary (1°) carbons	carbocation
alkanes	secondary (2°) carbons	carbanion
conformation	tertiary (3°) carbons	halogenation
conformer	quaternary (4°) carbon	combustion
homologous series	alkyl group, R	substitution reaction
homolog	IUPAC system of	reaction mechanism
isomers	nomenclature	free-radical chain
isomerism	common (trivial) names	mechanism
structural (constitutional)	octane number	geometric isomerism
isomers	free radical	configuration

Exercises

Structure and Nomenclature of Alkanes and Cycloalkanes [Secs. 2.5, 2.6, 2.15, 2.16]

2.1 Write structural formulas for the following compounds.

(a) 3-Methylheptane

(b) 2,3-Dimethylpentane

(c) 2,3-Dimethyl-4-ethylhexane

(d) 2,3,5-Trimethylhexane

(e) 2-Chloro-3-methylpentane

(f) 2-Bromo-2,3-dichlorobutane

(g) 1,1,2,2-Tetrabromopropane

(h) Methylcyclobutane

(i) trans-1,2-Dibromocyclopentane

(j) cis-1-Chloro-2-ethylcyclopropane

2.2 Give IUPAC names for the following compounds.

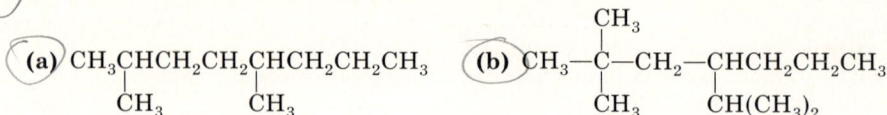

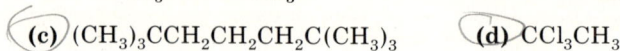

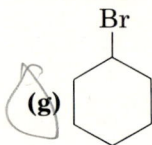

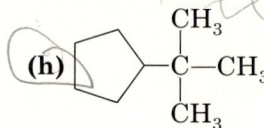

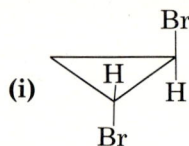

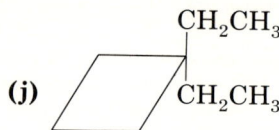

(a) $CH_3CHCH_2CH_2CHCH_2CH_2CH_3$
$\quad\quad\quad |\quad\quad\quad\quad\quad |$
$\quad\quad\quad CH_3\quad\quad\quad\quad CH_3$

(b)
$\quad\quad\quad\quad\quad\quad CH_3$
$\quad\quad\quad\quad\quad\quad |$
$CH_3-C-CH_2-CHCH_2CH_2CH_3$
$\quad\quad\quad\quad |\quad\quad\quad\quad |$
$\quad\quad\quad\quad CH_3\quad\quad CH(CH_3)_2$

(c) $(CH_3)_3CCH_2CH_2CH_2C(CH_3)_3$

(d) CCl_3CH_3

(e) $CHBr_2CHBr_2$

(f) CF_4

(g)

(h)

(i)

(j)

2.3 Write the structure for each of the incorrectly named compounds listed below. Explain why the given name is incorrect and give a correct name in each case.
(a) 3-Methylbutane
(b) 2-Ethylpropane
(c) 2,3-Dibromopropane
(d) *cis*-1,3-Dimethylcyclopropane
(e) 3,4-Dichloropentane
(f) 1,1,3-Trimethylbutane
(g) 2-Bromo-3-ethylbutane
(h) *trans*-1,6-Dimethylcyclohexane

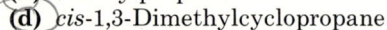

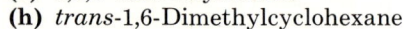

Structural Isomerism and Geometric Isomerism [Secs. 2.3, 2.16]

2.4 Draw the indicated number of structural isomers of compounds for each molecular formula.
(a) C_3H_7Cl (two isomers)
(b) C_2H_7N (two isomers)
(c) C_3H_8O (three isomers)
(d) C_3H_9N (four isomers)
(e) C_4H_9Cl (four isomers)
(f) C_3H_6BrCl (five isomers)

2.5 There are seven isomeric dibromocyclohexanes, including *cis-trans* isomers, having molecular formula $C_6H_{10}Br_2$. Draw the structures of the seven compounds, using condensed formulas for the rings.

2.6 For each group of structures, identify the structural isomers, the geometric isomers, and the structures that represent the same compound.

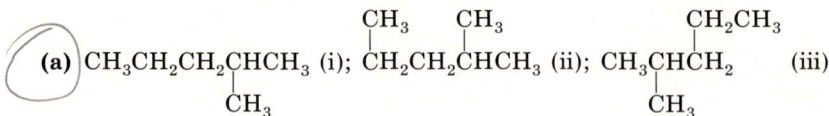

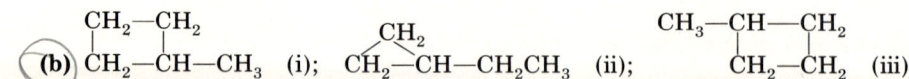

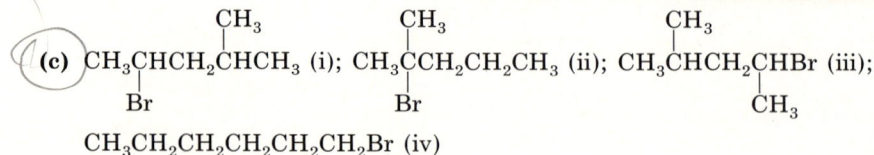

(a) $CH_3CH_2CH_2CHCH_3$ (i);
$\quad\quad\quad\quad\quad\quad\quad\quad |$
$\quad\quad\quad\quad\quad\quad\quad\quad CH_3$

CH_3
$|$
$CH_2CH_2CHCH_3$ (ii);

CH_2CH_3
$|$
CH_3CHCH_2 (iii)
$\quad\quad\quad\quad |$
$\quad\quad\quad\quad CH_3$

(b)
CH_2-CH_2
$|\quad\quad\quad\quad|$
$CH_2-CH-CH_3$ (i);

$\quad\quad CH_2$
$\quad\quad /\quad\backslash$
$CH_2-CH-CH_2CH_3$ (ii);

$CH_3-CH-CH_2$
$\quad\quad\quad\quad|\quad\quad|$
$\quad\quad\quad\quad CH_2-CH_2$ (iii)

(c)
$\quad\quad\quad\quad CH_3$
$\quad\quad\quad\quad |$
$CH_3CHCH_2CHCH_3$ (i);
$|$
Br

$\quad\quad CH_3$
$\quad\quad |$
$CH_3CCH_2CH_2CH_3$ (ii);
$\quad\quad |$
$\quad\quad Br$

$\quad\quad CH_3$
$\quad\quad |$
CH_3CHCH_2CHBr (iii);
$\quad\quad\quad\quad\quad\quad |$
$\quad\quad\quad\quad\quad\quad CH_3$

$CH_3CH_2CH_2CH_2CH_2CH_2Br$ (iv)

(d) (i); (ii); (iii); (iv); (v)

(e) (i); (ii); (iii); (iv); (v)

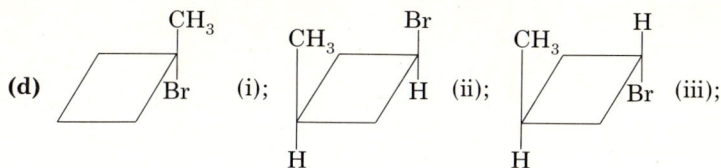

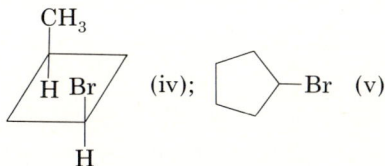

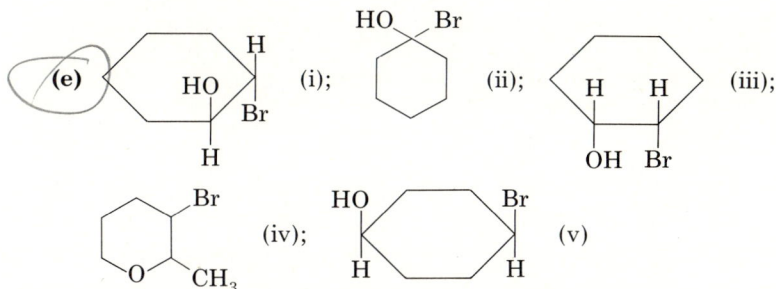

2.7 Four possible isomers can be obtained upon monochlorination of 2,2,4-trimethyl-pentane. Draw partially condensed formulas for each of these compounds and name each isomer according to the IUPAC system.

Classes of Carbons and Hydrogens [Sec. 2.4]
2.8 How many 1°, 2°, 3°, and 4° carbons and hydrogens (if any) are there in 2,2,4-trimethylpentane (the alkane with an octane rating of 100)?

Physical Properties of Alkanes [Sec. 2.8]
2.9 Without referring to tables, arrange each series of compounds in order of increasing boiling point.
   (a) *n*-Pentane; *n*-hexane, *n*-butane
   (b) *n*-Hexane; *n*-pentane, 2-methylpentane
   (c) *n*-Octane; 2,2,3-trimethylpentane, 2-methylheptane

Reactions of Alkanes and Cycloalkanes [Secs. 2.11–2.16]
2.10 Complete each of the following reactions by writing the structure of the product. If no reaction occurs, state so.

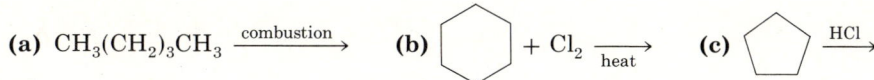

   (a) $CH_3(CH_2)_3CH_3 \xrightarrow{\text{combustion}}$ **(b)** ⬡ + $Cl_2 \xrightarrow[\text{heat}]{}$ **(c)** ⬠ $\xrightarrow{\text{HCl}}$

2.11 Draw structures for all possible monochlorinated and polychlorinated compounds that can be formed upon chlorination of ethane.
2.12 Bromination of ethane proceeds by the same mechanism as chlorination of ethane. Give the complete mechanism for the reaction, showing the initiation, propagation, and termination steps.
2.13 Isomeric compounds can sometimes be distinguished by observing the number of monochlorinated compounds that each one forms.
   (a) If a compound forms three monochloro compounds, is it *n*-pentane or 2-methylbutane?

(b) If a compound forms three monochloro compounds, is it *n*-hexane or 2,3-dimethylbutane?

2.14 Draw structures, including *cis-trans* isomers, for all possible monobrominated compounds that can be formed upon bromination of methylcyclopentane.

2.15 In the chlorination of ethane, C_2H_6, the product we want is C_2H_5Cl. Which one of the following experimental conditions is likely to give the best yield of monochlorinated product? Explain why.

(a) $C_2H_6 + Cl_2 \xrightarrow{\text{uv light}}$ **(b)** $C_2H_6 + Cl_2 \xrightarrow{\text{dark, room temp}}$

(c) $C_2H_6 + Cl_2(\text{excess}) \xrightarrow{\text{uv light}}$ **(d)** $C_2H_6(\text{excess}) + Cl_2 \xrightarrow{\text{uv light}}$

2.16 In the monochlorination of ethane, trace amounts of chlorinated butanes are also found, suggesting that at some point in the reaction *n*-butane was formed. Refer to the mechanism of free-radical halogenation of alkanes in Section 2.13 and explain how *n*-butane could be formed during the chlorination of ethane.

2.17 The heat of combustion of a straight-chain alkane is 850 ($\pm$5) kcal/mole. Which alkane has undergone complete combustion?

Unsaturated Hydrocarbons I: Alkenes

In Chapter 2 we discussed the chemistry of alkanes, the saturated hydrocarbons. Many hydrocarbons contain fewer hydrogens than do alkanes having the same number of carbon atoms. Because these compounds are deficient in hydrogen, we say that they are **unsaturated.** Unsaturated hydrocarbons *must* contain *multiple* bonds between carbon atoms. One family, distinguished by the presence of a carbon–carbon *double bond,* is called the **alkenes** or **olefins** and has the general formula C_nH_{2n}. A second class of compounds, the **alkynes,** is characterized by the presence of a carbon–carbon *triple bond.* The general formula of alkynes is C_nH_{2n-2}. **Dienes** are unsaturated hydrocarbons that contain *two* carbon–carbon double bonds. **Polyenes** are unsaturated hydrocarbons that contain more than two carbon–carbon double bonds.

This chapter treats the alkenes. The alkynes, dienes, and polyenes are discussed in Chapter 4.

Nomenclature of Alkenes 3.1

The simplest members of the alkene series (C_2 and C_3) are usually called by their common names, which are derived from the corresponding alkanes by replacing the *-ane* ending by *-ylene*. Larger alkenes are usually called by their IUPAC names. Except for a few additions, the IUPAC rules for naming alkenes are similar to those used for naming alkanes. Briefly, the rules are

1. The longest continuous carbon chain *containing the double bond* is selected as the parent chain.
2. The name of the parent carbon chain is obtained by replacing the *-ane* ending of the corresponding alkane by *-ene*.
3. The parent carbon chain is numbered in a manner that will give the doubly bonded carbon atoms the *lowest* numbers even if it results in the substituents getting higher numbers.
4. The position of the double bond is indicated by the number of the *lower*-numbered doubly bonded carbon.
5. In cycloalkenes, the double bond is always found between carbon 1 and carbon 2. It is therefore not necessary to specify the position of the double bond with a number. If substituents are present, the ring must be numbered, starting from the double bond, in the direction that gives the substituents the *lowest* number(s).

Thus,

$$CH_2\!=\!CH_2 \qquad\qquad CH_3CH\!=\!CH_2$$

Common name:	Ethylene	Propylene
IUPAC name:	Ethene	Propene

$$CH_3CH_2CH\!=\!CH_2 \qquad CH_3CH\!=\!CHCH_3 \qquad \overset{\displaystyle CH_3}{\underset{\textstyle |}{CH_3CHCH_2CH\!=\!CH_2}}$$

1-Butene   2-Butene   4-Methyl-1-pentene
(*not* 3-Butene)   (*not* 2-Methyl-4-pentene)

3-Methylcyclohexene   3-Chloro-4-ethylcyclobutene
(*not* 1-Methyl-2-cyclohexene)   (*not* 1-Chloro-2-ethyl-3-cyclobutene)

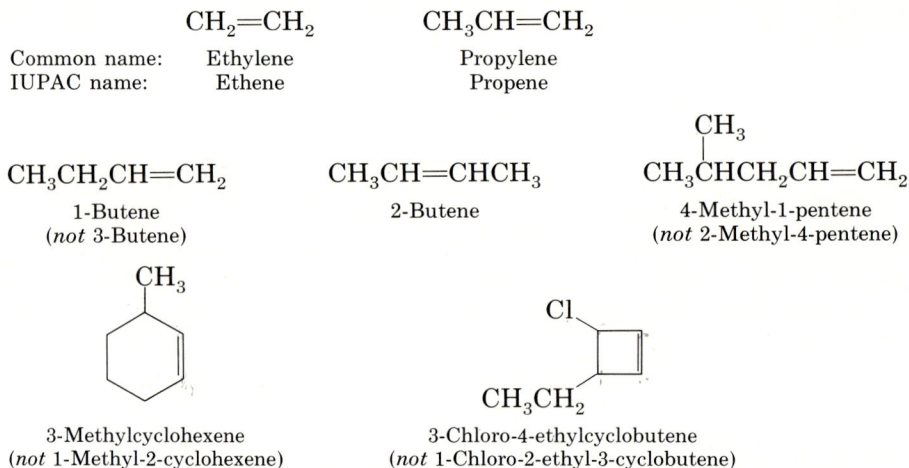

Compounds derived from ethylene and propylene are occasionally given special names. The $CH_2\!=\!CH-$ group is called the **vinyl group,** and $CH_2\!=\!CHCH_2-$ is called the **allyl group.**

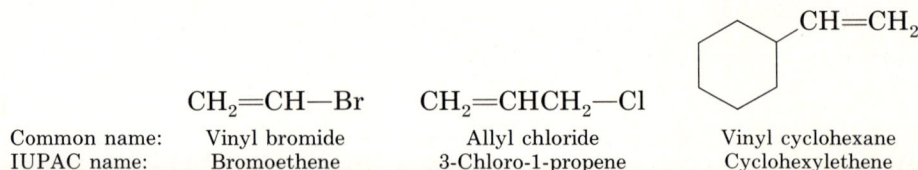

$$CH_2\!=\!CH-Br \qquad CH_2\!=\!CHCH_2-Cl$$

Common name:	Vinyl bromide	Allyl chloride	Vinyl cyclohexane
IUPAC name:	Bromoethene	3-Chloro-1-propene	Cyclohexylethene

To reinforce the rules of nomenclature let us work out a simple example.

Example 3.1 Write the structural formula of 3,5-dimethyl-4-isopropyl-2-octene.

Solution (1) The parent carbon chain is an octene, and the double bond is located between the second and third carbons. Write the parent carbon chain without bothering yet with the hydrogen atoms.

$$C-C=C-C-C-C-C-C$$
$$1 \quad 2 \quad 3 \quad 4 \quad 5 \quad 6 \quad 7 \quad 8$$

(2) Two methyl groups are attached on the parent carbon chain, one on carbon 3 and the other on carbon 5.

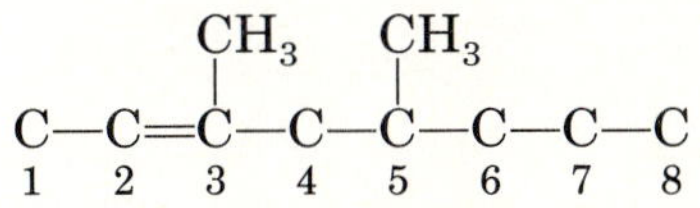

$$\begin{array}{ccccccc} & & CH_3 & & CH_3 & & & \\ & & | & & | & & & \\ C-C=&C&-C&-C&-C&-C&-C \\ 1 & 2 & 3 & 4 & 5 & 6 & 7 & 8 \end{array}$$

(3) An isopropyl group is attached on carbon 4.

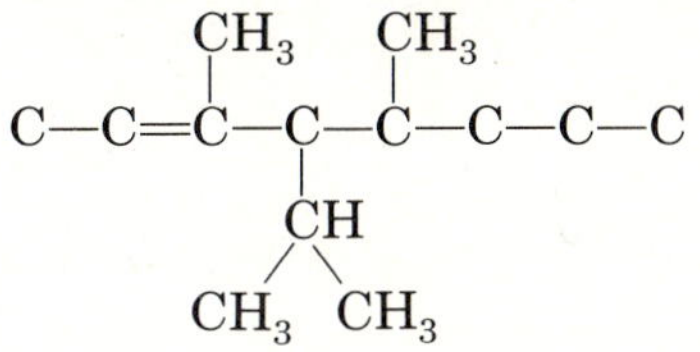

$$\begin{array}{ccccc} & & CH_3 & & CH_3 \\ & & | & & | \\ C-C=&C&-C&-C&-C-C-C \\ & & & | & \\ & & & CH & \\ & & CH_3 & & CH_3 \end{array}$$

(4) Put in the missing hydrogens to get the correct structure.

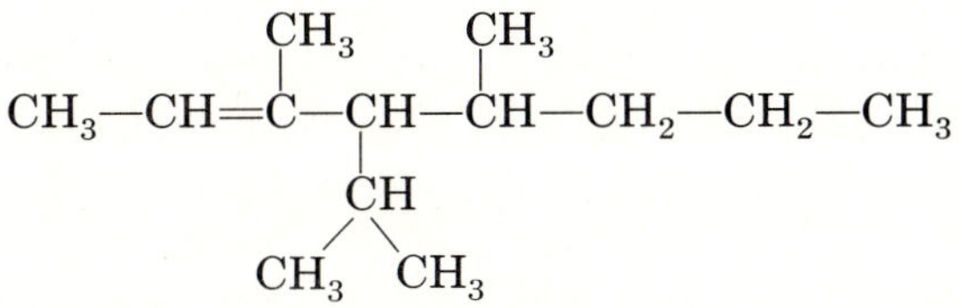

$$\begin{array}{cccc} & CH_3 & & CH_3 \\ & | & & | \\ CH_3-CH=&C&-CH&-CH-CH_2-CH_2-CH_3 \\ & | & & \\ & CH & & \\ & CH_3 & CH_3 \end{array}$$

Problem 3.1 Write the structural formulas for these compounds.
(a) 2-Methyl-2-butene
(b) 4,5-Dimethyl-3-isopropyl-2-hexene
(c) 2-Chloro-4-methyl-2-pentene
(d) 3-Bromo-2-chloro-3-methyl-1-pentene
(e) 1,3-Dimethylcyclohexene
(f) 1-Bromo-3-chlorocyclobutene

Problem 3.2 Name the following compounds.

$$\begin{array}{cccc} & CH_3 & & CH_3 \\ & | & & | \\ \textbf{(a)}\ CH_3CH_2-&C&-CH=CH-&C&-CH_2CH_3 \\ & | & & | \\ & CH_3 & & CH_2 \\ & & & | \\ & & & CH_2 \\ & & & | \\ & & & CH_3 \end{array}$$

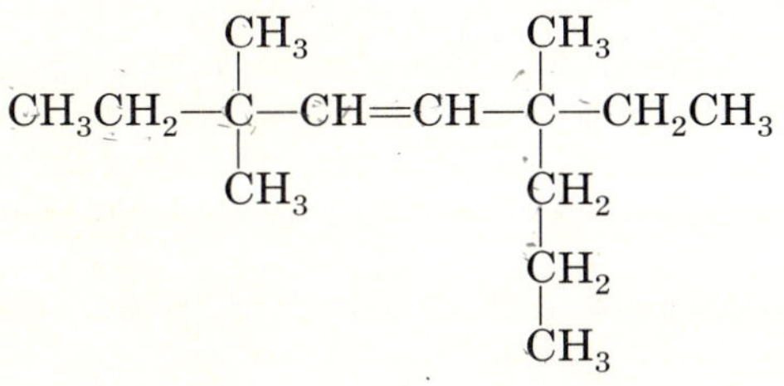

(b)
$$
\begin{array}{c}
\overset{\displaystyle CH_3}{\underset{\displaystyle |}{}}\overset{\displaystyle CH_3}{\underset{\displaystyle |}{}}\\
CH \\
CH_3CH_2-\overset{|}{C}=CHCH_2-\overset{|}{\underset{|}{C}}-CH_3 \\
CH_3-\overset{|}{\underset{|}{C}}-CH_3 \\
CH_3
\end{array}
$$

(c)
$$
\begin{array}{c}
\overset{\displaystyle CH_3}{\underset{\displaystyle |}{}} \\
\overset{\displaystyle CH_3}{}\quad \overset{\displaystyle CH_2}{\underset{\displaystyle |}{}} \\
CH_3C=C-CH_2CH_2CH-CH_3 \\
\underset{\displaystyle Br}{|}
\end{array}
$$

(d)
$$
\begin{array}{c}
\overset{\displaystyle F}{\underset{\displaystyle |}{}} \\
CH_3CH_2CH=CHCH-\overset{|}{\underset{|}{C}}-CH_3 \\
\underset{\displaystyle Br}{|}\ \ \underset{\displaystyle F}{|}
\end{array}
$$

(e)

(f)

(g)

—CH_2CH=CH_2

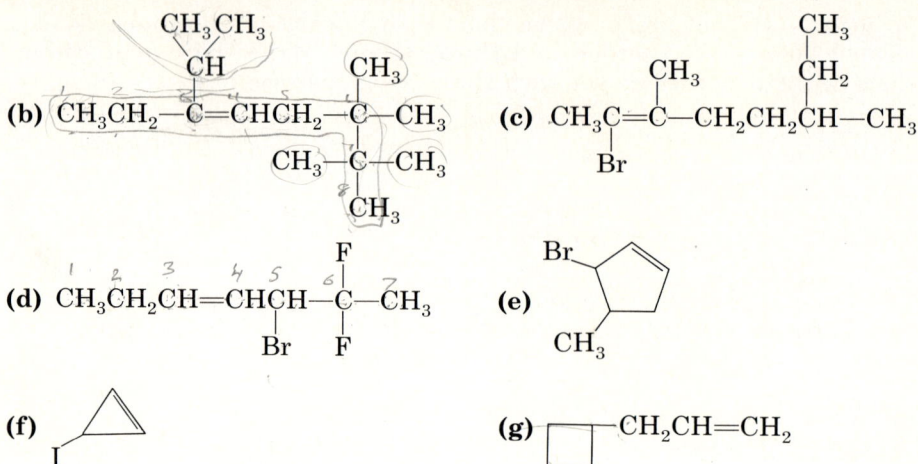

Problem 3.3 The names in this list are incorrect. Give the correct name and structure for each compound listed.
(a) 2,2-Dimethyl-4-pentene
(b) 4,5-Dimethyl-2-isopropyl-2-hexene
(c) 4-*n*-Butyl-5-hexene
(d) 2-Chloro-6-methylcyclohexene
(e) 1,2-Dimethyl-3-cyclopentene

Now that we know how to recognize and name alkenes, let us look into how a carbon–carbon double bond is formed, and also into its geometry.

3.2 Geometry of the Carbon–Carbon Double Bond: sp^2 Hybridization

In alkanes carbon is always bonded to four atoms. As a consequence, the singly bonded carbon uses sp^3-hybridized orbitals that are directed toward the corners of a regular tetrahedron. In alkenes, on the other hand, the doubly bonded carbon is always attached to only three other atoms. As a consequence, the doubly bonded carbon *must* use a different kind of hybridization and *must* assume a different shape. Let us look at the type of hybrid orbitals in and the shape of the simplest alkene, ethylene. In ethylene the carbons are sp^2 **hybridized** ($\frac{1}{3}s$ and $\frac{2}{3}p$ characteristics). These orbitals are formed in the following manner: As with sp^3 hybridization (Sec. 1.10), the ground-state carbon proceeds to its excited state, but this time the $2s$ orbital and only two of the three $2p$ orbitals hybridize. The result is three equivalent sp^2 hybrid orbitals and one unhybridized $2p_z$ orbital (Fig. 3.1). The three sp^2 orbitals get as far away from each other as possible by assuming a *planar* arrangement with an angle of 120° between the hybrid orbitals. This type of arrangement is also known as trigonal planar geometry. The remaining unhybridized $2p_z$ orbital is perpendicular to the plane of the sp^2 orbitals (Fig. 3.2).

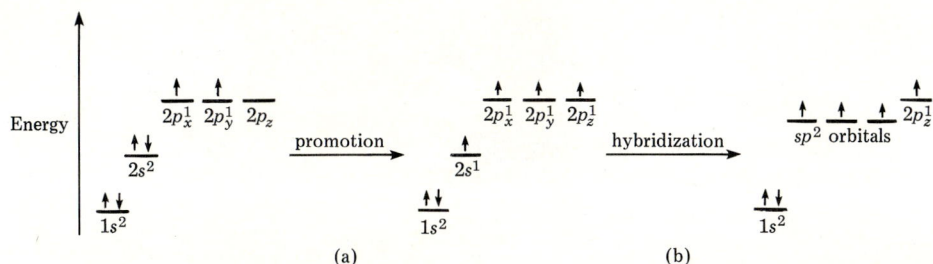

Figure 3.1 Formation of three sp^2 hybrid orbitals by **(a)** promotion of ground state carbon to excited state carbon followed by **(b)** hybridization of the $2s$ orbital with the $2p_x$ and $2p_y$ orbitals. The $2p_z$ orbital remains unhybridized.

Each trigonal carbon of ethylene overlaps two of its sp^2 orbitals with the s orbitals of two hydrogen atoms forming sp^2–s σ (sigma) bonds (shown by heavy lines and dotted lines in Fig. 3.3). The two carbons are connected by the end-on overlap of the remaining third sp^2 orbitals forming an sp^2–sp^2 σ bond. Now each carbon still has a $2p_z$ orbital containing one electron. These $2p_z$ orbitals, located above and below the plane of the six atoms, are capable of a new kind of overlap

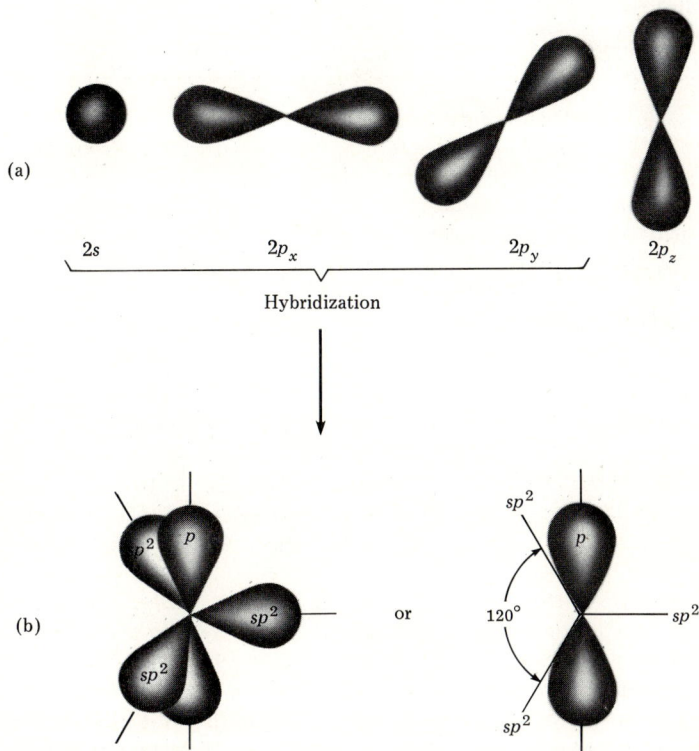

Figure 3.2 **(a)** Hybridization of $2s$ and $2p_x$ and $2p_y$ orbitals to form **(b)** three planar sp^2-hybridized orbitals with bond angles of 120° and a $2p_z$ orbital perpendicular to the plane.

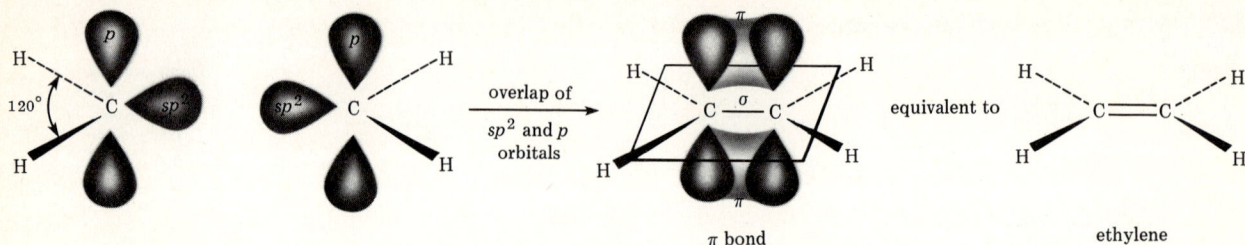

Figure 3.3 End-on overlap of two sp^2 orbitals to form a σ bond between the two carbons of ethylene and side-side overlap of the $2p_z$ orbitals to form a π bond.

called *side-side* overlap. The bond formed in this sidewise overlap is called a **π (pi) bond.** Thus the two carbons of ethylene are attached together by an sp^2–sp^2 σ bond and a π bond (from $2p_z$–$2p_z$ overlap) resulting in a carbon–carbon double bond (C=C). The π bond, a weaker bond than the σ bond (65 kcal/mole vs 83 kcal/mole), is the one that breaks in the course of a chemical reaction. The formation of these bonds is illustrated in Figure 3.3.

The carbon–carbon double bond, shown in Figure 3.3, accounts for several properties of alkenes. First, the carbon–carbon double bond is shorter by about 0.2 Å than the carbon–carbon single bond because two pairs of electrons pull the two nuclei closer together than does only one pair.

Second, the high electron density due to the pair of electrons in the π bond explains the reactivity of alkenes (Sec. 3.12).

Third, rotation about a carbon–carbon double bond is restricted because rotation of one carbon with respect to the other requires that the π bond be broken (see Fig. 3.4). Ordinarily this does not occur unless sufficient energy (about 65 kcal/mole) is supplied to the molecule in the form of heat or ultraviolet radiation.

Fourth, restricted rotation about the carbon–carbon double bond requires

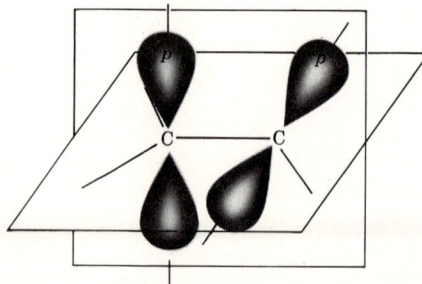

Figure 3.4 Rotation of one carbon with respect to the other diminishes the overlap of their $2p_z$ orbitals to the extent that if rotation is allowed to proceed to a full 90° the π bond would break.

that the two carbon atoms and the four other atoms attached to them lie in the same plane.

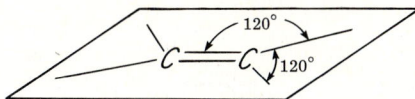

Finally, the restricted rotation about the carbon–carbon double bond and the planar geometry that results from it give rise to a type of isomerism, geometric isomerism, which is discussed in the next section.

Geometric Isomerism in Alkenes *3.3*

Alkenes of the type WXC=CYZ, where W differs from X and Y from Z, can exist as **geometric isomers.** This type of isomerism was encountered earlier in connection with cycloalkanes (Sec. 2.16). In cycloalkanes geometric isomerism is due to restricted rotation about the carbon–carbon single bond in a ring. In alkenes geometric isomerism is due to hindered rotation about the carbon–carbon double bond.

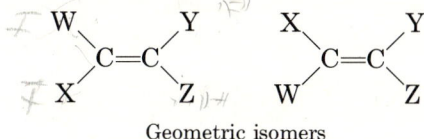

Geometric isomers

When two similar groups are on the same side of the double bond, the compound is called the *cis* isomer; when they are on the opposite sides of the double bond, the compound is called the *trans* isomer. For example,

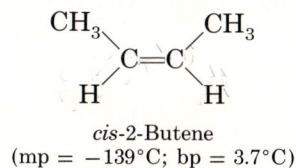

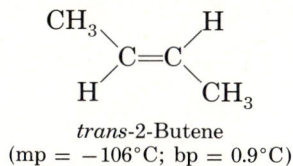

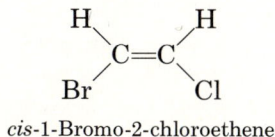

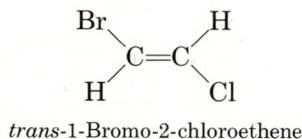

cis-2-Butene
(mp = −139°C; bp = 3.7°C)

trans-2-Butene
(mp = −106°C; bp = 0.9°C)

cis-1-Bromo-2-chloroethene

trans-1-Bromo-2-chloroethene

Geometric isomers differ from one another only in the way in which the substituents are arranged in space relative to the plane of the C=C bond. Since they can be interconverted only by the breaking and making of bonds, *cis* and *trans* isomers are stable molecules capable of independent existence. They have *different* physical properties and can be separated by fractional crystallization or distillation. Geometric isomers have *similar* chemical properties, since they are members of the same class of organic compounds. If two identical groups are attached to any one of the doubly bonded carbons (W = X or Y = Z), geometric isomerism is not possible. For example, the following compounds have no geometric isomers.

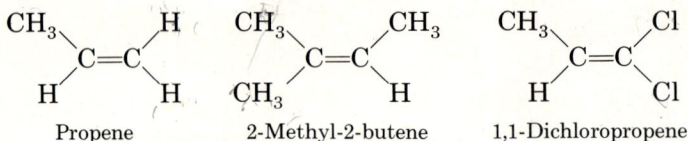

Propene   2-Methyl-2-butene   1,1-Dichloropropene

Problem 3.4 Which of the following compounds can exist as *cis-trans* isomers? Draw the structures of the geometric isomers.

(a) $CH_2=CHCH_2CH_2CH_3$ **(b)** $CH_3CH_2CH=CCH_2CH_3$
$$\overset{|}{CH_3}$$

(c) $CH_3CH_2CH=CHCH_2CH_3$ **(d)** $CH_3CH_2CH=CBr_2$

Problem 3.5 Draw the structures of the following compounds.
(a) *trans*-4-Octene   **(b)** *cis*-1,2-Dichloropropene
(c) *trans*-3-Methyl-2-pentene   **(d)** *cis*-3,4-Dibromocyclopentene

For alkenes with four different substituents such as

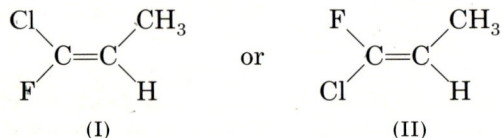

(I)   or   (II)

the terms *cis* and *trans* are ambiguous for distinguishing the two geometric isomers. As a result, another system, the **E,Z system,** has been devised to describe such compounds unambiguously.

Basically, the *E,Z* system works as follows. Arrange the groups on *each* carbon of the C=C bond in order of priority. The priority depends on atomic number: the higher the atomic number of the atom directly attached to the double-bonded carbon, the higher the priority.

Thus, in structure (I), the priority at one end of the C=C bond is Cl > F, and at the other end the priority is CH_3 > H. If the two groups of higher priority are on the same side of the C=C plane, the isomer is labeled *Z* (from the German *zusammen,* together). If the two groups of higher priority are on opposite sides of the C=C plane, the isomer is labeled *E* (from the German *entgegen,* opposite). Thus, for I and II we have

Priority: Cl > F, CH_3 > H

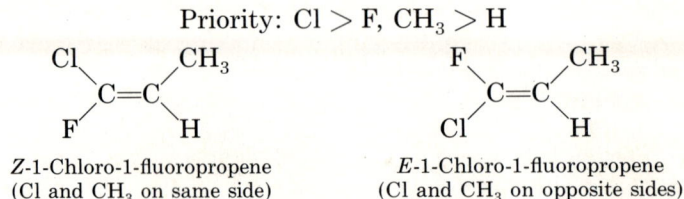

Z-1-Chloro-1-fluoropropene   E-1-Chloro-1-fluoropropene
(Cl and CH_3 on same side)   (Cl and CH_3 on opposite sides)

Further examples are

Priority: Br > H, I > CH$_3$

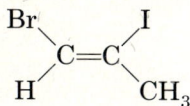

Br   I
  \   /
   C=C
  /   \
 H   CH$_3$

Z-1-Bromo-2-iodopropene

Priority: Br > H, Cl > F

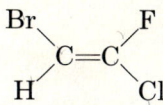

Br   F
  \   /
   C=C
  /   \
 H   Cl

E-1-Bromo-2-chloro-2-fluoroethene

Problem 3.6 Name the following compounds using the *E,Z* nomenclature.

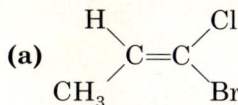

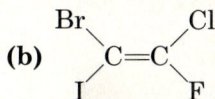

(a)
H   Cl
  \   /
   C=C
  /   \
CH$_3$ Br

(b)
Br   Cl
  \   /
   C=C
  /   \
 I   F

The *E,Z* system is less ambiguous and more versatile than the *cis-trans* system and is being used more and more in the literature. Nevertheless, *cis* and *trans* have been in common use for so many years that they are unlikely to disappear.

Physical Properties of Alkenes *3.4*

In general, the physical properties of alkenes are much the same as those of corresponding alkanes. At room temperature the C$_2$ to C$_4$ alkenes are gases; the C$_5$ to C$_{18}$ alkenes are liquids, and those above C$_{18}$ are solids.

Like alkanes, alkenes are insoluble in water and soluble in nonpolar organic solvents such as benzene or in carbon tetrachloride.

Preparation of Alkenes *3.5*

Alkenes are prepared in the laboratory by one of two general methods, each of which involves the **elimination** of an atom or group of atoms from adjacent carbons and the subsequent formation of a carbon–carbon double bond.

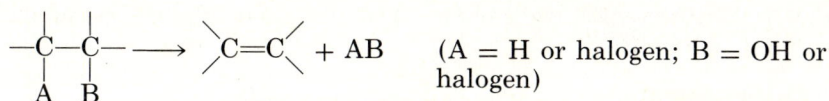

$$-\overset{|}{\underset{A}{C}}-\overset{|}{\underset{B}{C}}- \longrightarrow \ \ \overset{}{>}C=C\overset{}{<} \ + \ AB \qquad \text{(A = H or halogen; B = OH or halogen)}$$

The two preparative methods are briefly illustrated here and discussed in greater detail in the sections that follow.

1. Dehydration of alcohols (ROH)

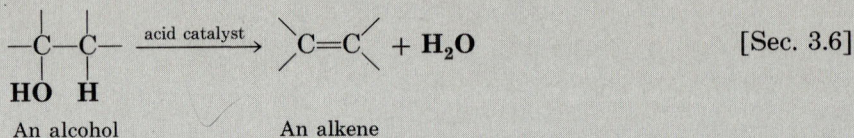

$$-\overset{\displaystyle|}{\underset{\displaystyle HO}{C}}-\overset{\displaystyle|}{\underset{\displaystyle H}{C}}-\ \xrightarrow{\text{acid catalyst}}\ \overset{\diagdown}{\diagup}C\!\!=\!\!C\overset{\diagup}{\diagdown}\ +\ \mathbf{H_2O}$$

An alcohol   An alkene   [Sec. 3.6]

2. Dehydrohalogenation of alkyl halides (RX)

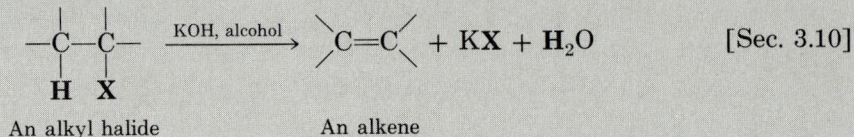

$$-\overset{\displaystyle|}{\underset{\displaystyle H}{C}}-\overset{\displaystyle|}{\underset{\displaystyle X}{C}}-\ \xrightarrow{\text{KOH, alcohol}}\ \overset{\diagdown}{\diagup}C\!\!=\!\!C\overset{\diagup}{\diagdown}\ +\ \mathbf{KX}\ +\ \mathbf{H_2O}$$

An alkyl halide   An alkene   [Sec. 3.10]

3.6 Dehydration of Alcohols

When an alcohol is heated in the presence of a mineral acid catalyst, it readily loses a molecule of water to give an alkene. The acid catalysts most commonly used are sulfuric acid, H_2SO_4, and phosphoric acid, H_3PO_4. In these equations H^+ represents the acid catalyst and Δ represents the heat applied.

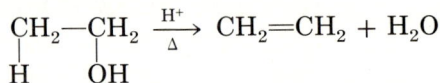

$$\underset{\underset{\displaystyle H}{|}}{CH_2}\!\!-\!\!\underset{\underset{\displaystyle OH}{|}}{CH_2}\ \xrightarrow[\Delta]{H^+}\ CH_2\!\!=\!\!CH_2\ +\ H_2O$$

IUPAC name: Ethanol   Ethene
Common name: Ethyl alcohol   Ethylene

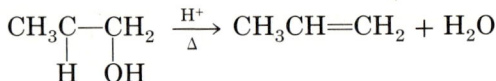

$$\underset{\underset{\displaystyle H}{|}}{CH_3C}\!\!-\!\!\underset{\underset{\displaystyle OH}{|}}{CH_2}\ \xrightarrow[\Delta]{H^+}\ CH_3CH\!\!=\!\!CH_2\ +\ H_2O$$

IUPAC name: 1-Propanol   Propene
Common name: *n*-Propyl alcohol   Propylene

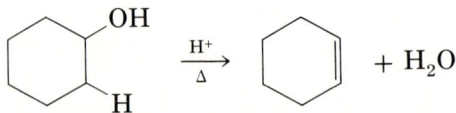

$$\xrightarrow[\Delta]{H^+}\ +\ H_2O$$

IUPAC name: Cyclohexanol   Cyclohexene
Common name: Cyclohexyl alcohol

Problem 3.7 Write the structure of the product of the dehydration of each of the following alcohols.

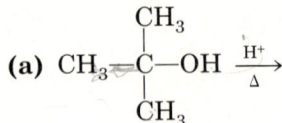

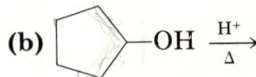

(a) $CH_3\!-\!\!\overset{\displaystyle CH_3}{\underset{\displaystyle CH_3}{\overset{|}{\underset{|}{C}}}}\!\!-\!\!OH\ \xrightarrow[\Delta]{H^+}$ **(b)** $-\!OH\ \xrightarrow[\Delta]{H^+}$

Which Alkene Predominates? Saytzeff's Rule 3.7

The examples of dehydration of alcohols discussed in the previous section produced a single alkene as the possible product. Suppose, however, that the loss of water from adjacent carbon atoms can give rise to more than one alkene, as in the dehydration of 2-butanol.

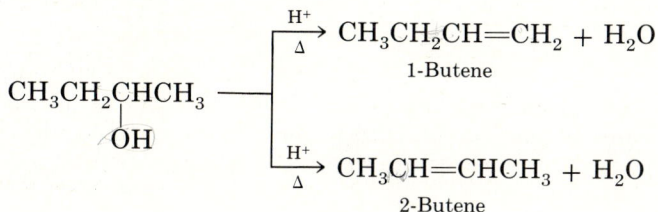

$$CH_3CH_2CHCH_3 \underset{OH}{|} \quad \begin{cases} \xrightarrow[\Delta]{H^+} CH_3CH_2CH{=}CH_2 + H_2O \\ \text{1-Butene} \\[2em] \xrightarrow[\Delta]{H^+} CH_3CH{=}CHCH_3 + H_2O \\ \text{2-Butene} \end{cases}$$

Which alkene predominates, 1-butene or 2-butene? A generalization known as **Saytzeff's rule** applies: *In every instance in which more than one alkene can be formed, the major product is always the alkene with the most alkyl substituents attached on the double-bonded carbons.*

Applying Saytzeff's rule to the dehydration of 2-butanol enables us to predict that 2-butene, with two alkyl substituents attached to $C{=}C$, is the major product, and 1-butene, with only one alkyl group, is the minor product. (Bear in mind that application of Saytzeff's rule requires you to count the number of alkyl groups bonded to $C{=}C$; the size of the alkyl group is immaterial.)

Problem 3.8 Give the major and minor products of the dehydration of

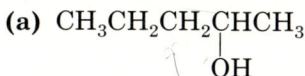

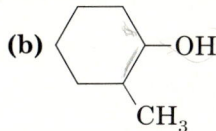

(a) $CH_3CH_2CH_2CHCH_3$
     $\qquad\qquad |$
     $\qquad\qquad OH$

(b) (cyclohexene ring with —OH and —CH$_3$ substituents)

Now that you have seen examples of the dehydration of some specific alcohols, let us look into the mechanism of the reaction.

Mechanism of Dehydration of Alcohols 3.8

The dehydration of alcohols involves a **carbocation,** a reaction intermediate mentioned earlier in Section 2.10. A carbocation is a highly reactive, unstable intermediate in which the carbon carries a positive charge. Once formed, carbocations do not persist for any length of time. Their lack of stability causes them to quickly react further to form uncharged products. When the reaction is the dehydration of an alcohol, the uncharged product is an alkene.

The mechanism of dehydration basically involves the following steps.*

Step 1. Protonation of the alcohol: The proton of the acid catalyst is transferred to the oxygen atom, which functions as a Lewis base. This step is identical to the protonation of water to form hydronium ions.

$$-\overset{|}{\underset{H}{C}}-\overset{|}{\underset{:\ddot{O}H}{C}}-\ +\ H^{+} \rightleftharpoons -\overset{|}{\underset{H}{C}}-\overset{|}{\underset{\underset{\overset{|}{H}}{+:\ddot{O}H}}{C}}-$$

Step 2. Formation of a carbocation: The highly electronegative oxygen atom pulls electrons away from carbon, breaking the bond. A water molecule is released, and a carbocation with its positively charged carbon is generated.

$$-\overset{|}{\underset{H}{C}}-\overset{|}{\underset{\underset{\overset{|}{H}}{+:\ddot{O}H}}{C}}- \rightleftharpoons -\overset{|}{\underset{H}{C}}-\overset{|}{\underset{+}{C}}-\ +\ H_2O$$

Step 3. Loss of a proton from the carbocation regenerates the acid catalyst and simultaneously forms the alkene.

$$-\overset{|}{\underset{\overset{|}{H}}{C}}\overset{|}{\underset{+}{C}}- \rightleftharpoons \ \overset{\diagup}{\diagdown}C=C\overset{\diagup}{\diagdown}\ +\ H^{+}$$

$$\text{An alkene}$$

Problem 3.9 Write the mechanism for the dehydration of 1-propanol to propene.

$$CH_3CH_2CH_2OH \xrightarrow[\text{heat}]{H^{+}} CH_3CH=CH_2 + H_2O$$
$$\text{1-Propanol} \qquad\qquad \text{Propene}$$

3.9 Classes of Carbocations and Ease of Dehydration of Alcohols

In the previous section it was established that the dehydration of alcohols involves a carbocation intermediate. Carbocations are classified as primary (1°), secondary (2°), or tertiary (3°) according to the number of carbon atoms attached to the positively charged carbon.

* To be consistent with the convention adopted in Section 2.10, a curved double-headed arrow ⌢ is used to indicate the making or breaking of a bond involving a pair of electrons. In writing mechanisms that involve movements of a pair of electrons, the curved arrow *always* goes from the electron-rich to the electron-deficient atom.

$$R-\overset{+}{C}H_2 \qquad R-\overset{+}{\underset{\underset{R}{|}}{C}}-H \qquad R-\overset{+}{\underset{\underset{R}{|}}{C}}-R$$

1° carbocation $\qquad$ 2° carbocation $\qquad$ 3° carbocation

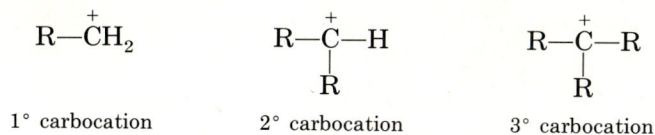

The ease of formation and the stabilities of carbocations follow the order

$$3° > 2° > 1°$$

← ————————————————————————————

Ease of formation and stabilities of carbocations

Thus, in a reaction that involves the formation of a carbocation intermediate, the more stable tertiary carbocation is the more easily formed. This means that when comparing the relative rates of dehydration among alcohols, we can expect a tertiary alcohol (one in which the C—OH group is directly attached to three carbons) to react faster than a secondary alcohol (one in which the C—OH is bonded to two carbons), and the latter faster than a primary alcohol (one in which the C—OH is attached to one carbon).

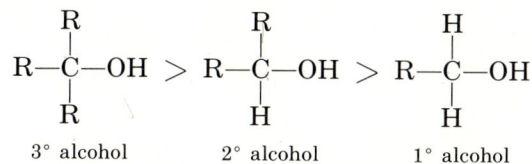

$$R-\overset{\overset{R}{|}}{\underset{\underset{R}{|}}{C}}-OH \; > \; R-\overset{\overset{R}{|}}{\underset{\underset{H}{|}}{C}}-OH \; > \; R-\overset{\overset{H}{|}}{\underset{\underset{H}{|}}{C}}-OH$$

3° alcohol $\qquad$ 2° alcohol $\qquad$ 1° alcohol

← ————————————————————————————

Ease of dehydration or rate of dehydration

The expectation is borne out by the fact that tertiary alcohols often undergo dehydration upon treatment with dilute acid at room temperature, whereas concentrated acid and elevated temperatures are usually required to convert primary alcohols into alkenes.

Problem 3.10 Arrange these carbocations in order of increasing stabilities.

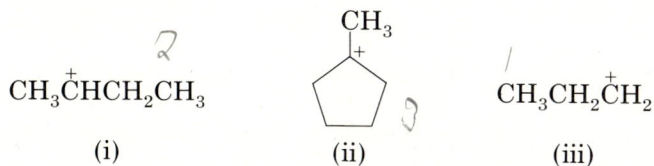

$$CH_3\overset{+}{C}HCH_2CH_3$$

(i)

$$CH_3CH_2\overset{+}{C}H_2$$

(ii) $\qquad$ (iii)

Problem 3.11 Arrange these alcohols in order of ease of dehydration.

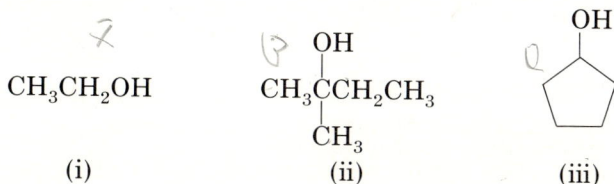

$$CH_3CH_2OH$$

(i)

$$CH_3\overset{\overset{OH}{|}}{\underset{\underset{CH_3}{|}}{C}}CH_2CH_3$$

(ii) $\qquad$ (iii)

Before discussing the second method of preparing alkenes, let us summarize what you have learned thus far in the first method of preparation, the dehydration of alcohols.

1. The dehydration of alcohols requires an acid catalyst.
2. The predominant alkene formed follows Saytzeff's rule, which states that the major product is the alkene with the most alkyl substituents attached on the double-bonded carbons.
3. The reaction proceeds via a carbocation intermediate.
4. The stabilities of carbocations and the ease of dehydration of alcohols follows the order $3° > 2° > 1°$.

3.10 Dehydrohalogenation of Alkyl Halides

Alkenes can also be prepared under alkaline conditions, in which case an alkyl halide, RX, is required as the starting material. Thus, heating an alkyl halide with a solution of potassium hydroxide, KOH, in alcohol, yields an alkene. This reaction is known as **dehydrohalogenation** because it involves the elimination of H and of X from adjacent carbon atoms.

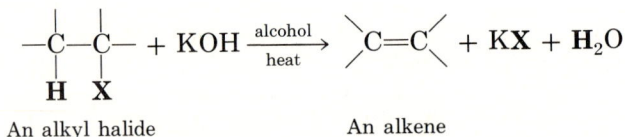

$$-\overset{|}{\underset{\underset{H}{|}}{C}}-\overset{|}{\underset{\underset{X}{|}}{C}}- + KOH \xrightarrow[heat]{alcohol} \underset{/}{\overset{\backslash}{}}C{=}C\overset{/}{\underset{\backslash}{}} + KX + H_2O$$

An alkyl halide   An alkene

As with the dehydration of an alcohol, the dehydrohalogenation of an alkyl halide may form more than one alkene. In such case Saytzeff's rule again applies; that is, the alkene with the most alkyl substituents on the double-bonded carbons predominates. For example,

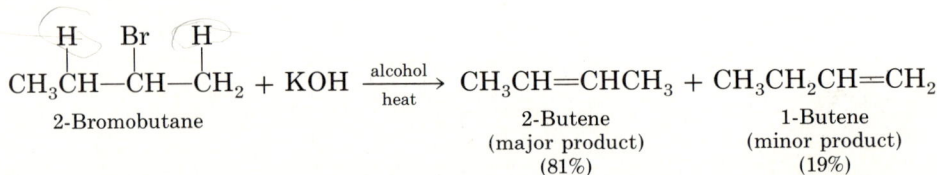

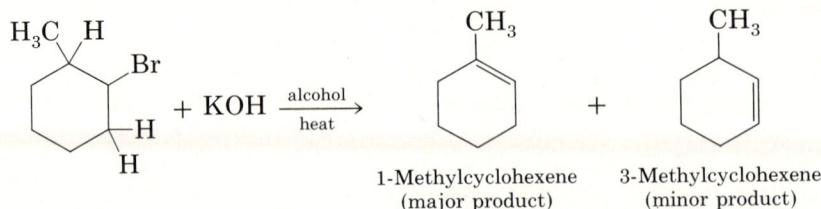

$$CH_3\overset{\overset{H}{|}}{CH}-\overset{\overset{Br}{|}}{CH}-\overset{\overset{H}{|}}{CH_2} + KOH \xrightarrow[heat]{alcohol} CH_3CH{=}CHCH_3 + CH_3CH_2CH{=}CH_2$$

2-Bromobutane   2-Butene (major product) (81%)   1-Butene (minor product) (19%)

1-Methylcyclohexene (major product)   3-Methylcyclohexene (minor product)

Problem 3.12 Give the structure of the major and minor products of the dehydrohalogenation of

(a) $CH_3-CH=CH-CH_3$ **(b)** ⬡$-CH_2=CH-CH_3$
$\quad\quad\quad\quad\ \ CH_3$ Cl

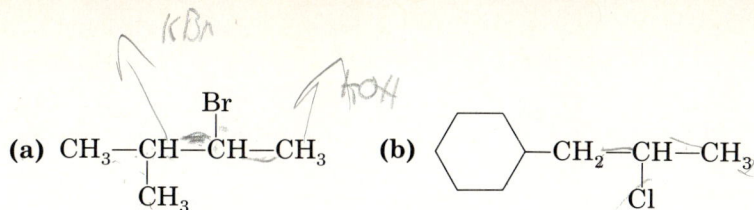

The mechanisms of dehydrohalogenation depend on the type of alkyl halide used (whether 1°, 2°, or 3° alkyl halide). They are discussed in Chapter 8.

Reactions of Alkenes   *3.11*

All alkenes contain a carbon–carbon double bond, and all except ethylene also contain a saturated alkyl chain as part of the molecule. The chemistry of alkenes can therefore be divided into two general types of reactions: (1) addition reactions that involve the carbon–carbon double bond and (2) substitution reactions that usually involve the saturated alkyl chain. Let us first examine the various kinds of addition reactions.

Additions to the Carbon-Carbon Double Bond   *3.12*

The typical reaction of alkenes is addition to the C=C group. An **addition reaction** is one in which a reagent, A—B, adds across the double bond to give a saturated product that contains all the atoms of both reactants.

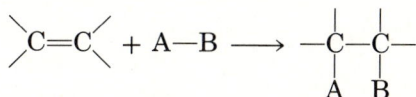

$$\text{\Large$>$}C=C\text{\Large$<$} \ + \ A-B \ \longrightarrow \ -\overset{|}{C}-\overset{|}{C}-$$
$$\quad\quad\quad\quad\quad\quad\quad\quad\quad\quad A \ \ B$$

The addition reactions we will encounter in this chapter are summarized as follows.

1. Addition of hydrogen: catalytic hydrogenation

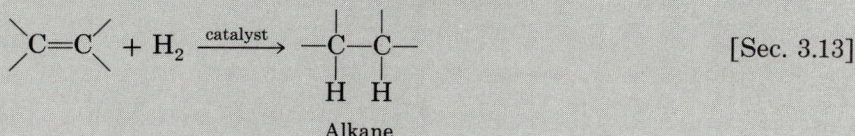

$$\text{\Large$>$}C=C\text{\Large$<$} \ + \ H_2 \ \xrightarrow{\text{catalyst}} \ -\overset{|}{\underset{H}{C}}-\overset{|}{\underset{H}{C}}- \quad\quad\quad \text{[Sec. 3.13]}$$

$$\text{Alkane}$$

2. Addition of halogens: halogenation

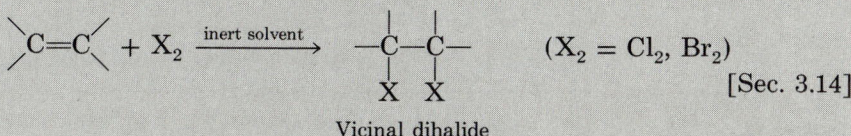

$$\text{\Large$>$}C=C\text{\Large$<$} \ + \ X_2 \ \xrightarrow{\text{inert solvent}} \ -\overset{|}{\underset{X}{C}}-\overset{|}{\underset{X}{C}}- \quad (X_2 = Cl_2, \ Br_2)$$
$$\quad\quad\quad\quad\quad\quad\quad\quad\quad\quad\quad\quad\quad\quad\quad\quad\quad\quad\quad\quad\quad\quad \text{[Sec. 3.14]}$$

$$\text{Vicinal dihalide}$$

3. Electrophilic addition to alkenes: addition of acids (H—A)

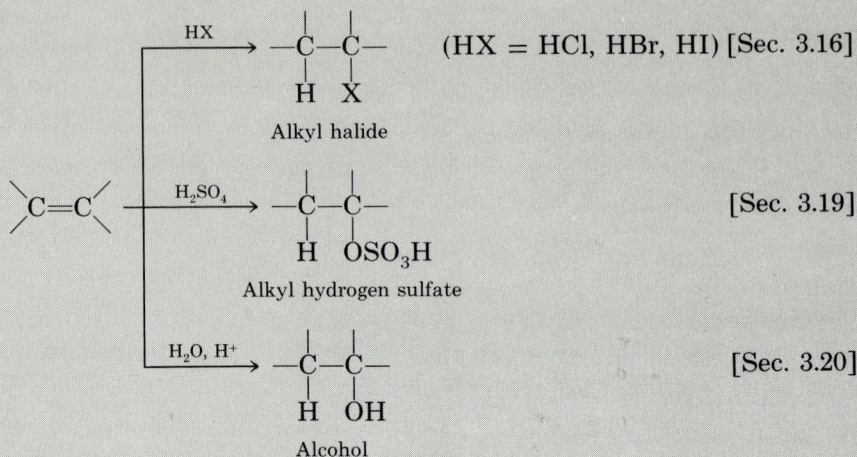

$(HX = HCl, HBr, HI)$ [Sec. 3.16]

Alkyl halide

[Sec. 3.19]

Alkyl hydrogen sulfate

[Sec. 3.20]

Alcohol

4. Addition of HOX: halohydrin formation

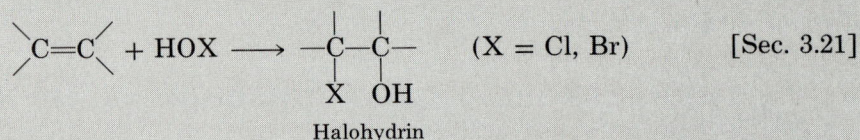

$(X = Cl, Br)$ [Sec. 3.21]

Halohydrin

5. Ozonolysis

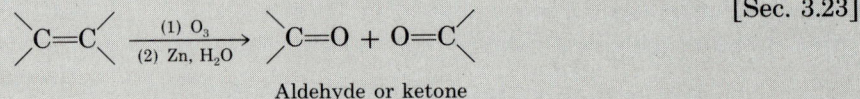

[Sec. 3.23]

Aldehyde or ketone

6. Polymerization (addition of >C=C< to another >C=C<)

[Sec. 3.25]

Polymer

3.13 Addition of Hydrogen: Catalytic Hydrogenation

The carbon–carbon double bond of alkenes can add a mole of hydrogen in the presence of suitable catalysts to give an alkane. This method, called the **catalytic hydrogenation** of alkenes, is carried out as follows. The alkene is dissolved in a suitable solvent in the presence of finely divided platinum (Pt), nickel (Ni), or palladium (Pd) catalyst. Hydrogen gas is then bubbled at low pressure into the reaction vessel with constant stirring. The yield of alkane by this method is quite good.

General equation

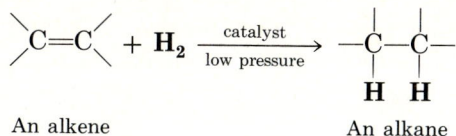

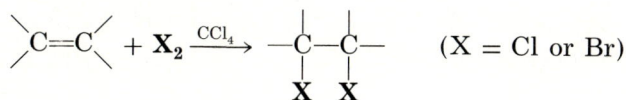

$$\text{C=C} + \mathbf{H_2} \xrightarrow[\text{low pressure}]{\text{catalyst}} \underset{\underset{\text{H} \quad \text{H}}{|}}{-}\text{C}-\text{C}-$$

An alkene   An alkane

Specific examples

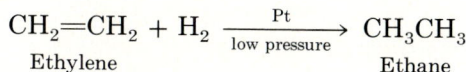

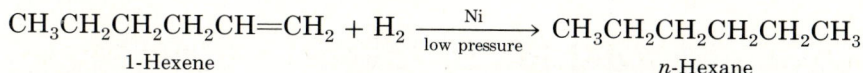

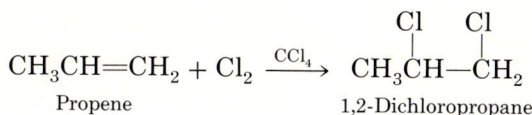

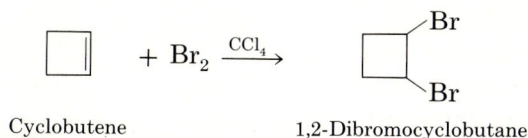

$$CH_2{=}CH_2 + H_2 \xrightarrow[\text{low pressure}]{\text{Pt}} CH_3CH_3$$
     Ethylene   Ethane

$$CH_3CH_2CH_2CH_2CH{=}CH_2 + H_2 \xrightarrow[\text{low pressure}]{\text{Ni}} CH_3CH_2CH_2CH_2CH_2CH_3$$
         1-Hexene   *n*-Hexane

Hydrogenation has important industrial applications in the manufacture of high-octane automobile and aviation fuels and in the preparation of synthetic detergents. The reaction can also be used to analyze the extent of unsaturation in a molecule.

Addition of Halogens: Halogenation   *3.14*

When an alkene is treated at room temperature with a solution of bromine or chlorine in carbon tetrachloride or some other inert solvent, the halogen adds rapidly to the double bond of the alkene to give the corresponding **vicinal dihalide** (two halogens attached to adjacent carbons).

General equation

$$\text{C=C} + \mathbf{X_2} \xrightarrow{\text{CCl}_4} \underset{\underset{\text{X} \quad \text{X}}{|}}{-}\text{C}-\text{C}- \qquad (X = Cl \text{ or } Br)$$

Specific examples

$$CH_3CH{=}CH_2 + Cl_2 \xrightarrow{\text{CCl}_4} \overset{\overset{\text{Cl} \quad \text{Cl}}{|}}{CH_3CH{-}CH_2}$$
     Propene   1,2-Dichloropropane

$$\square + Br_2 \xrightarrow{\text{CCl}_4} \square\!\!\!\begin{smallmatrix}\text{Br}\\ \\\text{Br}\end{smallmatrix}$$
    Cyclobutene   1,2-Dibromocyclobutane

Iodine is too unreactive and will not add to the double bond of an alkene. Fluorine is too reactive and reacts explosively with an alkene, making it an unsuitable reactant.

The addition of bromine to an alkene is useful as a practical test for detecting unsaturation (see Sec. 3.22).

The addition of halogens to an alkene proceeds by a mechanism similar to (but somewhat more complicated than) the addition of acids discussed in the next sections.

Problem 3.13 Write the structure of the product expected from the reaction at room temperature of **(a)** 2-butene with Br_2 in CCl_4 and **(b)** cyclohexene with Cl_2 in CCl_4.

3.15 Electrophilic Addition to Alkenes: Addition of Acids

The next few reactions of alkenes to be discussed involve the addition to the double bond of reagents whose general formulas are represented by the symbol H—A. These may be hydrogen halides (H—Cl, H—Br, H—I), sulfuric acid (H—OSO$_3$H), or water (H—OH). The addition of all these reagents to the double bond follows the same mechanism. For this reason, the general mechanism of their addition will be discussed before their specific reactions are given.

Hydrogen halides, sulfuric acid, and water all contain an ionizable hydrogen.

$$H—A \longrightarrow H^+ + A:^-$$

The positively charged hydrogen ion is an electron-deficient species. Any electron-deficient species is called an **electrophile,** and any electron-rich species is called a **nucleophile.** The addition of H—A to an alkene is believed to be a two-step process.

Step 1. The hydrogen ion (the electrophile) attacks the π electrons of the alkene, forming a C—H bond and a carbocation.

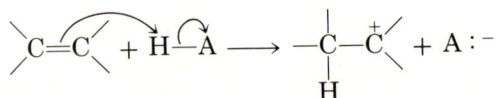

Step 2. The negatively charged species A:$^-$ (a nucleophile) attacks the carbocation and forms a new C—A bond.

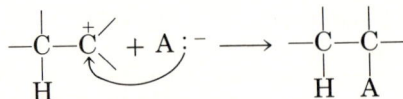

The formation of the carbocation in step 1 is the more difficult and consequently the slower step in the mechanism. The slowest step in any reaction mechanism is called the **rate-determining step.** The type of reaction just described, which involves the attack by an electrophilic reagent on the π electrons, falls in a general category called **electrophilic addition** reactions.

Now that you have seen the general mechanism of electrophilic addition, some specific examples will be discussed.

Addition of Hydrogen Halides *3.16*

Alkenes react with hydrogen chloride, HCl, hydrogen bromide, HBr, and hydrogen iodide, HI, to form alkyl halides, RX. This reaction is also known as **hydrohalogenation** (in contrast to dehydrohalogenation) because H and X are added to the double-bonded carbons.

General equation

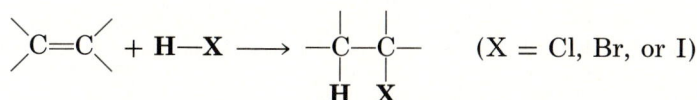

$$\text{C=C} + \text{H—X} \longrightarrow \underset{\text{H} \quad \text{X}}{-\text{C}-\text{C}-} \qquad (\text{X = Cl, Br, or I})$$

Specific examples

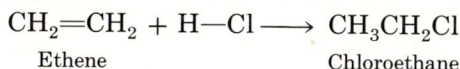

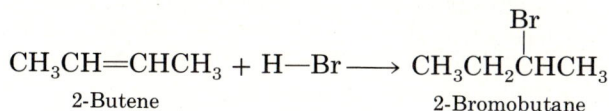

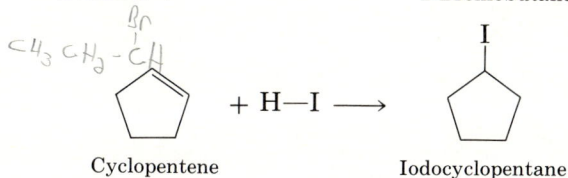

$$\text{CH}_2\text{=CH}_2 + \text{H—Cl} \longrightarrow \text{CH}_3\text{CH}_2\text{Cl}$$
Ethene   Chloroethane

$$\text{CH}_3\text{CH=CHCH}_3 + \text{H—Br} \longrightarrow \underset{}{\text{CH}_3\text{CH}_2\overset{\text{Br}}{\text{CHCH}_3}}$$
2-Butene   2-Bromobutane

$CH_3CH_2-\overset{Br}{CH}$

 + H—I ⟶

Cyclopentene   Iodocyclopentane

Problem 3.14   Write the mechanism of the addition of HCl to ethene.

Markovnikov's Rule *3.17*

When hydrogen halide is added to a symmetrical alkene such as RCH=CHR, there is only one possible product because the two double-bonded carbons are equivalent. This is illustrated by the reaction of 2-butene with HBr.

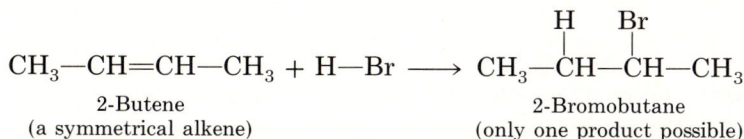

$$\text{CH}_3\text{—CH=CH—CH}_3 + \text{H—Br} \longrightarrow \underset{\text{H} \quad \text{Br}}{\text{CH}_3\text{—CH—CH—CH}_3}$$
2-Butene   2-Bromobutane
(a symmetrical alkene)   (only one product possible)

With unsymmetrical alkenes, however, such as those of the type RCH=CHR′ (R′ ≠ R), the possibility exists for the hydrogen halide to add in two ways and thus give two isomeric products. The addition of HBr to propene, for example, could yield 1-bromopropane, 2-bromopropane, or both.

$$CH_3CH{=}CH_2 + HBr \longrightarrow$$

Propene
(an unsymmetrical
alkene)

$$\begin{array}{c} CH_3CHCH_3 \\ | \\ Br \end{array}$$

2-Bromopropane
(major product)

$$CH_3CH_2CH_2Br$$

1-Bromopropane
(minor product)

Statistically, the two products should be formed in equal amounts. In fact, the major product is 2-bromopropane. Based on the results of many observations of alkene reactions, the Russian chemist Vladimir Markovnikov in 1869 summarized his findings in a statement now known as **Markovnikov's rule:** *In electrophilic addition of H—X to unsymmetrical alkenes the hydrogen of the hydrogen halide adds to the double-bonded carbon that bears the greater number of hydrogen atoms and the negative halide ion adds to the other double-bonded carbon.*

According to this rule, the addition of HCl to 2-methylpropene, for example, yields 2-chloro-2-methylpropane as the major product and 1-chloro-2-methylpropane as the minor product.

$$\underset{\underset{CH_3}{|}}{CH_3C}{=}CH_2 + HCl \longrightarrow \quad \underset{\underset{Cl}{|}}{\overset{\overset{CH_3}{|}}{CH_3CCH_3}} \quad + \quad \overset{\overset{CH_3}{|}}{CH_3CHCH_2Cl}$$

2-Chloro-2-methylpropane   1-Chloro-2-methylpropane
(major product)   (minor product)

In some unsymmetrical alkenes the two double-bonded carbons may be equivalent (may contain the same number of hydrogen atoms), but the addition of H—X still gives two possible products. In such cases an approximately equimolar mixture of the two possible addition products is obtained. For example, the addition of HCl to 2-pentene yields an equal mixture of 2-chloropentane and 3-chloropentane.

$$CH_3CH_2CH{=}CHCH_3 + HCl \longrightarrow \underset{}{\overset{\overset{Cl}{|}}{CH_3CH_2CH_2CHCH_3}} + \overset{\overset{Cl}{|}}{CH_3CH_2CHCH_2CH_3}$$

2-Pentene
(an unsymmetrical alkene
with equivalent
double-bonded carbons)

2-Chloropentane   3-Chloropentane

Equimolar mixture

3.18 Explanation for Markovnikov's Rule

The explanation for Markovnikov's rule may be given in light of the general mechanism of electrophilic addition. The addition of HX to an alkene involves the formation of a carbocation intermediate. We should expect the more stable carbocation to be preferentially formed. Recall from Section 3.9 that the stability of a carbocation follows the order $3° > 2° > 1°$.

Returning to the previous example of the addition of HBr to propene, we can see that two different carbocation intermediates are possible. The formation of each depends on which double-bonded carbon the hydrogen ion of HBr adds to. Addition of the hydrogen to C-2 gives a primary carbocation, and addition at C-1 gives the more stable secondary carbocation.

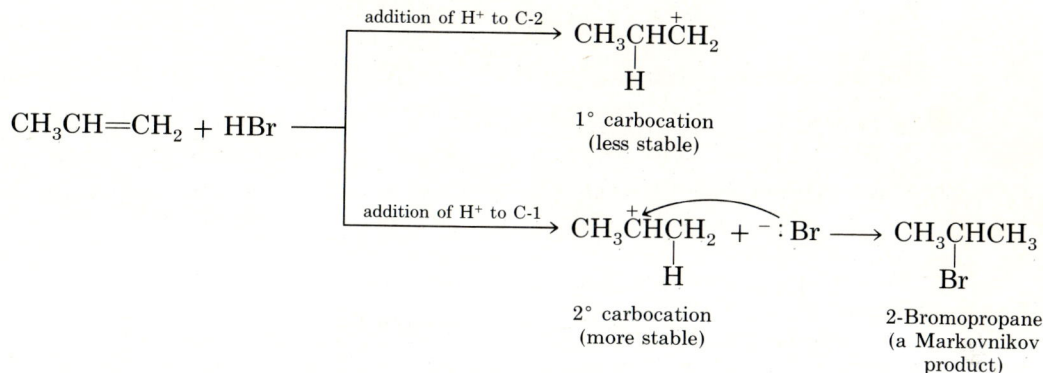

$$CH_3CH{=}CH_2 + HBr$$

addition of H^+ to C-2

$$CH_3CH\overset{+}{CH_2}$$
$$|$$
$$H$$

1° carbocation
(less stable)

addition of H^+ to C-1

$$CH_3\overset{+}{CH}CH_2 + {}^-{:}Br \longrightarrow CH_3CHCH_3$$
$$\qquad\quad |\qquad\qquad\qquad\qquad |$$
$$\qquad\quad H\qquad\qquad\qquad\qquad Br$$

2° carbocation
(more stable)

2-Bromopropane
(a Markovnikov
product)

In modern terms Markovnikov's rule can be restated: *The addition of an unsymmetrical reagent HX to an unsymmetrical alkene proceeds in such a direction as to produce the more stable carbocation.*

Problem 3.15 Write the structures of the major products of the following reactions.

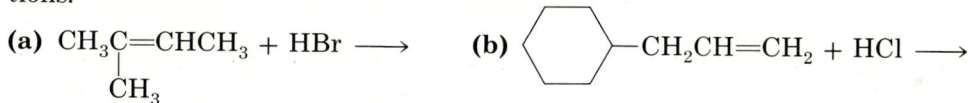

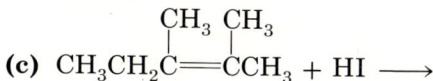

(a) $CH_3C{=}CHCH_3 + HBr \longrightarrow$
$\quad\quad\quad |$
$\quad\quad\quad CH_3$

(b) ⬡—$CH_2CH{=}CH_2 + HCl \longrightarrow$

(c) $CH_3CH_2C{=}CCH_3 + HI \longrightarrow$
$\quad\quad\quad\quad\quad\; |\quad\; |$
$\quad\quad\quad\quad\quad CH_3\; CH_3$

Addition of Sulfuric Acid *3.19*

Upon thorough mixing, cold concentrated sulfuric acid adds across the double bond of alkenes to give alkyl hydrogen sulfate. The general equation for this reaction is

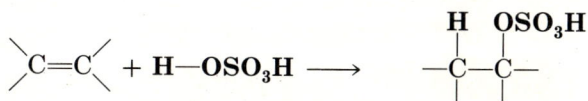

$$\diagdown C{=}C\diagup + H{-}OSO_3H \longrightarrow$$

$$\begin{matrix} H & OSO_3H \\ | & | \\ {-}C{-}C{-} \\ | & | \end{matrix}$$

An alkyl hydrogen sulfate

Addition of sulfuric acid to alkenes also follows Markovnikov's rule, as illustrated in the reaction with propene.

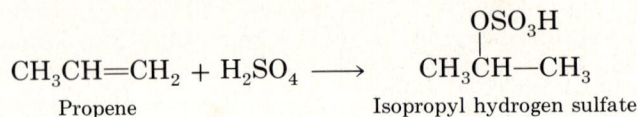

$$CH_3CH{=}CH_2 + H_2SO_4 \longrightarrow \overset{\displaystyle OSO_3H}{\underset{\text{Isopropyl hydrogen sulfate}}{CH_3CH{-}CH_3}}$$

Propene

Problem 3.16 Write the mechanism of the addition of H_2SO_4 to propene.

Problem 3.17 Write the structure of the major product of the addition of H_2SO_4 to **(a)** 1-butene; **(b)** 2-butene; and **(c)** 1-methylcyclohexene.

3.20 Addition of Water: Hydration

When heated with water in the presence of an acid catalyst, alkenes yield alcohols (ROH). The process is called **hydration** of alkenes because it involves the addition of water across the double bond.

The addition of HOH across the double bond is in accordance with Markovnikov's rule: The hydrogen goes to the double-bonded carbon that bears the greater number of hydrogen atoms and the hydroxyl group goes to the other double-bonded carbon.

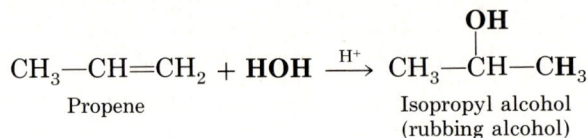

$$CH_3{-}CH{=}CH_2 + \textbf{HOH} \xrightarrow{\ H^+\ } \overset{\displaystyle \textbf{OH}}{CH_3{-}CH{-}CH_3}$$

Propene   Isopropyl alcohol (rubbing alcohol)

Problem 3.18 Write the structure of the major product of the hydration of **(a)** ethylene; **(b)** 1-butene; and **(c)** 1-methylcyclopentene.

3.21 Addition of HOX: Halohydrin Formation

When an alkene is treated with aqueous chlorine or aqueous bromine, the addition product is a **halohydrin.** (When Cl_2 is used, the product is a chlorohydrin; when Br_2 is used, the product is a bromohydrin.)

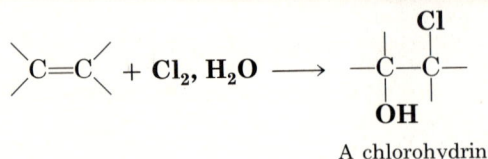

$$\text{C=C} + Cl_2, H_2O \longrightarrow \overset{\displaystyle Cl}{\underset{\displaystyle OH}{-C-C-}}$$

A chlorohydrin

$$\text{C}=\text{C} + \textbf{Br}_2, \textbf{H}_2\textbf{O} \longrightarrow \underset{\overset{\displaystyle |}{\text{OH}}}{-}\text{C}\overset{\overset{\displaystyle \text{Br}}{|}}{-}\text{C}-$$

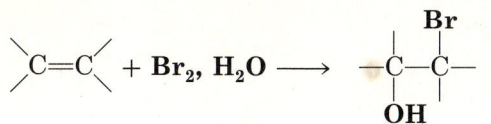

A bromohydrin

The reaction proceeds as if hypochlorous acid, HO—Cl, or hypobromous acid, HO—Br, were the adding reagent. The chloronium ion, Cl^+, or bromonium ion, Br^+, is the electrophile, and the hydroxide ion, OH^-, the nucleophile. Addition of HOX also follows Markovnikov's rule, as illustrated with propene.

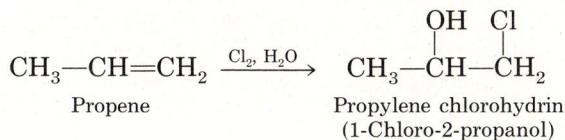

$$\text{CH}_3-\text{CH}=\text{CH}_2 \xrightarrow{\text{Cl}_2,\ \text{H}_2\text{O}} \text{CH}_3-\underset{\underset{\displaystyle \text{OH}}{|}}{\text{CH}}-\underset{\underset{\displaystyle \text{Cl}}{|}}{\text{CH}}_2$$

Propene   Propylene chlorohydrin
(1-Chloro-2-propanol)

Problem 3.19 Write the mechanism for the addition of HOCl to propene.

Problem 3.20 Write the structure of the major product expected on reaction of **(a)** ethylene with aqueous chlorine; **(b)** 1-butene with aqueous chlorine; **(c)** 2-methylpropene with aqueous bromine; and **(d)** 1-methylcyclopentene with aqueous bromine.

Visual Tests for Unsaturation *3.22*

The functional group of alkenes is the C=C bond. What simple visual test can we use to detect unsaturation? Of the reactions discussed so far the addition of bromine is frequently used. Bromine itself is a dark red liquid, whereas both the alkene and the addition product are colorless. Thus, to test whether a substance is an alkene, we add to it a solution of bromine in an inert solvent (CCl_4). If the substance is an alkene (or an alkyne), the bromine solution will be rapidly decolorized.

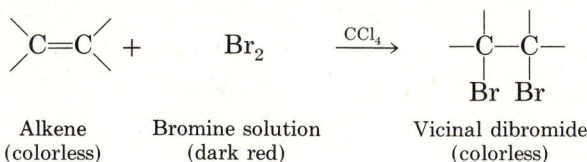

$$\text{C}=\text{C} + \text{Br}_2 \xrightarrow{\text{CCl}_4} \underset{\underset{\displaystyle \text{Br}\ \ \text{Br}}{|\quad|}}{-}\text{C}-\text{C}-$$

Alkene   Bromine solution   Vicinal dibromide
(colorless)   (dark red)   (colorless)

Another visual test for unsaturation is the **Baeyer test.** In this test, a dilute, alkaline solution of purple potassium permanganate, $KMnO_4$, is added to an alkene (or alkyne) at room temperature. A *glycol,* a compound with two OH groups, and a brown precipitate of manganese dioxide, MnO_2, are formed.

$$\text{C=C} + KMnO_4 \xrightarrow[\text{alkaline}]{\text{cold, dilute,}} \underset{\underset{\text{HO} \quad \text{OH}}{|}}{-\overset{|}{C}-\overset{|}{C}-} + MnO_2\downarrow$$

(purple)   A glycol   (brown
precipitate)

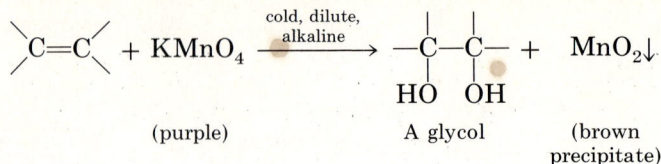

Replacement of a purple-colored solution by a brown precipitate indicates a positive Baeyer test.

The addition of H_2SO_4 to alkenes (or alkynes) is yet another visual test for unsaturation. Alkenes will dissolve in concentrated H_2SO_4, with evolution of heat and form homogeneous mixtures. Alkanes do not react with concentrated H_2SO_4 and will separate from the acid; two distinct layers form. *A word of caution:* Other types of compounds also react with and dissolve in H_2SO_4 (oxygen-containing compounds, many nitrogen-containing compounds, and some aromatic compounds).

Problem 3.21 Compound **A**, C_6H_{12}, gave the following results.

(1) **A** + Br_2 $\xrightarrow{CCl_4}$ solution remains dark red
(2) **A** + cold, concentrated H_2SO_4 $\longrightarrow$ two distinct layers
(3) **A** + dilute $KMnO_4$ $\longrightarrow$ solution remains purple
(4) **A** + Br_2 $\xrightarrow{\text{heat}}$ $C_6H_{11}Br$ (*a single isomer*)

What is the structure of **A**?

3.23 Ozonolysis

When ozone-rich oxygen gas is passed through a solution of an alkene dissolved in an inert solvent, the carbon–carbon double bond is broken and an ozonide is produced. Ozonides, like other compounds containing the peroxide group, —O—O—, are quite unstable and may explode violently and unpredictably. Consequently, they are not isolated.

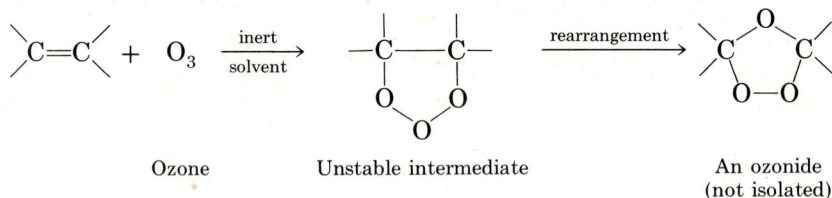

$$\text{C=C} + O_3 \xrightarrow[\text{solvent}]{\text{inert}} \underset{\underset{O}{\underset{\diagdown \diagup}{O \quad O}}}{-\overset{|}{C}-\overset{|}{C}-} \xrightarrow{\text{rearrangement}} \underset{O-O}{\overset{O}{C \diagup \diagdown C}}$$

Ozone   Unstable intermediate   An ozonide
(not isolated)

Further addition of water in the presence of a zinc catalyst results in the formation of two smaller products, each of which contains a carbonyl group, C=O, at the position where the carbon–carbon double bond was. These products may be aldehydes, $R-\overset{\overset{\displaystyle H}{|}}{C}=O$, or ketones, $R-\overset{\overset{\displaystyle R}{|}}{C}=O$, depending on the structure of the starting alkene.

$$\underset{\text{Aldehyde or ketone}}{\overset{\text{O}}{\underset{\text{O}-\text{O}}{\text{C}\diagdown\text{C}}}} + H_2O \xrightarrow{\text{Zn}} \diagup C=O + O=C\diagdown$$

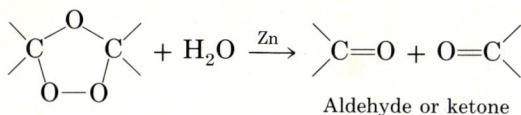

Ozonolysis is a valuable degradative reaction, for it can be used to locate the position of a double bond on the parent alkene chain. This is done by identifying the structure(s) of the product(s) obtained on treatment of an alkene with ozone. For example, suppose we wanted to know whether a given compound is 1-butene or its isomer, 2-butene. Both compounds have the same molecular formula, C_4H_8; both decolorize bromine in carbon tetrachloride; and both give a positive Baeyer test. One way we can distinguish between them is by ozonolysis, because they both yield different identifiable products. The ozonolysis of 1-butene yields formaldehyde, $H_2C=O$, and propionaldehyde, $CH_3CH_2CH=O$, whereas the ozonolysis of 2-butene yields only one product, acetaldehyde, $CH_3CH=O$.

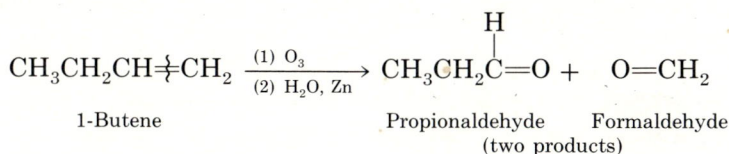

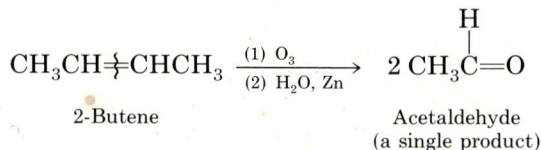

$$\underset{\text{1-Butene}}{CH_3CH_2CH\!\!\dashv\!\!CH_2} \xrightarrow[\text{(2) H}_2\text{O, Zn}]{\text{(1) O}_3} \underset{\substack{\text{Propionaldehyde} \\ \text{(two products)}}}{CH_3CH_2\overset{\text{H}}{\underset{|}{C}}=O} + \underset{\text{Formaldehyde}}{O=CH_2}$$

$$\underset{\text{2-Butene}}{CH_3CH\!\!\dashv\!\!CHCH_3} \xrightarrow[\text{(2) H}_2\text{O, Zn}]{\text{(1) O}_3} \underset{\substack{\text{Acetaldehyde} \\ \text{(a single product)}}}{2\ CH_3\overset{\text{H}}{\underset{|}{C}}=O}$$

To be sure that you understand the value of ozonolysis in determining the structure of an alkene, you should work through Example 3.2.

Example 3.2 The products of the ozonolysis of an alkene were identified as acetone, $(CH_3)_2C=O$, and formaldehyde, $CH_2=O$. What is the structure of the parent alkene?

Solution If you recall, the carbonyl carbons were originally double-bonded carbons. The structure of the parent alkene can therefore be determined by eliminating the oxygen atoms from the two carbonyl carbons and joining them by a double bond. Thus, the parent alkene is 2-methyl-propene.

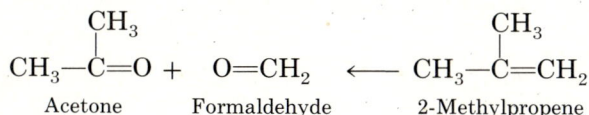

$$\underset{\text{Acetone}}{CH_3\!-\!\overset{\text{CH}_3}{\underset{|}{C}}\!=\!O} + \underset{\text{Formaldehyde}}{O=CH_2} \longleftarrow \underset{\text{2-Methylpropene}}{CH_3\!-\!\overset{\text{CH}_3}{\underset{|}{C}}\!=\!CH_2}$$

Problem 3.22 Write the structure of the product expected from the ozonolysis of **(a)** 2-methyl-2-pentene; **(b)** cyclopentene; and **(c)** 2-hexene.

Problem 3.23 Write the structure of the parent alkene given the following products of ozonolysis.

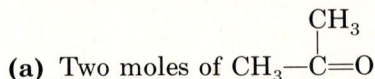

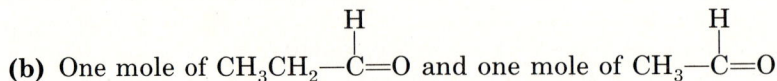

(a) Two moles of $CH_3 - \overset{\overset{\displaystyle CH_3}{|}}{C} = O$

(b) One mole of $CH_3CH_2 - \overset{\overset{\displaystyle H}{|}}{C} = O$ and one mole of $CH_3 - \overset{\overset{\displaystyle H}{|}}{C} = O$

Problem 3.24 Compound **B**, C_6H_{12}, gave the following results.

(1) **B** + Br_2 $\xrightarrow{CCl_4}$ colorless solution
(2) **B** + dilute $KMnO_4$ $\longrightarrow$ brown precipitate
(3) **B** + O_3 followed by treatment with Zn, H_2O yielded CH_3CH_2CHO as the only product.

What are the two possible structures for **B**? (*Hint:* The two structures are geometric isomers.)

3.24 Substitution Reactions: Halogenation at High Temperatures

So far discussion has focused on addition reactions that take place exclusively at the carbon–carbon double bond. With the exception of ethylene, all alkenes contain saturated alkyl groups as part of their structures. What kind of reactions can we expect to occur there? Since these alkyl groups are saturated, we can expect them to undergo the typical alkane reaction, which is free-radical substitution. Let us use halogenation as an example of a substitution reaction. We already know that halogen can attack the double bond as well as the saturated alkyl site (Sec. 2.12). Can we control the experimental conditions so that the halogen attacks only one of these sites? The answer is yes. Alkenes are more reactive than alkanes. Their reactions therefore take place under milder experimental conditions. Alkenes undergo addition of halogen at low temperature, even in the dark, and generally in the liquid phase. On the other hand, alkanes undergo substitution by halogen only at elevated temperatures or under the influence of ultraviolet light (Sec. 2.12) and generally in the gas phase. The latter conditions favor a free-radical substitution reaction, whereas the former favor an addition reaction.

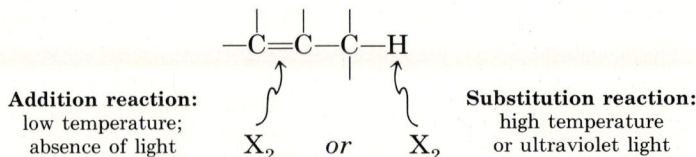

$$-\overset{|}{C}=\overset{|}{C}-\overset{|}{C}-H$$

Addition reaction:
low temperature;
absence of light
X_2 *or* X_2

Substitution reaction:
high temperature
or ultraviolet light

For example, the treatment of propene with chlorine at high temperatures yields chiefly the substitution product 3-chloropropene (allyl chloride). Reaction of chlorine with propene in the liquid phase at low temperature, in the dark, yields only 1,2-dichloropropane, an addition product.

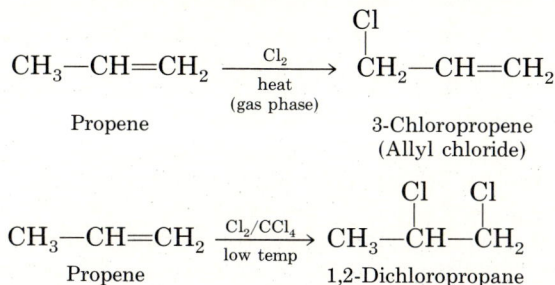

$$CH_3-CH=CH_2 \xrightarrow[\substack{heat \\ (gas\ phase)}]{Cl_2} \overset{\displaystyle Cl}{\overset{\displaystyle |}{CH_2}}-CH=CH_2$$

Propene

3-Chloropropene
(Allyl chloride)

$$CH_3-CH=CH_2 \xrightarrow[low\ temp]{Cl_2/CCl_4} CH_3-\overset{\displaystyle Cl}{\overset{\displaystyle |}{CH}}-\overset{\displaystyle Cl}{\overset{\displaystyle |}{CH_2}}$$

Propene

1,2-Dichloropropane

Problem 3.25 Write the structure of the major product expected in each reaction.

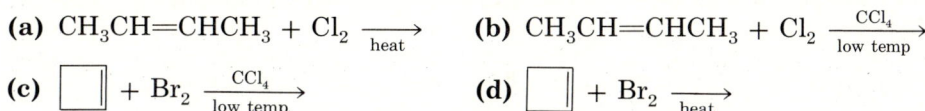

(a) $CH_3CH=CHCH_3 + Cl_2 \xrightarrow{heat}$ **(b)** $CH_3CH=CHCH_3 + Cl_2 \xrightarrow[low\ temp]{CCl_4}$

(c) $\square + Br_2 \xrightarrow[low\ temp]{CCl_4}$ **(d)** $\square + Br_2 \xrightarrow{heat}$

Polymerization 3.25

One important property of alkenes is their ability to form giant molecules called **polymers.** Polymers are prepared by the polymerization process, which involves the reaction of alkene units with themselves (from the Greek *poly,* many; *meros,* unit). The basic alkene unit is referred to as a **monomer.** Under the influence of various catalysts alkene monomers add to one another indefinitely in a process called **addition polymerization.** Alkene polymers form useful products such as plastics and rubber. For example, the polymerization of ethylene, which takes places when ethylene is heated in the presence of a suitable catalyst, produces a waxy polymer called polyethylene. The overall reaction is

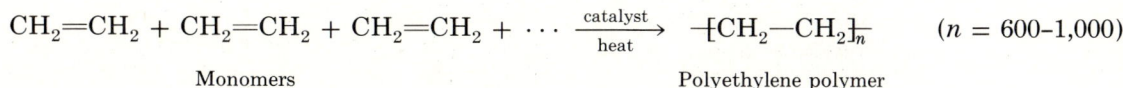

$$CH_2=CH_2 + CH_2=CH_2 + CH_2=CH_2 + \cdots \xrightarrow[heat]{catalyst} -\!\!\left[CH_2-CH_2\right]_n \qquad (n = 600\text{--}1{,}000)$$

Monomers

Polyethylene polymer

Polyethylene is widely used for the manufacture of plastic bottles, wire insulation, toys, packaging, and wearing apparel. Ethylene is one of the most important alkenes produced in the United States. In 1980 the United States production of ethylene was 28.3 billion pounds.

The substitution of one of the hydrogens of ethylene by some other group produces a vinyl compound ($CH_2=CH-$, a vinyl group; $CH_2=CH-G$, a vinyl compound). These polymerize in much the same manner as ethylene, and produce **vinyl polymers.** The general equation for the polymerization of vinyl compounds is

$$n \; CH_2\!=\!\underset{\underset{G}{|}}{\overset{\overset{R}{|}}{C}} \xrightarrow[\text{catalyst}]{\text{peroxide}} \left[\!-CH_2\!-\!\underset{\underset{G}{|}}{\overset{\overset{R}{|}}{C}}\!-\!\right]_n \qquad (R = H, \text{ alkyl, or Cl})$$

A vinyl compound   A vinyl polymer

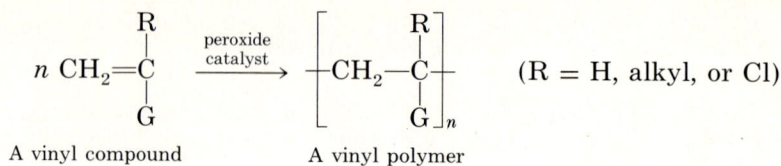

Vinyl polymers are important because they constitute the raw materials used by the plastics industry for the manufacture of many useful finished products. For example, the polymerization of vinyl chloride produces polyvinyl chloride, which is used as a rubber substitute, for pipes, and as an insulator. The polymerization of tetrafluoroethylene produces the tough plastic known as Teflon. Teflon is a chemically inert substance widely used in insulation materials and as a coating in "nonstick" cooking utensils. Table 3.1 lists some important vinyl polymers and their uses in the plastic industry.

Table 3.1 Some Important Vinyl Polymers and Their Uses in the Plastics Industry

$$n \; CH_2\!=\!\underset{\underset{G}{|}}{\overset{\overset{R}{|}}{C}} \xrightarrow[\text{catalyst}]{\text{peroxide}} \left[\!-CH_2\!-\!\underset{\underset{G}{|}}{\overset{\overset{R}{|}}{C}}\!-\!\right]_n$$

R	G	Monomer	Polymer	Name	Uses
—H	—Cl	$CH_2\!=\!\underset{Cl}{\overset{H}{C}}$	$\left[\!-CH_2\!-\!\underset{Cl}{\overset{H}{C}}\!-\!\right]_n$	polyvinyl chloride	electrical insulators; pipes
—H	—C≡N	$CH_2\!=\!\underset{C\equiv N}{\overset{H}{C}}$	$\left[\!-CH_2\!-\!\underset{C\equiv N}{\overset{H}{C}}\!-\!\right]_n$	polyacrylonitrile (Orlon)	fibers for clothing
—H	(phenyl ring)	$CH_2\!=\!\overset{H}{C}$ (phenyl)	$\left[\!-CH_2\!-\!\overset{H}{C}\!-\!\right]_n$ (phenyl)	polystyrene	electrical insulators; foamed plastic fabrication
—H	$-O\!-\!\overset{\overset{O}{\|}}{C}\!-\!CH_3$	$CH_2\!=\!\underset{\underset{O}{\|}}{\overset{H}{\underset{O-C-CH_3}{C}}}$	$\left[\!-CH_2\!-\!\underset{\underset{O}{\|}}{\overset{H}{\underset{O-C-CH_3}{C}}}\!-\!\right]_n$	polyvinyl acetate	plastic sheets, films, and fibers
—Cl	—Cl	$CH_2\!=\!\underset{Cl}{\overset{Cl}{C}}$	$\left[\!-CH_2\!-\!\underset{Cl}{\overset{Cl}{C}}\!-\!\right]_n$	polyvinylidene chloride (Saran)	seat covers; self-adhering food wrappers
—CH₃	$-\overset{\overset{O}{\|}}{C}\!-\!OCH_3$	$CH_2\!=\!\underset{\underset{O}{\|}}{\overset{CH_3}{\underset{C-OCH_3}{C}}}$	$\left[\!-CH_2\!-\!\underset{\underset{O}{\|}}{\overset{CH_3}{\underset{C-OCH_3}{C}}}\!-\!\right]_n$	polymethyl methacrylate (Plexiglas; Lucite)	transparent sheets; unbreakable substitute for glass; paints

Now that you have learned the chemistry of alkanes and alkenes, let us see how that knowledge can be put to practice in synthesizing a compound in the best way possible.

In general, the best approach to a problem involving the synthesis of a compound is working *backward,* that is, from the product to the starting material. To illustrate, let us work out the following example.

Example 3.3 Suppose you wish to prepare ethanol, CH_3CH_2OH, starting from ethane. (You may use any inorganic reagent and any reaction conditions).

Solution First draw the structures of the starting material and of the product.

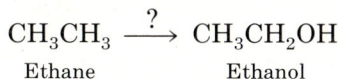

$$CH_3CH_3 \xrightarrow{\ ?\ } CH_3CH_2OH$$
Ethane   Ethanol

Working backward, ethanol can be obtained via hydration of an alkene, ethylene.

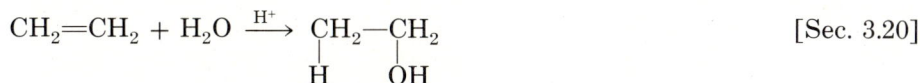

$$CH_2{=}CH_2 + H_2O \xrightarrow{H^+} \underset{\underset{H\qquad OH}{\mid\qquad\mid}}{CH_2{-}CH_2} \qquad\qquad\text{[Sec. 3.20]}$$

Ethylene, in turn, can be obtained via dehydrohalogenation of an ethyl halide, for example, bromoethane.

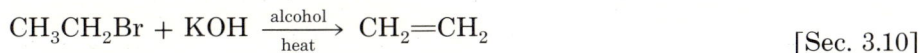

$$CH_3CH_2Br + KOH \xrightarrow[\text{heat}]{\text{alcohol}} CH_2{=}CH_2 \qquad\qquad\text{[Sec. 3.10]}$$

Bromoethane, in turn, is obtained from ethane via free-radical bromination.

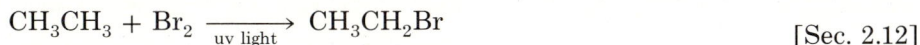

$$CH_3CH_3 + Br_2 \xrightarrow[\text{uv light}]{} CH_3CH_2Br \qquad\qquad\text{[Sec. 2.12]}$$

The overall series of reactions is therefore

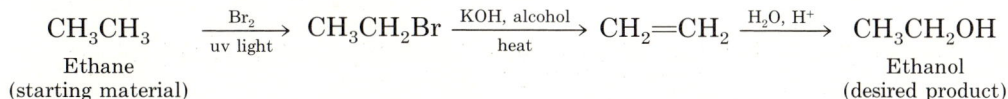

$$CH_3CH_3 \xrightarrow[\text{uv light}]{Br_2} CH_3CH_2Br \xrightarrow[\text{heat}]{\text{KOH, alcohol}} CH_2{=}CH_2 \xrightarrow{H_2O,\ H^+} CH_3CH_2OH$$
  Ethane   Ethanol
(starting material)   (desired product)

Now work out the following synthesis problem.

Problem 3.26

Starting from	Synthesize
(a) CH_3CH_2OH	CH_3CH_3
(b) CH_3CH_2Br	$HOCH_2CH_2Cl$
(c) CH_3CH_2Cl	$HOCH_2CH_2OH$
(d) $CH_3CH_2CH_2OH$	$CH_3CHBrCH_3$
(e) $CH_3CH{=}CH_2$	$BrCH_2CHBrCH_2Br$
	(*Hint:* First do a reaction at the saturated carbon.)

Summary of Concepts and Reactions

Alkenes are unsaturated hydrocarbons that contain a carbon–carbon double bond,
$\diagdown C{=}C \diagup$. [Sec. 3.1]

Alkenes are named in the IUPAC system by replacement of the ending -ane of alkanes
by -ene. The position of the double bond is indicated by the lowest possible number.
[Sec. 3.1]

The groups attached to the sp^2-hybridized carbons in an alkene lie in the same plane.
[Sec. 3.2]

Because of the lack of rotation about the carbon–carbon double bond, geometric isomers
can exist when different groups are attached to both carbons. [Sec. 3.3]

Alkenes may be prepared by the dehydration of alcohols

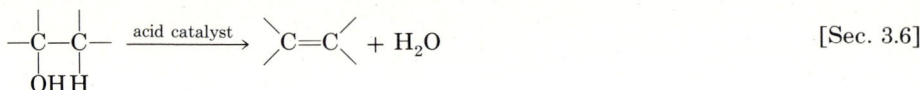

$$-\underset{\underset{OH}{|}}{C}-\underset{\underset{H}{|}}{C}- \xrightarrow{\text{acid catalyst}} \diagdown C{=}C \diagup + H_2O \qquad \text{[Sec. 3.6]}$$

or by the dehydrohalogenation of alkyl halides

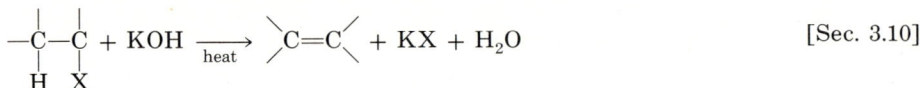

$$-\underset{\underset{H}{|}}{C}-\underset{\underset{X}{|}}{C} + KOH \xrightarrow[\text{heat}]{} \diagdown C{=}C \diagup + KX + H_2O \qquad \text{[Sec. 3.10]}$$

The typical reaction of alkenes is addition to $\diagdown C{=}C \diagup$. [Sec. 3.12]

Most reactions of alkenes are electrophilic additions

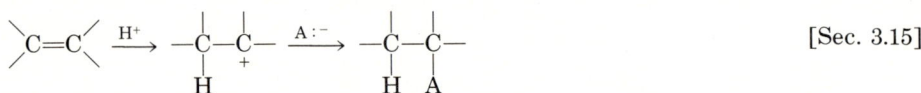

$$\diagdown C{=}C \diagup \xrightarrow{H^+} -\underset{\underset{H}{|}}{C}-\overset{+}{C}- \xrightarrow{A:^-} -\underset{\underset{H}{|}}{C}-\underset{\underset{A}{|}}{C}- \qquad \text{[Sec. 3.15]}$$

All additions of H—A to unsymmetrical alkenes follow Markovnikov's rule.
[Sec. 3.17]

Addition of H—A involves the formation of the most stable carbocation;
$R_3\overset{+}{C} > R_2\overset{+}{C}H > R\overset{+}{C}H_2$. [Sec. 3.18]

Alkenes decolorize (a) Br_2 in CCl_4 and (b) dilute solution of $KMnO_4$ (Baeyer's test).
[Sec. 3.22]

Ozonolysis

$$\diagdown C{=}C \diagup \xrightarrow[\text{(2) Zn, H}_2\text{O}]{\text{(1) O}_3} -\underset{|}{C}{=}O + O{=}\underset{|}{C}-$$

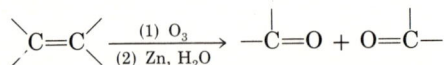

is a reaction used to locate the position of the double bond. [Sec. 3.23]

The saturated part of an alkene reacts like an alkane. [Sec. 3.24]

Alkenes can form giant molecules called polymers. The basic unit of a polymer is called a
monomer. [Sec. 3.25]

Alkene monomers add to one another in a process called addition polymerization.

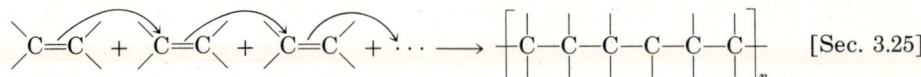

$$\diagdown C{=}C \diagup + \diagdown C{=}C \diagup + \diagdown C{=}C \diagup + \cdots \longrightarrow {\Big[}\overset{|}{C}-\overset{|}{C}-\overset{|}{C}-\overset{|}{C}-\overset{|}{C}-\overset{|}{C}{\Big]}_n \qquad \text{[Sec. 3.25]}$$

The substitution of one hydrogen of a doubly bonded carbon by another group produces
a vinyl compound ($CH_2{=}CH-G$). Polymerization of a vinyl compound produces a
vinyl polymer. [Sec. 3.25]

The best approach to a problem involving the synthesis of a compound is to work
backward, that is, from the product back to the starting material. [Sec. 3.26]

unsaturated	geometric isomers	electrophilic addition
alkenes (olefins)	*E,Z* system	hydrohalogenation
C_nH_{2n}	elimination	Markovnikov's rule
alkynes	Saytzeff's rule	hydration
C_nH_{2n-2}	carbocation	vicinal dihalide
dienes	dehydrohalogenation	halohydrin
polyenes	addition reaction	Baeyer test
vinyl group	catalytic hydrogenation	polymers
allyl group	electrophile	monomer
sp^2 hybridized	nucleophile	addition polymerization
π (pi) bond	rate-determining step	vinyl polymers

Exercises

Structure, Nomenclature, and Geometric Isomerism   [Secs. 3.1, 3.3]

3.1 Give the IUPAC names for the following structures. Use *cis* and *trans* designations where pertinent for geometric isomers.

(a) 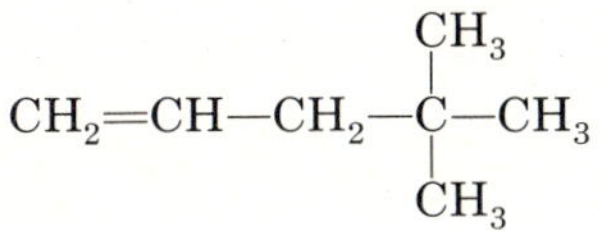 $CH_2{=}CH{-}CH_2{-}\underset{\underset{CH_3}{|}}{\overset{\overset{CH_3}{|}}{C}}{-}CH_3$

(b) ▢

(c) 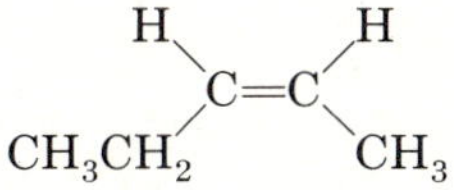 $\underset{CH_3CH_2}{\overset{H}{}}C{=}C\underset{CH_3}{\overset{H}{}}$

(d) 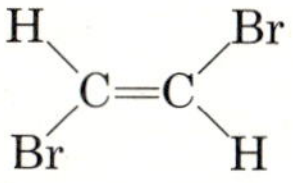 $\underset{Br}{\overset{H}{}}C{=}C\underset{H}{\overset{Br}{}}$

(e) 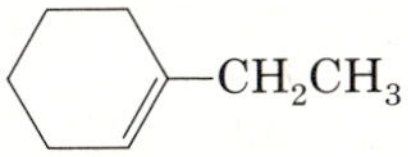 (cyclohexene)—CH_2CH_3

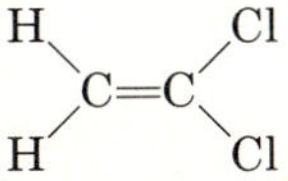

(f) 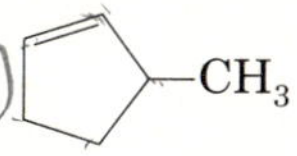 (cyclopentene)—CH_3

(g) $\underset{H}{\overset{H}{}}C{=}C\underset{Cl}{\overset{Cl}{}}$

(h) $H_2C{=}C\underset{CH_2CH_3}{\overset{Cl}{}}$

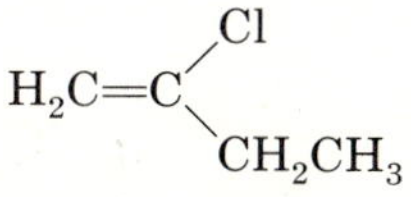

3.2 Write formulas for the following named compounds.
(a) 3-Methyl-1-butene   **(b)** 5-Bromo-2-methyl-2-hexene
(c) 4-Chlorocyclohexene   **(d)** Vinylcyclopropane
(e) Allylcyclopentane   **(f)** *cis*-3-Hexene
(g) *trans*-2-Heptene   **(h)** *trans*-1,2-Dicyclopropylethene
(i) *Z*-2-Bromo-1-chloropropene   **(j)** *E*-1-Bromo-1-chloro-2-fluoroethene

3.3 State what is wrong with the following names and give the correct name for each molecule.
(a) 2-Ethyl-2-pentene   **(b)** 2-*n*-Propyl-2-butene
(c) 3-Methyl-2-butene   **(d)** 3-*n*-Butyl-1-hexene
(e) 1-Methyl-2-cyclobutene   **(f)** 2,5-Dimethylcyclohexene

3.4 Which of the following compounds can exist as *cis* and *trans* isomers? Draw the structures of the geometric isomers.
(a) 1-Butene   **(b)** 2-Pentene

(c) 1-Bromo-3-hexene　　　　　　　**(d)** 1-Chloro-2-methylpropene
(e) 1,1-Dichloroethene　　　　　　 **(f)** 2-Methyl-2-pentene

3.5 There are three compounds with molecular formula $C_2H_2Cl_2$. Draw the structure of each compound and indicate which isomer(s) is (are) structural (constitutional) and which is (are) geometric (*cis-trans*).

3.6 Including geometric isomers, there are sixteen alkenes of formula C_6H_{12}. Draw and name each structure, and indicate which are geometric isomers.

3.7 Name the following structures using the *E,Z* system of nomenclature.

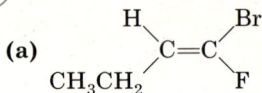

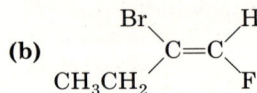

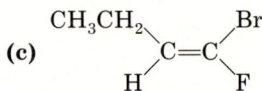

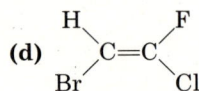

(a)
$$\underset{CH_3CH_2}{\overset{H}{\diagup}}C=C\underset{F}{\overset{Br}{\diagdown}}$$

(b)
$$\underset{CH_3CH_2}{\overset{Br}{\diagup}}C=C\underset{F}{\overset{H}{\diagdown}}$$

(c)
$$\underset{H}{\overset{CH_3CH_2}{\diagup}}C=C\underset{F}{\overset{Br}{\diagdown}}$$

(d)
$$\underset{Br}{\overset{H}{\diagup}}C=C\underset{Cl}{\overset{F}{\diagdown}}$$

Hybridization and Shapes of Molecules [Secs. 1.10, 3.2]

3.8 Indicate the type of hybridization and the shape about each carbon in the following structures.

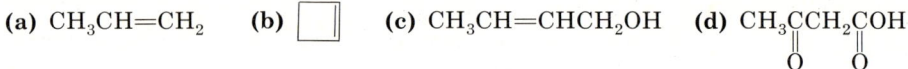

(a) $CH_3CH=CH_2$　　**(b)** □　　**(c)** $CH_3CH=CHCH_2OH$　　**(d)** $CH_3\underset{\underset{O}{\|}}{C}CH_2\underset{\underset{O}{\|}}{C}OH$

Preparation of Alkenes [Secs. 3.6–3.9]

3.9 Which of the products named for each reaction is the major product according to Saytzeff's rule?

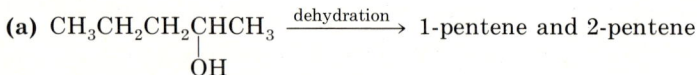

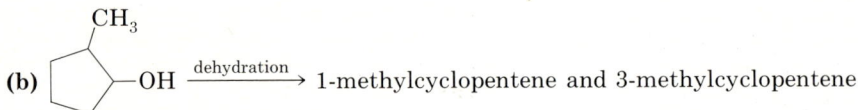

(a) $CH_3CH_2CH_2\underset{\underset{OH}{|}}{C}HCH_3 \xrightarrow{\text{dehydration}}$ 1-pentene and 2-pentene

(b) [cyclopentane with CH_3 and $-OH$ substituents] $\xrightarrow{\text{dehydration}}$ 1-methylcyclopentene and 3-methylcyclopentene

3.10 Using Saytzeff's rule write the structure of the *main* product obtained on dehydration of each of the following alcohols.

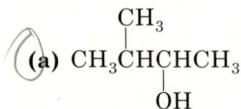

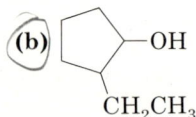

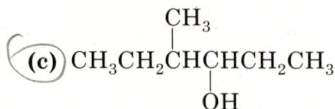

(a) $CH_3\underset{\underset{OH}{|}}{\overset{\overset{CH_3}{|}}{C}}HCHCH_3$

(b) [cyclopentane ring with $-OH$ and CH_2CH_3 substituents]

(c) $CH_3CH_2\underset{\underset{OH}{|}}{\overset{\overset{CH_3}{|}}{C}}HCHCH_2CH_3$

(d) [cyclohexane ring]$-CH_2\underset{\underset{OH}{|}}{C}HCH_3$

3.11 Dehydration of 2-butanol, $CH_3\underset{\underset{OH}{|}}{C}HCH_2CH_3$, yields 2-butene (major product) and 1-butene (minor product). Write a mechanism that accounts for the formation of these two products.

3.12 Arrange the alcohols in each group in order of ease of dehydration.

(a) $CH_3CH_2CH_2CH_2OH$ (i); $CH_3CH_2\overset{\underset{|}{OH}}{C}HCH_3$ (ii); $CH_3\overset{\overset{\displaystyle CH_3}{|}}{\underset{\underset{\displaystyle OH}{|}}{C}}CH_3$ (iii)

(b) ⬠—CH_2OH (i); ⬠ (ii); ⬠—OH (iii)

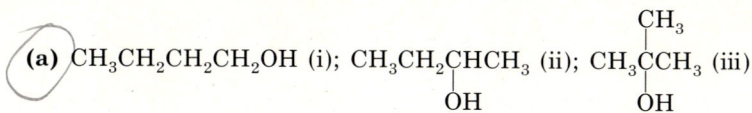

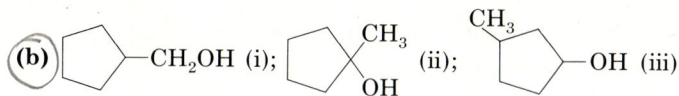

3.13 Which of the products named for each reaction is the major product according to Saytzeff's rule?

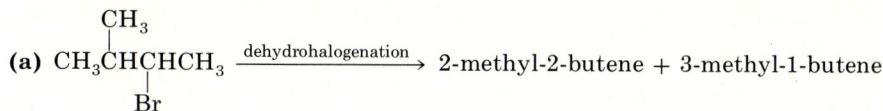

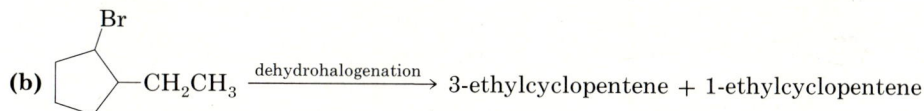

(a) $CH_3\overset{\overset{\displaystyle CH_3}{|}}{C}H\overset{\underset{|}{Br}}{C}HCH_3$ $\xrightarrow{\text{dehydrohalogenation}}$ 2-methyl-2-butene + 3-methyl-1-butene

(b) ⬠(—Br)—CH_2CH_3 $\xrightarrow{\text{dehydrohalogenation}}$ 3-ethylcyclopentene + 1-ethylcyclopentene

3.14 Using Saytzeff's rule write the structure of the *main* product obtained on dehydrohalogenation of each of the following compounds.
(a) 2-Bromopentane   **(b)** 2-Bromo-2-methylpentane
(c) 3-Bromo-2-methylpentane   **(d)** 3-Bromo-2,3-dimethylpentane

Reactions of Alkenes [Secs. 3.11–3.24]
3.15 Draw the structure of the product expected on treatment of 2-butene with each of the following.
(a) H_2/Pd   **(b)** HCl
(c) Cold, concentrated H_2SO_4 **(d)** H_2O, H^+
(e) Br_2, H_2O **(f)** Br_2 in CCl_4
(g) Cold, dilute $KMnO_4$ **(h)** O_3 followed by Zn, H_2O
3.16 Draw the structures of the major product(s) expected on treatment of 1-butene with the reagents listed in Exercise 3.15.
3.17 Draw the structures of the products, **A** through **N**.

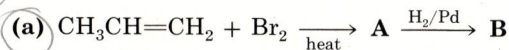

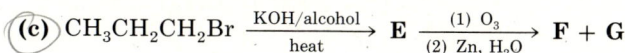

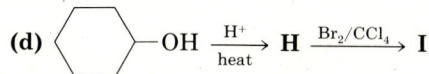

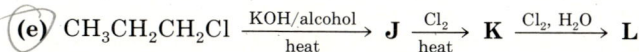

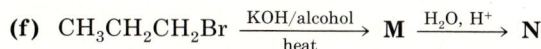

(a) $CH_3CH{=}CH_2 + Br_2 \xrightarrow{\text{heat}}$ **A** $\xrightarrow{H_2/Pd}$ **B**

(b) $CH_3CH_2CH_2OH \xrightarrow[\text{heat}]{H^+}$ **C** $\xrightarrow{HCl}$ **D**

(c) $CH_3CH_2CH_2Br \xrightarrow[\text{heat}]{\text{KOH/alcohol}}$ **E** $\xrightarrow[\text{(2) Zn, H}_2\text{O}]{\text{(1) O}_3}$ **F** + **G**

(d) ⬡—OH $\xrightarrow[\text{heat}]{H^+}$ **H** $\xrightarrow{Br_2/CCl_4}$ **I**

(e) $CH_3CH_2CH_2Cl \xrightarrow[\text{heat}]{\text{KOH/alcohol}}$ **J** $\xrightarrow[\text{heat}]{Cl_2}$ **K** $\xrightarrow{Cl_2, H_2O}$ **L**

(f) $CH_3CH_2CH_2Br \xrightarrow[\text{heat}]{\text{KOH/alcohol}}$ **M** $\xrightarrow{H_2O, H^+}$ **N**

3.18 For each reaction, fill in the missing reagent(s) and, where pertinent, the reaction conditions.

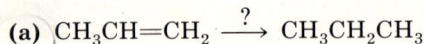

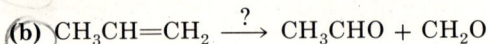

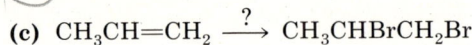

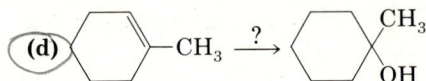

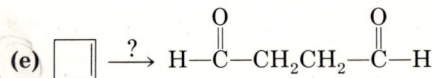

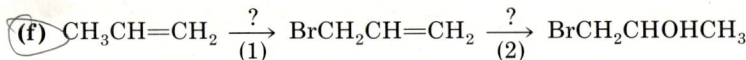

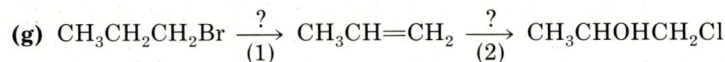

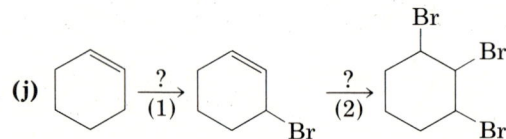

(a) $CH_3CH{=}CH_2 \xrightarrow{?} CH_3CH_2CH_3$

(b) $CH_3CH{=}CH_2 \xrightarrow{?} CH_3CHO + CH_2O$

(c) $CH_3CH{=}CH_2 \xrightarrow{?} CH_3CHBrCH_2Br$

(d)

(e) $\square \xrightarrow{?} \underset{\text{O}}{H{-}\overset{\displaystyle\text{O}}{\overset{\|}{C}}{-}CH_2CH_2{-}\overset{\displaystyle\text{O}}{\overset{\|}{C}}{-}H}$

(f) $CH_3CH{=}CH_2 \xrightarrow[(1)]{?} BrCH_2CH{=}CH_2 \xrightarrow[(2)]{?} BrCH_2CHOHCH_3$

(g) $CH_3CH_2CH_2Br \xrightarrow[(1)]{?} CH_3CH{=}CH_2 \xrightarrow[(2)]{?} CH_3CHOHCH_2Cl$

(h) $CH_3CH_2CH_2OH \xrightarrow[(1)]{?} CH_3CH{=}CH_2 \xrightarrow[(2)]{?} CH_3CHBrCH_3$

(i) $CH_2{=}CH_2 \xrightarrow[(1)]{?} CH_3CH_3 \xrightarrow[(2)]{?} CH_3CH_2Cl$

(j)

Identification and Structural Determination [Secs. 3.22, 3.23]

3.19 *n*-Hexane and 1-hexene are both colorless liquids with similar boiling points. What three simple tests would distinguish the two compounds? Indicate what you would see.

3.20 Give the structure of an alkene that would give each of the following product(s) on ozonolysis.

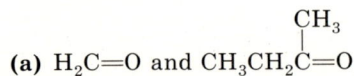

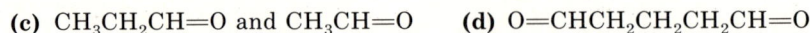

(a) $H_2C{=}O$ and $CH_3CH_2\overset{\displaystyle CH_3}{\overset{|}{C}}{=}O$ 　　(b) $2\ CH_3CH_2\overset{\displaystyle CH_3}{\overset{|}{C}}{=}O$

(c) $CH_3CH_2CH{=}O$ and $CH_3CH{=}O$ 　　(d) $O{=}CHCH_2CH_2CH_2CH{=}O$

3.21 Compound **A**, C_4H_9Br, on treatment with alcoholic KOH, gave compound **B**, C_4H_8. Treatment of **B** with ozone followed by zinc and water yields $CH_3CH_2CH{=}O$ and $O{=}CH_2$. What are the structures of **A** and **B**?

3.22 A compound C_4H_9Br **(A)**, on treatment with alcoholic KOH, gave compound **B**, C_4H_8. Treatment of **B** with ozone followed by zinc and water yielded $CH_3CH{=}O$ as the only product. What two possible structures for **B** account for these facts? What is the structure for **A**?

3.23 A compound **A**, C_4H_9Br, on treatment with alcoholic KOH, gave compound **B**, C_4H_8. Treatment of **B** with ozone followed by zinc and water yielded $(CH_3)_2C{=}O$ and $O{=}CH_2$. What two possible structures for **B** account for these facts? What is the structure for **A**?

3.24 By means of equations show how you would carry out the synthetic conversion indicated. For each step, indicate the reagents needed and the reaction conditions.

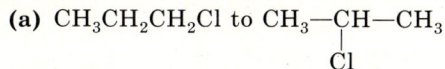

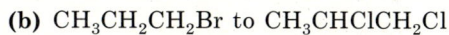

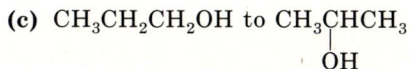

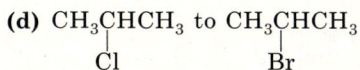

(a) $CH_3CH_2CH_2Cl$ to $CH_3-CH-CH_3$ with Cl on the central carbon

(b) $CH_3CH_2CH_2Br$ to $CH_3CHClCH_2Cl$

(c) $CH_3CH_2CH_2OH$ to CH_3CHCH_3 with OH

(d) CH_3CHCH_3 (with Cl) to CH_3CHCH_3 (with Br)

3.25 Beginning with ethanol, CH_3CH_2OH, as your only organic starting material, show how you would prepare each of the compounds listed. (You may use any other needed reagent.) Where more than one step is required, show each reaction clearly.

(a) Ethene
(b) Bromoethane
(c) Ethane
(d) 1,2-Dichloroethane
(e) 2-Chloroethanol, $HOCH_2CH_2Cl$
(f) Ethylene glycol, $HOCH_2CH_2OH$

4

Unsaturated Hydrocarbons II: Dienes, Polyenes, and Alkynes

In Chapter 3 we discussed the chemistry of alkenes. In this chapter we will study the chemistry of hydrocarbons that are even more unsaturated than simple alkenes. These include compounds that contain two or more double bonds as well as those that contain triple bonds.

4.1 Structure and Nomenclature of Dienes

Alkenes that contain two double bonds are called **dienes** (that is, they have two *ene* functions). If the double bonds are separated by only *one* single bond,

$-\overset{|}{C}=\overset{|}{C}-\overset{|}{C}=\overset{|}{C}-$, the diene is said to be **conjugated.**

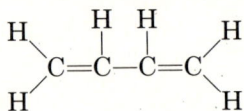

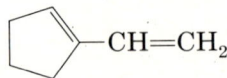

$$\underset{H}{\overset{H}{\diagdown}}C=\underset{H}{\overset{H}{\underset{|}{C}}}-\underset{}{\overset{H}{\underset{|}{C}}}=C\underset{H}{\overset{H}{\diagup}} \qquad \bigcirc\!\!-CH=CH_2$$

If the double bonds are separated by *more* than one single bond, the diene is called a **nonconjugated** diene, and the double bonds are said to be **isolated.**

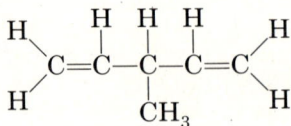

$$\bigcirc \qquad \underset{H}{\overset{H}{\diagdown}}C=\underset{H}{\overset{H}{\underset{|}{C}}}-\underset{CH_3}{\overset{H}{\underset{|}{C}}}-\underset{}{\overset{H}{\underset{|}{C}}}=C\underset{H}{\overset{H}{\diagup}}$$

Dienes are named by the IUPAC system in essentially the same way as alkenes except that the suffix *-adiene* replaces the ending *-ene* of the alkene. Also, *two* numbers are needed to indicate the locations of the double bonds in the chain. In cyclic dienes one of the double bonds is always assigned the number 1, and the other is given the lowest possible number. This system can be extended to compounds that contain three double bonds (**trienes**), four double bonds (**tetraenes**), or many double bonds (**polyenes**).

$$CH_2=CH-CH_2-CH=CH_2 \qquad\qquad CH_2=CH-CH=CH-CH=CH_2$$

1,4-Pentadiene   1,3,5-Hexatriene
(nonconjugated)   (conjugated)

1,3-Cyclohexadiene   1,3,5,7-Cyclooctatetraene
(conjugated)   (conjugated)

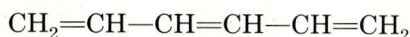

Problem 4.1   Write the structures of the following compounds.
(a) 1,3-Butadiene   **(b)** 1,4-Cyclohexadiene
(c) 2-Methyl-2,5-heptadiene   **(d)** 2-Methyl-1,3-cyclopentadiene
(e) 1-Vinylcyclobutene   **(f)** 1-Allylcyclopentene

Problem 4.2   Indicate which dienes in Problem 4.1 are conjugated and which contain isolated double bonds.

Problem 4.3   Name the following compounds.

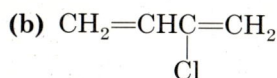

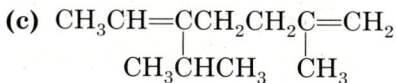

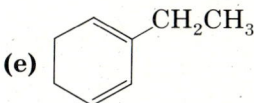

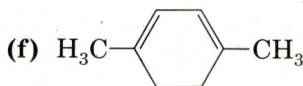

(a) $CH_3CH=CHCH=CH_2$

(b) $CH_2=CHC=CH_2$
            |
            Cl

(c) $CH_3CH=CCH_2CH_2C=CH_2$
            | |
        CH_3CHCH_3 CH_3

(d)

(e)

(f) $H_3C-$$-CH_3$

Preparation and Properties of Dienes *4.2*

Dienes may be prepared by applying any of the appropriate methods used to prepare simple alkenes (Sec. 3.5). Of course, the starting molecule must have the right kind of functional groups in the proper positions. For example, a conven-

ient method for preparing 1,3-butadiene in the laboratory is to heat 1,4-dichloro-
butane with an alcoholic solution of potassium hydroxide, a dehydrohalogenation reaction (Sec. 3.10).

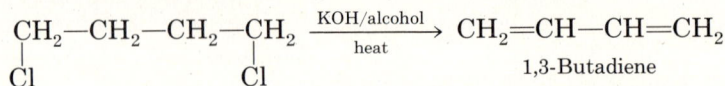

$$\underset{\substack{| \\ Cl}}{CH_2}-CH_2-\underset{\substack{| \\ Cl}}{CH_2}-CH_2 \xrightarrow[\text{heat}]{\text{KOH/alcohol}} \underset{\text{1,3-Butadiene}}{CH_2{=}CH{-}CH{=}CH_2}$$

1,3-Butadiene is a raw material used in the manufacture of synthetic rubber
(Sec. 4.6A).

As far as the chemical properties are concerned, dienes undergo the typical
reactions of unsaturated hydrocarbons—addition reactions. In nonconjugated
dienes the isolated double bonds have little effect on each other, and they therefore behave as if they were two separate alkenes. Consequently, the chemical
properties of nonconjugated dienes are the same as those of simple alkenes except for the fact that they consume twice as much reagent.

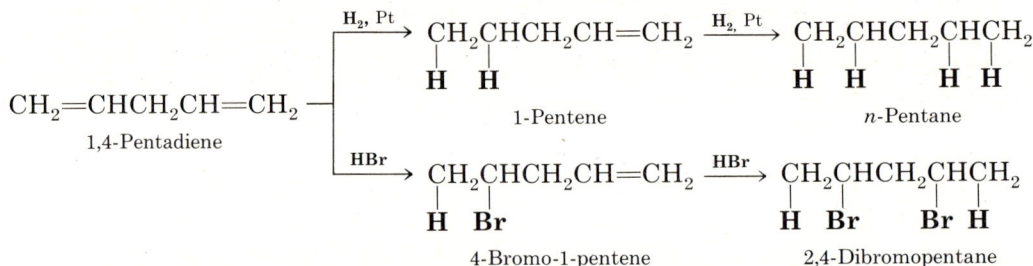

$$CH_2{=}CHCH_2CH{=}CH_2$$
1,4-Pentadiene

$\xrightarrow{H_2, Pt}$ $\underset{\substack{| \quad | \\ H \quad H}}{CH_2CHCH_2CH{=}CH_2}$ $\xrightarrow{H_2, Pt}$ $\underset{\substack{| \quad | \quad\quad | \quad | \\ H \quad H \quad\quad H \quad H}}{CH_2CHCH_2CHCH_2}$
1-Pentene · · · · · · · · · · · · · · n-Pentane

$\xrightarrow{HBr}$ $\underset{\substack{| \quad | \\ H \quad Br}}{CH_2CHCH_2CH{=}CH_2}$ $\xrightarrow{HBr}$ $\underset{\substack{| \quad | \quad\quad | \quad | \\ H \quad Br \quad\quad Br \quad H}}{CH_2CHCH_2CHCH_2}$
4-Bromo-1-pentene · · · · · · · · · · · 2,4-Dibromopentane

In conjugated dienes, on the other hand, the proximity of the double bonds
affects their chemical properties in ways that make them different from
nonconjugated dienes. This difference is evidenced by the fact that conjugated
dienes are *more stable* than nonconjugated dienes and that they undergo an
unexpected reaction, *1,4-addition*.

4.3 Stability of Conjugated Dienes: Heat of Hydrogenation

Experimentally, one of the methods used to determine the stability of a
diene is to measure the **heat of hydrogenation.** The heat of hydrogenation,
ΔH, is the amount of energy released (in kilocalories per mole) when the diene is
reduced to an alkane.

For example, if we compare the amounts of heat released when 1,4-pentadi-
ene and 1,3-pentadiene are converted to pentane, we find that hydrogenation of
the nonconjugated diene releases 60.8 kcal/mole, whereas the heat of hydrogenation of 1,3-pentadiene (a conjugated diene) is 54.1 kcal/mole (Fig. 4.1). The
product, pentane, has the same energy regardless of the starting diene. The fact
that 1,3-pentadiene releases less energy on hydrogenation than 1,4-pentadiene
means that the conjugated diene contains less energy, or is more stable, than the
nonconjugated diene.

$$CH_2{=}CH{-}CH_2{-}CH{=}CH_2 \text{ (nonconjugated)}$$

$$H_2, Pt$$

$$\Delta H = 60.8 \text{ kcal/mole}$$

$$CH_2{=}CH{-}CH{=}CH{-}CH_3 \text{ (conjugated)}$$

$$H_2, Pt$$

$$\Delta H = 54.1 \text{ kcal/mole}$$

$$CH_3CH_2CH_2CH_2CH_3$$
Pentane

Energy ⟶

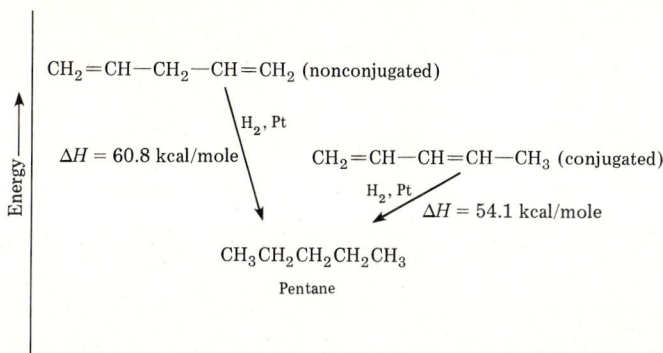

Figure 4.1 Conversion of two dienes to the same alkane releases different amounts of energy; 1,3-pentadiene (conjugated) is more stable than 1,4-pentadiene (nonconjugated).

Numerous experiments, similar to the one just described, have established that whenever two dienes, one conjugated and the other isolated, are converted to the same alkane, *invariably* the conjugated diene is more stable than the diene with the isolated double bonds.

Before we discuss *why* conjugated dienes are more stable, let us point out how they also differ from nonconjugated dienes in their addition reactions.

Electrophilic Addition to Conjugated Dienes: *4.4*
1,4-Addition

When a mole of a reagent capable of adding to an alkene reacts with a nonconjugated diene, the expected addition to *adjacent* double-bonded carbons is obtained. The addition of a reagent to a pair of adjacent carbons is called **1,2-addition.** For example, treatment of 1,4-pentadiene with just enough of a solution of bromine in carbon tetrachloride to form the dihalide gives the expected 1,2-addition product, 4,5-dibromo-1-pentene. Addition of another mole of bromine gives another 1,2-addition product, 1,2,4,5-tetrabromopentane.

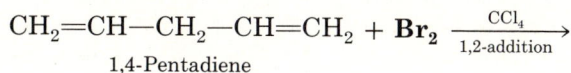

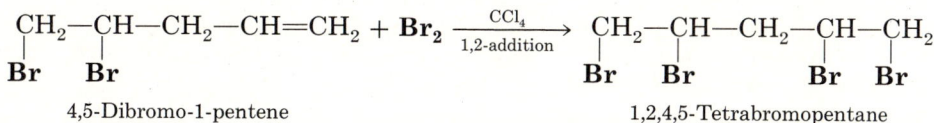

$$CH_2{=}CH{-}CH_2{-}CH{=}CH_2 + Br_2 \xrightarrow[\text{1,2-addition}]{CCl_4}$$
1,4-Pentadiene

$$\underset{\underset{Br}{|}}{CH_2}{-}\underset{\underset{Br}{|}}{CH}{-}CH_2{-}CH{=}CH_2 + Br_2 \xrightarrow[\text{1,2-addition}]{CCl_4} \underset{\underset{Br}{|}}{CH_2}{-}\underset{\underset{Br}{|}}{CH}{-}CH_2{-}\underset{\underset{Br}{|}}{CH}{-}\underset{\underset{Br}{|}}{CH_2}$$
4,5-Dibromo-1-pentene   1,2,4,5-Tetrabromopentane

Treatment of a conjugated diene with bromine under similar conditions gives, in addition to the expected 1,2-addition product, an unexpected **1,4-addition** product.

$$-\overset{|}{\underset{1}{C}}=\overset{|}{\underset{2}{C}}-\overset{|}{\underset{3}{C}}=\overset{|}{\underset{4}{C}}- \;+\; A-B \longrightarrow \; -\overset{|}{C}-\overset{|}{\underset{\underset{B}{|}}{\underset{A}{C}}}\underset{}{\overset{}{}}$$

1,2-Addition
(an expected product)

and

1,4-Addition
(an unexpected product)

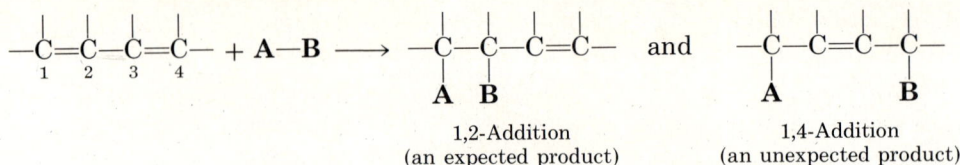

1,4-Addition occurs when a reagent attaches itself to the carbons at the two ends of a conjugated system.

Thus, treatment of 1,3-butadiene with bromine yields not only the expected 3,4-dibromo-1-butene (1,2-addition product) but also the unexpected 1,4-dibromo-2-butene (1,4-addition product). Treatment with hydrogen bromide, HBr, yields not only the expected product 3-bromo-1-butene (1,2-addition) but also an unexpected product, 1-bromo-2-butene (1,4-addition). Treatment with hydrogen in the presence of a suitable catalyst yields not only the expected 1-butene (1,2-addition) but also 2-butene (1,4-addition). These reactions are represented in Figure 4.2. *Frequently, the unexpected 1,4-addition product is the major one* (especially when reactions are conducted at 25°C and above).

We have just seen that conjugated dienes undergo an unexpected reaction, 1,4-addition. We have also observed that they are more stable than their nonconjugated isomers (Sec. 4.3). How do we account for these facts? The explanation is provided by a concept in structural theory called resonance.

Problem 4.4 Draw the structures of the product(s) that can result from the addition of 1 mole each of

(a) HBr to 2,4-hexadiene

(b) Br_2 in CCl_4 to 2,4-hexadiene

(c) H_2O, H^+ to 2,4-hexadiene

(d) HCl to 1,3-cyclohexadiene

(e) H_2/Pt to 1,4 pentadiene

(f) HBr to 1,4-cyclohexadiene

4.5 Resonance

To account for 1,4-addition, we must examine the mechanism of the reaction. As with simple alkenes, addition occurs by a two-step process. The first step involves the attack on the π bond by an electrophile to form a carbocation. Thus,

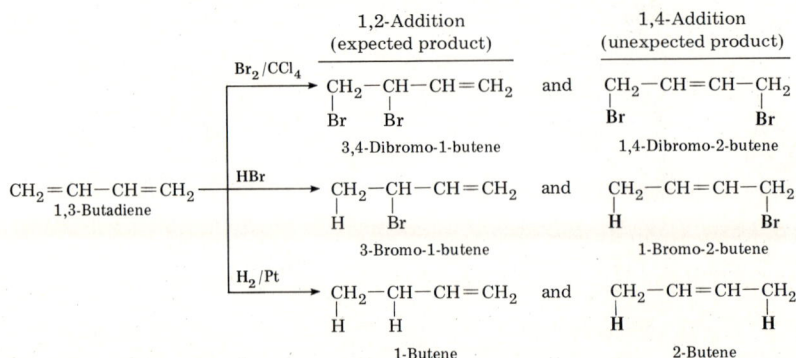

Figure 4.2 Addition of one mole of a reagent to a conjugated diene gives rise to an expected 1,2-addition product and an unexpected 1,4-addition product.

addition of bromine to 1,3-butadiene results in a secondary carbocation. This carbocation is also called an **allylic carbocation** because it is part of an allylic structure ($-CH_2-CH=CH_2$, the allyl group; $-\overset{+}{C}H-CH=CH_2$, the allylic carbocation).

$$CH_2=CH-CH=CH_2 \longrightarrow CH_2-\overset{+}{C}H-CH=CH_2 + {}^-:\textbf{Br}$$
$$\underset{\textbf{Br}}{\overset{\curvearrowleft}{}} \qquad\qquad \underset{\textbf{Br}}{|}$$

Allylic carbocation
(I)

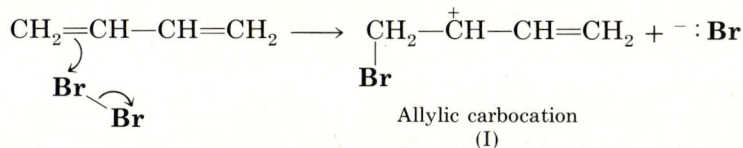

Actually, our representation of the allylic carbocation is inadequate. Figure 4.3 illustrates why this is so. As can be seen, the representation of the allylic carbocation (I) shows that the positive charge is on a carbon having a vacant p orbital. This p orbital is next to a π bond. The proximity of π electrons to the positive charge will cause them to be attracted to it, resulting in structure II, also an allylic carbocation. In II, as in I, there is a positively charged carbon with a vacant p orbital, and next to it there is a π bond. The π electrons in II are also attracted to the positive charge, resulting in allylic carbocation (I). In effect the π electrons are not bonded specifically to two atoms but are dispersed over three atoms. The term **delocalization** is used whenever π electrons *simultaneously* form a bond between more than a single pair of atoms. The phenomenon of delocalization of electrons is called **resonance.**

Note that neither structure I nor II truly represents the allylic carbocation. The allylic carbocation is actually a new type of species called a **resonance hybrid.** A resonance hybrid is a *single* species that combines the characteristics of two or more structures that differ only in their electronic arrangements. The individual structures are called **contributing structures** that have no real existence. The symbol $\longleftrightarrow$ is used to represent resonance between contributing structures of a resonance hybrid. The double-headed arrow should not be confused with $\rightleftharpoons$, which indicates an equilibrium between real structures. The resonance hybrid is illustrated in Figure 4.4.

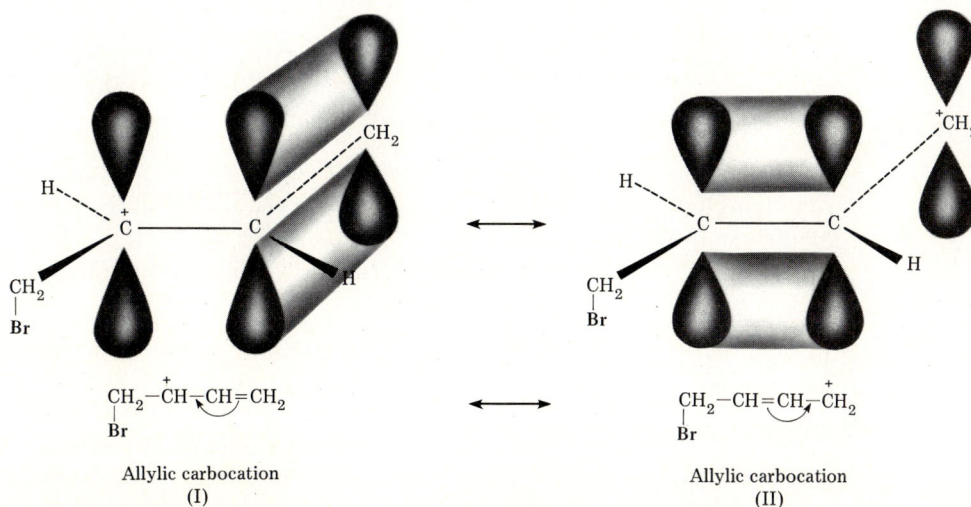

Allylic carbocation
(I)

Allylic carbocation
(II)

Figure 4.3 Two contributing structures of an allylic carbocation, neither of which exists in reality.

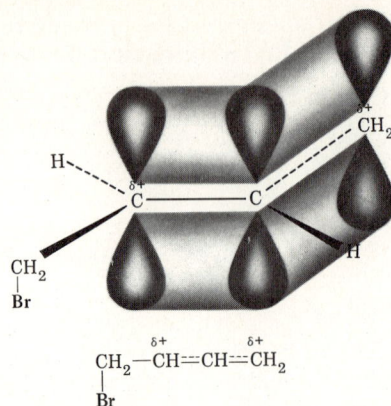

$$CH_2-CH\overset{\delta+}{=\!=\!=}CH\overset{\delta+}{=\!=\!=}CH_2$$
$$|$$
$$Br$$

Figure 4.4 Two representations of an allylic carbocation resonance hybrid. Below the orbital picture is a single structure containing broken bond lines to indicate the distribution of the positive charge between the C-2 and C-4 positions.

Because of the delocalization of the positive charge over two atoms, C-2 and C-4, the resonance hybrid is more stable (i.e., contains less energy) than any of the contributing structures where the positive charge is localized on one carbon atom only. The extra stability of the resonance hybrid is called the **resonance energy.**

With this more accurate picture of the allylic carbocation we can now account for the formation of the unexpected 1,4-addition product. The attack by $^-$:Br, the nucleophile, in the second step of the addition mechanism can take place at either C-2 or C-4. Attack at C-2 gives the 1,2-addition product, and attack at C-4 gives the 1,4-addition product (Fig. 4.5).

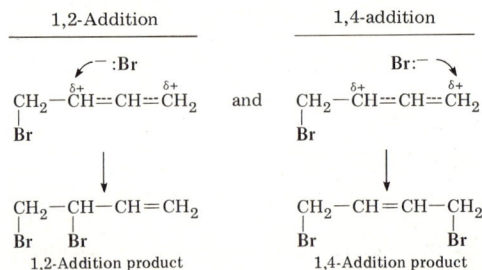

Figure 4.5 Addition of the bromide ion to C-2 and C-4 giving a 1,2-addition and a 1,4-addition product, respectively.

Problem 4.5 The carboxylate anion, R—C(=O)(O:$^-$), is stabilized by resonance since the negative charge can be delocalized over both oxygen atoms. Write formulas for **(a)** the two contributing resonance structures and **(b)** the resonance hybrid using dashed bond lines to indicate the distribution of the negative charge over the two oxygen atoms.

Polymerization of Conjugated Dienes: Polyenes *4.6*

A Natural and Synthetic Rubber

Like ethylene or substituted ethylenes (Sec. 3.25), conjugated dienes may polymerize into compounds that still contain many double bonds. Compounds that contain several double bonds are generally referred to as polyenes. The polymerization of conjugated dienes is commercially important because both natural and synthetic rubber are polymers of conjugated dienes. Natural rubber is a polymer of the conjugated diene 2-methyl-1,3-butadiene or **isoprene.**

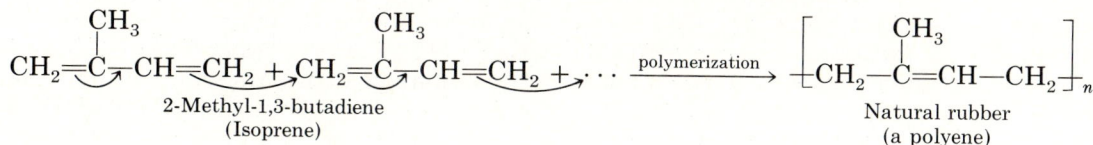

$$CH_2=\overset{\underset{\displaystyle CH_3}{|}}{C}-CH=CH_2 + CH_2=\overset{\underset{\displaystyle CH_3}{|}}{C}-CH=CH_2 + \cdots \xrightarrow{\text{polymerization}} \left[CH_2-\overset{\underset{\displaystyle CH_3}{|}}{C}=CH-CH_2 \right]_n$$

2-Methyl-1,3-butadiene
(Isoprene)

Natural rubber
(a polyene)

Synthetic rubbers are made through polymerization of different conjugated dienes. For example, the polymerization of 1,3-butadiene produces a soft rubbery polymer manufactured since 1927 under the trade name of *Buna rubber.*

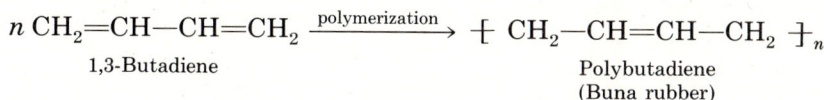

$$n\ CH_2=CH-CH=CH_2 \xrightarrow{\text{polymerization}} \left[CH_2-CH=CH-CH_2 \right]_n$$

1,3-Butadiene

Polybutadiene
(Buna rubber)

The similar polymerization of its derivative 2-chloro-1,3-butadiene, or *chloroprene,* forms a rubber substitute called *neoprene.* Neoprene is superior to natural rubber in its resistance to oil, gasoline, and other organic solvents.

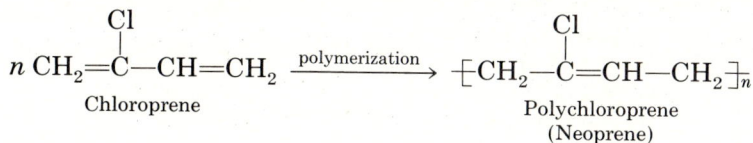

$$n\ CH_2=\overset{\underset{\displaystyle Cl}{|}}{C}-CH=CH_2 \xrightarrow{\text{polymerization}} \left[CH_2-\overset{\underset{\displaystyle Cl}{|}}{C}=CH-CH_2 \right]_n$$

Chloroprene

Polychloroprene
(Neoprene)

Natural or synthetic rubber is too soft to be of much use in industry. Its hardness and durability can be improved by **vulcanization,** a technique that was discovered accidentally in 1834 by Charles Goodyear. He found that when rubber is treated with sulfur (vulcanized) a tougher polymer is produced. The sulfur atoms form bridges between different chains in the rubber molecule. These cross-links make the rubber harder and stronger than the original unvulcanized rubber.

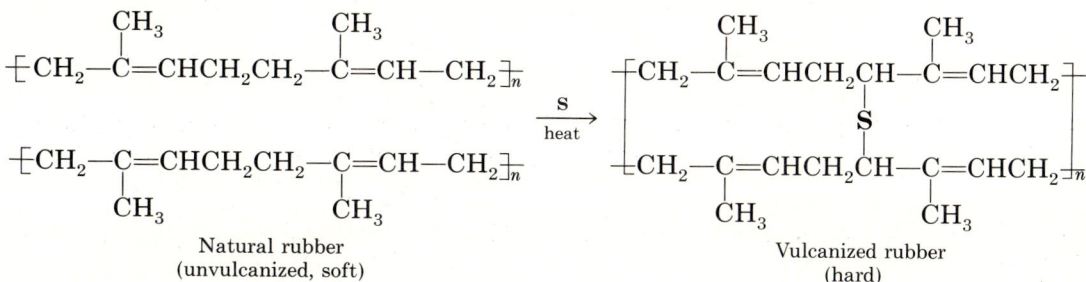

Natural rubber
(unvulcanized, soft)

Vulcanized rubber
(hard)

The process of vulcanization made possible the manufacture of tough rubber tires for the automobile industry. We frequently observe that old tires lose their flexibility, become stiff and brittle, and develop small cracks. This is because, like simple alkenes, the rubber undergoes ozonolysis with the ozone and moisture present in the atmosphere, resulting in the cleavage of the long polymer chain into smaller chains.

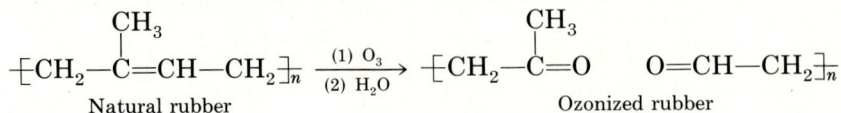

$$\underset{\text{Natural rubber}}{\left[CH_2-\underset{\underset{CH_3}{|}}{C}=CH-CH_2\right]_n} \xrightarrow[\text{(2) } H_2O]{\text{(1) } O_3} \underset{\text{Ozonized rubber}}{\left[CH_2-\underset{\underset{CH_3}{|}}{C}=O \quad O=CH-CH_2\right]_n}$$

Other synthetic rubber polymers may also be prepared by the process of copolymerization. **Copolymerization** occurs when two or more different unsaturated monomers are mixed together and are allowed to polymerize. In the polymer thus formed the monomeric units may be distributed in a random manner or they may alternate along the polymer chain. During World War II, when natural rubber became unavailable to this country, an excellent synthetic substitute called SBR (<u>s</u>tyrene-<u>b</u>utadiene <u>r</u>ubber) was developed from the copolymerization of 1,3-butadiene and styrene, $CH_2=CHC_6H_5$. SBR, which has since been used extensively in the automobile industry, consists of about three parts butadiene and one part styrene.

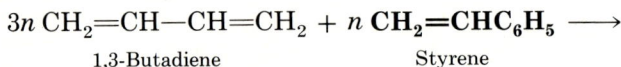

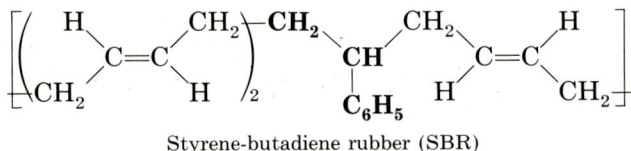

$$3n \ \textbf{CH}_2=\textbf{CH}-\textbf{CH}=\textbf{CH}_2 + n \ \textbf{CH}_2=\textbf{CHC}_6\textbf{H}_5 \longrightarrow$$

1,3-Butadiene   Styrene

Styrene-butadiene rubber (SBR)

B   Natural Polyenes

Some natural polyenes, called **terpenes,** are found in many plants. Terpenes may be viewed as consisting of **isoprene units** joined together in groups of two, three, four, six, or eight. An isoprene unit is a sequence of five carbon atoms that resembles isoprene.

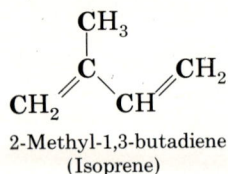

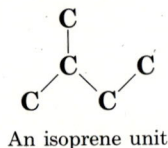

2-Methyl-1,3-butadiene
(Isoprene)

An isoprene unit

Terpenes are classified according to the number of isoprene units, as indicated in Table 4.1.

Geraniol, which has a sweet rose odor, and limonene, which gives lemon and

Table 4.1   Classes of Terpenes

	Number of carbon atoms	Number of isoprene units
Monoterpenes	10	2
Sesquiterpenes	15	3
Diterpenes	20	4
Triterpenes	30	6
Tetraterpenes	40	8

orange peels their characteristic odors, are examples of monoterpenes. The isoprene units are shaded.

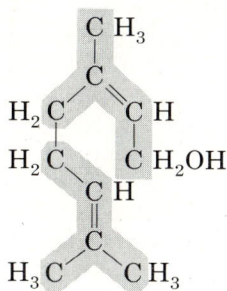

Geraniol
(rose)

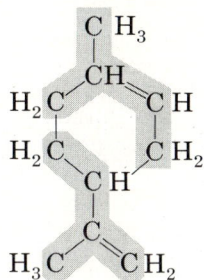

Limonene
(lemon, orange)

Vitamin A, a fat-soluble vitamin essential for resisting infections and for proper vision, is an example of a diterpene.

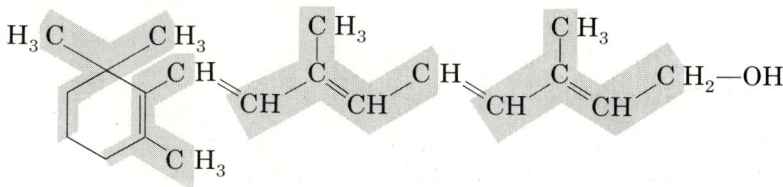

Vitamin A
(Retinol)

Problem 4.6   The reddish yellow pigments found in tomatoes, carrots, and other fruits and vegetables are terpenes. Lycopene occurs in ripe tomatoes and watermelon, and β-carotene is found in carrots, tomatoes, and spinach. From the given structures determine how many isoprene units each has.
(a) Lycopene

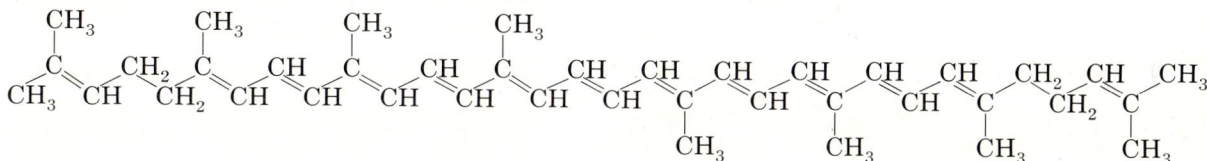

(b) β-Carotene

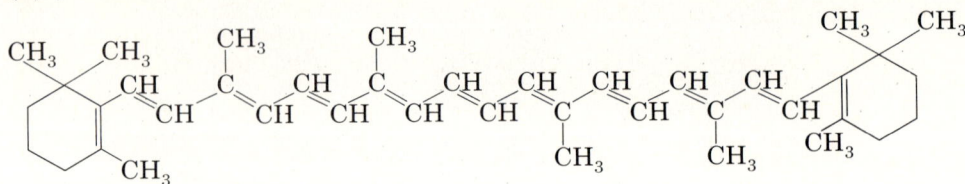

$CH_3 \quad CH_3 \qquad CH_3 \qquad CH_3 \qquad\qquad\qquad CH_3 \quad CH_3$

$CH \quad C \quad CH \quad C \quad CH \quad CH \quad CH \quad CH \quad CH$
$CH \qquad CH \quad CH \quad CH \quad CH \quad C \qquad CH \quad C \qquad CH$

$CH_3 \qquad\qquad\qquad\qquad CH_3 \qquad CH_3 \qquad CH_3$

4.7 Alkynes

An **alkyne** is an unsaturated hydrocarbon that contains a **carbon–carbon triple bond.** Like alkanes and alkenes, alkynes also form a homologous series, the increment again being —CH_2—, a methylene group. The general formula that characterizes the alkyne series is C_nH_{2n-2}. As the formula indicates, alkynes are even more unsaturated than alkenes, and we would expect them to undergo the same reactions as alkenes; namely, electrophilic addition. This is in fact the case, except that twice as much reagent can be added to alkynes as can be added to alkenes. Alkynes are also synthesized in much the same manner as alkenes, except that twice as many atoms are eliminated from adjacent carbon atoms.

4.8 Geometry of the Carbon–Carbon Triple Bond: *sp* Hybridization

The simplest member of the alkyne series is acetylene, C_2H_2.

$$H—C\equiv C—H$$

Acetylene

From the structure of acetylene we see that each carbon is bonded to two other atoms. Acetylene is also a linear molecule with a bond angle of 180°. This geometry can be explained by a third type of hybridization called *sp* hybridization ($\frac{1}{2}s$ and $\frac{1}{2}p$ characteristics). As with alkanes and alkenes, carbon proceeds first to its excited state, but in the hybridization step only one *s* and one *p* orbital mix together to form two equivalent *sp* orbitals. This leaves two unhybridized *p* orbitals, which are perpendicularly oriented to each other and to the plane of the hybrid *sp* orbitals (Fig. 4.6).

The union of the two carbons in acetylene occurs by the end-on overlap of their *sp* orbitals to form a σ bond. The remaining *sp* orbitals bond with the *s* orbital of hydrogen and also form σ bonds. The two unhybridized *p* orbitals overlap in a sidewise fashion to form two π bonds. Both carbons thus are joined by one σ bond and two π bonds, which together form a triple bond (Fig. 4.7).

Not only is acetylene a linear molecule, but substituted acetylenes R—C≡C—H or R—C≡C—R are also linear, and for this reason *cis-trans* isomerism is impossible in this class of compounds. The carbon–carbon triple bond is

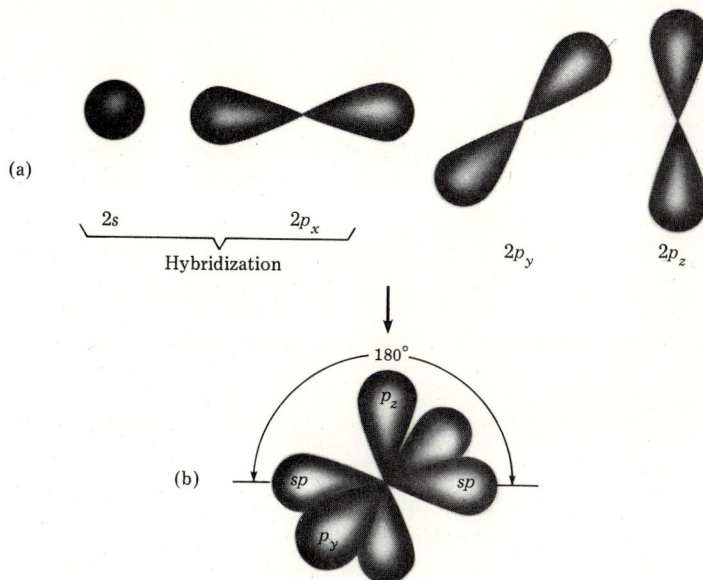

Figure 4.6 **(a)** Hybridization of a 2s orbital and a 2p orbital to form **(b)** two linear sp-hybridized orbitals with bond angle of 180° and two unhybridized p orbitals perpendicularly oriented to each other and to the plane of the hybrid sp orbitals.

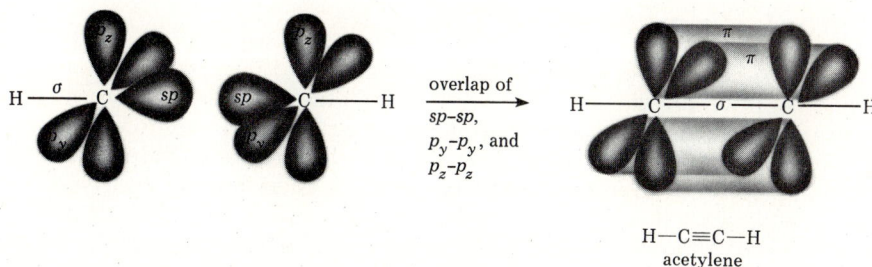

Figure 4.7 End-on overlap of two sp orbitals to form a σ bond between the two carbons of acetylene and side-side overlap of the p_y orbitals and the p_z orbitals to form two π bonds.

made up of a strong σ bond and two weaker π bonds. Consequently, it is both shorter (1.20 Å) and stronger (200 kcal/mole) than the carbon–carbon double bond of alkenes (1.34 Å and 83 kcal/mole).

Nomenclature of Alkynes 4.9

Although alkynes can be named by both common and IUPAC nomenclatures, all alkynes except acetylene are generally named by the IUPAC nomenclature. The IUPAC rules for alkynes are the same as those used for alkenes, except that the ending -yne replaces -ene.

	HC≡CH	CH₃C≡CH	CH₃CH₂C≡CH	CH₃C≡CCH₃
IUPAC name:	Ethyne	Propyne	1-Butyne	2-Butyne
Common name:	Acetylene			

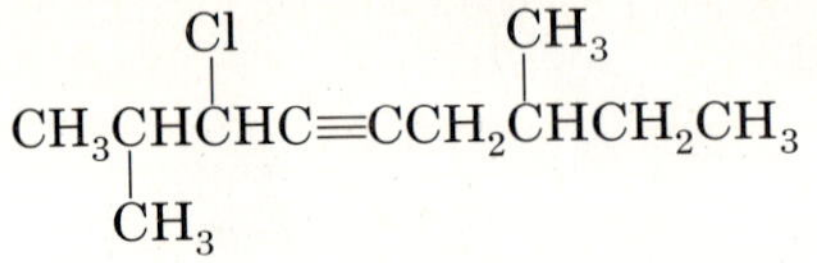

$$CH_3CHCHC{\equiv}CCH_2CHCH_2CH_3$$

with Cl and CH₃ and CH₃ substituents

3-Chloro-2,7-dimethyl-4-nonyne

Problem 4.7 Write the condensed structural formula for
(a) 1-Pentyne
(b) 3-Hexyne
(c) 3,3-Dimethyl-1-butyne
(d) 1-Cyclohexyl-2-butyne

Problem 4.8 The names given for the compounds listed here are incorrect. Draw their structures and give their correct name.
(a) 4-Pentyne
(b) 2-Chloro-2-*n*-propyl-3-butyne
(c) 2,2-Dibromo-5-methyl-3-pentyne
(d) 4,4-Dimethyl-2-butyne

4.10 Physical Properties of Alkynes

Alkynes have physical properties that are essentially the same as those of alkenes and alkanes. They are insoluble in water but quite soluble in the usual organic solvents such as benzene, ether, and carbon tetrachloride. Like alkanes and alkenes, alkynes are less dense than water; their boiling points show the usual increase with increasing molecular weight; and their boiling points are close to those of alkanes or alkenes with the same carbon skeleton. Like alkanes or alkenes the C_2–C_4 alkynes are gases, the C_5–C_{18} alkynes are liquids, and those with more than eighteen carbons are solids.

4.11 Preparation of Alkynes

Basically, alkynes are synthesized by two methods.

1. Dehydrohalogenation of alkyl dihalides using a strong base

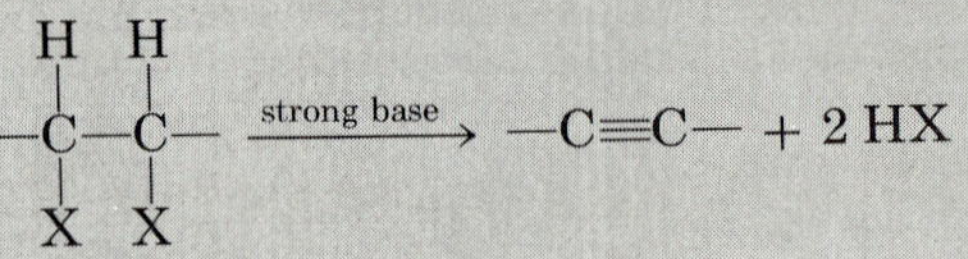

$$-\underset{\underset{X}{|}}{\overset{\overset{H}{|}}{C}}-\underset{\underset{X}{|}}{\overset{\overset{H}{|}}{C}}- \xrightarrow{\text{strong base}} -C{\equiv}C- + 2\,HX$$

2. Reaction of sodium acetylide with 1° alkyl halides

$$-C\equiv C:^- Na^+ + \quad R-X \quad \longrightarrow \quad -C\equiv C-R + NaX$$

Sodium acetylide   1° alkyl halide

A Dehydrohalogenation of Alkyl Dihalides

Treatment of vicinal dihalides (compounds that contain halogen atoms on adjacent carbon atoms) with strong bases, alcoholic KOH followed by sodium amide, $NaNH_2$, results in the formation of alkynes. This is a useful preparation since the vicinal dihalides themselves are readily obtained from the addition of halogen to corresponding alkenes (Sec. 3.14). This reaction therefore provides a general method for the conversion of alkenes to alkynes. The following example illustrates this point.

$$CH_3-CH_2-CH=CH_2 + Br_2 \xrightarrow{CCl_4} CH_3-CH_2-\underset{\underset{\textstyle Br}{|}}{CH}-\underset{\underset{\textstyle Br}{|}}{CH_2}$$

1-Butene   1,2-Dibromobutane
(a vicinal dihalide)

$$CH_3-CH_2-\underset{\underset{\textstyle Br}{|}}{CH}-\underset{\underset{\textstyle Br}{|}}{CH_2} \xrightarrow[\text{heat}]{KOH/alcohol} CH_3-CH_2-CH=\overset{\overset{\textstyle H}{|}}{C}-Br$$

1-Bromo-1-butene
(a vinyl bromide)

$$CH_3-CH_2-CH=\overset{\overset{\textstyle H}{|}}{C}-Br \xrightarrow[\text{heat}]{NaNH_2} CH_3-CH_2-C\equiv CH$$

1-Butyne

1-Bromo-1-butene, the compound obtained from dehydrohalogenation of 1,2-dibromobutane with KOH in alcohol, is a *vinylic* bromide. Vinyl halides and all halides where the halogen is attached to a double-bonded carbon atom (sp^2-hybridized carbon) are more difficult to dehydrohalogenate than are alkyl halides. This is why a stronger base than alcoholic KOH, such as sodium amide, is needed to eliminate the second mole of hydrogen halide to give the alkyne.

Problem 4.9 Starting with 1-pentene, show how you would synthesize **(a)** 1-pentyne and **(b)** 2-pentyne.

B Reaction of Sodium Acetylide with Primary Alkyl Halides

Acetylene and monosubstituted acetylenes, $R-C\equiv C-H$, contain a hydrogen atom attached to a triple-bonded carbon atom. Such *acetylenic* hydrogens are somewhat acidic and may be replaced by certain metals to form salts known as **metal acetylides.** For example, sodium in liquid ammonia reacts with acetylene to form a sodium acetylide salt and hydrogen.

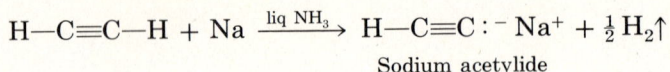

$$H-C\equiv C-H + Na \xrightarrow{\text{liq } NH_3} H-C\equiv C:^- Na^+ + \tfrac{1}{2}H_2\uparrow$$

Sodium acetylide

This sodium salt can react with primary alkyl halides to form higher alkynes with the triple bond at the end of the chain (that is, terminal alkynes).

General equation

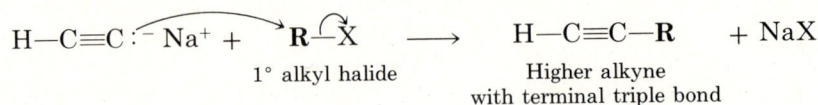

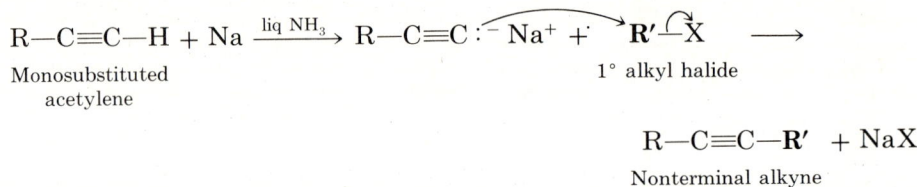

$$H-C\equiv C:^- Na^+ + \quad R-X \longrightarrow H-C\equiv C-R \quad + NaX$$

1° alkyl halide   Higher alkyne with terminal triple bond

Specific examples

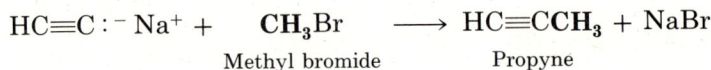

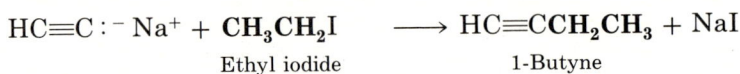

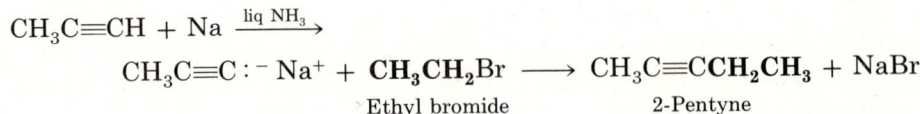

$$HC\equiv C:^- Na^+ + \quad CH_3Br \longrightarrow HC\equiv CCH_3 + NaBr$$

Methyl bromide   Propyne

$$HC\equiv C:^- Na^+ + CH_3CH_2I \longrightarrow HC\equiv CCH_2CH_3 + NaI$$

Ethyl iodide   1-Butyne

Monosubstituted acetylenes can react in a similar fashion, but give higher alkynes with nonterminal triple bonds.

General equation

$$R-C\equiv C-H + Na \xrightarrow{\text{liq } NH_3} R-C\equiv C:^- Na^+ + \quad R'-X \longrightarrow$$

Monosubstituted acetylene   1° alkyl halide

$$R-C\equiv C-R' \quad + NaX$$

Nonterminal alkyne

Specific example

$$CH_3C\equiv CH + Na \xrightarrow{\text{liq } NH_3}$$

$$CH_3C\equiv C:^- Na^+ + CH_3CH_2Br \longrightarrow CH_3C\equiv CCH_2CH_3 + NaBr$$

Ethyl bromide   2-Pentyne

With secondary and tertiary alkyl halides, sodium acetylides generally bring about dehydrohalogenation to give alkenes rather then the desired alkylation product.

Problem 4.10 Starting with acetylene, show how you would synthesize the following compounds (you may use any other needed reagents).
(a) 2-Butyne   **(b)** 1-Pentyne   **(c)** 2-Pentyne

Alkynes undergo reactions that are similar to those of alkenes, except that they are capable of adding two molecules of a reagent for each triple bond present. Addition of reagents occurs in two stages, producing first an alkene and, upon further addition, a saturated compound. By proper selection of experimental conditions, it is possible to stop the reaction at the alkene stage, but it is usually difficult to do so. We shall focus our attention on the following reactions.

1. Addition of hydrogen

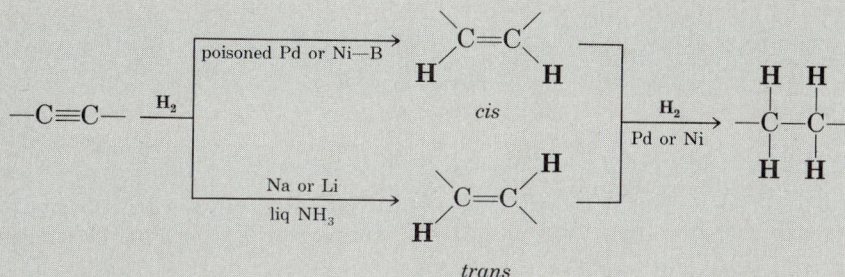

$$-C \equiv C- \xrightarrow{H_2}
\begin{cases}
\xrightarrow{\text{poisoned Pd or Ni—B}} & \underset{H}{\overset{\diagdown}{C}} = \underset{H}{\overset{\diagup}{C}} \quad cis \\
\\
\xrightarrow[\text{liq NH}_3]{\text{Na or Li}} & \underset{H}{\overset{\diagdown}{C}} = \underset{H}{\overset{\diagup}{C}} \quad trans
\end{cases}
\xrightarrow[\text{Pd or Ni}]{H_2} \underset{H \ H}{\overset{H \ H}{-C-C-}}$$

2. Addition of halogen

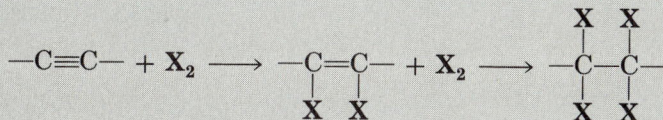

$$-C \equiv C- + X_2 \longrightarrow \underset{X \ X}{-C=C-} + X_2 \longrightarrow \underset{X \ X}{\overset{X \ X}{-C-C-}}$$

3. Addition of hydrogen halide

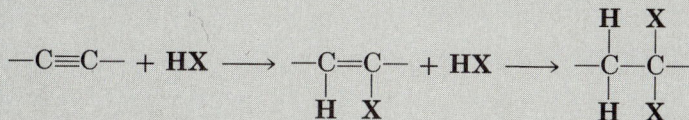

$$-C \equiv C- + HX \longrightarrow \underset{H \ X}{-C=C-} + HX \longrightarrow \underset{H \ X}{\overset{H \ X}{-C-C-}}$$

4. Addition of water: hydration

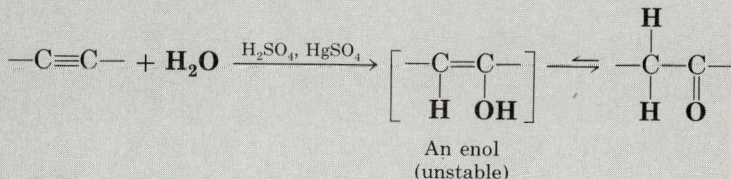

$$-C \equiv C- + H_2O \xrightarrow{H_2SO_4, \ HgSO_4} \left[\underset{H \ OH}{-C=C-}\right] \rightleftharpoons \underset{H \ O}{\overset{H}{-C-C-}}$$

An enol
(unstable)

A   Addition of Hydrogen

Like alkenes, alkynes may add hydrogen in the presence of suitable catalysts, such as platinum, palladium, or nickel. Under these conditions, however, the hydrogenation cannot be stopped at the alkene stage, and the final product is always an alkane.

$$-C{\equiv}C- \xrightarrow{\text{H}_2/\text{Pd (or Ni)}} \left[\underset{\text{H}}{\overset{}{>}}C{=}C\underset{\text{H}}{\overset{}{<}} \right] \xrightarrow{\text{H}_2/\text{Pd (or Ni)}} -\underset{\underset{\text{H}}{|}}{\overset{\overset{\text{H}}{|}}{C}}-\underset{\underset{\text{H}}{|}}{\overset{\overset{\text{H}}{|}}{C}}-$$

An alkane

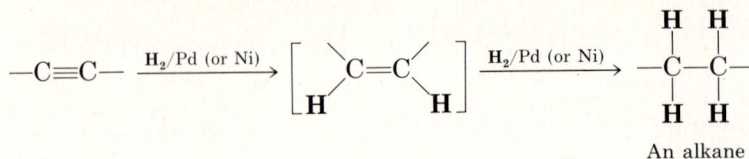

Conversion of alkynes to the alkene stage can be accomplished only by using special catalysts. Depending on the choice of catalyst, the product may be a *cis*-alkene or a *trans*-alkene, where such isomerism is possible. Addition of a controlled amount of hydrogen either in the presence of deactivated or *poisoned* palladium (which is prepared by adding lead acetate to palladium) or using a nickel boride catalyst, Ni—B, gives almost entirely a *cis*-alkene.

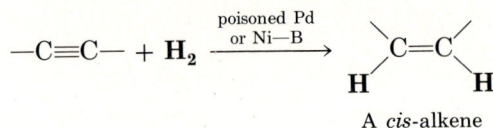

$$-C{\equiv}C- + \text{H}_2 \xrightarrow[\text{or Ni--B}]{\text{poisoned Pd}} \underset{\text{H}}{\overset{}{>}}C{=}C\underset{\text{H}}{\overset{}{<}}$$

A *cis*-alkene

Treatment of alkynes with sodium or lithium in liquid ammonia yields predominantly *trans*-alkenes.

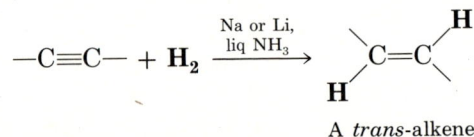

$$-C{\equiv}C- + \text{H}_2 \xrightarrow[\text{liq NH}_3]{\text{Na or Li,}} \underset{\text{H}}{\overset{}{>}}C{=}C\overset{\text{H}}{\underset{}{<}}$$

A *trans*-alkene

B   Addition of Halogen

Alkynes react with equivalents of halogen to give tetrahalides. As with alkenes, this reaction is restricted to chlorine and bromine.

General equation

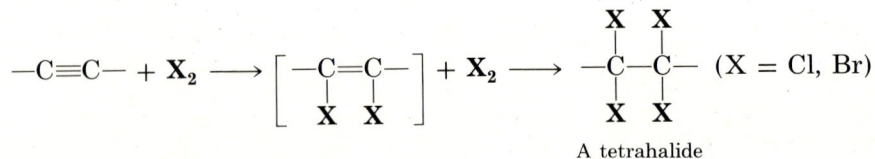

$$-C{\equiv}C- + \text{X}_2 \longrightarrow \left[-\underset{\underset{\text{X}}{|}}{C}{=}\underset{\underset{\text{X}}{|}}{C}- \right] + \text{X}_2 \longrightarrow -\underset{\underset{\text{X}}{|}}{\overset{\overset{\text{X}}{|}}{C}}-\underset{\underset{\text{X}}{|}}{\overset{\overset{\text{X}}{|}}{C}}- \quad (\text{X = Cl, Br})$$

A tetrahalide

Specific example

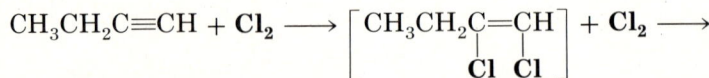

$$CH_3CH_2C{\equiv}CH + Cl_2 \longrightarrow \left[CH_3CH_2\underset{\underset{\text{Cl}}{|}}{C}{=}\underset{\underset{\text{Cl}}{|}}{CH} \right] + Cl_2 \longrightarrow$$

1-Butyne

$$CH_3CH_2\underset{\underset{\text{Cl}}{|}}{\overset{\overset{\text{Cl}}{|}}{C}}-\underset{\underset{\text{Cl}}{|}}{\overset{\overset{\text{Cl}}{|}}{CH}}$$

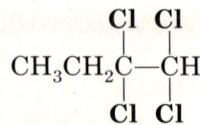

1,1,2,2-Tetrachlorobutane

C Addition of Hydrogen Halide

The addition of hydrogen chloride, bromide, or iodide to alkynes follows Markovnikov's rule (Sec. 3.17). The reaction proceeds in two steps and may be stopped at the haloalkene (or vinyl halide) stage or, if allowed to react further with another mole of HX, to the *gem*-dihalide stage. The term *gem*-dihalide (from the Latin *geminus,* twin) signifies that both halogens are on the same carbon atom.

General equation

$$-C\equiv C- + HX \longrightarrow \underset{\substack{| \quad | \\ H \quad X}}{-C=C-} + HX \longrightarrow \underset{\substack{| \quad | \\ H \quad X}}{\overset{\substack{H \quad X \\ | \quad |}}{-C-C-}}$$

A haloalkene
(A vinyl halide)   A *gem*-dihalide

Specific example

$$CH_3CH_2C\equiv CH + HBr \longrightarrow \underset{\substack{| \quad | \\ Br \quad H}}{CH_3CH_2C=CH} + HBr \longrightarrow \underset{\substack{| \quad | \\ Br \quad H}}{\overset{\substack{Br \quad H \\ | \quad |}}{CH_3CH_2C-CH}}$$

1-Butyne   2-Bromo-1-butene   2,2-Dibromobutane

Problem 4.11 Starting with acetylene, show how you would synthesize these compounds.
(a) 1,1-Dibromoethane   **(b)** 1,2-Dibromoethane
(c) 2,2-Dibromobutane   **(d)** 2,3-Dibromobutane

D Addition of Water: Hydration

Water adds to alkynes in the presence of dilute sulfuric acid and mercuric sulfate catalyst. The addition of water follows Markovnikov's rule (Sec. 3.17) to give initially an adduct called an enol.

$$-C\equiv C- + H-OH \xrightarrow{H_2SO_4,\ HgSO_4} \left[\underset{}{\overset{H}{\diagdown}}C=C\diagup_{OH} \right]$$

Initial adduct: an enol
(unstable)

The term **enol** is derived from the combination of the suffixes -*ene*, characteristic of alkenes, and -*ol,* characteristic of alcohols. Structurally, an enol is any compound that contains a carbon–carbon double bond (alkenes) to which is attached a hydroxyl group (alcohols). Enols are unstable compounds and are converted immediately to more stable products that contain a C=O or carbonyl group. For example, the addition of water to acetylene gives first vinyl alcohol, an unstable enol,

$$H-C\equiv C-H + H-OH \xrightarrow{H_2SO_4,\ HgSO_4} \left[\begin{matrix} H \\ \\ H \end{matrix} C=C \begin{matrix} H \\ \\ OH \end{matrix} \right]$$

Vinyl alcohol
(an unstable enol)

which rearranges to a carbonyl compound, acetaldehyde.

Acetaldehyde
(a stable carbonyl compound)

Acetaldehyde is but a single member of a larger class of compounds known as aldehydes ($R-\overset{H}{\underset{}{C}}=O$), which will be discussed in Chapter 10. Acetylene is the only alkyne that will give an aldehyde upon hydration. Substituted acetylenes also undergo hydration, but yield only ketones ($R-\overset{O}{\overset{\|}{C}}-R'$). For example, the hydration of propyne yields the ketone acetone.

$$CH_3-C\equiv C-H + H-OH \longrightarrow \left[CH_3-\underset{O-H}{C}=CH_2 \right] \rightleftharpoons CH_3-\overset{}{\underset{O}{C}}-CH_3$$

Propyne   An enol   Acetone
(unstable)

Because all enols, except vinyl alcohol, are converted to ketones, the spontaneous conversion of the unstable enol into a carbonyl, or **keto,** form is referred to as **enol–keto tautomerism,** an isomerization reaction. Note that the enol and keto forms of a compound, also called **tautomers,** are *distinct molecules.* They should not be confused with the resonance forms (Sec. 4.5), which have no real independent existence. Enol–keto tautomerism is an *equilibrium reaction* that involves the shift of a proton and of an electron pair. In the hydration of alkynes, this equilibrium strongly favors the keto side.

Problem 4.12 Draw the structures of the product formed from the hydration of the following compounds. Draw the enol and keto forms of each product.

(a) $CH_3CH_2CH_2C\equiv CH$ **(b)** $CH_3C\equiv CCH_3$ **(c)**

Acetylene is the only alkyne of significant industrial importance. Acetylene is produced industrially by the pyrolysis of methane.

$$6 \text{ CH}_4 + \text{O}_2 \xrightarrow[1500°C]{} 2 \text{ HC}\equiv\text{CH} + 2 \text{ CO}\uparrow + 10 \text{ H}_2\uparrow$$

Acetylene

One of the earlier uses for acetylene was as an illuminant because, when mixed with the proper amount of air, it burns with a bright, intense white flame. "Carbide" lamps, as they were called at the time, were widely used as a source of light prior to the invention of electric bulbs. Today, about half of the acetylene consumed in the United States is used for welding, cutting, and cleaning iron and steel by means of the oxyacetylene torch. Because of its low cost and high chemical reactivity, acetylene is also used as the starting material for a wide variety of important organic compounds, such as acetaldehyde, acetic acid, plastics, and rubber compounds. For example, catalytic addition of acetic acid to acetylene yields vinyl acetate, which is used to manufacture plastics such as polyvinyl acetate and polyvinyl alcohol (PVA), from which floor coverings, upholstering materials, and shoe soles are made.

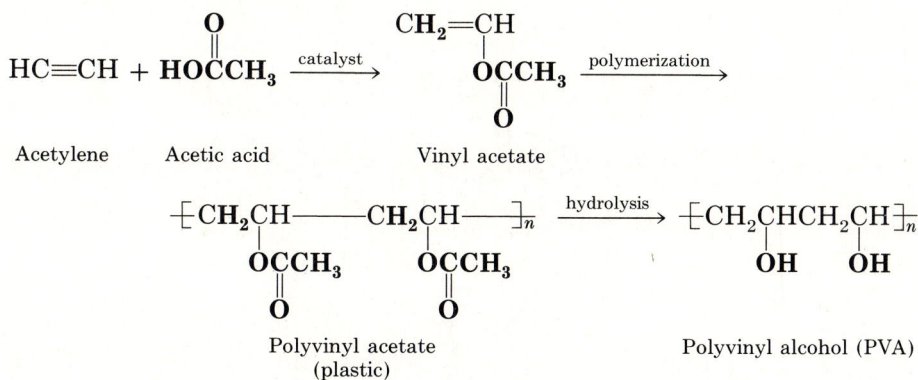

Addition of hydrogen cyanide, $HC\equiv N$, yields an important compound, acrylonitrile, used in the production of the synthetic fiber polyacrylonitrile, or *Orlon,* and the synthetic rubber *Buna N rubber* (a copolymer of 1,3-butadiene and acrylonitrile).

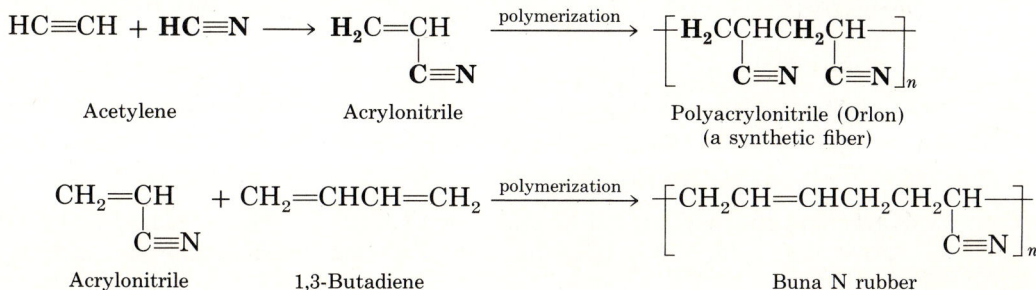

Under the influence of cuprous chloride, Cu_2Cl_2, and ammonium chloride, NH_4Cl, acetylene can be made to dimerize into vinylacetylene, $CH_2=CH-C\equiv CH$. Careful addition of hydrogen chloride to vinylacetylene gives the important conjugated diene 2-chloro-1,3-butadiene, or *chloroprene*. Recall from Section 4.6 that polymerization of this conjugated diene monomer gives the commercially important synthetic rubber *Neoprene*. This valuable synthesis was discovered by J. A. Niewland in 1929.

$$2\ HC\equiv CH \xrightarrow[\text{NH}_4\text{Cl}]{\text{Cu}_2\text{Cl}_2} CH_2=CHC\equiv CH$$

Acetylene   Vinylacetylene

$$CH_2=CHC\equiv CH + HCl \longrightarrow CH_2=CH\overset{\overset{\displaystyle Cl}{|}}{C}=CH_2 \xrightarrow{\text{polymerization}}$$

Chloroprene

$$\left[-CH_2CH=\overset{\overset{\displaystyle Cl}{|}}{C}CH_2 - \right]_n$$

Neoprene

Acetylene is a reactive substance. When being liquefied under pressure, it is liable to explode violently if subjected to heat or shock. At atmospheric pressure, however, acetylene is soluble in acetone (25 volumes of acetylene for each volume of acetone) and forms a stable mixture. Therefore, for commercial purposes acetylene can be safely transported in pressure cylinders saturated with acetone. At higher pressure, acetylene can be stored safely in the presence of an inert gas such as nitrogen.

4.14 Visual Tests for Alkynes

Alkynes, like alkenes, are unsaturated hydrocarbons. Therefore, the same tests that were used to distinguish between alkenes and alkanes apply to alkynes. Alkynes give positive tests with Br_2 in carbon tetrachloride, with dilute $KMnO_4$ (Baeyer test) and with H_2SO_4.

Terminal alkynes, $RC\equiv CH$, can be differentiated from nonterminal alkynes, $RC\equiv CR$, by means of reactions involving heavy metal ions (Ag^+ or Cu^+). Terminal alkynes form insoluble precipitates; nonterminal alkynes do not react and leave a clear solution.

$$CH_3C\equiv CH \xrightarrow{\text{AgNO}_3} CH_3C\equiv C:^-\ Ag^+\downarrow$$

Propyne   Insoluble precipitate

$$CH_3C\equiv CCH_3 \xrightarrow{\text{AgNO}_3} \text{No reaction}$$

2-Butyne

Heavy metal acetylides are unstable and, if allowed to dry, are likely to explode. Consequently, they should be destroyed while still wet by treating them with warm nitric acid. The strong mineral acid regenerates the alkyne.

$$\text{CH}_3\text{C}\equiv\text{C}:^- \text{Ag}^+ \xrightarrow[\text{warm}]{\text{HNO}_3} \text{CH}_3\text{C}\equiv\text{CH} + \text{Ag}^+$$

Propyne

Problem 4.13 Indicate how you could differentiate between the following pairs of compounds by means of simple chemical tests.
(a) *n*-Butane and 1-butyne   **(b)** 1-Butene and propyne
(c) 1-Butyne and 2-butyne

Problem 4.14 Compound **A**, C_4H_6, gave the following tests.
  (1) **A** + Br_2/CCl_4 (excess) → **B** ($C_4H_6Br_4$)
  (2) **A** + H_2/Pt (excess) → **C** (C_4H_{10})
  (3) **A** + $AgNO_3$ → **D** (precipitate)
What are the structures of **A, B, C,** and **D**?

Summary of Concepts and Reactions

Dienes are alkenes that contain two double bonds. A diene may contain isolated or conjugated double bonds.   [Sec. 4.1]

Dienes are named by the IUPAC system from replacement of the ending -ene of alkenes by -adiene. The positions of the double bonds are indicated by two numbers.
                                                                         [Sec 4.1]

Dienes are prepared by the same reactions used to prepare simple alkenes.
                                                                         [Sec. 4.2]

Conjugated dienes are more stable than nonconjugated dienes.   [Sec. 4.3]

Conjugated dienes undergo both an expected 1,2-addition and an unexpected 1,4-addition.   [Sec. 4.4]

Conjugated dienes undergo 1,2- and 1,4-addition because of allylic resonance.

[Sec. 4.5]

Resonance is the result of the delocalization of electrons over more than a single pair of atoms.   [Sec. 4.5]

Like simple alkenes, dienes form large molecules called polymers consisting of repeating diene monomers.   [Sec. 4.6A]

Natural rubber is a polymer of 2-methyl-1,3-butadiene (isoprene).   [Sec. 4.6A]

Some natural polyenes, called terpenes, consist of isoprene units joined together in groups of two, three, four, six, or eight.   [Sec. 4.6B]

Alkynes are unsaturated hydrocarbons that contain a carbon–carbon triple bond, —C≡C—. They have the general formula C_nH_{2n-2}.   [Sec. 4.7]

The carbon–carbon triple bond and the two groups attached to the two *sp*-hybridized carbons lie in the same plane with a bond angle of 180°.   [Sec. 4.8]

Alkynes are named by the IUPAC system from replacement of the ending -ene of alkenes by -yne.   [Sec. 4.9]

Alkynes have physical properties similar to those of alkenes. [Sec. 4.10]
Terminal alkynes react with certain metals to form salts called metal acetylides.
[Sec. 4.11B]

Alkynes may be synthesized by dehydrohalogenation of vicinal dihalides

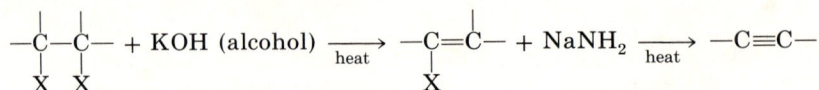

$$-\overset{\displaystyle |}{\underset{\displaystyle X}{C}}-\overset{\displaystyle |}{\underset{\displaystyle X}{C}}- + \text{ KOH (alcohol)} \xrightarrow[\text{heat}]{} -\overset{\displaystyle |}{C}=\overset{\displaystyle }{\underset{\displaystyle X}{C}}- + \text{ NaNH}_2 \xrightarrow[\text{heat}]{} -C\equiv C-$$

or by reaction of sodium acetylide with a 1° alkyl halide

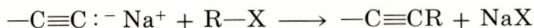

$$-C\equiv C:^- \text{ Na}^+ + \text{R}-\text{X} \longrightarrow -C\equiv CR + \text{NaX}$$

[Sec. 4.11A, B]

Alkynes undergo reactions similar to those of alkenes. [Sec. 4.12]
Addition of H_2O in the presence of dilute H_2SO_4 and an $HgSO_4$ catalyst produces first an
   unstable enol that is converted into a keto group, a process called enol–keto tautomerism. [Sec. 4.12D]
Acetylene is the only alkyne of significant industrial importance. [Sec. 4.13]
Alkynes decolorize Br_2 in CCl_4 and also give a positive Baeyer test. Only terminal alkynes form precipitates with heavy metal ions such as Ag^+ and Cu^+. [Sec. 4.14]

Key Terms

dienes	allylic carbocation	isoprene units
conjugated	delocalization	alkyne
nonconjugated	resonance	carbon–carbon triple bond
isolated	resonance hybrid	C_nH_{2n-2}
trienes	contributing structures	metal acetylides
tetraenes	resonance energy	enol
polyenes	isoprene	keto
heat of hydrogenation	vulcanization	enol–keto tautomerism
1,2-addition	copolymerization	tautomers
1,4-addition	terpenes	

Exercises

Structure and Nomenclature of Dienes, Polyenes, and Alkynes [Secs. 4.1, 4.9]

4.1 Draw structures for each of the following.
   (a) 2,4-Heptadiene
   (b) 3,4-Dimethyl-1,4-octadiene
   (c) 2-Methyl-1,3-cyclohexadiene
   (d) Isoprene
   (e) 1,3,5,7-Cyclooctatetraene
   (f) 2-Heptyne
   (g) 1,4-Heptadiyne
   (h) Cyclooctyne
   (i) Vinylacetylene
   (j) Cyclopropylacetylene
   (k) 3,3-Dimethyl-1-hexyne
   (l) Allylacetylene

4.2 Classify the dienes in Exercise 4.1 as conjugated or isolated.

4.3 Name the following compounds according to the IUPAC rules.
   (a) $(CH_3)_2CHCH=CHCH_2CH=CH_2$
   (b) $CH_2=CBr—CH=CH_2$

(c) ⬠—CH_2CH_3

(d) $CH_3C{\equiv}C{-}CH_2{-}\underset{\underset{\displaystyle CH_3}{|}}{\overset{\overset{\displaystyle CH_3}{|}}{C}}{-}CH_3$

(e) $CH_3C{\equiv}C{-}\underset{\underset{\displaystyle Cl}{|}}{CH}{-}CH_3$

Wait, Cl is above.

(e) $CH_3C{\equiv}C{-}\overset{\overset{\displaystyle Cl}{|}}{CH}{-}CH_3$

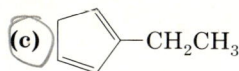

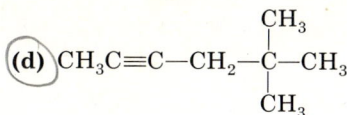

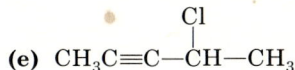

4.4 Write the structural formulas and give the IUPAC names for all dienes with the indicated molecular formulas.
   (a) C_5H_8 (5 structural isomers)   **(b)** C_6H_{10} (16 structural isomers)

4.5 Draw the structural formulas and give the IUPAC names for all alkynes with the molecular formulas shown in Exercise 4.4.

Resonance and Resonance Contributing Structures [Sec. 4.5]

4.6 When an electrophile (H^+, for example) attacks 1,3-butadiene, it always does so at C-1 rather than at C-2. How do you account for this fact? (*Hint:* Draw the structures of the carbocations formed in each case.)

4.7 Write a resonance contributing form for each structure (the delocalization of electrons is indicated by the arrows).

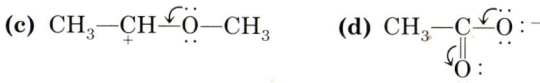

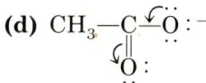

   (a) $CH_2{-}CH{=}CH{-}CH_3$

   (b) ⬡

   (c) $CH_3{-}CH{-}\ddot{O}{-}CH_3$

   (d) $CH_3{-}\underset{\underset{\displaystyle \ddot{O}:}{||}}{C}{-}\ddot{O}:^-$

Preparations and Reactions of Dienes and Alkynes [Secs. 4.2, 4.4, 4.11, 4.12]

4.8 Draw the structure of the product for each reactions. If no reaction occurs, state so.)
   (a) $CH_2{=}CHCH_2CH_2CH{=}CH_2 + Cl_2$ (excess) $\rightarrow$
   (b) $CH_2{=}\underset{\underset{\displaystyle CH_3}{|}}{C}{-}CH_2CH_2CH{=}CH_2 + HI$ (excess) $\rightarrow$

   (c) $BrCH_2CH_2CH_2CH_2Br + KOH \xrightarrow[\text{heat}]{\text{ethanol}}$

   (d) $CH_3CH{=}CHBr + NaNH_2 \xrightarrow[\text{heat}]{}$

   (e) $CH_3CH{=}CH{-}CH{=}CH{-}CH_3 + Br_2 \rightarrow$ (1,4-addition product)
   (f) $CH_3C{\equiv}CH + HCl$ (excess) $\rightarrow$
   (g) $CH_3C{\equiv}CCH_3 + Cl_2$ (excess) $\rightarrow$

   (h) $CH_3C{\equiv}CCH_3 + Na \xrightarrow{\text{liq } NH_3}$

   (i) ⬠—$C{\equiv}CH + Na \xrightarrow{\text{liq } NH_3}$

   (j) Product (i) $+ CH_3CH_2Br \rightarrow$

   (k) ⬠—$C{\equiv}CH + H_2O \xrightarrow{H_2SO_4, HgSO_4}$

   (l) $CH_3CH_2C{\equiv}C{-}CH_3 + H_2 \xrightarrow{\text{Ni}{-}B}$

(m) $CH_3CH_2C{\equiv}C{-}CH_3 + H_2 \xrightarrow{\text{Na, liq } NH_3}$

(n) [structure] $-CH_2C{\equiv}CH + Ag^+ \longrightarrow$ *Insoluble Salts (Don't Try to Draw Molecule)*

(o) [structure] $-C{\equiv}CCH_3 + Ag^+ \longrightarrow$ *No Reaction*

4.9 Starting from acetylene write equations for the preparation of the compounds indicated here. If more than one step is required, show each step clearly.

(a) Ethane   **(b)** Ethyl chloride   **(c)** 1,1-Dichloroethane
(d) 1,1,2,2-Tetrabromoethane   **(e)** 1-Butyne   **(f)** 3-Hexyne
(g) *cis*-3-Hexene   **(h)** *trans*-3-Hexene

4.10 Starting with 1-pentyne in each case, show how you would convert it to the products indicated. You may use any other required organic or inorganic chemicals. If more than one step is necessary, show each step clearly.

(a) Pentane   **(b)** 1,1,2,2-Tetrabromopentane
(c) 2-Pentanone, $CH_3{-}\underset{\underset{O}{\|}}{C}{-}CH_2CH_2CH_3$ **(d)** 1-Pentene

(e) 2,2-Dichloropentane   **(f)** 3-Heptyne

4.11 How can **(a)** 1-hexene be converted to 1-hexyne and **(b)** 1-bromopentane be converted to 1-pentyne?

Enol-Keto Tautomerism [Sec. 4.12D]

4.12 Write the structure of the keto tautomer corresponding to each of the following enol forms.

(a) $CH_3{-}\underset{\underset{OH}{|}}{C}{=}CHCH_3$ **(b)** [structure] $-OH$ **(c)** $CH_3{-}\underset{\underset{O}{\|}}{C}{-}CH{=}\underset{\underset{OH}{|}}{C}{-}CH_3$

Visual Tests and Identification of Structures [Secs. 4.12, 4.13]

4.13 An unknown compound **A** has the molecular formula C_5H_8. Suggest a possible structure for **A** from these data.

     (1) One mole of **A** takes up two moles of bromine.
     (2) **A** does not react with either sodium metal or heavy metal ions (Ag^+ or Cu^+).
     (3) Ozonolysis of one mole of **A** yields two moles of $H{-}\underset{\underset{H}{|}}{C}{=}O$ and one mole of $O{=}\underset{\underset{CH_3}{|}}{C}{-}\underset{\underset{H}{|}}{C}{=}O$.

4.14 An unknown compound **B** has the molecular formula C_5H_8. Suggest a possible structure for **B** from these data.

     (1) One mole of **B** takes up two moles of hydrogen.
     (2) **B** forms a precipitate with Ag^+ or Cu^+.
     (3) Treatment of **B** with H_2O in the presence of dilute H_2SO_4 and mercuric sulfate catalyst yields $CH_3CH_2CH_2\underset{\underset{O}{\|}}{C}CH_3$ as the main product.

4.15 An unknown compound **C** has the molecular formula C_5H_8. Suggest a possible structure for **C** from these data.

     (1) One mole of **C** takes up two moles of chlorine.
     (2) Compound **C** decolorizes a solution of Br_2 in CCl_4.
     (3) Compound **C** forms no precipitate with either Ag^+ of Cu^+.

(4) Treatment of **C** with H_2O in the presence of dilute sulfuric acid and mercuric sulfate catalyst yields an equimolar mixture of $CH_3\overset{\displaystyle O}{\underset{\displaystyle \|}{C}}CH_2CH_2CH_3$ and $CH_3CH_2\overset{\displaystyle O}{\underset{\displaystyle \|}{C}}CH_2CH_3$.

Definitions of Terms

4.16 Define or give an example of **(a)** conjugated diene, **(b)** resonance contributing structures, **(c)** allyl carbocation, **(d)** 1,2-addition, **(e)** 1,4-addition, **(f)** sesquiterpene, **(g)** copolymerization, **(h)** vulcanization, **(i)** acetylide ion, **(j)** enol–keto tautomerism.

5

Benzene and Aromatic Compounds

The term *aromatic* was used originally to designate compounds with spicy or sweet-smelling odors derived from plants. Often these pleasantly fragrant substances contained a variety of groups, such as $-OCH_3$, $-CH{=}CH-COOH$,

and $-\overset{\overset{\displaystyle H}{|}}{C}{=}O$, attached to a C_6H_5 unit. This same unit was found also among products obtained by distillation of coal tar. One such product, phenol, has the formula C_6H_5OH, and another, **benzene,** has the formula C_6H_6 or C_6H_5-H. With time, as odorless and vile-smelling substances that contained the C_6H_5 unit were discovered, the original meaning of the term aromatic was abandoned. The expression *aromatic compounds* came to mean benzene and derivatives of benzene. Today a compound is said to be **aromatic** if it is *benzene-like in its properties*. This definition includes benzene and benzene derivatives as well as other substances that, although they bear no resemblance to benzene superficially, nevertheless behave like benzene. Obviously, we must begin our study of aromatic compounds with a discussion of benzene itself.

5.1 Structure of Benzene: Resonance Description

Benzene, C_6H_6, is a planar, cyclic compound.

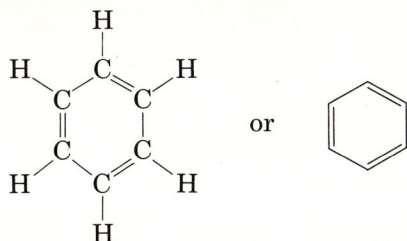

Each carbon in benzene is attached to three atoms and is therefore sp^2 hybridized. As expected, the bond angles in benzene are 120°. For two principal reasons, however, the description just given is unsatisfactory.

1. If the structure we have shown were correct, we would expect benzene to be "1,3,5-cyclohexatriene." If benzene were 1,3,5-cyclohexatriene, we would expect the molecule to have the shape of an irregular hexagon with three C—C bond lengths of 1.54 Å (typical of single bonds) and three C=C bonds of 1.34 Å (typical of double bonds). However, evidence from x-ray diffraction experiments reveals that *all* carbon–carbon bonds in benzene are of equal length, 1.39 Å.
2. If the structure of benzene we have shown were correct, we would expect benzene to undergo addition reactions. In fact, the typical reaction of benzene (and other aromatic compounds) is **substitution,** rather than addition.

The true structure of benzene can be explained by the concept of resonance. Recall that according to the resonance theory (Sec. 4.5), whenever a substance can be represented by two or more equivalent or nearly equivalent structures that differ only in the position of valence electrons, the actual molecule does not correspond to any of the *contributing structures* but is a *resonance hybrid* of all of them. Thus, benzene is actually a resonance hybrid (III) of the two imaginary contributing structures (I and II).

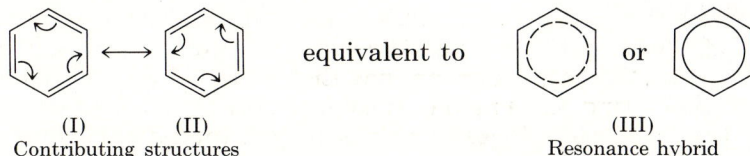

(I)   (II)   (III)
Contributing structures   Resonance hybrid

In the resonance hybrid, the broken and solid circles within the hexagon represent the even distribution of the valence electrons over the six carbon atoms.

Structure of Benzene: 5.2
Molecular Orbital Description

The delocalization of electrons over the six carbons may easily be seen in the molecular orbital picture of benzene. It should be emphasized that the results of the molecular orbital approach are identical to those of the resonance method.

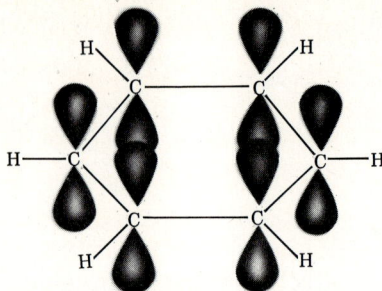

Figure 5.1 Molecular orbital picture of benzene without *p* orbital overlap.

Figure 5.1 shows the molecular orbital picture of benzene without *p*-orbital overlap.

To complete the picture, it should be remembered that any pair of adjacent *p* orbitals is capable of side-side overlap to form a π bond. Three double bonds can be formed by overlap of the *p* orbitals of C-1 and C-2, C-3 and C-4, and C-5 and C-6 (Fig. 5.2a) or by overlap of *p* orbitals of C-2 and C-3, C-4 and C-5, and C-6 and C-1 (Fig. 5.2b).

Which of these structures is the correct one? Neither. They both are contributing forms that have no physical reality. The true structure of benzene is a

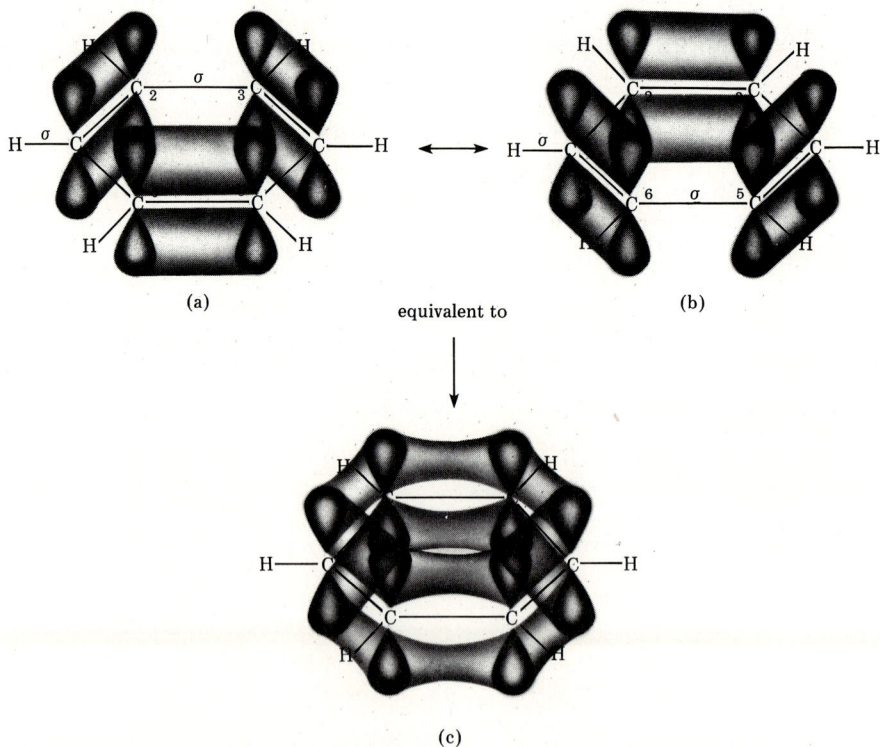

(a)   (b)

equivalent to

(c)

Figure 5.2 (a, b) Imaginary contributing structures of the resonance hybrid **(c)** that represents the true structure of benzene.

hybrid, not a mixture of (a) and (b). In the hybrid, the π-electron cloud is delocalized all over the ring (Fig. 5.2c).

An immediate benefit of the resonance or the molecular orbital description of benzene is that it agrees with the physical measurements of the molecule. As a resonance hybrid, we expect benzene to be a planar molecule having the shape of a regular hexagon, with bond angles of 120°. We also expect *all* the carbon–carbon bond distances to be intermediate (1.39 Å) between the usual C—C single bond (1.54 Å) and the typical C=C double bond (1.34 Å). All the experimental data confirm our expectations.

The resonance picture helps also to explain the lack of reactivity of benzene toward addition. The reasoning is that the delocalization of π electrons confers a great degree of stability on benzene. This degree of stability is so great that the π bonds of the molecule will normally resist breaking. The typical reaction of benzene is substitution, which leaves the π-electron cloud around the ring intact, rather than addition, which disrupts the π cloud around the ring.*

One line of experimental evidence showing that the stability of benzene is due to the π cloud around the ring is discussed in the next section.

Stability of Benzene: Resonance Energy   **5.3**

The heat of hydrogenation of cyclohexene is 28.6 kcal/mole. If a benzene molecule has three localized double bonds—that is, if it were 1,3,5-cyclohexatriene—we would expect its heat of hydrogenation to be 3×28.6 or 85.8 kcal/mole.

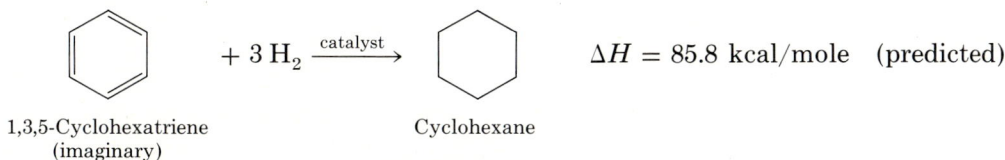

Cyclohexene   + H$_2$ $\xrightarrow{\text{catalyst}}$ Cyclohexane $\Delta H = 28.6$ kcal/mole

1,3,5-Cyclohexatriene (imaginary)   + 3 H$_2$ $\xrightarrow{\text{catalyst}}$ Cyclohexane $\Delta H = 85.8$ kcal/mole   (predicted)

Actually, when benzene is hydrogenated to cyclohexane, *only 49.8 kcal/mole of energy is liberated.*

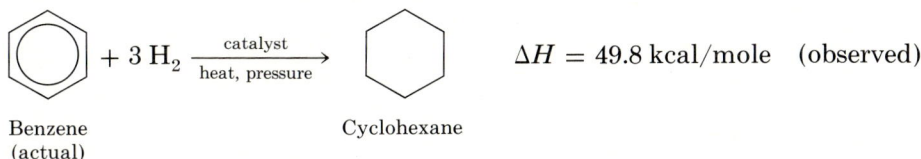

Benzene (actual)   + 3 H$_2$ $\xrightarrow[\text{heat, pressure}]{\text{catalyst}}$ Cyclohexane $\Delta H = 49.8$ kcal/mole   (observed)

---

* Addition reactions (hydrogenation, bromination, etc.) can take place, but vigorous conditions, such as high heat or high pressure, are needed.

The amount of energy released on hydrogenation of benzene is 36.0 kcal/mole less ($85.8 - 49.8 = 36.0$ kcal/mole) than that predicted for the hypothetical 1,3,5-cyclohexatriene. The fact that benzene liberates 36 kcal/mole less energy than predicted can only mean that benzene contains 36 kcal/mole *less* energy or, in other words, that benzene is more *stable* by 36 kcal/mole than predicted. Obviously, this greater stability of the molecule must come from the fact that the π electrons in benzene are delocalized. We refer to this amount of stabilization between the energy liberated by the actual structure of benzene, with its delocalized π electrons, and the energy liberated by the hypothetical 1,3,5-cyclohexatriene, with its localized π electrons, as the **resonance energy.** It is the preservation of this large resonance energy that is the source of benzene's tendency to undergo substitution rather than addition reactions and gives it its aromatic character.

Problem 5.1 Hydrogenation of one mole of naphthalene, $C_{10}H_8$, with five moles of hydrogen releases 82.0 kcal of heat. Assuming the heat of hydrogenation of a normal double bond to be 28.6 kcal/mole, calculate the resonance energy of naphthalene.

Recall that we defined aromatic compounds as those with benzene-like properties. Like benzene, aromatic compounds tend to undergo substitution rather than addition reactions and, for the same reason, to retain their resonance energy. Can we predict which compound is aromatic? The answer is yes, as will be shown in the next section.

5.4 Aromatic Character: The $(4n + 2)\ \pi$ Rule

Extensive studies have revealed that aromatic character, or **aromaticity,** is associated with several structural requirements. First, aromatic compounds are *cyclic* structures that contain what *looks like* a continuous system of alternating double and single bonds. Actually, the π electrons are delocalized through overlap of adjacent p orbitals. To permit such delocalization, a second structural requirement must be met: aromatic compounds must be *planar*. Finally, aromaticity is possible only if the number of π electrons in the compound is $(4n + 2)$, where n is zero or a positive integer ($n = 0, 1, 2, 3$, and so on). Examples of aromatic compounds that obey this $(4n + 2)\ \pi$ rule, also known as **Hückel's rule,** are shown in Table 5.1.

A glance at the table reveals that some aromatic compounds, such as naphthalene, anthracene, or phenanthrene, resemble benzene structurally: they consist of two or more benzene rings fused together. Others, such as pyrrole, bear no structural resemblance to benzene. Nevertheless, pyrrole and all other compounds listed in Table 5.1 show aromatic properties because they satisfy the $(4n + 2)\ \pi$ rule.*

---

* The two nonbonded electrons *within* the ring of pyrrole, furan, and other molecules shown in the table are part of the π system and should be counted when applying Hückel's rule.

n	$4n + 2$	Structure and name of aromatic compound

1	6	Benzene Pyridine Pyrrole Furan Thiophene
2	10	Naphthalene Quinoline Indole
3	14	Anthracene Phenanthrene

Aromaticity according to the $(4n + 2)$ π rule shows up even more strikingly in certain organic ions. All carbocations encountered up to this point have been very reactive, short-lived intermediates. Tropylium bromide, on the other hand, is a stable, high-melting salt that is soluble in water.

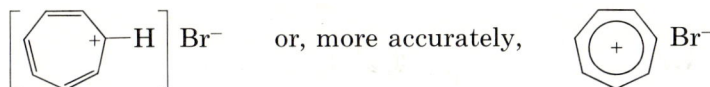

$[\quad$ —H $]$ Br$^-$ or, more accurately, $(+)$ Br$^-$

Tropylium bromide

The tropylium ion is a seven-membered ring that contains three double bonds. Since the ring has a total of six π electrons, it is therefore aromatic. Similarly, cyclopentadiene, in contrast to nearly all other hydrocarbons, is quite acidic. It is readily converted to its sodium salt.

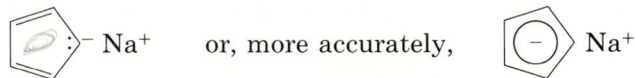

:$^-$ Na$^+$ or, more accurately, $(-)$ Na$^+$

Sodium cyclopentadienide

The cyclopentadienide anion, $C_5H_5^-$, is also aromatic and particularly stable because the five-membered ring has a sextet or π electrons, two double bonds from the ring and a pair of nonbonded electrons.

What does it mean if the sum of the π electrons in a cyclic, conjugated system does not follow the $(4n + 2)$ π rule? It means, simply, that the compound is not aromatic. Cyclooctatetraene, for example, contains a π system of electrons. Because no integral value of n can give the number 8 according to the $(4n + 2)$ π rule, cyclooctatetraene is not aromatic. As predicted, cyclooctatetraene reacts readily with bromine in the dark and with cold dilute permanganate, just as an ordinary alkene does. Furthermore, x-ray diffraction studies have shown that cyclooctatetraene has a "tub" shape rather than the planar shape required for an aromatic compound.

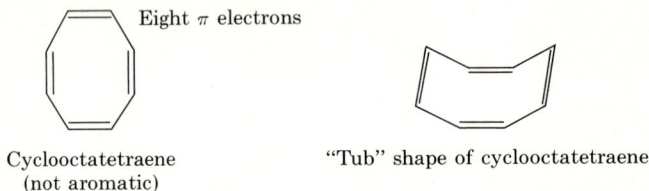

Eight π electrons

Cyclooctatetraene
(not aromatic)

"Tub" shape of cyclooctatetraene

In our discussion of aromaticity we have represented benzene and other compounds as having alternating double and single bonds. This was done as a matter of convenience to illustrate the application of the $(4n + 2) \pi$ rule. Keep in mind that in aromatic compounds such bonds do not really exist; in the actual structure the π bonds are delocalized.

Problem 5.2 Predict which of the following structures might be expected to possess aromatic character. Explain the basis for your choice.

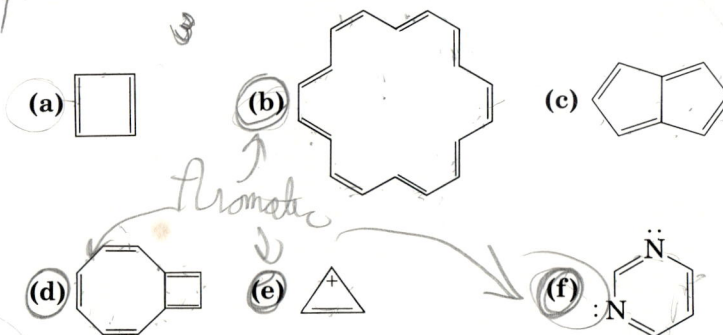

(a)   (b)   (c)

(d)   (e)   (f)

5.5   Nomenclature of Aromatic Compounds

Now that we are familiar with the criteria for aromaticity, we should learn how to name members of the various families of aromatic compounds, especially the derivatives of benzene. We will then be ready to study their reactions.

A   Monosubstituted Benzenes

Because all six positions in benzene are equivalent, there is no need to specify by a number the position of a substituent for monosubstituted benzenes. All that is required is to indicate the nature of the substituents present. For example, replacement of one of the hydrogens by bromine yields a single product, bromobenzene.

Br

= Br

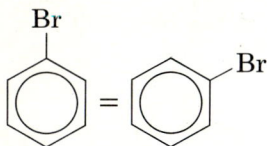

Bromobenzene
The two structures represent the same molecule because all positions are equivalent.

Other monosubstituted benzenes that can be named in the same manner, by placing the name of the substituent in front of the word "benzene," are

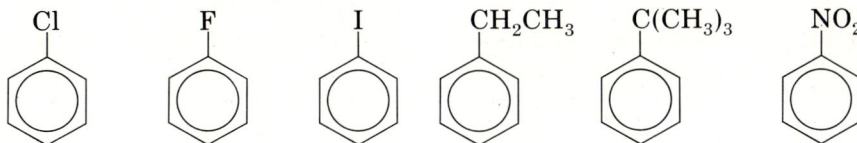

Cl $\qquad$ F $\qquad$ I $\qquad$ CH_2CH_3 $\qquad$ $C(CH_3)_3$ $\qquad$ NO_2

Chlorobenzene Fluorobenzene Iodobenzene Ethylbenzene t-Butylbenzene Nitrobenzene

Some monosubstituted benzenes are considered parent compounds because they occur so frequently. For that reason they are given common names that you should memorize.

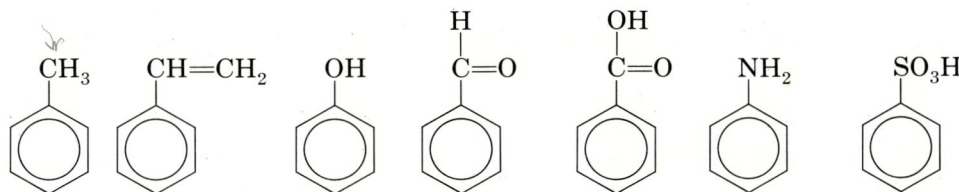

CH_3 $\qquad$ $CH{=}CH_2$ $\qquad$ OH $\qquad$ H–C=O $\qquad$ OH–C=O $\qquad$ NH_2 $\qquad$ SO_3H

Toluene Styrene Phenol Benzaldehyde Benzoic acid Aniline Benzenesulfonic acid

Sometimes it is more convenient to name the benzene ring as a substituent, as in the following structures.

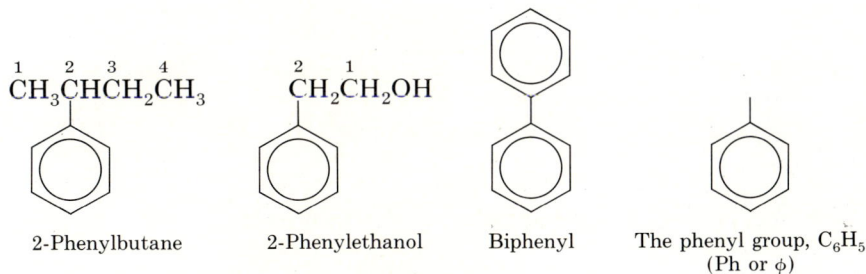

$\overset{1}{C}H_3\overset{2}{C}H\overset{3}{C}H_2\overset{4}{C}H_3$ $\qquad$ $\overset{2}{C}H_2\overset{1}{C}H_2OH$ $\qquad\qquad\qquad$

2-Phenylbutane 2-Phenylethanol Biphenyl The phenyl group, C_6H_5
(Ph or ϕ)

The **phenyl** group, symbolized by Ph or ϕ, is a benzene from which one hydrogen has been removed. Another common aromatic substituent is the **benzyl** group, derived by removing a hydrogen from the CH_3 group in toluene.

For example,

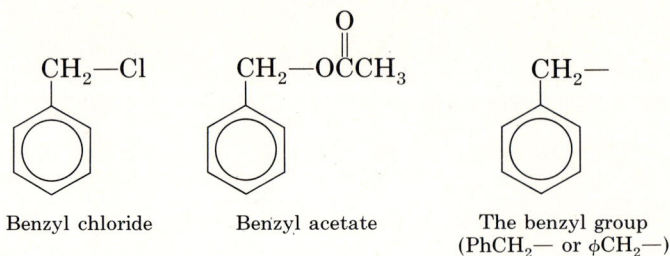

Benzyl chloride   Benzyl acetate   The benzyl group
                                                 (PhCH$_2$— or ϕCH$_2$—)

The benzyl group is often abbreviated as PhCH$_2$— or ϕCH$_2$—.

B   Disubstituted Benzenes

All disubstituted benzenes, no matter what the substituents, can give rise to three possible isomers. To differentiate between the isomers, the relative positions of the substituents are designated by numbers or, more commonly, by the prefixes **ortho (*o*-), meta (*m*-)**, or **para (*p*-).**

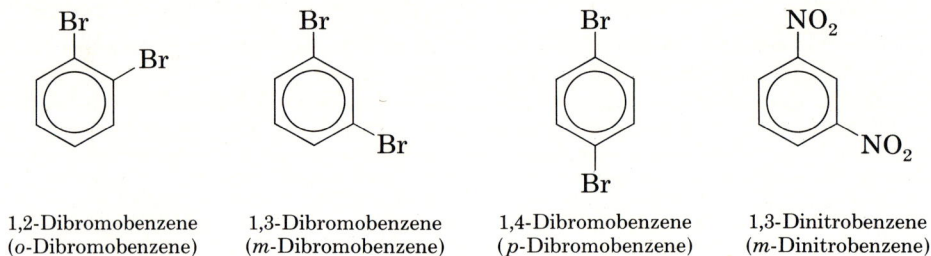

1,2-Dibromobenzene   1,3-Dibromobenzene   1,4-Dibromobenzene   1,3-Dinitrobenzene
(*o*-Dibromobenzene)   (*m*-Dibromobenzene)   (*p*-Dibromobenzene)   (*m*-Dinitrobenzene)

When the substituents are different, they are listed in alphabetical order. For example,

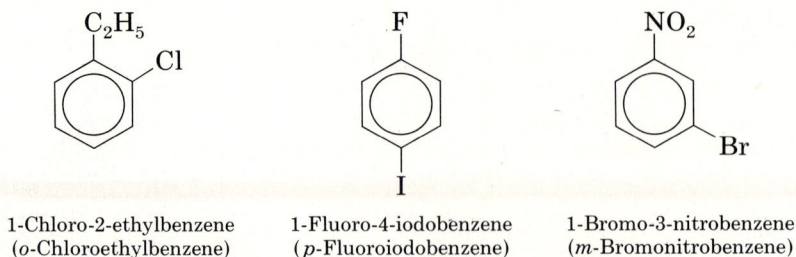

1-Chloro-2-ethylbenzene   1-Fluoro-4-iodobenzene   1-Bromo-3-nitrobenzene
(*o*-Chloroethylbenzene)   (*p*-Fluoroiodobenzene)   (*m*-Bromonitrobenzene)

If one of the substituents is part of a parent compound, then the disubstituted benzene is named as a derivative of the parent compound.

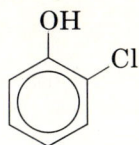

2-Chlorophenol
(*o*-Chlorophenol)

4-Nitrotoluene
(*p*-Nitrotoluene)

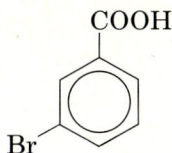

3-Bromobenzoic acid
(*m*-Bromobenzoic acid)

4-Aminophenol
(*p*-Aminophenol)
or
4-Hydroxyaniline
(*p*-Hydroxyaniline)

Like monosubstituted benzenes, certain disubstituted benzenes are referred to by their common names. The dimethylbenzenes, for example, are known as xylenes.

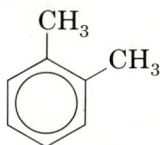

o-Xylene

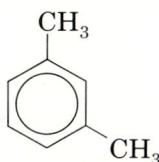

m-Xylene

p-Xylene

Problem 5.3 Name each of the following structures using ortho, meta, and para designations for disubstituted compounds.

(a)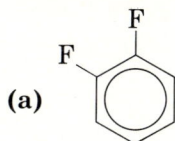

(b)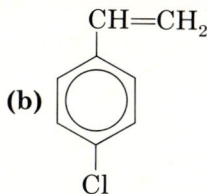

(c)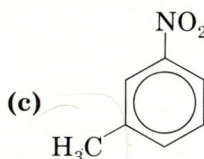

(d)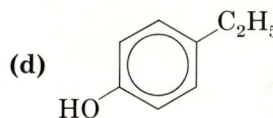

(e)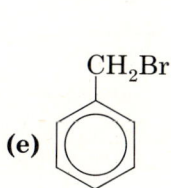

(f)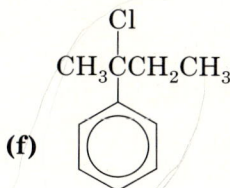

(g)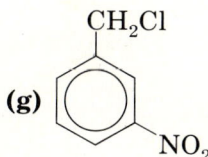

(h)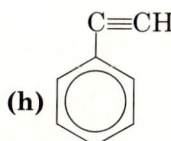

Problem 5.4 Draw the structure of each of the compounds named.
(a) 2-Phenylpentane
(b) *p*-Nitroethylbenzene
(c) *m*-Chlorobenzaldehyde
(d) *o*-Ethylaniline
(e) *m*-Bromobenzenesulfonic acid
(f) 2,2-Dimethyl-1-phenylbutane
(g) *p*-Bromotoluene
(h) *m*-Nitrobenzoic acid

C Polysubstituted Benzenes

When three or more substituents are present, the ring must be numbered and the positions of the substituents must be specified by numbers. As with other classes of compounds previously encountered, the numbering starts with one substituent and continues around the ring so as to use the lowest possible numbers for the other substituents. When the substituents are different the last one named is understood to be in position 1. As with disubstituted benzenes, if one of the group is part of the parent compound, then the carbon that bears the functional group of the parent compound is assigned the number 1.

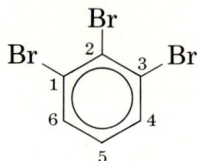

1,2,3-Tribromobenzene
(*not* 1,2,6-Tribromobenzene)

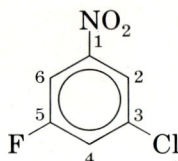

3-Chloro-5-fluoronitrobenzene
(nitro on carbon 1)

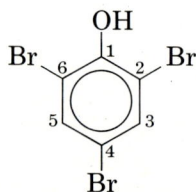

2,4,6-Tribromophenol
(OH on carbon 1)

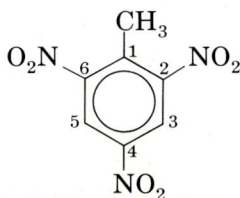

2,4,6-Trinitrotoluene (TNT)
(CH_3 on carbon 1)

Problem 5.5 Write the structural formulas that correspond to the following names.

(a) 2,6-Dibromotoluene
(c) 2-Benzyl-3-nitro-5-bromophenol
(e) 3,5-Dinitrobenzoic acid
(g) 2,4-Dinitrofluorobenzene

(b) 1,2,4,5-Tetrachlorobenzene
(d) 2,4,6-Tribromoaniline
(f) 2,3-Dichlorobenzaldehyde
(h) 3,5-Dibromobenzenesulfonic acid

Problem 5.6 Give names for the following structures.

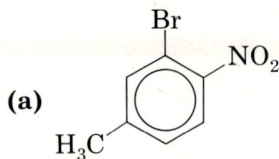

(a)

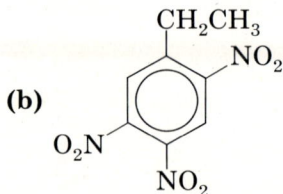

(b)

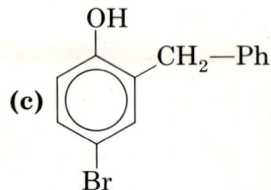

(c)

(d) H₃C, CH₃ / H₃C, CH₃ (e) O₂N, Br, COOH (f) Ph, Br, Ph, Ph, Ph, Br

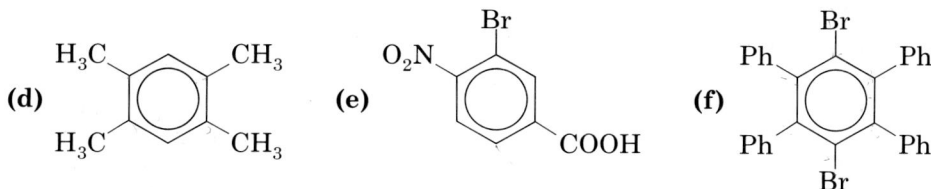

D Polynuclear Aromatic Hydrocarbons

Benzene rings may be fused together to form *polynuclear aromatic hydrocarbons*. Examples of polynuclear aromatic hydrocarbons containing two, three, and four rings are

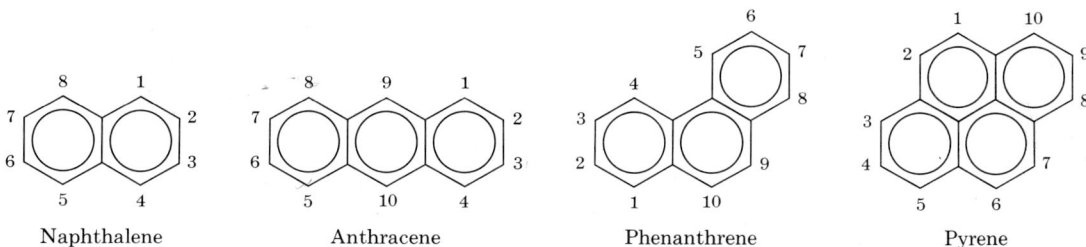

Naphthalene   Anthracene   Phenanthrene   Pyrene

In naphthalene, positions 1, 4, 5, and 8 are equivalent, as are positions 2, 3, 6, and 7. Often, the 1-position in naphthalene is called the *alpha* (α) position and C-2 the *beta* (β) position. Thus,

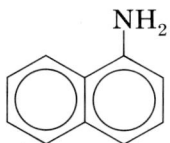

IUPAC name: 1-Aminonaphthalene   2-Aminonaphthalene
Common name: α-Naphthylamine   β-Naphthylamine
       (a weak carcinogen)   (a strong carcinogen)

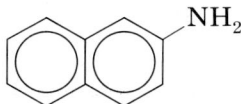

Problem 5.7 Write the structures of
(a) 2-Methylnaphthalene   (b) 1-Chloronaphthalene
(c) α-Nitronaphthalene   (d) β-Fluoronaphthalene
(e) 1,5-Dimethylnaphthalene   (f) 1,3,6,8-Tetranitronaphthalene
(g) 1-Bromoanthracene   (h) 9-Bromoanthracene
(i) 1,5-Dinitroanthracene   (j) 1-Bromophenanthrene
(k) 3-Bromophenanthrene   (l) 4,5-Dimethylphenanthrene

Problem 5.8 Name the following structures.

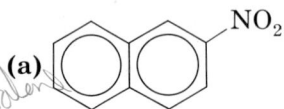 (a)

2 Nitro Naphthalene

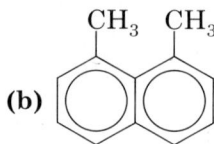 (b)

1,8 Dimethyl Napthalene

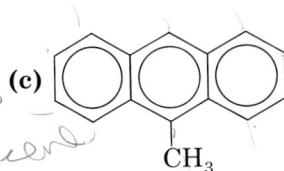 (c)

9 Methyl Anthracene

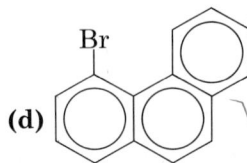 (d)

4 Bromo Phena Anthrene

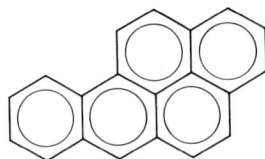

(a)

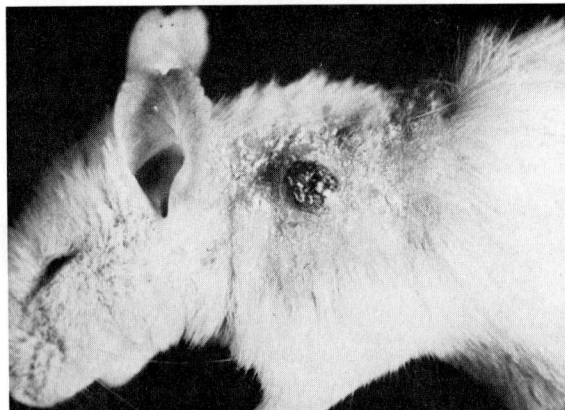

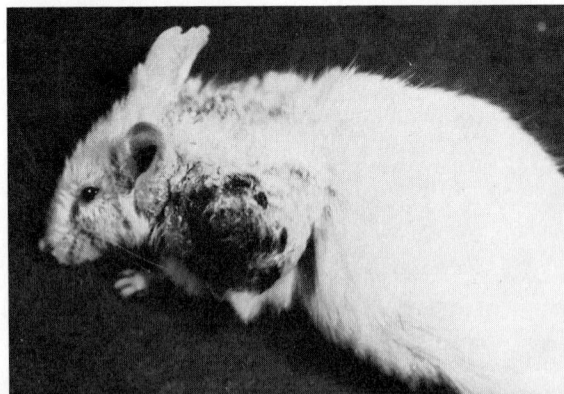

(b)

Figure 5.3 **(a)** 3,4-Benzopyrene, a potent carcinogen. **(b)** Two stages in the development of a skin tumor induced in the mouse using 3,4-benzopyrene. [From J. C. Arcos, "Cancer: chemical factors in environment," part 1. *American Laboratory, 10*(6), 65 (1978). Reprinted with permission.]

Problem 5.9 Which other position(s) in monosubstituted anthracenes are equivalent to **(a)** position 1, **(b)** position 2, and **(c)** position 9?

A number of polycyclic aromatic hydrocarbons are carcinogens. 3,4-Benzopyrene (Fig. 5.3a) has the dubious distinction of being the most widely distributed hydrocarbon carcinogen in the environment. The compound is the main hydrocarbon carcinogen in cigarette smoke, auto emissions, and soot.

Sources of Aromatic Compounds **5.6**

Although compounds that contain the benzene ring are widely dispersed in nature, the two major sources of aromatic compounds are coal tar and petroleum. Until about 1940, benzene, toluene, xylenes, naphthalene, anthracene, and various other aromatic compounds were obtained by the distillation of coal tar, a by-product in the manufacture of coke for steel. In the past 40 years, the demand for aromatic compounds has far outstripped the amount available from coal tar, so that the major source of the commercially significant aromatic compounds (benzene, toluene, and xylenes) is now the petroleum industry. Alkanes and cycloalkanes found in petroleum are converted to benzene and alkylbenzenes by catalytic dehydrogenation. Benzene, for example, is synthesized on a large scale by dehydrogenation of cyclohexane.

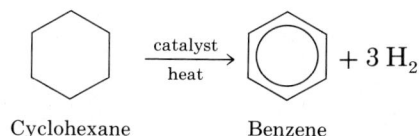

Cyclohexane   Benzene

A cyclic precursor is not always required, since cyclization of an aliphatic chain occurs in the same process. Thus, toluene can be obtained by the dehydrogenation of methylcyclohexane or of *n*-heptane, both of which are found in petroleum.

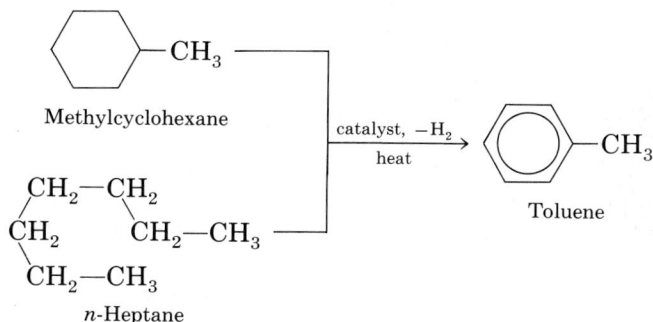

Methylcyclohexane

n-Heptane

Toluene

Xylenes are also made industrially in the same manner from dimethylcyclohexanes or from the appropriate octanes found in petroleum.

Benzene, toluene, and xylenes are the basic raw materials for most other aromatic compounds. Derivatives are obtained from these hydrocarbons by reactions that are discussed in the sections that follow.

Problem 5.10 Assuming that no rearrangement occurs during dehydrogenation, predict the products obtained from the dehydrogenation of **(a)** 1,2-dimethylcyclohexane, **(b)** 1,3-dimethylcyclohexane, and **(c)** 1,4-dimethylcyclohexane.

5.7 Electrophilic Aromatic Substitution: General Mechanism

Because the aromatic ring is extremely stable, a highly reactive agent is required to react with the benzene ring. The aromatic ring, with its delocalized π electrons, is also an electron-rich system. We might therefore expect that attack on the ring takes place by means of an electron-deficient species, or an **electrophile.** This is indeed the case. The reaction of benzene with a typical electrophile, E$^+$, is similar to the addition of an electrophile to an alkene, and is shown as step (1) of the reaction mechanism.

Step 1. The electrophile E$^+$ approaches the π cloud of the aromatic ring and forms a bond to carbon, creating a positive charge in the ring.

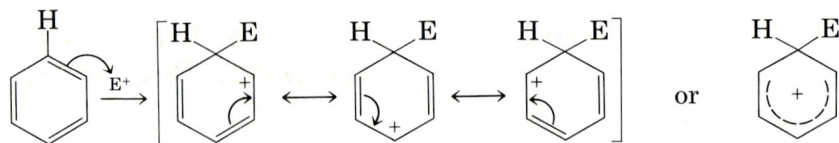

The resulting carbocation has three important contributing structures that spread the positive charge over the other five carbon atoms, although more so on the carbons that are ortho and para to the position attacked by the electrophile.

The intermediate carbocation has two possible fates. It could react with an electron-rich species, a nucleophile, :Nu, to give an addition product. With alkenes, this is the normally observed process. If this were to happen to aromatic compounds, however, the product that would result, a cyclohexadiene, would no longer have the great amount of resonance energy characteristic of the aromatic system.

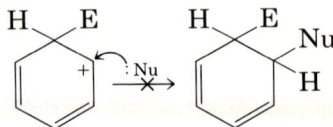

Addition of a nucleophile to the carbocation intermediate would destroy the aromatic system

To retain the aromatic system, the carbocation does *not* combine with the nucleophile. Instead, the nucleophile abstracts a proton from the carbocation ion intermediate. The loss of the proton allows the electrons from the carbon–hydrogen bond to go back into the ring, thus restoring the aromatic system.

135

5.8 The Role of
Catalysts in
Electrophilic
Aromatic
Substitution

Step 2. The removal of the proton by the nucleophile, which leads to the restoration of the aromatic ring, is the second step in the reaction mechanism.

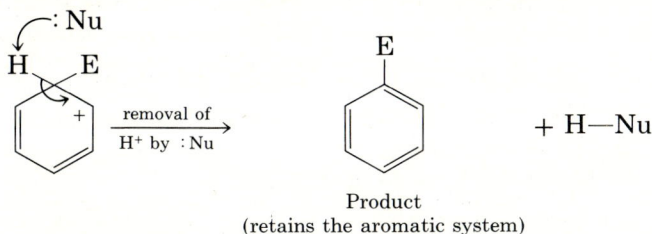

Product
(retains the aromatic system)

The net overall result of this two-step mechanism is the *substitution of the group* E^+ *for a proton* (H^+). Hence the name given to this mechanism is **electrophilic aromatic substitution.**

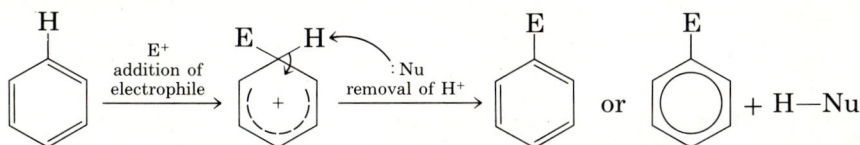

Now that we are familiar with the general aspects of electrophilic aromatic substitution, we are almost ready to study the reactions of specific electrophiles with benzene. Before doing so, however, let us see how electrophiles sufficiently reactive to attack the aromatic ring are generated.

The Role of Catalysts in Electrophilic **5.8** Aromatic Substitution

In almost all electrophilic aromatic substitutions a catalyst is needed for a reaction to take place. The catalysts are usually Lewis acids or a protonic acid. *The purpose of these catalysts is to generate powerful electrophiles.* Contrast, for example, the bromination of cyclohexene with the bromination of benzene. Bromination of cyclohexene (an alkene) takes place rapidly in the presence of molecular bromine, Br_2, and no catalyst is needed. The attacking electrophile in this case is molecular bromine itself.

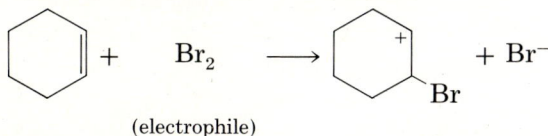

(electrophile)

Bromine alone, however, is not a reactive enough electrophile to attack the benzene ring; a catalyst is needed to generate a more powerful electrophile, **Br^+**. The positively charged bromine ion, or **bromonium ion,** is formed when molecular bromine, Br_2, reacts with a Lewis acid, such as ferric bromide.

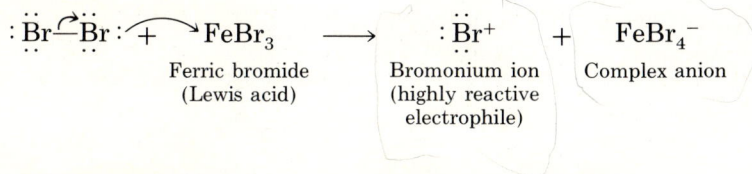

| Ferric bromide | Bromonium ion | Complex anion |
| (Lewis acid) | (highly reactive electrophile) | |

Ferric bromide, the Lewis acid, is by definition deficient in electrons. It will, therefore, combine with any species that is capable of donating an electron pair to it, which in this case is the bromine molecule. In the process, a complex anion, $FeBr_4^-$, and the bromonium ion, Br^+, are simultaneously formed. The latter is a more reactive electrophile than Br_2 and is able to attack the benzene ring to eventually yield bromobenzene.

Similarly, chlorination of benzene requires the presence of a Lewis acid catalyst, usually ferric chloride. Again, the function of ferric chloride is to generate a highly reactive electrophile, the **chloronium ion (Cl^+).**

$$:\ddot{C}l\frown\ddot{C}l: + FeCl_3 \longrightarrow :\ddot{C}l^+ + FeCl_4^-$$

Ferric chloride	Chloronium ion	Complex anion
(Lewis acid)	(highly reactive electrophile)	

Aluminum chloride, $AlCl_3$, is another Lewis acid catalyst frequently used to generate highly reactive electrophiles. For example, the reaction of an alkyl chloride, R—Cl, with aluminum chloride yields a complex anion, $AlCl_4^-$, and a **carbocation, R^+,** a powerful electrophile.

$$R\frown\ddot{C}l: + AlCl_3 \longrightarrow R^+ + AlCl_4^-$$

Alkyl chloride	Aluminum chloride	Carbocation	Complex anion
	(Lewis acid)	(highly reactive electrophile)	

The carbocation, once formed, can attack the aromatic ring, a process known as *alkylation* (see Section 5.9).

Nitration of benzene, the substitution of NO_2 for H, and *sulfonation* of benzene, the substitution of SO_3H for H, usually also require the addition of a catalyst to help produce reactive electrophiles. The attacking electrophile in nitration is the **nitronium ion, NO_2^+,** whose formation is facilitated by the presence of sulfuric acid (the catalyst).

$$HO\!-\!NO_2 + H_2SO_4 \rightleftharpoons HSO_4^- + \left[HO\overset{+}{\frown}NO_2 \atop H \right] \longrightarrow$$

Nitric acid	Sulfuric acid	Protonated nitric acid
	(strong acid)	

$$NO_2^+ + H_2O$$

Nitronium ion
(highly reactive
electrophile)

Sulfuric acid, the stronger acid, donates a proton to nitric acid, the weaker acid, to yield a protonated nitric acid intermediate and hydrogen sulfate. Upon losing water, the intermediate generates the nitronium ion, NO_2^+, a powerful electrophile.

Sulfonation of benzene is usually carried out in fuming sulfuric acid, which is a solution of SO_3 in H_2SO_4. In sulfonation, an active electrophile is the **SO_3H^+ ion** whose formation is brought about according to

$$SO_3 + H_2SO_4 \rightleftharpoons SO_3H^+ + HSO_4^-$$

Acid catalyst	Electrophile	Hydrogen sulfate
		(anion)

137

5.9 Specific
Electrophilic
Aromatic
Substitution
Reactions

Table 5.2 The Electrophiles in Common Aromatic Substitution Reactions

Electrophile	Name of reaction
Cl^+ or Br^+	halogenation
R^+	alkylation
NO_2^+	nitration
SO_3H^+	sulfonation

As in nitration, sulfuric acid acts as a catalyst by providing a hydrogen ion to SO_3, thus creating the active electrophile SO_3H^+.

To recapitulate, acid catalysts are needed in electrophilic aromatic substitution reactions to generate reactive electrophiles. The electrophiles in common aromatic substitutions are shown in Table 5.2. Each of these electrophilic agents plays the same role as E^+ does in the general mechanism of electrophilic aromatic substitution shown on page 135.

The common electrophilic aromatic substitutions of benzene are illustrated in the next section.

Specific Electrophilic Aromatic Substitution Reactions   5.9

Halogenation, alkylation,[*] **nitration,** and **sulfonation** are the typical electrophilic aromatic substitution reactions. These reactions are shown in Figure 5.4. Note that in each reaction the net result is the replacement of a hydro-

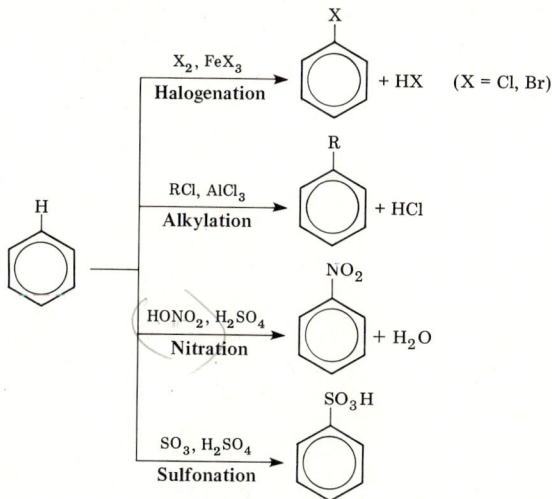

Figure 5.4 Common electrophilic aromatic substitution reactions.

---

[*] The reaction, known as Friedel–Crafts alkylation, was discovered in 1877 by a French chemist, Charles Friedel, and an American, M. Crafts.

gen in benzene by one of the electrophiles listed in Table 5.2. The specific mechanism for each reaction also follows the general pattern described in Section 5.7.

Problem 5.11 Based on the general mechanism of electrophilic aromatic substitution shown in Section 5.7, and from the discussion in Section 5.8, write the mechanism for **(a)** the chlorination of benzene and **(b)** the nitration of benzene.

5.10 Side-Chain Reactions of Aromatic Compounds

In the previous section methods for introducing a substituent directly into an aromatic ring via electrophilic substitution were examined. In this section we consider reactions involving alkyl side chains.

A Halogenation of an Alkyl Side Chain

An alkylbenzene contains both an aliphatic and an aromatic portion. We might, therefore, expect the aliphatic portion (the alkyl side chain) to undergo free-radical substitution characteristic of alkanes (see Sec. 2.12) and the ring to undergo electrophilic substitution characteristic of benzene. This is indeed the case. For example, bromination of toluene in the presence of uv light (free-radical conditions) yields exclusively benzyl bromide; there is no substitution in the aromatic ring.

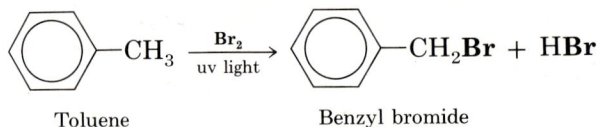

Toluene   Benzyl bromide

Similarly, when ethylbenzene is chlorinated under free-radical conditions, substitution occurs in the alkyl portion only, and the product formed is almost exclusively 1-chloro-1-phenylethane, with small amounts of 1-chloro-2-phenylethane.

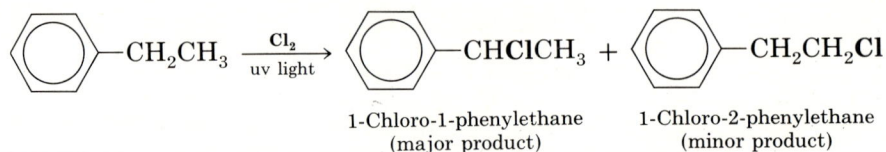

1-Chloro-1-phenylethane   1-Chloro-2-phenylethane
(major product)   (minor product)

To account for the almost exclusive formation of 1-chloro-1-phenylethane we must look at the free-radical intermediates. Removal of a hydrogen from the carbon adjacent to the aromatic ring gives a benzyl type of free radical that is relatively easy to form because it can be stabilized by resonance.

Delocalization of the odd electron
stabilizes the free radical,
facilitating its formation.

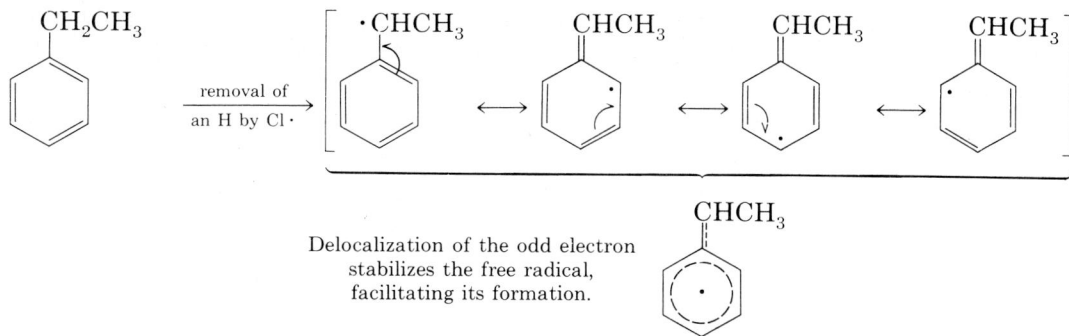

Removal of a hydrogen from the methyl group of ethylbenzene gives a free radical in which the odd electron cannot be stabilized by resonance, thus minimizing the formation of 1-chloro-2-phenylethane.

The odd electron cannot be delocalized; therefore the free radical is more difficult to form.

Problem 5.12 Predict the major organic products formed in these reactions.

(a) *n*-Propylbenzene $\xrightarrow[\text{uv light}]{\text{Br}_2}$ **(b)** Isopropylbenzene $\xrightarrow[\text{uv light}]{\text{Cl}_2}$

Halogenation of alkylbenzenes under electrophilic conditions (that is, treatment of an alkylbenzene with bromine or chlorine in the presence of ferric halide) leads to substitution in the ring. The exact position(s) of the incoming halogen will be dealt with later (Sec. 5.11).

B   Oxidation of an Alkyl Side Chain

Another reaction of the alkyl side chain is its conversion into a *carboxyl group,* COOH, by treatment with hot potassium permanganate. Regardless of the length of the alkyl chain, the product is always the same—benzoic acid, C_6H_5COOH. For example,

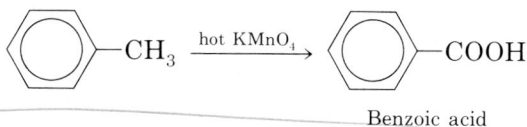

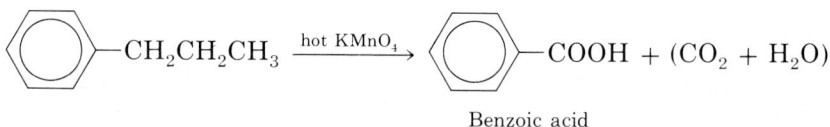

Benzoic acid

Benzoic acid

139

Rings with more than one alkyl chain are similarly oxidized; each alkyl chain, regardless of its length, is converted to a COOH group. For example,

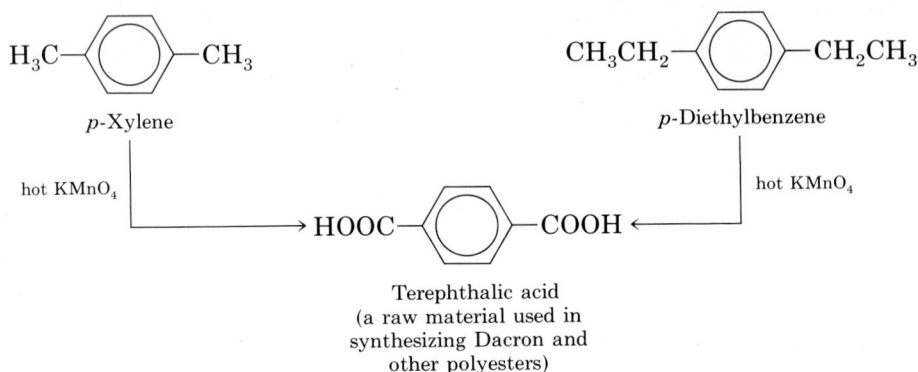

H_3C—⟨◯⟩—CH_3 CH_3CH_2—⟨◯⟩—CH_2CH_3

p-Xylene   *p*-Diethylbenzene

hot $KMnO_4$ hot $KMnO_4$

$HOOC$—⟨◯⟩—$COOH$

Terephthalic acid
(a raw material used in
synthesizing Dacron and
other polyesters)

5.11 Disubstituted Benzenes: Orientation

In previous sections the discussion centered on methods of introducing one substituent on a benzene ring. As we have noted, certain monosubstituted benzenes are made via direct electrophilic substitution (Sec. 5.9), whereas others are prepared indirectly by converting a substituent already present in the ring into another (Sec. 5.10). Like benzene, monosubstituted benzenes are capable of undergoing further substitution. When a second group, G, is introduced into a monosubstituted benzene, C_6H_5—Y, three possible isomeric compounds can theoretically be formed.

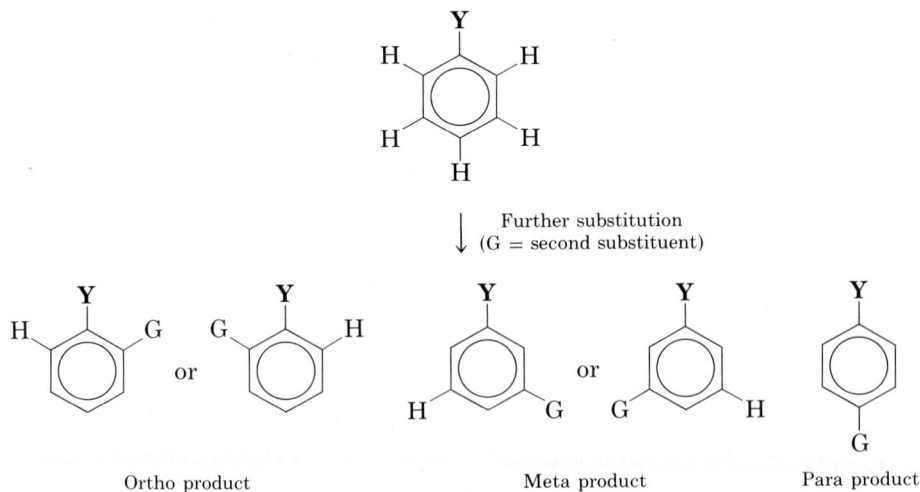

Further substitution
(G = second substituent)

Ortho product   Meta product   Para product

If the second substituent were to join the ring on a purely statistical basis, the relative proportions of ortho, meta, and para products could be calculated easily. The monosubstituted benzene, C_6H_5—Y, has five replaceable hydrogens. Among these five hydrogens, the two that are ortho to Y are equivalent to each other, as are the two hydrogens in the meta position. Therefore, if the entering substituent G were to join the ring on a purely statistical basis, the proportions of each isomer would be

ortho = $\frac{2}{5}$ of the total, or 40%

meta = $\frac{2}{5}$ of the total, or 40%

para = $\frac{1}{5}$ of the total, or 20%

In fact, such distribution is never observed. *The actual distribution depends on the nature of the first substituent, Y,* and falls into two and only two categories. Certain Ys direct the second substituent G, regardless of what G is, into the ortho and para positions. Other Ys direct the second substituent G, regardless of the nature of G, primarily into the meta position. For example, when phenol is nitrated, the only products formed are *o*-nitrophenol and *p*-nitrophenol; no meta isomer is produced. The OH group (the original substituent) is said to be an **ortho, para director.**

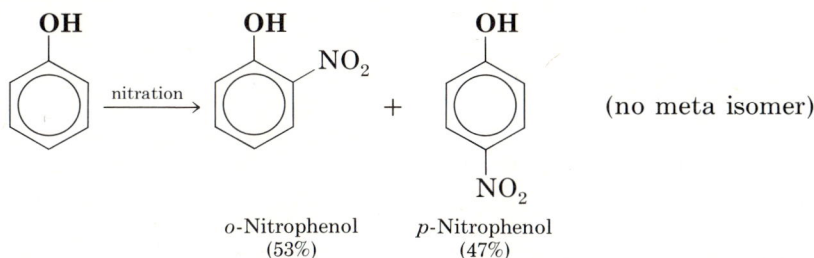

o-Nitrophenol (53%)   *p*-Nitrophenol (47%)   (no meta isomer)

Nitration of nitrobenzene, on the other hand, yields almost exclusively *m*-dinitrobenzene. The NO_2 group (the original substituent) is therefore referred to as a **meta director.**

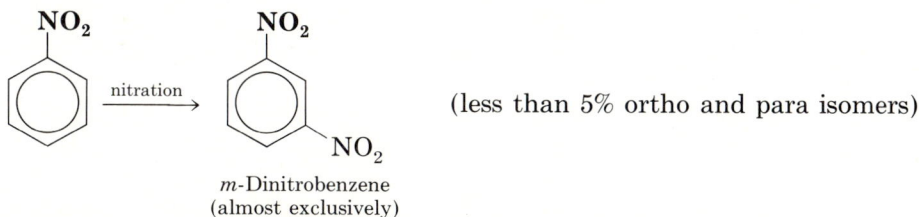

m-Dinitrobenzene (almost exclusively)   (less than 5% ortho and para isomers)

In Table 5.3 you will find a summary of the orienting effects of a number of substituents. Remember, the orientation of the second substituent, regardless of its nature, is dependent on Y only.

Table 5.3 Orientation Effects of Substituents Y in Electrophilic Aromatic Substitution

Ortho,para directors	Meta directors
—OH, —OR	—NO_2
—NH_2, —NHR, —NR_2	—SO_3H
—C_6H_5	—COOH, —COOR
—CH_3, —R (alkyl)	—CHO, —COR
—F, —Cl, —Br, —I	—CN

Problem 5.13 Consult Table 5.3 and predict the major products in the follow-ing reactions. If the product is a mixture of ortho and para isomers, show both isomers.

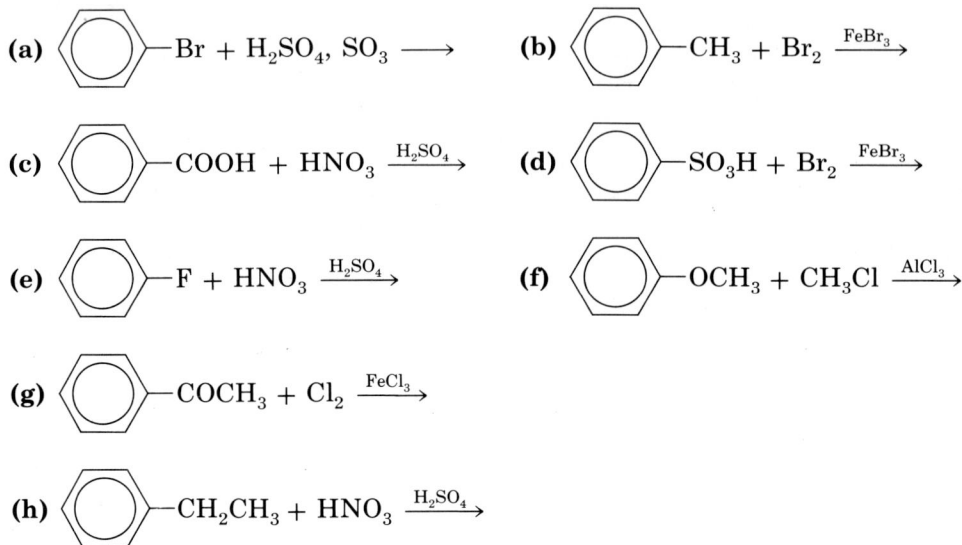

(a) $\bigcirc$—Br + H_2SO_4, SO_3 ⟶   (b) $\bigcirc$—CH_3 + Br_2 $\xrightarrow{FeBr_3}$

(c) $\bigcirc$—COOH + HNO_3 $\xrightarrow{H_2SO_4}$ (d) $\bigcirc$—SO_3H + Br_2 $\xrightarrow{FeBr_3}$

(e) $\bigcirc$—F + HNO_3 $\xrightarrow{H_2SO_4}$ (f) $\bigcirc$—OCH_3 + CH_3Cl $\xrightarrow{AlCl_3}$

(g) $\bigcirc$—$COCH_3$ + Cl_2 $\xrightarrow{FeCl_3}$

(h) $\bigcirc$—CH_2CH_3 + HNO_3 $\xrightarrow{H_2SO_4}$

5.12 Disubstitution: Reactivity

In addition to the orientation effect, the presence of a substituent also af-fects the *rate* of electrophilic substitution reactions. Intuitively, it is reasonable to assume that those substituents that release electrons to the ring more effec-tively than the hydrogen they replace will *activate* the ring toward electrophilic substitution. Such substituents will cause the reaction to go *faster* than with benzene. Conversely, those substituents that withdraw electrons from the ring more effectively than the hydrogen they replace will *deactivate* the ring toward electrophilic substitution. Such substituents will cause the reaction to go *slower* than with benzene.

The order of reactivity predicted for electrophilic substitution is shown in Figure 5.5. Experimental observations confirm these predictions: ortho,para di-

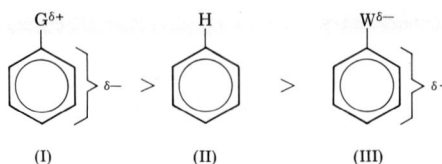

Figure 5.5 Substituent G donates electrons to the ring, thus activating it, relative to benzene. Substituent W withdraws electrons from the ring, thus deactivating it.

Table 5.4 Orientation and Reactivity Effects of Substituents in Electrophilic Aromatic Substitution

Substituent	Effect on reactivity
Ortho,para directors	
—OH, —OR, —NH$_2$, —NHR, —NR$_2$	strongly activating
—C$_6$H$_5$, —CH$_3$, —R (alkyl)	moderately activating
—F, —Cl, —Br, —I	deactivating
Meta directors	
—NO$_2$, —SO$_3$H, —COOH, —COOR	strongly deactivating
—CHO, —COR, —CN	

rectors, which donate electrons to a ring, activate the ring toward electrophilic substitution, whereas meta-directing groups, which withdraw electrons from a ring, deactivate the ring toward electrophilic substitution. The only important exception to this generalization occurs in the halobenzenes, C$_6$H$_5$X (X = F, Cl, Br, I). Although fluorine, chlorine, bromine, and iodine are ortho,para directors, these substituents deactivate an aromatic ring in electrophilic substitutions. The orientation and reactivity effects of substituents are summarized in Table 5.4.

Planning an Aromatic Synthesis 5.13

Now that you have learned a great deal of aromatic chemistry, let us see how that knowledge can be used to prepare a particular aromatic compound in the best way possible.

In any aromatic synthesis two kinds of information must be kept in mind. First, you must know the different methods by which a given substituent is introduced into the ring. Second, you must be aware of how the group(s) already present will influence the orientation of an incoming substituent. These points are illustrated in the following examples.

Example 5.1 Starting from benzene, synthesize *m*-bromonitrobenzene.

Solution   (1) Draw the structure of the starting material and the structure of the product.

Benzene   *m*-Bromonitrobenzene

Obviously, two substitutions have taken place. Therefore two steps are required: bromination of the ring and nitration of the ring.
(2) Working backward, decide what reaction leads to the product.

$$\xrightarrow{\ ?\ } \text{product}$$

Since the —NO$_2$ group is a meta director (whereas —Br is an ortho,para-directing group), the last reaction is bromination of nitrobenzene.

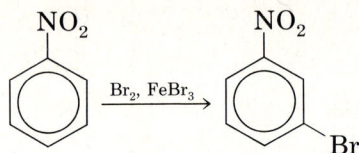

(3) The reactants leading to nitrobenzene are

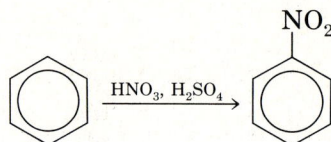

(4) Write the overall synthesis in the right order.

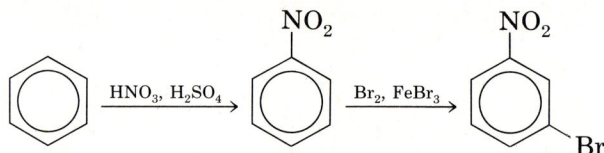

The alternate sequence (bromination followed by nitration) would have given a mixture of o-bromonitrobenzene and p-bromonitrobenzene, the wrong isomers.

Example 5.2 Suppose that you wish to synthesize p-nitrobenzoic acid, starting from toluene.

Solution First, draw the structure of the starting material and the structure of the product.

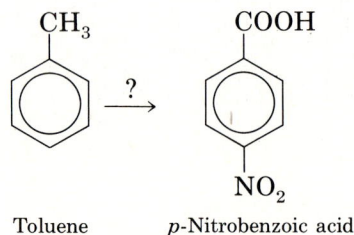

Toluene   p-Nitrobenzoic acid

Obviously, two changes have occurred. The side chain (CH$_3$) has been oxidized to COOH, and a nitro group has been introduced. Therefore two steps are required: a side-chain oxidation and a nitration. Which step should be carried out first?

Note that the para isomer of nitrobenzoic acid is required. Toluene *must therefore be nitrated first* because the methyl group is an ortho,para director.

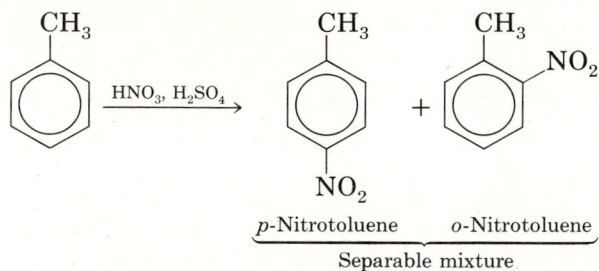

p-Nitrotoluene   o-Nitrotoluene

Separable mixture

2. Oxidation of the side chain

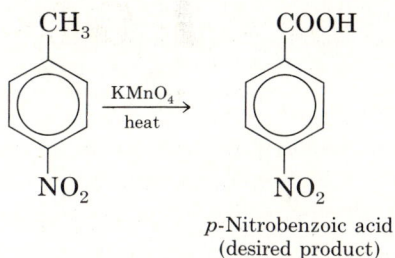

p-Nitrobenzoic acid
(desired product)

The correct sequence of steps to arrive at the desired product is therefore nitration followed by oxidation.

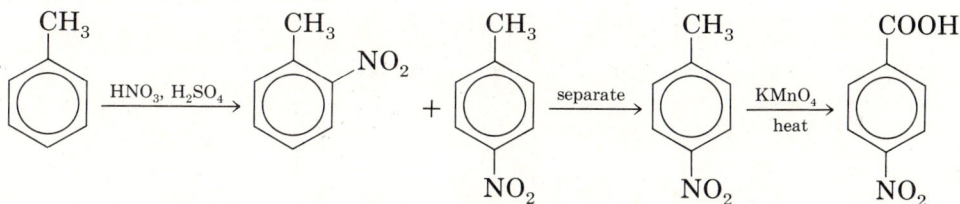

If the reverse were done (that is, if toluene were oxidized first, followed by the nitration of benzoic acid), the product obtained would be exclusively *m*-nitrobenzoic acid, an unwanted isomer.

3. Oxidation followed by nitration yields the wrong isomer.

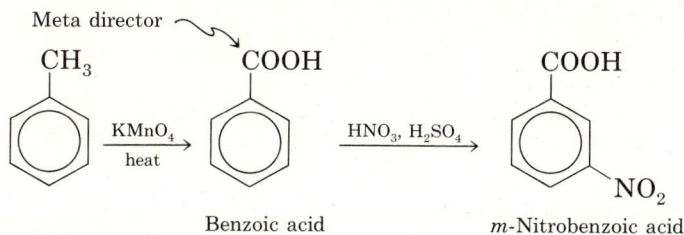

Benzoic acid   *m*-Nitrobenzoic acid

As you see, it is important to use the correct sequence of steps if the desired product is to be obtained.

Problem 5.14 Starting from benzene and any other reagents needed, show how you would synthesize these compounds. Assume that ortho,para mixtures can be separated.

(a) *o*-Nitrotoluene

(b) *m*-Chlorobenzenesulfonic acid

(c) Benzoic acid

(d) Benzyl chloride

Summary of Concepts and Reactions

Aromatic compounds are compounds that are benzene-like in their properties. Benzene itself is a cyclic, planar structure of formula C_6H_6.   [Sec. 5.1]

In benzene all carbons are sp^2 hybridized and all carbon–carbon bonds are of equal length.   [Sec. 5.1]

Benzene is actually a resonance hybrid of two imaginary contributing structures.   [Secs. 5.1, 5.2]

Benzene and other aromatic hydrocarbons are stabilized by large resonance energies. It is the preservation of these large resonance energies (36 kcal/mole for benzene) that is the source of aromatic hybrocarbons' tendency to undergo substitution rather than addition.   [Sec. 5.3]

All aromatic compounds must obey the $(4n + 2)$ π rule, also known as Hückel's rule.   [Sec. 5.4]

The two major sources of aromatic compounds are coal tar and petroleum. [Sec. 5.6]

Reactions of aromatic compounds proceed via an electrophilic aromatic substitution mechanism. In almost all electrophilic aromatic substitutions a catalyst is needed for reaction to take place. The purpose of the catalyst is to generate powerful electrophiles.   [Secs. 5.7, 5.8]

The most common electrophilic aromatic substitution reactions are

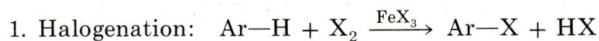

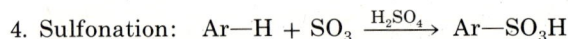

1. Halogenation: $Ar—H + X_2 \xrightarrow{FeX_3} Ar—X + HX$

2. Alkylation: $Ar—H + RX \xrightarrow{AlX_3} Ar—R + HX$

3. Nitration: $Ar—H + HONO_2 \xrightarrow{H_2SO_4} Ar—NO_2 + H_2O$

4. Sulfonation: $Ar—H + SO_3 \xrightarrow{H_2SO_4} Ar—SO_3H$

[Sec. 5.9]

The aliphatic portion of an alkylbenzene can undergo two kinds of reactions: (1) free-radical halogenation

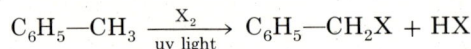

$$C_6H_5—CH_3 \xrightarrow[\text{uv light}]{X_2} C_6H_5—CH_2X + HX$$

and (2) oxidation

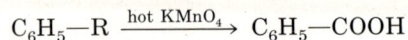

[Sec. 5.10A]

$$C_6H_5—R \xrightarrow{\text{hot } KMnO_4} C_6H_5—COOH$$

[Sec. 5.10B]

When a substituent is already present on the ring, it affects the ring in two ways:

1. It can direct a second substituent to an ortho or para position (ortho,para director) or to a meta position (meta director).   [Sec. 5.11]

2. It can activate or deactivate the ring.   [Sec. 5.12]

benzene
aromatic
substitution
resonance energy
aromaticity
Hückel's rule
phenyl
benzyl
ortho (*o-*)

meta (*m-*)
para (*p-*)
electrophile
electrophilic aromatic
   substitution
bromonium ion (Br^+)
chloronium ion (Cl^+)
carbocation (R^+)

nitronium ion ($NO_2{}^+$)
SO_3H^+ ion
halogenation
Friedel-Crafts alkylation
nitration
sulfonation
ortho,para director
meta director

Exercises

Resonance, Resonance Energy, Aromaticity [Secs. 5.1–5.4]

5.1 Refer to the diagram on page 121 and draw

(a) A resonance-contributing structure of pyridine

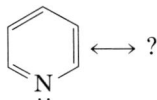

(b) Two resonance-contributing structures of naphthalene

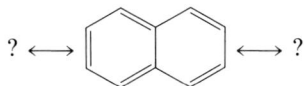

5.2 Hydrogenation of one mole of anthracene, $C_{14}H_{10}$, with seven moles of hydrogen releases 116.2 kcal of heat. Assuming the heat of hydrogenation of a normal double bond to be 28.6 kcal/mole, calculate the resonance energy of anthracene.

5.3 Which of these structures will be aromatic according to the $(4n + 2)$ π rule?

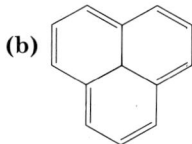

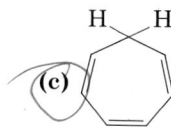

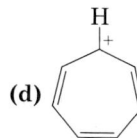

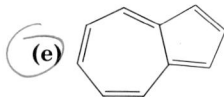

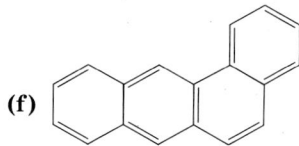

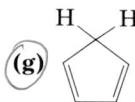

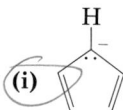

Structure and Nomenclature [Sec. 5.5]

5.4 Draw structures for these compounds.

(a) Aniline	**(b)** Toluene
(c) Benzoic acid	**(d)** Phenol
(e) Isopropylbenzene (cumene)	**(f)** *o*-Diethylbenzene
(g) Triphenylmethane	**(h)** *p*-Xylene
(i) *p*-Chlorobenzyl bromide	**(j)** *m*-Nitrobenzenesulfonic acid
(k) 3,5-Dibromostyrene	**(l)** 1,2-Diphenylethane
(m) *m*-Chloroaniline	**(n)** Pentachlorophenol
(o) Hexamethylbenzene	**(p)** 2-Aminonaphthalene
(q) 2,6-Dichloronaphthalene	**(r)** 2-Bromoanthracene
(s) 9,10-Diphenylanthracene	**(t)** 1-Nitrophenanthrene
(u) 3,5-Dimethylphenanthrene	**(v)** 2,4,6-Tribromophenanthrene

5.5 Name these compounds.

(a)

(b)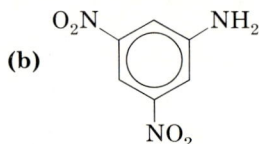

(c)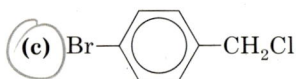

(d)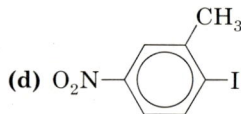

(e)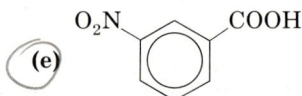

(f)

(g)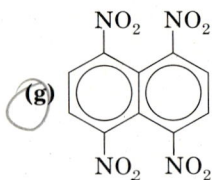

(h)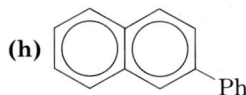

(i)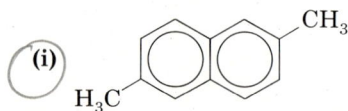

(j)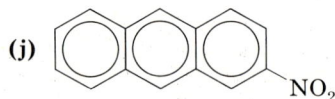

(k)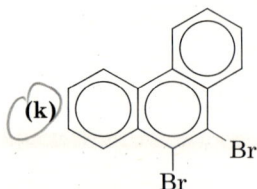

(l)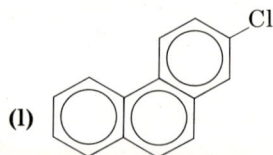

5.6 Write the structures and give names for four aromatic compounds with the formula C_7H_7Br.

5.7 Write structures and give names for all possible isomeric:
(a) Dichloroanilines (six isomers)   **(b)** Chloronitrotoluenes (ten isomers)
(c) Bromonitrophenols (ten isomers)   **(d)** Dibromonaphthalenes (ten isomers)

5.8 Treatment of benzene with CH_3Cl catalyzed by $AlCl_3$ yields toluene. Write the mechanism for the reaction and show all contributing forms in the reaction intermediate.

5.9 When a mixture of benzene and 2-methylpropene is heated in the presence of sulfuric acid catalyst, *t*-butylbenzene is formed. Write the complete mechanism to account for the formation of *t*-butylbenzene.

Orientation and Reactivity [Secs. 5.11, 5.12]

5.10 Arrange the following series of compounds in decreasing order of reactivity toward bromination.

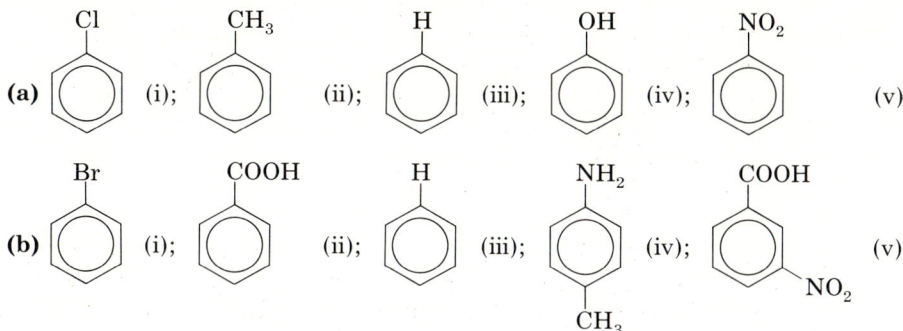

(a) Cl (i); CH$_3$ (ii); H (iii); OH (iv); NO$_2$ (v)

(b) Br (i); COOH (ii); H (iii); NH$_2$ ... CH$_3$ (iv); COOH ... NO$_2$ (v)

5.11 Predict the orientation and reactivity effects of these substituents.

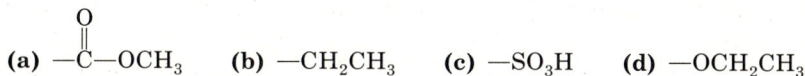

(a) $-\overset{\overset{\textstyle O}{\|}}{C}-OCH_3$ **(b)** $-CH_2CH_3$ **(c)** $-SO_3H$ **(d)** $-OCH_2CH_3$

Reactions and Syntheses [Secs. 5.9–5.13]

5.12 Draw the structures and give the names, using ortho, meta, and para designations, for the major products of the monochlorination when each of these compounds is reacted with chlorine and ferric chloride.

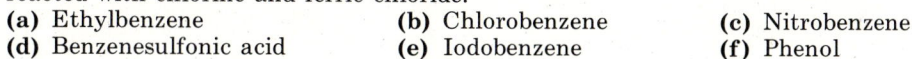

    (a) Ethylbenzene   **(b)** Chlorobenzene   **(c)** Nitrobenzene

    (d) Benzenesulfonic acid   **(e)** Iodobenzene   **(f)** Phenol

5.13 Complete the following reactions. When a mixture of ortho and para isomers is expected, show both isomers.

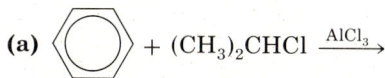

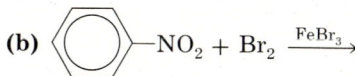

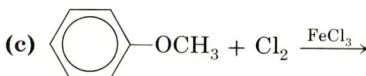

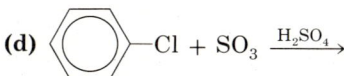

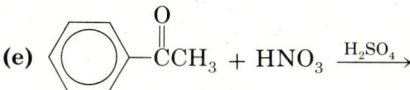

(a) $+ (CH_3)_2CHCl \xrightarrow{AlCl_3}$

(b) $-NO_2 + Br_2 \xrightarrow{FeBr_3}$

(c) $-OCH_3 + Cl_2 \xrightarrow{FeCl_3}$

(d) $-Cl + SO_3 \xrightarrow{H_2SO_4}$

(e) $-\overset{\overset{\textstyle O}{\|}}{C}CH_3 + HNO_3 \xrightarrow{H_2SO_4}$

(f) H_3C—⟨○⟩—NO_2 + Br_2 (1 mole) $\xrightarrow[\text{uv light}]{}$

(g) [structure: benzene ring with CH_3 at top and CH_3 at bottom] + $KMnO_4$ (excess) $\xrightarrow[\text{heat}]{}$

(h) [structure: benzene ring with CH_3 at top and NO_2 at bottom] + $KMnO_4$ $\xrightarrow[\text{heat}]{}$

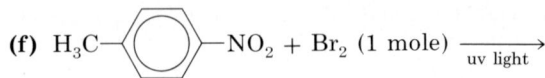

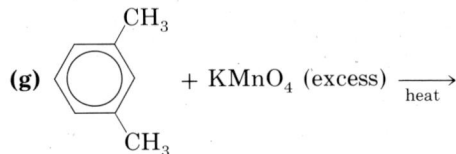

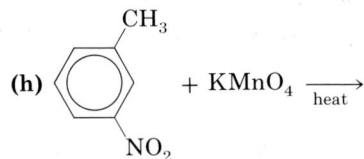

5.14 Draw the structures of the reactant and the product for the transformations shown here and supply the necessary reagents **A** through **I**. (Assume that substitutions on naphthalene, anthracene, and phenanthrene require the same reagents as benzene.)

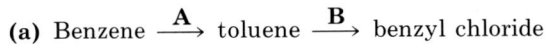

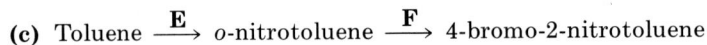

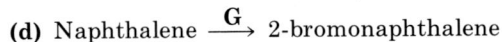

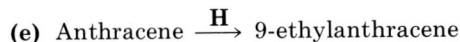

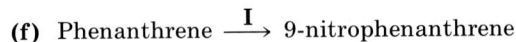

(a) Benzene $\xrightarrow{\textbf{A}}$ toluene $\xrightarrow{\textbf{B}}$ benzyl chloride

(b) Toluene $\xrightarrow{\textbf{C}}$ p-chlorotoluene $\xrightarrow{\textbf{D}}$ p-chlorobenzoic acid

(c) Toluene $\xrightarrow{\textbf{E}}$ o-nitrotoluene $\xrightarrow{\textbf{F}}$ 4-bromo-2-nitrotoluene

(d) Naphthalene $\xrightarrow{\textbf{G}}$ 2-bromonaphthalene

(e) Anthracene $\xrightarrow{\textbf{H}}$ 9-ethylanthracene

(f) Phenanthrene $\xrightarrow{\textbf{I}}$ 9-nitrophenanthrene

5.15 Starting from benzene and any other reagents needed, show how you would synthesize the following products. Assume that ortho,para mixtures can be separated.

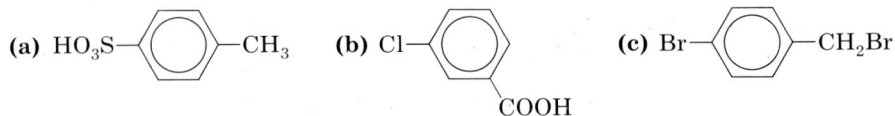

(a) HO_3S—⟨○⟩—CH_3 **(b)** Cl—⟨○⟩ with COOH   **(c)** Br—⟨○⟩—CH_2Br

5.16 There are three isomeric diethylbenzenes, **A, B,** and **C.** On nitration, isomer **A** gave *three* mononitro derivatives, isomer **B** gave *one* nitro compound, and isomer **C** gave *two* nitro compounds. What are the structures of **A, B,** and **C?**

6

Stereochemistry

Stereochemistry is a branch of chemistry that deals with the description of the spatial arrangements of atoms in molecules and with the consequences of molecular shape on the physical and chemical properties of molecules. It is also the study of the three-dimensional orientations of the atoms in molecules as they undergo a chemical reaction. As you can see, the topic of stereochemistry is enormous and varied in scope. This chapter will concentrate on one aspect of stereochemistry, that of **stereoisomerism.** In particular, the significance of stereoisomerism as it relates to biological processes will be emphasized.

Stereoisomerism 6.1

Two compounds that have the same molecular formulas but different structural formulas were defined as **structural,** or *constitutional,* isomers (Sec. 2.3). Methyl ether, CH_3OCH_3, and ethanol, CH_3CH_2OH, are classical examples of structural isomers. Stereoisomerism is a more subtle form of isomerism. Stereoisomers have not only the same molecular formulas but also the *identical sequence of bonding of the atoms in the molecule.* They differ in one respect only: the spatial orientation, or *configuration,* of their constituent atoms.

In Chapter 3, you learned that certain alkenes, such as 2-butene, can exist as *cis* or *trans* isomers and that *cis*-2-butene and *trans*-2-butene, for example, can be isolated as separate entities. The two compounds were referred to as **geometric isomers.**

151

If you recall, these geometric isomers exist because of the restricted rotation about the carbon–carbon double bond. Geometric isomers differ from one another only in the spatial orientation of the groups attached to the C=C bond. We can therefore say that geometric isomerism is one kind of stereoisomerism.

Another type of stereoisomerism is **optical isomerism.** This type of stereoisomerism manifests itself by its effect on *plane-polarized light.* In order to fully appreciate optical isomerism, you must first understand the nature of plane-polarized light.

6.2 Plane-Polarized Light

A beam of ordinary white light may be considered as a bundle of rays of different wavelengths (400–800 nm) vibrating in all possible planes perpendicular to the direction in which they propagate (Fig. 6.1a). When a suitably colored filter is placed in front of the white light, all the rays of the transmitted light are of the same color and therefore of the same wavelength. The transmitted light is called *ordinary monochromatic light.* It also vibrates in all possible planes (Fig. 6.1b).

If a sheet of Polaroid film is placed in front of ordinary monochromatic light, the emerging light vibrates in *one plane only.* This light is called **plane-polarized light** (Fig. 6.2). The Polaroid film, which consists of a transparent

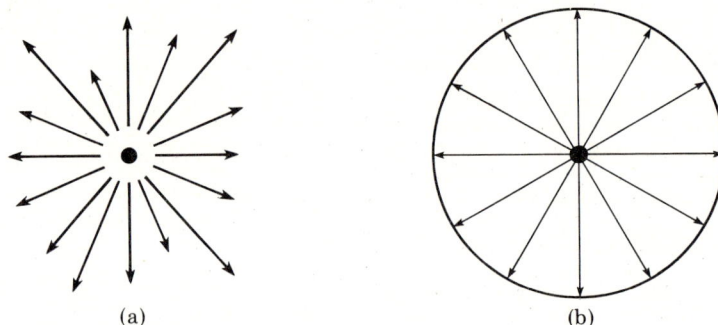

(a)   (b)

Figure 6.1 Ordinary light **(a)** vibrates in all the planes perpendicular to the page. Monochromatic light **(b)** also vibrates in all possible planes perpendicular to the page, but its vibrations are uniform.

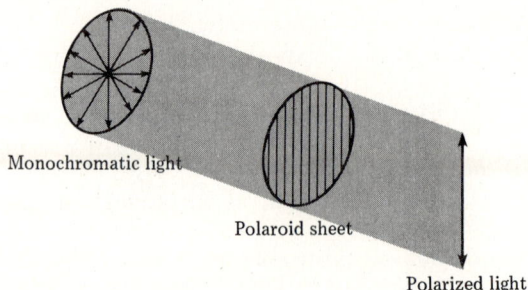

Monochromatic light

Polaroid sheet

Polarized light

Figure 6.2 When a polaroid film is placed in front of monochromatic light, the emerging light vibrates in only one plane.

sheet of oriented polyvinyl alcohol embedded with iodine crystals, acts as a *polarizer*. Other polarizers include tourmaline and a Nicol prism, a pair of calcium carbonate crystals cemented by Canadian balsam.

The next step in the discussion is to learn how to determine whether or not a compound has any effect on plane-polarized light.

Optical Activity and Its Measurement 6.3

Any compound that has the ability to change the direction of plane-polarized light, or to *rotate* it, is said to be *optically active*. The rotation itself is called **optical activity.**

How is optical activity measured? In principle, it can be measured by placing a solution of the compound suspected of being optically active in front of the plane-polarized light and seeing what happens to the light transmitted by the solution. (If the compound to be studied is a liquid, the pure compound may be used. In such case, the measurement is said to be carried out on a *neat* sample.) To see how the light transmitted by a solution is affected by plane-polarized light, refer to Figure 6.3. You can see that the plane of the light emerging from the solution has been rotated by a certain number of degrees, alpha (α), to the right of the original plane. The compound is therefore *optically active*. The experimental number of degrees of rotation, α, is called the *observed rotation*.

In practice, the apparatus used to measure optical activity is called a *polarimeter* (Fig. 6.4). If, in looking into the polarimeter, the viewer finds the observed rotation to be to the right (clockwise rotation), the optically active compound is

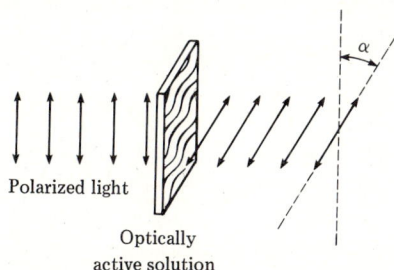

Figure 6.3 Plane-polarized light passing through an optically active solution is rotated by a certain number of degrees, α, to the right of the original plane.

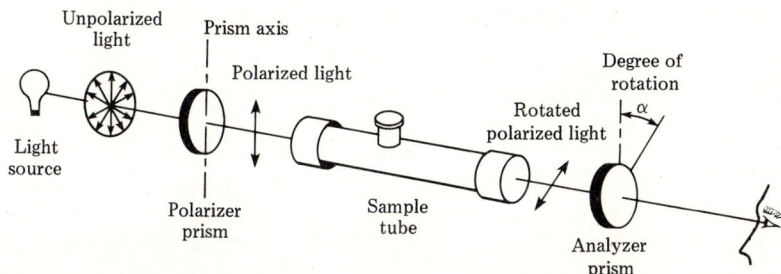

Figure 6.4 Diagrammatic sketch of a polarimeter.

Table 6.1 Physical Properties of Two Optical Isomers

	(+)-2-Butanol	(−)-2-Butanol
Boiling point (°C)	99.5	99.5
Density[a] (g/ml)	0.8080	0.8080
Refractive index[a]	1.395	1.395
Specific rotation[a]	+13.9°	−13.9°

[a] Values at 20°C.

designated as **dextrorotatory.** The symbol (+) is used to indicate an optically active compound that is dextrorotatory. For example, (+)-2-butanol means we are dealing with an optically active compound that rotates plane-polarized light to the right.

By the same token, an optically active compound that rotates plane-polarized light to the left (counterclockwise rotation) is termed **levorotatory.** The symbol (−) is used to indicate levorotation. Thus, (−)-2-butanol is an optical isomer of (+)-2-butanol. The two compounds are identical in all their physical properties except the signs of their optical rotation, which are exactly opposite, as shown in Table 6.1.

All optically active compounds are either dextrorotatory (+) or levorotatory (−). This tells us only the direction or the sign of the rotation. It indicates nothing about the number of degrees, or the *magnitude,* of the optical rotation.

The value of the experimentally observed rotation for a compound is not a fixed number. It varies proportionately with the *concentration* of the solution and with the *length* of the polarimeter tube for a given concentration. The observed rotation is also a function of the *temperature* at which the measurement is made, the *wavelength* of the light, and the *solvent* (if any). An example will serve to illustrate these points.

Example 6.1 A solution of 2.0 g/ml of (+)-2-butanol in a 10 cm polarimeter tube has an observed rotation of +27.8°. Calculate the observed rotation
(a) If the length of the polarimeter tube is 5 cm and the concentration of the solution remains the same.
(b) If the concentration is doubled and the length of the polarimeter tube remains the same (10 cm).

Solution **(a)** If the length of the tube is halved, then the observed rotation is also halved, because there is a direct relationship between the length of the polarimeter tube and the observed rotation.

+27.8°/2 = +13.9°

(b) If the concentration is doubled, the observed rotation is also doubled, because there is direct relationship between the two parameters.

+27.8° × 2 = +55.6°

The values of the observed rotation are therefore dependent upon experimental conditions.

Specific Rotation **6.4**

To arrive at a fixed value for the optical rotation of a compound, a specified set of conditions must be established, just as the statement that the boiling point of water is 100°C implies a specified condition of pressure (namely, 760 torr). The value of the optical rotation of a compound under standard conditions is called the **specific rotation, [α]**, and is defined by the equation

$$[\alpha]_{\lambda}^{t} = \frac{\alpha}{l \times c} \qquad \text{(solvent specified)}$$

where [α] = specific rotation (in degrees)

t = temperature (in °C)

λ = wavelength of light used (in nanometers); usually sodium D line, 589 nm

α = experimental observed rotation (in degrees)

l = length of polarimeter tube (in decimeters; 1 dm = 10 cm)

c = concentration of the solution (in g/ml)

Example 6.2 Calculate the specific rotation [α] for (+)-2-butanol based on the data given in Example 6.1.

Solution In Example 6.1, the original conditions were

$\alpha = +27.8°$ $l = 1$ dm $c = 2.0$ g/ml

Therefore

$$[\alpha] = \frac{\alpha}{l \times c} = \frac{+27.8°}{(1)(2)} = +13.9°$$

In 6.1(a) the following conditions prevailed.

$\alpha = +13.9°$ $l = 0.5$ dm $c = 2.0$ g/ml

Therefore

$$[\alpha] = \frac{\alpha}{l \times c} = \frac{+13.9°}{(0.5)(2)} = +13.9°$$

In 6.1(b) the conditions were as follows.

$\alpha = +55.6°$ $l = 1$ dm $c = 4.0$ g/ml

Therefore

$$[\alpha] = \frac{\alpha}{l \times c} = \frac{+55.6°}{(1)(4)} = +13.9°$$

As you can see, as long as the wavelength, temperature, and solvent used remain the same, we get the same value for the specific rotation. The specific rotation is therefore a physical constant characteristic of a compound.

Problem 6.1 Calculate the specific rotation of cholesterol if the concentration of the solution is 5.20 g/100 ml, the length of the polarimeter tube is 5 cm, and the observed rotation is −2.5°.

The value of the specific rotation for an optically active compound is a physical property, just as its melting point and boiling point are. It is therefore important as an aid in identifying a compound or in determining its degree of purity. For example, in looking up the physical constants of heroin in a handbook of physics and chemistry you would find its specific rotation listed as $[\alpha]_D^{15} = -166°$ (methanol). This means that heroin is an optically active compound. It is levorotatory, and the value of its specific rotation is $-166°$ when measured at $15°$C with plane-polarized light at the wavelength of the sodium D line in methanol as the solvent.

6.5 Optical Activity and Structures of Compounds

Many organic compounds, but not the majority, are optically active. The question arises, Can we predict whether or not a compound will be optically active based on its structure? The answer is yes. In order to exhibit optical activity, a molecule must be **chiral** (from the Greek *cheir,* hand); that is, *it cannot be superimposed on its mirror image.*

As an illustration of what is meant by this statement, suppose the compound in question were your right hand. If you place the hand in front of a mirror, you will have a mirror image of it (that is, an image of your right hand that looks like your left hand).

Imagine it were possible to take the mirror image out of the plane of the mirror and bring it in front of the mirror next to the original right hand. Try now to answer the following question: Will your right hand (original) and its mirror image (the "left" hand) superimpose on each other? (Does your left hand fit a right-hand glove?)

The answer in our example is obviously no: the right hand and its mirror image are not superimposable. Because the mirror image of the right hand cannot be superimposed on it, the right hand in this example is chiral. The same holds true for the left hand: the right hand is a nonsuperimposable mirror image of the left hand, so the left hand is also a chiral object. Had the right and left hands been chemical compounds, they would have been optically active.

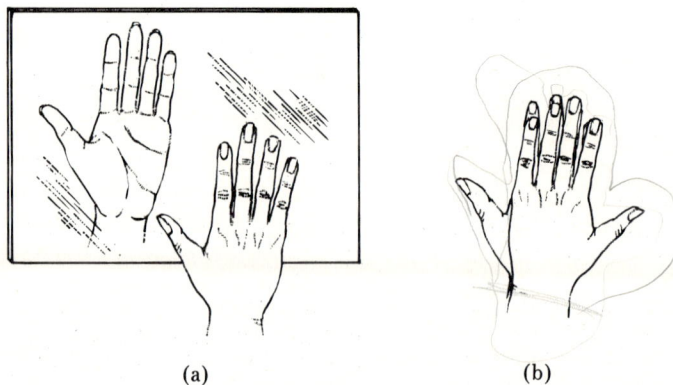

(a)   (b)

Figure 6.5 The mirror image (the "left" hand) is not superimposable on the right hand.

Compounds that are mirror images of each other and are not superimposable are called **enantiomers** (from the Greek *enantio,* opposite; *meros,* part). Enantiomers have the same physical properties except that they rotate plane-polarized light in opposite directions (but in equal amounts). If, for example, the "right hand" compound had a specific rotation of $+120°$, its enantiomer, the "left hand" compound, would have a specific rotation of $-120°$.

Problem 6.2 Which of the following objects are chiral? (Assume the objects have no markings.)

(a) A sphere   **(b)** A tennis racket   **(c)** A bicycle

(d) An automobile   **(e)** A screw   **(f)** A baseball glove

(g) A paper clip   **(h)** A spoon

Let us return to chemical compounds and decide which structures are chiral and which are *achiral* (achiral = not chiral).*

Most chiral chemical compounds have a **chiral center,** which is an atom (usually carbon) bonded to *four different atoms or groups.* For example, methane, CH_4, is achiral because it has no chiral center (four identical atoms are attached to the carbon atom) and its mirror image is superimposable. Therefore methane will not exhibit optical activity, and there are no enantiomers of methane.

Bromochlorofluoromethane, $CHBrClF$, on the other hand, is chiral. It has a chiral center (four different atoms are attached to the central carbon atom), and therefore the original and its mirror image are not superimposable (see Fig. 6.6). Thus there are two optical isomers (enantiomers) that will rotate plane-polarized light by equal amounts but in opposite directions. The chiral center is usually indicated by an asterisk.

Now that you have studied optical activity from a structural standpoint, return to the previous example of the two optically active forms of 2-butanol. If you draw their structures, you will note the presence of a chiral carbon atom

* Note to students: A set of molecular models would help you greatly in mastering the material in the next few sections.

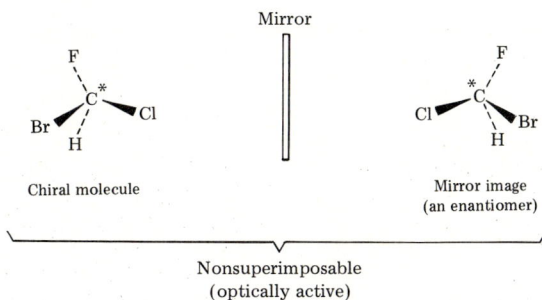

Figure 6.6 A chiral molecule is optically active because it contains a chiral center and is not superimposable on its mirror image. Note that the tetrahedral geometry is represented by the use of wedged lines to indicate bonds coming out of the plane of the paper and dashed lines to indicate bonds directed behind the plane of the paper. Straight lines are used for bonds that are in the plane of the paper.

Mirror

2-Butanol   2-Butanol

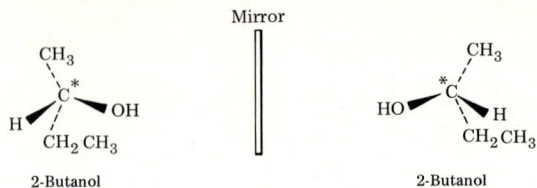

Nonsuperimposable enantiomers

Figure 6.7 Perspective formulas for the enantiomers of 2-butanol.

(the one with the asterisk in Fig. 6.7), and you will see that the mirror images are not superimposable. The two compounds are therefore enantiomers.

Problem 6.3 Indicate each chiral carbon with an asterisk in these formulas.

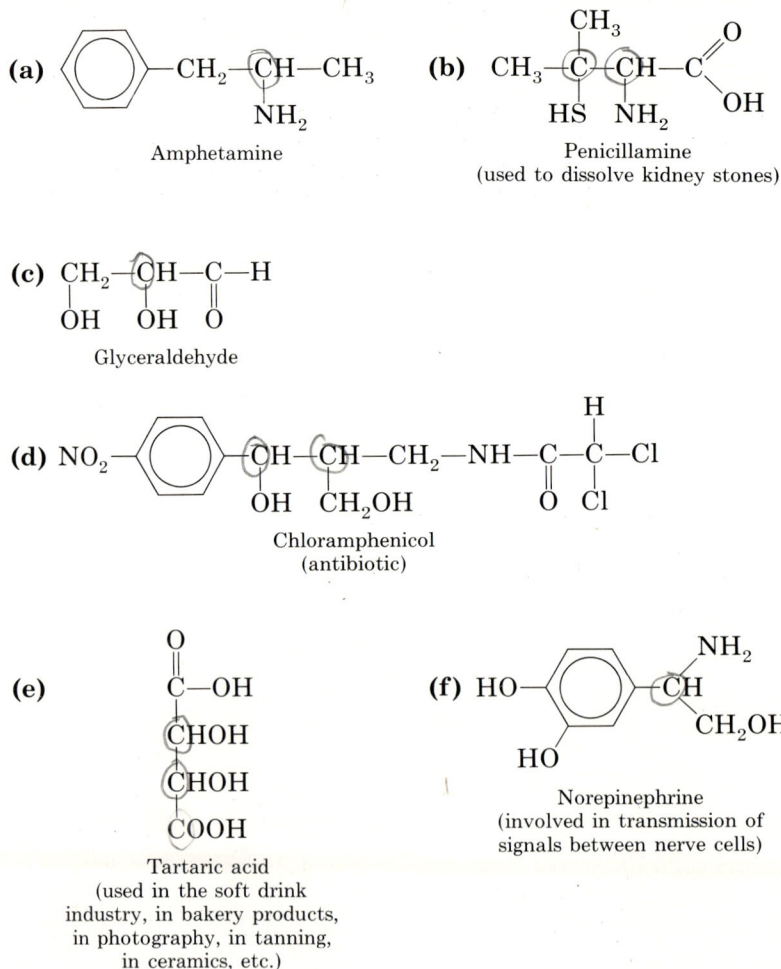

(a)

$\bigcirc$—CH_2—CH—CH_3
$\quad\quad\quad\quad\quad$ |
$\quad\quad\quad\quad\quad NH_2$

Amphetamine

(b)

$\quad\quad\quad CH_3\quad\quad O$
$\quad\quad\quad\quad |$
CH_3—C—CH—C
$\quad\quad\quad |\quad\quad |\quad\quad\quad\quad OH$
$\quad\quad\quad HS\quad NH_2$

Penicillamine
(used to dissolve kidney stones)

(c)

CH_2—CH—C—H
$\quad |\quad\quad |\quad\quad ||$
$\quad OH\quad OH\quad O$

Glyceraldehyde

(d)

$\quad\quad\quad\quad\quad\quad\quad\quad\quad\quad\quad\quad H$
$\quad\quad\quad\quad\quad\quad\quad\quad\quad\quad\quad\quad |$
NO_2—$\bigcirc$—CH—CH—CH_2—NH—C—C—Cl
$\quad\quad\quad\quad\quad |\quad\quad |\quad\quad\quad\quad\quad\quad ||\quad |$
$\quad\quad\quad\quad\quad OH\quad CH_2OH\quad\quad\quad O\quad Cl$

Chloramphenicol
(antibiotic)

(e)

$\quad O$
$\quad ||$
$\quad C$—OH
$\quad |$
$CHOH$
$\quad |$
$CHOH$
$\quad |$
$COOH$

Tartaric acid
(used in the soft drink
industry, in bakery products,
in photography, in tanning,
in ceramics, etc.)

(f)

$\quad\quad\quad\quad\quad\quad\quad\quad\quad NH_2$
$\quad\quad\quad\quad\quad\quad\quad\quad\quad |$
HO—$\bigcirc$—CH
$\quad\quad\quad\quad\quad\quad\quad\quad\quad CH_2OH$
HO

Norepinephrine
(involved in transmission of
signals between nerve cells)

Suppose we dissolve an unknown sample in a polarimeter tube, attempt to measure its optical activity, and find that the solution is *optically inactive.* Can we necessarily conclude that the substance in question is achiral? The answer is, not necessarily. We may very well have attempted to measure the optical activity of *an equal mixture of enantiomers,* called a **racemic mixture.** A racemic mixture of compounds is optically inactive. The reason for this is not that the molecules are achiral, but that the optical rotation caused by one enantiomer is exactly canceled by the optical rotation, in the opposite direction, of the other enantiomer. For example, if a given amount of ($+$)-amphetamine causes the observed rotation to be $+6.25°$, then an equal amount of ($-$)-amphetamine, under the same experimental conditions, will result in an observed rotation of $-6.25°$. Thus, in mixing equal weights of ($+$)- and ($-$)-amphetamine (a racemic mixture) we find that the net observed rotation is zero ($+6.25° + -6.25° = 0°$). The solution is therefore optically inactive. The symbol ($\pm$) is used to denote a racemic mixture. For example, ($\pm$)-amphetamine indicates a racemic mixture of amphetamine.

Predicting the Number of Stereoisomers: 2^n Rule **6.7**

In the examples discussed so far we saw that the presence of one chiral center gave rise to two enantiomers. For each additional chiral center, there exists another pair of stereoisomers. We can therefore state that *the number of stereoisomers for a structure with n different chiral centers is 2^n.* This statement is known as **van't Hoff's rule.**

Consider 2,3,4-trihydroxybutanal, $HOCH_2-\overset{*}{C}HOH-\overset{*}{C}HOH-CH=O$. It has two unlike chiral carbon atoms and can therefore exist as four possible isomers ($2^2 = 4$). Let us write them down and consider the relationships among them.

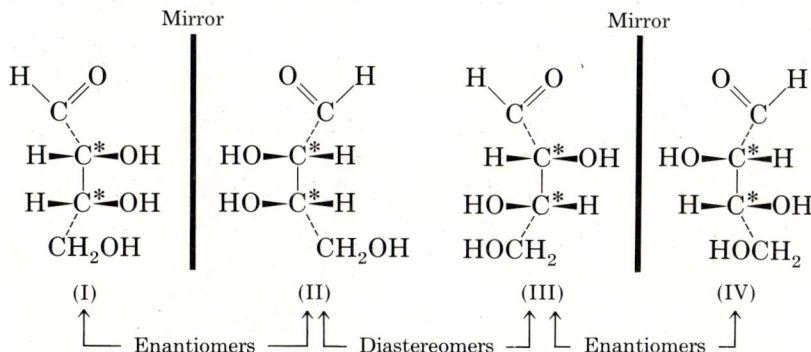

You can see that stereoisomers I and II are enantiomers and that stereoisomers III and IV are also enantiomers. Compounds II and III are stereoisomers, but are *not* enantiomers; they are known as **diastereomers.**

Diastereomers are stereoisomers that are not mirror images of each other.
Unlike enantiomers, which have the same physical properties (except toward
polarized light), diastereomers have different boiling points, melting points, refractive indices, solubilities, and so on. Diastereomers can therefore be separated
from each other by fractional crystallization or distillation.

For example, if we have an equimolar mixture of I, II, III, and IV and distill,
we get two fractions. One fraction contains I and II. The second fraction contains the diastereomers of I and II, compounds III and IV. Each fraction is
optically inactive because it is made up of a racemic mixture.

Problem 6.4 Looking at the preceding structures (I–IV), state which other
pairs are diastereomers besides II and III.
(a) Structures I and III   **(b)** Structures I and IV
(c) Structures II and IV

6.8 Meso Compounds

Tartaric acid, $HOOC-\overset{*}{C}HOH-\overset{*}{C}HOH-COOH$, is another example of a
compound that has two chiral centers. If we apply van't Hoff's rule, we could
expect a maximum of four (2^2) stereoisomers. Let us draw all the possible stereoisomers of tartaric acid.

An examination of structures I–IV shows the following. Structures I and II are
enantiomers. The pair of compounds I and III are diastereomers, as are the II
and III pair. What about III and IV? A close examination reveals that the
structures are mirror images of each other but also that they are superimposable. Structures III and IV, then, do not represent enantiomers; rather, they are
the same molecule. A 180° rotation of IV makes it identical to III. Because III
and IV are identical, and therefore superimposable, we can say that the molecule
as a whole is achiral even through it possesses two chiral centers. We refer to III
as a **meso compound** (mesotartaric acid). A meso compound may thus be defined as an *optically inactive compound even though it possesses more than one
chiral center.* One simple way of recognizing a meso compound is to note that
the molecule possesses a *plane of symmetry.* A substance is said to have a plane

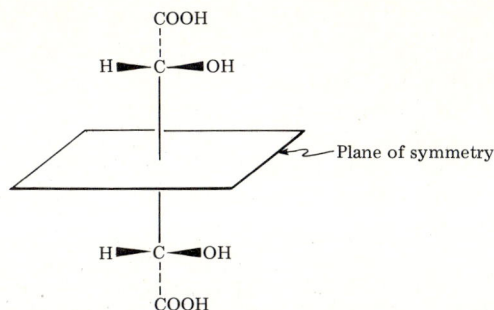

Figure 6.8 Internal symmetry of mesotartaric acid.

of symmetry if it can be divided by a plane, acting as a mirror, in such a way that all parts on one side of the plane are identical to all parts on the other side of the plane. The upper half of mesotartaric acid, for example, is the mirror image of its lower half (Fig. 6.8). There is internal compensation of optical activity, making the molecule as a whole optically inactive.

Because of the presence of a meso isomer, there are only three stereoisomers of tartaric acid and not four. Table 6.2 summarizes some physical properties of the stereoisomers of tartaric acid. You can see from the table that if we had an equimolar mixture of the three tartaric acids and did a fractional crystallization, we would get two fractions. One would contain racemic ($\pm$)-tartaric acid; the other would contain the diastereomer mesotartaric acid. Each fraction would be optically inactive (why?)

Problem 6.5 Determine the number of possible stereoisomers in each of the compounds listed in Problem 6.3.

Table 6.2 Physical Properties of the Tartaric Acids

	Dextro	**Levo**	**Meso**
Melting point (°C)	170	170	140
Specific gravity	1.76	1.76	1.67
Solubility (g/100 ml at 20°C)	139	139	125
$[\alpha]_D^{20}$ in H_2O	$+12°$	$-12°$	$0°$

Resolution 6.9

In Sections 6.7 and 6.8 we saw that stereoisomers can be separated by fractional crystallization or distillation providing they are diastereomers. This is because diastereomers possess different physical properties. But is it also possible to separate a pair of enantiomers? You will recall that, except for their signs of optical rotation, which are opposite, enantiomers have identical physical properties. Thus, we would not expect to be able to separate a pair of enantio-

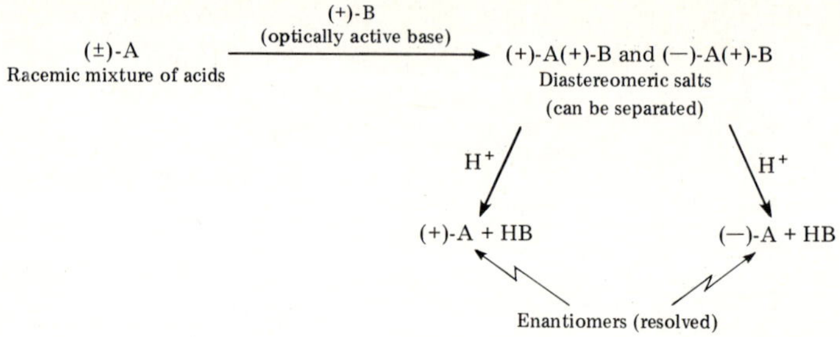

$(\pm)$-A
Racemic mixture of acids

$(+)$-B
(optically active base)
$\longrightarrow$
$(+)$-A$(+)$-B and $(-)$-A$(+)$-B
Diastereomeric salts
(can be separated)

H$^+$ H$^+$

$(+)$-A + HB $(-)$-A + HB

Enantiomers (resolved)

Figure 6.9 Resolution of a racemic acid, $(\pm)-$A, with an optically active base, $(+)-$B.

mers by fractional distillation, because their boiling points, for example, are the same. Nevertheless, the answer to the question is, yes, it is possible to separate a pair of enantiomers into its $(+)$ and $(-)$ isomers. The process by which this can be done is called **resolution.** The first instance of resolution of a racemic mixture was accomplished by Louis Pasteur in 1848. Pasteur painstakingly hand-picked and separated crystals of the $(+)$ form of sodium ammonium tartrate from crystals of the $(-)$ isomer. (Crystals of the two forms are actually physically different.) He then showed that the $(+)$ form rotated the plane of polarized light to the right, whereas the $(-)$ isomer caused the light to be rotated in a counterclockwise manner. Pasteur further showed that when he mixed equal amounts of the $(+)$ and $(-)$ crystals he had just separated and dissolved them in water, the resulting solution, like the starting material, was optically inactive.

Pasteur's method of resolution of $(\pm)$-sodium ammonium tartrate is primarily one of historical interest. The general utility of the method is limited, because very few racemic mixtures yield crystals that make manual separation of one enantiomer from another possible.

The most useful method of resolution involves the conversion of a pair of enantiomers into a diastereomeric pair. The pair of diastereomers, once formed, is separated by fractional crystallization or distillation. After such separation, each diasteromer is treated in such a way as to liberate one enantiomer uncontaminated with the other. As you can see, *the key step in a chemical resolution is the conversion of a pair of enantiomers into a pair of diastereomers.*

Formation of a pair of diastereomers is usually accomplished in the following way. If the $(\pm)$ pair is an acid, such as $(\pm)$-tartaric acid, it is treated with a pure form of an optically active base, say the $(+)$ form. The salts that are formed are diastereomers, and they can therefore be separated. Once the diastereomeric salts are separated, each may be treated with a mineral acid to liberate pure $(+)$ acid, in this case $(+)$-tartaric acid, and pure $(-)$ acid, or $(-)$-tartaric acid. The resolution of a racemic acid, $(\pm)$-A, with an optically active base, $(+)$-B, is summarized in Figure 6.9.

If the starting racemic mixture is a base, then an optically active acid can be used to resolve it. Compounds containing other functional groups can be resolved by variation of the same general method.

Problem 6.6 Before proceeding to the next section, test yourself by defining the following terms introduced thus far in the chapter.
(a) Plane-polarized light   **(b)** Optical activity

(c) Specific rotation   **(d)** Chiral compound   *163*
(e) Chiral center   **(f)** Enantiomers
(g) Diastereomers   **(h)** Meso compound   6.10 Fischer
(i) Racemic mixture   **(j)** Resolution   Projections

Fischer Projections *6.10*

Up to now we made use of perspective formulas in considering the stereochemical relationship between enantiomers, diastereomers, and meso compounds. Since it is difficult to represent a three-dimensional molecule on a two-dimensional surface (the paper on which this is printed), chemists have adopted the use of **Fischer projection*** formulas to represent in an easy manner molecules whose stereochemistry is important. For example,

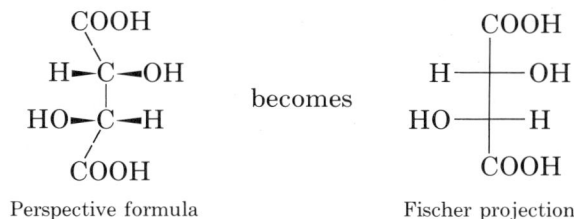

F
\
Br—C—Cl   becomes   Br——|——Cl
/
H

Perspective formula   Fischer projection

COOH   COOH
H—C—OH   becomes   H——|——OH
HO—C—H   HO——|——H
COOH   COOH

Perspective formula   Fischer projection

In writing a Fischer projection of a compound whose perspective formula is given, the following points must be kept in mind.

1. The intersection of the horizontal and vertical lines represents the chiral center.
2. The horizontal lines attached to the chiral center represent bonds directed toward the reader.
3. The vertical lines attached to the chiral center represent bonds directed away from the reader.

* Emil Fischer (1852–1919) was professor of organic chemistry at the University of Berlin and the outstanding director of organic chemical research. He is considered the father of carbohydrate chemistry. Fischer made monumental contributions to our understanding of the chemistry of amino acids, proteins, and purines. He also made important contributions in many other fields, such as stereochemistry, enzymes, dyes, hydrazines (used in rocket fuels), and indoles. In 1902 he became the second recipient after van't Hoff (1901) of the Nobel prize in chemistry.

6.11 Manipulating Fischer Projections

It is sometimes necessary to compare two Fischer projection formulas that are written in different orientations. Consider, for example, these two representations of 2-butanol.

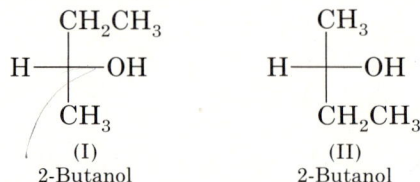

$$
\begin{array}{c}
\text{CH}_2\text{CH}_3 \\
\text{H} \!-\!\!\!-\!\!\!-\!\!\! \text{OH} \\
\text{CH}_3 \\
\text{(I)} \\
\text{2-Butanol}
\end{array}
\qquad\qquad
\begin{array}{c}
\text{CH}_3 \\
\text{H} \!-\!\!\!-\!\!\!-\!\!\! \text{OH} \\
\text{CH}_2\text{CH}_3 \\
\text{(II)} \\
\text{2-Butanol}
\end{array}
$$

What relation is there between 2-butanol I and 2-butanol II? Are they identical molecules or enantiomers?

To make the comparison possible between the two structures, one or both molecules must be redrawn. In doing so, we must be very careful not to change their original configurations. Although perspective formulas may be turned over or rotated in any direction without causing a change in the original configuration, this is *not* the case with Fischer projection formulas. *The only permitted manipulation that will not change the configuration of a Fischer projection formula is to rotate it by 180° within the plane of the paper.*

Let us now go back to structures I and II and decide what relationship exists between them. If we leave I alone and rotate II by 180° within the plane of the paper, we obtain

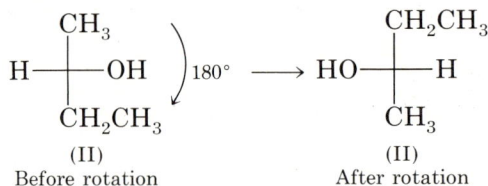

$$
\begin{array}{c}
\text{CH}_3 \\
\text{H} \!-\!\!\!-\!\!\!-\!\!\! \text{OH} \\
\text{CH}_2\text{CH}_3 \\
\text{(II)} \\
\text{Before rotation}
\end{array}
\quad
\xrightarrow{180°}
\quad
\begin{array}{c}
\text{CH}_2\text{CH}_3 \\
\text{HO} \!-\!\!\!-\!\!\!-\!\!\! \text{H} \\
\text{CH}_3 \\
\text{(II)} \\
\text{After rotation}
\end{array}
$$

If we compare II after 180° rotation with I, it is obvious that I and II are enantiomers.

Now try to solve the following problems.

Problem 6.7 What relationship (identical, enantiomers) is there between the following pairs of structures?

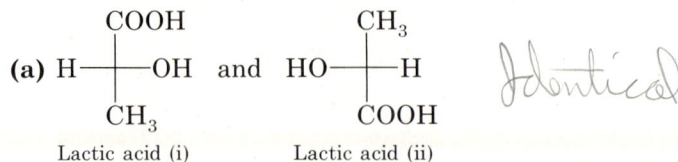

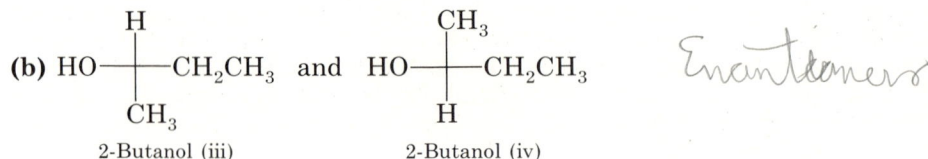

(a)
$$
\begin{array}{c}
\text{COOH} \\
\text{H} \!-\!\!\!-\!\!\!-\!\!\! \text{OH} \\
\text{CH}_3 \\
\text{Lactic acid (i)}
\end{array}
\quad \text{and} \quad
\begin{array}{c}
\text{CH}_3 \\
\text{HO} \!-\!\!\!-\!\!\!-\!\!\! \text{H} \\
\text{COOH} \\
\text{Lactic acid (ii)}
\end{array}
\qquad \textit{Identical}
$$

(b)
$$
\begin{array}{c}
\text{H} \\
\text{HO} \!-\!\!\!-\!\!\!-\!\!\! \text{CH}_2\text{CH}_3 \\
\text{CH}_3 \\
\text{2-Butanol (iii)}
\end{array}
\quad \text{and} \quad
\begin{array}{c}
\text{CH}_3 \\
\text{HO} \!-\!\!\!-\!\!\!-\!\!\! \text{CH}_2\text{CH}_3 \\
\text{H} \\
\text{2-Butanol (iv)}
\end{array}
\qquad \textit{Enantiomers}
$$

164

(c)

$$\underset{\text{Alanine (v)}}{H_3C-\overset{\displaystyle NH_2}{\underset{\displaystyle COOH}{|}}-H}$$ and $$\underset{\text{Alanine (vi)}}{CH_3-\overset{\displaystyle COOH}{\underset{\displaystyle NH_2}{|}}-H}$$ *Enantiomers*

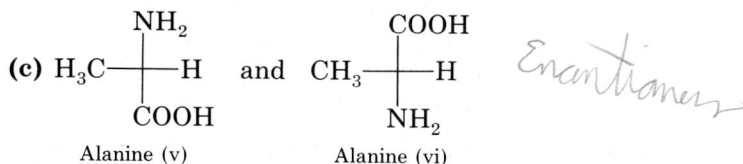

Problem 6.8 Given this Fischer projection formula,

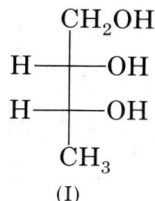

$$\overset{\displaystyle CH_2OH}{\underset{\displaystyle CH_3}{\overset{\displaystyle |}{\underset{\displaystyle |}{H-\!\!\!-OH}}}}$$
$$H-\!\!\!-OH$$
(I)

indicate what relationship (identical, enantiomer, or diastereomer) there is between it and each of the following Fischer projections.

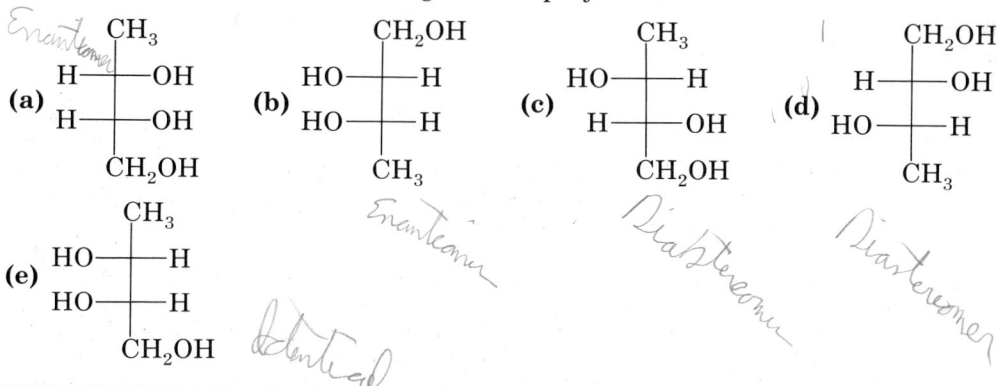

(a) *Enantiomer*

$$\overset{CH_3}{\underset{CH_2OH}{\overset{|}{\underset{|}{\begin{array}{c}H-\!\!-OH\\H-\!\!-OH\end{array}}}}}$$

(b)

$$\overset{CH_2OH}{\underset{CH_3}{\begin{array}{c}HO-\!\!-H\\HO-\!\!-H\end{array}}}$$ *Enantiomer*

(c)

$$\overset{CH_3}{\underset{CH_2OH}{\begin{array}{c}HO-\!\!-H\\H-\!\!-OH\end{array}}}$$ *Diastereomer*

(d)

$$\overset{CH_2OH}{\underset{CH_3}{\begin{array}{c}H-\!\!-OH\\HO-\!\!-H\end{array}}}$$ *Diastereomer*

(e)

$$\overset{CH_3}{\underset{CH_2OH}{\begin{array}{c}HO-\!\!-H\\HO-\!\!-H\end{array}}}$$ *Identical*

Problem 6.9 What is the relationship between the following pairs of Fischer
projections in Problem 6.8?

(a) and **(b)**   **(c)** and **(d)**   **(d)** and **(e)**   **(a)** and **(c)**   **(a)** and **(d)**   **(b)** and **(e)**

Identical   Enantiomer   Diastereomers   Diastereomers   Diastereomer   Diastereomers

Specification of Configuration **6.12**

The configuration of a molecule, if you recall, was defined as the orientation
in space of the atoms that can be changed only by the breaking and making of
bonds. It is often desirable to know the actual, or **absolute configuration,** of a
molecule. In the case of geometric isomers, the *E,Z* notation was used to describe
unambiguously the configuration of the stereoisomers (Sec. 3.3). With regard to
optical isomers, the notations (+) and (−) tell us only that a compound is either

dextrorotatory or levorotatory. Unfortunately, there is no simple correlation between the sign of optical rotation of a compound and its configuration. Therefore, methods have been devised to specify the absolute configuration of chiral molecules. The two most widely used systems of notation are the **D,L system** and the **R,S system.**

A   The D,L System

In the early 1900s M. A. Rosanoff introduced a method for designating configurations by the use of the letters D and L. In this system, the simplest known sugar, glyceraldehyde, was chosen as a standard to which other configurations were related. Glyceraldehyde, with one chiral center, can exist in two enantiomeric forms. Traditionally, the two enantiomers (I and II) are written so that the CHO is at the top of the Fischer projection and the CH_2OH at the bottom.

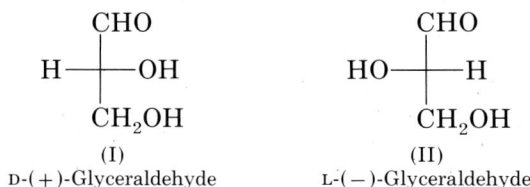

CHO	CHO
H——OH	HO——H
CH_2OH	CH_2OH
(I)	(II)
D-(+)-Glyceraldehyde	L-(−)-Glyceraldehyde

One enantiomer is dextrorotatory (+) and the other levorotatory (−). However, which of the two enantiomers, I or II, is actually dextrorotatory and which is levorotatory? For a long time no one could answer the question. Rosanoff made an arbitrary decision to call the enantiomer with the —OH group to the right of the chiral carbon (structure I) the D form of glyceraldehyde. He then named the enantiomer with the —OH group to the left of the chiral center (structure II) the L configuration of glyceraldehyde. At the time Rosanoff proposed this D and L nomenclature, the odds in favor of (or against) his choice were 50:50. It was not until 1951 that, by means of x-ray techniques, the absolute configuration of D-glyceraldehyde was established; it did, in fact, have the structure chosen by Rosanoff.

The configurations of many other compounds related to D-(+)-glyceraldehyde or L-(−)-glyceraldehyde have been established. Any transformations of D-(+)-glyceraldehyde that do not involve breaking a bond directly attached to the chiral carbon yield products with D configuration. For example,

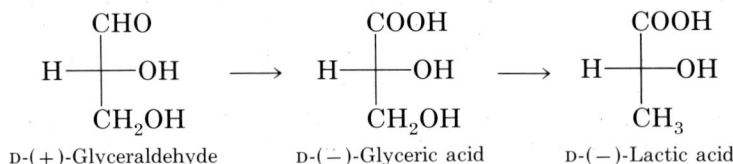

CHO	COOH	COOH
H——OH	H——OH	H——OH
CH_2OH	CH_2OH	CH_3
D-(+)-Glyceraldehyde	D-(−)-Glyceric acid	D-(−)-Lactic acid

Note that D-glyceric acid and D-lactic acid, although they both have the D configuration, do *not* rotate plane-polarized light to the right as does D-(+)-glyceraldehyde. The direction of optical rotation must always be determined experimentally by the use of a polarimeter.

The D,L system works well for compounds that can be related to D- or L-glyceraldehyde, such as carbohydrates having more than one chiral center. In the next structures, for instance, if the —OH attached to the chiral carbon farthest away from the aldehyde group lies to the right, the compound belongs to the D series. If it lies to the left, the compound belongs to the L family (see also Chapter 11).

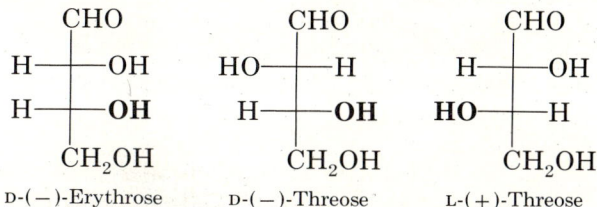

$$
\begin{array}{ccc}
\text{CHO} & \text{CHO} & \text{CHO} \\
\text{H}\!-\!\!\!-\!\text{OH} & \text{HO}\!-\!\!\!-\!\text{H} & \text{H}\!-\!\!\!-\!\text{OH} \\
\text{H}\!-\!\!\!-\!\textbf{OH} & \text{H}\!-\!\!\!-\!\textbf{OH} & \text{HO}\!-\!\!\!-\!\text{H} \\
\text{CH}_2\text{OH} & \text{CH}_2\text{OH} & \text{CH}_2\text{OH} \\
\text{D-}(-)\text{-Erythrose} & \text{D-}(-)\text{-Threose} & \text{L-}(+)\text{-Threose}
\end{array}
$$

The D,L system works well also for amino acids (see also Chapter 15). For example,

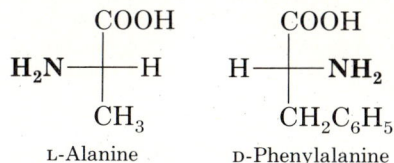

$$
\begin{array}{cc}
\text{COOH} & \text{COOH} \\
\textbf{H}_2\textbf{N}\!-\!\!\!-\!\text{H} & \text{H}\!-\!\!\!-\!\textbf{NH}_2 \\
\text{CH}_3 & \text{CH}_2\text{C}_6\text{H}_5 \\
\text{L-Alanine} & \text{D-Phenylalanine}
\end{array}
$$

Problem 6.10 Which of these compounds belong to the D configuration and which belong to the L configuration?

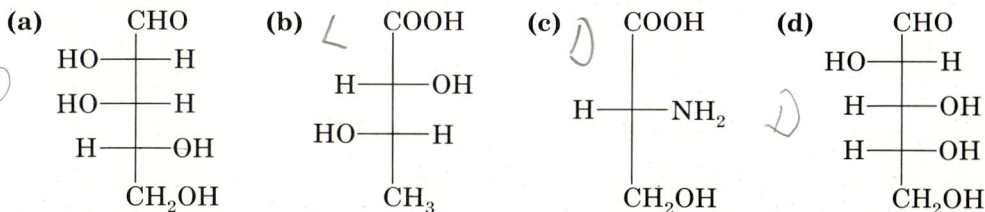

$$
\begin{array}{cccc}
\textbf{(a)}\;\text{CHO} & \textbf{(b)}\;\text{COOH} & \textbf{(c)}\;\text{COOH} & \textbf{(d)}\;\text{CHO} \\
\text{HO}\!-\!\!\!-\!\text{H} & \text{H}\!-\!\!\!-\!\text{OH} & & \text{HO}\!-\!\!\!-\!\text{H} \\
\text{HO}\!-\!\!\!-\!\text{H} & \text{HO}\!-\!\!\!-\!\text{H} & \text{H}\!-\!\!\!-\!\text{NH}_2 & \text{H}\!-\!\!\!-\!\text{OH} \\
\text{H}\!-\!\!\!-\!\text{OH} & \text{CH}_3 & \text{CH}_2\text{OH} & \text{H}\!-\!\!\!-\!\text{OH} \\
\text{CH}_2\text{OH} & & & \text{CH}_2\text{OH}
\end{array}
$$

The D,L terminology is used extensively in carbohydrate and protein chemistry. The system works well for designating compounds that resemble those classes of compounds. Moreover, it has the advantage of being easy to master in such cases. The D,L system is less readily applicable, however, to molecules that bear no resemblance and are totally unrelated to carbohydrates or amino acids. For this reason another system of notation has been devised.

B The R,S System

In the early 1960s, the R,S system was introduced to specify absolute configurations of chiral molecules. The method is applicable to all kinds of compounds and presents no ambiguity once it has been mastered. To specify configuration in the R,S system, two steps need to be taken.

1. Assign priorities to the four substituents bonded to a chiral carbon.
2. Designate the chiral carbon as either R or S depending on the priority sequence established in the first step.

Briefly, the most important rules used to establish the priority sequence are

Rule 1. Priority is based on *atomic numbers*. The higher the atomic number of the atom directly attached to the chiral carbon, the higher the priority. For example, the order of priority of atoms bonded to a chiral carbon is

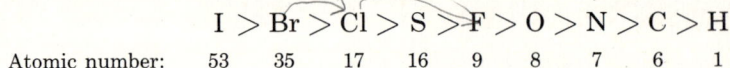

$$I > Br > Cl > S > F > O > N > C > H$$

Atomic number: 53 35 17 16 9 8 7 6 1

Rule 2. If two identical atoms are attached to the chiral carbon, then the priority is determined by comparing the atomic numbers of next set of atoms, two atoms away from the chiral carbon. If this fails, the third set should be used, and so on. For example, according to rule 2, the substituent $-C(CH_3)_3$ has priority over $-CH(CH_3)_2$, which has priority over $-CH_2CH_3$, which in turn has priority over $-CH_3$. Similarly, the order of priority for the substituents $-CH_2NH_2$, $-CH_2OH$, and $-CH_2Cl$ is $-CH_2Cl > -CH_2OH > -CH_2NH_2$.

Rule 3. For double or triple bonds, the atom at the remote end is counted two or three times, respectively, in succession. For example,

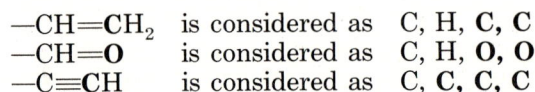

$-CH=CH_2$	is considered as	**C, H, C, C**
$-CH=O$	is considered as	**C, H, O, O**
$-C\equiv CH$	is considered as	**C, C, C, C**

Thus, $-CH=O$ has priority over $-C\equiv CH$, which has priority over $-CH=CH_2$, which in turn has priority over $-CH_2-CH_3$.

Let us apply the R,S system to the 2-butanol molecule.

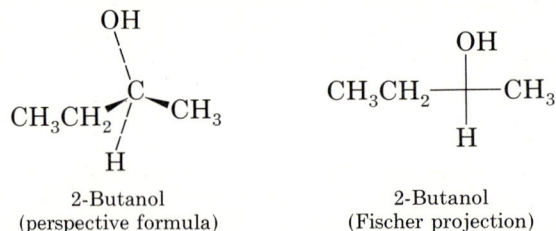

2-Butanol
(perspective formula)

2-Butanol
(Fischer projection)

First, establish the priority sequence of the groups attached to the chiral carbon, which is

$$HO > CH_2CH_3 > CH_3 > H$$

Second, designate the molecule as R-2-butanol or S-2-butanol. To do so, we must orient the molecule so that the group of lowest priority (H) is directed *away* from us. (In the example under consideration, the molecule is already oriented properly.) Now, if your eyes travel in a clockwise direction from the substituent of highest priority to the one of second priority and on to the one of third priority, the configuration assigned to the chiral carbon is R (from the Latin *rectus,* right).

Conversely, if your eyes travel in a counterclockwise direction, the configuration is specified as S (from the Latin *sinister,* left). Thus, the 2-butanol in our example is S-2-butanol.

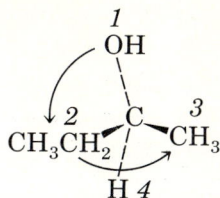
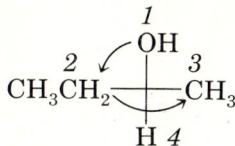

Priority *1 → 2 → 3* counterclockwise = *S*

Note again that the letters *R* and *S* specify configurations only; the direction of optical rotation must be determined experimentally by the use of a polarimeter.

The same general procedure is followed to specify the configuration, *R* or *S*, of *each* chiral carbon in a molecule that contains several chiral centers. For example, the tartaric acid isomers represented here are

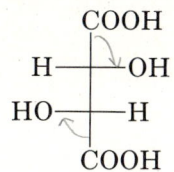   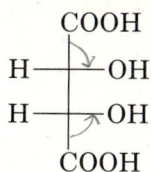

R,R-Tartaric acid   *R,S*-Tartaric acid

Problem 6.11   Designate each compound as *R* or *S*.

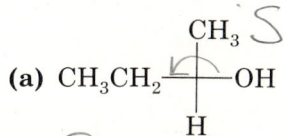   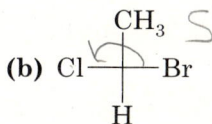

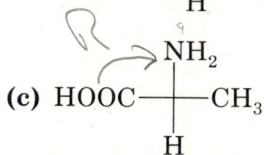   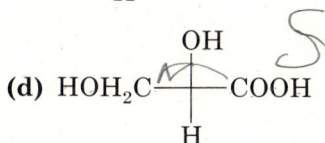

Problem 6.12   Complete the following structures by placing the missing substituents in the correct positions.

(a) *R*-2-Chlorobutane, $CH_3CH_2CHClCH_3$

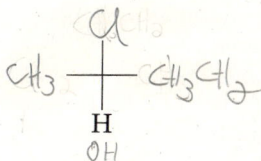

(b) *S*-2-Butanol, $CH_3CH_2CHOHCH_3$

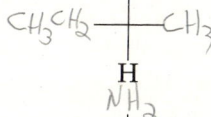

(c) *S*-Alanine, CH_3CHNH_2COOH

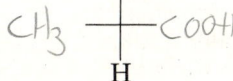

(d) *S*-Glyceric acid, $HOCH_2CHOHCOOH$

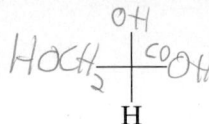

$HOCH_2 \overset{\displaystyle OH}{\underset{\displaystyle H}{\underset{\rule{1.5cm}{0.4pt}}{\overline{}}}} COOH$

(e) *R*-3-Methyl-1-pentyne, $HC \equiv CCH(CH_3)CH_2CH_3$

$CH_3 \underset{\displaystyle H}{\underset{\rule{1.5cm}{0.4pt}}{\overline{}}} =$

(f) *S*-3-Chloro-1-butene, $H_2C = CHCHClCH_3$

$\underset{\displaystyle H}{\underset{\rule{1.5cm}{0.4pt}}{\overline{}}}$

6.13 Significance of Stereoisomerism

We have dealt at some length with the topic of stereoisomerism and have noted the similarities and differences in physical properties that may be expected between one stereoisomer and another. A question that may come to mind is, Why is there so much emphasis, so much detail, on the subject? Is this just a matter of idle curiosity or a means for torturing students? There is ample justification for the emphasis placed on the subject even though some students may have difficulties with it.

A   Biochemical Applications

As a practical consideration, the importance of stereochemistry in biological processes cannot be overemphasized. Many of the compounds occurring in living organisms are chiral, and in every such case the organism makes use of only one of the enantiomers of the compound, just as a right hand fits comfortably only in a right-hand glove. For example, L-lysine, an amino acid, is essential in human nutrition. It is often added as a supplement in wheat-based foods to improve their protein quality. The enantiomer, on the other hand, cannot be utilized by the body until it has been transformed by enzymatic action into the proper stereoisomer.

Time and again we will encounter the dramatic effects produced by stereochemical changes on the properties of certain compounds. To cite a few examples, one form of dopamine, L-dopa, is an effective drug recently introduced in the treatment of Parkinson's disease, whereas the enantiomer D-dopa, is inactive. One form of asparagine is sweet tasting; its enantiomer has a bitter taste. Maltose, a sugar, has a slightly sweet taste, whereas the diastereomer, cellobiose, is neutral in taste. Human beings are incapable of digesting cellulose, but they can easily metabolize its stereoisomer, starch, to provide the necessary energy for proper body functioning. (Termites or cows, unlike humans, possess microorganisms that produce the enzymes needed to break down cellulose into glucose.) One form of androsterone is an active hormone, while a stereoisomer is inactive. Of the various stereoisomeric prostaglandins, some are under active investigation as potential birth-control agents, whereas others are under consideration as birth-fertility pills. Still others are being considered for use in the treatment of such diverse diseases as hypertension, arthritis, and peptic ulcers. L-Thyroxine is a drug used in the treatment of thyroid deficiency, whereas D-thyroxine is used to lower the cholesterol level in the blood.

For a number of years, the control of disease-carrying and crop-ravaging insects was carried out by using DDT or Dieldrin. With our growing concern for protecting the environment from the long-lasting effects of these insecticides, major efforts are being made in the search for alternative methods of pest control. One alternative is the use of compounds that act as specific sex attractants for the male or female of a destructive species of insects (sex pheromones). Recently, it has been found that the European corn borer, which is responsible for a great deal of crop destruction, does not respond to pure *cis*-11-tetradecenyl acetate and is only weakly attracted by pure *trans* isomer. But a mixture of 4% *trans* and 96% *cis* isomer acts as a powerful sex attractant for the males. This mixture may be used to trap the male corn borers and prevent them from mating.

Another discovery of potential economic importance is the recent finding that D-disparlure is a thousand times more effective than the L form in attracting the male gypsy moth, an insect responsible for extensive crop damage.

B Theoretical Organic Applications

Historically, the discovery of optical activity (by Biot, 1815) in certain molecules contributed to the theory of the tetrahedral model for the sp^3-hybridized carbon atoms (Le Bel and van't Hoff). This concept, which is universally accepted today, is fundamental to the understanding of the properties of organic compounds.

Stereochemistry has proved significant in one more area: the elucidation of reaction mechanisms. For example, it is often desirable to compare the configuration of the starting material and that of the product when reactions take place at a chiral carbon. Does the product have the same or the opposite configuration as the starting material, or does it consist of a racemic mixture? The answers to such questions enable us to rationalize how certain reactions take place. The most obvious examples are the nucleophilic substitution reactions of alkyl halides, which are discussed in Chapter 8.

Summary of Concepts and Reactions

Stereochemistry is a branch of chemistry that deals with the description of the spatial arrangements of atoms in molecules and with the consequences of molecular shape on the physical and chemical properties of molecules. It is also the study of the three-dimensional orientations of the atoms in molecules as they undergo a chemical reaction.   [Sec. 6.1]

Optical isomerism is a type of stereoisomerism that manifests itself by its effect on plane-polarized light.   [Sec. 6.1]

Plane-polarized light is light that vibrates in one plane only.   [Sec. 6.2]

Any compound that has the ability to change the direction of plane-polarized light or to rotate it is said to be optically active.   [Sec. 6.3]

Optical activity is measured with an instrument called a polarimeter.   [Sec. 6.3]

An optically active compound is said to be dextrorotatory, (+), if it rotates plane-polarized light to the right and levorotatory, (−), if it rotates plane-polarized light to the left.   [Sec. 6.3]

The value of the optical rotation of a compound under standard conditions is called the specific rotation, [α], and is

$$[\alpha]_\lambda^t = \frac{\alpha}{l \times c}$$ [Sec. 6.4]

To exhibit optical activity a compound must be chiral.   [Sec. 6.5]

A compound is chiral if it cannot be superimposed, atom for atom, on its mirror image.   [Sec. 6.5]

Compounds that are mirror images of each other and are not superimposable are called enantiomers.   [Sec. 6.5]

Most chiral compounds have a chiral center, which is an atom bonded to four different atoms or groups.   [Sec. 6.5]

An equal mixture of enantiomers is called a racemic mixture and is optically inactive.   [Sec. 6.6]

The maximum number of stereoisomers for a structure with n different chiral centers is 2^n. This is also known as van't Hoff's rule.   [Sec. 6.7]

Stereoisomers that are not mirror images of each other are called diastereomers.   [Sec. 6.7]

A meso compound is an optically inactive compound even though it possesses more than one chiral center. A meso compound always has a plane of symmetry.   [Sec. 6.8]

A pair of enantiomers (a racemic mixture) can be separated by a process called resolution.   [Sec. 6.9]

Fischer projections are simple representations of three-dimensional molecules.   [Sec. 6.10]

The absolute configuration of chiral molecules are specified by means of two widely used systems of notation: the D,L and R,S systems.   [Sec. 6.12]

Stereochemistry is important in biological processes and also in the elucidation of reaction mechanisms.   [Sec. 6.13]

Key Terms

stereoisomerism
structural isomers
geometric isomers
optical isomerism
plane-polarized light
optical activity
dextrorotatory

levorotatory
specific rotation
chiral
enantiomers
chiral center
racemic mixture
van't Hoff's rule

diastereomers
meso compound
resolution
Fischer projections
absolute configuration
D,L system
R,S system

Exercises

Optical Activity, Specific Rotation, Chiral Compounds (Secs. 6.3–6.5)

6.1 The compounds (+)- and (−)-alanine will show differences in which of the following properties?

(a) Melting point   **(b)** Density   **(c)** Solubility in water

(d) Optical rotation   **(e)** Refractive index

6.2 The specific rotation of (+)-menthol in ethanol is $+48.3°$. What would be the observed rotation of a 10% (wt/vol) menthol solution when measured in a 2 dm polarimeter tube?

6.3 Calculate the specific rotation of cortisone from these data. When 500 mg of cortisone was dissolved in 100 ml of ethanol and the solution was placed in a 25 cm polarimeter tube, the observed rotation of the solution was $+2.61°$.

6.4 Which of these substances are capable of existing in an optically active form?
(a) 2-Bromobutane
(b) 1-Chloro-1-phenylethane
(c) 2-Chloro-1-phenylethane
(d) Methylcyclohexane
(e) 1-Bromo-2-methylcyclohexane
(f) 2-Amino-3-methylbutane
(g) 2-Chloro-2-methylpropane
(h) 2-Fluoro-3-methylbutane

6.5 Identify any chiral carbon atoms in these structures:
(a) $CH_3CHBrCH_2CH_3$
(b) $CH_3CH(NH_2)COOH$
(c) $C_6H_5CH(OH)COOH$
(d) CH_3CHOH with CH_3
(e) $CH_3CH(OH)CH(OH)CH_3$
(f) cyclopentane with $-CH_3$ and OH
(g) $CH_3CHClCH(OH)CH_3$
(h) $CH_3CHCH_2CH_2CHCH_3$ with Br, Br

6.6 Draw the structural formula for an optically active compound whose molecular formula is **(a)** $C_4H_{10}O$; **(b)** C_8H_9Br; **(c)** C_5H_9N; **(d)** $C_4H_{10}S$.

Enantiomers, Diastereomers, Meso Compounds, van't Hoff's Rule [Secs. 6.5–6.9]

6.7 How many stereoisomers does each structure in Exercise 6.5 have? Draw the structure of any meso compound.

6.8 Which of the following compounds has a plane of symmetry?
(a) cis-1,2-Dibromocyclobutane
(b) trans-1,2-Dibromocyclobutane
(c) CH_2OH / $H-C-OH$ / $H-C-OH$ / CH_2OH
(d) $COOH$ / $H-C-Br$ / $Br-C-H$ / CH_3

6.9 Draw the Fischer projection formulas for all the stereoisomers of 2,3-dichloropentane. Indicate which isomers are enantiomers and which are diastereomers. Predict the maximum number of stereoisomers for this compound. What is the actual number of stereoisomers?

6.10 Repeat Exercise 6.9 for 2,3-dichlorobutane.

Manipulating Fischer Projections [Sec. 6.11]

6.11 A Fischer projection representation of lactic acid is

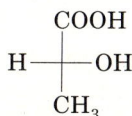

$COOH$ / $H-OH$ / CH_3

Indicate what relationship (identical or enantiomer) there is between the above structure and each of these Fischer projections.

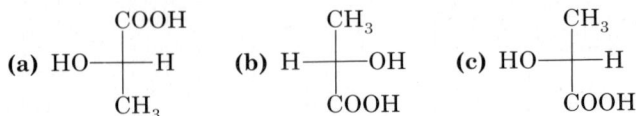

(a) HO—|—H
  COOH (top), CH₃ (bottom)

(b) H—|—OH
  CH₃ (top), COOH (bottom)

(c) HO—|—H
  CH₃ (top), COOH (bottom)

6.12 What is the relationship (identical, enantiomers, or diastereomers) between each of the following pairs of structures?

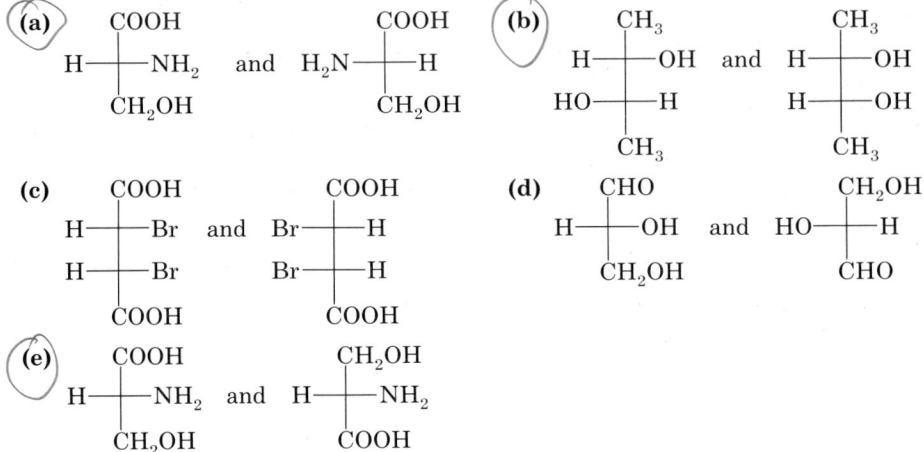

(a)

COOH
H—|—NH₂   and   H₂N—|—H
CH₂OH   CH₂OH

(b)

CH₃
H—|—OH   and   H—|—OH
HO—|—H   H—|—OH
CH₃   CH₃

(c)

COOH
H—|—Br   and   Br—|—H
H—|—Br   Br—|—H
COOH   COOH

(d)

CHO   CH₂OH
H—|—OH   and   HO—|—H
CH₂OH   CHO

(e)

COOH   CH₂OH
H—|—NH₂   and   H—|—NH₂
CH₂OH   COOH

D,L and R,S Methods of Specifying Configurations [Sec. 6.12]
6.13 Use D or L to specify the configuration of each structure.

(a)

CHO
H—|—OH
CH₃

(b)

CHO
HO—|—H
H—|—OH
CH₂OH

(c)

COOH
HO—|—H
CH₃

(d)

COOH
H—|—OH
HO—|—H
CH₂OH

(e)

COOH
H₂N—|—H
CH₂SH

(f)

COOH
H—|—NH₂
CH₂OH

6.14 Assign priority sequence to the following substituents according to the R,S system of notation: H, —CH₃, Cl, Br, F, CHO, COOH, —NH₂, —CH=CH₂, —C≡CH.

Highest priority   Lowest priority

⟶

6.15 Use R or S to specify the configuration of each structure.

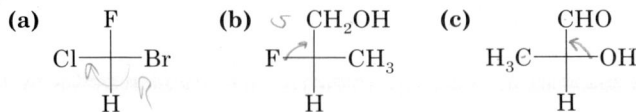

(a)

F
Cl—|—Br
H

(b)

CH₂OH
F—|—CH₃
H

(c)

CHO
H₃C—|—OH
H

6.16 Complete the following structures by placing the missing substituents in the correct positions.

(a) R-2-Aminobutane, CH₃CH₂CHNH₂CH₃

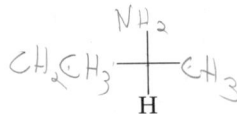

(b) *S*-Serine, HOCH₂CHNH₂COOH

$$HOCH_2 \underset{H}{\overset{NH_2}{\mid}} COOH$$

(c) *S*-1-Bromo-1-chloro-1-fluoroethane, CH₃CBrClF

$$F \underset{CH_3}{\overset{Cl}{\mid}} Br$$

(d) *R*-3-Chloro-3-methyl-1-pentene, CH₂=CH—CClCH₂CH₃
  |
  CH₃

$$\underset{CH_3}{\overset{}{\mid}}$$

7

Alcohols, Phenols, and Thiols

Structurally, alcohols and phenols may be viewed as organic derivatives of water.

$$H{-}O{-}H \qquad R{-}O{-}H \qquad Ar{-}O{-}H$$

Water   Alcohol   Phenol

Alcohols and phenols have a common functional group, the **hydroxyl group, —OH.** In alcohols the hydroxyl group is attached to an alkyl group, —R. In phenols the hydroxyl function is bonded directly to an aromatic ring, Ar.

7.1 Classification and Nomenclature of Alcohols

Alcohols are subdivided into three classes: primary (1°), secondary (2°), and tertiary (3°). The classification depends on the number of alkyl groups bonded to the carbon bearing the —OH, the so-called **carbinol carbon.** Primary alcohols have *one* alkyl group attached to the carbinol carbon, secondary alcohols have *two,* and tertiary alcohols have *three.*

$$-\overset{|}{\underset{|}{C}}{-}OH \qquad R{-}\overset{H}{\underset{H}{C}}{-}OH \qquad R{-}\overset{H}{\underset{R}{C}}{-}OH \qquad R{-}\overset{R}{\underset{R}{C}}{-}OH$$

Carbinol group   Primary alcohol   Secondary alcohol   Tertiary alcohol

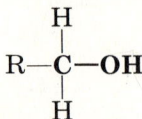

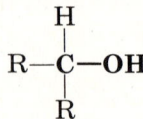

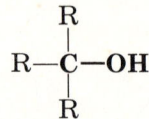

This classification is used because it is helpful in dealing with the many reactions of alcohols.

Alcohols can be designated by common names or by the IUPAC system. Common names are often used for the simplest alcohols or for very complicated ones, such as cholesterol.

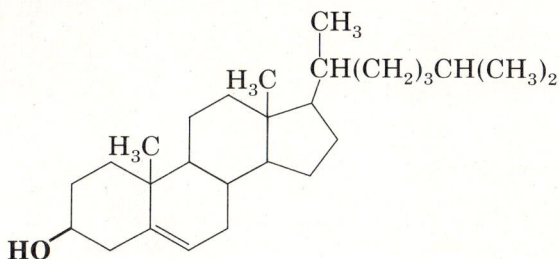

Cholesterol—a complex alcohol with a common name

The common names for the simplest alcohols consist of the name of the alkyl group attached to the hydroxyl function followed by the word *alcohol*. Examples of the three classes of alcohols and their common names are

Primary alcohols

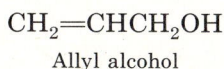

CH_3OH CH_3CH_2OH $CH_2\!\!=\!\!CHCH_2OH$

Methyl alcohol   Ethyl alcohol   Allyl alcohol

Secondary alcohols

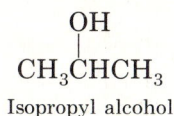

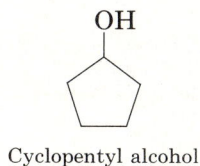

Isopropyl alcohol   Cyclopentyl alcohol

Tertiary alcohol

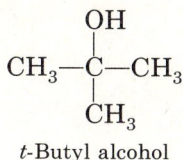

t-Butyl alcohol

Problem 7.1 Give a common name to each of the following alcohols. Classify each structure as a 1°, 2°, or 3° alcohol.

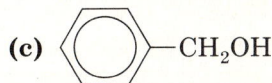

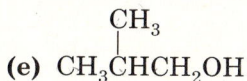

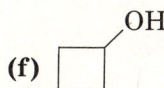

(a) $CH_3CH_2CH_2OH$ **(b)** $CH_3CH_2CH_2CH_2OH$ **(c)**

(d)   **(e)** $CH_3\overset{\displaystyle CH_3}{\underset{\displaystyle |}{C}}HCH_2OH$ **(f)**

Another common method of nomenclature is based on naming alcohols as derivatives of the carbinol group, $-\overset{|}{\underset{|}{C}}-OH$. The groups attached to the carbinol carbon are named. For example, ethyl alcohol, CH_3-CH_2OH, is named methyl-carbinol in the **carbinol system** of nomenclature. This system is seldom used except for phenyl-substituted methyl alcohols. For example,

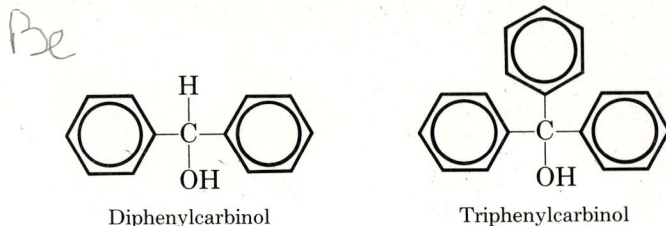

Diphenylcarbinol

Triphenylcarbinol

Problem 7.2 Name the following alcohols using the carbinol system.

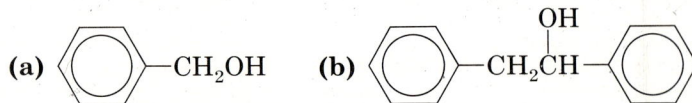

(a) ⟨◯⟩—CH_2OH (b) ⟨◯⟩—$CH_2\overset{OH}{\overset{|}{C}}H$—⟨◯⟩

In the IUPAC system, alcohols are named according to the following rules.

1. Select the longest continuous carbon chain that *contains the* —OH *group.* Drop the *-e* ending of the parent alkane and replace it by the suffix *-ol*. For example,

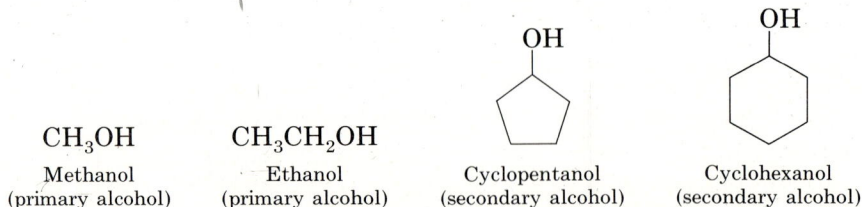

CH_3OH

Methanol
(primary alcohol)

CH_3CH_2OH

Ethanol
(primary alcohol)

Cyclopentanol
(secondary alcohol)

Cyclohexanol
(secondary alcohol)

2. When isomers are possible, the chain is numbered so as to give the functional group (—OH) the *lowest possible number*. For example,

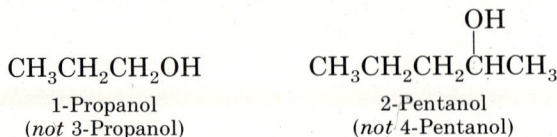

$CH_3CH_2CH_2OH$

1-Propanol
(*not* 3-Propanol)

$CH_3CH_2CH_2\overset{OH}{\overset{|}{C}}HCH_3$

2-Pentanol
(*not* 4-Pentanol)

3. When alkyl side chains or other groups are present, they are named alphabetically and their positions are indicated by a number. The position of the functional group (—OH) is always given the lowest possible number at the end of the name. For cyclic alcohols, numbering always starts from the carbon bearing the —OH group. The following are examples.

$$CH_3CH_2CHCH_2CHCH_3$$

with ethyl group CH_2CH_3 and OH

4-Ethyl-2-hexanol
(*not* 3-Ethyl-5-hexanol)

$$CH_3CH_2CHCH_2CH_2CHCH_2OH$$

with Cl and CH_3 substituents

5-Chloro-2-methyl-1-heptanol
(*not* 3-Chloro-6-methyl-7-heptanol)

3-Methylcyclohexanol
(*not* 1-Methyl-3-cyclohexanol)

3-Bromo-2-phenylcyclopentanol
(*not* 1-Bromo-2-phenyl-3-cyclopentanol)

4. If a molecule contains both an —OH group and a C=C or C≡C bond, the name should include (if possible) both the hydroxyl and the unsaturated groups, even if *this does not make the longest chain the parent hydrocarbon.* As a main functional group of alcohols, the —OH group takes preference before the double or triple bonds in getting the lower number. For example,

$$CH_2=CHCHCH_3$$ with OH

3-Buten-2-ol
(*not* 1-Buten-3-ol)

$$HC\equiv CCH_2CH_2OH$$

3-Butyn-1-ol
(*not* 1-Butyn-4-ol)

$$CH_3CHC=CH_2$$ with CH_2CH_3 and OH

3-Ethyl-3-buten-2-ol
(longest chain including C=C)

Problem 7.3 Give an IUPAC name for each of the commonly named alcohols.
(a) Ethyl alcohol
(b) *n*-Propyl alcohol
(c) Isopropyl alcohol
(d) *t*-Butyl alcohol
(e) Cyclopropyl alcohol
(f) Cyclohexyl alcohol
(g) Benzyl alcohol
(h) Allyl alcohol
(i) Diphenylcarbinol
(j) Benzylcarbinol

Problem 7.4 Name the following alcohols by the IUPAC system.

(a) $CH_3CH_2CH_2CHCH_2CH_3$ with OH

(b)

(c) CH_3CHCH_2OH with phenyl ring

(d) $(CH_3)_2CHCH_2CHCH_3$ with OH

(e)
$$\underset{\underset{\text{Br}\ \ \ \text{OH}}{|\qquad|}}{\overset{\overset{\text{CH}_2\text{CH}_3}{|}}{\text{CH}_3\text{CHCHCHCH}_3}}$$

(f) [structure: cyclopentane ring with OH and CH$_3$]

(g) [structure: cyclohexane ring with H$_3$C, —OH, and Cl substituents]

(h) $CH_3CH{=}CHCH_2OH$

(i)
$$\underset{\underset{\text{CH}_2\text{CH}_3}{|}}{\text{CH}_3\text{CHCH}{=}\text{CHCH}_2\text{OH}}$$

(j) $CH_3C{\equiv}CCH_2CH_2OH$

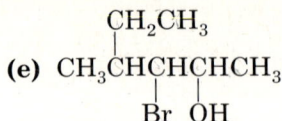

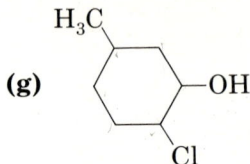

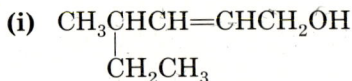

Problem 7.5 Classify each structure in Problem 7.4 as a 1°, 2°, or 3° alcohol.

Problem 7.6 What is the correct IUPAC name for each of the following incorrectly named alcohols?

(a) 4-Hexanol
(c) 2-Ethyl-2-hexanol
(e) 1-Methyl-2-cyclobutanol
(g) 1-Chloro-2-methyl-3-cyclohexanol
(i) 2-Methyl-2-penten-4-ol

(b) 3-Methyl-5-pentanol
(d) 4-Chloro-5-phenyl-5-heptanol
(f) 4-Bromocyclopentanol
(h) 2-Hexen-4-ol
(j) 2-Pentyn-4-ol

Some alcohols contain more than one hydroxyl group. In general, to be stable such alcohols must have the —OH groups on different carbons. In the IUPAC system, the suffix **-diol** is added to the name of the parent hydrocarbon when two hydroxyl groups are present, and the suffix **-triol** is added when there are three —OH groups. Common names are often used for the simplest polyhydroxy alcohols. For example, those containing two —OH groups on adjacent carbons are known as 1,2-**glycols.**

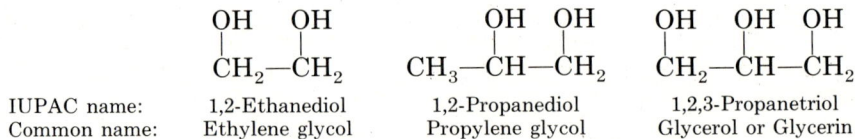

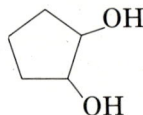

$$\underset{}{\overset{\overset{\text{OH}\quad\text{OH}}{|\qquad|}}{\text{CH}_2{-}\text{CH}_2}}\qquad\underset{}{\overset{\overset{\text{OH}\quad\text{OH}}{|\qquad|}}{\text{CH}_3{-}\text{CH}{-}\text{CH}_2}}\qquad\underset{}{\overset{\overset{\text{OH}\quad\text{OH}\quad\text{OH}}{|\qquad|\qquad|}}{\text{CH}_2{-}\text{CH}{-}\text{CH}_2}}$$

IUPAC name:	1,2-Ethanediol	1,2-Propanediol	1,2,3-Propanetriol
Common name:	Ethylene glycol	Propylene glycol	Glycerol or Glycerin

[structure: cyclopentane ring with two OH groups]

IUPAC name: 1,2-Cyclopentanediol

Alcohols that contain more than three hydroxyl groups per molecule are known as **polyols.** Polyols are constituents of, or are derived from, sugars (Chapter 11).

Problem 7.7 Name the following compounds by the IUPAC system.

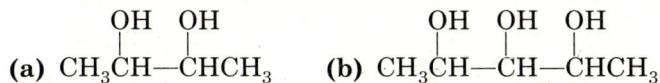

(a) $CH_3\overset{\underset{|}{OH}}{CH}-\overset{\underset{|}{OH}}{CH}CH_3$ (b) $CH_3\overset{\underset{|}{OH}}{CH}-\overset{\underset{|}{OH}}{CH}-\overset{\underset{|}{OH}}{CH}CH_3$

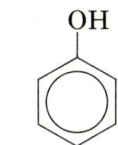

(c)   (d) HO—⟨ ⟩—OH

Nomenclature of Phenols **7.2**

Phenols are generally named as derivatives of the simplest member of the family, phenol. For example,

 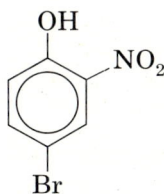 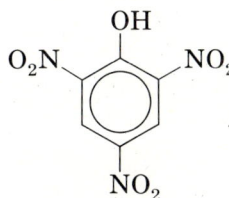 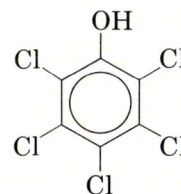

Phenol   *p*-Aminophenol   4-Bromo-2-nitrophenol   2,4,6-Trinitrophenol   Pentachlorophenol
(Carbolic acid)   (Picric acid)

However, because a great amount of work on phenols was done a long time ago, many common names are still prevalent. *Carbolic acid,* for example, is the name given by pharmacists to phenol itself. *Picric acid* is the common name for 2,4,6-trinitrophenol. These names are indicative of the acidic properties of phenols, a subject we consider in Section 7.4. Examples of other phenols with common names are

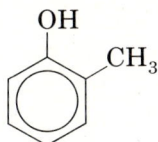 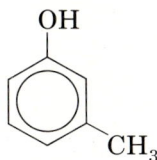 

o-Cresol   *m*-Cresol   *p*-Cresol

The mixture of cresols is used as a wood preservative, known as creosote.

Several important hydroxysubstituted phenols also have common names.

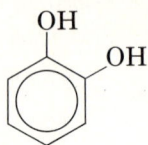 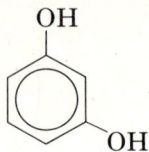 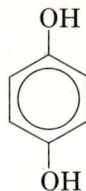 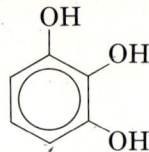

Catechol   Resorcinol   Hydroquinone   Pyrogallol
(germicide)   (antiseptic)   (photographic film developers)

Compounds in which the —OH group is bonded to the naphthalene ring are called *naphthols*.

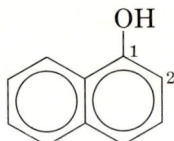   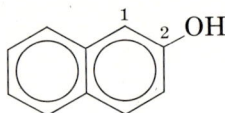

1-Naphthol   2-Naphthol
(α-Naphthol)   (β-Naphthol)
(insecticide intermediate)   (dye intermediate)

Problem 7.8 Name each compound as a derivative of phenol, 1-naphthol, or 2-naphthol.

(a) *o*-Cresol   **(b)** *m*-Cresol   **(c)** *p*-Cresol
(d) Catechol   **(e)** Resorcinol   **(f)** Hydroquinone

(g) Pyrogallol   **(h)** 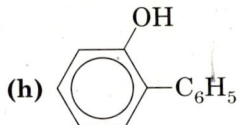   **(i)**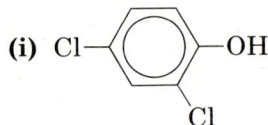

(j) 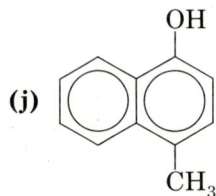   **(k)**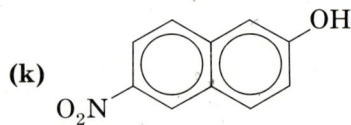

7.3 Physical Properties of Alcohols and Phenols

A Monohydroxy Alcohols

Table 7.1 gives the physical properties of some of the lower members of the **monohydroxy alcohol** series. The data in Table 7.1 show that the introduction of a single hydroxyl group into a hydrocarbon has a marked effect on the physi-

Table 7.1 Physical Properties of Straight-Chain Primary Alcohols

Name	Structure	Mp (°C)	Bp (°C)	Solubility in H_2O at 20°C
Methanol	CH_3OH	−97	64.5	completely soluble
Ethanol	CH_3CH_2OH	−115	78	completely soluble
1-Propanol	$CH_3CH_2CH_2OH$	−126	97	completely soluble
1-Butanol	$CH_3(CH_2)_2CH_2OH$	−90	118	8.0 g/100 g H_2O
1-Pentanol	$CH_3(CH_2)_3CH_2OH$	−78.5	138	2.3 g/100 g H_2O
1-Hexanol	$CH_3(CH_2)_4CH_2OH$	−52	157	0.6 g/100 g H_2O

cal properties of the compound. For example, even the simplest alcohol, methanol, is a liquid at room temperature. In contrast, alkanes from methane to butane are gases.

The influence of the —OH group is shown also in the solubility properties of alcohols. In contrast to alkanes, which are insoluble in water, the lower alcohols are completely miscible with water. As the number of carbons in the alcohol increases, the solubility in water decreases.

Table 7.1 reveals also that within a homologous series of normal alcohols, the boiling points increase with increase in molecular weights.

A comparison of boiling points among isomeric alcohols shows that the boiling points decrease as the number of alkyl branches from the carbinol group increases. For example,

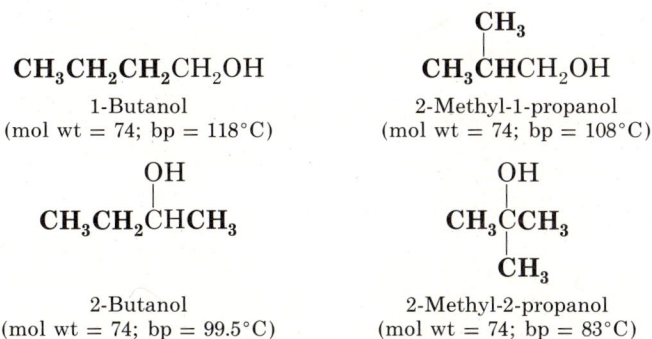

$$CH_3CH_2CH_2CH_2OH$$

1-Butanol
(mol wt = 74; bp = 118°C)

$$\overset{\displaystyle CH_3}{\overset{|}{CH_3CHCH_2OH}}$$

2-Methyl-1-propanol
(mol wt = 74; bp = 108°C)

$$\overset{\displaystyle OH}{\overset{|}{CH_3CH_2CHCH_3}}$$

2-Butanol
(mol wt = 74; bp = 99.5°C)

$$\overset{\displaystyle OH}{\overset{|}{\underset{\underset{\displaystyle CH_3}{|}}{CH_3CCH_3}}}$$

2-Methyl-2-propanol
(mol wt = 74; bp = 83°C)

Such orders of boiling points are not unexpected: we encountered similar trends among alkanes (Sec. 2.8B). The straight-chain isomers have larger molecular surfaces than do their branched-chain isomers. Therefore, they have higher intermolecular interactions and, as a result, higher boiling points.

What may be surprising at first are the high values of the boiling points of alcohols compared with alkanes of similar molecular weights. For example, the boiling point of ethanol is more than 100°C higher than the boiling point of propane.

$$CH_3CH_2CH_3$$

Propane
(mol wt = 44; bp = −42°C)

$$CH_3CH_2OH$$

Ethanol
(mol wt = 46; bp = 78°C)

That this large increase in boiling point is due to the O—H bond and not to the presence of oxygen can be proved by comparing the boiling point of ethanol with that of its structural isomer, methyl ether:

$$CH_3CH_2OH \qquad\qquad CH_3OCH_3$$

Ethanol	Methyl ether
(mol wt = 46; bp = 78°C)	(mol wt = 46; bp = −24°C)

The boiling point of methyl ether, −24°C, is much closer to the boiling point of propane than it is to the boiling point of ethanol. Therefore, the large increase in boiling points of alcohols compared with alkanes (or ethers) also must be due to the O—H group.

B Hydrogen Bonding

The effect of the OH group on the physical properties of alcohols can be explained as follows. The O—H bond is *highly polar* (Sec. 1.4B). The oxygen, a highly electronegative atom carries a partial negative charge ($\delta-$), and the hydrogen a partial positive charge ($\delta+$). The polarity of the O—H bond gives rise to forces of attraction between a partially positive H in one molecule and a partially negative O in a neighboring molecule. These forces of attraction, called **hydrogen bonding** or **H bonding,** are sizable and unique whenever *hydrogen is bonded to oxygen, nitrogen, or fluorine,* the three most electronegative atoms.

The reason, then, that alcohols have high boiling points is that a great deal of energy (in the form of heat) is required to overcome the intermolecular attractive forces due to hydrogen bonding between neighboring alcohol molecules. The absence of OH groups in alkanes precludes H bonding; as a result alkanes have lower boiling points than do alcohols of comparable molecular weights. Figure 7.1 illustrates H bonding between alcohol molecules.

Hydrogen bonding accounts also for the differences in physical states between lower alcohols and lower alkanes. Methanol, even though it has a lower molecular weight (32) than, say, butane (58), is a liquid at room temperature because the attractive forces due to hydrogen bonding bring methanol molecules close together. The lower alkanes are gases because there are no such strong forces of attraction between the molecules.

Association through hydrogen bonding also explains the solubility of the lower alcohols in water (Fig. 7.2). The short-chain alcohols are miscible with water because they resemble water more than they do hydrocarbons. In other words, the effect of the polar OH group predominates over that of the relatively small nonpolar hydrocarbon portion. When those alcohols are mixed with water, they dissolve because of hydrogen bonding between the OH bond of the alcohol and the hydroxyl group of the water.

As the number of carbons in an alcohol increases, the nonpolar alkyl group becomes more and more important and the polar OH group becomes less important. In other words, higher alcohols resemble hydrocarbons more than they do

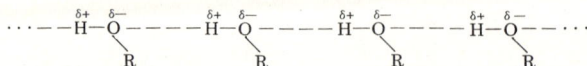

Figure 7.1 Hydrogen bonding, denoted by dashed lines, between alcohol molecules. Energy is needed to overcome these intermolecular attractive forces, hence the high boiling points of alcohols.

$$\overset{\delta+}{H}-\overset{\delta-}{\underset{\underset{R}{|}}{O}}----\overset{\delta+}{H}-\overset{\delta-}{\underset{\underset{H}{|}}{O}}----\overset{\delta+}{H}-\overset{\delta-}{\underset{\underset{R}{|}}{O}}----\overset{\delta+}{H}-\overset{\delta-}{\underset{\underset{H}{|}}{O}}----\overset{\delta+}{H}-\overset{\delta-}{\underset{\underset{R}{|}}{O}}--\cdots$$

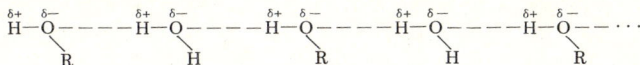

Figure 7.2 Hydrogen bonding (dashed lines) between alcohol and water molecules explains the solubility of lower alcohols in water.

water. As a result, the solubility of higher alcohols in water diminishes until 1-nonanol is reached, at which point the solubility in water is decreased to zero. Conversely, the solubility of higher alcohols increases in nonpolar solvents. The solubility characteristics of alcohols is another application of the "like dissolves like" rule discussed previously (Sec. 2.8A).

C Diols and Triols

The physical properties of diols and triols reflect even more strikingly the influence of H bonding. Diols with two OH groups available for H bonding, and triols with three, have even higher boiling points than do monohydroxy alcohols of comparable molecular weights. Diols and triols are also miscible with water, but are only slightly soluble in alkanes. Table 7.2 illustrates how the number of OH bonds available for hydrogen bonding more than molecular weight affects the physical properties of alcohols.

Problem 7.9 Arrange each group of compounds in order of increasing boiling points. Explain your reasoning.
(a) Butane; 1-propanol; 2-propanol; 2-methylpropane
(b) Ethanol; 1-propanol; 2-butanol; 1-butanol
(c) Glycerol; 1-pentanol; 2-pentanol; 2-methyl-2-butanol

Problem 7.10 Arrange each group of compounds in order of increasing solubility in water. Explain your reasoning.
(a) 1-Butanol; pentane; 1-hexanol; 1-pentanol
(b) 1-Hexanol; glycerol; 1-heptanol; 1,2-cyclohexanediol
(c) 1-Hexanol; ethylene glycol; 1,2-hexanediol; hexane

Table 7.2 Physical Properties of Some Alcohols and Number of OH Groups

Alcohol	Mol wt	Bp (°C)	Solubility in H_2O at 20°C	OH groups per molecule
1-Propanol	60	97	Soluble	1
1-Butanol	74	118.5	Soluble	1
Ethylene glycol (1,2-ethanediol)	62	197	Soluble	2
1-Pentanol	88	138	2.3 g/100 g H_2O	1
1-Hexanol	102	157	0.6 g/100 g H_2O	1
Glycerol (1,2,3-propanetriol)	92	290	Soluble	3

D Phenols

Phenol is a colorless, crystalline, low-melting solid, with a high boiling point, that is moderately soluble in water. Most other phenols also are solids, with slight solubility in water and high boiling points.

Problem 7.11 The solubility of phenol (mol wt = 94) is 9.0 g/100 g H_2O at 20°C, and its boiling point is 182°C, whereas toluene (mol wt = 92) is insoluble in water and boils at 111°C. Give a reasonable explanation for the differences in physical properties of the two compounds.

Perhaps the most significant physical property that distinguishes alcohols from phenols is the *acidity* of phenols.

7.4 Acidities of Phenols and Alcohols Compared

The acidity of a compound may be expressed quantitatively in terms of its pK_a value. We shall examine the exact significance of pK_a in Chapter 12. For the purpose of this discussion, all you need to know is that (1) the lower the pK_a value for a compound, the greater its acidity, and (2) if, for example, compound A has a pK_a of 11 and compound B has a pK_a of 12, A is 10 times more acidic than B. In other words, *a change in one pK_a unit represents a tenfold change in acidity.* Correspondingly, a two-unit change in pK_a represents a 10^2, or hundred-fold, change in acidity.

The pK_a of phenol is approximately 10. Alcohols have pK_as of about 16, and water has a pK_a of 14. This means that phenol, whose pK_a is 6 units lower than alcohols, is 10^6, or 1 million times more acidic than alcohols. The pK_as of alcohols (16) and water (14) indicate that alcohols are weaker acids than water.

One way to explain the great difference in acidities between phenol and alcohols is illustrated in Figure 7.3, using cyclohexanol (whose structure is somewhat similar to phenol) as an example of an alcohol. E_1 represents the internal energy of cyclohexanol and phenol in their acid forms. We place the two compounds on almost the same energy level because their stabilities are nearly identical. The free energies of the conjugate bases of each of the two compounds are symbolized by E_2. From the figure, you can see that ΔE, the difference between

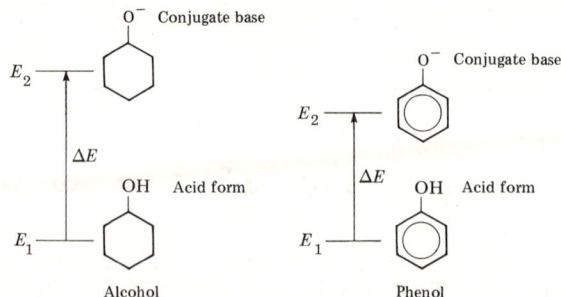

Figure 7.3 Phenol is a stronger acid than cyclohexanol because ΔE ($= E_2 - E_1$) is smaller for phenol than for cyclohexanol.

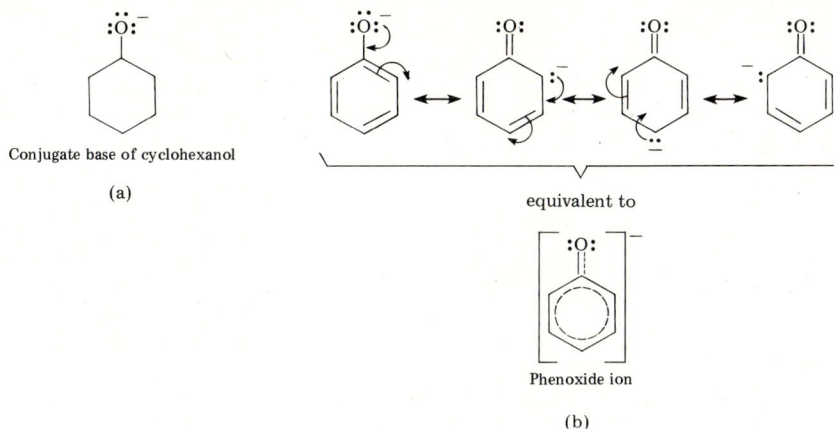

Conjugate base of cyclohexanol

(a)

equivalent to

Phenoxide ion

(b)

Figure 7.4 **(a)** The negative charge on oxygen is localized; the ion is not stabilized (high E_2 value). **(b)** The negative charge is delocalized via resonance; the ion is stabilized (low E_2 value).

E_2 and E_1, is smaller for phenol and its conjugate base, the **phenoxide ion,** than for cyclohexanol and its conjugate base. *The smaller value of ΔE for the phenol-phenoxide pair makes it a stronger acid.* The lower position of E_2 for the phenoxide ion as compared with the conjugate base of cyclohexanol indicates that the phenoxide ion is more stable (it contains less internal energy).

Why is the phenoxide ion more stable? Because the negative charge on the oxygen is *dispersed* by *resonance* through the benzene ring. No such delocalization of charge can take place for the conjugate base of cyclohexanol (see Fig. 7.4).

Most phenols have a pK_a in the vicinity of 10. However, introduction of electron-withdrawing groups, such as NO_2 or CN, on the ring increases the acidity of phenols dramatically. For example, *p*-nitrophenol (pK_a = 7.6) is almost 1000 times more acidic than phenol itself, and picric acid (2,4,6-trinitrophenol) is as strong an acid as HCl.

Problem 7.12 Predict the order of acidity of these compounds.

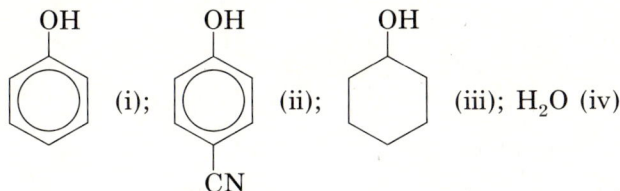

(i); (ii); (iii); H_2O (iv)

Reactions of Alcohols and Phenols 7.5

Alcohols undergo two kinds of reactions: those that involve the breaking of the oxygen–hydrogen bond (CO⟍H) and those that involve the rupture of the carbon–oxygen bond (C⟍OH).

Phenols do not participate in reactions where the C⟍OH bond is broken.

7.6 Alcohols and Phenols as Acids: Salt Formation

Alcohols and phenols can act as acids whenever they donate a proton to a base. The base must be stronger than the conjugate base of the alcohol (the alkoxide ion) or the phenol (the phenoxide ion), respectively.

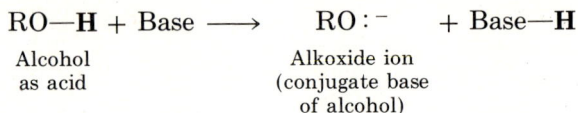

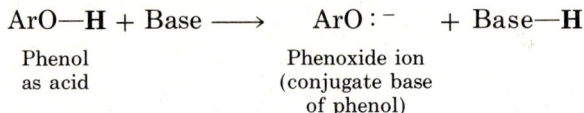

$$RO—H + Base \longrightarrow \quad RO:^- \quad + Base—H$$

Alcohol   Alkoxide ion
as acid   (conjugate base
            of alcohol)

$$ArO—H + Base \longrightarrow \quad ArO:^- \quad + Base—H$$

Phenol — Phenoxide ion (conjugate base of phenol)
as acid

Since alcohols are weaker acids than water (Sec. 7.4), it is not possible to form the salt of an alcohol in aqueous alkaline solutions. Thus,

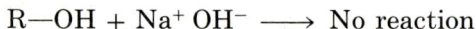

$$R—OH + Na^+ OH^- \longrightarrow \text{No reaction}$$

Although the salt of an alcohol cannot be obtained in aqueous solutions, it is possible to form the salt by the action of active metals such as Na or K.

General equation

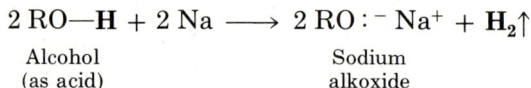

$$2\,RO—H + 2\,Na \longrightarrow 2\,RO:^- Na^+ + H_2\uparrow$$

Alcohol   Sodium
(as acid)   alkoxide

Specific example

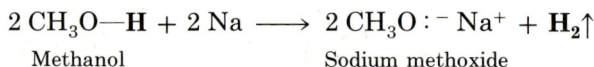

$$2\,CH_3O—H + 2\,Na \longrightarrow 2\,CH_3O:^- Na^+ + H_2\uparrow$$

Methanol   Sodium methoxide

The action of metallic sodium on an alcohol is similar to the reaction of metallic sodium with water.

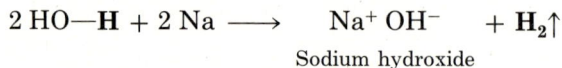

$$2\,HO—H + 2\,Na \longrightarrow \quad Na^+ OH^- \quad + H_2\uparrow$$

Sodium hydroxide

Alkoxides are strong bases, even stronger than hydroxides. Alkoxides are used extensively in the synthesis of ethers (Sec. 9.6) and wherever strong basic solutions, free of water, are needed.

The liberation of hydrogen gas, seen as bubbles, from an unknown liquid is used sometimes as a test for alcohols. Of course, any compound containing an acidic hydrogen will also give a positive test.

Phenols are far more acidic than either water or alcohols (Sec. 7.4), and it is possible to obtain **phenoxide** salts from aqueous solutions of alkali. For example,

$$\underset{\substack{\text{Phenol} \\ \text{(as an acid)}}}{\underset{\text{O—H}}{\bigcirc}} + \underset{\substack{\text{Sodium hydroxide} \\ \text{(a base)}}}{Na^+\,OH^-} \longrightarrow \underset{\substack{\text{Sodium phenoxide} \\ \text{(a salt)}}}{\underset{O:^-\,Na^+}{\bigcirc}} + H—OH$$

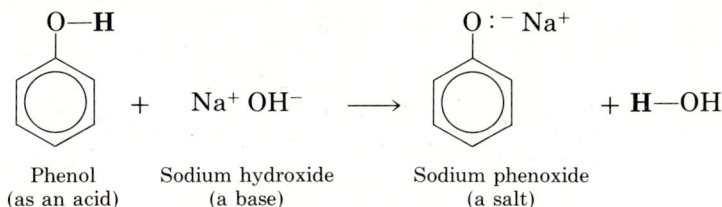

   Phenoxide salts are water soluble, in contrast to phenols, which are only slightly soluble. It is therefore possible to extract phenols from neutral organic mixtures by treatment with dilute aqueous sodium hydroxide. The aqueous extract will contain the phenoxide salt. The free phenol can be regenerated from the aqueous extract by acidifying it with mineral acid (HCl).

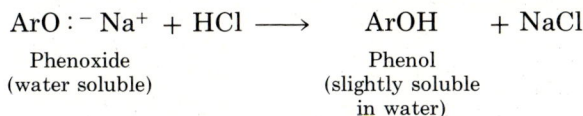

$$\underset{\substack{\text{Phenoxide} \\ \text{(water soluble)}}}{ArO:^-\,Na^+} + HCl \longrightarrow \underset{\substack{\text{Phenol} \\ \text{(slightly soluble} \\ \text{in water)}}}{ArOH} + NaCl$$

Problem 7.13 Indicate how best to separate a mixture of cyclohexanol and phenol by means other than fractional distillation.

_____

_____

Problem 7.14 Write the equation for each reaction described.
(a) Ethanol reacts with metallic sodium with evolution of hydrogen to form sodium ethoxide.
(b) *p*-Nitrophenol treated with dilute aqueous sodium hydroxide forms sodium *p*-nitrophenoxide, which is soluble in water.

_____

_____

Problem 7.15 Given three vials labeled **A, B,** and **C,** each containing one of these compounds—methanol, 1-hexene, and 3-methylpentane—identify the contents of each vial based on the following data.

	Test	Result	Conclusion
A	$\xrightarrow{Br_2/CCl_4}$	no discoloration	Vial **A** contains _____
	$\xrightarrow{Na}$	no bubbles	
B	$\xrightarrow{Br_2/CCl_4}$	no discoloration	Vial **B** contains _____
	$\xrightarrow{Na}$	vigorous bubbling	
C	$\xrightarrow{Br_2/CCl_4}$	rapid discoloration	Vial **C** contains _____
	$\xrightarrow{Na}$	no bubbles	

7.7 Oxidation

We shall use the term **oxidation** to signify the removal of H from a compound and/or the addition of O to a compound. Conversely, the addition of H to a compound and/or the removal of O from a compound will be described as **reduction.**

Examples of oxidation and reduction reactions are

Oxidation
                                                            Change

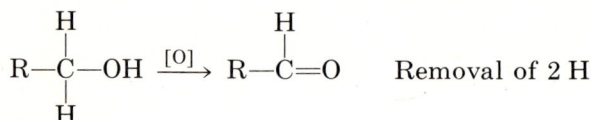   Removal of 2 H

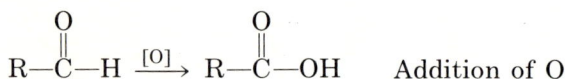   Addition of O

Reduction

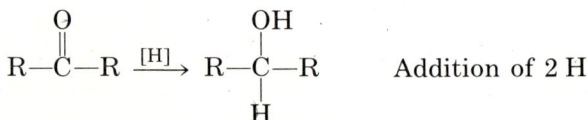   Addition of 2 H

$$Ar—NO_2 \xrightarrow{[H]} Ar—NH_2 \qquad \text{Addition of 2 H and removal of 2 O}$$

In oxidation reactions, the symbol [O] is used to represent the **oxidizing agent,** without specifying its exact nature. An oxidizing agent is the chemical reagent that does the oxidation. The general process of reduction is symbolized by [H] when no specific **reducing agent** is mentioned. A reducing agent is a substance that does the reduction.

Problem 7.16 State which of the following changes represent either an oxidation or a reduction.

(a) $CH_3—COOH \longrightarrow CH_3CH_2OH$

(b) $R—S—S—R \longrightarrow 2\,R—SH$

(c)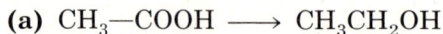

(d)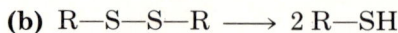

(e)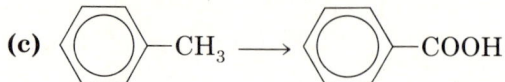

(f) $HOCH_2CH_2OH \longrightarrow HOOC—COOH$

Oxidation of Alcohols *7.8*

A   Laboratory Methods

Oxidation of alcohols gives different products depending on the *class of alcohols that is oxidized* and on the *kind of oxidizing agent that is used.*

Primary alcohols yield **aldehydes** when treated with *mild* oxidizing agents such as hot metallic copper or CrO_3 in pyridine.

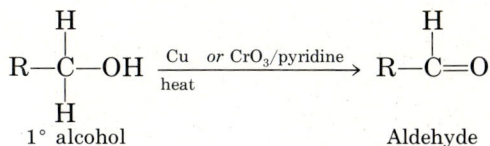

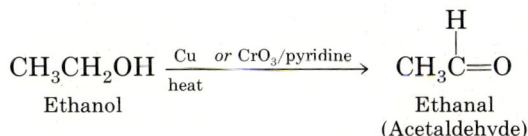

$$
\begin{array}{c}
\text{H} \\
| \\
\text{R}-\text{C}-\text{OH} \\
| \\
\text{H} \\
\text{1° alcohol}
\end{array}
\xrightarrow[\text{heat}]{\text{Cu } \textit{or } \text{CrO}_3/\text{pyridine}}
\begin{array}{c}
\text{H} \\
| \\
\text{R}-\text{C}=\text{O} \\
\\
\text{Aldehyde}
\end{array}
$$

$$
\begin{array}{c}
\text{CH}_3\text{CH}_2\text{OH} \\
\text{Ethanol}
\end{array}
\xrightarrow[\text{heat}]{\text{Cu } \textit{or } \text{CrO}_3/\text{pyridine}}
\begin{array}{c}
\text{H} \\
| \\
\text{CH}_3\text{C}=\text{O} \\
\\
\text{Ethanal} \\
\text{(Acetaldehyde)}
\end{array}
$$

When primary alcohols are allowed to react with stronger oxidizing agents, such as chromic acid, $H_2Cr_2O_7$, or neutral potassium permanganate, $KMnO_4$, the intermediate aldehydes formed initially are oxidized further to carboxylic acids, RCOOH. For example,

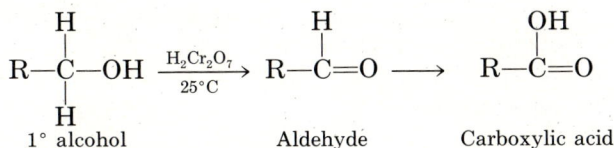

$$
\begin{array}{c}
\text{H} \\
| \\
\text{R}-\text{C}-\text{OH} \\
| \\
\text{H} \\
\text{1° alcohol}
\end{array}
\xrightarrow[\text{25°C}]{\text{H}_2\text{Cr}_2\text{O}_7}
\begin{array}{c}
\text{H} \\
| \\
\text{R}-\text{C}=\text{O} \\
\\
\text{Aldehyde}
\end{array}
\longrightarrow
\begin{array}{c}
\text{OH} \\
| \\
\text{R}-\text{C}=\text{O} \\
\\
\text{Carboxylic acid}
\end{array}
$$

Chromic acid is prepared by adding sulfuric acid to potassium dichromate, giving an orange solution. Visual evidence of a reaction with chromic acid is shown by a change in color of the solution from orange to green. For example, the oxidation of ethyl alcohol with chromic acid

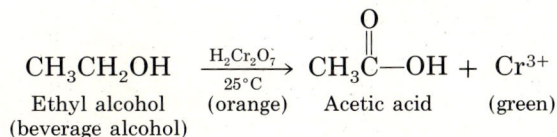

$$
\begin{array}{c}
\text{CH}_3\text{CH}_2\text{OH} \\
\text{Ethyl alcohol} \\
\text{(beverage alcohol)}
\end{array}
\xrightarrow[\text{25°C}]{\text{H}_2\text{Cr}_2\text{O}_7^-}
\begin{array}{c}
\text{O} \\
\| \\
\text{CH}_3\text{C}-\text{OH} \\
\text{Acetic acid}
\end{array}
+ \begin{array}{c}
\text{Cr}^{3+} \\
\\
\text{(green)}
\end{array}
$$

(orange)

is the basis of *roadside breath test* used by police to detect potential drunken drivers. A sample of the suspect's breath is passed over a Breathalyzer, a device containing chromic acid. If a green color develops and persists beyond a certain point in the apparatus, this is taken as a probable indication that the motorist's blood contains more than the legal level of alcohol (0.10%) to allow further driving. The suspect is then taken to the police station for further tests.

Secondary alcohols, when treated with *any* of the oxidizing agents mentioned previously, yield **ketones.**

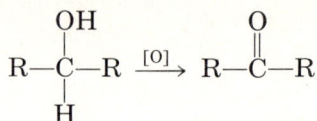

2° alcohol   Ketone

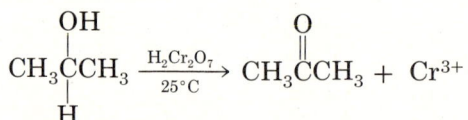

2-Propanol (orange)   Acetone   (green)

Tertiary alcohols do not react with oxidizing agents. If chromic acid is used as an oxidizing agent, the lack of reaction can be ascertained visually by noting that the orange color of the solution persists.

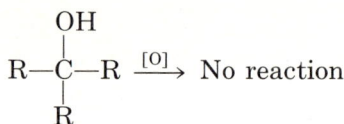

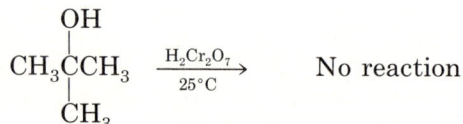

t-Butyl alcohol (orange)   (solution remains orange)

Oxidation of alcohols is an important route to the synthesis of aldehydes and ketones (see Chapter 10) and of acids (Chapter 12). Chromic acid oxidation is also useful for distinguishing 1° and 2° alcohols (positive test) from 3° alcohols (negative test). Neutral potassium permanganate, $KMnO_4$, oxidizes alcohols in the same manner as does chromic acid.

Problem 7.17 Devise simple chemical tests that would differentiate 1-hexene from ethanol and from 2-methyl-2-butanol. Tell exactly what you would see.

B   Biological Oxidation of Alcohols

The oxidation of alcohols in the laboratory has its counterpart in living organisms. Obviously, the human body does not require hot copper or chromic acid to oxidize ethanol (drinking alcohol) to acetaldehyde, CH_3CHO, or to acetic acid, CH_3COOH. The oxidative process, which takes place primarily in the liver, is much more sophisticated and is mediated by several chemical steps catalyzed by enzymes (enzymes are biological catalysts). What we should bear in mind is that the *same* chemical principles apply within our living environment as those that we enunciated for reactions in the laboratory. That ethanol is transformed by the liver first to acetaldehyde and immediately thereafter to acetic acid should therefore come as no surprise. That our bodies can tolerate ethanol up to

(a)

$$CH_3CH_2OH \xrightarrow{\text{enzyme}_1} CH_3-\overset{\displaystyle O}{\overset{\|}{C}}-H \ + \ \text{Enzyme}_1-2H$$

Ethanol   Acetaldehyde

$$\xrightarrow{\text{enzyme}_2} CH_3-\overset{\displaystyle O}{\overset{\|}{C}}-OH \ + \ \text{Enzyme}_2-2H$$

Acetic acid
(used by body cells)

(b)

$$CH_3CH_2OH \ + \ \text{Antabuse} \xrightarrow{\text{enzyme}_1} CH_3-\overset{\displaystyle O}{\overset{\|}{C}}-H \ + \ \text{Enzyme}_1-H$$

enzyme$_2$ ↓   ↓ no enzyme$_2$ to oxidize acetaldehyde

Antabuse–enzyme$_2$ complex

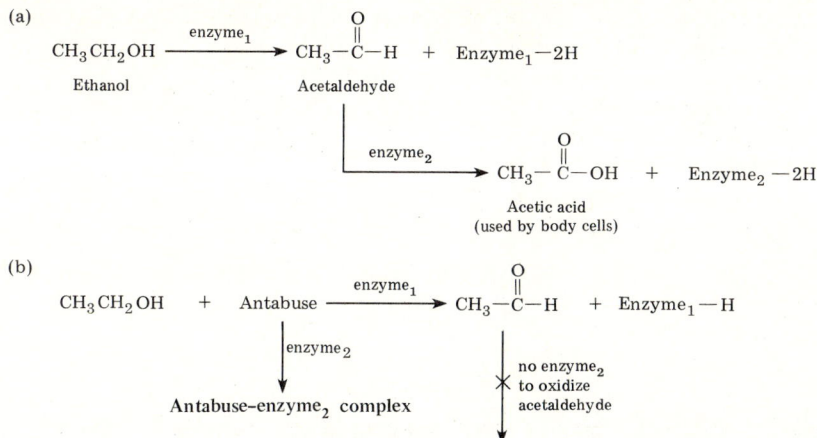

Figure 7.5 Oxidation products of ethanol: **(a)** normal and **(b)** in the presence of Antabuse. The end product in **(a)**, acetic acid, is used by the body cells. In **(b)** the buildup of acetaldehyde leads to violent pains.

a certain dose again is not surprising: after all, the acetic acid produced can be utilized by almost every cell of the body. Any excess alcohol, the overload that the liver cannot transform, continues to circulate in the blood, eventually causing intoxication. The treatment of alcoholics with the drug Antabuse (disulfiram), even though not the answer to the problem of alcoholism, can be understood from a chemical point of view. Antabuse inhibits the ability of the liver to transform acetaldehyde into acetic acid, thereby quickly causing a toxic pileup of acetaldehyde. This is why anyone who has taken disulfiram cannot drink even a single glass of alcohol without becoming violently ill. This process is described in Figure 7.5.

Methanol, when taken internally, is poisonous. Small doses can cause blindness by degeneration of the optic nerve. Large amounts may be fatal, undoubtedly because the products of oxidation, formaldehyde and formic acid, cannot be assimilated quickly enough by the body.

Ethylene glycol is toxic when swallowed, whereas 1,2-propylene glycol is harmless. If we consider the expected products of oxidation for each alcohol, it becomes clear why the two alcohols behave so differently.

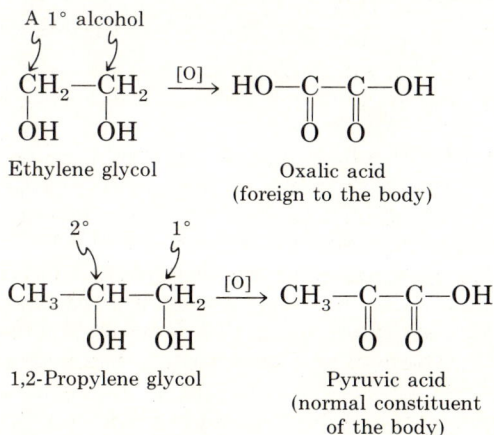

A 1° alcohol

$$\underset{\underset{OH}{|}}{CH_2}-\underset{\underset{OH}{|}}{CH_2} \xrightarrow{[O]} HO-\overset{\overset{\displaystyle O}{\|}}{C}-\overset{\overset{\displaystyle O}{\|}}{C}-OH$$

Ethylene glycol   Oxalic acid
(foreign to the body)

2°   1°

$$CH_3-\underset{\underset{OH}{|}}{CH}-\underset{\underset{OH}{|}}{CH_2} \xrightarrow{[O]} CH_3-\overset{\overset{\displaystyle O}{\|}}{C}-\overset{\overset{\displaystyle O}{\|}}{C}-OH$$

1,2-Propylene glycol   Pyruvic acid
(normal constituent
of the body)

Pyruvic acid, derived from the oxidation of 1,2-propylene glycol, is a normal constituent of body metabolism, making 1,2-propylene glycol harmless. Ethylene glycol is a poison because the oxalic acid formed in oxidation is a constituent foreign to the human body. It depresses the central nervous system and causes kidney malfunction. Anyone who is excessively fond of rhubarb may have suffered discomfort, since rhubarb contains substantial amounts of oxalic acid, which also gives rhubarb its tart taste.

Problem 7.18 Draw the structure of the product expected in each of the following oxidation reactions (if no reaction, state so).

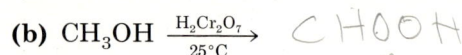

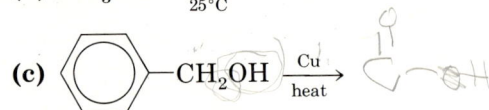

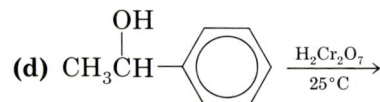

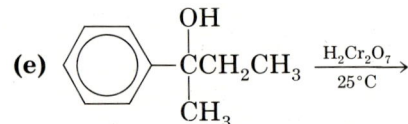

(a) $CH_3CH_2CH_2OH$ $\xrightarrow{CrO_3/pyridine}$ *NR — CH3 CH2 COOH*

(b) CH_3OH $\xrightarrow[25°C]{H_2Cr_2O_7}$ *CHOOH*

(c) ⬡—CH_2OH $\xrightarrow[heat]{Cu}$ *C—O.H*

(d) $CH_3\overset{\displaystyle OH}{\underset{\displaystyle |}{CH}}$—⬡ $\xrightarrow[25°C]{H_2Cr_2O_7}$

(e) ⬡—$\overset{\displaystyle OH}{\underset{\displaystyle \underset{\displaystyle CH_3}{|}}{\overset{|}{C}}}CH_2CH_3$ $\xrightarrow[25°C]{H_2Cr_2O_7}$

7.9 Oxidation of Phenols

Phenols are easily oxidized, but often the oxidation products are ill-defined, complex substances. We can oxidize phenols with the same oxidizing agents that were used to oxidize 1° and 2° alcohols (CrO_3, $H_2Cr_2O_7$, for example). Some phenols are oxidized slowly by oxygen in the air alone; in that case, we say that the phenol has undergone *autoxidation*. The food, petroleum, rubber, and plastic industries take advantage of the autoxidation of certain phenols to incorporate them as *antioxidants* in their products. Just look at the label on any package of potato chips, for example, and you will almost certainly find among the listed ingredients BHT, BHA, or *n*-propyl gallate. These are phenolic additives that act as antioxidants. The spoilage of the chips, which is due to oxidation, is retarded because these antioxidants are autoxidized preferentially.

In the special cases of hydroquinone and catechol, the oxidation products are well-defined, colored substances called *quinones*. Hydroquinone is oxidized to the yellow, stable *p*-benzoquinone. The reaction may be reversed with a reducing agent.

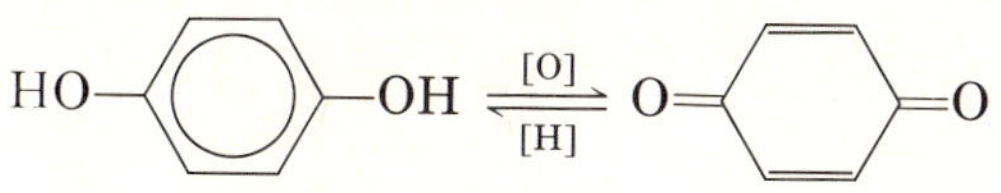

Hydroquinone is the primary reducing agent used today as a photographic film developer. The developing involves essentially the reduction of silver ions, Ag^+, to metallic silver, Ag^0. Substituted p-quinones are the coloring materials in many fungi and molds. *Ubiquinone,* or coenzyme Q, is found in biological systems, where it is associated with the vital respiratory cycle.

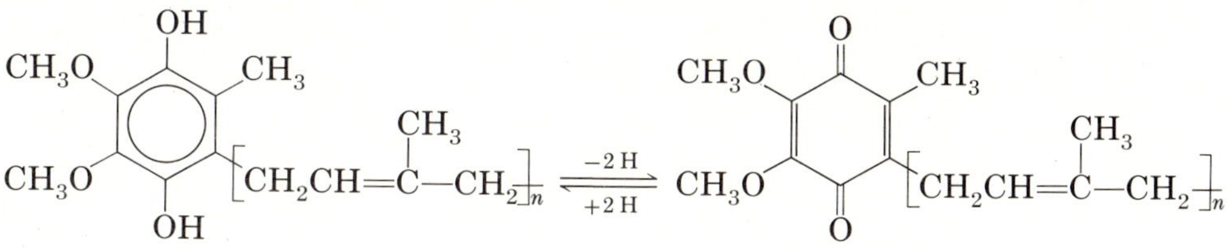

Catechol is oxidized in a similar manner to o-benzoquinone. o-Quinones are not as stable as the para isomer.

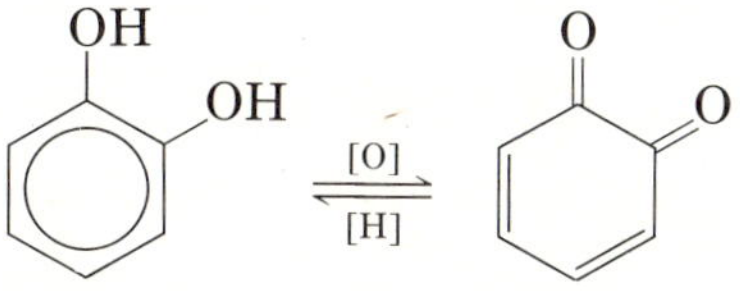

In intermediary metabolism in human beings, certain substances obtained from proteins contain catechol as part of the molecule. These are oxidized into *melanin pigments* (Fig. 7.6). Melanins are the substances that impart color to the skin, eyes, and hair. The exact structure of the melanins is not known.

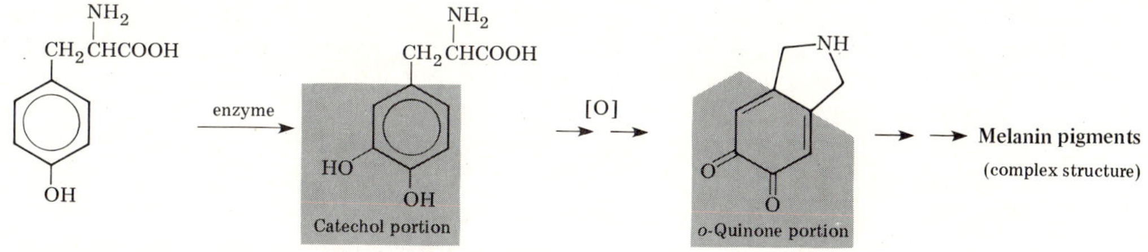

Figure 7.6 Formation of melanin pigments, one of the metabolic routes taken by tyrosine.

7.10 Reactions Involving Carbon-Hydroxyl Bond Breaking

The cleavage of the C$\dashv$OH bond is applicable to alcohols only.

A   Inorganic Esters

Treatment of alcohols with inorganic oxyacids, HOX, produces **inorganic esters.** The reaction involves the replacement of one or more of the hydrogens in the oxyacid by alkyl groups.

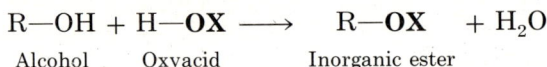

The general formulas of inorganic esters formed upon treatment of alcohols with specific oxyacids are shown in Figure 7.7. Examples of inorganic esters of interest are

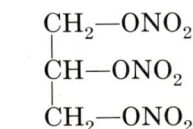

Glyceryl trinitrate
(Nitroglycerin)
(explosive, vasodilator)

CH$_3$CH$_2$—ONO

Ethyl nitrite
(vasodilator)

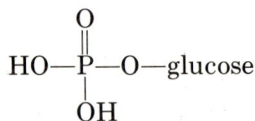

Glucose 6-phosphate
(intermediate in
carbohydrate metabolism)

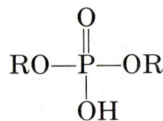

Dialkyl hydrogen phosphate
(integral part of nucleic acids)

Organic esters, derived from a carboxylic acid and an alcohol, are discussed in detail in Chapter 12.

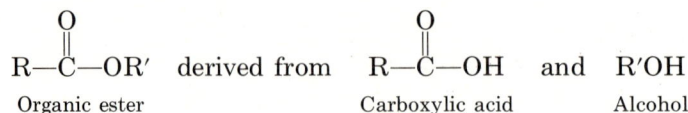

Organic ester   Carboxylic acid   Alcohol

Problem 7.19   Write the structure of the inorganic ester resulting from the reaction of methanol (1 mole) with each acid (see Fig. 7.7).
(a) Nitric acid   **(b)** Nitrous acid   **(c)** Sulfuric acid   **(d)** Phosphoric acid

---

Problem 7.20   Write the structures of these compounds (see Fig. 7.7).
(a) Ethyl nitrate   **(b)** *n*-Butyl nitrite

(c) *n*-Butyl hydrogen sulfate   **(d)** Ethyl dihydrogen phosphate
(e) Diethyl hydrogen phosphate   **(f)** Triethyl phosphate

B   Replacement of the OH Group by Halide: Alkyl Halides

The hydroxyl group of alcohols can be replaced by halide to form **alkyl halides.** The net equation is

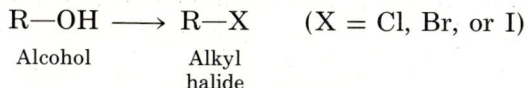

$$R\text{—}OH \longrightarrow R\text{—}X \qquad (X = Cl,\ Br,\ or\ I)$$

Alcohol   Alkyl halide

The replacement of OH by halogen is applicable only to chlorides, bromides, and iodides. Alkyl fluorides, R—F, are not prepared from alcohols. In this section we discuss only the various halogen-containing compounds used in preparing alkyl halides from alcohols. The mechanisms of alkyl halide formation are presented in Section 8.6.

Reaction with hydrogen halides: Lucas test. Alcohols when treated with aqueous concentrated hydrogen halides H—X, or with anhydrous H—X, give alkyl halides. The general equation is

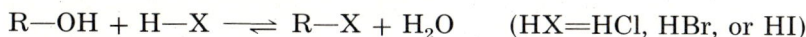

$$R\text{—}OH + H\text{—}X \;\rightleftharpoons\; R\text{—}X + H_2O \qquad (HX = HCl,\ HBr,\ or\ HI)$$

The reaction of halogen acids with various alcohols proceeds at different rates. In general, tertiary alcohols react rapidly with hydrogen halides; secondary alcohols react somewhat slower; and primary alcohols, even more slowly.

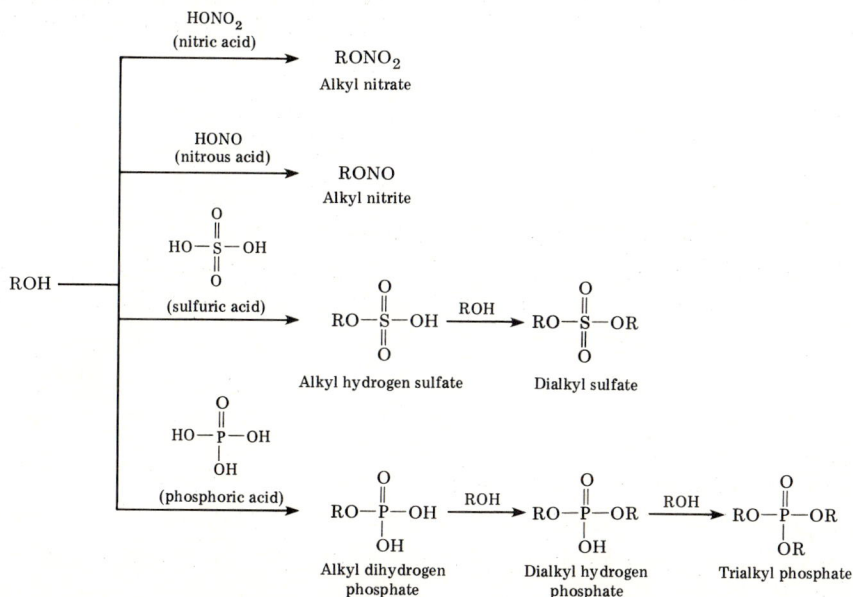

Figure 7.7 Names and structures of some inorganic esters formed by the reaction of an alcohol with several oxyacids.

These differences in reaction rates of $1°$, $2°$, and $3°$ alcohols form the basis of the **Lucas test** for distinguishing among the three classes of alcohols.

In practice, the Lucas test is carried out as follows. An alcohol is mixed, at room temperature, with concentrated HCl and $ZnCl_2$. The latter serves as a catalyst. The alkyl chloride, as it is formed, is insoluble in the medium and causes the solution to become cloudy before it separates as a distinct layer.

Tertiary alcohols react almost immediately; with secondary alcohols, the cloudiness appears after about five minutes. Primary alcohols do not react under these conditions, and the solution remains clear. The results expected in the Lucas test are summarized in the following equations.

$$R_3COH + HCl \xrightarrow[25°C]{ZnCl_2} R_3CCl + H_2O$$

A 3° alcohol (immediate cloudiness)

$$R_2CHOH + HCl \xrightarrow[25°C]{ZnCl_2} R_2CHCl + H_2O$$

A 2° alcohol (cloudiness in 5 min)

$$RCH_2OH + HCl \xrightarrow[25°C]{ZnCl_2} \text{No visible reaction}$$

A 1° alcohol (solution remains clear)

From a practical viewpoint, alkyl halides often are prepared as intermediates for obtaining other desired compounds, not because they are needed per se. In such cases, the method that gives access to an alkyl halide most readily, at a minimum of cost, is normally the one chosen. Among the halogen halides, HBr meets the criteria best, and alkyl bromides are usually prepared by the reaction

$$R-OH + HBr \underset{\rightleftharpoons}{\xrightarrow{heat}} R-Br + HOH$$

Because the reaction of alcohols with hydrogen halides is reversible, alternate methods of preparing alkyl halides from alcohols have been devised.

Reaction with thionyl chloride. Thionyl chloride, $(Cl)_2S=O$, is a useful reagent for synthesizing alkyl chlorides from alcohols. An advantage of this method over the previous one is that the reaction is irreversible because the by-products, HCl and SO_2, are gases. The overall reaction is

$$\underset{\text{Thionyl chloride}}{ROH + Cl-\overset{O}{\overset{\|}{S}}-Cl} \longrightarrow HCl\uparrow + \underset{\text{Alkyl chlorosulfite}}{RO-\overset{O}{\overset{\|}{S}}-Cl} \xrightarrow{heat}$$

$$\underset{\text{Alkyl chloride}}{R-Cl} + SO_2\uparrow$$

The intermediate alkyl chlorosulfite decomposes spontaneously on heating to give SO_2 and alkyl chloride.

Reaction with phosphorus trihalides, PX_3. Alcohols react with phosphorus trihalides as follows.

$$3\,R-OH + PX_3 \xrightarrow{heat} 3\,R-X + H_3PO_3 \qquad (PX_3 = PCl_3,\ PBr_3,\ or\ PI_3)$$

The reaction goes to completion (is irreversible) because the alkyl halide, as it is formed, can be distilled off from the reaction mixture. In practice, the reaction is limited to phosphorus trichloride, bromide, and iodide, respectively. Phosphorus trifluoride, PF_3, does not react with alcohols to form R—F.

Problem 7.21 Arrange the following alcohols in order of increasing reactivity toward Lucas reagent ($HCl + ZnCl_2$).

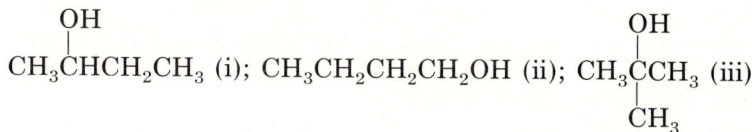

$$CH_3\overset{\underset{|}{OH}}{C}HCH_2CH_3 \text{ (i); } CH_3CH_2CH_2CH_2OH \text{ (ii); } CH_3\overset{\underset{|}{OH}}{\underset{\underset{CH_3}{|}}{C}}CH_3 \text{ (iii)}$$

Problem 7.22 Draw the structure of the organic products expected from the following reactions (if no reaction is expected, state so).

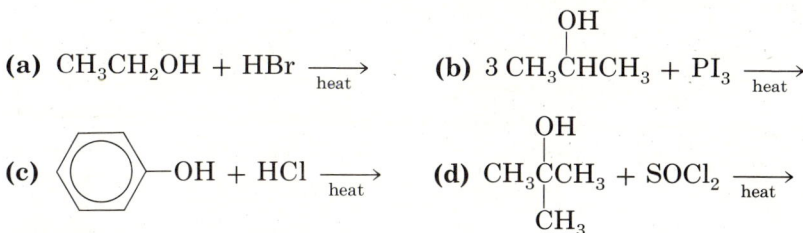

(a) $CH_3CH_2OH + HBr \xrightarrow[heat]{}$

(b) $3\ CH_3\overset{\underset{|}{OH}}{C}HCH_3 + PI_3 \xrightarrow[heat]{}$

(c) ⬡—OH + HCl $\xrightarrow[heat]{}$

(d) $CH_3\overset{\underset{|}{OH}}{\underset{\underset{CH_3}{|}}{C}}CH_3 + SOCl_2 \xrightarrow[heat]{}$

C   Dehydration of Alcohols: Formation of Alkenes

The elimination of H_2O from alcohols to give alkenes has been dealt with in Chapter 3.

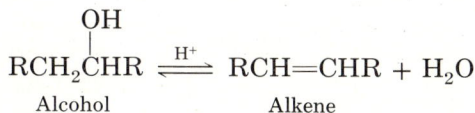

$$RCH_2\overset{\underset{|}{OH}}{C}HR \underset{}{\overset{H^+}{\rightleftharpoons}} RCH=CHR + H_2O$$

Alcohol   Alkene

As a reminder, let us recapitulate some important features of the reaction.

1. Dehydration of alcohols is catalyzed by acids.
2. Elimination of H_2O involves a carbocation as an intermediate.
3. The ease of dehydration of alcohols follows the order $3° > 2° > 1° >$ methanol, which is also the order of ease of formation and stability of carbocations.
4. The carbocation loses a proton to give an alkene.
5. If more than one alkene can be formed, the major one is the one that has the greater number of alkyl groups bonded to C=C (*Saytzeff's rule*).

Problem 7.23 Draw the structure of the major alkene obtained on dehydration of the following alcohols.

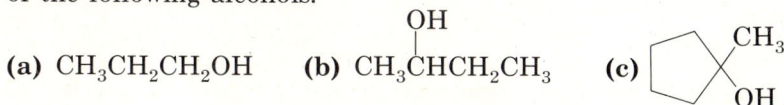

(a) $CH_3CH_2CH_2OH$ **(b)** $CH_3\overset{\underset{|}{OH}}{C}HCH_2CH_3$ **(c)** ⬠ with CH_3 and OH

7.11 Reactions on the Aromatic Ring of Phenols

The hydroxyl group in phenols is a strong ortho,para director and a strong activator in aromatic substitution reactions (Secs. 5.11, 5.12). For this reason, phenol undergoes reactions on the ring much more easily than does benzene. For example, nitration of phenol goes so rapidly that *dilute nitric acid* at room temperature must be used. (Remember that nitration of benzene requires concentrated nitric acid and sulfuric acid as a catalyst.)

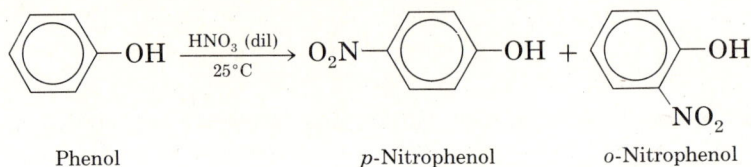

Phenol   *p*-Nitrophenol   *o*-Nitrophenol

As another example of the high reactivity of phenols, halogenation takes place *without catalyst*. The products depend on the solvent used. In *aprotic solvents*—solvents that do not release protons (CCl_4, CS_2)—bromination, for example, gives a mixture of *o*- and *p*-bromophenol. In protic solvents—solvents that can release protons (H_2O)—halogenation of phenols takes place with such ease that it is impossible to stop at monohalogenation: a trisubstituted phenol is produced.

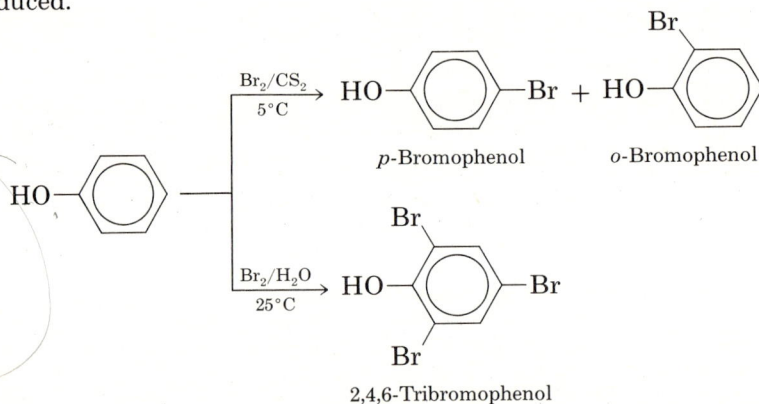

p-Bromophenol   *o*-Bromophenol

2,4,6-Tribromophenol

7.12 Preparation of Alcohols

This section presents methods for preparing alcohols starting from alkenes only. Synthesis of alcohols from alkyl halides is discussed in Chapter 8; preparations of alcohols from aldehydes and ketones and from acids are discussed in Chapters 10 and 12, respectively.

A Hydration of Alkenes

The addition of water to alkenes proceeds in the reverse manner as the dehydration of alcohols. We have encountered the reaction before in Chapter 3 (Sec. 3.20). Let us just point out the highlights of the reaction here.

1. The overall reaction consists of the addition of water to a double bond in the presence of an acid catalyst, H^+.

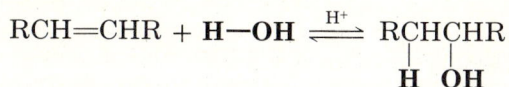

$$RCH{=}CHR + H{-}OH \xrightleftharpoons{H^+} \underset{\underset{H \ \ OH}{\mid \ \ \mid}}{RCHCHR}$$

2. The addition follows *Markovnikov's rule,* which enables us to predict the major product when more than one can be formed. For example,

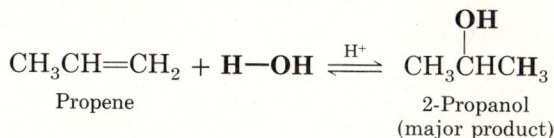

$$CH_3CH{=}CH_2 + H{-}OH \xrightleftharpoons{H^+} \underset{\underset{\mid}{CH_3CHCH_3}}{\overset{\overset{OH}{\mid}}{}}$$

Propene   2-Propanol (major product)

3. *It is not possible to prepare primary alcohols* (ethanol is the only exception) *via hydration of alkenes.* The method is nevertheless extremely useful for preparing a number of alcohols, because the reactants are inexpensive and easy to obtain.

B Hydroboration-Oxidation

A second method for synthesizing alcohols from alkenes involves the addition of diborane, B_2H_6, followed by oxidation with alkaline hydrogen peroxide. The net result is the addition of H_2O across $C{=}C$ in an **anti-Markovnikov** manner.

General equation

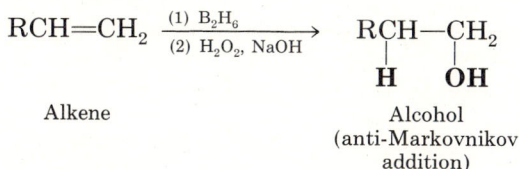

$$RCH{=}CH_2 \xrightarrow[\text{(2) } H_2O_2,\text{ NaOH}]{\text{(1) } B_2H_6} \underset{\underset{H \quad OH}{\mid \quad \ \mid}}{RCH{-}CH_2}$$

Alkene   Alcohol (anti-Markovnikov addition)

Specific example

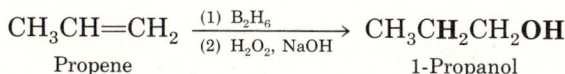

$$CH_3CH{=}CH_2 \xrightarrow[\text{(2) } H_2O_2,\text{ NaOH}]{\text{(1) } B_2H_6} CH_3CH_2CH_2OH$$

Propene   1-Propanol

As you can see, this method complements the preceding one.

Problem 7.24 Draw the structure of the major product expected in each reaction.

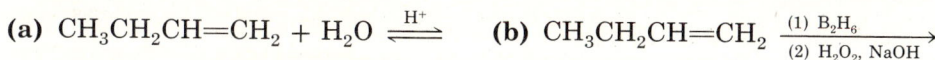

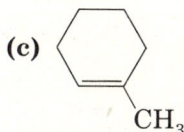

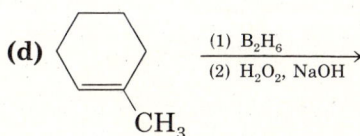

(a) $CH_3CH_2CH{=}CH_2 + H_2O \xrightleftharpoons{H^+}$ (b) $CH_3CH_2CH{=}CH_2 \xrightarrow[\text{(2) } H_2O_2,\text{ NaOH}]{\text{(1) } B_2H_6}$

(c) [cyclohexene with CH₃ substituent] $+ H_2O \xrightleftharpoons{H^+}$ (d) [cyclohexene with CH₃ substituent] $\xrightarrow[\text{(2) } H_2O_2,\text{ NaOH}]{\text{(1) } B_2H_6}$

7.13 Sources and Uses of Important Alcohols

Methanol is also known as *wood alcohol* because it was formerly obtained from the destructive distillation of wood. (Destructive distillation is the heating of a substance in the absence of air.) Today, methanol is manufactured industrially by treating a mixture of carbon monoxide and hydrogen, known as *synthesis gas*, at elevated temperature and pressure in the presence of catalysts.

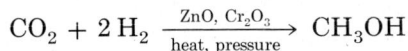

$$CO_2 + 2H_2 \xrightarrow[\text{heat, pressure}]{\text{ZnO, Cr}_2\text{O}_3} CH_3OH$$

In 1980, 7.2 billion pounds of methanol was produced in the United States. A large fraction (40–50%) goes into the production of formaldehyde, a starting material for many plastics. The rest is used as a solvent or as a nonpermanent antifreeze or is converted to a variety of other compounds. A 20% mixture of methanol and gasoline, *gasohol,* makes a good motor fuel, a fact that should receive serious consideration in the search for a way to alleviate the fuel problem. The taste and poisonous properties of methanol make it a good **denaturant** for ethyl alcohol. A denaturant is a substance that when added to ethyl alcohol makes it unfit for drinking.

Ethanol, or *grain alcohol,* used in alcoholic beverages, is obtained from the fermentation of sugars and starches (see Chapter 11). The fermentation solution contains from 5 to 13% ethanol and a number of other products, such as aldehydes, ketones, *n*-propyl alcohol, isobutyl alcohol, and pentanols. The mixture of alcohols other than ethanol is known as *fusel oil.* Ethanol is separated from aldehydes, ketones, and fusel oil by fractional distillation. The ethanol recovered from the distillation is 95.5% pure; it still contains 4.5% water. The 95.5% ethanol–4.5% water mixture cannot be separated further by distillation. We call such a solution an **azeotropic solution,** or an **azeotrope.** For most purposes, the azeotrope is suitable for use as such. When pure anhydrous (without water) ethanol, or **absolute ethanol,** is needed, the azeotrope is treated chemically (with lime, for example) to remove the last trace of water.

In the United States, most of the ethanol produced comes not from fermentation but from the hydration of ethylene. In 1980, about 1.2 billion pounds of ethyl alcohol was synthesized from ethylene for industrial and medicinal uses. Medicinally, ethanol per se has little therapeutic value. When imbibed, it is not a stimulant, as many of us may believe, but a central nervous system depressant. Some of the effects of various concentrations of alcohol in the blood are shown in Table 7.3.

Ethyl alcohol is the solvent used in many medicinal preparations labeled "tincture of ———" and in perfumes or flavoring extracts. A 70% solution of ethanol is an effective antiseptic.

Hydration of propylene with 80% sulfuric acid below 40°C gives *isopropyl alcohol.*

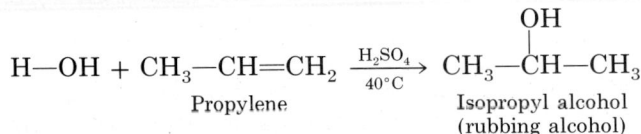

$$H\text{—}OH + CH_3\text{—}CH{=}CH_2 \xrightarrow[40°C]{H_2SO_4} CH_3\text{—}\overset{\overset{\displaystyle OH}{|}}{C}H\text{—}CH_3$$

Propylene   Isopropyl alcohol
(rubbing alcohol)

This alcohol finds application as a solvent and as an intermediate for the produc-

Table 7.3 Effects of Various Concentrations of Alcohol in the Blood

Alcohol in blood (%)[a]	Observable effect
0.01	clearing of the head, freer breathing through nasal passages, mild tingling of mucous membranes of the mouth and throat
0.02	mild throbbing at the back of the head, dizziness
0.03	mild euphoria, no sense of worry, exaggerated feeling of confidence; time seems to pass quickly
0.04	excess energy, boisterousness, flippancy
0.10	perceptible staggering
0.30	stuporous condition
0.40	deep anesthesia; may be fatal

[a] For every 0.01% of alcohol in the blood, 5–10 ml of alcohol is ingested.

tion of acetone, $(CH_3)_2C{=}O$. It is also the rubbing alcohol sold in the neighborhood drugstore.

Many complex alcohols occur in nature, and a number of them have been synthesized. Examples are cholesterol (page 177) and vitamins A and D, which are discussed in Section 13.15.

All the alcohols considered so far have been monohydroxy alcohols. The two most important simple polyhydroxy alcohols are ethylene glycol and glycerol.

Ethylene glycol is the main constituent of permanent antifreeze used in automobiles. Zerex and Prestone consist mainly of ethylene glycol plus small amounts of a dye and antioxidant. A solution of 60% glycol and 40% water freezes at $-49°C$ ($-56°F$) and is an effective antifreeze. Ethylene glycol is also one of the two materials used in manufacturing Dacron polyester (Sec. 12.16).

Glycerol is the most important triol. Glycerol is a viscous, sweet-tasting, high-boiling liquid, soluble in all proportions in water and alcohol. Besides serving as a raw material in the synthesis of nitroglycerin (Sec. 7.10), glycerol is used also in the manufacture of certain plastics. Because it is nontoxic, glycerol is often used as a solvent in liquid medications. Because of its moisturizing properties, glycerol is an ingredient in cosmetic formulations, in tobacco products, and in foods. It is also the main constituent of suppositories.

Glycerol can be obtained as a by-product of the manufacture of soap from fats and oils (Chapter 13). However, much more glycerol is needed than can be supplied from fats and oils, and large amounts are produced synthetically starting from propylene.

Preparation of Phenols *7.14*

The preparation of phenols requires special methods because there is no direct way to introduce an —OH group onto an aromatic ring.

In the laboratory we prepare phenols by one of two methods: (1) alkali fusion of sulfonates or (2) hydrolysis of diazonium salts. In this chapter only the first method is discussed. The second is presented when we study the chemistry of aromatic amines (Chapter 14).

The alkali fusion of sulfonates involves the following steps.

1. Sulfonation of an aromatic ring.
2. Melting (fusion) of the aromatic sulfonic acid with sodium hydroxide to give a phenoxide salt.
3. Acidification of the phenoxide with HCl to produce the phenol.

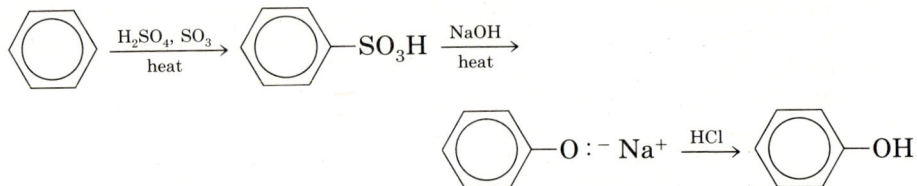

Most of the phenol produced industrially in the United States (over 80%) is via the **cumene process.** Cumene (isopropylbenzene) is obtained by the Friedel–Crafts alkylation of benzene with propylene.

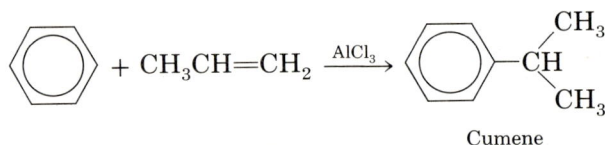

Cumene

When air is bubbled through cumene, cumene hydroperoxide is formed. Reaction with a strong acid converts the hydroperoxide to phenol and acetone, a valuable co-product.

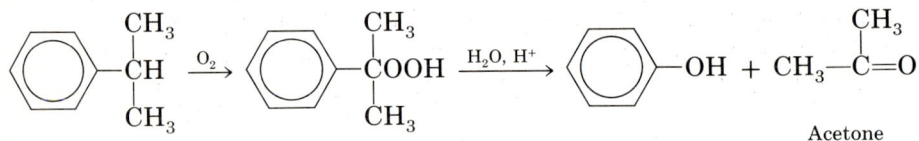

Acetone

As you can see, the manufacture of phenol depends ultimately on the availability of two petrochemicals, benzene and propylene.

7.15 Uses of Phenols

Phenol is the most important member of this class of compound. In 1980, 2.6 billion pounds of phenol was produced in the United States. Most of it is used in the manufacture of polymers, with formaldehyde and other aldehydes. These polymers, known as *phenolic resins,* have a variety of applications; they are used as molding and laminating materials, as protective coatings, as adhesives in plywood and other wood-based products, and as electrical insulators, to name a few. Phenol and its derivatives are also the starting point of a myriad of other items: aspirin, detergents, dyes, nylon, gasoline additives, herbicides, hormones, wood preservatives, explosives, and countless other products.

Sulfur Analogs of Alcohols and Phenols: 7.16
Thiols

Sulfur is in the same group in the periodic table as oxygen, and **thiols** are the sulfur analogs of alcohols and phenols. The prefix *thio* means sulfur. Thiols are also called **mercaptans,** meaning capturing mercury, because this class of compounds reacts and forms precipitates with mercuric ions. The functional group of thiols is —SH, called a **sulfhydryl group.**

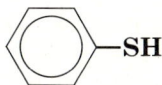

$$CH_3CH_2-SH \qquad \langle\bigcirc\rangle-SH$$

Common names:	Ethyl mercaptan	Phenyl mercaptan
	(Thioethanol)	(Thiophenol)
IUPAC name:	Ethanethiol	Benzenethiol

The formula of thiols suggests that they are hydrocarbon derivatives of hydrogen sulfide. As with H_2S, perhaps the most characteristic property of thiols is their unpleasant odor. For example, the main constituent of the offensive odor given off by skunks when they are attacked is a mixture of sulfur compounds.

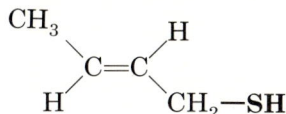

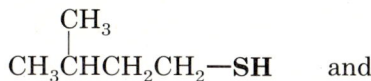

$$\underset{H}{\overset{CH_3}{\diagdown}}C=C\underset{CH_2-SH}{\overset{H}{\diagup}} \qquad CH_3\overset{CH_3}{\overset{|}{C}}HCH_2CH_2-SH \qquad and \qquad \underset{H}{\overset{CH_3}{\diagdown}}C=C\underset{CH_2-S-S-CH_3}{\overset{H}{\diagup}}$$

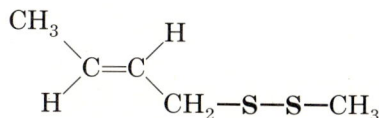

The sulfhydryl group is found among many proteins and enzymes. Because the —SH group is slightly acidic, many proteins and enzymes form salts with mercuric ions (as well as with other heavy metal ions), causing the proteins to precipitate and the enzyme to be deactivated. The formation of a mercury salt is

$$2\,R-SH + HgCl_2 \longrightarrow (RS)_2Hg\downarrow + 2\,HCl$$

One important reaction of thiols is *oxidation*. Even mild oxidizing agents, such as dilute hydrogen peroxide or cupric ions, convert thiols to *disulfides* by the loss of hydrogen.

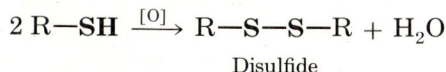

$$2\,R-SH \overset{[O]}{\longrightarrow} \underset{Disulfide}{R-S-S-R} + H_2O$$

Disulfides are analogous to peroxides, but are more stable. Disulfides are easily reduced to mercaptans by mild reducing agents.

$$R-S-S-R \overset{[H]}{\longrightarrow} 2\,R-SH$$

These two reactions, interconversions of sulfhydryl and disulfide groups, are important in protein chemistry. For example

$$
2 \begin{array}{c} COOH \\ | \\ H—C—NH_2 \\ | \\ CH_2 \\ | \\ \textbf{SH} \end{array}
\underset{[H]}{\overset{[O]}{\rightleftharpoons}}
\begin{array}{c} COOH \\ | \\ H—C—NH_2 \\ | \\ CH_2 \\ | \\ S———— \end{array}
\begin{array}{c} COOH \\ | \\ H—C—NH_2 \\ | \\ CH_2 \\ | \\ S \end{array}
$$

Cysteine   Cystine

(a sulfhydryl compound)   (a disulfide compound)

Lipoic acid, which plays a considerable role as an oxidizing agent in cellular metabolism, also contains a disulfide group. The decarboxylation (loss of CO_2) of pyruvic acid is catalyzed by lipoic acid. The latter is transformed into dihydro-lipoic acid, a sulfhydryl compound.

$$
\underset{\substack{\text{Lipoic acid} \\ \text{(a disulfide compound)}}}{\overset{S—S}{\diagup\!\!\diagdown}}—(CH_2)_4COOH + \underset{\text{Pyruvic acid}}{CH_3\overset{O}{\overset{||}{C}}COOH} + H_2O \xrightarrow{\text{enzyme}}
$$

$$
CH_3COOH + CO_2 + \underset{\substack{\text{Dihydrolipoic acid} \\ \text{(a sulfhydryl compound)}}}{H_2C\overset{\overset{\textbf{HS}}{|}}{\diagup}\underset{\underset{CH_2}{|}}{\diagdown}CH\overset{\overset{\textbf{SH}}{|}}{}(CH_2)_4COOH}
$$

Like alcohols, thiols form thioesters when treated with carboxylic acids. A thioester of biological importance is acetyl coenzyme A. Acetyl coenzyme A is involved in the metabolism of carbohydrates, fats, and proteins. It is formed when coenzyme A, a complex molecule containing an SH group, reacts with acetic acid.

$$
\underset{\substack{\text{Coenzyme A} \\ \text{(abbreviated structure)}}}{CoA—\textbf{SH}} + \underset{\text{Acetic acid}}{HO—\overset{O}{\overset{||}{C}}—CH_3} \xrightarrow{\text{enzyme}} \underset{\substack{\text{Acetyl coenzyme A} \\ \text{(a thioester)}}}{CH_3—\overset{O}{\overset{||}{C}}—S—CoA} + H_2O
$$

Problem 7.25 Write the structures of
(a) 1-Propanethiol   **(b)** Isobutyl mercaptan
(c) Ethyl disulfide   **(d)** Methyl phenyl disulfide

Summary of Concepts and Reactions

Alcohols and phenols have a common functional group, the —OH group. In alcohols, the —OH group is attached to an alkyl group —R. In phenols, the —OH group is bonded directly to an aromatic ring —Ar.   [Sec. 7.1]

Alcohols are subdivided into three classes: primary (1°), secondary (2°), and tertiary (3°). The classification depends on the number of alkyl groups bonded to the carbon bearing the —OH, the carbinol carbon.   [Sec. 7.1]

The common names of alcohols consist of the name of the alkyl group attached to the hydroxyl function plus the word alcohol. In the IUPAC system the -e ending of the parent alkane is replaced by -ol. For phenyl-substituted methyl alcohols, the carbinol system is used.   [Sec. 7.1]

Alcohols that contain two —OH groups are called diols or glycols, and those that contain three —OH groups are called triols. Alcohols that contain more than three —OH groups are known as polyols.   [Sec. 7.1]

Phenols are usually named as derivatives of the simplest member of the family, phenol, which is also known as carbolic acid.   [Sec. 7.2]

The boiling points of alcohols are much higher than are the boiling points of similar-molecular-weight alkanes because of the existence of hydrogen bonding. Hydrogen bonding also accounts for the differences in physical states between lower alcohols and lower alkanes.   [Sec. 7.3B]

Phenols are much more acidic than alcohols because the phenoxide ion is resonance stabilized whereas the alkoxide ion is not.   [Sec. 7.4]

Alcohols undergo two kinds of reactions: (1) those that involve the breaking of the oxygen–hydrogen bond (CO$\rightarrow$H)

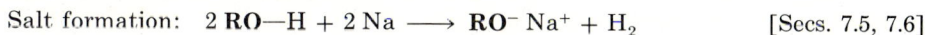

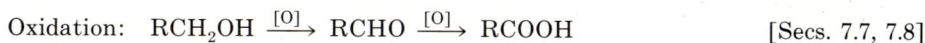

Salt formation: $2\ \mathbf{RO}\text{—H} + 2\,\text{Na} \longrightarrow \mathbf{RO}^-\,\text{Na}^+ + \text{H}_2$ [Secs. 7.5, 7.6]

Oxidation: $\text{RCH}_2\text{OH} \xrightarrow{[O]} \text{RCHO} \xrightarrow{[O]} \text{RCOOH}$ [Secs. 7.7, 7.8]

and (2) those that involve the breaking of the carbon–oxygen bond (C$\rightarrow$OH):

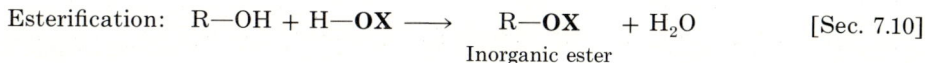

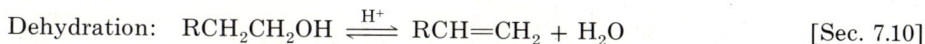

Esterification: $\text{R—OH} + \text{H—}\mathbf{OX} \longrightarrow \quad \text{R—}\mathbf{OX} \quad + \text{H}_2\text{O}$ [Sec. 7.10]
Inorganic ester

Formation of alkyl halides: $\text{R—OH} + \text{H}\mathbf{X} \rightleftharpoons \text{R}\mathbf{X} + \text{H}_2\text{O}$ [Sec. 7.10]

Dehydration: $\text{RCH}_2\text{CH}_2\text{OH} \xrightleftharpoons{\text{H}^+} \text{RCH}\text{=}\text{CH}_2 + \text{H}_2\text{O}$ [Sec. 7.10]

Phenols undergo only reactions where the CO$\rightarrow$H bond is broken.

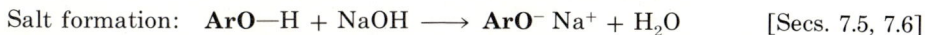

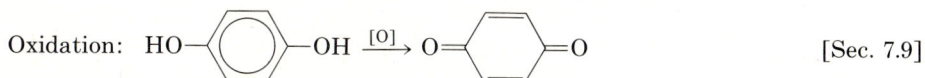

Salt formation: $\mathbf{ArO}\text{—H} + \text{NaOH} \longrightarrow \mathbf{ArO}^-\,\text{Na}^+ + \text{H}_2\text{O}$ [Secs. 7.5, 7.6]

Oxidation: HO—⟨◯⟩—OH $\xrightarrow{[O]}$ O=⟨ ⟩=O   [Sec. 7.9]

The —OH group in phenols is a strong ortho,para director and a strong activator.   [Sec. 7.11]

Two methods for preparing alcohols are considered in this chapter: (1) hydration of alkenes

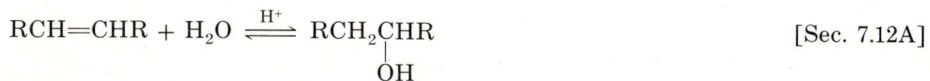

$$\text{RCH}\text{=}\text{CHR} + \text{H}_2\text{O} \xrightleftharpoons{\text{H}^+} \underset{\overset{|}{\text{OH}}}{\text{RCH}_2\text{CHR}}$$ [Sec. 7.12A]

and (2) hydroboration-oxidation

$$\text{RCH}\text{=}\text{CH}_2 \xrightarrow[\text{(2) H}_2\text{O}_2,\ \text{NaOH}]{\text{(1) B}_2\text{H}_6} \text{RCH}_2\text{CH}_2\text{OH}$$ [Sec. 7.12B]

A 95.5% ethanol and 4.5% water mixture cannot be separated further by distillation. Such a solution is called an azeotropic solution or an azeotrope.   [Sec. 7.13]

Absolute ethanol is ethanol without water (anhydrous ethanol).   [Sec. 7.13]

One laboratory method of preparing phenols is called the alkali fusion of sulfonates.

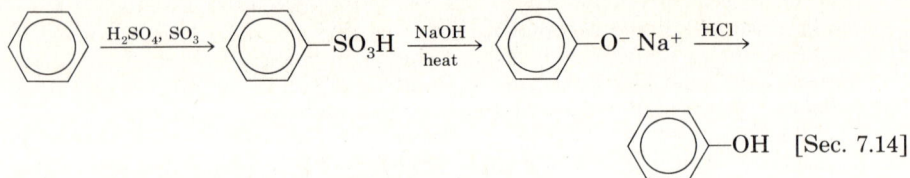

$$\bigcirc \xrightarrow{\text{H}_2\text{SO}_4,\ \text{SO}_3} \bigcirc -\text{SO}_3\text{H} \xrightarrow[\text{heat}]{\text{NaOH}} \bigcirc -\text{O}^-\ \text{Na}^+ \xrightarrow{\text{HCl}}$$

$$\bigcirc -\text{OH} \quad \text{[Sec. 7.14]}$$

In industry, phenol is prepared by the cumene process.

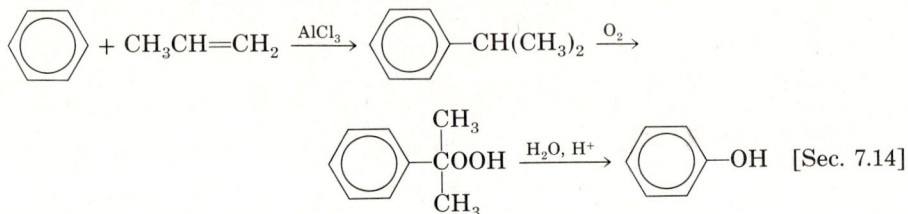

$$\bigcirc + \text{CH}_3\text{CH}{=}\text{CH}_2 \xrightarrow{\text{AlCl}_3} \bigcirc -\text{CH(CH}_3)_2 \xrightarrow{\text{O}_2}$$

$$\bigcirc -\underset{\underset{\text{CH}_3}{|}}{\overset{\overset{\text{CH}_3}{|}}{\text{C}}}\text{OOH} \xrightarrow{\text{H}_2\text{O, H}^+} \bigcirc -\text{OH} \quad \text{[Sec. 7.14]}$$

Most phenols are used industrially for the production of phenolic resins, which have a
variety of commercial uses. [Sec. 7.15]
The sulfur analogs of alcohols and phenols are called thiols or mercaptans.
[Sec. 7.16]

Key Terms

hydroxyl group, —OH
carbinol carbon
carbinol system
diol
triol
glycols
polyols
monohydroxy alcohols
hydrogen (H) bonding
phenol
phenoxide ion
alkoxides

phenoxide
oxidation
reduction
oxidizing agent
reducing agent
primary alcohols
aldehydes
secondary alcohols
ketones
tertiary alcohols
inorganic esters
alkyl halides

Lucas test
hydroboration-oxidation
anti-Markovnikov
denaturant
azeotropic solution
azeotrope
absolute ethanol
cumene process
phenolic resins
thiols
mercaptans
sulfhydryl group

Exercises

Structure, Nomenclature of Alcohols and Phenols. Classification of Alcohols
[Secs. 7.1, 7.2]
7.1 Write the structural formula for each compound.
    (a) *n*-Hexyl alcohol   **(b)** 2-Methyl-2-pentanol
    (c) 2,2-Dimethyl-1-butanol   **(d)** Benzyl alcohol
    (e) Grain alcohol   **(f)** Wood alcohol
    (g) Rubbing alcohol   **(h)** 1-Phenylethanol

(i)   2-Phenylethanol
(k)   1,5-Pentanediol
(m)   *trans*-3-Bromocyclohexanol
(o)   Ethylmethylphenylcarbinol

(j)   3-Chloro-5-phenyl-1-hexanol
(l)   Cyclopentanol
(n)   *cis*-1,3-Cyclobutanediol
(p)   3-Penten-2-ol

7.2   Give the IUPAC name of each compound.

(a) $CH_3CHOHCH_2CH_2CH_3$

(b) $CH_3CH_2\overset{\overset{\displaystyle CH_3}{|}}{\underset{\underset{\displaystyle OH}{|}}{C}}CH_2CH_2CH_3$

(c) $CH_3CH_2\overset{}{\underset{\underset{\displaystyle Cl}{|}}{C}HCH_2CH_2OH}$

(d) $HOCH_2CH_2CH_2CH_2OH$

(e) $(CH_3)_2CH\overset{}{\underset{\underset{\displaystyle OH}{|}}{C}HCH_3}$

(f) $CH_3\overset{\overset{\displaystyle Cl}{|}}{C}HCH_2\overset{}{\underset{\underset{\displaystyle CH_2CH_3}{|}}{C}HCH_2OH}$

(g)   (ring structure with CH_3 and —OH)

(h) $CH_2{=}\overset{}{\underset{\underset{\displaystyle CH_2CH_3}{|}}{C}CH_2CH_2OH}$

7.3   Classify the alcohols in Exercise 7.2 as 1°, 2°, or 3° alcohols.

7.4   There is something wrong with each of the following names. Write the structure and give the correct name for each compound.
(a)   3-Butanol
(b)   3-Hexen-5-ol
(c)   2,2-Dimethyl-4-hexanol
(d)   3-Methyl-4-pentanol
(e)   3,5-Pentanediol
(f)   *cis*-4-Bromocyclopentanol
(g)   *trans*-1,5-Cyclohexanediol
(h)   4-Cyclohexen-1-ol

7.5   Draw the formulas.
(a)   *o*-Bromophenol
(b)   Resorcinol
(c)   Picric acid
(d)   2,6-Dibromo-4-methylphenol
(e)   *m*-Phenylphenol
(f)   2,6-Di-*t*-butylphenol
(g)   1-Chloro-2-naphthol
(h)   Sodium *p*-nitrophenoxide

Boiling Point, Solubility, and Hydrogen Bonding   [Sec. 7.3]
7.6   Arrange each set of compounds in order of increasing boiling points. Give reasons for the order you choose.
(a)   Ethanol; propane; 1-pentanol.
(b)   Butane; 1,2,3-propanetriol; 1,2-propanediol.
(c)   1-Pentanol; 1-hexanol; 1,2-hexanediol.
7.7   Arrange each set of compounds in Exercise 7.6 in order of increasing solubility in water and give reasons for the order you choose.

Prediction of Acidity   [Sec. 7.4]
7.8   Arrange the compounds in each set in order of increasing acidity.
(a)   Ethanol; phenol; water; *p*-nitrophenol.
(b)   Cyclohexane; cyclohexanol; *p*-chlorophenol; phenol.

Recognition of Oxidation-Reduction Reactions   [Sec. 7.7]
7.9   Which transformation is an oxidation and which is a reduction?

(a) $CH_3NO_2 \longrightarrow CH_3NH_2$

(b) $CH_3\overset{\overset{\displaystyle O}{||}}{C}CH_3 \longrightarrow CH_3CH_2CH_3$

(c) $H_2C=O \longrightarrow HCOOH$

(d) ⬡—$CH_2OH \longrightarrow$ ⬡—$\overset{\displaystyle O}{\overset{\|}{C}}-H$

(e) $CH_3CN \longrightarrow CH_3CH_2NH_2$

(f) $2\ CH_3{-}SH \longrightarrow CH_3{-}S{-}S{-}CH_3$

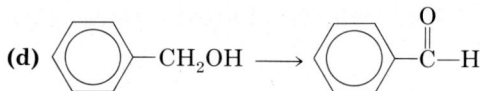

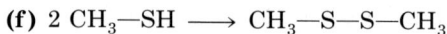

Preparations and Reactions of Alcohols and Phenols [Secs. 7.5–7.12]

7.10 Write the structures of the reactant and of the major product for each reaction. If no reaction occurs, state so.

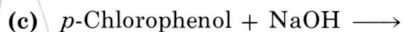

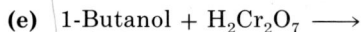

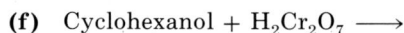

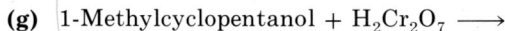

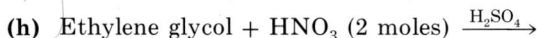

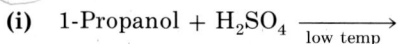

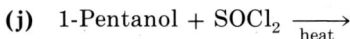

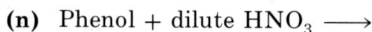

(a) 1-Propanol + Na $\longrightarrow$

(b) 1-Propanol + NaOH $\longrightarrow$

(c) *p*-Chlorophenol + NaOH $\longrightarrow$

(d) 1-Butanol + Cu $\xrightarrow{\text{heat}}$

(e) 1-Butanol + $H_2Cr_2O_7 \longrightarrow$

(f) Cyclohexanol + $H_2Cr_2O_7 \longrightarrow$

(g) 1-Methylcyclopentanol + $H_2Cr_2O_7 \longrightarrow$

(h) Ethylene glycol + HNO_3 (2 moles) $\xrightarrow{H_2SO_4}$

(i) 1-Propanol + $H_2SO_4 \xrightarrow{\text{low temp}}$

(j) 1-Pentanol + $SOCl_2 \xrightarrow{\text{heat}}$

(k) 1-Pentanol + HBr $\xrightarrow{\text{heat}}$

(l) 1-Pentanol + $PI_3 \xrightarrow{\text{heat}}$

(m) Cyclopentanol $\underset{}{\overset{H^{\cdot}}{\rightleftharpoons}}$

(n) Phenol + dilute $HNO_3 \longrightarrow$

(o) Phenol + $Br_2 \xrightarrow[25°C]{H_2O}$

7.11 Write equations, giving structures of reactants and major products, for the following preparations.

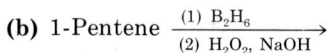

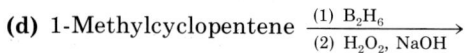

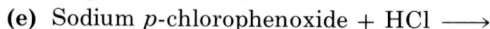

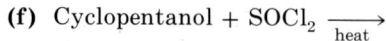

(a) 1-Pentene + $H_2O \xrightarrow{H^+}$

(b) 1-Pentene $\xrightarrow[\text{(2) } H_2O_2,\ NaOH]{\text{(1) } B_2H_6}$

(c) 1-Methylcyclopentene + $H_2O \xrightarrow{H^+}$

(d) 1-Methylcyclopentene $\xrightarrow[\text{(2) } H_2O_2,\ NaOH]{\text{(1) } B_2H_6}$

(e) Sodium *p*-chlorophenoxide + HCl $\longrightarrow$

(f) Cyclopentanol + $SOCl_2 \xrightarrow{\text{heat}}$

7.12 Write equations, showing necessary reagents and conditions, for the following preparations. (*Hint:* You may have to use reactions from earlier chapters.)
(a) Ethanol from ethylene
(b) 1-Butanol from 1-butene
(c) 2-Butanol from 1-butene
(d) Phenol from benzenesulfonic acid
(e) 1-Chloro-2-propanol from propene
(f) Allyl alcohol from 3-chloro-1-propanol
(g) Sodium ethoxide from ethanol
(h) Sodium propoxide ($CH_3CH_2CH_2O^-\ Na^+$) from propene
(i) Ethyl dihydrogen phosphate from ethanol

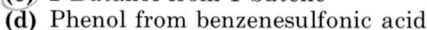

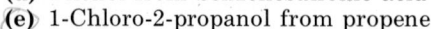

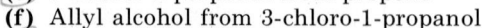

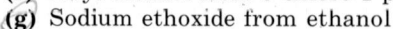

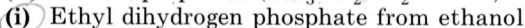

7.13 Arrange each set of alcohols in order of increasing reactivity toward Lucas reagent.

   (a) 1-Pentanol; 2-methyl-2-butanol; 2-pentanol.

   (b) 1-Methylcyclopentanol; cyclopentanol; cyclopentylcarbinol.

7.14 Identify compound **A**, $C_4H_{10}O$, from the following data.

   (1) $A + Br_2/CCl_4 \longrightarrow$ No reaction

   (2) $A + Na \longrightarrow$ bubbles

   (3) $A + ZnCl_2$, HCl $\longrightarrow$ immediate cloudiness

7.15 Identify compounds **A** and **B** from the following data.

   (1) $A + Na \longrightarrow$ bubbles

   (2) $A + HCl \xrightarrow{\text{heat}} B$ $(C_6H_{13}Cl)$

   (3) $B + KOH \xrightarrow[\text{heat}]{\text{alcohol}}$ 1-hexene (the only product)

7.16 Compound **A**, C_7H_8O, is a solid that is slightly soluble in water but readily soluble in aqueous sodium hydroxide. When **A** is treated with bromine water, the solid **B**, $C_7H_6Br_2O$, is formed. Draw possible structures for **A** and **B**.

7.17 What simple chemical tests would distinguish between the compounds in each of the following pairs.

   (a) 1-Butanol and 2-hexyne   **(b)** 1-Butanol and 1-hexyne

   (c) 1-Butanol and 2-butanol   **(d)** 1-Butanol and 1-butene

   (e) 2-Butanol and *t*-butyl alcohol   **(f)** *p*-Cresol and benzyl alcohol

Structures of Thiols and Disulfides; Boiling Points of Mercaptans [Sec. 7.16]

7.18 Write structures for these compounds.

   (a) Methyl mercaptan   **(b)** *t*-Butyl mercaptan

   (c) 1-Butanethiol   **(d)** 2-Methylcyclohexanethiol

   (e) Methyl disulfide   **(f)** Phenyl disulfide

7.19 In general, mercaptans have lower boiling points than the corresponding alcohols. Give a possible explanation for this fact.

Definitions of Terms

7.20 Define **(a)** oxidation, **(b)** reduction, **(c)** absolute ethyl alcohol, **(d)** Lucas test, **(e)** hydrogen bonding, **(f)** azeotrope, **(g)** 1,2-glycol, **(h)** phenoxide ion, **(i)** alkoxide ion.

8

Organic Halogen Compounds

Organic halogen compounds are rare in nature, although chloride ions abound in all living cells. The few organic chlorides found in nature are produced by microorganisms. Examples are the antibiotics aureomycin and chloromycetin and the fungicide griseofulvin.

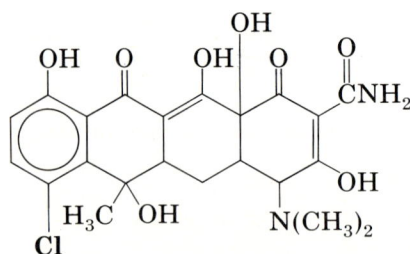

Aureomycin
(antibiotic)

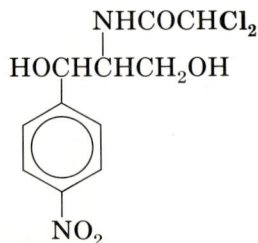

Chloromycetin
(antibiotic)

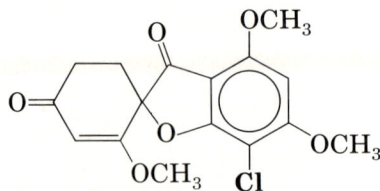

Griseofulvin
(fungicide)

212

Another organic halogen compound of natural origin is thyroxine, which contains iodine. Thyroxine is a hormone produced by the thyroid gland; it regulates the rates of many reactions in human metabolism.

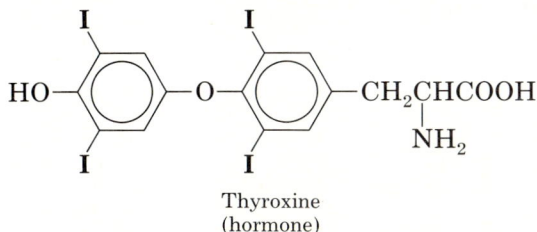

Thyroxine
(hormone)

A recently discovered additional natural source of organic halides is seawater red algae. It has been found that these marine organisms are capable of producing low-molecular-weight, volatile organic bromo compounds such as

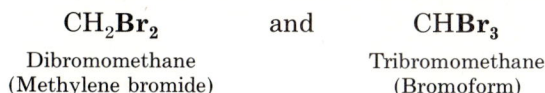

CH_2Br_2 and $CHBr_3$

Dibromomethane   Tribromomethane
(Methylene bromide)   (Bromoform)

There is strong evidence that chloro and iodo compounds can also be formed by certain algae. If confirmed, these findings are extremely interesting in light of current environmental concern over halocarbon-catalyzed depletion of ozone from the stratosphere (Sec. 8.4).

Although few organic halogen compounds are natural in origin, vast quantities are manufactured every year. The synthetic halogen compounds, particularly the polyhalogen compounds, are used as cleaning solvents, pesticides, anesthetics, aerosol propellants, refrigerants, polymers, and so on. The chemistry of organic halogen compounds is also important because these compounds are very useful intermediates for the synthesis of a wide range of other substances.

Classes and Names of Halogen Compounds *8.1*

Most of the material presented in this section has been introduced in previous chapters. Here we shall review, correlate, and amplify the topics.

One method of classifying organic halogen compounds is according to the halogen, that is, as fluoro, chloro, bromo, and iodo compounds. A more useful method—more useful because it organizes halides into groups whose members display similar chemical properties—is to classify them according to the hydrocarbon group.

1. Alkyl halides, R—X   (X may be F, Cl, Br, or I). Alkyl halides are subdivided into primary (1°), secondary (2°), or tertiary (3°), depending on the type of carbon to which the halogen is attached. Examples of the three types of halides and their names are

$$CH_3Cl \qquad CH_3CH_2CH_2Br \qquad \overset{\overset{\textstyle F}{|}}{CH_3CHCH_3}$$

	CH₃Cl	CH₃CH₂CH₂Br	CH₃CHCH₃
Common name:	Methyl chloride	n-Propyl bromide	Isopropyl fluoride
IUPAC name:	Chloromethane	1-Bromopropane	2-Fluoropropane
Class:	1°	1°	2°

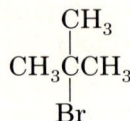

	Cyclohexyl iodide	*t*-Butyl bromide
Common name:	Cyclohexyl iodide	*t*-Butyl bromide
IUPAC name:	Iodocyclohexane	2-Bromo-2-methylpropane
Class:	2°	3°

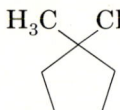

Common name:	Methylcyclopentyl chloride
IUPAC name:	1-Chloro-1-methylcyclopentane
Class:	3°

2. Allylic halides. The key structural feature is a halogen attached to a carbon next to a doubly bonded carbon, as in

$$CH_2{=}CHCH_2Br \qquad CH_3CH{=}CHCH_2Cl$$

Common name:	Allyl bromide	
IUPAC name:	3-Bromo-1-propene	1-Chloro-2-butene

IUPAC name:	3-Chlorocyclopentene

3. Benzylic halides, Ar—C̣—X. The key structural feature is a halogen *one carbon away* from an aromatic ring, as in

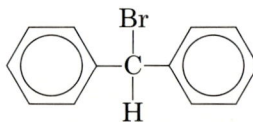

Benzyl chloride   Diphenylmethyl bromide

4. Aryl halides, Ar—X. The halogen is *directly* attached to an aromatic ring, as in

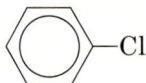  

Chlorobenzene   *p*-Bromotoluene   2-Chloronaphthalene

5. Vinylic halides. The key structural feature is a halogen attached *directly* to a doubly bonded carbon, as in

$$CH_2{=}CHCl$$
Vinyl chloride

1-Chlorocyclobutene

Polyhalogen compounds are classified in the same way. Although we will concentrate on the study of monohalogen compounds, you should memorize the common names of the di-, tri-, and tetrahalogen derivatives of methane.

CH_2X_2 = **methylene halides,** as in methylene chloride, CH_2Cl_2
CHX_3 = **haloforms,** as in chloroform, $CHCl_3$
CX_4 = **carbon tetrahalides,** as in carbon tetrachloride, CCl_4

Problem 8.1 Identify each organic halide as alkyl (1°, 2°, or 3°), allyl, benzyl, aryl, or vinyl.

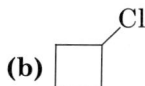

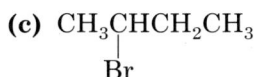

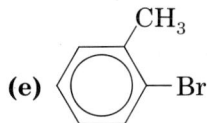

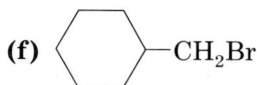

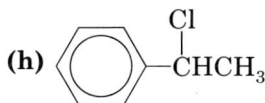

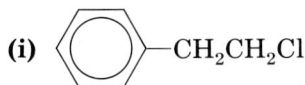

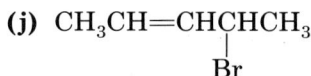

(a) $CH_3CH_2CH_2F$

(b)

(c) $CH_3CHCH_2CH_3$
$\quad\quad |$
$\quad\quad Br$

(d) —Cl

(e) with CH_3 and —Br

(f) —CH_2Br

(g) with CH_2CH_3 and Br

(h) —$\overset{\displaystyle Cl}{\underset{\displaystyle |}{C}}HCH_3$

(i) —CH_2CH_2Cl

(j) $CH_3CH{=}CHCHCH_3$
$\quad\quad\quad\quad\quad |$
$\quad\quad\quad\quad\quad Br$

(k) —$CH{=}CHCl$

(l) —Cl

Problem 8.2 Name the following compounds according to the IUPAC system.

(a) Methylene chloride
(b) Chloroform
(c) Bromoform
(d) Iodoform
(e) Carbon tetrachloride
(f) Isopropyl bromide
(g) Vinyl bromide
(h) Allyl chloride
(i) *t*-Butyl iodide
(j) Cyclobutyl chloride

Problem 8.3 Draw all eight structural isomers that have the molecular formula $C_5H_{11}Br$. Classify each as a primary, secondary, or tertiary bromide. Name each isomer according to the IUPAC system.

8.2 Physical Properties of Organic Halides

All organic halides are insoluble in water and soluble in common organic solvents (benzene, ether, etc.). The simple monofluoro and monochloro compounds are less dense than water, but the monobromo and monoiodo derivatives have densities greater than water. As the number of halogen atoms increases, so does the density. As the halogen content increases, the flammability decreases. Because they are good solvents for fats and oils and because they do not burn easily, polychloro compounds, such as trichloroethylene and tetrachloroethylene, are widely used as solvents for drycleaning.

Within a series of halides, the boiling points increase with increasing molecular weights. Therefore, the boiling points increase in the order $F < Cl < Br < I$. For example,

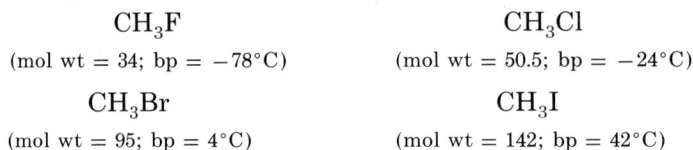

CH_3F	CH_3Cl
(mol wt = 34; bp = $-78°C$)	(mol wt = 50.5; bp = $-24°C$)
CH_3Br	CH_3I
(mol wt = 95; bp = $4°C$)	(mol wt = 142; bp = $42°C$)

Within a homologous series, the boiling points also increase regularly with molecular weights. For example,

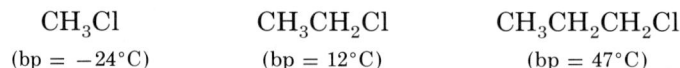

CH_3Cl	CH_3CH_2Cl	$CH_3CH_2CH_2Cl$
(bp = $-24°C$)	(bp = $12°C$)	(bp = $47°C$)

As expected, within a series of isomers, the straight-chain compound has the highest boiling point, and the most branched isomer the lowest boiling point.

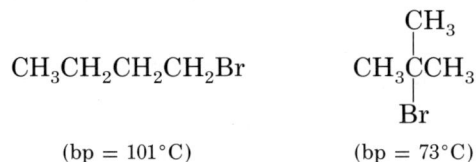

$$CH_3CH_2CH_2CH_2Br \qquad CH_3\overset{\overset{\displaystyle CH_3}{|}}{\underset{\underset{\displaystyle Br}{|}}{C}}CH_3$$

(bp = $101°C$)   (bp = $73°C$)

It should be evident that the solubility and boiling-point characteristics of organic halides are similar to those of alkanes—and for the same reason: in neither case is there possibility for hydrogen bonding.

Problem 8.4 Arrange each set of compounds in order of increasing boiling point.
(a) Iodobenzene; chlorobenzene; fluorobenzene; bromobenzene.
(b) Methyl bromide; bromoform; methyl chloride; methylene bromide.

8.3 Preparation of Halogen Compounds

A   Chloro, Bromo, and Iodo Compounds

The main methods for preparing organic chlorides, bromides, and iodides are summarized here. All these reactions have been discussed in previous chapters.

1. Direct halogenation of hydrocarbons
   (a) Halogenation of alkanes: alkyl halides   [Sec. 2.12]

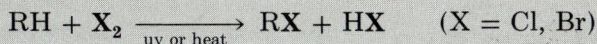

$$RH + X_2 \xrightarrow{\text{uv or heat}} RX + HX \qquad (X = Cl, Br)$$

   (b) Halogenation of alkenes: allyl halides   [Sec. 3.24]

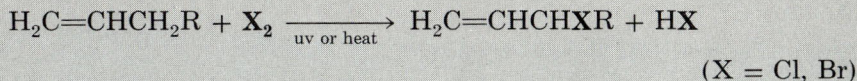

$$H_2C{=}CHCH_2R + X_2 \xrightarrow{\text{uv or heat}} H_2C{=}CHCHXR + HX$$

$$(X = Cl, Br)$$

   (c) Halogenation of alkyl benzenes: benzyl halides   [Sec. 5.10A]

$$ArCH_2R + X_2 \xrightarrow{\text{uv or heat}} ArCHXR + HX \qquad (X = Cl, Br)$$

   (d) Halogenation of aromatic ring: aryl halides   [Sec. 5.9]

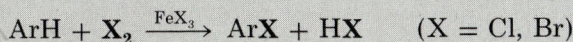

$$ArH + X_2 \xrightarrow{FeX_3} ArX + HX \qquad (X = Cl, Br)$$

2. Addition of HX to unsaturated hydrocarbons
   (a) Addition of HX to alkenes: alkyl halides   [Sec. 3.16]

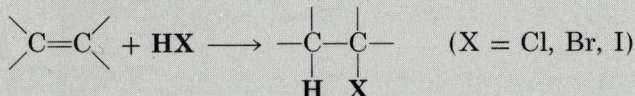

$$\text{C=C} + HX \longrightarrow -\overset{|}{\underset{H}{C}}-\overset{|}{\underset{X}{C}}- \qquad (X = Cl, Br, I)$$

   (b) Addition of HX to alkynes: vinyl halides   [Sec. 4.12C]

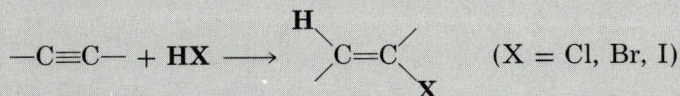

$$-C{\equiv}C- + HX \longrightarrow \overset{H}{\underset{X}{C=C}} \qquad (X = Cl, Br, I)$$

3. Conversion of alcohols: alkyl halides   [Sec. 7.10B]

$$ROH + HX \text{ (or } PX_3 \text{ or } SOX_2) \xrightarrow{\text{heat}} RX \qquad (X = Cl, Br, I)$$

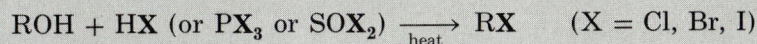

B   Halogen Exchange: A Way to Alkyl Fluorides

None of the methods just mentioned is suitable for preparing alkyl fluorides. Direct fluorination of hydrocarbons is unsatisfactory because elemental fluorine is too reactive and the reactions are hard to control.

The best way to prepare alkyl fluorides is by **halogen exchange.** An alkyl chloride or bromide is heated in the presence of a metallic fluoride such as AgF, Hg_2F_2, or SbF_3. For example,

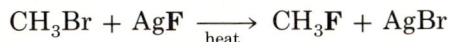

$$CH_3Br + AgF \xrightarrow{\text{heat}} CH_3F + AgBr$$

The halogen exchange reaction is used in the manufacture of chlorofluoro compounds, known as **Freons.** The simplest Freon, CCl_2F_2, is made as follows.

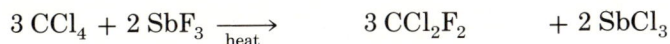

$$3\,CCl_4 + 2\,SbF_3 \xrightarrow{\text{heat}} 3\,CCl_2F_2 + 2\,SbCl_3$$

Dichlorodifluoromethane
(Freon-12)

The introduction of fluorine (and iodine) into an aromatic ring requires special methods (Sec. 14.11).

Problem 8.5 Draw the structure of the major monohalogen compound expected from each reaction.

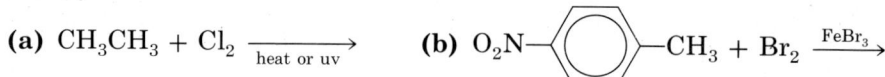

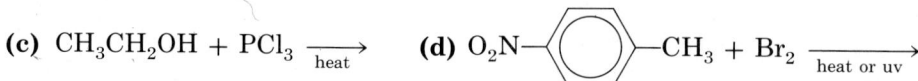

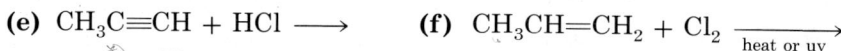

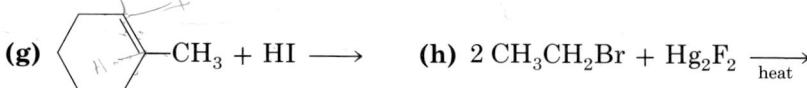

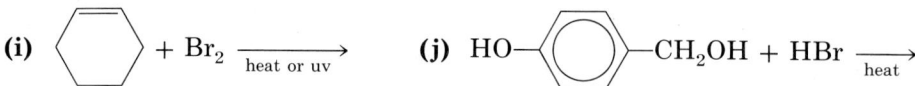

(a) $CH_3CH_3 + Cl_2 \xrightarrow{\text{heat or uv}}$

(b) $O_2N-\langle ring \rangle-CH_3 + Br_2 \xrightarrow{\text{FeBr}_3}$

(c) $CH_3CH_2OH + PCl_3 \xrightarrow{\text{heat}}$

(d) $O_2N-\langle ring \rangle-CH_3 + Br_2 \xrightarrow{\text{heat or uv}}$

(e) $CH_3C\equiv CH + HCl \longrightarrow$

(f) $CH_3CH=CH_2 + Cl_2 \xrightarrow{\text{heat or uv}}$

(g) $\langle ring \rangle-CH_3 + HI \longrightarrow$

(h) $2 CH_3CH_2Br + Hg_2F_2 \xrightarrow{\text{heat}}$

(i) $\langle ring \rangle + Br_2 \xrightarrow{\text{heat or uv}}$

(j) $HO-\langle ring \rangle-CH_2OH + HBr \xrightarrow{\text{heat}}$

8.4 Uses of Organic Halogen Compounds

Many organic halogen compounds play an important role in the chemical industry and in our daily lives. They include some of the most toxic man-made compounds* (pesticides) and some of the most inert and safe substances (Teflon). Almost invariably, they are polyhalogenated compounds.

The simplest polychloro compounds are liquids and are used as solvents. Methylene chloride, CH_2Cl_2, because of its low boiling point and low flammability, is an effective extraction solvent used in the pharmaceutical and food industries. The use of trichloroethylene and tetrachloroethylene as drycleaning agents has been mentioned previously (Sec. 8.2). Tetrachloroethylene serves also as a drug in the treatment of hookworm.

Some useful plastics derived from organic chloro compounds include polyvinyl chloride (PVC) and polyvinylidene chloride (Saran).

---

* 2,3,7,8-Tetrachlorodibenzo-p-dioxin (2,3,7,8-TCDD), a contaminant in the powerful herbicide Agent Orange, causes death at levels as low as 0.6 μg/kg of body weight. This level of toxicity is comparable to some of the most toxic natural substances, such as mold toxins.

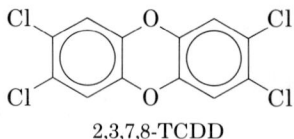

2,3,7,8-TCDD

$$\text{+CH}_2\text{—CHCl+}_n$$

Polyvinyl chloride
(PVC)

$$\text{+CH}_2\text{—CCl}_2\text{+}_n$$

Polyvinylidene chloride
(Saran)

Among the most controversial organic chlorine compounds are those used as pesticides. They include DDT, lindane, chlordane, dieldrin, and *p*-dichlorobenzene (see Table 8.1). DDT, for example, when first introduced in the 1940s was considered a miraculous substance. The use of DDT to delouse refugees after World War II is believed to have saved millions of people from dying of typhus. DDT can also be credited with the almost complete eradication of malaria-carrying mosquitoes and plague-infested flies. It has been estimated that over 75 million people have been protected from death and disease by DDT in the first 30 years of its use. Economically, billions of dollars worth of crops have been saved from destruction annually.

Despite its remarkable achievements in protecting health and agriculture, the indiscriminate use of DDT (and other polychlorinated compounds) has been severely criticized in countries, such as the United States, where environmental

Table 8.1 Some Common Chlorinated Pesticides

Structure	Name	Effective against
	DDT	Malaria mosquitoes and flies
	Lindane	Roaches and flies
	Chlordane	Mosquitoes and flies
	Dieldrin	Roaches, mosquitoes, and flies
	p-Dichlorobenzene (mothball)	Moths

concerns are of major importance. A problem with DDT is its exceptional stability, which causes it to persist in the environment for long periods of time. A second problem is that insects have developed a resistance to it. Insect resistance means that larger and more frequent doses of the insecticide must be applied to produce the same protective effect from year to year. The combination of DDT stability and accumulation in the fields kills not only harmful insects but also bees and other beneficial insects as well. When washed out of the soil into lakes and rivers, DDT can kill fish. Birds who eat fish can in turn accumulate DDT in their tissues. Thus DDT persists along the food chain and eventually reaches mammals, including humans.

Although there is no direct evidence to indicate that DDT has harmed humans, its devastating effects on the reproductive cycle of several species of birds and fish have been confirmed.

The American bald eagle, for example, was facing extinction because DDT affected its calcium metabolism to the extent that its eggshells were too thin to survive their incubation period. In the United States concern about the environment has been great enough to have DDT banned for most agricultural uses since 1972. In countries, such as India, where protection from famine and disease outweighs environmental considerations, DDT continues to be the most widely used insecticide.

In contrast to polychlorinated compounds, many of which are toxic, most polyfluoro molecules, even those that contain chlorine, are usually nontoxic. Halothane, $CF_3CHClBr$, is a nontoxic, nonflammable, widely used inhalation anesthetic. In recent years, scientists have discovered that certain fluorinated compounds are excellent transporters of oxygen and that they act as possible blood substitutes. In their experiments, the scientists submerged mice for extended periods in liquid fluorocarbons containing dissolved oxygen. The mice survived and showed no ill effects (Fig. 8.1).

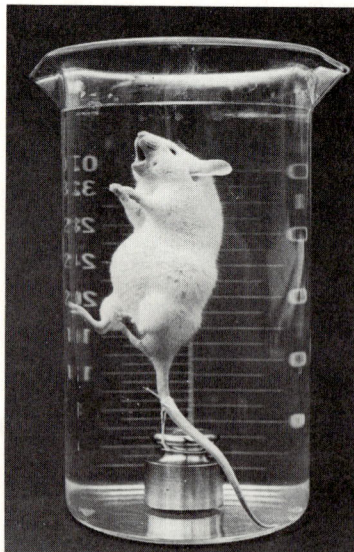

Figure 8.1 After an hour's submersion in fluorocarbon this mouse remained alive and well. To drain the liquid from his lungs he was held upside down. [Photo by William E. Schneider; courtesy of Leland C. Clark, Jr.]

The most widely used fluorine-containing molecules are the **Freons,** such as CCl_2F_2 and $CClF_3$. These chlorofluoro compounds are nontoxic, noncorrosive, low-boiling liquids and gases. Freons are used as refrigerants (hence the name) in household and commercial refrigerators and in air conditioners. Freons have also extensively been used as propellants in aerosol sprays. It appears that the Freons, since they are stable molecules, remain unchanged in the atmosphere for long periods of time until they reach the stratosphere. There, ultraviolet light decomposes the Freons, releasing chlorine atoms. The chlorine atoms catalyze the decomposition of the vital ozone layer surrounding the earth. The ozone layer acts as a screen against the ultraviolet rays of the sun. Without the screening effect of the ozone layer the chances of contracting skin cancer are greatly increased. As a result, the use of Freons in aerosol sprays has been banned.

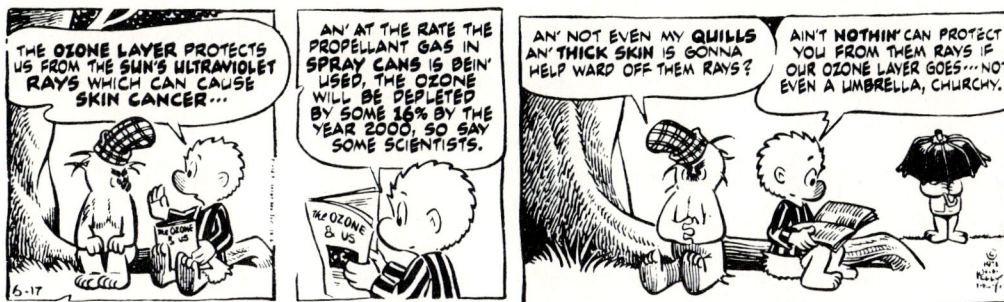

POGO by Walt Kelly. © 1975 by Walt Kelly. Courtesy of Selby Kelly.

Tetrafluoroethylene is polymerized to form a white, extremely unreactive polymer, Teflon, which is familiar as a nonstick coating in cooking utensils.

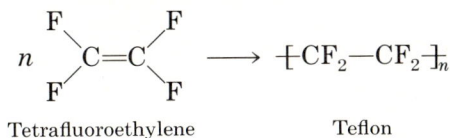

$$n \; {}^{F}_{F}\!\!>\!\!C\!\!=\!\!C\!\!<\!\!{}^{F}_{F} \longrightarrow -\!\!(CF_2\!\!-\!\!CF_2)\!\!-_n$$

Tetrafluoroethylene   Teflon

Reactions of Organic Halides **8.5**

The reactions of organic halides we shall discuss fall into three categories.

1. Those in which the halogen is replaced by some other atom or group **(nucleophilic substitution, or S_N, reactions).**
2. Those that involve the loss of HX from the halide **(elimination, or E, reactions).**
3. Those that involve reaction with certain metals to form **organometallic** compounds.

Under normal conditions, nucleophilic substitution and elimination reactions (categories 1 and 2) are limited to compounds in which the halogen is attached to an sp^3-hybridized carbon. These include alkyl, allyl, and benzyl halides. Aryl halides, in which halogen is bonded to an sp^2-hybridized carbon, are *very unreac-*

tive toward S_N reactions under normal conditions. Vinyl halides, likewise, are inert toward substitution reactions. Formation of organometallic compounds (category 3) takes place with both alkyl and aryl halides. We shall discuss each of these three types of reactions and show how S_N and E reactions are related to one another.

8.6 Nucleophilic Substitution Reactions

A   General Considerations

Before discussing specifically the nucleophilic substitution, S_N, reactions of alkyl halides, let us look at nucleophilic reactions in general. The overall process describing any S_N reaction is

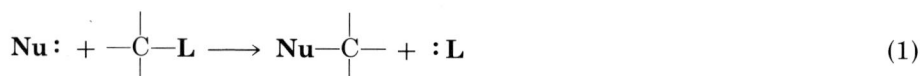

$$\text{Nu:} + -\overset{|}{\underset{|}{C}}-L \longrightarrow \text{Nu}-\overset{|}{\underset{|}{C}}- + \text{:L} \tag{1}$$

In this equation the incoming group, Nu :, is the nucleophile. The substance on which the nucleophile acts is the **substrate,** and the group displaced, : L, is the **leaving group.** In most S_N reactions the C—L bond of the substrate is polarized, with the negative end of the dipole close to the leaving group and the positive end near the carbon.

The nucleophile, Nu :, has an unshared electron pair available for bonding and is, in most cases, basic in character. Examples of nucleophiles are shown in Table 8.2.

The leaving group, : L, is also a nucleophile since it departs with an unshared pair of electrons. Therefore, if we wish an S_N reaction to go in the direction shown by equation (1) and to minimize the reverse reaction, the incoming nucleophile must be stronger than the departing one. Good leaving groups include H_2O and *anions* (the conjugate bases) of strong acids, such as Cl^-, Br^-, I^-, and HSO_4^-. Strong nucleophiles, but poor leaving groups, are HO^-, RO^-, and CN^-, which are conjugate bases of weak acids. The iodide ion is both a good nucleophilic reagent and a good leaving group.

Let us now apply these concepts to S_N reactions of alkyl halides. (Remember, however, that nucleophilic substitutions can take place with many other classes of compounds.)

B S_N Reactions of Alkyl Halides: Applications

The general equation describing nucleophilic substitution of alkyl halides is

$$\text{Nu:} + RX \longrightarrow \text{NuR} + X^-$$

The reaction (no matter what the mechanism) always involves breakage of the carbon–halogen bond. It is logical, therefore, to expect that the rate of substitution will be fastest for the compound with the weakest carbon–halogen bond and

**Table 8.2 Examples of
Common Nucleophiles**

HO:$^-$	Br:$^-$	:NH$_3$
RO:$^-$	Cl:$^-$	:NH$_2$R
H$_2$N:$^-$	:CN$^-$	:NHR$_2$
HS:$^-$	H$_2$Ö:	:NR$_3$
I:$^-$	RÖH	

slowest for the halide with the strongest carbon–halogen bond. This is indeed the case. The rates of S$_N$ reactions for a series of halides always follow the order

$$R\text{—I} > R\text{—Br} > R\text{—Cl} \gg R\text{—F}$$

C—X bond dissociation
energy (kcal/mole):   ∼55   ∼67   ∼80   ∼107

In fact, the C—F bond is so strong that organic fluorides do not undergo nucleophilic substitutions under ordinary conditions.

A partial list of nucleophilic reagents (Nu:) that can be used to attack an alkyl halide is given in Figure 8.2, together with the classes of compounds that can be formed. From this figure it should be obvious why organic halides are such useful intermediates for synthesizing other classes of organic compounds. The common sources of the nucleophiles are shown in parentheses.

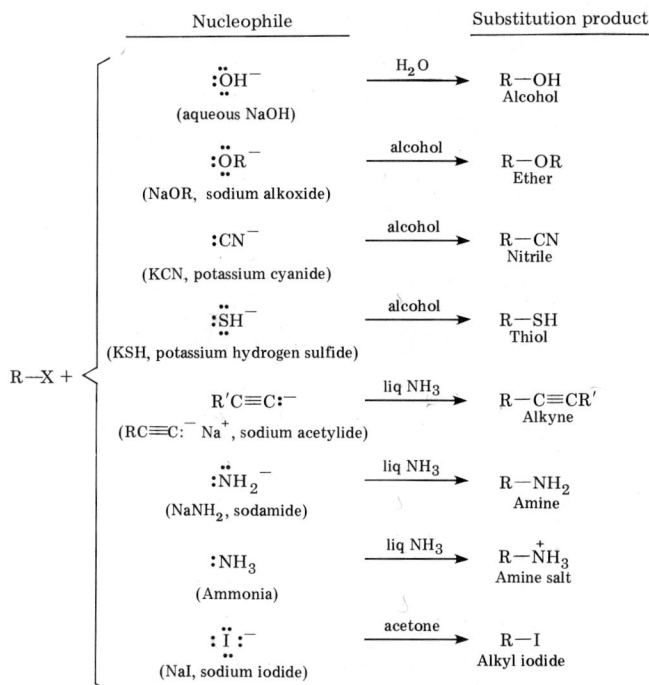

Figure 8.2 Some nucleophilic reactions of alkyl halides   R—X + Nu:→ R—Nu + X$^-$

Problem 8.6 Refer to Figure 8.2 and give the structure of the organic product in each of the following nucleophilic substitution reactions. Name the class of compound formed in each case.

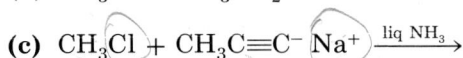

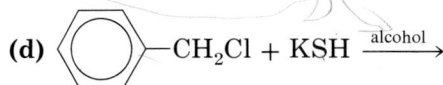

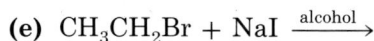

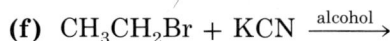

(a) $CH_3CH_2CH_2Br + NaOH \xrightarrow{H_2O}$

(b) $CH_3Br + CH_3CH_2O^- Na^+ \xrightarrow{alcohol}$

(c) $CH_3Cl + CH_3C{\equiv}C^- Na^+ \xrightarrow{liq\ NH_3}$

(d) ⟨⌬⟩$-CH_2Cl + KSH \xrightarrow{alcohol}$

(e) $CH_3CH_2Br + NaI \xrightarrow{alcohol}$

(f) $CH_3CH_2Br + KCN \xrightarrow{alcohol}$

C S_N Mechanism

Alkyl halides may undergo nucleophilic substitutions in two different ways: by a one-step mechanism and by a two-step mechanism. Which route is taken depends on the structure of the halide, the nature of the solvent, and the strength of the nucleophile, Nu:. In general, primary alkyl halides undergo nucleophilic substitutions by the one-step mechanism; the two-step mechanism is most common for tertiary halides. Secondary halides react by either mechanism. Let us consider the conversion of an alkyl halide to an alcohol by hydroxide,

$$R{-}X + {^-OH} \longrightarrow R{-}OH + X^-$$

to illustrate each mechanism.

1. One-Step Mechanism: S_N2 Reactions

The mechanism of nucleophilic substitution of hydroxide ions on primary alkyl halides has been studied extensively. Let us list some of the known facts about the reaction.

1. The reaction rate depends on the concentrations of both reactants, the alkyl halide and the nucleophile.
2. For a given concentration of alkyl halide, the rate of the reaction increases proportionally with the concentration of hydroxide ion. This means that if we double the concentration of hydroxide ion, but maintain the same concentration of alkyl halide, the S_N reaction proceeds twice as fast.
3. The nucleophile attacks the halogen-bearing carbon from the side opposite the leaving group, a so-called backside attack.
4. At the same time that the nucleophile attacks from the back side and the leaving group departs, the other bonds to the carbon invert (or "flip over"). This flipping of the bonds, known as the **Walden inversion** after its discoverer, is much like the inversion of an umbrella in a strong wind.

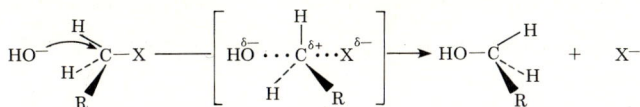

Figure 8.3 S_N2 reaction: nucleophilic attack by OH^- on a primary alkyl halide. The activated complex, shown in brackets, cannot be isolated.

A mechanism that conforms with these facts is shown in Figure 8.3. It is known as an **S_N2 reaction** (S = substitution; N = nucleophilic; 2 = bimolecular, indicating that there are two molecular species involved in the formation of the activated complex).

Many nucleophilic substitution reactions proceed according to the S_N2 mechanism. Under certain conditions the S_N2 mechanism described for the hydrolysis of primary alkyl halides operates also for secondary and tertiary alkyl halides. When this is the case, the rates of hydrolysis follow the order

$$1° \ RX > 2° \ RX > 3° \ RX$$

Rate in S_N2 reactions

Tertiary alkyl halides react slowest by the S_N2 mechanism because the three alkyl groups on the carbon that bears the halogen hinder the backside approach of the attacking nucleophile (Fig. 8.4). This is known as **steric hindrance.**

To summarize,

1. The S_N2 reaction is a single, concerted process.
2. The rate of an S_N2 reaction depends on the concentrations of both the alkyl halide and the nucleophile.
3. The reaction is fastest for primary halides and slowest for tertiary halides.
4. All S_N2 reactions proceed with complete inversion of configuration.

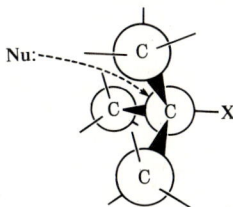

Figure 8.4 Steric hindrance. The presence of three alkyl groups on the carbon bonded to the leaving group X makes it difficult for the nucleophile to attack that carbon.

Problem 8.7 The reaction $CH_3OH + HBr \rightarrow CH_3Br + H_2O$ is an S_N2 reaction, in which the bromide ion (the nucleophile) simultaneously displaces water (the leaving group) from the protonated alcohol. Write the mechanism for this reaction.

Problem 8.8 Predict the order of reactivity of the following compounds in an S_N2 displacement by KCN: CH_3Cl (i); $C_6H_5CH_2Cl$ (ii); $(CH_3)_2CHCl$ (iii); $(CH_3)_3CCl$ (iv).

Problem 8.9 Write the structure of the product that would be obtained in an S_N2 reaction of *cis*-1-bromo-4-methylcyclohexane with KSH.

Problem 8.10 Write the structure and give the name (including R or S configuration) of the product that would be obtained in an S_N2 reaction of R-2-bromobutane with OH^-.

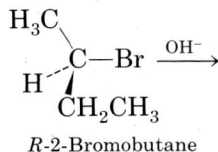

$$H_3C \atop H \diagdown C{-}Br \xrightarrow{OH^-} \atop CH_2CH_3$$

R-2-Bromobutane

2. Two-Step Mechanism: S_N1 Reaction

As it is usually conducted, the reaction of tertiary alkyl halides with aqueous base occurs at a faster rate than does the hydrolysis of primary alkyl halides. We find also that the rate of substitution is independent of the OH^- ion concentration and depends *only* on the concentration of the alkyl halide. A substitution reaction whose rate depends on the concentration of *one* molecular species is called an **S_N1 reaction** (S = substitution; N = nucleophilic; 1 = unimolecular).

In contrast to S_N2 reactions, which proceed with complete inversion of configuration, S_N1 reactions take place with both *inversion* and *retention* of configuration. A mechanism that accounts for these facts involves two steps.

Step 1. The tertiary alkyl halide ionizes to form a carbocation. This is the slow, *rate-determining step.*

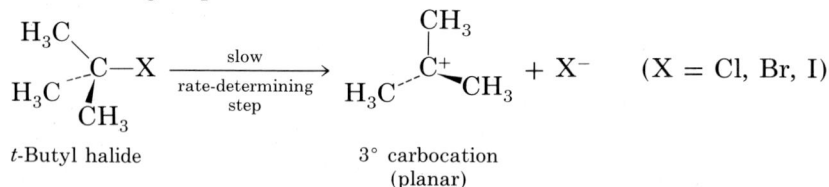

$$H_3C \atop H_3C \diagdown C{-}X \atop CH_3 \xrightarrow[\text{rate-determining step}]{\text{slow}} H_3C \diagdown \overset{CH_3}{\underset{}{C^+}} {-}CH_3 + X^- \qquad (X = Cl, Br, I)$$

t-Butyl halide   3° carbocation
                                     (planar)

Note that the rate-determining step does not involve the OH^- ion.

Step 2. A covalent bond forms between the carbocation and the OH^- nucleophile. This step occurs rapidly. Because the carbocation is planar (sp^2-hybridized carbon), the nucleophile has as much chance to bond with the positively charged carbon on the same side as the departed leaving group (giving *retention* of configuration) as on the back side (leading to *inversion*), as shown in Figure 8.5. If the starting alkyl halide were optically active, the product would be a racemic mixture* (Sec. 6.6).

Since the rate-determining step in an S_N1 reaction is the formation of a carbocation, the halide that can form the most stable carbocation reacts the

---

* The discussion is somewhat simplified. Often, in an S_N1 reaction, the product contains a larger amount of the inverted isomer, giving a *partially* racemized mixture. The reason why there is more of the inverted isomer is that the leaving group shields the front side of the molecules from the incoming nucleophile.

(a)

Retained configuration

(b) HO:⁻

Inverted configuration
(Walden inversion)

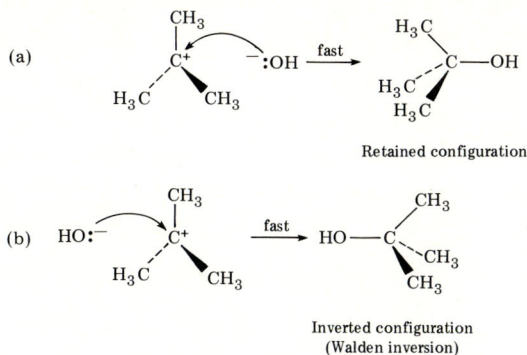

Figure 8.5 Nucleophilic attack **(a)** on same side as leaving group and **(b)** on side opposite leaving group.

fastest. The reactivity order in S_N1 reactions is therefore

Benzylic $>$ allylic $\approx 3° > 2° > 1°$ halide

To summarize;

1. The S_N1 reaction is a two-step process.
2. The rate-determining step is the formation of a carbocation.
3. The rates of S_N1 reactions follow the order of stability of carbocation: benzylic $>$ allylic $\approx 3° > 2° > 1°$ halide.
4. S_N1 reactions proceed with racemization.

Problem 8.11 The reaction $(CH_3)_3COH + HBr \rightarrow (CH_3)_3CBr + H_2O$ is an S_N1 reaction. Write the mechanism of this reaction.

Problem 8.12 Predict the order of reactivity of the following compounds in an S_N1 reaction: CH_3CH_2Br (i); $C_6H_5CH_2Br$ (ii); $(CH_3)_2CHBr$ (iii); $(CH_3)_3CBr$ (iv).

Problem 8.13 Write the structures of the products that would be obtained in an S_N1 reaction of *cis*-1-bromo-4-methylcyclohexane with aqueous sodium hydroxide.

Problem 8.14 Write the structures and give the names (including stereochemical configurations, R or S) of the products obtained in an S_N1 reaction of R-2-bromobutane with dilute aqueous hydroxide.

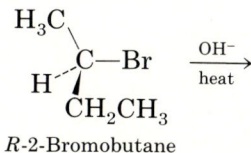

R-2-Bromobutane

8.7 Visual Tests Based on Nucleophilic Substitution Reactions

It is possible to distinguish among organic halogen compounds by any reaction that occurs by way of an intermediate carbocation, that is, an S_N1 **reaction.** One such visual reaction is that with silver nitrate in ethanol solution. Benzyl, allyl, and tertiary halides give an immediate precipitate of silver halide. A secondary halide reacts only slowly (within 5 min), primary halides do not react (the solution remains clear), and aryl or vinyl halides do not react.

The color of the silver halide precipitate is also helpful in distinguishing chlorides from bromides and iodides. Silver chloride is a white precipitate, and silver bromide and silver iodide are pale yellow.

A second visual chemical test used to distinguish among bromides is based on an S_N2 reaction: the displacement of bromide by iodide in acetone. To carry out this test, the bromide is dissolved in acetone containing sodium iodide.

$$R-Br + NaI \xrightarrow{\text{acetone}} R-I + NaBr\downarrow$$

Primary bromides react immediately to give an alkyl iodide and a precipitate of sodium bromide. Secondary bromides react only slowly; tertiary, aryl, and vinyl bromides do not react (the solution remains clear).

To make certain that you understand how to apply the visual tests just described, work out the following problem.

Problem 8.15 Compound **A,** a halide with the molecular formula C_4H_9X, gave the following results.

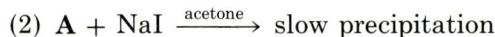

(1) $A + AgNO_3 \xrightarrow{\text{ethanol}}$ yellow precipitate within 5 minutes

(2) $A + NaI \xrightarrow{\text{acetone}}$ slow precipitation

Give the structure of **A** and explain the significance of each test.

8.8 Elimination Reactions

A General Consideration

In Section 3.10 you learned that alkyl halides can lose H and X from adjacent carbons to form alkenes by an elimination process.

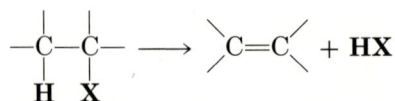

$$-\overset{|}{\underset{\underset{H}{|}}{C}}-\overset{|}{\underset{\underset{X}{|}}{C}}- \longrightarrow \overset{\diagdown}{\diagup}C=C\overset{\diagup}{\diagdown} + HX$$

In this section we will consider the mechanisms of elimination of alkyl halides. Like substitutions, eliminations are of two types: the **E1** (elimination unimolecular) **reaction** and the **E2** (elimination bimolecular) **reaction.** Generally, the E1 process accompanies the S_N1 reaction, and the E_2 reaction accompanies the S_N2 reaction.

B   E1 Mechanism

The E1, like the S_N1 mechanism, involves the formation of a carbocation in the rate-determining step. In S_N1 the carbocation forms a bond with the nucleophile to give a substitution product. In E1, the formation of the carbocation is followed by removal of a proton from an adjacent carbon by a base (B :) in a fast step. The overall process is

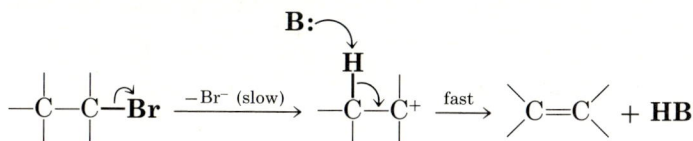

$$-\overset{|}{\underset{|}{C}}-\overset{|}{\underset{|}{C}}-\overset{\frown}{\mathbf{Br}} \xrightarrow[\text{}]{-Br^- \text{ (slow)}} \quad \overset{\mathbf{B:}\frown}{\underset{-\overset{|}{\underset{|}{C}}-\overset{\mathbf{H}}{\underset{|}{C}}{}^+}{}} \xrightarrow{\text{fast}} \quad \overset{}{\underset{}{C}}{=}\overset{}{\underset{}{C}} + \mathbf{HB}$$

Since the rate-determining step is the formation of the carbocation, it follows that the order of reactivity of E1 reactions is

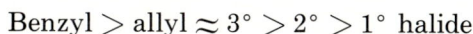

Benzyl > allyl $\approx$ 3° > 2° > 1° halide

This is the same as the order of reactivities of S_N1 reactions.

The product of an E1 reaction is an alkene. If more than one alkene can be formed, then the major product is the alkene with the most alkyl groups on C=C (Saytzeff's rule, Sec. 3.7). For example,

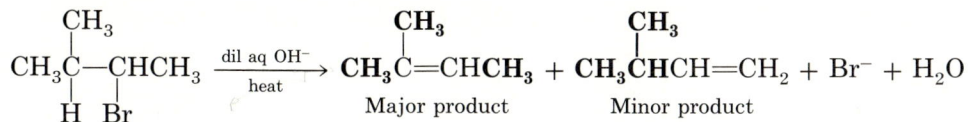

$$\underset{\underset{\mathbf{H}\ \mathbf{Br}}{\overset{\mathbf{CH_3}}{|}}}{CH_3\overset{|}{\underset{|}{C}}-CHCH_3} \xrightarrow[\text{heat}]{\text{dil aq } OH^-} \underset{\text{Major product}}{\overset{\mathbf{CH_3}}{|}{CH_3\overset{|}{C}}=CHCH_3} + \underset{\text{Minor product}}{\overset{\mathbf{CH_3}}{|}{CH_3\overset{|}{C}HCH}{=}CH_2} + Br^- + H_2O$$

C   E2 Mechanism

E2 and S_N2 reactions compete with each other and proceed in related pathways. Both reactions occur in a single step. In eliminations, however, the attack by the nucleophile (acting as a base, B :) is on the hydrogen atom attached to the carbon next to the one bearing the halogen (the β hydrogen). The E2 mechanism involves simultaneous bond breaking and bond formation and *occurs most readily when the β hydrogen and the halogen atom are trans to one another.*

The single-step E2 elimination

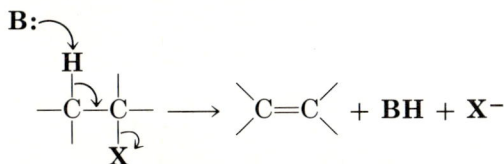

$$\overset{\mathbf{B:}\frown}{\underset{\underset{\mathbf{X}}{|}}{-\overset{\mathbf{H}}{\underset{|}{C}}\overset{}{\underset{|}{C}}-}} \longrightarrow \overset{}{\underset{}{C}}{=}\overset{}{\underset{}{C}} + \mathbf{BH} + \mathbf{X}^-$$

The order of reactivity of E2 elimination is

3° RX > 2° RX > 1° RX

which is the reverse of the order for S_N2 reactions. If more than one alkene can be formed, a mixture of products usually results. The major alkene is the one with the most alkyl substituents on C=C (Saytzeff's rule, Sec. 3.7).

To summarize,

1. E1 reactions compete with S_N1 reactions.
2. E2 reactions compete with S_N2 reactions.
3. No matter which elimination mechanism operates, the rates of halide elimination always follow the order $3° > 2° > 1°$.
4. E1 and E2 eliminations follow Saytzeff's rule.

Problem 8.16 One of the products of the reaction of 2-bromobutane with hot, dilute, aqueous sodium hydroxide is 2-butene. Write the E1 mechanism for this reaction.

Problem 8.17 The dehydrohalogenation of *cis*-1-bromo-2-methylcyclohexane yields two alkenes, 1-methylcyclohexene and 3-methylcyclohexene.
(a) Which alkene is the major product?
(b) Write the E2 mechanism that accounts for the formation of the two alkenes.

D   Elimination Versus Substitution

Since eliminations and substitutions compete, it is important to adjust the reaction conditions to obtain the highest yields of desired product. In general, elimination is favored over substitution when strongly basic solutions and high temperatures are used. This is why, for example, if we wish to prepare propene from 1-chloropropane, the reaction conditions are hot, alcoholic KOH. If we wish to obtain 1-propanol (substitution product), we maximize the yield by treating the halide with dilute aqueous hydroxide at moderate temperature.

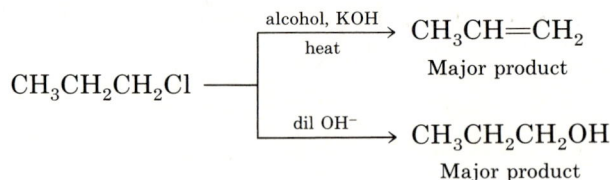

$$CH_3CH_2CH_2Cl \xrightarrow{\begin{array}{c}\text{alcohol, KOH}\\ \text{heat}\end{array}} CH_3CH=CH_2$$

Major product

$$\xrightarrow{\text{dil } OH^-} CH_3CH_2CH_2OH$$

Major product

8.9   Formation and Uses of Some Organometallic Compounds

Most organic chlorides, bromides, and iodides react with certain metals to give **organometallic compounds,** molecules with carbon–metal bonds. For organic synthesis, one of the most important classes of organometallic com-

pounds is the **Grignard reagents,** named after their discoverer, Victor Grignard, who received the Nobel prize (1912) for the discovery. Grignard reagents
are obtained by the reaction of alkyl or aryl halides with metallic magnesium in
dry ether as the solvent.

General reaction Na Water

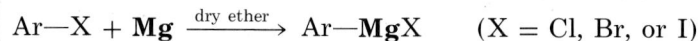

R—X + **Mg** $\xrightarrow{\text{dry ether}}$ R—**Mg**X   (X = Cl, Br, or I)

Ar—X + **Mg** $\xrightarrow{\text{dry ether}}$ Ar—**Mg**X   (X = Cl, Br, or I)

Specific example

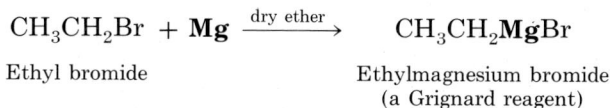

CH_3CH_2Br + **Mg** $\xrightarrow{\text{dry ether}}$ CH_3CH_2**Mg**Br

Ethyl bromide   Ethylmagnesium bromide
                                         (a Grignard reagent)

Although the exact structure of Grignard reagents is not known, the best way to
interpret their reactions is to consider the compounds as if the carbon–magnesium bond were ionic, that is, $R^-Mg^{2+}X^-$.

Grignard reagents react readily with any source of protons to give hydrocarbons. Even water or alcohols are sufficiently acidic to convert a Grignard reagent
to a hydrocarbon.

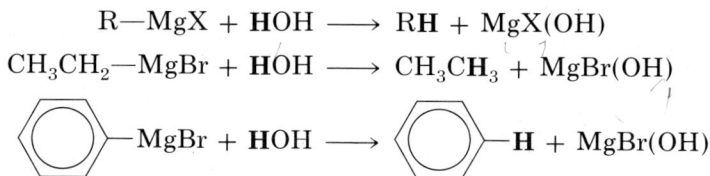

R—MgX + **HOH** $\longrightarrow$ R**H** + MgX(OH)

CH_3CH_2—MgBr + **HOH** $\longrightarrow$ CH_3CH_3 + MgBr(OH)

—MgBr + **HOH** $\longrightarrow$ —**H** + MgBr(OH)

It is therefore absolutely necessary to avoid even traces of moisture to form a
Grignard reagent.

Grignard reagents are among the most important compounds in organic
chemistry because they react with many functional groups. In later chapters we
discuss the reactions of Grignard reagents with aldehydes, ketones, and esters.

Organic halides form organometallic compounds when treated with metals
other than magnesium. Organometallic compounds that are made by the reaction of halides with group IA metals (Li, Na, or K), such as

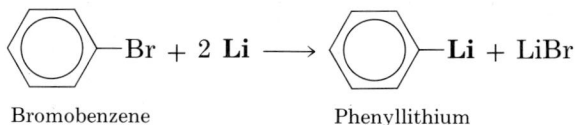

—Br + 2 **Li** $\longrightarrow$ —**Li** + LiBr

Bromobenzene   Phenyllithium

behave like Grignard reagents: they are extremely sensitive to moisture and
react with a great many functional groups.

Problem 8.18 Write the structures of the reactant and of the organic prod-
uct(s) in each of the following equations.

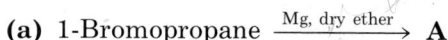

(a) 1-Bromopropane $\xrightarrow{\text{Mg, dry ether}}$ **A**

(b) 1-Bromonaphthalene $\xrightarrow{\text{2 Li}}$ **B**

(c) 2-Bromobutane $\xrightarrow{\text{Mg, dry ether}}$ **C** $\xrightarrow{\text{H}_2\text{O}}$ **D**

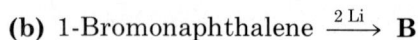

Summary of Concepts and Reactions

Organic halogen compounds can be classified by either one of two methods: (1) according to the kind of halogen present or (2) according to the hydrocarbon group attached to the halogen atom.   [Sec. 8.1]

All organic halides are insoluble in water and soluble in common organic solvents. Within a series of halides, the boiling points increase with increasing molecular weights.   [Sec. 8.2]

The main methods for preparing organic chlorides, bromides, and iodides are (1) direct halogenation of hydrocarbons, (2) addition of HX to unsaturated hydrocarbons, and (3) conversion of alcohols.   [Sec. 8.3A]

The best way to prepare alkyl fluorides is by halogen exchange.   [Sec. 8.3B]

Many organic halogen compounds play an important role in the chemical industry and in our daily lives.   [Sec. 8.4]

In nucleophilic substitution or S_N reactions a nucleophile, Nu:, attacks a compound, the substrate, and displaces a group, :L, called the leaving group. The overall process is

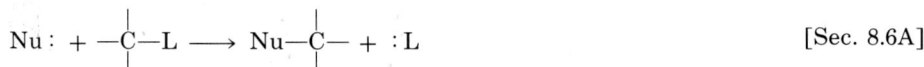

$$\text{Nu:} + -\overset{|}{\underset{|}{\text{C}}}-\text{L} \longrightarrow \text{Nu}-\overset{|}{\underset{|}{\text{C}}}- + \text{:L} \qquad\qquad \text{[Sec. 8.6A]}$$

Chlorides, bromides, and iodides are good leaving groups.   [Sec. 8.6A]

The rates of S_N reactions for a series of halides always follow the order RI > RBr > RCl $\gg$ RF.   [Sec. 8.6B]

Alkyl halides may undergo nucleophilic substitution by two different mechanisms: (1) one-step S_N2 reactions and (2) two-step S_N1 reactions.   [Sec. 8.6C]

The S_N2 reaction is a single, concerted process that involves complete inversion of the configuration (Walden inversion) and works best with primary alkyl halides, least well with tertiary alkyl halides.   [Sec. 8.6C]

The S_N1 reaction is a two-step process that involves a carbocation intermediate and works best with tertiary halides, least well with primary halides. The S_N1 reaction proceeds with equal amounts of inversion and retention of configuration, thus forming a racemic mixture.   [Sec. 8.6C]

Visual tests based on S_N1 or S_N2 reactions are used to distinguish among organic halogen compounds and among organic bromides, respectively.   [Sec. 8.7]

Alkyl halides can lose HX from adjacent carbons to form alkenes by an elimination, or E, reaction.

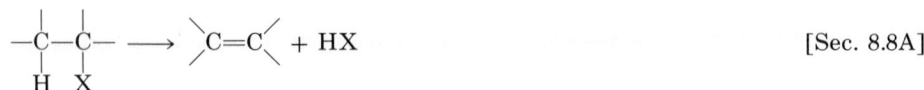

$$-\overset{|}{\underset{\underset{\text{H}}{|}}{\text{C}}}-\overset{|}{\underset{\underset{\text{X}}{|}}{\text{C}}}- \longrightarrow \overset{\diagdown}{}\text{C}=\text{C}\overset{\diagup}{} + \text{HX} \qquad\qquad \text{[Sec. 8.8A]}$$

If more than one alkene can be formed, the major product is the alkene with the most alkyl groups on C=C (Saytzeff's rule).   [Sec. 8.8B]

There are two types of elimination mechanisms: the E1 (elimination unimolecular) mechanism and the E2 (elimination bimolecular) mechanism. The former usually accompanies the S_N1 reaction and the latter the S_N2 reaction.   [Sec. 8.8A]

The E1, like the S_N1 mechanism, is a two-step process that involves the formation of a carbocation in the rate-determining step. The overall mechanism is

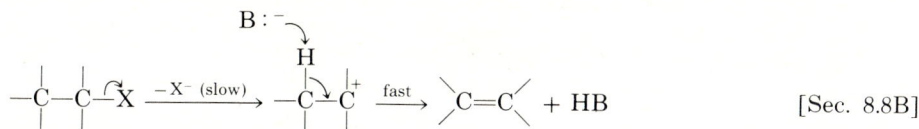

$$-\overset{|}{\underset{|}{C}}-\overset{|}{\underset{|}{C}}\overset{\frown}{X} \xrightarrow{-X^- \text{ (slow)}} \overset{B:^-}{\overset{\searrow}{\underset{H}{-\overset{|}{\underset{|}{C}}}}}\overset{|^+}{\overset{\frown}{C}} \xrightarrow{\text{fast}} \overset{\diagdown}{C}=\overset{\diagup}{C}\diagdown + HB \qquad [\text{Sec. 8.8B}]$$

The order of reactivity of E1 reactions is benzylic > allylic $\simeq$ $3°$ > $2°$ > $1°$ halide.
[Sec. 8.8B]

The E2, like the S_N2 mechanism, is a one-step process that involves simultaneous bond breaking and bond formation and occurs most readily when the β hydrogen and the halogen are *trans* to one another.

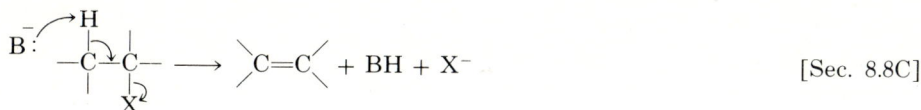

$$B:^-\overset{\curvearrowright H}{\overset{|}{\underset{|}{-C}}}\overset{|}{\underset{\underset{X}{\underset{\downarrow}{\curvearrowright}}}{C-}} \longrightarrow \overset{\diagdown}{C}=\overset{\diagup}{C}\diagdown + BH + X^- \qquad [\text{Sec. 8.8C}]$$

Since eliminations and substitutions compete, it is important to adjust the reaction conditions to obtain the highest yield of desired product. [Sec. 8.8D]

Most organic chlorides, bromides, and iodides react with certain metals to give organometallic compounds, molecules with carbon–metal bonds. [Sec. 8.9]

One of the most important classes of organometallic compounds is the Grignard reagents, RMgX. [Sec. 8.9]

Key Terms

alkyl halides, R—X	Freons	S_N2 reaction
allylic halides	nucleophilic, S_N,	steric hindrance
benzylic halides	substitution reactions	rate-determining step
aryl halides, Ar—X	elimination, E, reactions	S_N1 reaction
vinylic halides	nucleophile	E1 reaction
methylene halides	substrate	E2 reaction
haloforms	leaving group	organometallic compounds
carbon tetrahalides	Walden inversion	Grignard reagents
halogen exchange		

Exercises

Structure, Nomenclature, and Classification [Sec. 8.1]

8.1 Write the structures of the following compounds.
(a) Vinyl bromide
(b) Allyl chloride
(c) Bromoform
(d) Methylene iodide
(e) *p*-Nitrobenzyl bromide
(f) Triphenylmethyl chloride
(g) 3-Bromo-1-pentene
(h) Hexafluoroethane
(i) *t*-Butyl iodide
(j) *cis*-2-Methylcyclohexyl bromide

8.2 Give the correct IUPAC name for each of the following compounds.

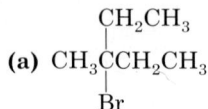

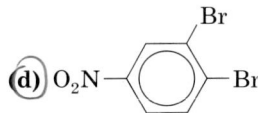

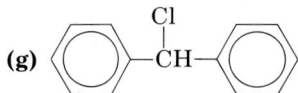

(a) $CH_3\underset{\underset{Br}{|}}{\overset{\overset{CH_2CH_3}{|}}{C}}CH_2CH_3$

(b)

(c) —I

(d) O_2N— —Br

(e) —Br

(f) $CH_3CH_2\underset{\underset{}{\overset{\overset{CH_3}{|}}{C}}}HCH_2F$

(g) —$\underset{\overset{|}{CH}}{\overset{Cl}{|}}$—

(h) Cl_3C—$CHCl_2$

(i) CH_3CH=$CHCH_2CH_2Br$

(j) CHI_3

8.3 Which compound(s) in Exercise 8.2 fit the following descriptions?
(a) 1° alkyl halide   (b) 2° alkyl halide   (c) 3° alkyl halide
(d) Allylic halide   (e) Benzylic halide   (f) Aryl halide
(g) Vinylic halide

Physical Properties [Sec. 8.2]
8.4 Arrange the compounds in each group in order of increasing boiling point.
(a) CBr_4 (i); $CHBr_3$ (ii); CH_3Br (iii); CH_2Br_2 (iv)

(b) (i);   (ii);   (iii);   (iv)

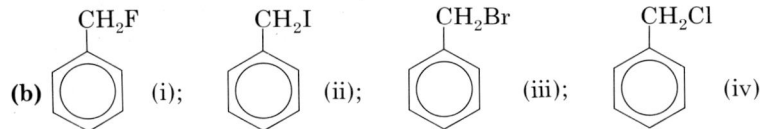

Preparations of Organic Halides [Sec. 8.3]
8.5 Give the structure of the major monohalogen compound expected in each of the following reactions.

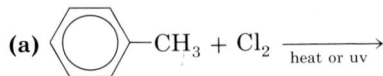

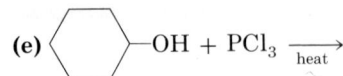

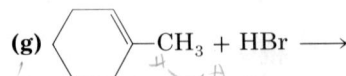

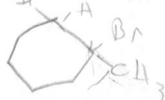

(a) —$CH_3 + Cl_2 \xrightarrow[\text{heat or uv}]{}$

(b) $CH_3\underset{\underset{CH_3}{|}}{C}$=$CHCH_3 + HI \longrightarrow$

(c) $CH_3CH_2CH_2OH + SOCl_2 \xrightarrow[\text{heat}]{}$

(d) $HOOC$—$+ Cl_2 \xrightarrow{\text{FeCl}_3}$

(e) —$OH + PCl_3 \xrightarrow[\text{heat}]{}$

(f) CH_3CH_2C≡$CH + HBr \longrightarrow$

(g) —$CH_3 + HBr \longrightarrow$

(h) $CH_3CH_2Cl + AgF \xrightarrow[\text{heat}]{}$

(i) HO$_3$S—⟨benzene ring with CH$_3$⟩—CH$_3$ + Br$_2$ $\xrightarrow{\text{FeBr}_3}$

(j) HO—⟨benzene ring⟩—CH$_2$CH$_2$OH + HCl $\xrightarrow{\text{heat}}$

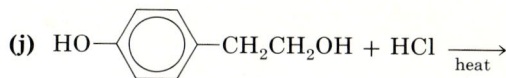

Reactions of Organic Halogen Compounds [Secs. 8.5, 8.9]

8.6 Write the structure of the organic product of each of the following nucleophilic substitution reactions.

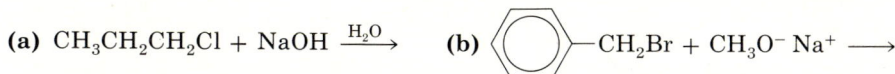

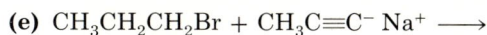

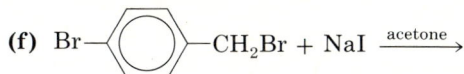

(a) CH$_3$CH$_2$CH$_2$Cl + NaOH $\xrightarrow{\text{H}_2\text{O}}$

(b) ⟨benzene ring⟩—CH$_2$Br + CH$_3$O$^-$ Na$^+$ $\longrightarrow$

(c) CH$_3$CH$_2$CH$_2$Br + KCN $\longrightarrow$

(d) ⟨cyclopentane ring⟩—CH$_2$Cl + KSH $\longrightarrow$

(e) CH$_3$CH$_2$CH$_2$Br + CH$_3$C≡C$^-$ Na$^+$ $\longrightarrow$

(f) Br—⟨benzene ring⟩—CH$_2$Br + NaI $\xrightarrow{\text{acetone}}$

8.7 Give the reagent and/or reaction conditions necessary to cause the transformations shown.

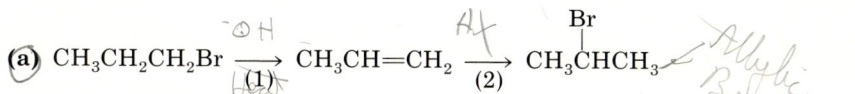

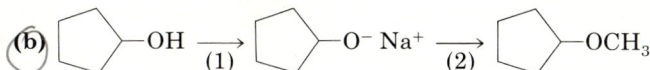

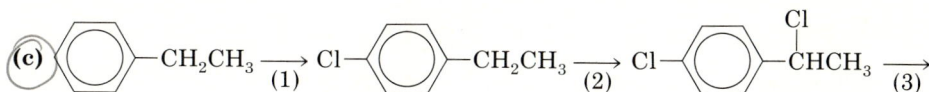

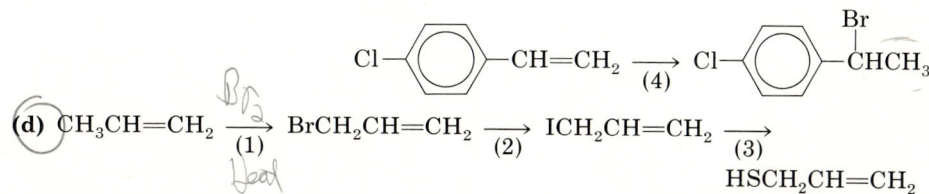

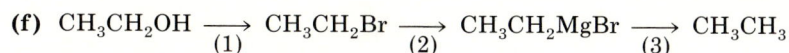

(a) CH$_3$CH$_2$CH$_2$Br $\xrightarrow[(1)]{}$ CH$_3$CH=CH$_2$ $\xrightarrow[(2)]{}$ CH$_3$CHCHCH$_3$ (with Br)

(b) ⟨cyclopentane ring⟩—OH $\xrightarrow[(1)]{}$ ⟨cyclopentane ring⟩—O$^-$ Na$^+$ $\xrightarrow[(2)]{}$ ⟨cyclopentane ring⟩—OCH$_3$

(c) ⟨benzene ring⟩—CH$_2$CH$_3$ $\xrightarrow[(1)]{}$ Cl—⟨benzene ring⟩—CH$_2$CH$_3$ $\xrightarrow[(2)]{}$ Cl—⟨benzene ring⟩—CHCH$_3$ (with Cl) $\xrightarrow[(3)]{}$

Cl—⟨benzene ring⟩—CH=CH$_2$ $\xrightarrow[(4)]{}$ Cl—⟨benzene ring⟩—CHCH$_3$ (with Br)

(d) CH$_3$CH=CH$_2$ $\xrightarrow[(1)]{}$ BrCH$_2$CH=CH$_2$ $\xrightarrow[(2)]{}$ ICH$_2$CH=CH$_2$ $\xrightarrow[(3)]{}$ HSCH$_2$CH=CH$_2$

(e) CH$_3$—⟨benzene ring⟩ $\xrightarrow[(1)]{}$ CH$_3$—⟨benzene ring⟩—Br $\xrightarrow[(2)]{}$ CH$_3$—⟨benzene ring⟩—MgBr

(f) CH$_3$CH$_2$OH $\xrightarrow[(1)]{}$ CH$_3$CH$_2$Br $\xrightarrow[(2)]{}$ CH$_3$CH$_2$MgBr $\xrightarrow[(3)]{}$ CH$_3$CH$_3$

Nucleophilic Substitution Reactions [Sec. 8.6]

8.8 Predict which compound in each pair would be more reactive in the S$_N$2 reaction and explain why.
  (a) 1-Chloropropane or 1-iodopropane
  (b) 1-Chlorobutane or 2-chlorobutane

(c) 1-Bromo-2-butene or 2-bromo-2-butene

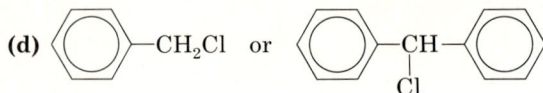

(d) —CH$_2$Cl   or   —CH—
           |
          Cl

8.9 1-Chloro-2,2-dimethylpropane is a primary alkyl chloride. Nevertheless, it shows very low reactivity in S$_N$2 reactions. Suggest an explanation.

8.10 Predict which compound in each pair would be more reactive in the S$_N$1 reaction and explain why.

(a) 1-Chlorobutane or 2-chlorobutane

(b) 1-Bromo-2-butene or 2-bromo-2-butene

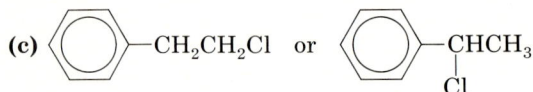

(c) —CH$_2$CH$_2$Cl   or   —CHCH$_3$
                                                        |
                                                       Cl

(d) (CH$_3$)$_3$CBr or (C$_6$H$_5$)$_3$CBr

8.11 When *cis*-1-bromo-4-methylcyclohexane is heated with aqueous NaOH, only *trans*-4-methylcyclohexanol is formed.

(a) Draw structures (including stereochemistry) for the starting material and the product.

(b) Did this displacement proceed by the S$_N$1 or by the S$_N$2 mechanism? Explain.

8.12 Write structures for the products (including stereochemical configurations) that would be obtained in the following S$_N$2 reactions.

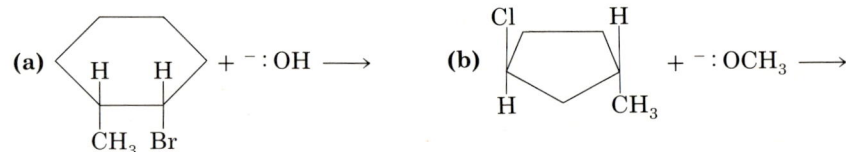

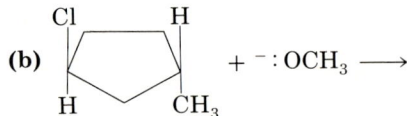

(a) + $^-$:OH $\longrightarrow$ **(b)** + $^-$:OCH$_3$ $\longrightarrow$

(c) *R*-2-Chlorobutane + $^-$SH $\longrightarrow$ **(d)** *S*-2-Bromobutane + $^-$CN $\longrightarrow$

8.13 Write structures for the products (including stereochemical configurations) that would be obtained in the S$_N$1 reactions of the starting compounds in Exercise 8.12 with dilute aqueous NaOH.

Elimination Reactions [Sec. 8.8]

8.14 Write structures for the products that would be obtained in E1 eliminations of the following compounds. If more than one product is expected, indicate which is the major one according to Saytzeff's rule.

(a) Bromocyclohexane   **(b)** *trans*-1-Chloro-2-methylcyclopentane
(c) 2-Bromo-2-methylbutane   **(d)** 2-Chloro-1-phenylpropane

8.15 Write structures for the products that would be obtained in E2 elimination of the compounds in Exercise 8.14.

S$_N$ Versus E Reactions [Sec. 8.8]

8.16 A student treated 2-bromo-3-methylbutane with KCN in alcohol and obtained three products. One had formula C$_6$H$_{11}$N, the other two had the formula C$_5$H$_{10}$. Write structures for the three compounds and explain how they were formed. (*Hint:* Consider competing S$_N$2 and E2 reactions.)

Structure Identification by Visual Tests [Sec. 8.7]

8.17 What simple visual chemical test(s) could be used to distinguish between each pair of compounds?

(a) *t*-Butyl chloride and *n*-butyl chloride

(b) Allyl bromide and 1-bromopropane
(c) *p*-Bromotoluene and benzyl bromide
(d) Bromobenzene and phenol
(e) 2-Bromoheptane and 2-heptanol

8.18 A halide with the formula C_3H_7X gave an immediate pale yellow precipitate when treated with silver nitrate in alcohol. Immediate precipitation took place also when the halide was treated with sodium iodide in acetone. Write the structure of the halide.

8.19 Compound **A,** C_8H_9Br, is optically active. When treated with alcoholic silver nitrate solution, **A** precipitated AgBr immediately, and on oxidation **A** gave benzoic acid, C_6H_5COOH. What is the structure of **A?**

8.20 A white precipitate formed slowly when silver nitrate in alcohol was added to compound **A,** with the formula $C_6H_{13}Cl$. When **A** was treated with hot alcoholic KOH, a mixture of two alkenes, **B** and **C,** with the formula C_6H_{12} was obtained. Compounds **B** and **C** could not be separated easily, so the mixture of the two was subjected to ozonolysis. Four compounds

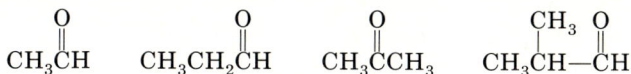

$$\underset{CH_3CH}{\overset{O}{\parallel}} \qquad \underset{CH_3CH_2CH}{\overset{O}{\parallel}} \qquad \underset{CH_3CCH_3}{\overset{O}{\parallel}} \qquad \underset{CH_3CH-CH}{\overset{CH_3\ \ O}{\parallel}}$$

were obtained. Write the structures of **A, B,** and **C.**

9

Ethers and Epoxides

In 1842 Crawford W. Long, a physician from Georgia, removed a tumor from the neck of a friend who had been rendered unconscious by inhaling ether. The patient felt no pain. This was the first successful use of surgical anesthesia. Even today, the word ether is associated in the layperson's mind with anesthesia. To the chemist, on the other hand, the anesthetic ether is only one member of a class of compounds named ethers. All ethers share a common structural feature: *they are molecules in which an oxygen atom is linked by single bonds to two organic groups.* Ether, the familiar anesthetic, has two ethyl groups attached to the oxygen. Other ethers may be used as herbicides, refrigerants, solvents, and flavoring agents, or in a variety of other applications, depending on the groupings that are bonded to the oxygen. Chemically, ethers are relatively unreactive substances, and for this reason they are used as solvents in many reactions.

9.1 Structure and Nomenclature of Ethers

Any compound with the general formula R—O—R, Ar—O—R, or Ar—O—Ar is called an ether. In these formulas, R represents any alkyl group, Ar any aryl group. We may think of ethers as water molecules in which each hydrogen has been replaced by an organic group. In fact, the geometry of simple ethers is similar to that of water. The H—O—H bond angle in water is about 105°; the R—O—R bond angle in methyl ether (R = CH_3) is 110°.

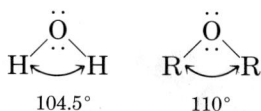

$\overset{\cdot\cdot}{\underset{}{O}}$
H⌣H   R⌣R
104.5°   110°

When the organic groups attached to the oxygen are identical, the ether is classified as a **symmetrical ether;** if the two groups are different, the ether is said to be **unsymmetrical.** Unsymmetrical ethers are also referred to as **mixed ethers.**

Simple ethers are given common names. Usually, the organic groups are named first in alphabetical order; then the word *ether* is added. For symmetrical ethers, only one group need be named. Table 9.1 shows the names and structures of some symmetrical and unsymmetrical ethers.

Methyl phenyl ether has the special name *anisole* (oil of anis). Several fragrant ethers that occur in nature contain the anisole structure as part of the total molecule.

More complicated ethers are named according to the IUPAC system. The ethers are considered as **alkoxy** (RO—) derivatives of a parent compound. As usual, the position of the alkoxy group is indicated by a number along the chain of the parent compound. Specific alkoxy groups (*methoxy* for CH_3O-, *ethoxy* for CH_3CH_2O-, etc.) are named in the same manner as the sodium salts of alcohols (Sec. 7.6). The names of a few ethers according to the IUPAC system are

$CH_3CH_2CH_2\underset{\underset{OCH_3}{|}}{C}HCH_3$ $CH_3CH_2CH_2\underset{\underset{OCH_2CH_3}{|}}{C}HCH_2CH_3$ $H_3CO-CH_2CH_2OH$

2-Methoxypentane   3-Ethoxyhexane   2-Methoxyethanol

Problem 9.1   Give a correct name for each of the following compounds.

(a) $CH_3-O-CH_2CH_2CH_3$ **(b)** 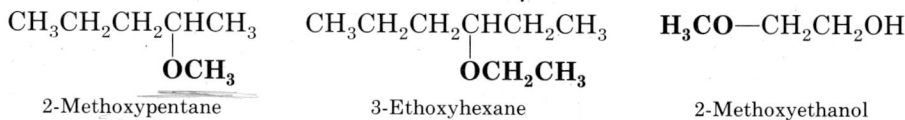—$CH_2-O-CH_2CH_3$

(c) $CH_3\underset{\underset{OCH_3}{|}}{C}HCH_2CH_3$ **(d)** $CH_2{=}CHCH_2-O-CH_2CH{=}CH_2$

(e) $CH_3CH_2\underset{\underset{OCH_2CH_3}{|}}{C}HCH{=}CH_2$ **(f)** CH_3-O-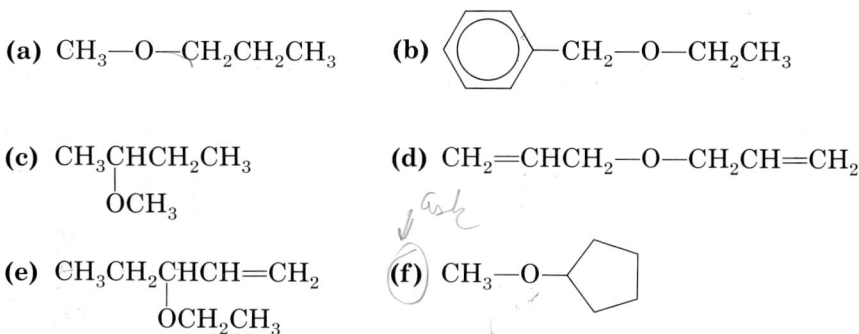

Problem 9.2   Draw the structure corresponding to each of the following compounds.
(a) Isopropyl ether   **(b)** 3-Ethoxyheptane
(c) Benzyl ether   **(d)** *p*-Chloroanisole
(e) 4-Methoxy-2-hexene   **(f)** 2-Chloro-4-ethoxyphenol

Table 9.1   Nomenclature and Structure of Some Simple and Mixed Ethers

Symmetrical ethers		Unsymmetrical ethers	
Name	**Structure**	**Name**	**Structure**
Methyl ether	CH_3—O—CH_3	Ethyl methyl ether	CH_3—O—CH_2CH_3
Ethyl ether	CH_3CH_2—O—CH_2CH_3	Ethyl-n-propyl ether	CH_3CH_2—O—$CH_2CH_2CH_3$
Vinyl ether	CH_2=CH—O—CH=CH_2	t-Butyl methyl ether	$(CH_3)_3$C—O—CH_3
Phenyl ether	⬡—O—⬡	Methyl phenyl ether (anisole)	⬡—O—CH_3

Problem 9.3   Which ether(s) in Problems 9.1 and 9.2 is (are) **(a)** symmetrical and **(b)** unsymmetrical?

9.2 Physical Properties of Ethers

The boiling points and solubility characteristics (Table 9.2) of ethers may be understood from their structures. Ethers, unlike water or alcohols, have no hydrogens directly attached to the oxygen atom. As a result there is no hydrogen bonding between one ether molecule and another, as there is between alcohol molecules. We can therefore predict that the boiling points of ethers will be considerably lower than those of alcohols, and about the same as those of alkanes, of comparable molecular weights.

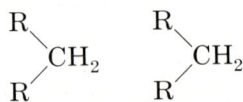

Alkanes: No hydrogen bonding between molecules; low boiling points

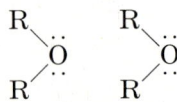

Ethers: No hydrogen bonding between molecules; low boiling points

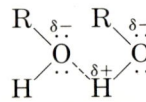

Alcohols: Hydrogen bonding between molecules; high boiling points

With regard to solubility in water, we may expect ethers to behave much like alcohols of similar molecular weights. This is because the oxygen atom in ethers, as in alcohols, is capable of hydrogen bonding with water molecules. Alkanes, of course, are insoluble because no hydrogen bonding with water is possible.

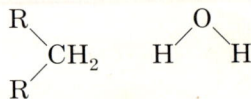

Alkanes: No hydrogen bonding with water; insoluble

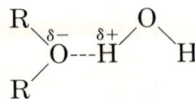

Ethers: Hydrogen bonding with water; soluble

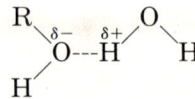

Alcohols: Hydrogen bonding with water; soluble

Table 9.2 A Comparison of Boiling Points and Solubilities of Alkanes, Ethers, and Alcohols

Structure	Name	Mol wt	Bp (°C)	Solubility in H$_2$O at 20°C
CH$_3$CH$_2$CH$_3$	propane	44	−42	insoluble
CH$_3$OCH$_3$	methyl ether	46	−24	soluble
CH$_3$CH$_2$OH	ethanol	46	78	soluble
CH$_3$CH$_2$CH$_2$CH$_3$	n-butane	58	−0.5	insoluble
CH$_3$CH$_2$OCH$_3$	ethyl methyl ether	60	8	soluble
CH$_3$CH$_2$CH$_2$OH	1-propanol	60	97	soluble
CH$_3$(CH$_2$)$_3$CH$_3$	n-pentane	72	35	insoluble
CH$_3$CH$_2$OCH$_2$CH$_3$	ethyl ether	74	36	7.5 g/100 g
CH$_3$(CH$_2$)$_2$CH$_2$OH	1-butanol	74	118	7.9 g/100 g
CH$_3$(CH$_2$)$_5$CH$_3$	n-heptane	100	98	insoluble
CH$_3$(CH$_2$)$_2$O(CH$_2$)$_2$CH$_3$	n-propyl ether	102	91	0.2 g/100 g
CH$_3$(CH$_2$)$_4$CH$_2$OH	1-hexanol	102	157	0.6 g/100 g

Hazards of Ethers *9.3*

We shall see that ethers are quite stable chemically. This is true as long as they are handled with the proper precautions. When exposed to air, most liquid aliphatic ethers (R—O—R) react slowly to form **peroxides.** These peroxides are unstable and, even in small concentrations, can cause violent explosions. It is imperative, therefore, to ascertain the absence of peroxides before using an ether. Peroxides can be detected by shaking the ether with an aqueous solution of iron(II) ammonium sulfate plus potassium thiocyanate. A blood-red color is produced if peroxides are present. Peroxides can be removed from ethers by washing with iron(II) ammonium sulfate.

The formation of peroxides is one hazard encountered in leaving open a container of ether. Another danger comes from the volatility of the ether itself. Certain air–ether vapor mixtures detonate violently. A spark can set off the reaction of ethyl ether. It is therefore important that ethyl ether be used only in a well-ventilated laboratory or operating room. Obviously, open flames are forbidden in the vicinity of ethers.

Preparation of Ethers *9.4*

There are two general methods for synthesizing ethers. One, the dehydration of alcohols, is used commercially and in the laboratory to make certain symmetrical ethers. The second, the Williamson synthesis, is a general laboratory method used to prepare all kinds of ethers, symmetrical and unsymmetrical. Both methods depend ultimately on alcohols as starting materials, and both involve nucleophilic substitution (S$_N$) reactions.

The two reactions are illustrated here in a general way and are discussed in greater detail in the next two sections.

---

1. Dehydration of alcohols

$$2\ R{-}OH \xrightarrow[\text{heat}]{H^+} R{-}O{-}R + H_2O \qquad\qquad \text{[Sec. 9.5]}$$

2. The Williamson synthesis

   a. $R{-}O^-\ Na^+ + R'{-}X \longrightarrow R{-}O{-}R' + NaX$

   b. $Ar{-}O^-\ Na^+ + R'{-}X \longrightarrow Ar{-}O{-}R' + NaX \qquad \text{[Sec. 9.6]}$

9.5   Ethers by Dehydration of Alcohols

The *dehydration of alcohols* takes place in the presence of acid catalysts (H_2SO_4, H_3PO_4) under controlled temperatures. The general reaction for ether formation is

$$R{-}\mathbf{OH} + \mathbf{H}{-}OR \xrightarrow[\text{heat}]{H^+} R{-}O{-}R + \mathbf{H_2O}$$

A   Mechanism

The formation of ethers by dehydration of alcohols proceeds by an S_N mechanism. With primary alcohols the reaction takes the S_N2 route; with secondary and tertiary alcohols, the S_N1 mechanism prevails. The formation of ethyl ether by the S_N2 mechanism proceeds according to the following steps.

Step 1. Protonation of the alcohol. The alcohol acting as a base, is rapidly protonated by sulfuric acid (H^+) to give a protonated alcohol.

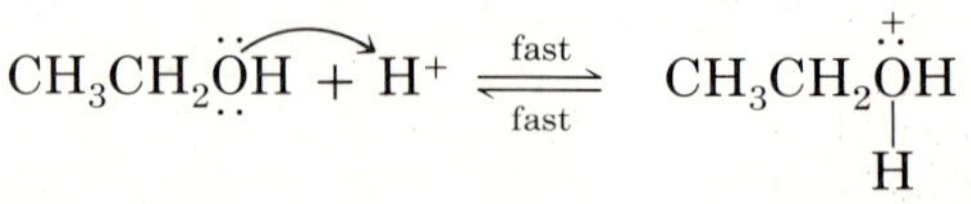

$$CH_3CH_2\overset{\cdot\cdot}{\underset{\cdot\cdot}{O}}H + H^+ \underset{\text{fast}}{\overset{\text{fast}}{\rightleftharpoons}} \quad CH_3CH_2\overset{+}{\underset{\underset{H}{|}}{\overset{\cdot\cdot}{O}}}H$$

      Ethyl alcohol   Protonated alcohol

Step 2. Nucleophilic attack on protonated alcohol (S_N2 reaction). The protonated alcohol is attacked from the backside by an unprotonated alcohol acting as a nucleophile. At the same time the leaving group (H_2O) departs. The products are water and a protonated ether.

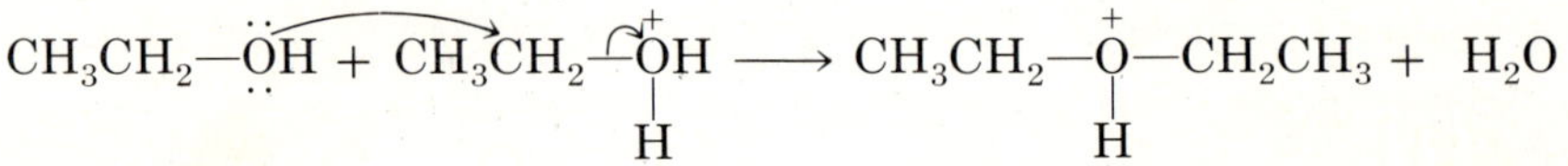

$$CH_3CH_2{-}\overset{\cdot\cdot}{\underset{\cdot\cdot}{O}}H + CH_3CH_2{-}\overset{+}{O}H \longrightarrow CH_3CH_2{-}\overset{+}{\underset{\underset{H}{|}}{O}}{-}CH_2CH_3 + H_2O$$

      Ethyl alcohol   Protonated ether   Leaving
      (nucleophile)   group

Step 3. Regeneration of the sulfuric acid catalyst. The protonated ether from step 2 loses its proton, giving the ether and regenerating the catalyst. The catalyst, H^+, can be reused in step 1.

$$CH_3CH_2-\overset{+}{\underset{|}{O}}-CH_2CH_3 \rightleftharpoons CH_3CH_2-\ddot{O}-CH_2CH_3 + H^+$$
$$H$$

Ethyl ether   Catalyst
regenerated

Problem 9.4 Illustrate the formation of ethers from *t*-butyl alcohol by an S_N1 mechanism. (*Hint:* See Sec. 8.6C.)

B Scope and Limitations

Formation of ethers by dehydration of alcohols involves the loss of water between two alcohol molecules, an intermolecular process. We should not forget, however, that heating an alcohol with acid may lead also to dehydration to an alkene, an intramolecular process. Thus the reaction conditions should be controlled if the desired product is to be obtained. When ethyl alcohol is dehydrated by sulfuric acid at 180°C, the dominant product is ethylene.

$$\underset{\text{Ethyl alcohol}}{\underset{\text{H}\quad\text{OH}}{CH_2CH_2}} \xrightarrow[180°C]{H_2SO_4} \underset{\text{Ethylene}}{CH_2{=}CH_2} + H_2O$$

To prepare ethyl ether, we first dissolve ethyl alcohol in sulfuric acid at ambient temperature and then heat the solution to 140°C while adding more alcohol. Under these moderate conditions ethyl ether is obtained as the major product.

$$\underset{\text{Ethyl alcohol}}{2\ CH_3CH_2OH} \xrightarrow[140°C]{H_2SO_4} \underset{\text{Ethyl ether}}{CH_3CH_2-O-CH_2CH_3} + H_2O$$

Generally, the dehydration of alcohols is limited to the preparation of symmetrical ethers. If you attempt to synthesize an unsymmetrical ether by this method, you will obtain a mixture of at least three products. An example will show why this is so.

Example 9.1 Assume you needed ethyl *n*-propyl ether and tried to prepare it by dehydrating a mixture of ethyl alcohol and *n*-propyl alcohol. The following would happen.

$$\underset{\text{Ethyl alcohol}}{CH_3CH_2OH} + \underset{\text{*n*-Propyl alcohol}}{HOCH_2CH_2CH_3} \xrightarrow[\text{heat}]{H_2SO_4} \underset{\substack{\text{Ethyl *n*-propyl ether}\\(\text{desired})}}{CH_3CH_2-O-CH_2CH_2CH_3} + H_2O$$

$$CH_3CH_2OH + HOCH_2CH_3 \xrightarrow[\text{heat}]{H_2SO_4} CH_3CH_2-O-CH_2CH_3 + H_2O$$

Ethyl ether
(undesired)

$$CH_3CH_2CH_2OH + HOCH_2CH_2CH_3 \xrightarrow[\text{heat}]{H_2SO_4}$$

$$CH_3CH_2CH_2-O-CH_2CH_2CH_3 + H_2O$$

n-Propyl ether
(undesired)

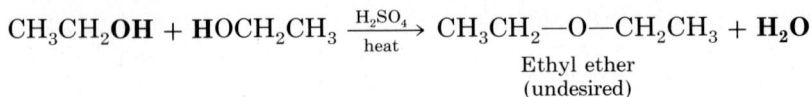

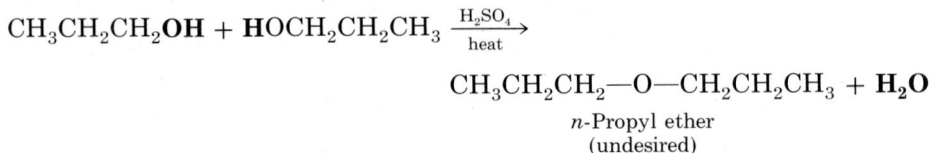

You would get not only the desired compound but at least two other unwanted ethers, making the method impractical.

Despite its limitations, the formation of ethers by dehydration of alcohols is an important reaction: the reaction conditions are simple and the raw materials are inexpensive and readily available.

Problem 9.5 Write equations for these conversions.
(a) *n*-Propyl alcohol to *n*-propyl ether
(b) Benzyl alcohol to benzyl ether

9.6 The Williamson Synthesis of Ethers

The reaction of a sodium alkoxide or a sodium phenoxide with an alkyl halide to form an ether is known as the **Williamson synthesis.**

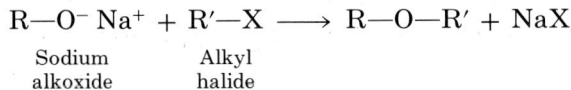

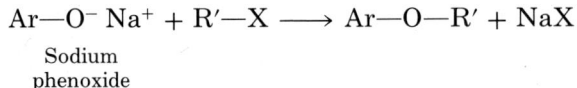

$$R-O^- Na^+ + R'-X \longrightarrow R-O-R' + NaX$$

Sodium   Alkyl
alkoxide   halide

$$Ar-O^- Na^+ + R'-X \longrightarrow Ar-O-R' + NaX$$

Sodium
phenoxide

Because of its versatility, the Williamson synthesis is the preferred method for making ethers in the laboratory; it can be used to prepare symmetrical as well as mixed ethers. The reaction goes by an S_N2 mechanism. The alkoxide (or phenoxide) ion serves as a strong nucleophile and displaces the halide ion from the alkyl halide.

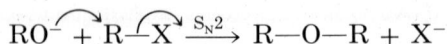

$$RO^- + R-X \xrightarrow{S_N2} R-O-R + X^-$$

You will recall that the S_N2 mechanism does not work well (1) if the approach of the nucleophile is hindered by bulky groups in the alkyl halide or (2) if the halogen that is being displaced is directly attached to an aromatic ring. For these reasons it is important to choose the right combination of reagents if the Williamson synthesis is to succeed. The following examples will illustrate each condition.

Example 9.2   Suppose you wish to prepare *t*-butyl methyl ether, $(CH_3)_3C$—O—CH_3, in the laboratory.* In theory, this could be done by either of two reactions. You could react sodium methoxide, $CH_3O^- Na^+$, with *t*-butyl chloride, $(CH_3)_3C$—Cl. A second possible combination is to react sodium *t*-butoxide, $(CH_3)_3C$—O$^-$ Na$^+$, with methyl chloride, CH_3Cl. *Only the second route gives the desired ether by substitution;* the first combination leads to dehydrohalogenation to an alkene, an elimination reaction.

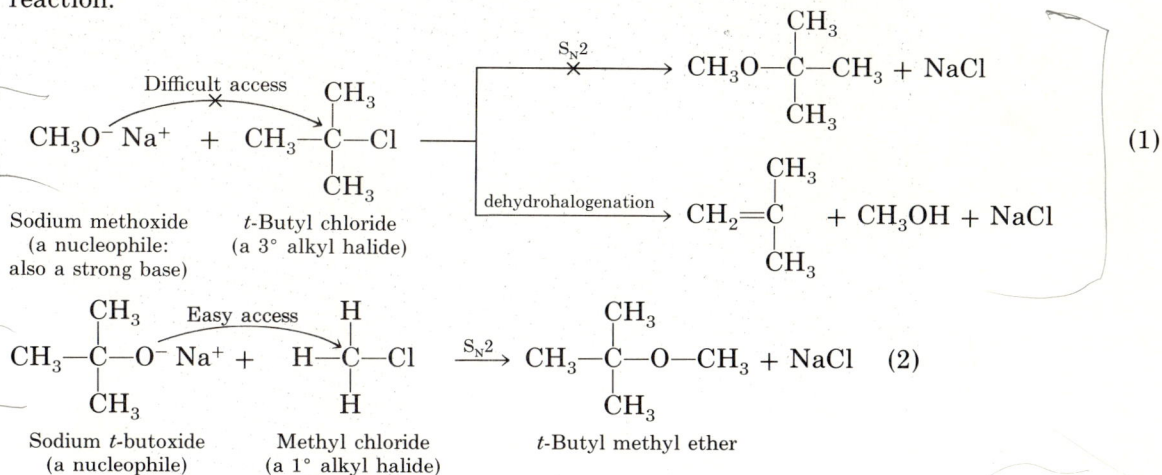

Difficult access

$$CH_3O^- Na^+ \;+\; CH_3{-}\underset{\underset{\displaystyle CH_3}{|}}{\overset{\overset{\displaystyle CH_3}{|}}{C}}{-}Cl$$

Sodium methoxide
(a nucleophile:
also a strong base)

t-Butyl chloride
(a 3° alkyl halide)

$\xrightarrow{S_N2}$ $CH_3O{-}\underset{\underset{\displaystyle CH_3}{|}}{\overset{\overset{\displaystyle CH_3}{|}}{C}}{-}CH_3 + NaCl$

$\xrightarrow{\text{dehydrohalogenation}}$ $CH_2{=}\underset{\underset{\displaystyle CH_3}{|}}{C} + CH_3OH + NaCl$

(1)

$$CH_3{-}\underset{\underset{\displaystyle CH_3}{|}}{\overset{\overset{\displaystyle CH_3}{|}}{C}}{-}O^- Na^+ + H{-}\underset{\underset{\displaystyle H}{|}}{\overset{\overset{\displaystyle H}{|}}{C}}{-}Cl \xrightarrow{S_N2} CH_3{-}\underset{\underset{\displaystyle CH_3}{|}}{\overset{\overset{\displaystyle CH_3}{|}}{C}}{-}O{-}CH_3 + NaCl \quad (2)$$

Easy access

Sodium *t*-butoxide
(a nucleophile)

Methyl chloride
(a 1° alkyl halide)

t-Butyl methyl ether

No ether is formed in reaction (1) because the nucleophile, CH_3^-, has no easy access to the carbon that carries the halogen in the 3° alkyl halide. Instead, we have dehydrohalogenation (elimination) because the methoxide ion, like the OH^- ion, is not only a nucleophile but also a strong base.

Elimination favored when alkyl halide is tertiary

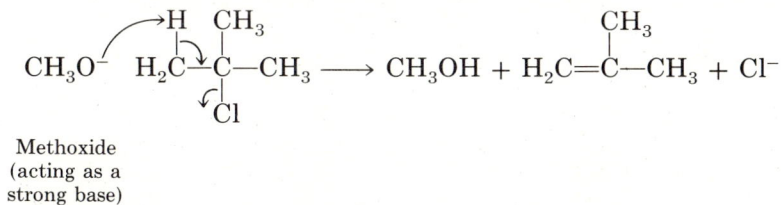

$$CH_3O^- \quad H_2\overset{\displaystyle \curvearrowright H}{C}{\overset{\displaystyle CH_3}{\underset{\underset{\displaystyle Cl}{|}}{C}}}{-}CH_3 \longrightarrow CH_3OH + H_2C{=}\underset{\overset{\displaystyle CH_3}{|}}{C}{-}CH_3 + Cl^-$$

Methoxide
(acting as a
strong base)

It is also necessary to plan carefully to prepare an alkyl aryl ether, R—O—Ar.

Example 9.3   Assume you need to synthesize methyl phenyl ether (anisole), CH_3—O—C_6H_5, by the Williamson method. In theory, you could obtain anisole in either of two ways.

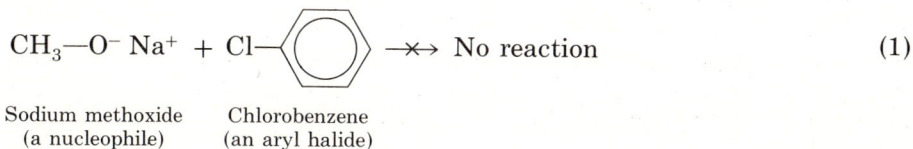

$$CH_3{-}O^- Na^+ + Cl{-}\langle\bigcirc\rangle \xrightarrow{\times} \text{No reaction} \qquad (1)$$

Sodium methoxide
(a nucleophile)

Chlorobenzene
(an aryl halide)

* This ether is prepared on a commercial scale by a different method (see Exercise 9.7).

$$\langle\bigcirc\rangle\!\!-\!\!O^-Na^+ + \ \overset{\curvearrowleft}{CH_3}\!\!-\!\!\overset{\curvearrowleft}{Cl} \ \xrightarrow{S_N2} \ \langle\bigcirc\rangle\!\!-\!\!O\!\!-\!\!CH_3 + NaCl \qquad (2)$$

Sodium phenoxide	Methyl chloride	Anisole
(a nucleophile)	(a 1° alkyl halide)	

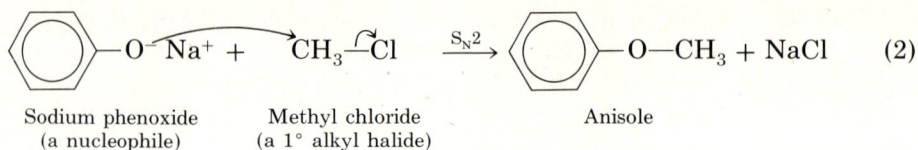

In fact, no reaction takes place with the first combination because it is extremely difficult to displace a halogen that is directly attached to a benzene ring (Sec. 8.5). The second route gives anisole by a straightforward S_N2 reaction.

In practice, anisole is often prepared by reacting sodium phenoxide with methyl sulfate, rather than with the more expensive methyl chloride. The reaction in this case also proceeds by an S_N2 mechanism.

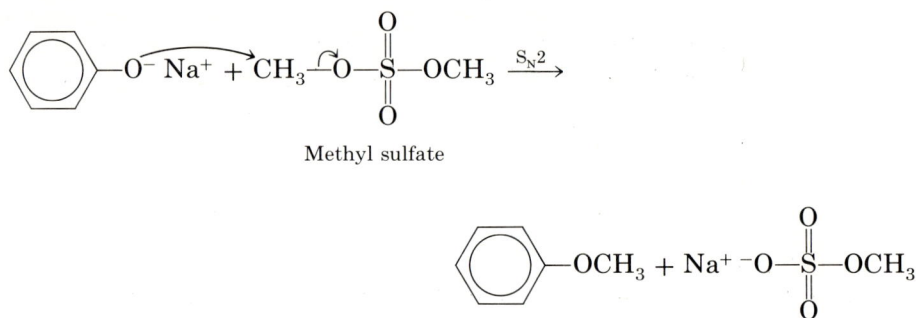

$$\langle\bigcirc\rangle\!\!-\!\!O^-Na^+ + \overset{\curvearrowleft}{CH_3}\!\!-\!\!\overset{\curvearrowleft}{O}\!\!-\!\!\overset{\displaystyle O}{\underset{\displaystyle O}{\overset{\|}{\underset{\|}{S}}}}\!\!-\!\!OCH_3 \ \xrightarrow{S_N2}$$

Methyl sulfate

$$\langle\bigcirc\rangle\!\!-\!\!OCH_3 + Na^+ \ ^-O\!\!-\!\!\overset{\displaystyle O}{\underset{\displaystyle O}{\overset{\|}{\underset{\|}{S}}}}\!\!-\!\!OCH_3$$

To recapitulate,

1. The Williamson synthesis is the preferred laboratory method for making ethers. The reaction goes by an S_N2 mechanism.
2. To prepare a mixed ether, it is necessary to choose the proper combination of reagents.
3. To obtain the best yields of mixed dialkyl ethers, we select a 1° rather than a 2° or 3° alkyl halide whenever possible and react it with a sodium alkoxide.
4. To prepare an alkyl aryl ether, we must be careful not to pick a combination in which one of the reagents has a halogen directly attached to an aromatic ring.

Problem 9.6 The compound 2,4-D

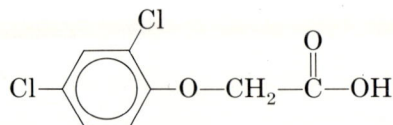

$$Cl\!\!-\!\!\langle\bigcirc\rangle\!\!-\!\!O\!\!-\!\!CH_2\!\!-\!\!\overset{\displaystyle O}{\overset{\|}{C}}\!\!-\!\!OH$$
$$\overset{\displaystyle Cl}{|}$$

is one of the most powerful weed killers. Assume that the following four compounds are available.

| (i) | (ii) | (iii) | (iv) |

where structures (iii) is $Cl—CH_2—\overset{\overset{\displaystyle O}{\|}}{C}—OH$ and (iv) is $HO—CH_2—\overset{\overset{\displaystyle O}{\|}}{C}—OH$

Show the combination you would use to make 2,4-D.

Reactions of Ethers *9.7*

Ethers are quite stable compounds. The ether linkage does not react with bases, reducing agents, oxidizing agents, or active metals. *Ethers react only under strongly acidic conditions.*

A   Reaction with Cold Concentrated Acids

The reaction of ethers with cold concentrated acids takes place as follows.

$$R—\overset{..}{\underset{..}{O}}—R + H^+X^- \rightleftharpoons R—\overset{+}{\underset{\underset{\displaystyle H}{|}}{\overset{..}{O}}}—R + X^-$$

Ether   Protonated ether
(as a base)

An ether, acting as a base, donates an electron pair to a hydrogen ion to form a protonated ether. Protonated ethers, once produced, are soluble in the cold concentrated acid solution. Since alkanes do not react with, and are insoluble in, cold concentrated acid solutions, it is possible to distinguish alkanes from ethers by this reaction. For example, n-pentane and ethyl ether have almost the same boiling points, but only the ether dissolves in cold concentrated sulfuric acid. Mixing n-pentane with cold concentrated H_2SO_4 produces two distinct layers.

$$CH_3CH_2—\overset{..}{\underset{..}{O}}—CH_2CH_3 + H_2SO_4 \xrightarrow{\text{(cold, conc)}} CH_3CH_2—\overset{+}{\underset{\underset{\displaystyle H}{|}}{O}}—CH_2CH_3 + HSO_4^-$$

Solution: one layer

$$CH_3CH_2CH_2CH_2CH_3 + H_2SO_4 \xrightarrow{\text{(cold, conc)}} \text{No reaction}$$

No solution: two distinct layers

Problem 9.7 An overworked laboratory assistant placed 1-propanol, n-propyl ether, and ethyl bromide in three different bottles, but forgot to label them. What simple chemical test could be performed on a sample from each of the bottles to identify them?

B   Cleavage of Ethers by Hot Concentrated Acids

When ethers are heated in concentrated acid solutions, the ether linkage is broken.

General equation

$$\text{R}-\overset{..}{\underset{..}{\text{O}}}-\text{R} + \mathbf{H-X} \xrightarrow[\text{heat}]{} \text{R}-\text{O}-\text{H} + \text{R}-\text{X}$$
$$\text{(conc)}$$

Specific example

$$\text{CH}_3\text{CH}_2-\overset{..}{\underset{..}{\text{O}}}-\text{CH}_2\text{CH}_3 + \underset{\text{(conc)}}{\mathbf{HI}} \xrightarrow[\text{heat}]{} \text{CH}_3\text{CH}_2\text{OH} + \text{CH}_3\text{CH}_2\text{I}$$

The acids most often used in this reaction are HI, HBr, and HCl.

The cleavage of ethers involves two steps: (1) a protonated ether is formed, and (2) when heated, the protonated ether undergoes cleavage. The cleavage occurs by an S_N1 (step 2a) or an S_N2 (2b) mechanism, depending on the structure of the ether.

Step 1

$$\text{R}-\overset{..}{\underset{..}{\text{O}}}-\text{R} + \text{HX} \rightleftharpoons \text{R}-\overset{+}{\underset{\underset{\text{H}}{|}}{\overset{..}{\text{O}}}}-\text{R} + \text{X}^-$$

Protonated ether   Nucleophile

Step 2

(a) $\text{R}-\overset{+}{\underset{\underset{\text{H}}{|}}{\overset{..}{\text{O}}}}-\text{R} \xrightarrow[\substack{\text{if Rs are 2°} \\ \text{and 3° alkyls}}]{S_N1} \text{R}^+ + \text{R}-\text{OH}$

$$\underset{\downarrow}{\overset{+}{\text{X}^-}}$$
$$\text{R}-\text{X}$$

(b) $\text{X}^- + \text{R}-\overset{+}{\underset{\underset{\text{H}}{|}}{\overset{..}{\text{O}}}}-\text{R} \xrightarrow[\text{if Rs are 1° alkyls}]{S_N2} \text{X}-\text{R} + \text{R}-\text{OH}$

If an *excess of acid* is present, the alcohol initially produced is converted into an alkyl halide by the reaction

$$\text{R}-\text{OH} + \text{HX} \longrightarrow \text{RX} + \text{H}_2\text{O}$$

For example,

$$\text{CH}_3\text{CH}_2-\text{O}-\text{CH}_2\text{CH}_3 + 2\,\text{HBr} \xrightarrow[\text{heat}]{} 2\,\text{CH}_3\text{CH}_2\text{Br} + \text{H}_2\text{O}$$
$$\text{(conc)}$$

Problem 9.8   Compound **A**, C_7H_8O, was found to give no reaction with metallic sodium or $SOCl_2$. When **A** was heated with HI, two products were formed. One was CH_3I; the other, **B**, had formula C_6H_6O. Compound **B** dissolved in aqueous sodium hydroxide. What are the structures of **A** and **B**? Write an equation for the conversion of **A** to **B**.

Cyclic Ethers *9.8*

All the ethers discussed thus far are open-chain compounds. In the process of dehydration of alcohols, they were made by reaction between two separate alcohol molecules, an *intermolecular* process. It is possible to obtain cyclic ethers by an *intramolecular* reaction, if we start with an alcohol that has two OH groups (a diol). For example, heating 1,4-butanediol in the presence of sulfuric acid gives the cyclic ether *tetrahydrofuran* (THF).

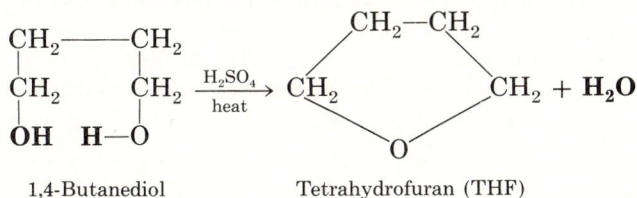

1,4-Butanediol   Tetrahydrofuran (THF)

In this reaction, one of the hydroxyl groups is protonated and undergoes nucleophilic attack by the other OH group to give the cyclic ether. The mechanism is exactly the same as that described for ethyl ether. THF is so named because it is related to the unsaturated cyclic ether *furan*.

Furan

Similarly, when 1,5-pentanediol is heated in acid, the six-membered ring ether *tetrahydropyran* is obtained.

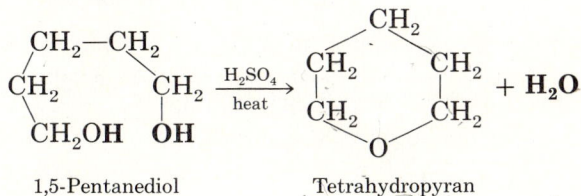

1,5-Pentanediol   Tetrahydropyran

Tetrahydropyran is so called because it is related to the unsaturated ether *pyran*.

Pyran

You will encounter structures resembling furan and pyran when you study carbohydrates (Chapter 11).

Chemically, five- and six-membered ring ethers behave much like their open-chain analogs. They are relatively inert except toward acids.

Three-membered ring ethers, $\overset{\diagdown}{}C\underset{O}{\text{——}}C\overset{\diagup}{}$, can also be made. These are reactive compounds, which will be discussed later (Sec. 9.10).

9.9 Uses of Ethers

Ethyl ether is by far the most important ether. We are all aware of its anesthetic properties. Despite its flammability and danger of explosion, ethyl ether has been used safely countless times in surgery. It is a relatively slow inhalation anesthetic that can be administered over a fairly wide range of concentrations. Because it dissolves many types of organic compounds, ethyl ether is often used to separate organic substances from inorganic material. In biochemical applications, lipids, which are soluble in ether, are separated from carbohydrates and proteins, which are usually insoluble. Ethyl ether is also a solvent in many chemical reactions because of its relative inertness. It is the preferred solvent in Grignard reactions; in this application, it is necessary to use *absolute ether* (ether that is 100% moisture free). Because ordinary ethyl ether is hygroscopic—that is, it picks up water from the air—any traces of water that may be present must be removed if absolute ether is to be obtained. This is done by treating the ether with metallic sodium. (What is formed when sodium reacts with water?) The ether is then ready for use in a Grignard reaction.

Vinyl ether, $CH_2{=}CH{-}O{-}CH{=}CH_2$, a volatile ether that boils at 31°C, has anesthetic properties similar to ethyl ether. It is known as *Vinethene* and is valuable for its rapid inhalation anesthetic action.

Methoxyflurane, $CHCl_2{-}CF_2{-}O{-}CH_3$, or 2,2-dichloro-1,1-difluoroethyl methyl ether, was introduced as an anesthetic in 1959. It is the most potent of the inhalation anesthetics and is not flammable. Methoxyflurane has been recommended for use in obstetrical anesthesia. It is now widely used in the United States.

The fragrant aromas of anise seeds, vanilla beans, and cloves come from compounds that contain the anisole nucleus. Eugenol mixed with zinc oxide forms a cement that is used by dentists as temporary fillings.

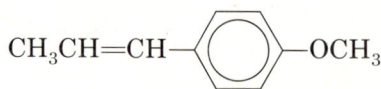

Anethole
(oil of anise)

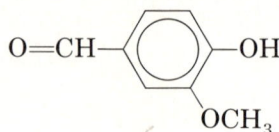

Vanillin
(vanilla beans)

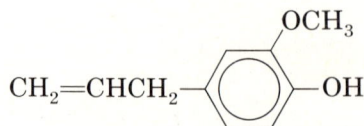

Eugenol
(oil of cloves)

Phenyl ether is used as a heat transfer medium because of its high boiling point (259°C). *Monobenzone,* or *p*-benzyloxyphenol, is an antioxidant in the manufacture of rubber. Monobenzone is also used as a depigmenting agent in cases of excessive formation of melanin, such as occurs in severe freckling and in Addison's disease.

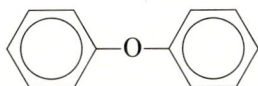
Phenyl ether

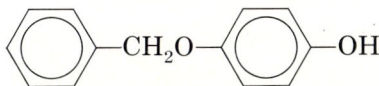
Monobenzone

The commercial herbicides *2,4-D* and *2,4,5-T* play major roles in selectively controlling broadleaf weeds in grasslands and in major crops such as wheat, barley, and sugar cane.

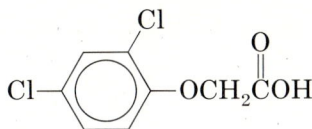

2,4-D

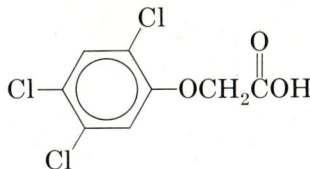

2,4,5-T

Epoxides *9.10*

Epoxides are cyclic ethers in which the ether oxygen is part of a three-membered ring.

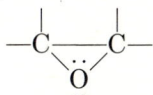

An epoxide

The simplest and most important epoxide is *ethylene oxide*. Ethylene oxide is prepared commercially by air oxidation of ethylene.

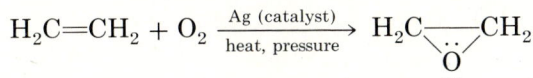

Ethylene oxide

Epoxides undergo many reactions that other ethers do not. Their reactivity is due to the strain in the three-membered ring, which is relieved when the epoxide ring is opened after a reaction has taken place. Examples of ring-opening reactions of ethylene oxide that form commercially important products are

Ethylene oxide

Ethylene glycol
(a component in the manufacture of Dacron;
main ingredient in automobile antifreeze)

$$n\ H_2C\underset{\underset{\cdot\cdot}{\overset{\cdot\cdot}{O}}}{\diagdown\!\!\diagdown}CH_2 \xrightarrow{\text{SrCO}_3} \left[O\!-\!CH_2\!-\!CH_2\right]_n$$

      Ethylene oxide   Polyethylene oxide

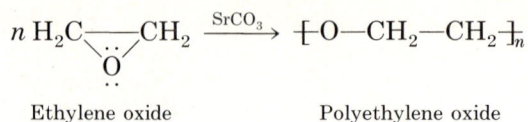

Polyethylene oxide is used as an additive by fire departments under the name Rapid-Water (Fig. 9.1). Apparently, the addition of small amounts of polyethylene oxide to water reduces the friction between water molecules so that flames are extinguished even when the water pressure is reduced.

Another important epoxide is *epichlorhydrin*.

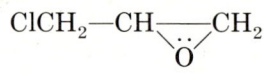

$$ClCH_2\!-\!CH\underset{\underset{\cdot\cdot}{\overset{\cdot\cdot}{O}}}{\diagdown\!\!\diagdown}CH_2$$

            Epichlorhydrin

Ring-opening reactions of epichlorhydrin with a number of reagents produce **epoxy resins.** Epoxy resins are polymers used as protective coatings because of their resistance to chemicals, their excellent electrical insulating properties, and their flexibility and hardness. Epoxy glues have outstanding adhesive properties and are used to bond ceramics, glass, and metals. The replacement of metal rivets by epoxy glues in airplanes is a novel application that promises to save costly fuel by reducing the weight of future aircraft.

Figure 9.1 A mixture of polyethylene oxide and water (Rapid-Water) can be sprayed farther and faster than conventional water. [Courtesy of Union Carbide.]

Summary of Concepts and Reactions

Any compound with the general formula R—O—R, Ar—O—R, or Ar—O—Ar is called an ether. [Sec. 9.1]

When the organic groups attached to the oxygen atom are identical, the ether is classified as a symmetrical ether; if the two groups are different, the ether is said to be unsymmetrical or a mixed ether. [Sec. 9.1]

The boiling points of ethers are much lower than the boiling points of alcohols having the same molecular weight because, unlike alcohols, ether molecules are incapable of forming hydrogen bonds with one another. The solubility of ethers in water is similar to that of alcohols because, like alcohols, ethers can be hydrogen-bonded to water molecules. [Sec. 9.2]

When exposed to air, most liquid aliphatic ethers react slowly to form unstable peroxides, which can cause violent explosions. [Sec. 9.3]

There are two general methods for synthesizing ethers: (1) dehydration of alcohols

$$2 \text{ ROH} \xrightarrow[\text{heat}]{\text{H}^+} \text{R—O—R} + \text{H}_2\text{O} \qquad \text{[Sec. 9.5]}$$

and (2) the Williamson synthesis

$$\text{RO}^- \text{Na}^+ + \text{R}'\text{X} \xrightarrow{\text{S}_\text{N}2} \text{R—O—R}' + \text{NaX} \qquad \text{[Sec. 9.6]}$$

Ethers do not react with bases, oxidizing agents, reducing agents, or active metals, but they do react with strong acids. [Sec. 9.7A, B]

Epoxides are cyclic ethers in which the ether oxygen is part of a three-membered ring. [Sec. 9.10]

Key Terms

symmetrical ether
unsymmetrical ether
mixed ethers
alkoxy

peroxides
dehydration of alcohols
Williamson synthesis

ethyl ether
epoxides
epoxy resins

Exercises

Structure and Nomenclature of Ethers [Sec. 9.1]

9.1 Write the formulas for these compounds.

(a) *n*-Propyl ether
(b) Vinyl ether
(c) Cyclohexyl methyl ether
(d) 1-Bromo-3-ethoxy-2-butanol
(e) 2,2-Dimethoxybutane
(f) *p*-Methoxyphenol
(g) Benzyl isopropyl ether
(h) *m*-Nitroanisole

9.2 Name the following structures, using any nomenclature system that is appropriate.

(a) $CH_3OCH(CH_3)_2$

(b) $(CH_3)_3CO$—⬡

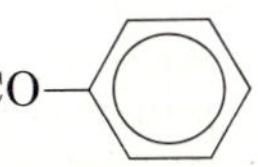

(c) 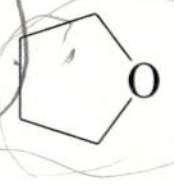 ⬠ O

(d) 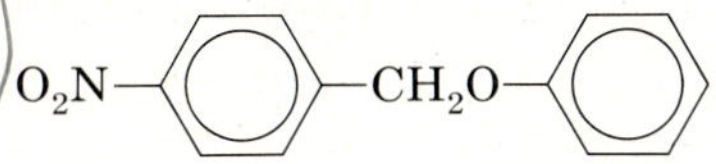 O_2N—◯—CH_2O—◯

(e) $CH_3CHCH{=}CH_2$
$\quad\quad\quad |$
$\quad\quad\quad OC_2H_5$

(f) (benzene ring)$-OCH_3$
$\quad\quad$ Br

(g) CH_3CHCH_2OH
$\quad\quad\quad |$
$\quad\quad\quad OCH_3$

(h) (cyclopentane ring)$-OCH_2CH_3$

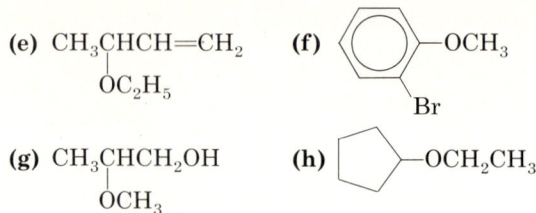

Physical Properties of Ethers [Sec. 9.2]

9.3 Arrange these compounds in order of increasing boiling point: $HOCH_2CH_2OH$ (i); $CH_3OCH_2CH_2OH$ (ii); $CH_3CH_2CH_2CH_3$ (iii); $CH_3OCH_2CH_2OCH_3$ (iv)

Preparations and Reactions of Ethers [Secs. 9.5–9.7]

9.4 Name the possible ethers that can be formed by heating a mixture of methanol and ethanol with sulfuric acid. Illustrate by writing equations.

9.5 The Williamson synthesis is an example of an S_N2 reaction. Draw a model of the transition state for the reaction $RO^- + R'{-}X \rightarrow R{-}O{-}R' + X^-$.

9.6 Give equations for two different combinations of reagents for preparing the following ethers by the Williamson synthesis.

(a) Methyl *n*-propyl ether
(b) Ethyl *n*-butyl ether

9.7 Methyl *t*-butyl ether (MTBE) is added to certain gasolines to extend their supply and to enhance their octane rating.

(a) In the laboratory MTBE is prepared by the Williamson synthesis. Write an equation for the reaction.

(b) Industrially MTBE is prepared by the reaction of isobutylene with methyl alcohol in the presence of an acid catalyst. Write a mechanism for the reaction.

9.8 Outline laboratory syntheses of the following compounds from alcohols and phenols, using any necessary inorganic reagents.

(a) Ethyl *t*-butyl ether
(b) Phenetole ($C_6H_5{-}O{-}C_2H_5$)
(c) *p*-Nitrobenzyl phenyl ether
(d) 1,3-Dimethoxybenzene

9.9 Draw the structures of the reactants and major products for each reaction. If no reaction occurs, state so.

(a) *t*-Butyl bromide + potassium ethoxide
(b) Potassium *t*-butoxide + ethyl bromide
(c) Bromobenzene + sodium ethoxide
(d) Sodium phenoxide + ethyl bromide
(e) Ethyl ether + boiling NaOH
(f) Ethyl ether + Na
(g) *p*-Nitroanisole + HI
(h) Methyl ether + HBr (excess)
(i) Ethyl ether + H_2SO_4 (cold)

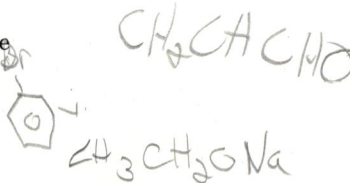

Structure Identification and Structure Determination [Secs. 9.5–9.11 and previous chapters]

9.10 Describe simple chemical tests that would distinguish between these pairs.

(a) *n*-Butyl ether and *n*-pentyl alcohol
(b) Methyl *n*-propyl ether and 2-pentene
(c) Ethyl ether and pentane

9.11 Compound **A**, $C_4H_{10}O_3$, shows the properties of both an alcohol and an ether. When heated with excess hydrogen iodide, **A** yields only ICH_2CH_2I. Draw a structural formula for **A**.

9.12 Compound **A**, $C_8H_{18}O$, does not react with metallic sodium or PCl_5. When heated with excess hydrogen bromide, **A** yields only one product, **B**, C_4H_9Br. When **B** is treated with ethanolic silver nitrate, the solution remains clear, but treatment of **B** with sodium iodide in acetone yields an immediate precipitate, C_4H_9I. Draw possible structural formulas for **A** and **B**, and write an equation for each reaction.

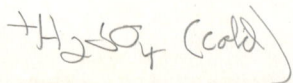

Aldehydes and Ketones

Aldehydes and **ketones** take us for the first time to the chemistry of several families of organic compounds that contain the **carbonyl group,** which consists of a carbon–oxygen double bond.

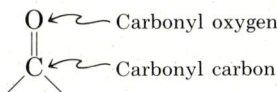

O ← Carbonyl oxygen
‖
C ← Carbonyl carbon

The carbonyl group

We can have an aldehyde, a ketone, an acid, an ester, an amide, and so on, depending on what other atoms are attached to the carbonyl carbon. This chapter presents aldehydes and ketones, in which the carbonyl carbon is bonded to hydrogen or to carbon. Aldehydes and ketones resemble each other closely enough in many of their properties to justify treating them in the same chapter. Acids and acid derivatives, which have oxygen, nitrogen, or another atom (other than carbon or hydrogen) attached directly to the carbonyl carbon, are discussed in later chapters because the chemistry of these compounds is different from that of aldehydes and ketones.

Aldehydes and ketones are widely distributed in nature. They are involved in many biological reactions and are used as starting materials or as intermediates in a great number of chemical syntheses. Most of the properties of aldehydes and ketones are due to the carbonyl group. Therefore, the significance of this functional group cannot be overemphasized when discussing the chemistry of aldehydes and ketones.

10.1 Aldehydes: Structure and Nomenclature

Aldehydes have the general formula

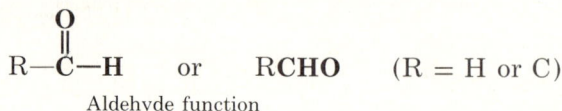

$$R-\overset{\overset{\displaystyle O}{\|}}{C}-H \quad \text{or} \quad RCHO \quad (R = H \text{ or } C)$$

Aldehyde function

Aldehydes may be referred to by their common names or by the IUPAC method. Aldehydes of up to four carbons are often called by their common names, which are derived from the common names of the acids to which they are related (see Sec. 12.1). In the IUPAC system, aliphatic aldehydes are named by dropping the suffix -e from the name of the hydrocarbon that has the same carbon skeleton as the aldehyde and replacing it with the suffix -al. Because the aldehyde group is always at the end of a chain, it automatically becomes carbon number 1, although this number does not appear in the name.

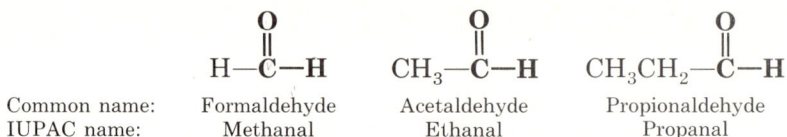

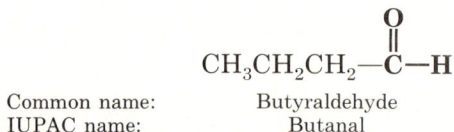

$$H-\overset{\overset{\displaystyle O}{\|}}{C}-H \qquad CH_3-\overset{\overset{\displaystyle O}{\|}}{C}-H \qquad CH_3CH_2-\overset{\overset{\displaystyle O}{\|}}{C}-H$$

Common name:	Formaldehyde	Acetaldehyde	Propionaldehyde
IUPAC name:	Methanal	Ethanal	Propanal

$$CH_3CH_2CH_2-\overset{\overset{\displaystyle O}{\|}}{C}-H$$

Common name: Butyraldehyde
IUPAC name: Butanal

Substituted aldehydes are named in the usual ways by numbering and alphabetically listing the substituents. The CHO group is assigned the number 1 position and takes precedence over other functional groups that may be present such as —OH, C=C, or C≡C. For example,

$$CH_3\overset{\overset{\displaystyle Cl}{|}}{C}H-\overset{\overset{\displaystyle O}{\|}}{C}-H \qquad HO-CH_2CH_2-\overset{\overset{\displaystyle O}{\|}}{C}-H \qquad CH_3CH=CH-\overset{\overset{\displaystyle O}{\|}}{C}-H$$

2-Chloropropanal   3-Hydroxypropanal   2-Butenal

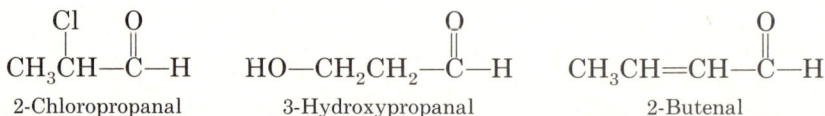

Aromatic aldehydes are usually designated as derivatives of the simplest aromatic aldehyde, **benzaldehyde.**

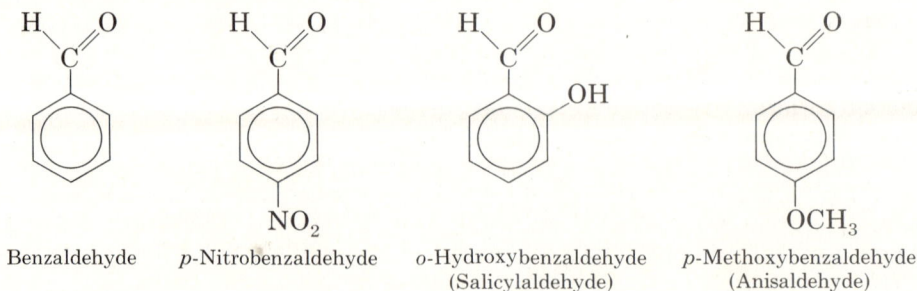

Benzaldehyde   p-Nitrobenzaldehyde   o-Hydroxybenzaldehyde   p-Methoxybenzaldehyde
                                      (Salicylaldehyde)   (Anisaldehyde)

Problem 10.1 Write the structures for these aldehydes.
(a) Hexanal   **(b)** *m*-Bromobenzaldehyde   **(c)** Propionaldehyde
(d) 3-Methylbutanal   **(e)** Acetaldehyde   **(f)** 4-Chloro-3-pentenal

Problem 10.2 Name these compounds.

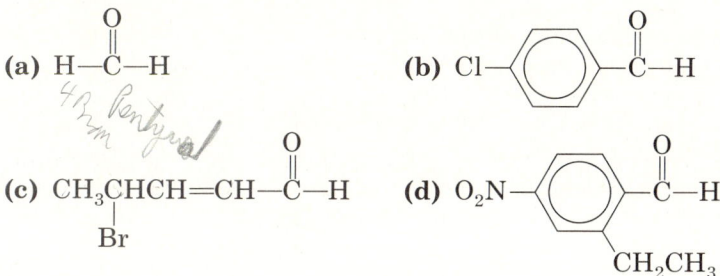

$$\text{(a)} \quad H-\overset{\displaystyle O}{\overset{\|}{C}}-H$$

4m Pentyral

$$\text{(b)} \quad Cl-\underset{}{\bigcirc}-\overset{\displaystyle O}{\overset{\|}{C}}-H$$

$$\text{(c)} \quad CH_3\underset{\underset{Br}{|}}{C}HCH{=}CH-\overset{\displaystyle O}{\overset{\|}{C}}-H$$

$$\text{(d)} \quad O_2N-\underset{\underset{CH_2CH_3}{}}{\bigcirc}-\overset{\displaystyle O}{\overset{\|}{C}}-H$$

Ketones: Structure and Nomenclature *10.2*

Ketones have the general formula

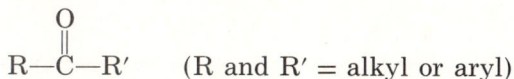

$$R-\overset{\displaystyle O}{\overset{\|}{C}}-R' \qquad \text{(R and R}' = \text{alkyl or aryl)}$$

In a ketone the carbonyl group is flanked on both sides by carbon atoms. For this reason, the carbonyl group may appear at any of various positions in the chain, *except at the end*. The R groups need not be the same; either or both may be aliphatic or aromatic. In ketones the carbonyl group may also be part of a cyclic structure. For example, these structures are all ketones.

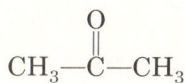

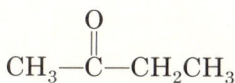

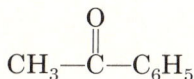

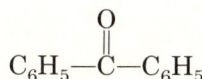

$$CH_3{-}\overset{O}{\overset{\|}{C}}{-}CH_3 \quad CH_3{-}\overset{O}{\overset{\|}{C}}{-}CH_2CH_3 \quad CH_3{-}\overset{O}{\overset{\|}{C}}{-}C_6H_5 \quad C_6H_5{-}\overset{O}{\overset{\|}{C}}{-}C_6H_5$$

Simple aliphatic ketones are named by listing the alkyl substituents attached to the carbonyl group, followed by the word *ketone*. The simplest aliphatic ketone, dimethyl ketone, is usually called **acetone.**

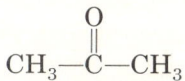

$$CH_3{-}\overset{O}{\overset{\|}{C}}{-}CH_3 \quad CH_3{-}\overset{O}{\overset{\|}{C}}{-}CH_2CH_3 \quad CH_3{-}\overset{O}{\overset{\|}{C}}{-}CH{=}CH_2 \quad CH_3CH_2{-}\overset{O}{\overset{\|}{C}}{-}\bigcirc$$

Acetone (Dimethyl ketone) — Methyl ethyl ketone — Methyl vinyl ketone — Ethyl cyclopentyl ketone

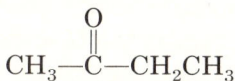

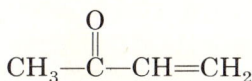

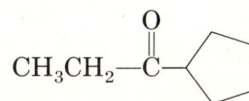

When the carbonyl group of a ketone is attached to a benzene ring, the ketone may be similarly named, or it may be given a special name.

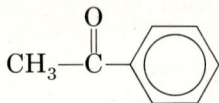

Methyl phenyl ketone
Acetophenone

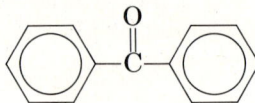

Diphenyl ketone
Benzophenone

In the IUPAC system, we name ketones in the usual manner. We find the longest continuous chain *carrying the carbonyl group* and name the parent structure by dropping the suffix -*e* from the corresponding hydrocarbon and replacing it with the ending -*one*. The chain is numbered in such a way as to give the lowest possible number to the C=O group. When the position of the C=O group is unambiguous, no number is necessary.

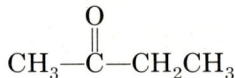

Butanone
(no number needed, the position of C=O is unambiguous)

Cyclopentanone

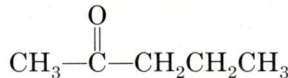

2-Pentanone
(*not* 4-Pentanone)

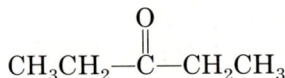

$CH_3CH_2—\overset{\displaystyle O}{\overset{\|}{C}}—CH_2CH_3$

3-Pentanone

Substituted ketones are named by numbering and listing the substituents alphabetically. The C=O group is always assigned the lowest possible number. For cyclic ketones, numbering always starts from the C=O group.

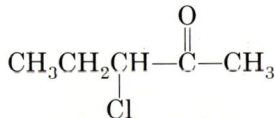

3-Chloro-2-pentanone
(*not* 3-Chloro-4-pentanone)

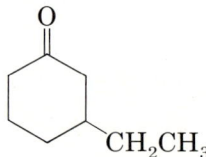

3-Ethylcyclohexanone
(*not* 5-Ethylcyclohexanone)

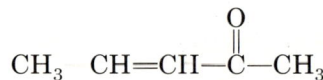

3-Penten-2-one
(*not* 2-Penten-4-one)

Problem 10.3   Write the structures for these ketones.
(a) Acetone
(b) 2-Hexanone
(c) Methyl *n*-propyl ketone
(d) Methyl-*p*-nitrophenyl ketone
(e) 4-Hexen-2-one
(f) Methyl allyl ketone

Problem 10.4   Name these compounds.

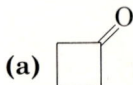

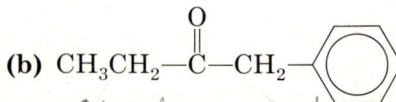

(a)

(b) $CH_3CH_2—\overset{\displaystyle O}{\overset{\|}{C}}—CH_2—$

↑ Phenyl-2-Butanone

(c) $CH_3CH_2-\overset{\overset{\displaystyle O}{\|}}{C}-CH_2CH_2CH_3$

(d)

(e) $CH_3CH_2-\overset{\overset{\displaystyle O}{\|}}{C}-$

(f) $CH_2=CH-\overset{\overset{\displaystyle O}{\|}}{C}-CH=CH_2$

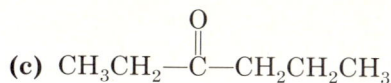

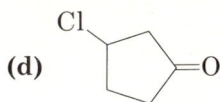

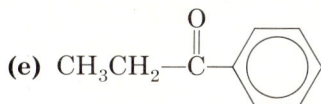

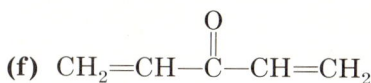

The Carbonyl Group 10.3

Because the role of the carbonyl group is central to an understanding of the physical and chemical properties of aldehydes and ketones, we need to review some of its features. In several respects the carbonyl function strongly resembles the carbon–carbon double bond of alkenes. As with ethylene, for example, the carbonyl carbon is joined to three other atoms. In forming bonds to the three atoms, the carbonyl carbon uses three equivalent sp^2-hybridized orbitals, as do the carbons in ethylene. Because of this factor, all the bonds to the carbonyl carbon lie in a single plane and are at 120° angles from one another, just as in ethylene (Fig. 10.1).

The electronic structure of the carbon–oxygen double bond also resembles strongly the carbon–carbon double bond. Here, as with ethylene, the carbon–oxygen double bond is a combination of a sigma (σ) bond and a pi (π) bond. The σ bond is formed from the overlap of an sp^2 orbital of carbon and a p orbital of oxygen; it is cylindrically symmetrical about the line between the carbon and oxygen nuclei (Fig. 10.2a). The π bond, which is formed from the overlap between the p orbital of the sp^2-hybridized carbonyl carbon and a p orbital of oxygen, represents two regions of electron density, one above and one below the plane of the bonds that join the carbonyl carbon.

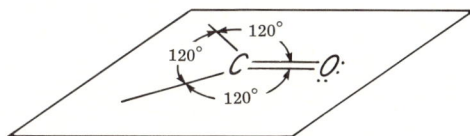

Figure 10.1 Geometry of the carbonyl group: planar, 120° angles.

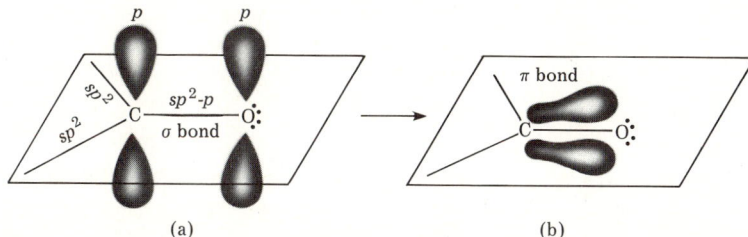

Figure 10.2 **(a)** Formation of the carbon–oxygen π bond from overlap between p orbitals of carbon and oxygen. **(b)** Once formed, the π bond is polarized, as shown by the greater electron density in the proximity of oxygen.

In one important respect, however, the carbonyl group differs markedly from the carbon–carbon double bond of ethylene: the π bond of the carbonyl group is *polarized*. Because of the greater electronegativity of oxygen relative to carbon, the electron density around the oxygen is larger near it than around the carbon. The situation is illustrated in Figure 10.2b. The polar character of the carbonyl group can be described by way of the two resonance-contributing structures.

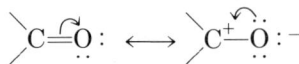

$$\overset{\diagdown}{\underset{\diagup}{C}}\!\!=\!\!\overset{\frown}{O}: \quad \longleftrightarrow \quad \overset{\diagdown}{\underset{\diagup}{C}}{}^{+}\!\!-\!\!\overset{\frown}{O}:{}^{-}$$

Resonance-contributing structures
of the polar C=O group

Alternatively, the partial negative charge on the carbonyl oxygen and the partial positive charge on the carbonyl carbon may be indicated as follows.

$$\overset{\diagdown}{\underset{\diagup}{C}}{}^{\delta+}\!\!=\!\!O^{\delta-}$$

Partial charge on the
polar carbonyl group

10.4 Physical Properties of Aldehydes and Ketones

A Boiling Points

Because of the polarity of the carbonyl group, aldehydes and ketones are polar compounds. The polar character of the molecules gives rise to intermolecular attractions. These attractive forces, called **dipole–dipole attractions,** occur between the partial negative charge on the carbonyl oxygen of one molecule and the partial positive charge on the carbonyl carbon of another molecule.

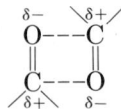

$$\begin{array}{c} \overset{\delta-}{O}\text{---}\overset{\delta+}{C}{\diagup} \\ \| \qquad \| \\ {\diagdown}\underset{\delta+}{C}\text{---}\underset{\delta-}{O} \end{array}$$

Dipole–dipole attractions among carbonyl compounds

Dipole–dipole attractions, although important, are not as strong as interactions due to hydrogen bonding. As a result, the boiling points of aldehydes and ketones are higher than those of nonpolar alkanes, but lower than those of alcohols, of comparable molecular weights. For example,

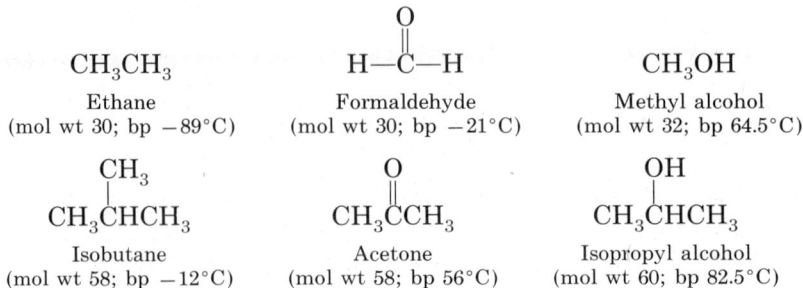

CH_3CH_3	$\overset{\displaystyle O}{\overset{\|}{H-C-H}}$	CH_3OH
Ethane	Formaldehyde	Methyl alcohol
(mol wt 30; bp $-89°C$)	(mol wt 30; bp $-21°C$)	(mol wt 32; bp $64.5°C$)
$\overset{\displaystyle CH_3}{\overset{\|}{CH_3CHCH_3}}$	$\overset{\displaystyle O}{\overset{\|}{CH_3CCH_3}}$	$\overset{\displaystyle OH}{\overset{\|}{CH_3CHCH_3}}$
Isobutane	Acetone	Isopropyl alcohol
(mol wt 58; bp $-12°C$)	(mol wt 58; bp $56°C$)	(mol wt 60; bp $82.5°C$)

Problem 10.5 Arrange these compounds in order of increasing boiling points.

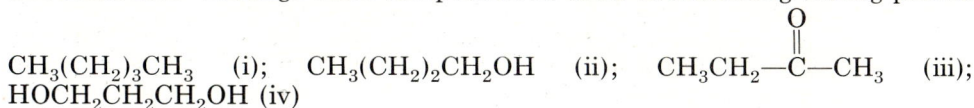

$CH_3(CH_2)_3CH_3$ (i); $CH_3(CH_2)_2CH_2OH$ (ii); $CH_3CH_2-\overset{\overset{\displaystyle O}{\|}}{C}-CH_3$ (iii);
$HOCH_2CH_2CH_2OH$ (iv)

B Solubility in Water

The lower aldehydes and ketones are soluble in water because aldehydes and ketones form hydrogen bonds with water, even though they are incapable of intermolecular hydrogen bonding with themselves.

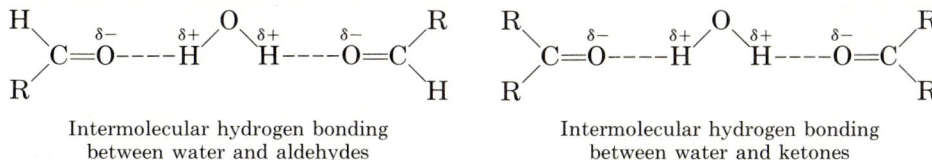

Intermolecular hydrogen bonding   Intermolecular hydrogen bonding
between water and aldehydes   between water and ketones

As the hydrocarbon portion of the molecule increases, the solubility in water decreases rapidly. Aldehydes and ketones with more than six carbons are essentially insoluble in water. However, the higher as well as the lower aldehydes and ketones are soluble in organic solvents such as benzene, ether, and carbon tetrachloride.

Problem 10.6 Arrange the following compounds in order of increasing solublity in water.
Methyl ethyl ketone; pentanal; hexanal; acetone.

Preparation of Aldehydes and Ketones *10.5*

The major methods for preparing aldehydes and ketones in the laboratory are summarized here and discussed in the following sections.

1. Oxidation of alcohols

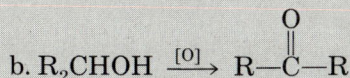

a. $RCH_2OH \xrightarrow{[O]} R-\overset{\overset{\displaystyle O}{\|}}{C}-H$ [Sec. 10.6]

b. $R_2CHOH \xrightarrow{[O]} R-\overset{\overset{\displaystyle O}{\|}}{C}-R$

2. Ozonolysis of alkenes

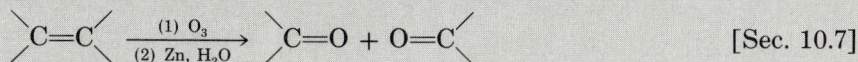

$\underset{/}{\overset{\backslash}{C}}=\underset{\backslash}{\overset{/}{C}} \xrightarrow[\text{(2) Zn, H}_2\text{O}]{\text{(1) O}_3} \underset{/}{\overset{\backslash}{C}}=O + O=\underset{\backslash}{\overset{/}{C}}$ [Sec. 10.7]

3. Hydration of alkynes

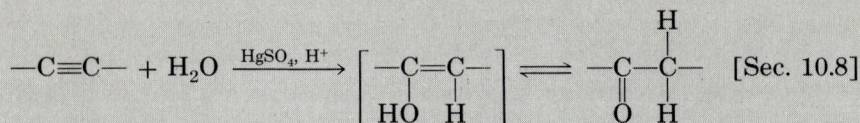

$$-C\equiv C- + H_2O \xrightarrow{HgSO_4,\ H^+} \left[\begin{matrix} -C=C- \\ HO \quad H \end{matrix} \right] \rightleftharpoons \begin{matrix} \quad\quad H \\ -C-C- \\ O \quad H \end{matrix}$$ [Sec. 10.8]

4. Friedel–Crafts acylation

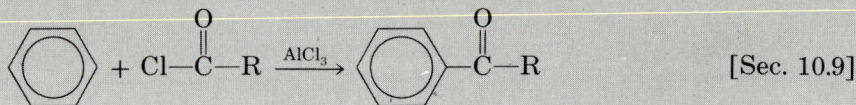

 [Sec. 10.9]

10.6 Oxidation of Primary and Secondary Alcohols

We have already seen (Sec. 7.8) that the oxidation of primary alcohols, under controlled conditions, yields aldehydes and that the oxidation of secondary alcohols yields ketones.

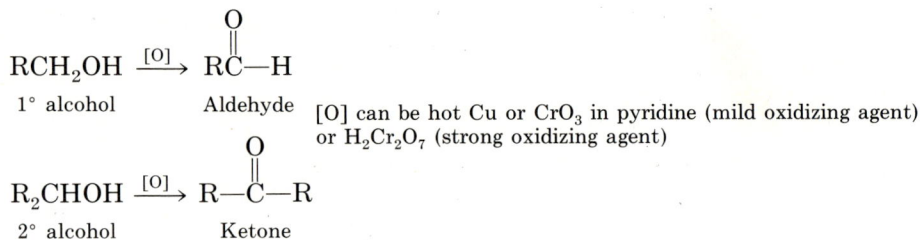

$$\underset{\text{1° alcohol}}{RCH_2OH} \xrightarrow{[O]} \underset{\text{Aldehyde}}{R\overset{O}{\underset{\|}{C}}-H}$$

[O] can be hot Cu or CrO_3 in pyridine (mild oxidizing agent) or $H_2Cr_2O_7$ (strong oxidizing agent)

$$\underset{\text{2° alcohol}}{R_2CHOH} \xrightarrow{[O]} \underset{\text{Ketone}}{R-\overset{O}{\underset{\|}{C}}-R}$$

As pointed out in Chapter 7, when strong oxidizing agents are used, aldehydes are very easily oxidized further to carboxylic acids.

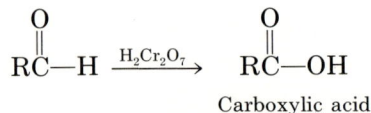

$$R\overset{O}{\underset{\|}{C}}-H \xrightarrow{H_2Cr_2O_7} \underset{\text{Carboxylic acid}}{R\overset{O}{\underset{\|}{C}}-OH}$$

To prevent this additional oxidation, precautions must be taken to remove aldehydes from contact with the oxidizing agent as soon as they are formed. For low-boiling aldehydes, this is accomplished by maintaining the reaction mixture at a temperature below the boiling point of the starting alcohol but above that of the aldehyde. In this way, the aldehyde distills out from the reaction mixture before oxidation to the acid occurs.

Problem 10.7   Write equations for these conversions.
(a) 1-Pentanol → pentanal
(b) Cyclohexanol → cyclohexanone
(c) *p*-Methylbenzyl alcohol → *p*-methylbenzaldehyde

Another oxidation method that yields aldehydes and ketones is the ozonolysis of alkenes. The reaction, which you have already seen (Sec. 3.23), proceeds in several stages. The net result, however, is the presence of two C=O bonds for every C=C that was initially present in the alkene. For example,

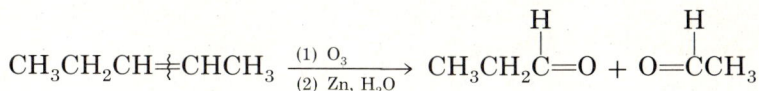

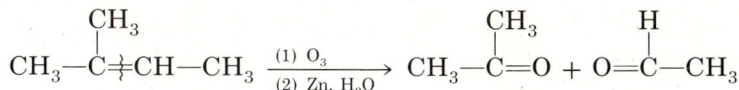

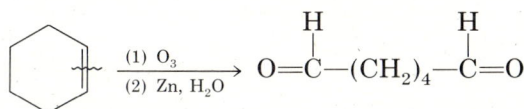

$$CH_3CH_2CH\!\!=\!\!CHCH_3 \xrightarrow[\text{(2) Zn, H}_2\text{O}]{\text{(1) O}_3} CH_3CH_2\overset{\overset{\displaystyle H}{|}}{C}\!\!=\!\!O + O\!\!=\!\!\overset{\overset{\displaystyle H}{|}}{C}CH_3$$

$$CH_3\!-\!\overset{\overset{\displaystyle CH_3}{|}}{C}\!\!=\!\!CH\!-\!CH_3 \xrightarrow[\text{(2) Zn, H}_2\text{O}]{\text{(1) O}_3} CH_3\!-\!\overset{\overset{\displaystyle CH_3}{|}}{C}\!\!=\!\!O + O\!\!=\!\!\overset{\overset{\displaystyle H}{|}}{C}\!-\!CH_3$$

$$\xrightarrow[\text{(2) Zn, H}_2\text{O}]{\text{(1) O}_3} O\!\!=\!\!\overset{\overset{\displaystyle H}{|}}{C}\!-\!(CH_2)_4\!-\!\overset{\overset{\displaystyle H}{|}}{C}\!\!=\!\!O$$

As pointed out in Chapter 3, the great utility of ozonolysis is that it provides an identification method to locate the position of a double bond in an alkene of unknown structure.

Problem 10.8 Write the structures of the product(s) expected on treatment of each alkene with ozone followed by Zn, H$_2$O.

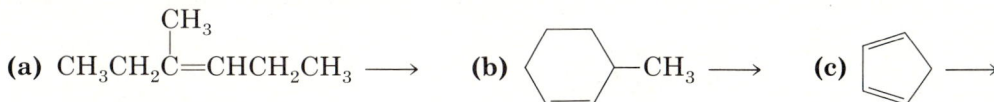

(a) $CH_3CH_2\overset{\overset{\displaystyle CH_3}{|}}{C}\!\!=\!\!CHCH_2CH_3 \longrightarrow$ (b) ⬡—CH$_3$ $\longrightarrow$ (c) ⬠ $\longrightarrow$

Hydration of alkynes is another reaction we encountered previously (Sec. 4.12D). Recall that addition of water to C≡C follows Markovnikov's rule. An unstable enol compound is formed initially, which rapidly rearranges to a stable carbonyl compound (keto form). The only aldehyde that can be obtained by this method is acetaldehyde, by the addition of water to acetylene. All other alkynes yield ketones on hydration.

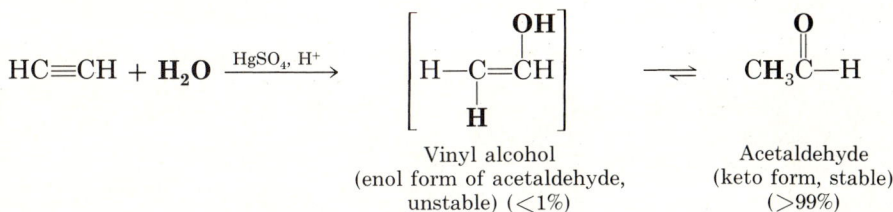

$$HC\!\!\equiv\!\!CH + H_2O \xrightarrow{\text{HgSO}_4,\ \text{H}^+} \left[H\!-\!\overset{\overset{\displaystyle OH}{|}}{\underset{\underset{\displaystyle H}{|}}{C}}\!\!=\!\!CH \right] \rightleftharpoons CH_3\overset{\overset{\displaystyle O}{||}}{C}\!-\!H$$

Vinyl alcohol
(enol form of acetaldehyde,
unstable) (<1%)

Acetaldehyde
(keto form, stable)
(>99%)

$$CH_3C{\equiv}CH + H_2O \xrightarrow{HgSO_4,\ H^+} \left[\underset{\text{OH}}{CH_3C{=}CH_2} \right] \rightleftharpoons \underset{\text{O}}{CH_3CCH_3}$$

(enol form, unstable) (keto form, stable)
(<1%) (>99%)

Formation of ketones by this method depends primarily on the availability of the starting alkyne.

Problem 10.9 Write structures of the carbonyl compounds obtained on hydration of these alkynes.

(a) $CH_3CH_2C{\equiv}CH \longrightarrow$ (b) 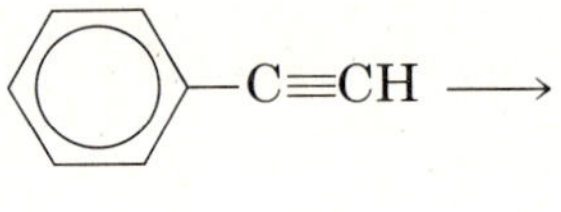 ⬡—C≡CH ⟶

10.9 Friedel–Crafts Acylation

A general method for preparing ketones that contain an aromatic ring is the **Friedel–Crafts acylation** reaction. The reaction involves treatment of an aromatic ring with an acyl chloride, $R{-}\overset{O}{\underset{\|}{C}}{-}Cl$, in the presence of $AlCl_3$, which acts as a catalyst.

General equation

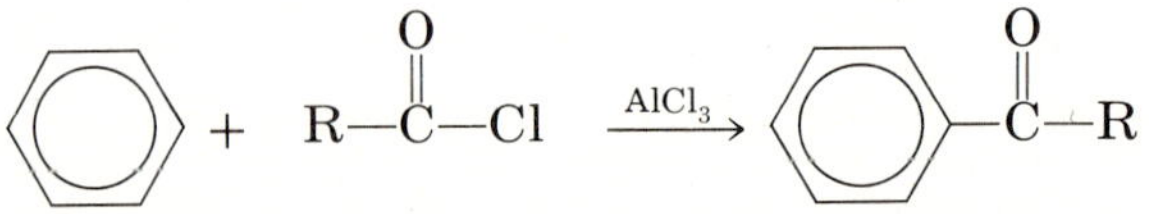

⬡ + R—C—Cl $\xrightarrow{AlCl_3}$ ⬡—C—R

An acyl chloride   A ketone

Specific examples

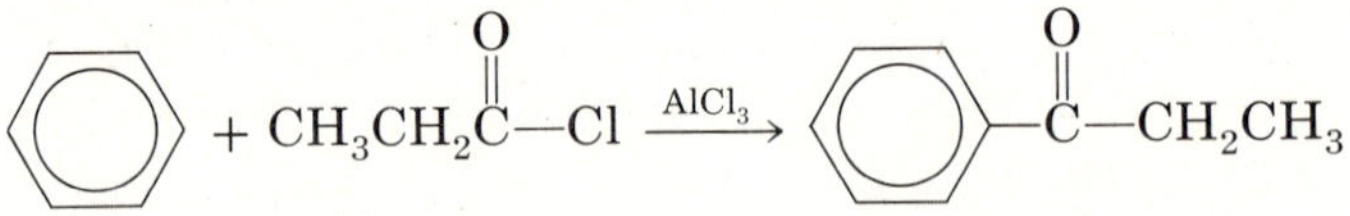

⬡ + $CH_3CH_2\overset{O}{\underset{\|}{C}}{-}Cl$ $\xrightarrow{AlCl_3}$ ⬡—C—CH_2CH_3

Propionyl chloride   Ethyl phenyl ketone
(Propiophenone)

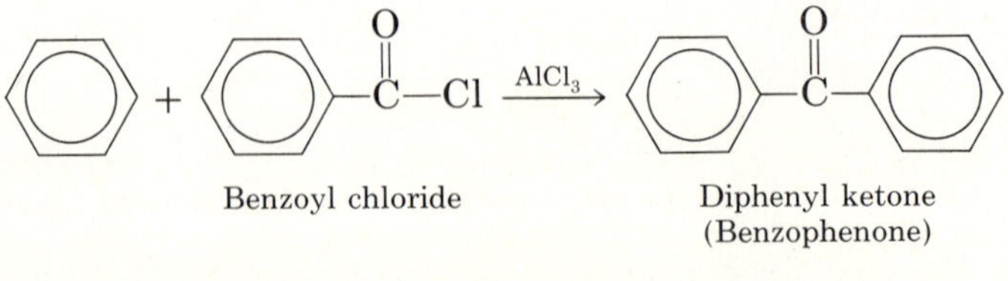

⬡ + ⬡—C—Cl $\xrightarrow{AlCl_3}$ ⬡—C—⬡

Benzoyl chloride   Diphenyl ketone
(Benzophenone)

An interesting application of Friedel–Crafts acylation is an intramolecular reaction resulting in ring closure.

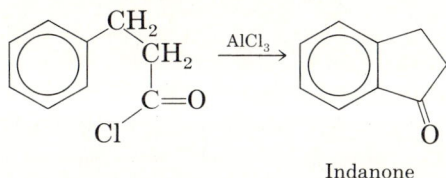

Indanone

The synthesis of acyl chlorides, which are needed in all these reactions, is discussed in Chapter 12.

Problem 10.10 Draw the structure(s) of the appropriate reagents needed to prepare these ketones.

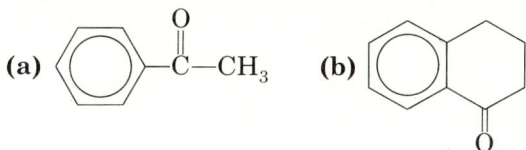

(a)

(b)

Industrial Preparations of Aldehydes *10.10*

The two simplest aldehydes, formaldehyde and acetaldehyde, are prepared on an industrial scale by special methods. **Formaldehyde** is produced by catalytic oxidation. In the commercial process, methanol is oxidized with air over a metal catalyst, such as Cu or Ag, in the temperature range 500–650°C.

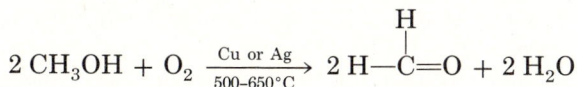

$$2\ CH_3OH + O_2 \xrightarrow[500-650°C]{Cu\ or\ Ag} 2\ \overset{\overset{\textstyle H}{|}}{H-C}=O + 2\ H_2O$$

Because of its extreme reactivity, formaldehyde cannot be isolated or handled readily in the pure form. It is often produced and marketed as an aqueous solution, known as *formalin,* which is formed by passing the formaldehyde–water mixture obtained in the reaction into water. The solution contains about 37% formaldehyde by weight. Formaldehyde reacts also with itself to form a solid polymer, *paraformaldehyde,* that is stable and easy to handle. Paraformaldehyde can be converted back to gaseous formaldehyde by heating.

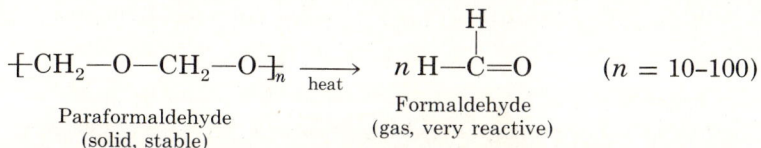

$$\left[CH_2-O-CH_2-O \right]_n \xrightarrow[heat]{} n\ \overset{\overset{\textstyle H}{|}}{H-C}=O \qquad (n = 10-100)$$

Paraformaldehyde
(solid, stable)

Formaldehyde
(gas, very reactive)

One industrial method for preparing **acetaldehyde** is by the direct oxidation of ethylene. Known as the **Wacker process,** the method uses palladium chloride as a catalyst and an air–water mixture as the oxidizing agent. The net overall reaction is

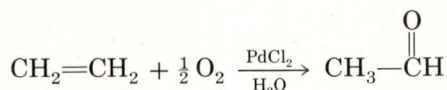

$$CH_2{=}CH_2 + \tfrac{1}{2}\,O_2 \xrightarrow[\text{H}_2\text{O}]{\text{PdCl}_2} CH_3{-}\overset{\overset{\textstyle O}{\|}}{C}H$$

The Wacker process has assumed increasing importance in recent years. In fact, it has displaced the older method of producing acetaldehyde by the hydration of acetylene.

Acetaldehyde is quite reactive and difficult to handle in its pure form (if you recall, it has a boiling point of 21°C). Often it is marketed in the liquid form as *paraldehyde,* which is a trimer of acetaldehyde that is more stable and easier to handle than acetaldehyde itself. On heating in the presence of acids, paraldehyde decomposes readily to yield three moles of acetaldehyde.

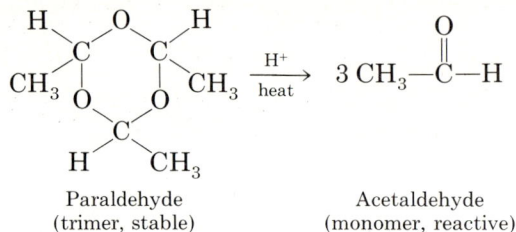

Paraldehyde	Acetaldehyde
(trimer, stable)	(monomer, reactive)

10.11   Uses of Important Aldehydes and Ketones

The structures of some important aldehydes and ketones are shown in Figure 10.3. The simpler carbonyl compounds, the first three entries in each column of Figure 10.3, are prepared commercially as synthetic intermediates or as solvents. The other compounds, whose structures are often somewhat complex, occur in a variety of natural products.

About 2 billion pounds of *formaldehyde* are produced annually. Most of it enters into the manufacture of resins (such as Bakelite and Melmac), which are used as protective coatings, for molded articles, and to impart "permanent press" and "wash and wear" qualities to clothing. The aqueous solution of formaldehyde, formalin, is an effective disinfectant used in hospitals to sterilize surgical instruments, gloves, and other contaminated articles. Although it is too irritating to be used on live tissues, formalin has been employed as an embalming fluid and as a preservative for anatomical specimens because it causes hardening of the skin.

Acetaldehyde is a synthetic intermediate used in the manufacture of acetic acid, ethyl acetate, vinyl acetate, and so on. These compounds go into the production of synthetic rubber and water-based paints, among other applications.

Chloral is a starting material for the manufacture of the insecticide DDT.

The lower aldehydes have sharp, acrid odors. But as the size of the molecule increases, the odor becomes less pungent and more fragrant. In fact, many naturally occurring aldehydes (and ketones) have been used in the blending of perfumes and as flavoring agents. *Butanal,* for example, is a constituent of the

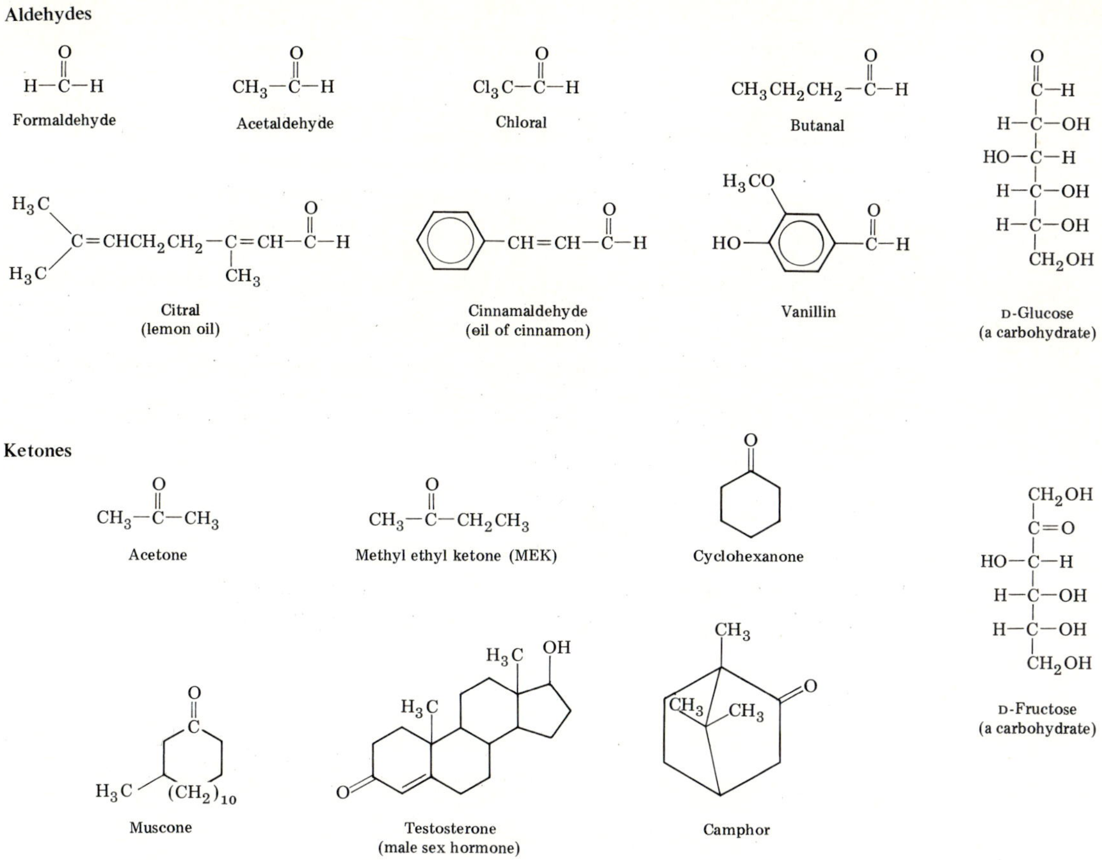

Aldehydes

$$H-\overset{\overset{\displaystyle O}{\|}}{C}-H$$

Formaldehyde

$$CH_3-\overset{\overset{\displaystyle O}{\|}}{C}-H$$

Acetaldehyde

$$Cl_3C-\overset{\overset{\displaystyle O}{\|}}{C}-H$$

Chloral

$$CH_3CH_2CH_2-\overset{\overset{\displaystyle O}{\|}}{C}-H$$

Butanal

$$\begin{array}{c} \overset{\overset{\displaystyle O}{\|}}{C}-H \\ H-C-OH \\ HO-C-H \\ H-C-OH \\ H-C-OH \\ CH_2OH \end{array}$$

D-Glucose
(a carbohydrate)

$$\underset{H_3C}{\overset{H_3C}{>}}C=CHCH_2CH_2-\overset{\overset{\displaystyle}{\underset{CH_3}{|}}}{C}=CH-\overset{\overset{\displaystyle O}{\|}}{C}-H$$

Citral
(lemon oil)

$$\text{C}_6\text{H}_5-CH=CH-\overset{\overset{\displaystyle O}{\|}}{C}-H$$

Cinnamaldehyde
(oil of cinnamon)

Vanillin

Ketones

$$CH_3-\overset{\overset{\displaystyle O}{\|}}{C}-CH_3$$

Acetone

$$CH_3-\overset{\overset{\displaystyle O}{\|}}{C}-CH_2CH_3$$

Methyl ethyl ketone (MEK)

Cyclohexanone

$$\begin{array}{c} CH_2OH \\ C=O \\ HO-C-H \\ H-C-OH \\ H-C-OH \\ CH_2OH \end{array}$$

D-Fructose
(a carbohydrate)

Muscone

Testosterone
(male sex hormone)

Camphor

Figure 10.3 Structures and names of some important aldehydes and ketones.

aroma of fresh bread. *Citral* is the major component of lemon grass oil. *Benzaldehyde,* once called "oil of bitter almond" because of its presence in almond seed, is used as a flavoring agent and as an intermediate in the manufacture of drugs, dyes, and other organic products. *Cinnamaldehyde* is the chief constituent of the oil of cinnamon bark. *Vanillin* is the fragrant component of the vanilla bean.

In terms of its use, *acetone* is the most important ketone. Acetone is manufactured either by the oxidation of isopropyl alcohol or as a by-product of the cumene process for phenols (Sec. 7.14). Because of its miscibility in water and in nonpolar solvents, acetone is used in large quantities as a solvent for such products as paints, lacquers, and cellulose acetate.

Acetone also plays a role in human health. Normally, the concentration of acetone in the blood is very low (less than 1 mg/100 ml of blood). Under abnormal conditions, such as carbohydrate starvation or diabetes, the energy needs of the body cannot be supplied by its normal source, the carbohydrates, and the metabolism of fats takes over. The accelerated breakdown of fats causes a buildup of ketone bodies in the blood, one of which is acetone. Because acetone is excreted in the urine or is exhaled in the breath, the detection of an acetone odor in the breath of a patient may be an indication of a diabetic condition. If excess acetone and other ketone bodies are allowed to accumulate, death can result.

Another ketone, *methyl ethyl ketone* (MEK), is used primarily as a solvent for lacquers, paints, and plastics. The petroleum industry also utilizes MEK as a dewaxer of lubricating oils.

The principal application of *cyclohexanone* is as an intermediate in the manufacture of nylon, but it is sometimes used also as a high-boiling solvent.

Camphor is obtained from the bark of the camphor tree and is characterized by a fragrant, penetrating odor. Once used as a plasticizer for celluloid, camphor currently is used sometimes as an expectorant in the treatment of colds, in some liniments for the treatment of bruises and sprains, and as a weak analgesic.

Muscone, a cyclic ketone containing fifteen carbons in the ring, is the odorous component of musk, a sex attractant secreted by the Himalayan male musk deer. Muscone is especially valued for its ability to enhance other fragrances even when present in minute amounts. Many of the expensive perfumes on the market contain muscone as a fixative.

Testosterone, the major constituent of the male sex hormone group, is produced in testes and is responsible for the development of the secondary sex characteristics. Testosterone belongs to a class of natural products known as *steroids,* which are discussed in Chapter 13.

The last entries in Figure 10.3, glucose and fructose, are representatives of *carbohydrates*. Because of their importance, carbohydrates will be treated separately in Chapter 11.

10.12 Nucleophilic Addition Reactions of Aldehydes and Ketones

Aldehydes and ketones undergo generally the same kinds of reactions. The most typical of these are **nucleophilic addition reactions** to the carbon–oxygen double bond. In nucleophilic addition reactions the partially positive carbonyl carbon undergoes attack by electron-rich reagents, or *nucleophiles* (Nu :), and the partially negative carbonyl oxygen is attacked by electron-deficient reagents, or *electrophiles* (E$^+$).

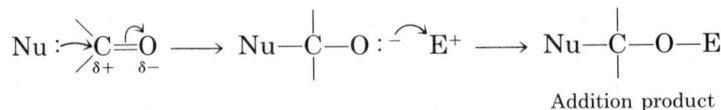

$$\text{Nu} : \longrightarrow \underset{\delta+}{\text{C}} = \underset{\delta-}{\text{O}} \longrightarrow \text{Nu}-\overset{|}{\underset{|}{\text{C}}}-\text{O} : ^- E^+ \longrightarrow \text{Nu}-\overset{|}{\underset{|}{\text{C}}}-\text{O}-\text{E}$$

Addition product

The typical nucleophilic addition reactions of carbonyl compounds are summarized here, and discussed in the sections that follow.

1. Addition of metal hydrides: formation of alcohols [Sec. 10.13]

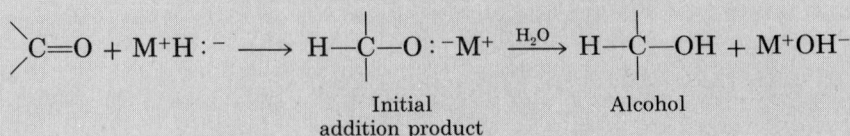

$$\diagdown \text{C}=\text{O} + \text{M}^+\text{H} : ^- \longrightarrow \text{H}-\overset{|}{\underset{|}{\text{C}}}-\text{O} : ^-\text{M}^+ \xrightarrow{\text{H}_2\text{O}} \text{H}-\overset{|}{\underset{|}{\text{C}}}-\text{OH} + \text{M}^+\text{OH}^-$$

Initial   Alcohol
addition product

2. Addition of Grignard reagents: formation of alcohols   [Sec. 10.14]

$$\overset{}{\underset{}{\diagdown}}C=O + \overset{\delta-}{R}-\overset{\delta+}{MgX} \longrightarrow R-\overset{|}{\underset{|}{C}}-\bar{O}\overset{+}{Mg}X \xrightarrow{H_2O,\ H^+}$$

Initial
addition product

$$R-\overset{|}{\underset{|}{C}}-OH + Mg(OH)X$$

Alcohol

3. Addition of hydrogen cyanide: formation of cyanohydrins [Sec. 10.15]

$$\overset{}{\underset{}{\diagdown}}C=O + {}^-:CN \longrightarrow NC-\overset{|}{\underset{|}{C}}-O:^- \xrightarrow{H^+} NC-\overset{|}{\underset{|}{C}}-OH$$

Initial addition product   Cyanohydrin

4. Addition of alcohols: formation of hemiacetals and acetals
[Sec. 10.16]

$$\overset{}{\underset{}{\diagdown}}C=O + R\ddot{O}H \xrightarrow{H^+} RO-\overset{|}{\underset{|}{C}}-OH + R'OH \xrightarrow{H^+}$$

Hemiacetal

$$RO-\overset{|}{\underset{|}{C}}-OR' + H_2O$$

Acetal

Addition of Metal Hydrides: *10.13*
Formation of Alcohols

Aldehydes can be reduced to primary alcohols, and ketones can be reduced to secondary alcohols when treated with **metal hydrides** (abbreviated $M^+H:^-$), such as sodium borohydride, $NaBH_4$. Sodium borohydride reduces the carbonyl group by nucleophilic addition to the carbonyl carbon. The initial addition product, after being hydrolyzed, yields an alcohol.

$$M^+H:^- + \underset{\delta+\ \ \delta-}{\diagdown C=O} \longrightarrow H-\overset{|}{\underset{|}{C}}-O:^-M^+ \xrightarrow{H_2O} H-\overset{|}{\underset{|}{C}}-OH + M^+OH^-$$

Initial addition product   Alcohol

Although the same result may be achieved by catalytic hydrogenation, $NaBH_4$ has the advantage of selectively reducing the carbonyl group of *non-conjugated* unsaturated aldehydes or ketones. Contrast, for example, the treatment of 3-pentenal with hydrogen and a catalyst and treatment of the same compound with $NaBH_4$ followed by water.

$$\text{CH}_3\text{CH}=\text{CHCH}_2\overset{\overset{\displaystyle H}{|}}{\text{C}}=\text{O} \quad\begin{cases} \xrightarrow{2\ \text{H}_2/\text{Pt}} \text{CH}_3\text{CH}_2\text{CH}_2\text{CH}_2\text{CH}_2\text{OH} \\ \qquad\qquad\qquad \text{1-Pentanol} \\ \\ \xrightarrow[\text{(2) H}_2\text{O}]{\text{(1) NaBH}_4} \text{CH}_3\text{CH}=\text{CHCH}_2\text{CH}_2\text{OH} \\ \qquad\qquad\qquad \text{3-Penten-1-ol} \end{cases}$$

3-Pentenal

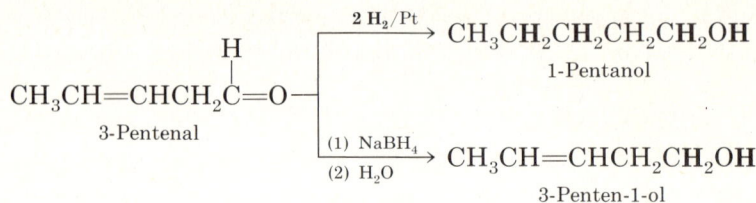

The reaction of 3-pentenal with NaBH_4 leaves the carbon–carbon double bond intact because these bonds are not usually susceptible to attack by nucleophiles.

Problem 10.11 Write the structures of the alcohols expected from these reactions.

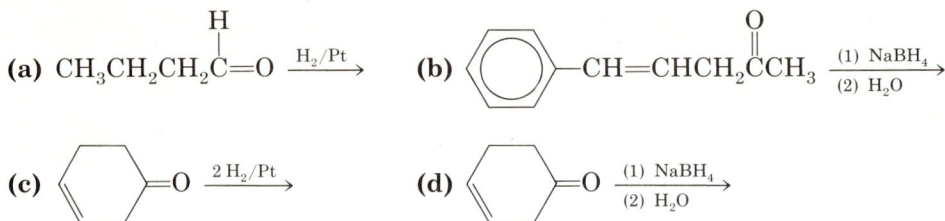

(a) $\text{CH}_3\text{CH}_2\text{CH}_2\overset{\overset{\displaystyle H}{|}}{\text{C}}=\text{O} \xrightarrow{\text{H}_2/\text{Pt}}$

(b) ⬡$-\text{CH}=\text{CHCH}_2\overset{\overset{\displaystyle O}{\|}}{\text{C}}\text{CH}_3 \xrightarrow[\text{(2) H}_2\text{O}]{\text{(1) NaBH}_4}$

(c) ⬡$=\text{O} \xrightarrow{2\ \text{H}_2/\text{Pt}}$

(d) ⬡$=\text{O} \xrightarrow[\text{(2) H}_2\text{O}]{\text{(1) NaBH}_4}$

10.14 Addition of Grignard Reagents: Formation of Alcohols

The reaction of **Grignard reagents, RMgX,** with aldehydes and ketones is one of the most versatile methods for synthesizing all kinds of alcohols. In its simplest form, a Grignard reagent may be regarded as a highly polar compound that has its negative charge on the carbon attached to the magnesium and its positive charge on the metal, thus: $\text{R}^-\text{Mg}^+\text{X}$. The reactions between carbonyl compounds and Grignard reagents thus follow the general pattern of addition reactions to the carbonyl group. The carbonyl carbon is attacked by the nucleophile, $\text{R}:^-$, whereas the carbonyl oxygen combines with the positive fragment, Mg^+X. The addition product, after it is formed, is hydrolyzed with aqueous acid to give an alcohol.

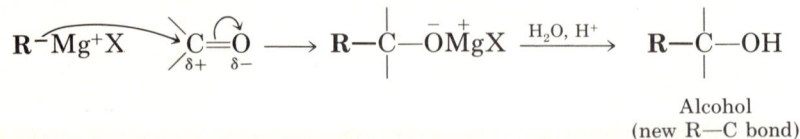

$$\text{R}^-\text{Mg}^+\text{X} \quad \overset{\curvearrowright}{\underset{\delta+\ \ \delta-}{\text{C}=\text{O}}} \longrightarrow \text{R}-\overset{|}{\underset{|}{\text{C}}}-\overset{-}{\text{O}}\overset{+}{\text{M}}\text{gX} \xrightarrow{\text{H}_2\text{O, H}^+} \text{R}-\overset{|}{\underset{|}{\text{C}}}-\text{OH}$$

Alcohol
(new R—C bond)

The addition of Grignard reagents to carbonyl compounds is a very useful reaction. First, the method allows us to synthesize complex alcohols from simpler reactants. This is because the alcohol that is produced contains a newly formed carbon–carbon bond. Second, the hydroxyl group in the newly formed alcohol provides a convenient handle for further transformations. Third, the method is versatile: it is possible to prepare 1°, 2°, or 3° alcohols, depending on what kind of carbonyl compound undergoes the addition reaction. For example, if *formaldehyde* is used as the starting carbonyl compound, addition of Grignard reagents to it always yields *primary alcohols*.

Grignard Reagents:
Formation of
Alcohols

General equation

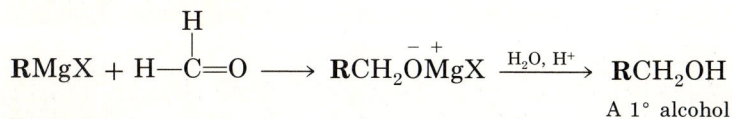

$$\begin{matrix} & H \\ & | \\ RMgX + H-C\!\!=\!\!O & \end{matrix} \longrightarrow RCH_2\overset{-}{O}\overset{+}{M}gX \xrightarrow{H_2O,\ H^+} RCH_2OH$$

A 1° alcohol

Specific example

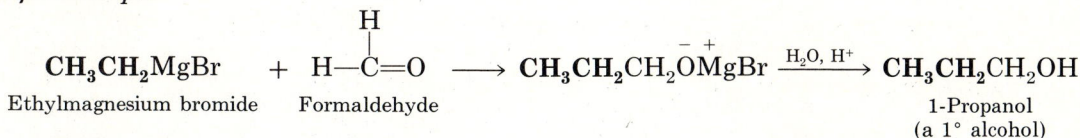

$$\begin{matrix} & H \\ & | \\ CH_3CH_2MgBr \quad + \quad H-C\!\!=\!\!O & \end{matrix} \longrightarrow CH_3CH_2CH_2\overset{-}{O}\overset{+}{M}gBr \xrightarrow{H_2O,\ H^+} CH_3CH_2CH_2OH$$

Ethylmagnesium bromide   Formaldehyde   1-Propanol
(a 1° alcohol)

If any aldehyde other than formaldehyde is used, the addition of Grignard reagent yields *secondary alcohols*.

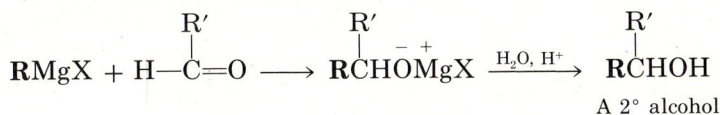

$$\begin{matrix} & R' & & R' & & R' \\ & | & & | & & | \\ RMgX + H-C\!\!=\!\!O & \longrightarrow & RCH\overset{-}{O}\overset{+}{M}gX & \xrightarrow{H_2O,\ H^+} & RCHOH \end{matrix}$$

A 2° alcohol

Tertiary alcohols are obtained on addition of Grignard reagents to ketones.

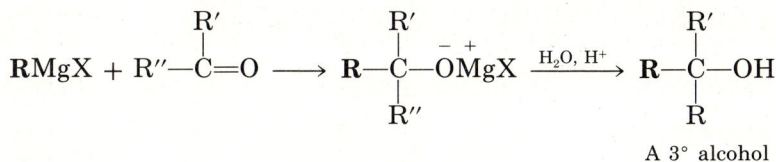

$$\begin{matrix} & R' & & R' & & R' \\ & | & & | & & | \\ RMgX + R''-C\!\!=\!\!O & \longrightarrow & R-C-\overset{-}{O}\overset{+}{M}gX & \xrightarrow{H_2O,\ H^+} & R-C-OH \\ & & & | & & | \\ & & & R'' & & R \end{matrix}$$

A 3° alcohol

For a better understanding of the synthetic utility of the reaction of Grignard reagents, let us work through an example.

> **Example 10.1** Suppose you were asked to prepare 2-methyl-2-pentanol using any organic compounds that contain no more than three carbon atoms.
> **Solution** As with many problems involving the synthesis of a compound, the best way to arrive at an answer is to work backward.
> (1) Write the structure of the desired alcohol.
>
> $$\begin{matrix} & & CH_3 \\ & & | \\ CH_3CH_2CH_2 & - & C-CH_3 \\ & & | \\ & & OH \end{matrix}$$
>
> (2) Determine what class of alcohol you are dealing with. In this case, it is a tertiary alcohol because the carbinol carbon is attached to three other carbon atoms.
> (3) Because 2-methyl-2-pentanol is a 3° alcohol, it can be obtained by addition of a Grignard reagent to a ketone. The question is, which Grignard reagent is to be added to which ketone?

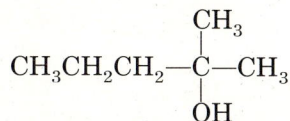

(4) Before answering the question specifically, you need to make the following generalizations. If the desired alcohol is a secondary one and the two R groups attached to C—O—H are the same, there is only one combination of Grignard reagent and aldehyde that gives the desired product. If the desired alcohol is a secondary one and the two R groups bonded to C—O—H are different, then two combinations of Grignard reagent and aldehyde are possible.

For tertiary alcohols the following generalizations are valid. If all three R groups attached to C—O—H are the same, there is one combination of Grignard reagent and ketone possible. If two of three R groups are the same, there are two possible combinations. And if all three R groups attached to C—O—H are different, then there are three possible combinations.

(5) Going back to the structure of 2-methyl-2-pentanol, you see that two of the three R groups attached to C—O—H are the same. There are therefore two possible combinations, (a) and (b), of Grignard reagent and ketone that will give the desired alcohol.

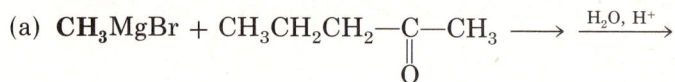

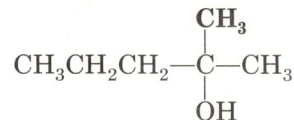

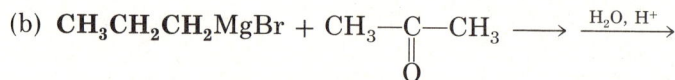

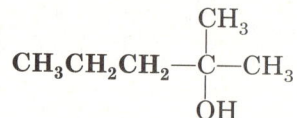

(a) $\text{CH}_3\text{MgBr} + \text{CH}_3\text{CH}_2\text{CH}_2\!-\!\overset{\displaystyle }{\underset{\displaystyle \underset{\|}{O}}{C}}\!-\!\text{CH}_3 \longrightarrow \xrightarrow{\text{H}_2\text{O, H}^+}$

$$\text{CH}_3\text{CH}_2\text{CH}_2\!-\!\overset{\displaystyle \text{CH}_3}{\underset{\displaystyle \text{OH}}{C}}\!-\!\text{CH}_3$$

(b) $\text{CH}_3\text{CH}_2\text{CH}_2\text{MgBr} + \text{CH}_3\!-\!\overset{\displaystyle }{\underset{\displaystyle \underset{\|}{O}}{C}}\!-\!\text{CH}_3 \longrightarrow \xrightarrow{\text{H}_2\text{O, H}^+}$

$$\text{CH}_3\text{CH}_2\text{CH}_2\!-\!\overset{\displaystyle \text{CH}_3}{\underset{\displaystyle \text{OH}}{C}}\!-\!\text{CH}_3$$

Normally, the combination that is most readily available and most economical is chosen. Here the choice of starting materials was limited to organic compounds that contain no more than three carbon atoms, so the answer must be (b), which happens to be readily available and inexpensive as well.

Problem 10.12 What combination(s) of Grignard reagent and carbonyl compound would be suitable for the preparation of these alcohols?

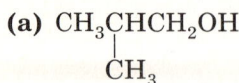

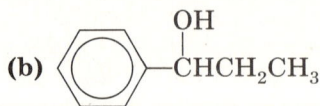

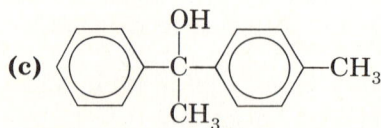

(a) $\text{CH}_3\overset{\displaystyle }{\underset{\displaystyle \text{CH}_3}{\text{CH}}}\text{CH}_2\text{OH}$

(b) $\overset{\displaystyle \text{OH}}{\underset{\displaystyle }{\text{C}_6\text{H}_5\!-\!\text{CHCH}_2\text{CH}_3}}$

(c) $\text{C}_6\text{H}_5\!-\!\overset{\displaystyle \text{OH}}{\underset{\displaystyle \text{CH}_3}{C}}\!-\!\text{C}_6\text{H}_4\!-\!\text{CH}_3$

Problem 10.13 Show how the following compounds could be prepared by addition of Grignard reagents to carbonyl compounds, followed by dehydration or oxidation of the alcohol. Use starting materials with no more than four carbons.
(a) 3-Heptanone   **(b)** 3-Methyl-1-butene   **(c)** 2-Methyl-2-pentene

Addition of Hydrogen Cyanide: *10.15*
Formation of Cyanohydrins

The addition of hydrogen cyanide to carbonyl compounds is another synthetically useful reaction. The addition is catalyzed by cyanide ion, $^-:C{\equiv}N$. The cyanide ion is a good nucleophile that readily attacks the carbonyl carbon. The intermediate that is formed abstracts a proton from HCN to give the final addition product, called a **cyanohydrin.**

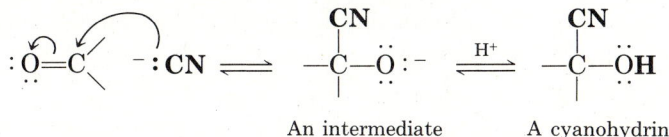

$$:\overset{..}{\underset{..}{O}}{=}C \qquad ^-:CN \rightleftharpoons \quad \underset{|}{\overset{CN}{\underset{|}{C}}}{-}\overset{..}{\underset{..}{O}}:^- \quad \overset{H^+}{\rightleftharpoons} \quad \underset{|}{\overset{CN}{\underset{|}{C}}}{-}\overset{..}{O}H$$

$$\text{An intermediate} \qquad \text{A cyanohydrin}$$

Mandelonitrile, the cyanohydrin obtained when hydrogen cyanide is added to benzaldehyde, can be found in the glands of some animals. If the animal is threatened, an enzyme is released that promotes the dissociation of mandelonitrile into benzaldehyde and hydrogen cyanide. The mixture of the two constitutes an effective repellent against attackers.

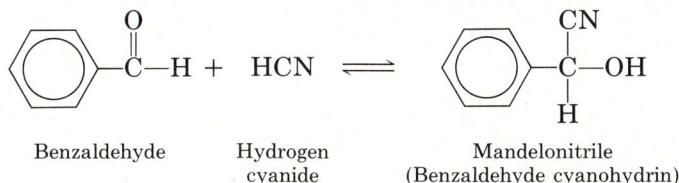

| Benzaldehyde | Hydrogen cyanide | Mandelonitrile (Benzaldehyde cyanohydrin) |

Cyanohydrins are valuable synthetic intermediates. On treatment with aqueous acid (hydrolysis), they yield hydroxy acids with *one more carbon than the starting aldehyde or ketone.*

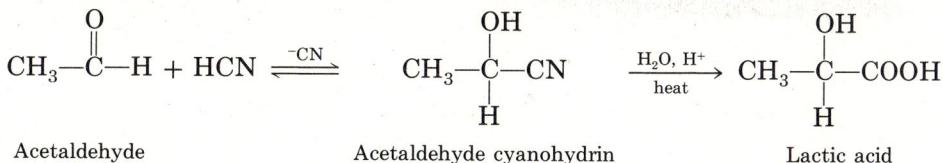

$$CH_3{-}\overset{O}{\overset{\|}{C}}{-}H + HCN \overset{^-CN}{\rightleftharpoons} \quad CH_3{-}\underset{\underset{H}{|}}{\overset{\overset{OH}{|}}{C}}{-}CN \quad \overset{H_2O,\ H^+}{\underset{heat}{\longrightarrow}} CH_3{-}\underset{\underset{H}{|}}{\overset{\overset{OH}{|}}{C}}{-}COOH$$

$$\text{Acetaldehyde} \qquad \text{Acetaldehyde cyanohydrin} \qquad \text{Lactic acid}$$

Problem 10.14 Draw the structural formulas for the products **A** through **F** in the following reactions.

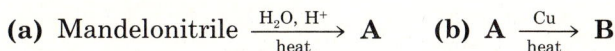

(a) Mandelonitrile $\overset{H_2O,\ H^+}{\underset{heat}{\longrightarrow}}$ **A**   **(b) A** $\overset{Cu}{\underset{heat}{\longrightarrow}}$ **B**

(c) Acetone + HCN $\underset{}{\overset{-CN}{\rightleftharpoons}}$ **C**   (d) **C** $\xrightarrow[\text{heat}]{H_2O, H^+}$ **D**

(e) **D** $\xrightarrow[\text{heat}]{H^+}$ **E** $(C_4H_6O_2)$ (f) **E** + Br_2/CCl_4 $\longrightarrow$ **F** $(C_4H_6Br_2O_2)$

10.16 Addition of Alcohols: Formation of Hemiacetals and Acetals

A   Hemiacetals (and Hemiketals)

Even weak nucleophiles, such as water and alcohols, can add to the carbonyl group of aldehydes and ketones. In such cases, an acid catalyst is required to speed the rate of reaction. The addition of one mole of an alcohol to the carbonyl group of an aldehyde yields a **hemiacetal,** and the addition of one mole of an alcohol to a ketone gives a **hemiketal.** Hemiacetals and hemiketals have an alkoxy group (OR) and a hydroxy group (OH) attached to the *same* carbon.

General equations

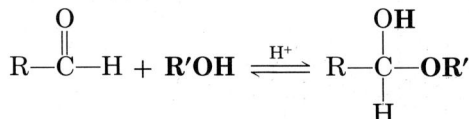

Aldehyde   Alcohol   Hemiacetal

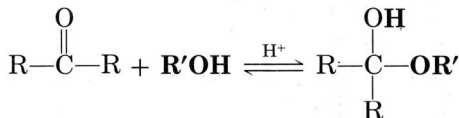

Ketone   Alcohol   Hemiketal

Specific examples

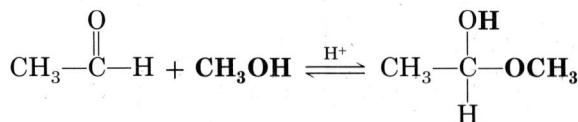

Ethanal   Methanol   1-Methoxyethanol
(a hemiacetal)

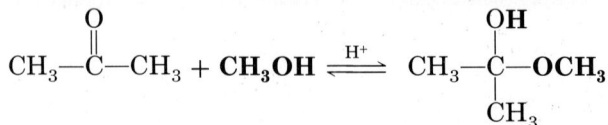

Propanone   Methanol   2-Methoxy-2-propanol
(a hemiketal)

The steps involved in the formation of 1-methoxyethanol, the specific example shown, are typical of hemiacetals in general.

Step 1. Protonation of the carbonyl oxygen: The carbonyl oxygen, acting as a Lewis base, attacks the proton. The resulting ion is stabilized by resonance to give a positive charge on the carbon, making the carbon more prone to attack by a nucleophile.

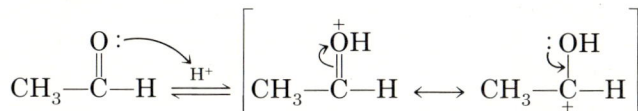

$$
CH_3-\overset{\overset{\displaystyle O:}{\|}}{C}-H \underset{H^+}{\rightleftharpoons} \left[CH_3-\overset{\overset{\displaystyle \overset{+}{O}H}{\|}}{C}-H \longleftrightarrow CH_3-\overset{\overset{\displaystyle :OH}{|}}{\underset{+}{C}}-H \right]
$$

Step 2. The positively charged carbon is attacked by the weak nucleophile methanol.

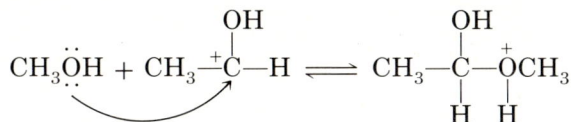

$$
CH_3\overset{..}{\underset{..}{O}}H + CH_3-\overset{\overset{\displaystyle OH}{|}}{\overset{+}{C}}-H \rightleftharpoons CH_3-\overset{\overset{\displaystyle OH}{|}}{\underset{\underset{\displaystyle H}{|}}{C}}-\overset{+}{O}CH_3
$$

Step 3. The loss of a proton regenerates the acid catalyst and simultaneously forms the hemiacetal.

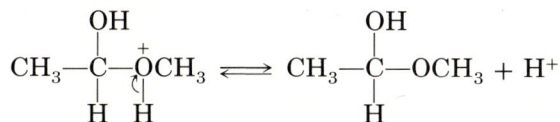

$$
CH_3-\overset{\overset{\displaystyle OH}{|}}{\underset{\underset{\displaystyle H\ \ H}{|}}{\overset{+}{C}}}-\overset{+}{O}CH_3 \rightleftharpoons CH_3-\overset{\overset{\displaystyle OH}{|}}{\underset{\underset{\displaystyle H}{|}}{C}}-OCH_3 + H^+
$$

Although most hemiacetals and hemiketals are too unstable to be isolated, cyclic hemiacetals and hemiketals are integral parts of the structure of simple sugars such as glucose and fructose that are discussed in the next chapter. A partial structure of glucose showing the hemiacetal group is

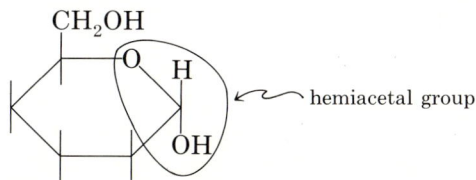

Partial structure of glucose

Problem 10.15 The mechanism of hemiketal formation follows the same steps as the mechanism of formation of hemiacetals. Show the steps that lead to the formation of 2-methoxy-2-propanol, the hemiketal of propanone and 1 mole of methanol.

B   Acetals

When hemiacetals are treated with an additional mole of alcohol in the presence of anhydrous acid, they are converted to **acetals.** Acetals have two alkoxy groups (OR) on the *same* carbon.

$$\text{R}-\underset{\underset{\text{H}}{|}}{\overset{\overset{\text{OH}}{|}}{\text{C}}}-\text{OR}' + \text{R}''\text{OH} \underset{}{\overset{\text{H}^+ \text{ (anhyd.)}}{\rightleftharpoons}} \text{R}-\underset{\underset{\text{H}}{|}}{\overset{\overset{\text{OR}''}{|}}{\text{C}}}-\text{OR}' + \text{H}_2\text{O}$$

Hemiacetal   Alcohol   Acetal

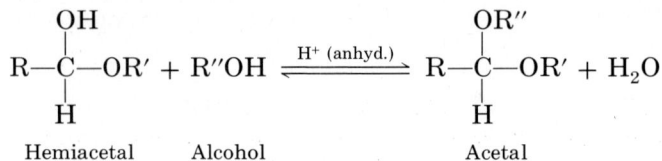

In contrast to hemiacetals, acetals are stable compounds that can be iso-lated easily. Chemically, acetals resemble ethers-they do not react with bases, oxidizing agents and reducing agents. Unlike ethers, which are stable to cold, dilute acids, *acetals are quite sensitive to aqueous acids*. Treatment of acetals with aqueous acids gives ultimately an aldehyde and 2 moles of alcohols. This reaction is just the reverse of acetal formation.

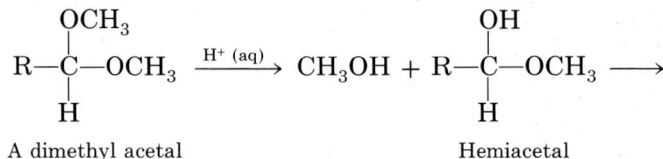

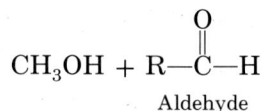

$$\text{R}-\underset{\underset{\text{H}}{|}}{\overset{\overset{\text{OCH}_3}{|}}{\text{C}}}-\text{OCH}_3 \xrightarrow{\text{H}^+ \text{ (aq)}} \text{CH}_3\text{OH} + \text{R}-\underset{\underset{\text{H}}{|}}{\overset{\overset{\text{OH}}{|}}{\text{C}}}-\text{OCH}_3 \longrightarrow$$

A dimethyl acetal   Hemiacetal

$$\text{CH}_3\text{OH} + \text{R}-\overset{\overset{\text{O}}{\|}}{\text{C}}-\text{H}$$

Aldehyde

Reaction of hemiketals with alcohols to form ketals seldom works and will not be discussed here.

Problem 10.16 Write structural formulas for **(a)** the hemiacetal derived by addition of ethanol to propanal and **(b)** the acetal derived by addition of a sec-ond mole of ethanol to the hemiacetal in (a).

10.17 Addition of Ammonia and Ammonia Derivatives

Nitrogen nucleophiles, such as ammonia ($:\text{NH}_3$) and substituted ammonias ($:\text{NH}_2—\text{Y}$), also add to the carbonyl group of aldehydes and ketones. The addi-tion reactions, like those of alcohols, are reversible and are acid catalyzed. For the sake of brevity the mechanism of nucleophilic addition of ammonia and ammonia derivatives to carbonyl is omitted. The net result of the addition is the conversion of a C=O group to a C=NY group.

$$\underset{/}{\overset{\backslash}{}}\text{C}=\text{O} + \text{H}_2\text{N}-\text{Y} \xrightarrow{\text{H}^+} \underset{/}{\overset{\backslash}{}}\text{C}=\text{N}-\text{Y} + \text{H}_2\text{O}$$

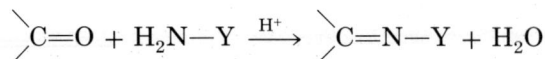

The nitrogen nucleophiles that react with carbonyl compounds are shown in Figure 10.4, along with the names and structures of the products. With most aldehydes and ketones, ammonia addition products are too unstable to give use-ful derivatives (first entry in Fig. 10.4). Again, formaldehyde is an exception. The product in this case is a crystalline solid of unusual structure, hexamethyl-enetetramine.

$$6 \text{ H}-\overset{\overset{\displaystyle O}{\|}}{C}-\text{H} + 4 \text{ NH}_3 \longrightarrow \quad + 6 \text{ H}_2\text{O}$$

Hexamethylenetetramine

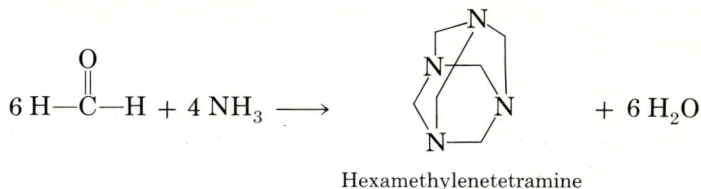

Hexamethylenetetramine has the distinction of being the first organic compound whose structure was confirmed by x-ray diffraction. The compound has been used as a urinary antiseptic and as an intermediate in the manufacture of the high explosive *cyclonite*.

The other nitrogen derivatives of carbonyl compounds, from oxime to semicarbazone, are used primarily to identify aldehydes and ketones. This is possible because these derivatives are mostly solid, crystalline compounds with characteristic melting points. The data in Table 10.1 illustrate how the conversion of carbonyl compounds into one or more nitrogen derivatives may be used for identification purposes.

2,4-Dinitrophenylhydrazones (2,4-DNPH) are especially useful derivatives. These compounds, which are yellow-orange or red solids, not only can be used to distinguish one carbonyl compound from another but also serve *as a test to differentiate carbonyl compounds from other classes of compounds*. Thus, treatment of an aldehyde or a ketone with 2,4-dinitrophenylhydrazine yields a 2,4-DNPH (yellow-orange or red precipitate). Alcohols, ethers, alkanes, alkenes, alkynes, and so on, give a negative test with 2,4-dinitrophenylhydrazine.

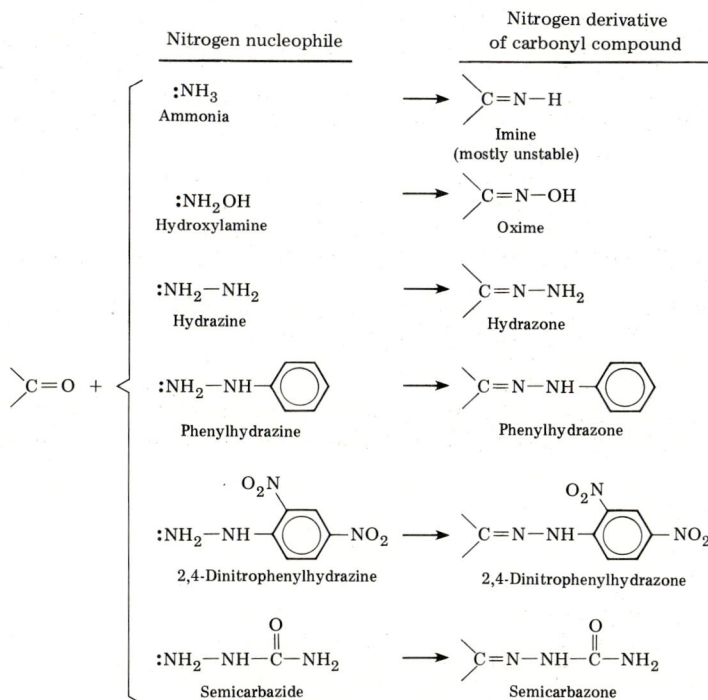

Figure 10.4 Structures and names of nitrogen nucleophiles that react with carbonyl compounds.

Table 10.1 Derivatives of Carbonyl Compounds Used for Identification Purposes

Carbonyl compound	Bp (°C)	Mp of derivative (°C)	
		2,4-Dinitrophenylhydrazone	Semicarbazone
Butanal	75	123	106
2,2-Dimethylpropanal	75	209	190
Hexanal	131	104	106
Cyclopentanone	131	146	203
4-Heptanone	145	75	133
5-Methyl-2-hexanone	145	95	147

Problem 10.17 Draw the structures of the products from the reactions of (1) acetone and (2) benzaldehyde with these compounds.
(a) Hydroxylamine   **(b)** Phenylhydrazine
(c) 2,4-Dinitrophenylhydrazine   **(d)** Semicarbazide

10.18 Reactions Involving the Enolate Anion

All the reactions discussed so far involve the C=O group directly. Let us now look at reactions that involve the carbon atom next to the carbonyl group.

A carbon atom next to the carbonyl group is called an **alpha carbon,** denoted **α carbon.** By the same token, any hydrogen attached to an α carbon is referred to as an **alpha hydrogen (α hydrogen).** The α hydrogens of aldehydes and ketones are somewhat special in that they are slightly acidic. This slight acidity is due to the fact that the anion, which results from the removal of an α hydrogen by a base B⁻, is stabilized by resonance. The resonance-stabilized anion is called the **enolate anion.**

Removal of H from an   Resonance-stabilized enolate anion
aldehyde or ketone by B⁻   (a nucleophile)

Because the α carbon of enolate anions possesses appreciable negative character, we may expect enolate anions to act as nucleophiles. Our predictions are borne out, as we shall see in the sections that follow.

Problem 10.18
(a) Draw the resonance-contributing structures of the enolate anion derived from propanal.
(b) Predict the product(s) formed on protonation of the enolate structures in (a). (*Hint:* Review enol ⇌ keto equilibrium in Sec. 4.12D.)

When enolate anions, acting as nucleophiles, add to the carbonyl group of aldehydes and ketones, a *larger* organic molecule is formed with loss of water, a **condensation process.** The simplest of these condensations is the addition of the enolate anion derived from acetaldehyde to the C=O bond of a nonionized acetaldehyde molecule. The reaction is known as the **aldol condensation.** Because the aldol condensation is the prototype of all such additions of enolate anions to aldehydes and ketones, let us consider its sequence of steps in detail.

Step 1. When acetaldehyde is treated with dilute sodium hydroxide, a small but significant number of molecules are converted to the enolate anion.

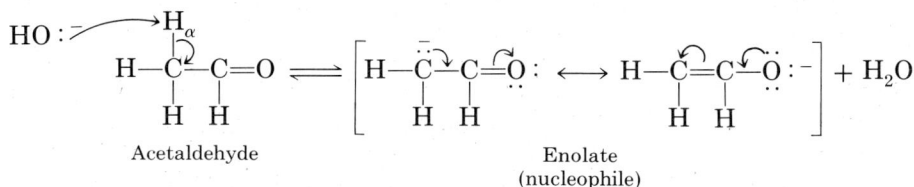

Acetaldehyde Enolate (nucleophile)

Step 2. The enolate anion, acting as a nucleophile, adds to the C=O bond of a nonionized acetaldehyde molecule left in solution to give an adduct that carries a negative charge on the oxygen. The adduct contains a newly formed carbon–carbon bond.

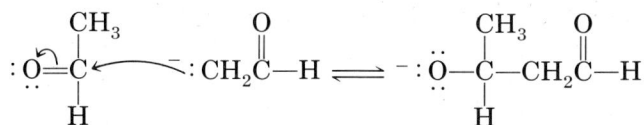

Step 3. The negatively charged oxygen abstracts a proton from water to give the final product, 3-hydroxybutanal, commonly known as *aldol.*

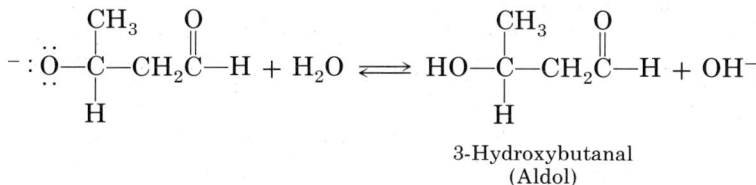

3-Hydroxybutanal (Aldol)

The term *aldol* is derived from the combination of the words *ald*ehyde and alcoh*ol,* the two functional groups present in the product.

The net overall reaction may be written as

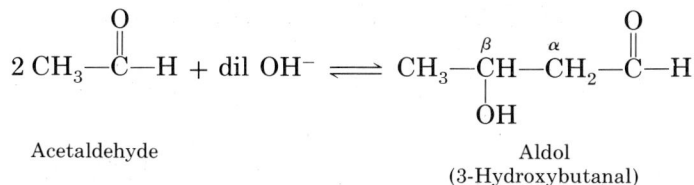

Acetaldehyde Aldol (3-Hydroxybutanal)

The same sequence of events—formation of enolate anion, addition to the C=O group of an nonionized molecule, and proton abstraction—takes place when other aldehydes containing α hydrogens are treated with dilute base. The general overall reaction may be written as

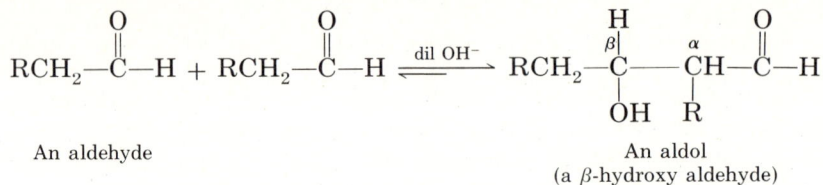

$$RCH_2-\overset{\overset{O}{\|}}{C}-H + RCH_2-\overset{\overset{O}{\|}}{C}-H \overset{\text{dil OH}^-}{\rightleftharpoons} RCH_2-\overset{\overset{\overset{H}{\underset{\beta}{|}}}{\underset{\underset{OH}{|}}{C}}}{}\overset{\overset{\alpha}{}}{\underset{\underset{R}{|}}{CH}}-\overset{\overset{O}{\|}}{C}-H$$

An aldehyde   An aldol
                                                             (a β-hydroxy aldehyde)

As you see, the product in all aldol condensations has several common structural characteristics.

1. The product is a larger molecule that contains a newly formed carbon–carbon bond.
2. The newly formed C—C bond occurs between the α carbon of one aldehyde molecule and what was originally the carbonyl carbon of the second aldehyde molecule.
3. The product contains two functional groups: one carbonyl group and one hydroxyl group.
4. The hydroxyl group is always attached to the beta carbon (β carbon), which is two carbons away from the carbonyl group.

Problem 10.19 Write the structure of the product of the aldol condensation that involves each of the following aldehydes.
(a) Propanal   **(b)** Butanal   **(c)** Phenylacetaldehyde

Aldol condensations are very useful in synthesis because they allow large molecules to be built from smaller ones. Moreover, the two functional groups present in an aldol, the aldehyde group and the β hydroxyl, are convenient handles for further transformations. Thus, aldols are easily dehydrated either by heating or by treatment with dilute acid. The facile loss of water is due to the fact that the resulting product contains a carbon–carbon double bond that is conjugated with the carbonyl group (an α,β-unsaturated aldehyde). For example,

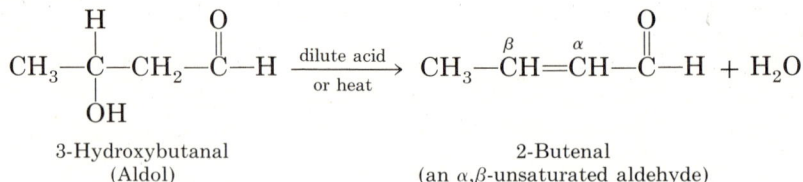

$$CH_3-\overset{\overset{H}{\underset{\underset{OH}{|}}{|}}}{C}-CH_2-\overset{\overset{O}{\|}}{C}-H \xrightarrow[\text{or heat}]{\text{dilute acid}} CH_3-\overset{\beta}{CH}=\overset{\alpha}{CH}-\overset{\overset{O}{\|}}{C}-H + H_2O$$

3-Hydroxybutanal   2-Butenal
(Aldol)   (an α,β-unsaturated aldehyde)

Problem 10.20 Write the structures of the α,β-unsaturated aldehydes expected from the dehydration of products **(a), (b),** and **(c)** in Problem 10.19.

Aldols can also be reduced to 1,3-diols. A useful application of the reaction is the synthesis of the mosquito repellent 6-12 (2-ethyl-1,3-hexanediol), which is obtained when the aldol condensation product from *n*-butyraldehyde is reduced catalytically with hydrogen and nickel.

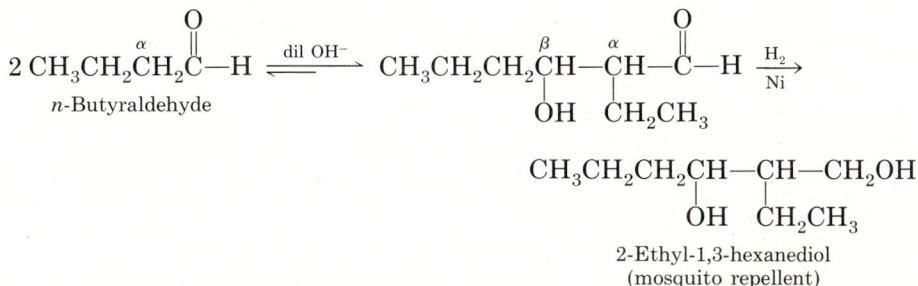

$$2 \ CH_3CH_2CH_2\overset{O}{\underset{}{\overset{\|}{C}}}-H \ \underset{}{\overset{dil \ OH^-}{\rightleftharpoons}} \ CH_3CH_2CH_2\overset{\beta}{\underset{\underset{OH}{|}}{C}H}-\overset{\alpha}{\underset{\underset{CH_2CH_3}{|}}{C}H}-\overset{O}{\overset{\|}{C}}-H \ \overset{H_2}{\underset{Ni}{\rightarrow}}$$

n-Butyraldehyde

$$CH_3CH_2CH_2\underset{\underset{OH}{|}}{C}H-\underset{\underset{CH_2CH_3}{|}}{C}H-CH_2OH$$

2-Ethyl-1,3-hexanediol
(mosquito repellent)

You may have noticed that all the steps in aldol condensations are reversible. With aldehydes, the overall equilibrium lies on the side of the condensation product, the aldol. The reverse is true with ketones: the equilibrium lies on the side of the reactants. Nevertheless, good yields of condensation products may be obtained from ketones if special techniques are used or if the initial condensation product is dehydrated. The aldol condensation of acetone, followed by dehydration, is shown as an example.

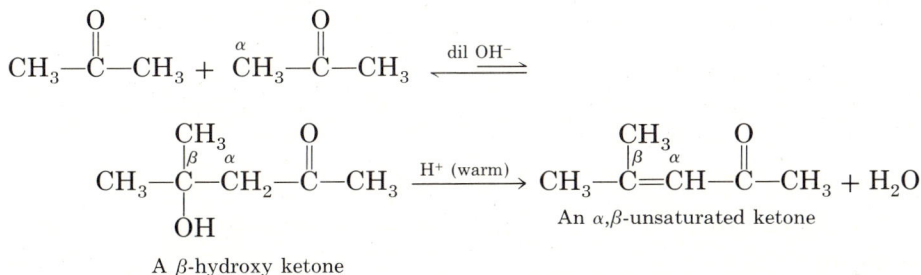

$$CH_3-\overset{O}{\overset{\|}{C}}-CH_3 + \overset{\alpha}{CH_3}-\overset{O}{\overset{\|}{C}}-CH_3 \ \underset{}{\overset{dil \ OH^-}{\rightleftharpoons}}$$

$$CH_3-\overset{CH_3}{\underset{\underset{OH}{|}}{\overset{|}{C}}}-\overset{\alpha}{C}H_2-\overset{O}{\overset{\|}{C}}-CH_3 \ \overset{H^+ \ (warm)}{\longrightarrow} \ CH_3-\overset{CH_3}{\overset{|}{C}}=\overset{\alpha}{C}H-\overset{O}{\overset{\|}{C}}-CH_3 + H_2O$$

A β-hydroxy ketone   An α,β-unsaturated ketone

Note again that the aldol condensation product with ketones, like that with aldehydes, contains a carbonyl group and a β OH group. Similarly, the dehydration of the aldol condensation product with ketones results in the formation of a C=C group conjugated with the C=O group (an α,β-unsaturated ketone).

Problem 10.21 Write the structures of (1) the product of the aldol condensation and (2) the α,β-unsaturated ketone resulting from dehydration of product (1) for each ketone.
(a) Methyl phenyl ketone   **(b)** Cyclohexanone

Crossed aldol condensation. If we subject a mixture of two different aldehydes, each containing an α hydrogen, to the aldol condensation, we get a mixture of condensation products that are difficult to separate. For this reason mixed aldol condensations are generally not useful for syntheses. However, if one of the aldehydes has *no* α hydrogen and therefore cannot form an enolate anion when treated with a base, mixed aldol condensations can be of value in synthesis. For example, if acetaldehyde is added slowly to an excess of benzaldehyde in the presence of dilute sodium hydroxide, it is possible to obtain good yields of **crossed aldol condensation** product. The enolate anion derived from acetaldehyde adds to the carbonyl group of benzaldehyde, which is present in excess. The crossed aldol condensation product that is formed

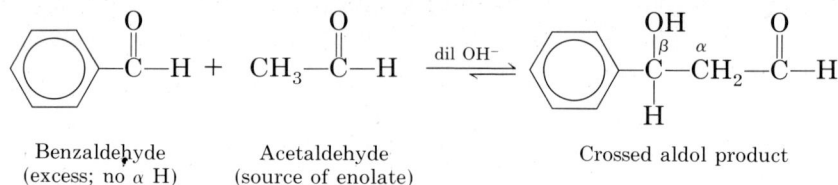

Benzaldehyde   Acetaldehyde   Crossed aldol product
(excess; no α H)   (source of enolate)

easily loses water to give cinnamaldehyde, the flavor constituent of cinnamon.

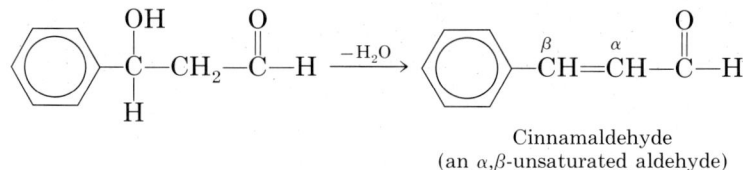

Cinnamaldehyde
(an α,β-unsaturated aldehyde)

Problem 10.22 Draw the structures of (1) the product of the crossed aldol condensation and (2) the dehydration product from (1) for each of the following pairs.
(a) Benzaldehyde + propionaldehyde
(b) Two moles of benzaldehyde and one mole of acetone

10.20 Visual Tests for Aldehydes and Ketones

Aldehydes and ketones can be differentiated from most other classes of compounds by treatment with 2,4-dinitrophenylhydrazine (2,4-DNPH). You will recall (Sec. 10.17) that both aldehydes and ketones form colored precipitates with 2,4-DNPH, whereas most other classes of compounds do not react.

A visual test that distinguishes aldehydes from ketones is the **Tollens' test.** Aldehydes when treated with Tollens' reagent, an alkaline solution of silver ammonia complex, give a positive test which consists of a bright silver mirror deposit on the walls of the reaction vessel. Ketones give a negative result in the Tollens' test.

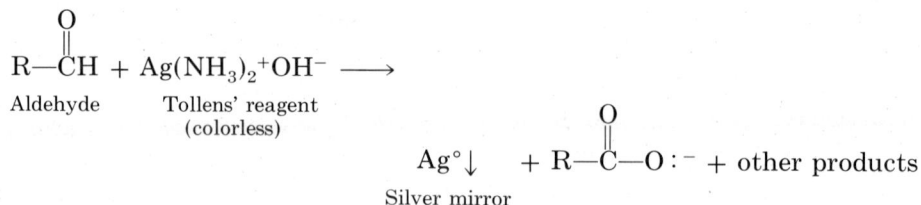

Aldehyde   Tollens' reagent
                     (colorless)

$Ag°\downarrow$ $+ R—C—O:^-$ + other products
Silver mirror

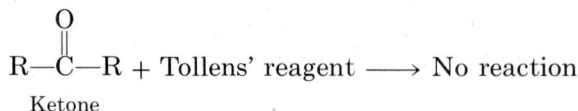

Ketone

Another visual test is the **iodoform test.** This test consists of treating a compound with a mixture of sodium hydroxide and iodine. A positive test is indicated by the formation of a yellow precipitate of iodoform, CHI_3. Any com-

$$CH_3\overset{\overset{\displaystyle O}{\|}}{C}-$$

pound that contains an **acetyl group, CH_3C-,** or a group that can be oxidized to an acetyl group will give a positive iodoform test. Thus, *all* methyl ketones give a positive test, as will alcohols with the structure

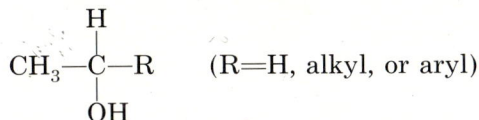

$$CH_3-\overset{\overset{\displaystyle H}{|}}{\underset{\underset{\displaystyle OH}{|}}{C}}-R \qquad (R=H, \text{ alkyl, or aryl})$$

Acetaldehyde, $CH_3CH=O$, is the *only* aldehyde that gives a positive iodoform test. The overall iodoform reaction is

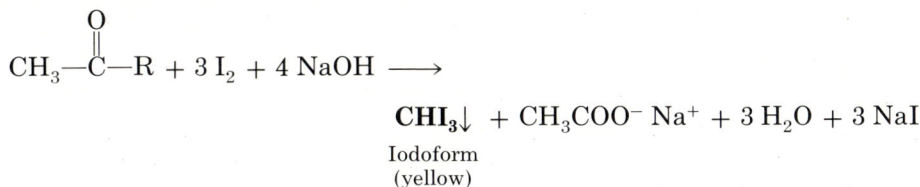

$$CH_3-\overset{\overset{\displaystyle O}{\|}}{C}-R + 3\,I_2 + 4\,NaOH \longrightarrow$$

$$\mathbf{CHI_3}\downarrow + CH_3COO^-\,Na^+ + 3\,H_2O + 3\,NaI$$
$$\text{Iodoform}$$
$$\text{(yellow)}$$

Problem 10.23 Which of the following compounds will give a positive iodoform test?

(a) Ethanol   (b) 1-Propanol   (c) Isopropyl alcohol
(d) 2-Pentanol   (e) 3-Pentanol   (f) Propionaldehyde
(g) Methyl phenyl ketone   (h) 2-Pentanone   (i) 3-Pentanone
(j) Diethyl ketone

Problem 10.24 A compound with the formula $C_5H_{10}O$ gave the following results.
(1) Treatment with 2,4-dinitrophenylhydrazine: a colored precipitate
(2) Tollens' test: negative
(3) Iodoform test: positive
Write the two structures that are consistent with these data.

Summary of Concepts and Reactions

Aldehydes have the general formula $R\overset{\overset{\displaystyle O}{\|}}{C}-H$ and ketones have the general formula

$R-\overset{\overset{\displaystyle O}{\|}}{C}-R'$.   [Secs. 10.1, 10.2]
The C=O group of aldehydes and ketones is called the carbonyl group.   [Sec. 10.1]

The carbonyl group is polar, and its polar character may be described by way of two resonance-contributing structures.

$$\left[\begin{array}{c} \diagup \\ \diagup \end{array}\!\!C\!\!=\!\!\ddot{O}: \longleftrightarrow \begin{array}{c} \diagup \\ \diagup \end{array}\!\!C^{+}\!\!-\!\!\ddot{O}:^{-} \right] \qquad\qquad \text{[Sec. 10.3]}$$

The polar character of the C=O group gives rise to intermolecular attractions called dipole–dipole attractions.

$$\begin{array}{c}
\overset{\delta-}{O}\text{--------}\overset{\delta+}{\diagup}C\diagup \\
\| \qquad\qquad \| \\
\diagup C\diagup\text{--------}\overset{}{O} \\
\overset{\delta+}{}\qquad\quad \overset{\delta-}{}
\end{array} \qquad\qquad \text{[Sec. 10.4A]}$$

Dipole–dipole attractions are not as strong as interactions due to hydrogen bonding. Consequently, the boiling points of aldehydes and ketones are higher than those of nonpolar alkanes of corresponding molecular weights but lower than those of alcohols of comparable weights. [Sec. 10.4A]

The lower aldehydes and ketones are soluble in water. [Sec. 10.4B]

There are basically four major methods for preparing aldehydes and ketones in the laboratory. [Secs. 10.5–10.9]

Aldehydes and ketones undergo generally the same kinds of reactions. The most typical of these are nucleophilic addition reactions to the C=O group. [Secs. 10.12–10.17]

Reactions of aldehydes and ketones may also take place on the carbon next to the carbonyl group, the α carbon. [Sec. 10.18]

The anion that results from the removal of an α hydrogen by a base is stabilized by resonance. The resonance-stabilized anion is called the enolate anion. [Sec. 10.18]

Enolate anions are nucleophiles and may add to the C=O group of aldehydes and ketones to form a larger molecule with loss of water, a condensation process. [Sec. 10.19]

The reaction of an enolate anion with an aldehyde or ketone is called the aldol condensation. [Sec. 10.19]

Aldehydes and ketones can be differentiated from other classes of compounds and also from each other by means of various visual tests. [Sec. 10.20]

Key Terms

aldehydes
ketones
carbonyl group
benzaldehyde
acetone
dipole–dipole
  attractions
Friedel–Crafts
  acylation
formaldehyde
acetaldehyde

Wacker process
nucleophilic addition
  reactions
nucleophiles
electrophiles
metal hydrides
Grignard reagents
  (RMgX)
cyanohydrin
hemiacetals
hemiketals

acetals
alpha carbon (α carbon)
alpha hydrogen
  (α hydrogen)
enolate anion
condensation process
aldol condensation
crossed aldol
  condensation
Tollens' test
iodoform test

Structure and Nomenclature of Aldehydes and Ketones [Secs. 10.1, 10.2]

10.1 Draw the structures of these compounds.

    (a) 4-Octanone   **(b)** 3-Methylcyclopentanone

    (c) Phenylacetaldehyde   **(d)** *m*-Methoxybenzaldehyde

    (e) 1,1,3-Tribromoacetone   **(f)** 4-Methyl-4-pentenal

    (g) Methyl *p*-nitrophenyl ketone   **(h)** 4-Bromo-1-phenyl-2-pentanone

    (i) 3,5-Dibromobenzaldehyde   **(j)** Methyl *t*-butyl ketone

10.2 Write names for the following compounds.

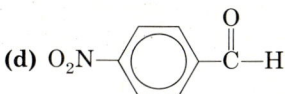

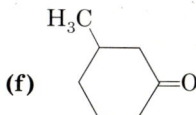

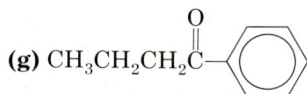

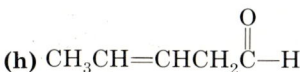

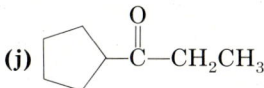

    (a) $CH_3CH_2CH_2CH_2CH_2\overset{\displaystyle O}{\overset{\|}{C}}-H$ **(b)** [cyclohexanone structure]

    (c) $CH_3CH_2CH_2\overset{\displaystyle O}{\overset{\|}{C}}CH_2CH_3$ **(d)** O_2N-[benzene ring]$-\overset{\displaystyle O}{\overset{\|}{C}}-H$

    (e) $CH_3CCl_2CH_2\overset{\displaystyle O}{\overset{\|}{C}}CH_3$ **(f)** [3-methylcyclohexanone structure]

    (g) $CH_3CH_2CH_2\overset{\displaystyle O}{\overset{\|}{C}}-$[benzene ring]   **(h)** $CH_3CH=CHCH_2\overset{\displaystyle O}{\overset{\|}{C}}-H$

    (i) $HOCH_2CH_2CH_2\overset{\displaystyle O}{\overset{\|}{C}}-H$ **(j)** [cyclopentyl]$-\overset{\displaystyle O}{\overset{\|}{C}}-CH_2CH_3$

10.3 Draw the structures and give the IUPAC names for all compounds of molecular formula $C_5H_{10}O$ that contain a carbonyl group.

Structures of Aldehyde and Ketone Derivatives [Secs. 10.15, 10.16, 10.17]

10.4 Draw the structures of the following compounds.

    (a) Acetaldehyde cyanohydrin

    (b) Benzaldehyde dimethyl acetal

    (c) Hexanal oxime

    (d) Diethyl ketone hydrazone

    (e) Acetaldehyde phenylhydrazone

    (f) Acetone-2,4-dinitrophenylhydrazone

    (g) Cyclohexanone semicarbazone

Prediction of Physical Properties [Sec. 10.4]

10.5 Predict the order of increasing boiling points for each group.

    (a) Acetaldehyde; propane; ethanol; 1-propanol.

    (b) Acetone; methyl ethyl ketone; 1-butanol; 1,2-butanediol.

10.6 Arrange the following compounds in order of increasing solubility in water: pentane; butanal; cyclohexanone; acetone.

Preparation of Aldehydes and Ketones [Secs. 10.5–10.9]

10.7 Draw the structure of each product. If no reaction is expected, state so.

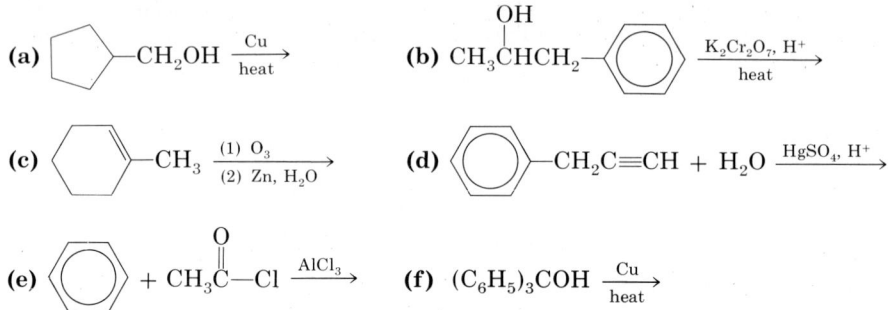

(a) ⬠—CH_2OH $\xrightarrow[\text{heat}]{\text{Cu}}$

(b) $CH_3\overset{\text{OH}}{\underset{|}{CH}}CH_2$—⬡ $\xrightarrow[\text{heat}]{K_2Cr_2O_7,\ H^+}$

(c) ⬡—CH_3 $\xrightarrow[\text{(2) Zn, H}_2\text{O}]{\text{(1) O}_3}$

(d) ⬡—$CH_2C\equiv CH + H_2O$ $\xrightarrow{\text{HgSO}_4,\ H^+}$

(e) ⬡ $+\ CH_3\overset{O}{\overset{\|}{C}}$—Cl $\xrightarrow{\text{AlCl}_3}$

(f) $(C_6H_5)_3COH$ $\xrightarrow[\text{heat}]{\text{Cu}}$

10.8 Draw the structures of the reactants and products in each sequence of reactions.

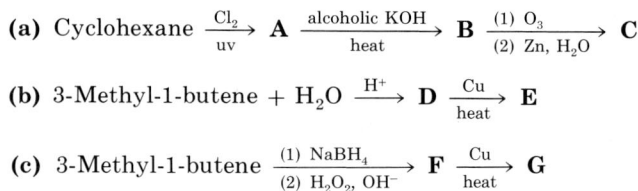

(a) Cyclohexane $\xrightarrow[\text{uv}]{\text{Cl}_2}$ **A** $\xrightarrow[\text{heat}]{\text{alcoholic KOH}}$ **B** $\xrightarrow[\text{(2) Zn, H}_2\text{O}]{\text{(1) O}_3}$ **C**

(b) 3-Methyl-1-butene $+ H_2O$ $\xrightarrow{H^+}$ **D** $\xrightarrow[\text{heat}]{\text{Cu}}$ **E**

(c) 3-Methyl-1-butene $\xrightarrow[\text{(2) H}_2\text{O}_2,\ \text{OH}^-]{\text{(1) NaBH}_4}$ **F** $\xrightarrow[\text{heat}]{\text{Cu}}$ **G**

10.9 Draw the structures of the alkenes or the dienes that give the following products on ozonolysis.

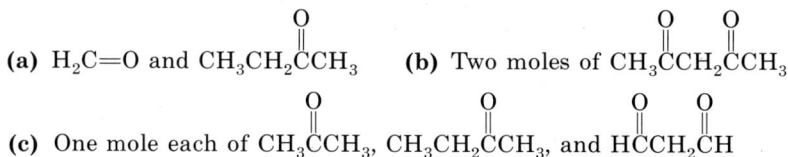

(a) $H_2C\!=\!O$ and $CH_3CH_2\overset{O}{\overset{\|}{C}}CH_3$

(b) Two moles of $CH_3\overset{O}{\overset{\|}{C}}CH_2\overset{O}{\overset{\|}{C}}CH_3$

(c) One mole each of $CH_3\overset{O}{\overset{\|}{C}}CH_3$, $CH_3CH_2\overset{O}{\overset{\|}{C}}CH_3$, and $H\overset{O}{\overset{\|}{C}}CH_2\overset{O}{\overset{\|}{C}}H$

Reactions of Aldehydes and Ketones [Secs. 10.12–10.20]

10.10 Write the structure of the major product(s), if any, of the reaction of acetone with each of the following reagents.

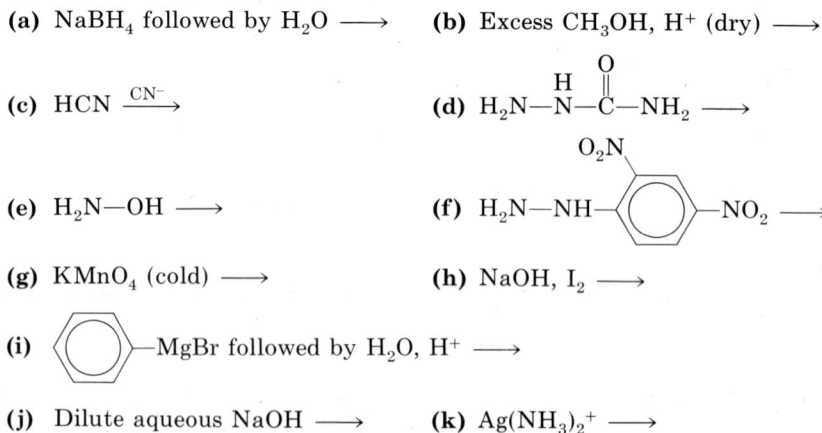

(a) $NaBH_4$ followed by H_2O $\longrightarrow$

(b) Excess CH_3OH, H^+ (dry) $\longrightarrow$

(c) HCN $\xrightarrow{\text{CN}^-}$

(d) H_2N—N—$\overset{O}{\overset{\|}{C}}$—$NH_2$ $\longrightarrow$ (with H on N)

(e) H_2N—OH $\longrightarrow$

(f) H_2N—NH—⬡(—O_2N)(—NO_2) $\longrightarrow$

(g) $KMnO_4$ (cold) $\longrightarrow$

(h) NaOH, I_2 $\longrightarrow$

(i) ⬡—MgBr followed by H_2O, H^+ $\longrightarrow$

(j) Dilute aqueous NaOH $\longrightarrow$

(k) $Ag(NH_3)_2^+$ $\longrightarrow$

10.11 Write the structure of the major product(s), if any, when benzaldehyde is treated with each of the reagents listed in Exercise 10.10.

10.12 Using a Grignard reagent and the appropriate aldehyde or ketone, show how each of the following alcohols could be prepared.

(a) 1-Butanol   **(b)** 2-Butanol   **(c)** 1-Methylcyclopentanol
(d) 3-Buten-2-ol   **(e)** Triphenylcarbinol

10.13 Complete the following reactions.

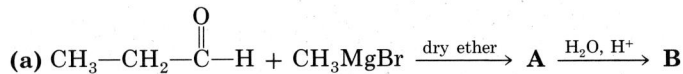

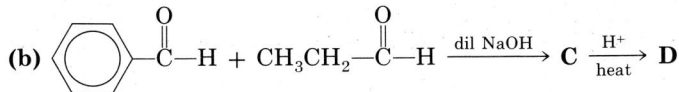

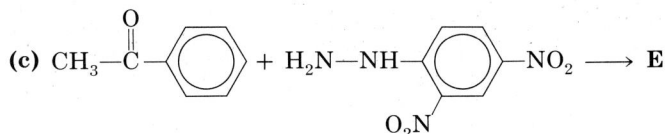

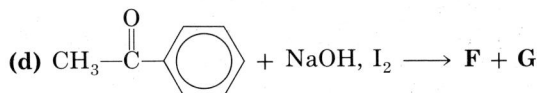

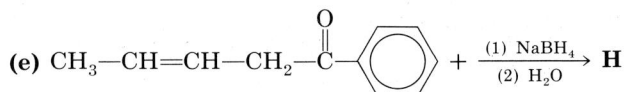

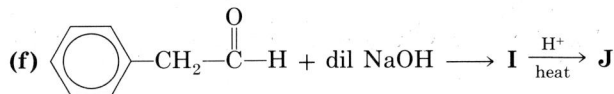

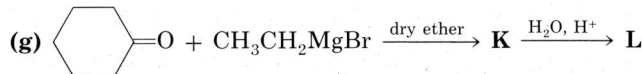

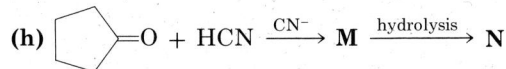

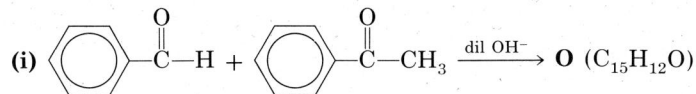

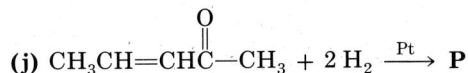

(a) $CH_3-CH_2-\overset{\displaystyle O}{\overset{\|}{C}}-H + CH_3MgBr \xrightarrow{\text{dry ether}} \textbf{A} \xrightarrow{H_2O,\ H^+} \textbf{B}$

(b) ⬡$-\overset{\displaystyle O}{\overset{\|}{C}}-H + CH_3CH_2-\overset{\displaystyle O}{\overset{\|}{C}}-H \xrightarrow{\text{dil NaOH}} \textbf{C} \xrightarrow[\text{heat}]{H^+} \textbf{D}$

(c) $CH_3-\overset{\displaystyle O}{\overset{\|}{C}}-$⬡$ + H_2N-NH-$⬡$-NO_2 \longrightarrow \textbf{E}$
     (with O_2N substituent)

(d) $CH_3-\overset{\displaystyle O}{\overset{\|}{C}}-$⬡$ + NaOH,\ I_2 \longrightarrow \textbf{F} + \textbf{G}$

(e) $CH_3-CH{=}CH-CH_2-\overset{\displaystyle O}{\overset{\|}{C}}-$⬡$ + \xrightarrow[\text{(2) } H_2O]{\text{(1) NaBH}_4} \textbf{H}$

(f) ⬡$-CH_2-\overset{\displaystyle O}{\overset{\|}{C}}-H + \text{dil NaOH} \longrightarrow \textbf{I} \xrightarrow[\text{heat}]{H^+} \textbf{J}$

(g) ⬡$={O} + CH_3CH_2MgBr \xrightarrow{\text{dry ether}} \textbf{K} \xrightarrow{H_2O,\ H^+} \textbf{L}$

(h) ⬠$={O} + HCN \xrightarrow{CN^-} \textbf{M} \xrightarrow{\text{hydrolysis}} \textbf{N}$

(i) ⬡$-\overset{\displaystyle O}{\overset{\|}{C}}-H + $⬡$-\overset{\displaystyle O}{\overset{\|}{C}}-CH_3 \xrightarrow{\text{dil OH}^-} \textbf{O}\ (C_{15}H_{12}O)$

(j) $CH_3CH{=}CH\overset{\displaystyle O}{\overset{\|}{C}}-CH_3 + 2\ H_2 \xrightarrow{Pt} \textbf{P}$

10.14 When an excess of formaldehyde is treated with acetaldehyde in the presence of base, the compound that is formed is

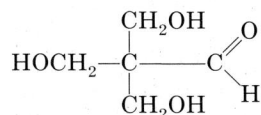

$$HOCH_2-\underset{\underset{\displaystyle CH_2OH}{|}}{\overset{\overset{\displaystyle CH_2OH}{|}}{C}}-\overset{\displaystyle O}{\overset{\diagdown}{C}}_H$$

Explain how this formation occurs. (*Hint:* There are three successive aldol condensations.)

10.15 Indicate what reagent(s) is (are) necessary to carry out each transformation. You may use any reagents you wish.

(a) Acetylene $\xrightarrow{?}$ acetaldehyde $\xrightarrow{?}$ ethanol

(b) Cyclopentanone $\xrightarrow{?}$ $\xrightarrow{?}$ 1-ethylcyclopentanol

(c) Cyclohexene $\xrightarrow{?}$ cyclohexanol $\xrightarrow{?}$ cyclohexanone

(d) 2-Pentanol $\xrightarrow{?}$ $\xrightarrow{?}$ 2-pentanone oxime

(e) 3-Pentanol $\xrightarrow{?}$ 3-pentanone $\xrightarrow{?}$ 3-methyl-3-pentanol

(f) 2-Hexanol $\xrightarrow{?}$ $\xrightarrow{?}$ 2-cyano-2-hexanol

10.16 Show how the following syntheses could be accomplished from the compounds indicated and from others that may be needed. Show all steps and reagents used.
(a) 2-Methylpentane-1,3-diol from propanal   (b) 4-Octanone from 1-butanol

Visual Tests and Structure Identification [Sec. 10.20 (and previous chapters)]

10.17 Explain how you could distinguish each of the following pairs of compounds by a chemical test. State what you would observe and write the equation(s) for any reaction(s) that would take place.
(a) Cyclohexanone and benzaldehyde

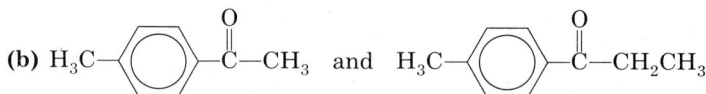

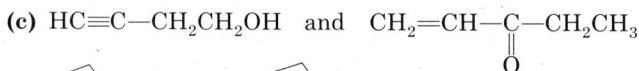

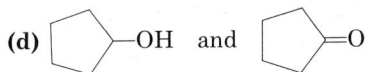

(b) H_3C—⟨⟩—$\overset{\overset{O}{\|}}{C}$—$CH_3$ and H_3C—⟨⟩—$\overset{\overset{O}{\|}}{C}$—$CH_2CH_3$

(c) $HC\equiv C$—CH_2CH_2OH and $CH_2{=}CH$—$\overset{}{\underset{\underset{O}{\|}}{C}}$—$CH_2CH_3$

(d) ⬠—OH and ⬠=O

10.18 Unknown **A** (C_5H_{10}) rapidly decolorizes a solution of bromine in carbon tetrachloride. When **A** is dissolved in cold concentrated sulfuric acid and then heated with water, **B** is formed. **B** ($C_5H_{12}O$) reacts with chromic acid to give **C** ($C_5H_{10}O$). Both **B** and **C** give positive iodoform tests. Each iodoform reaction mixture contains, in addition to CHI_3, the anion of butyric acid, $CH_3CH_2CH_2COOH$. What are the structures of **A**, **B**, and **C**?

10.19 An unknown of molecular formula C_8H_8O gives the following results.
Tollens' test: negative
2,4-Dinitrophenylhydrazine: positive
Iodoform: positive
Suggest a plausible structure for the compound.

10.20 Deduce the structure of a compound that is consistent with the following data.
Tollens' test: positive
2,4-Dinitrophenylhydrazine: yellow-orange product with the formula $C_9H_{10}N_4O_4$.

10.21 An unknown **A** ($C_6H_{12}O$) gives the results indicated.
(1) **A** + hydroxylamine: positive
(2) **A** + Tollens' reagent: negative
(3) Catalytic hydrogenation of **A** gives **B** ($C_6H_{14}O$).
(4) **B** + cold conc H_2SO_4 followed by treatment with water gives **C** (C_6H_{12}).
(5) **C** on ozonolysis gives two compounds, **D** (C_3H_6O) and **E** (C_3H_6O); **D** gives a negative Tollens' test and a positive iodoform test, whereas **E** gives a positive Tollens' test and a negative iodoform test.
What are the structures of **A**, **B**, **C**, **D**, and **E**?

Carbohydrates

The carbohydrates are the most widely distributed of the organic compounds. Among the familiar carbohydrates are sugars (such as glucose, fructose, and sucrose), starch, cellulose, and glycogen. Broadly speaking, carbohydrates represent a group of substances produced in green plants by means of **photosynthesis.** The leaf of a green plant combines water and carbon dioxide to produce the sugar glucose, a carbohydrate. Light provides the energy for the reaction, and chlorophyll, the green substance in a leaf, takes in the light and serves as a catalyst for photosynthesis. Once the glucose is produced, it is further synthesized by the plant into starch and cellulose.

When an animal eats a green plant, the starch in the plant is broken down into glucose molecules, from which it was previously synthesized. The animal's liver converts the glucose into glycogen, which is then stored in the muscle tissues. When the animal needs energy, the glycogen is broken down once again into glucose molecules. The glucose provides energy for the animal.

Almost all our foods can be traced to carbohydrates such as glucose. We make our clothes and many other materials from various forms of cellulose, such as cotton, rayon, and linen. Cellulose is also the basic component of wood, another useful product.

Definition *11.1*

The term carbohydrate stems from the fact that the earliest investigated compounds had the formula $C_n(H_2O)_n$, suggesting that carbohydrates are hydrates of carbon. Many carbohydrates are now known that do not correspond to this general formula.

In terms of the functional groups present, carbohydrates are defined as **polyhydroxy aldehydes** or **polyhydroxy ketones** or as *compounds that can easily be converted into one of them by hydrolysis.* We shall return to this second qualification when we take a closer look at the chemistry of carbohydrates.

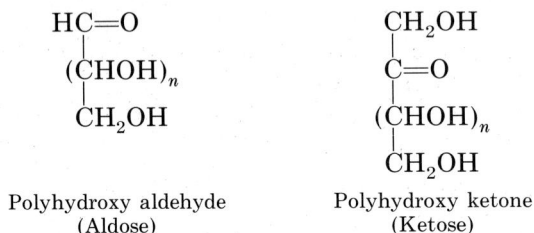

$$HC=O$$
$$(CHOH)_n$$
$$CH_2OH$$

$$CH_2OH$$
$$C=O$$
$$(CHOH)_n$$
$$CH_2OH$$

Polyhydroxy aldehyde   Polyhydroxy ketone
(Aldose)   (Ketose)

Polyhydroxy aldehydes are also known as **aldoses,** and polyhydroxy ketones are also called **ketoses.** These names are combinations of *ald-* for aldehyde and *ket-* for ketone, respectively, with the suffix *-ose* common to all sugars.

11.2 Classification of Carbohydrates

The classification of carbohydrates is based essentially on the following factors

1. The number of aldose and/or ketose units produced upon complete hydrolysis.
2. The number of carbon atoms in a given aldose or ketose.
3. The configuration of the hydroxyl group attached to the last chiral carbon atom in the aldose or ketose.

Let us consider each of these factors separately.

1. Hydrolysis products. Carbohydrates are classified according to the number of aldose and/or ketose units produced upon complete hydrolysis.* **Monosaccharides,** or simple sugars, consist of a single aldose or ketose unit and cannot be hydrolyzed. Examples of aldose monosaccharides are **ribose, glucose,** and **galactose.** The most common ketose monosaccharide is **fructose.**

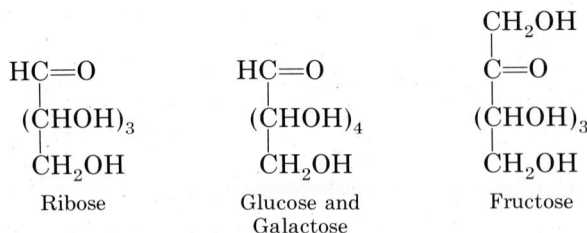

$$HC=O$$
$$(CHOH)_3$$
$$CH_2OH$$

$$HC=O$$
$$(CHOH)_4$$
$$CH_2OH$$

$$CH_2OH$$
$$C=O$$
$$(CHOH)_3$$
$$CH_2OH$$

Ribose   Glucose and   Fructose
Galactose

Sugars that yield two monosaccharides on hydrolysis are known as **disaccharides.** The monosaccharides produced may be the same or different. For

* Hydrolysis means splitting with water. A common method for hydrolyzing a carbohydrate is to treat it with hot water containing a small amount of sulfuric acid.

example, sucrose, the common table sugar, is a disaccharide that produces one mole of glucose and one mole of fructose on hydrolysis. The disaccharides maltose and cellobiose produce two moles of glucose on hydrolysis.

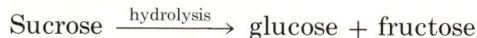

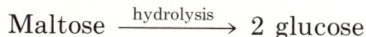

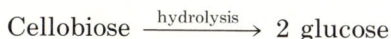

$$\text{Sucrose} \xrightarrow{\text{hydrolysis}} \text{glucose} + \text{fructose}$$

$$\text{Maltose} \xrightarrow{\text{hydrolysis}} 2 \text{ glucose}$$

$$\text{Cellobiose} \xrightarrow{\text{hydrolysis}} 2 \text{ glucose}$$

Oligosaccharides are carbohydrates that yield three to ten monosaccharides after hydrolysis. Carbohydrates that furnish ten or more monosaccharides when completely hydrolyzed are called **polysaccharides.**

To summarize,

1. Monosaccharides $\xrightarrow{\text{hydrolysis}}$ No reaction

2. Disaccharides $\xrightarrow{\text{hydrolysis}}$ 2 monosaccharides (same or different)

3. Oligosaccharides $\xrightarrow{\text{hydrolysis}}$ 3–10 monosaccharides

4. Polysaccharides $\xrightarrow{\text{hydrolysis}}$ 10 or more monosaccharides (usually glucose)

2. *Number of carbon atoms in monosaccharide.* Monosaccharides are specified further according to the number of carbons present in an aldose or a ketose. Aldoses are drawn with the aldehyde carbon, carbon number 1, at the top of the chain. All naturally occurring ketoses have the ketone group on the second carbon in the chain. For example,

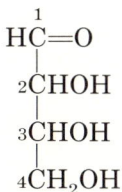

$$
\begin{array}{ll}
1\ \text{HC}=\text{O} & 1\ \text{CH}_2\text{OH} \\
2\ \text{CHOH} & 2\ \text{C}=\text{O} \\
3\ \text{CHOH} & 3\ \text{CHOH} \\
4\ \text{CH}_2\text{OH} & 4\ \text{CH}_2\text{OH}
\end{array}
$$

A four-carbon aldose (an aldotetrose)   A four-carbon ketose (a ketotetrose)

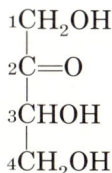

The simplest sugars have three carbon atoms and are called **trioses.** The addition of carbon atoms to a triose gives successively a **tetrose,** a **pentose,** a **hexose,** and so on. The simplest aldose is glyceraldehyde (*aldotriose*), and the simplest ketose is dihydroxyacetone (*ketotriose*).

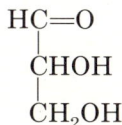

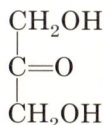

$$
\begin{array}{ll}
\text{HC}=\text{O} & \text{CH}_2\text{OH} \\
\text{CHOH} & \text{C}=\text{O} \\
\text{CH}_2\text{OH} & \text{CH}_2\text{OH}
\end{array}
$$

Glyceraldehyde   Dihydroxyacetone

Table 11.1 summarizes this aspect of classification of sugars.

Table 11.1 Classification of Monosaccharides

Number of carbon atoms	Type of monosaccharide	
	Aldose	Ketose
3	Aldotriose	Ketotriose
4	Aldotetrose	Ketotetrose
5	Aldopentose	Ketopentose
6	Aldohexose	Ketohexose
7	Aldoheptose	Ketoheptose

Aldohexoses

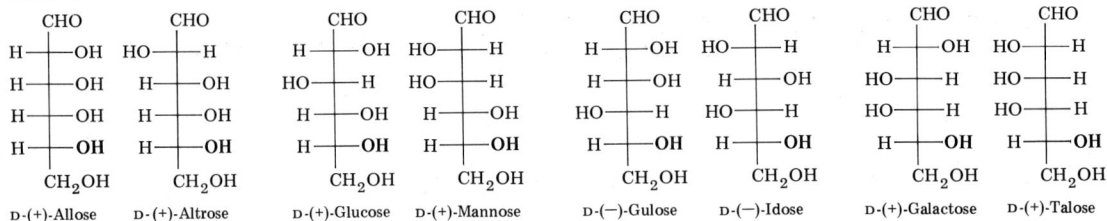

D-(+)-Allose D-(+)-Altrose D-(+)-Glucose D-(+)-Mannose D-(−)-Gulose D-(−)-Idose D-(+)-Galactose D-(+)-Talose

Aldopentoses

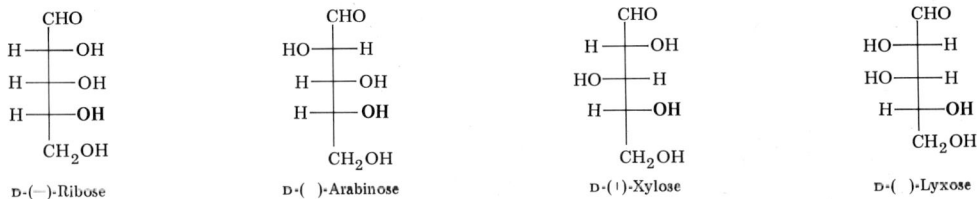

D-(−)-Ribose D-( )-Arabinose D-(⟩)-Xylose D-( )-Lyxose

Aldotetroses

D-(−)-Erythrose D-(−)-Threose

Aldotriose

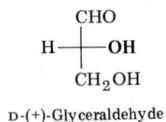

D-(+)-Glyceraldehyde

Figure 11.1 The series of D-aldoses up to six carbons.

Problem 11.1 Write **(a)** two different aldotetrose and **(b)** two different keto-pentose structures. In what respect do the structures in each pair differ?

3. Configurational isomers. Monosaccharides are also divided into D and L families. You will recall (Sec. 6.12A) that the letter D preceding the name of glyceraldehyde means that the OH on the chiral carbon points to the right when we draw the Fischer projection formula for this compound. Conversely, L-glycer-aldehyde means that the OH on the chiral carbon points to the left.

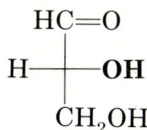

$$
\begin{array}{cc}
\text{HC}{=}\text{O} & \text{HC}{=}\text{O} \\
\text{H}\!-\!\!\!-\!\!\!-\!\text{OH} & \text{HO}\!-\!\!\!-\!\!\!-\!\text{H} \\
\text{CH}_2\text{OH} & \text{CH}_2\text{OH} \\
\text{D-(+)-Glyceraldehyde} & \text{L-(−)-Glyceraldehyde}
\end{array}
$$

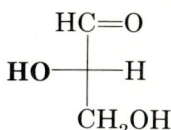

In classifying tetroses, pentoses, hexoses, and so on, into D or L families, it is only necessary to determine the position of the OH attached to the chiral carbon that is *farthest from the carbonyl group*. If that OH is to the right in the Fischer projection formula, the sugar belongs to the D family. If it is to the left, the sugar belongs to the L family. Most naturally occurring sugars belong to the D family, a fact that provides one more example of the selectivity found in nature where stereochemistry is involved. Figure 11.1 shows all the aldoses, up to six carbons, that belong to the D family.

Remember that assigning a given sugar to the D or L family tells us only the configuration about the last chiral center; it has nothing to do with the sign of optical rotation, which must be determined experimentally. For example, D-ribose is levorotatory, whereas D-glucose is dextrorotatory, but they both have the OH to the right of the last chiral center.

Physical Properties of Monosaccharides *11.3*

The physical properties of monosaccharides are in accord with our expectations for molecules that have several polar hydroxyl groups. For example, all the monosaccharides are highly soluble in water, a polar solvent, and insoluble in nonpolar solvents such as hexane (remember the "like dissolves like" rule).

The presence of several hydroxyl groups makes association between molecules by way of hydrogen bonding so strong that even the simplest sugars have high boiling points. Glyceraldehyde, for example, boils at about 150°C under reduced pressure of 0.8 mm. At atmospheric pressure (760 mm) decomposition of the molecule occurs before the boiling point is reached (which indicates that the boiling point is extremely high).

Problem 11.2 Show how hydrogen bonding exists between two glyceraldehyde molecules.

All monosaccharides are sweet tasting, and most are white solids.

11.4 Mutarotation

Before discussing the chemical properties of monosaccharides, it is necessary to explain the phenomenon of mutarotation.

As long ago as 1895, a pharmacist by the name of Tanret isolated two forms of D-glucose that had different physical properties. One form, which we shall call α-D-glucose, has a melting point of 146°C and a specific rotation of +112°. The other form, β-D-glucose, melts at 150°C and has a specific rotation of +19°. When either of these two forms of D-glucose is dissolved in water, a gradual change in specific rotation occurs until an equilibrium value of +53° is reached, after which the specific rotation of the solution remains constant. This change in optical rotation of a solution of either form of glucose until a constant value is obtained is known as **mutarotation.**

Problem 11.3
(a) What would you expect the specific rotation of an equal mixture of α- and β-D-glucose to be?
(b) The specific rotation after mutarotation is actually +53°. Which of the two forms predominates at equilibrium?

Many sugars other than glucose also exist in α and β forms and undergo mutarotation. Table 11.2 lists specific rotations of some common sugars, together with their equilibrium values after mutarotation.

Table 11.2   Specific Rotations of Some Sugars

Name of sugar	Pure α	Pure β	Equilibrium value after mutarotation
D-Glucose	+112°	+19°	+53°
D-Fructose	−21°	−133°	−92°
D-Galactose	+151°	−53°	+84°
D-Mannose	+30°	−17°	+14°
D-Lactose	+90°	+35°	+55°
D-Maltose	+168°	+112°	+136°

11.5 Cyclic Forms of Glucose

A   The Fischer Representation

In order to explain the phenomenon of mutarotation we must reconsider the open-chain form of D-glucose in light of one particular reaction between aldehydes (or ketones) and alcohols, the formation of **hemiacetals** (or **hemiketals**). The reaction is illustrated here for hemiacetals.

$$R-\overset{\overset{\displaystyle H}{|}}{C}=O \;+\; \mathbf{R'OH} \;\longrightarrow\; \underset{\overset{\displaystyle H}{\diagup}\;\overset{\displaystyle}{\diagdown}\mathbf{OR'}}{\overset{R\diagdown\;\overset{*}{\diagup}OH}{C}}$$

(* indicates chiral carbon)

Aldehyde   Alcohol   Hemiacetal

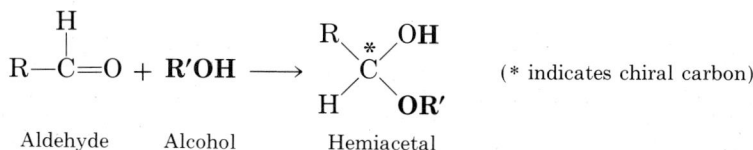

Usually, hemiacetal formation involves a reaction between two separate molecules, one molecule of alcohol and one of aldehyde, and is an *intermolecular process*. Note also that the carbonyl carbon (C=O) in the original aldehyde, which was sp^2 hybridized and achiral, has become sp^3 hybridized and chiral.

Problem 11.4 The preceding statement is true except when the starting aldehyde is formaldehyde. Explain why this is so.

In the case of glucose, however, hemiacetal formation is an *intramolecular process* because the carbonyl function and the alcohol groups are part of the *same* molecule. The result is a *cyclic hemiacetal*. Will any of the five different —OH groups react with the carbonyl group to form a cyclic hemiacetal? The answer is no. The ease of formation and the stability of the ring determine which —OH group reacts. We know that five- and six-membered rings are the most stable and the most easily prepared. In the case of glucose, it has been determined that it is the —OH bonded to the *fifth carbon* that reacts to form a *six-membered ring* (Fig. 11.2).

In the process of cyclization, a new chiral center is generated at carbon 1, giving two isomers (II and III in Fig. 11.2) that differ only in their configurations about the first carbon. This newly created chiral center is known as the **anomeric carbon.** By convention, in the Fischer representation of cyclic glucose (and other sugars as well), the α isomer has *the —OH to the right of the anomeric center,* the β isomer is the structure *with the —OH to the left of the anomeric carbon.*

Let us now consider the stereochemical relationship between α- and β-D-glucoses. Are they enantiomers? No; they are not mirror images of each other. They are therefore diastereomers. (Another way to differentiate between α and β forms of a sugar is to say that they are *anomers.*) As we would expect, the α and β forms of a sugar have different physical properties.

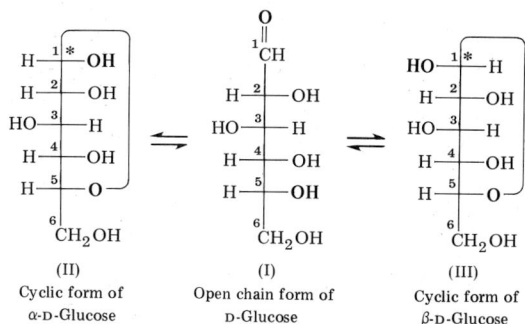

Figure 11.2 Open and cyclic forms of glucose (Fischer projections). Structures II and III are diastereomers.

From a structural viewpoint, mutarotation occurs because of the slow inter-conversion of α-D-glucose to β-D-glucose via the open-chain form, and vice versa, until equilibrium is established. The concentrations of the open-chain form (structure I in Fig. 11.2) and the cyclic forms (structures II and III) at equilibrium are 0.1%, 37%, and 63%, respectively.

Like glucose, the other sugars listed in Table 11.2 undergo mutarotation, also because of the slow interconversion of one cyclic form (α) to another (β) via the open-chain form, until an equilibrium is established.

Problem 11.5 Starting from the open-chain form of glucose, determine the sizes of the rings if the —OH attached to carbons 2, 3, 4, and 6, respectively, were involved in hemiacetal formation.

B   The Haworth Representation

Although Fischer projections of the cyclic forms of glucose helped in ex-plaining the mutarotation phenomenon and are easy to draw, the resulting structures are awkward because of the long bonds joining the heterocyclic oxy-gen to C-1 and C-5.

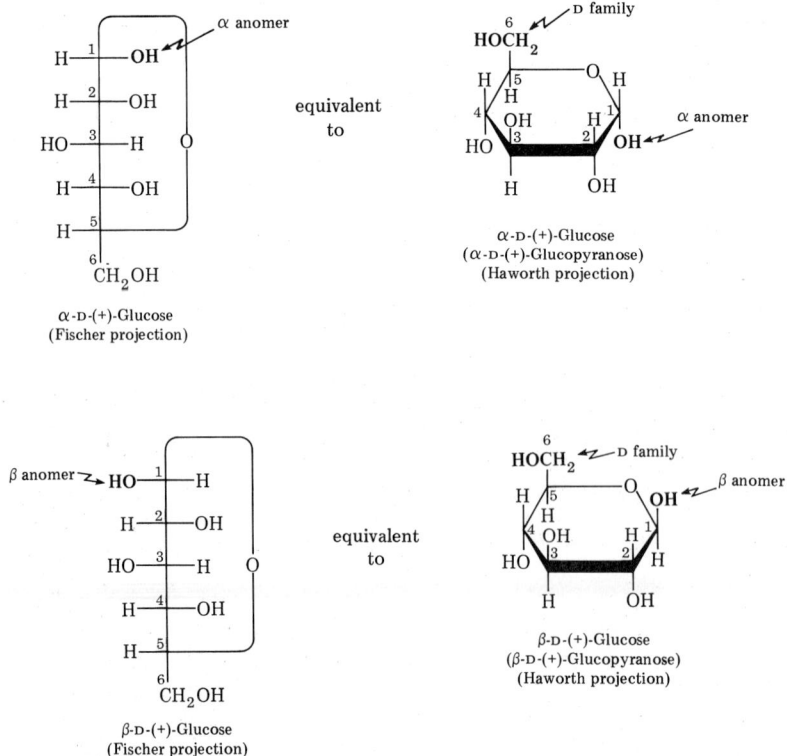

Figure 11.3 Fischer and Haworth representations of α-D-glucose and β-D-glucose.

More common representations of α-D-glucose and β-D-glucose, introduced by Haworth,* are shown in Figure 11.3. Note that in order to translate a Fischer projection into a Haworth projection, two simple rules must be followed.

Rule 1. All the hydroxyl groups attached to the right of the chiral centers in a Fischer formula should be directed *below* the plane of the ring in a Haworth representation. Conversely, the hydroxyl groups that appear to the left in a Fischer projection should be *above* the plane of the ring in a Haworth structure.

Rule 2. The terminal —CH$_2$OH unit in a Fischer formula should be projected *above* the plane of the ring in a Haworth structure if the sugar belongs to the D family. If the sugar belongs to the L family, the terminal —CH$_2$OH should be projected *below* the plane of the ring in a Haworth formula. Because most the naturally occurring sugars belong to the D family, the terminal —CH$_2$OH is most often seen above the plane of the ring.

Because of the similarity of Haworth projections of six-membered ring sugars to the heterocycle *pyran,* 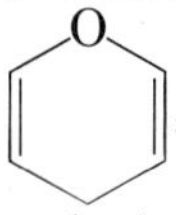 , these sugars have received the generic name **pyranose.** For example, α–D(+)-glucopyranose describes dextrorotatory (+)-D-glucose (D-gluco) in its α form as a six-membered ring (pyranose).

Problem 11.6 Starting from the open-chain form (see Fig. 11.1), draw the α form of D-galactose as **(a)** a Fischer projection six-membered ring and **(b)** a Haworth projection (pyranose form).

Five-Membered Ring Sugars *11.6*

Just as six-membered ring sugars may be drawn as Fischer projections or as Haworth structures (pyranose form), sugars that exist as five-membered cycles may be drawn in either of the two representations. The same rules enunciated previously for glucose apply. This is illustrated for D-ribose, a five-membered ring sugar, in Figure 11.4.

Because the Haworth structure is similar to the heterocycle furan,  ,

the five-membered ring sugars have received the generic name **furanose.** For example, α-D-ribofuranose describes D-ribose as a five-membered ring in its α form.

---

* Walter N. Haworth (1883–1950) was Professor of Chemistry at the University of Birmingham. His original field of interest was terpene chemistry, but while working under Purdie and Irvine his interest changed to the study of carbohydrates. As a result of his work, the ring structures and constitutions of most monosaccharides and disaccharides became established, as did the structures of many polysaccharides. He received the Nobel prize for his work in 1937.

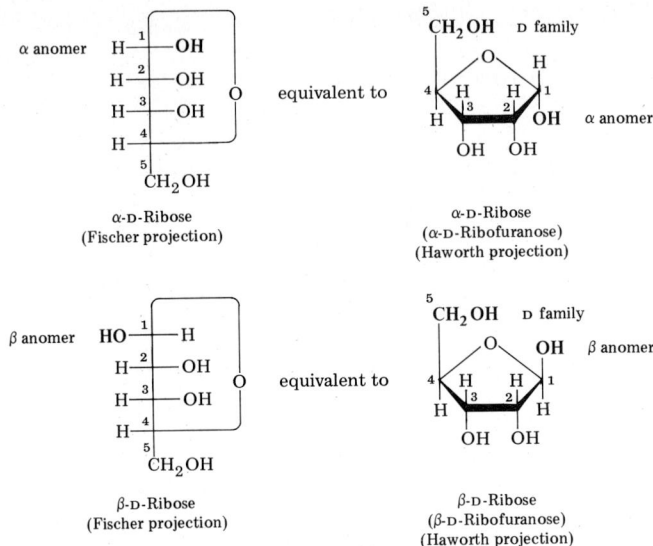

Figure 11.4 Fischer and Haworth representations of α-D-ribose and β-D-ribose.

Problem 11.7 Explain what is meant by α-D-2-deoxyribofuranose and draw its structure, starting from the open-chain form of D-2-deoxyribose given here.

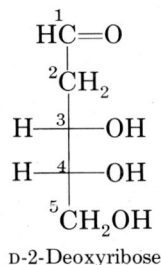

D-2-Deoxyribose

The most common ketose, fructose, also exists in cyclic forms. When fructose is a component of a saccharide, as in sucrose (Sec. 11.10), it usually occurs in its furanose form. The open-chain form and the Fischer and Haworth projections of fructose are shown in Figure 11.5.

Problem 11.8 In the uncombined form fructose exists as a six-membered ring structure. Draw **(a)** Fischer and **(b)** Haworth projections for both α and β forms of fructopyranose.

β anomer HO—2—1CH₂OH

(Fischer projection of β-D-(−)-Fructose)

β-D-(−)-Fructose
(Fischer projection)

D-(−)-Fructose
(open-chain form)

α-D-(−)-Fructose
(Fischer projection)

α anomer

or

or

β-D-(−)-Fructofuranose
(Haworth projection)

α-D-(−)-Fructofuranose
(Haworth projection)

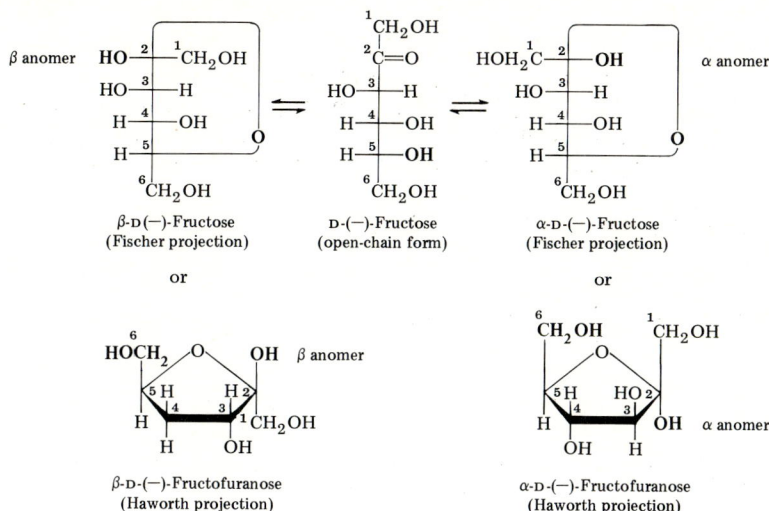

Figure 11.5 Open-chain form and Fischer and Haworth representations of fructose.

Expected Reactions of Monosaccharides *11.7*

Many of the reactions of monosaccharides are those we expect of their functional groups. Aldoses and ketoses contain the carbonyl function,* so they react with those reagents that normally attack the carbonyl group, $C=O$. The following reactions illustrate this point.

A Cyanohydrins, Oximes, Hydrazones, and Phenylhydrazones

Hydrogen cyanide reacts to form cyanohydrins.

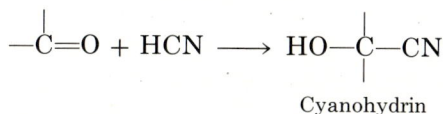

$$-C=O + HCN \longrightarrow HO-C-CN$$

Cyanohydrin

Hydroxylamine, hydrazine, phenylhydrazine, and derivatives condense to yield oximes, hydrazones, phenylhydrazones, and derivatives, respectively.

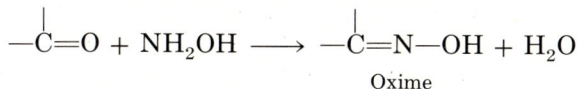

$$-C=O + NH_2OH \longrightarrow -C=N-OH + H_2O$$

Oxime

* Note that in explaining mutarotation, we indicated that sugars exist mainly in their cyclic forms, where the carbonyl function seemed to have disappeared. Remember, however, that a small concentration of the open-chain form, where the carbonyl group is evident, is always present. As soon as the carbonyl function has reacted with one of the attacking reagents, the concentration of the open-chain form is lower than that needed to maintain equilibrium. As a consequence, the cyclic structures open up to provide more of the open-chain form.

$$-\overset{|}{\underset{|}{C}}=O + H_2N-NH_2 \longrightarrow -\overset{|}{\underset{|}{C}}=N-NH_2 + H_2O$$

Hydrazone

$$-\overset{|}{\underset{|}{C}}=O + H_2N-NH-\text{⟨O⟩} \longrightarrow -\overset{|}{\underset{|}{C}}=N-NH-\text{⟨O⟩} + H_2O$$

Phenylhydrazone

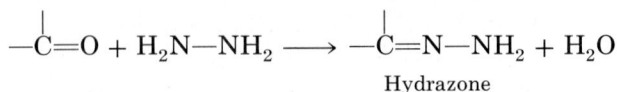

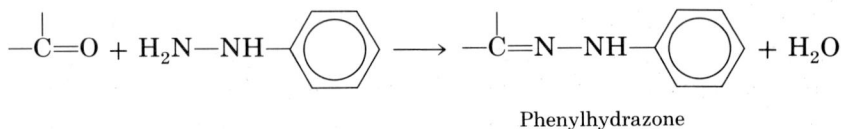

All these reactions were discussed previously (Secs. 10.15, 10.17), so that there is no need to elaborate here.

B   Glycosides: Sugar Acetals and Ketals

You have also seen previously that the internal condensation of the carbonyl group of aldoses and ketoses with one of the hydroxyl groups gives rise to cyclic hemiacetals and cyclic hemiketals, respectively. The furanose or pyranose structures can react further with alcohols or phenols, in the presence of catalytic amounts of dry acid, to give acetals or ketals.

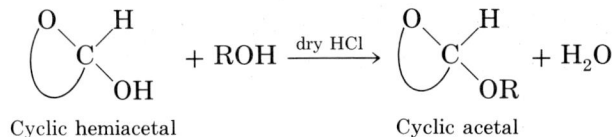

Cyclic hemiacetal   Cyclic acetal

The general term **glycoside** is used to denote any carbohydrate with an acetal or ketal group. The name of a specific glycoside consists of the radical of the alcohol (methyl for methyl alcohol, ethyl for ethyl alcohol, and so on) and the sugar from which it is derived (glucoside from glucose, fructoside from fructose, and so on). For example, the glycosides of D-glucose with methanol are called methyl α-D-glucoside and methyl β-D-glucoside.

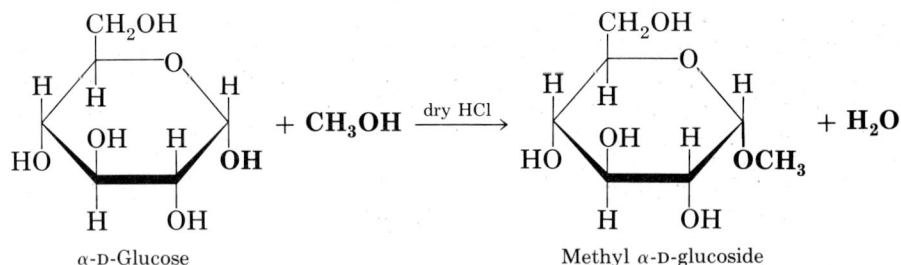

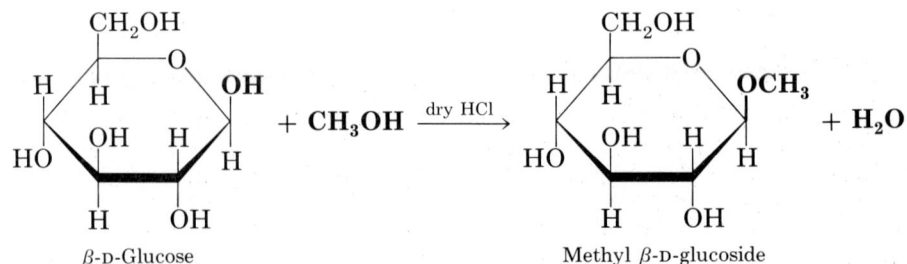

α-D-Glucose   Methyl α-D-glucoside

β-D-Glucose   Methyl β-D-glucoside

The carbon–oxygen–carbon linkage that bonds the two components of the acetal is called the **glycosidic linkage.** The glycosidic linkage is of great biological significance. Many natural products are glycosides. Among these products both the sugar fragment and the nonsugar moiety, called the *aglycone,* vary considerably. Examples are shown in Figure 11.6.

The glycosidic linkage is involved also in almost all interactions between one sugar and another. When one monosaccharide acts as a hemiacetal or a hemiketal and another plays the role of an alcohol, the resulting combination is a *disaccharide*. The formation of *maltose,* a disaccharide, illustrates this point. (Other disaccharides, as well as oligosaccharides and polysaccharides, are formed in the same manner; they are discussed in subsequent sections.)

α-D-Glucopyranose
(acts as a hemiacetal)   α-D-Glucopyranose
(acts as an alcohol)

Glycosidic linkage

Maltose

Glycosides, because they are acetals, are stable in neutral and basic solutions. Under acidic conditions, they are readily broken down to their sugar and alcohol components.

Nucleosides are glycosides in which the sugar fragments, D-ribose or D-2-deoxyribose, are joined through nitrogen to purine and pyrimidine bases, the aglycone portions. Nucleosides are partial building blocks of genetic material. Because of their great importance, the structure and the chemistry of these compounds are treated separately in Chapter 16.

C   Reduction (Glycitols)

The carbonyl group of aldoses and ketoses can be reduced, either by catalytic hydrogenation or with $NaBH_4$ (sodium borohydride), to the corresponding alcohol. The resulting polyhydric alcohols, or polyols, are given the class name of **glycitols.** For example, the reduction product of D-ribose is called D-ribitol, a constituent of vitamin B_2 (riboflavin); that of mannose is mannitol.

Sugar (hemiacetal)	Alcohol (aglycone)	Glycoside (acetal)

Salicin
(analgesic)

Digitoxin
(cardiac stimulant)

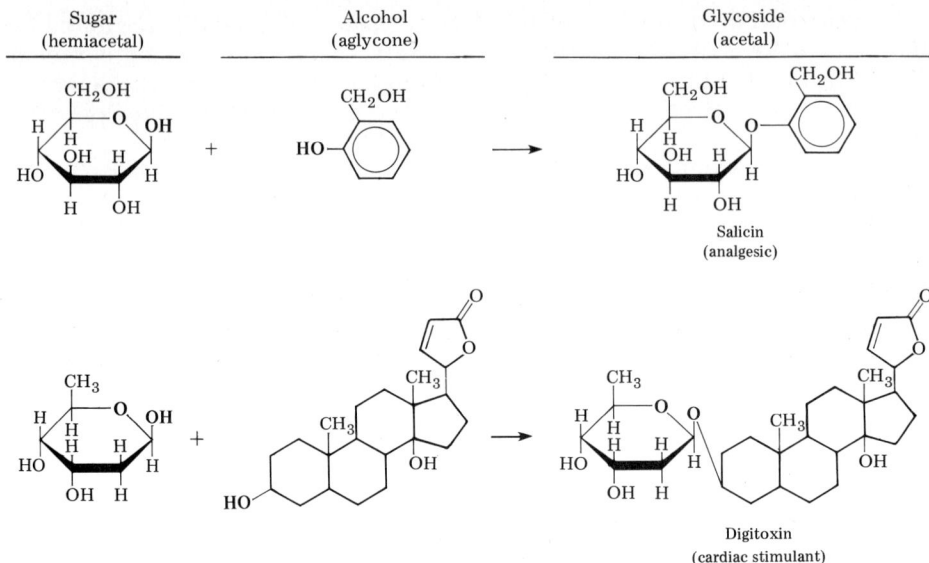

Figure 11.6 Salicin and digitoxin, two examples of glycosides of medicinal value, in which the glycosidic linkage is formed between a sugar acting as a hemiacetal and a nonsugar (aglycone) alcohol.

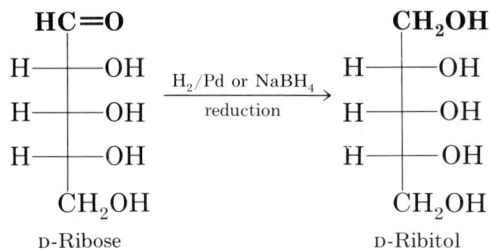

This system of nomenclature does not always hold, however. For example, the reduction product of D-glyceraldehyde is called glycerol, a principal constituent of fats and oils, and the product of reduction of D-glucose, glucitol, is more commonly called sorbitol.

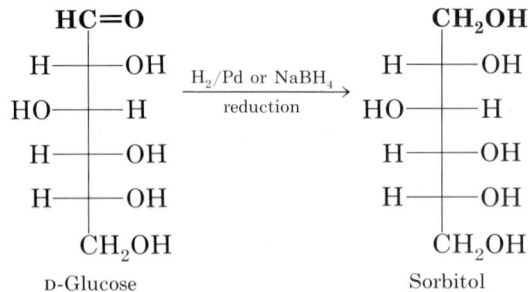

Problem 11.9 Fructose is reduced to a mixture of sorbitol and mannitol. Explain why.

Sorbitol and mannitol are often used as additives in foods, to which they impart sweetness and desirable texture qualities. Small amounts of sorbitol in wines produce a distinct smoothness and body, which is probably due to a combination of viscosity and complexing action. Sorbitol and mannitol are also used as rehydration agents in dehydrated foods where, because of their hygroscopic properties, they prevent the denaturation of the proteins. Sorbitol is also important as an intermediate in the manufacture of vitamin C (ascorbic acid).

Sorbitol has about the same caloric content as sugar (4 kcal/g), whereas mannitol produces only 2 kcal/g. The two are often used to replace sugars in dietetic foods. There are also indications that sorbitol, unlike sugars, has *noncariogenic* properties, which means that it causes no dental cavities. This is probably because it does not ferment as fast as sugar and is therefore washed off the teeth before the critical pH is reached.

D   Esters

Because aldoses and ketoses contain several hydroxyl groups, in addition to the carbonyl group, they react with reagents that normally attack alcohols as well as with reagents that normally attack the carbonyl group.

Reaction with acids produces esters. Phosphate esters, which are biologically the most important ones, are formed when phosphoric acid, H_3PO_4, condenses with a hydroxyl group in the sugar.

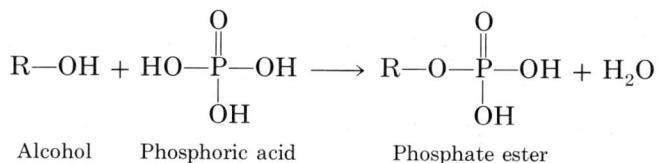

$$
\begin{array}{ccc}
& \quad\quad O & \quad\quad O \\
& \quad\quad \| & \quad\quad \| \\
R\text{—OH} + HO\text{—P—OH} & \longrightarrow & R\text{—O—P—OH} + H_2O \\
& \quad\quad | & \quad\quad | \\
& \quad\quad OH & \quad\quad OH
\end{array}
$$

Alcohol   Phosphoric acid   Phosphate ester

The simplest trioses, D-glyceraldehyde and dihydroxyacetone, do not occur in the free form to any extent in the human body. But the phosphate ester of each of these trioses is found in the metabolic breakdown of glucose. Their structures are

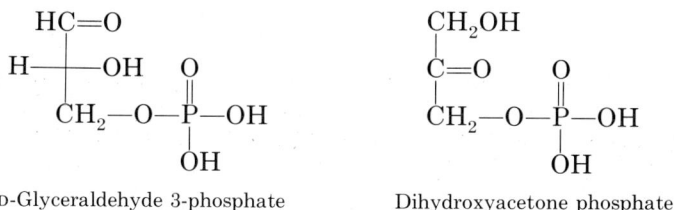

D-Glyceraldehyde 3-phosphate   Dihydroxyacetone phosphate

Phosphate esters of sugars also constitute the partial components of more complex molecules such as adenosine triphosphate (ATP), acetyl coenzyme A (acetyl CoA), nicotinamide adenine dinucleotide (NAD^+), and the nucleic acids, all of which are vitally important to the proper functioning of the body (see Chapter 16).

Problem 11.10   The structures of β-D-ribose (furanose form) and β-D-2-deoxy-ribose (furanose form) are

β-D-Ribofuranose

β-D-2-Deoxyribofuranose

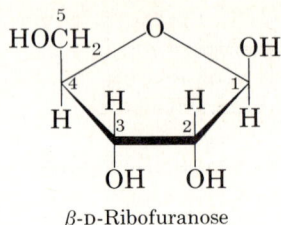

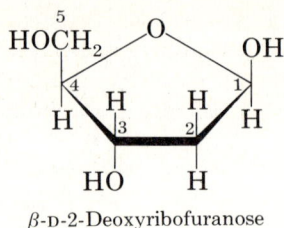

Draw the structures of **(a)** the 5-phosphate ester and **(b)** the 3-phosphate ester of each compound.

11.8 Unusual Reactions of Monosaccharides

In discussing the chemistry of monosaccharides, we have so far focused our attention first on the carbonyl function and then on the alcohol group. With one exception, the formation of cyclic hemiacetals, these functional groups were treated almost as if they were part of separate molecules. In fact, the two functionalities are in *close proximity* to each other. As a result, certain unusual reactions take place, two of which are presented in the following pages.

A Reaction with Phenylhydrazine (Osazones)

You have seen previously that it is possible to isolate the phenylhydrazone of an aldose or a ketose (Sec. 11.7A). More often, when aldoses or ketoses are treated with *an excess* of phenylhydrazine (three moles or more), they are converted into highly crystalline compounds called **osazones.** An osazone is a *phenylhydrazone derivative in which both the C-1 and the C-2 of an aldose or a ketose have reacted.* Because osazones have well-defined crystalline structures and are easy to isolate, they are often used for identification purposes. Moreover, because the osazone-producing reaction occurs only at C-1 and C-2, *sugars that have identical configurations at the other carbons give the same osazone.* For example, D-lyxose, D-xylose, and D-xylulose produce the same osazone (see Fig. 11.7). These sugars differ only in their stereochemistry or in the structure at the first two carbon atoms. Historically, osazone formation was one of the reactions used by Emil Fischer to determine the structure of all aldohexoses. To test your understanding of this concept, work out the following problem.

Problem 11.11 Refer to Figure 11.1 to answer these questions.
(a) D-Ribose and D-arabinose are converted to the same osazone. Explain why.
(b) Write the structure of a ketopentose that would give the same osazone as in (a).

B Oxidation: Reducing and Nonreducing Sugars

If you recall, the Tollens' reagent was used previously (Sec. 10.20) to differentiate between aldehydes and ketones. Simple aldehydes gave a positive test (silver mirror, Ag^0); simple ketones did not react.

D-Lyxose

$$
\begin{array}{c}
{}^{1}HC{=}O \\
HO{-}{}^{2}{-}H \\
HO{-}{}^{3}{-}H \\
H{-}{}^{4}{-}OH \\
{}^{5}CH_2OH
\end{array}
$$

D-Lyxose

$$
\begin{array}{c}
{}^{1}HC{=}O \\
H{-}{}^{2}{-}OH \\
HO{-}{}^{3}{-}H \\
H{-}{}^{4}{-}OH \\
{}^{5}CH_2OH
\end{array}
$$

D-Xylose

$\xrightarrow{\;3\ C_6H_5{-}NHNH_2\;}$ (phenylhydrazine)

$$
\begin{array}{c}
{}^{1}HC{=}N{-}\overset{H}{N}{-}C_6H_5 \\
{}^{2}C{=}N{-}\overset{H}{N}{-}C_6H_5 \\
HO{-}{}^{3}{-}H \\
H{-}{}^{4}{-}OH \\
{}^{5}CH_2OH
\end{array}
$$

An osazone

$+\quad (C_6H_5NH_2 + 2\ H_2O + NH_3)$

$$
\begin{array}{c}
{}^{1}CH_2OH \\
{}^{2}C{=}O \\
HO{-}{}^{3}{-}H \\
H{-}{}^{4}{-}OH \\
{}^{5}CH_2OH
\end{array}
$$

D-Xylulose

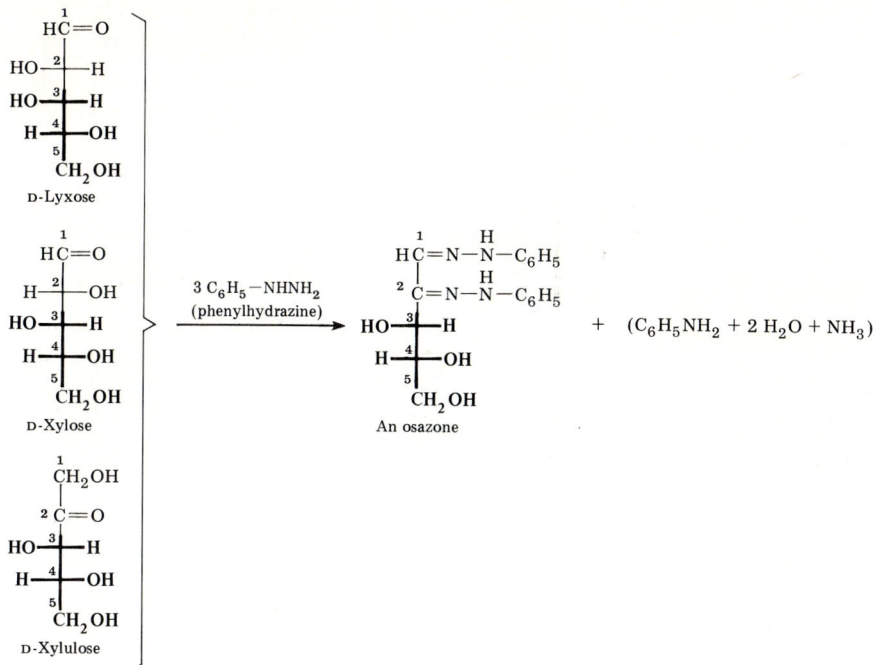

Figure 11.7 Three different sugars react with excess phenylhydrazine to produce the same osazone.

Because the carbonyl and hydroxyl groups are adjacent to each other, both aldoses and ketoses react positively with Tollens' reagent. In the process, silver ion (Ag^+) is reduced to metallic silver ($Ag^+ + e^- \rightarrow Ag^0\downarrow$), while the sugars are oxidized. Aldoses and ketoses also reduce **Benedict's** and **Fehling's reagents.** These reagents contain Cu^{2+} ions complexed with citrate and tartrate ions, respectively, in an alkaline medium. A positive Benedict or Fehling test is indicated when the original blue solution produces a yellow-red precipitate of copper(I) oxide, Cu_2O. Sugars that react with Tollens' and Benedict's (or Fehling's) reagents are classified as **reducing sugars.**

Reducing sugars are also those sugars that undergo mutarotation (see Sec. 11.4). Structurally, reducing sugars contain *a free α-hydroxy aldehyde, a free α-hydroxy ketone, a hemiacetal,* or *a hemiketal unit.* This includes, of course, *all monosaccharides.*

Disaccharides, oligosaccharides, and polysaccharides may or may not react with Tollens', Benedict's, and Fehling's reagents. Carbohydrates that give positive tests are, of course, reducing sugars. Those that do not are classified as **nonreducing sugars.**

Nonreducing sugars are also those sugars that do not mutarotate. Structurally, nonreducing sugars contain *acetal* or *ketal units* (glycosidic linkages). These functional groups are stable and do not hydrolyze under the alkaline conditions of Tollens', Benedict's, and Fehling's reagents. Table 11.3 summarizes the properties of reducing and nonreducing sugars.

Clinically, Benedict's test is used qualitatively and quantitatively to determine the presence of glucose in the blood or in the urine. An excess of glucose in the blood (hyperglycemia) may indicate a diabetic condition.

Table 11.3   Properties of Reducing and Nonreducing Sugars

Reducing sugars	Nonreducing sugars
(1) Give positive	(1) Give negative

(1) Give positive

    Tollens' test

       Ag^+ (ammonia complex) $\longrightarrow$ $Ag^0\downarrow$

         Clear solution   Silver mirror

    Benedict's test

       Cu^{2+} (citrate complex) $\longrightarrow$ $Cu_2O\downarrow$

         Clear blue solution   Yellow-red precipitate

    Fehling's test

       Cu^{2+} (tartrate complex) $\longrightarrow$ $Cu_2O\downarrow$

         Clear blue solution   Yellow-red precipitate

(1) Give negative

    Tollens' test

    Benedict's test

    Fehling's test

(2) Mutarotate

(3) Functional units present

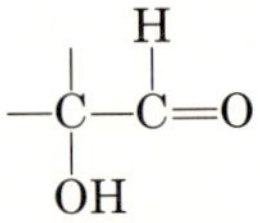

α-Hydroxyaldehyde

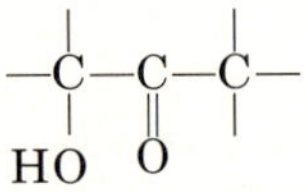

α-Hydroxyketone

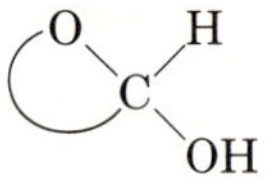

Hemiacetal

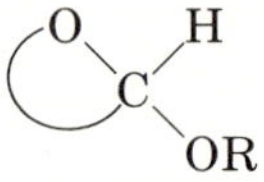

Hemiketal

(2) Do not mutarotate

(3) Functional units present

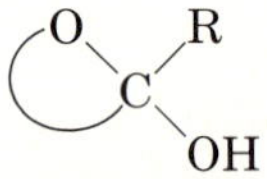

Acetal

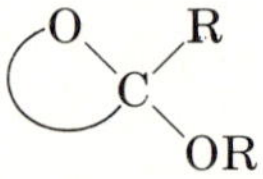

Ketal

11.9 Important Pentoses and Hexoses

Let us conclude our study of monosaccharides by singling out some individual pentoses and hexoses.

Two important pentoses are D-2-deoxyribose and D-ribose. Although they do not occur in the free state in nature, they play an extremely important role in metabolism. Deoxyribose is an integral component of deoxyribonucleic acid

(DNA), which is found in the nuclei of most living cells. DNA is essential for the transmission of hereditary traits. Ribose is found as a component of adenosine triphosphate (ATP) and of ribonucleic acid (RNA); RNA is involved in the biosynthesis of proteins and in the production of enzymes (see Chapter 16).

The most abundant and important aldohexose is glucose. Glucose, also known as *dextrose* because of its dextrorotatory characteristic, occurs in the free state and as a component of disaccharides and polysaccharides. Glucose is found in the human bloodstream (80–100 mg/100 ml of blood), which is why it is sometimes called *blood sugar*. Its presence is essential to human survival. Because glucose is readily absorbed and transported to tissues, where it provides energy, it is often administered intravenously as a 5% solution to hospitalized patients. A lower than normal concentration of blood sugar leads to a condition known as hypoglycemia. The symptoms of hypoglycemia are a general feeling of weakness, dizziness, and lethargy. A higher than normal concentration of glucose in the blood or its presence in the urine may indicate diabetes. Diabetes symptoms also involve a general feeling of dizziness, and to the untrained eye it may appear as if the diabetic individual is in a state of drunkenness.

Galactose, another aldohexose, does not occur in the free state. It is found in combination with glucose in lactose, also known as milk sugar (see Sec. 11.10). Galactose is also a component of cerebrosides and gangliosides, which are complex lipids present in the cerebral and neural tissues.

Fructose, also called *levulose* because it is highly levorotatory ($-92°$), is the most abundant ketohexose. It occurs both in the free state and in combination with other sugars (see Sec. 11.10). Fructose is the sugar in fruits, hence its name (from the Latin *fructus*, fruit).

Disaccharides   *11.10*

Disaccharides were defined as carbohydrates "that yield two monosaccharides upon hydrolysis" (Sec. 11.2). The two monosaccharide units in a disaccharide are held together by a glycosidic (acetal) linkage. The treatment of disaccharides in this section of the book will focus on three aspects.

1. The nature and structure of the monosaccharide components.
2. The kind of glycosodic linkage that joins one monosaccharide unit to the other.
3. Whether or not a given disaccharide is a reducing sugar, based on the structural criteria listed in Table 11.3.

Two of the disaccharides we shall consider, lactose and sucrose, are found in nature and are important because of their food value.

Lactose is the sugar found in milk. Cow's milk contains 4–6% lactose; its percentage in human milk varies from 5 to 8%. Upon hydrolysis (and digestion) lactose is broken down into D-glucose and D-galactose. In the body, galactose is isomerized normally to glucose in a series of steps catalyzed by enzymes. The glucose molecules that are formed can then be metabolized to provide energy to infants who drink the milk. Occasionally, the change of galactose to glucose does not take place because of a deficiency in one of the enzymes needed for the conversion, and galactose accumulates in the blood. This condition leads to a disease known as *galactosemia,* which can be fatal to the infants of the woman who has the disease.

Lactose may be visualized as being formed by the splitting off of water between β-D-galactose (pyranose form) and α-D-glucose (also in the pyranose structure). Galactose, acting as a hemiacetal, furnishes the β OH at the anomeric carbon (C-1), whereas glucose, acting as an alcohol, provides the hydroxylic OH at C-4. The two monosaccharides are linked to each other by a *glycosidic bond,* or, more specifically, by a β-1,4-glycosidic linkage. Lactose may also be referred to as β-D-galactose-(1,4)-α-D-glucose.

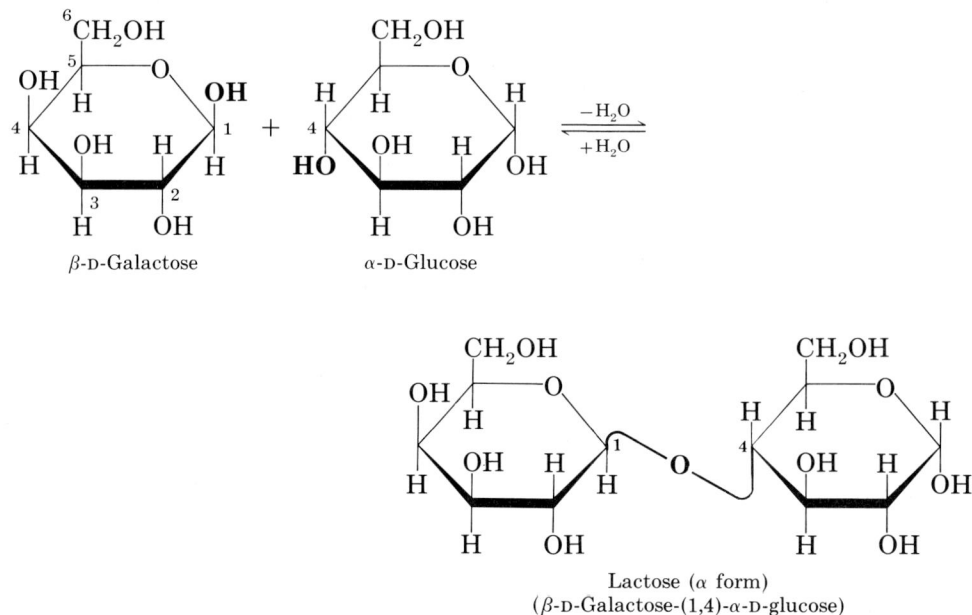

β-D-Galactose   α-D-Glucose

Lactose (α form)
(β-D-Galactose-(1,4)-α-D-glucose)

Is lactose a reducing or a nonreducing sugar? If we examine its structure, we find that the hemiacetal grouping of the glucose component remains intact. Lactose is therefore a reducing sugar.

Problem 11.12   Would you expect lactose to mutarotate? Explain.

Sucrose is the common table sugar. It is extracted from beets and from the juice of sugar cane. Hydrolysis (and digestion) of sucrose yields D-glucose and D-fructose in equimolar amounts. This equimolar mixture is called *invert sugar* because the initial dextrorotation of sucrose (+65.5°) is changed to a value of −19.9° (levorotatory) in the one-to-one mixture of glucose and fructose. Invert sugar is sweeter than sucrose; it is found in large amounts in honey.

α-D-Glucose

+

β-D-Fructose

$\xrightarrow[+H_2O]{-H_2O}$

Sucrose
(α-D-Glucose-(1,2)-β-D-fructose)

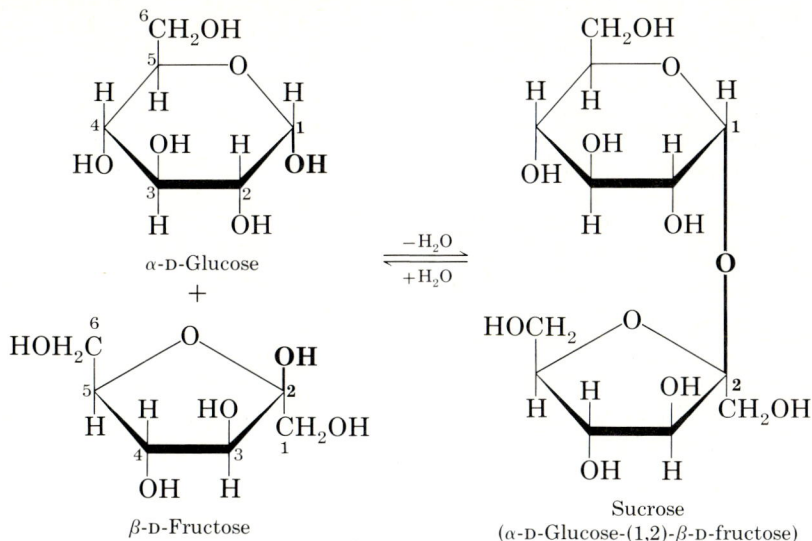

On examining the structure of sucrose you can see that the glucose unit exists in the pyranose form, whereas fructose is present as a five-membered ring. The two monosaccharides are again joined by a glycosidic linkage. In the example shown here, glucose plays the role of the hemiacetal by contributing the α OH at C-1, and fructose acts as the alcohol by providing the β OH at C-2. Sucrose may therefore be called α-D-glucose-(1,2)-β-D-fructose.

Because sucrose has an acetal group only, with no hemiacetal, it is a nonreducing sugar. Correspondingly, it does not undergo mutarotation.

Problem 11.13 Explain why sucrose may also be referred to as β-D-fructose-(2,1)-α-D-glucose.

Maltose, isomaltose, and **cellobiose** do not occur widely in the free state. Maltose and isomaltose are obtained usually on partial hydrolysis of starch; cellobiose results from partial hydrolytic breakdown of cellulose. All three disaccharides consist of diglucose units and yield glucose on hydrolysis. Structurally, maltose, isomaltose, and cellobiose differ only in the nature of their glycosidic linkages.

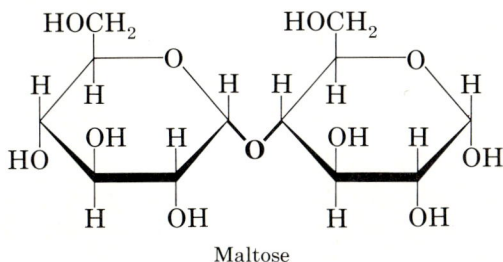

Maltose

Isomaltose

Cellobiose

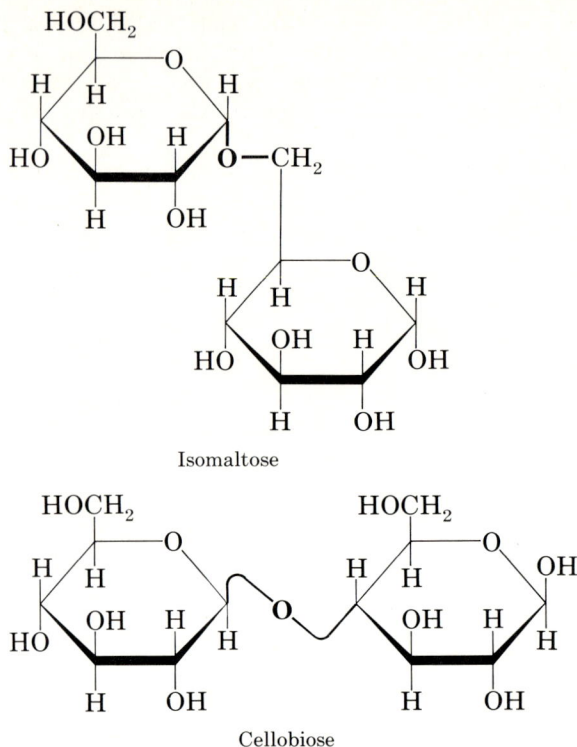

Problem 11.14 For each disaccharide (1) identify the monosaccharide units and state the form in which they exist; (2) circle and name the glycosidic linkage; and (3) indicate whether the sugar is reducing or nonreducing.
(a) Maltose   **(b)** Isomaltose   **(c)** Cellobiose

11.11 Polysaccharides

Polysaccharides were defined as "carbohydrates that furnish ten or more monosaccharides when completely hydrolyzed" (Sec. 11.2). We shall limit this discussion to **cellulose, starch,** and **glycogen.** Cellulose, starch, and glycogen are high-molecular-weight compounds made up entirely of D-glucose units. All three polysaccharides are nonreducing sugars. Structurally, cellulose, starch, and glycogen differ from one another only by the kind of glycosidic linkage joining one glucose unit to another.

Cellulose is the most abundant organic compound found in nature. It forms the fibrous tissue in the cell wall of all living plants. Wood is 40–50% cellulose. The paper this is printed on is made of cellulose, as are cotton and flax fibers. Because it is readily available from plants and undergoes various reactions, cellulose is also an important commercial raw material.

When cellulose is heated with dilute sulfuric acid, the glycosidic linkages are broken. A partial hydrolysis product is the disaccharide cellobiose, which contains a β-1,4-glycosidic linkage. Cellulose itself is an unbranched polymer of several thousand D-glucose units joined via β-1,4-glycosidic linkage. A partial structure of one cellulose chain is shown in Figure 11.8a.

Nutritionally, *starch* is the main contributor of carbohydrates in our diet. It exists exclusively in plants, especially in seeds, where it serves as a reserve carbohydrate. Starch is a mixture containing about 20% amylose and 80% amylopectin.

Structurally, amylose is similar to cellulose, except in its orientation of the glycosidic linkages. The glucose units in amylose are connected via α-1,4-glycosidic bonds rather than the β-1,4-glycosidic linkages that are present in cellulose. Compare Figures 1.8a and 1.8b. This difference in configuration between β and α in the glycosidic linkage is important. It allows our digestive enzymes to utilize starch as our only vegetal source of glucose (the human body cannot digest cellulose, if you recall). The resulting glucose is then metabolized further to supply energy and the carbon skeleton for our bodies.

Problem 11.15 What disaccharide is produced on partial hydrolysis of amylose? (*Hint:* See Sec. 11.10.)

Amylopectin, the other component of starch, is a branched polymer. The main portion of the molecule consists of several thousand glucose units connected by α-1,4-glycosidic linkages. Branching is due to the presence of the α-1,6-glycosidic linkages, which occur after about every 25–30 glucose residues.

Glycogen has a structure very like that of amylopectin. The main difference is in the frequency of branching. In glycogen the α-1,6-glycosidic bond occurs after about every 8–10 glucose units. The structures of amylopectin and glycogen are shown in Figure 11.9. Functionally, glycogen plays the same role in the animal kingdom as starch does in plants: it serves as a reserve carbohydrate.

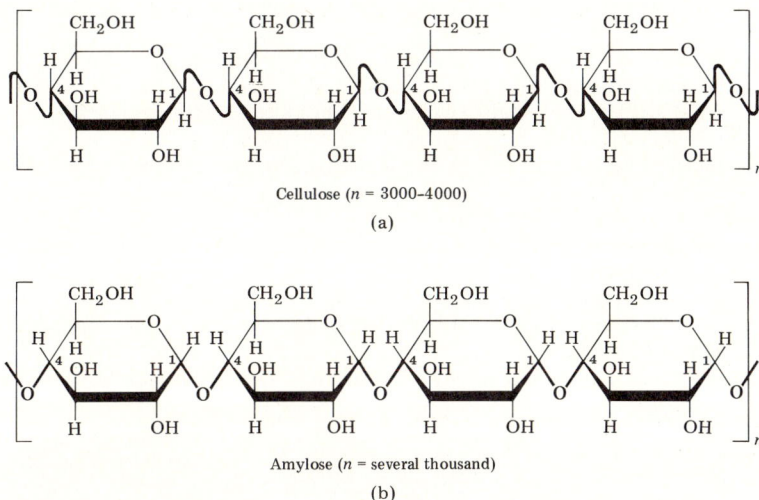

Figure 11.8 Partial structures of (a) cellulose and (b) amylose.

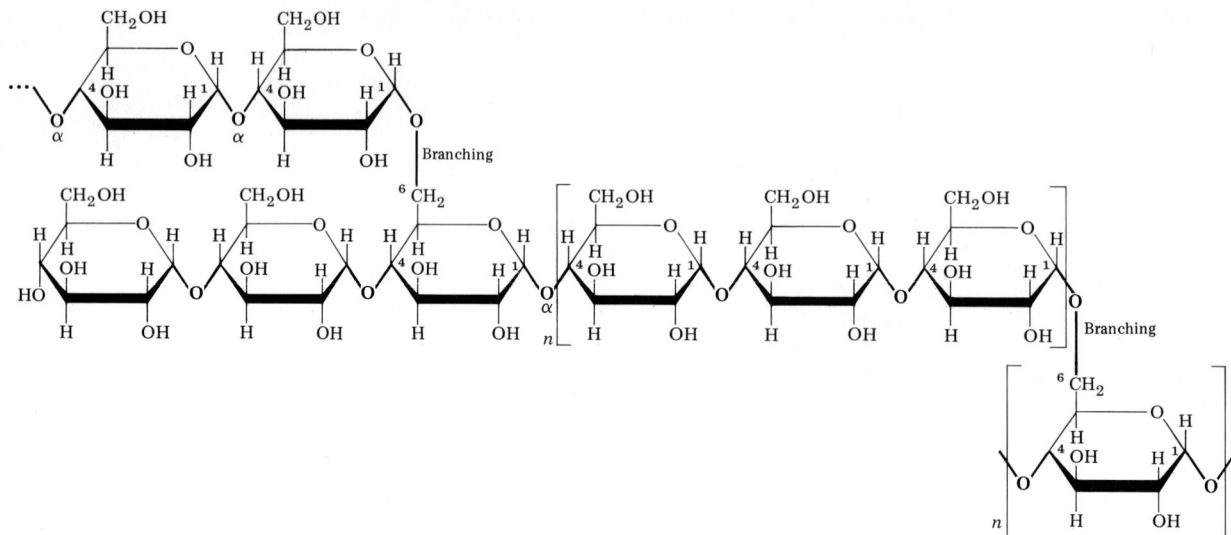

Figure 11.9 Partial structures of amylopectin and glycogen.

Amylopectin (n = 25–30)
Glycogen (n = 8–10)

Summary of Concepts and Reactions

Carbohydrates represent a group of substances produced in green plants by means of photosynthesis.   [Sec. 11.1]

Carbohydrates are polyhydroxyaldehydes or polyhydroxyketones or compounds that can easily be converted into one of them by hydrolysis.   [Sec. 11.1]

Polyhydroxyaldehydes are also known as aldoses. Polyhydroxyketones are also known as ketoses.   [Sec. 11.1]

Monosaccharides, or simple sugars, are carbohydrates that consist of a single aldose or ketose unit and cannot be hydrolyzed.   [Sec. 11.2]

Disaccharides are sugars that yield two monosaccharide units upon hydrolysis.
                                                                             [Sec. 11.2]

Oligosaccharides are carbohydrates that yield three to ten monosaccharides after hydrolysis. Polysaccharides are carbohydrates that yield ten or more monosaccharide units when completely hydrolyzed.   [Sec. 11.2]

Sugars that contain three, four, five, or six carbons are called trioses, tetroses, pentoses, and hexoses, respectively.   [Sec. 11.2]

Glyceraldehyde, the simplest aldose, contains three carbon atoms and is therefore the aldotriose. Dihydroxyacetone, the simplest ketose, also contains three carbon atoms and is the ketotriose.   [Sec. 11.2]

Monosaccharides are also divided into D or L families. The sugars found in nature belong to the D family.   [Sec. 11.2]

When the —OH group attached to the chiral carbon that is farthest from the carbonyl group points to the right in a Fischer projection formula, the sugar belongs to the D family. If the —OH points to the left, the sugar belongs to the L family.
                                                                             [Sec. 11.2]

312   All monosaccharides are soluble in water and have high boiling points.   [Sec. 11.3]

When the two forms of D-glucose are dissolved in water, their specific rotation changes until an equilibrium value of $+53°$ is reached. This change in optical rotation of a solution of the two forms of glucose until a constant value is obtained is called mutarotation. [Sec. 11.4]

Through an intramolecular process, glucose forms a six-membered cyclic hemiacetal. [Sec. 11.5A]

The cyclic hemiacetal form of glucose is produced by the reaction of the —OH bonded to the fifth carbon with the aldehyde group of the first carbon. [Sec. 11.5A]

As the cyclic structure is formed, a new chiral center, known as the anomeric carbon, is generated. [Sec. 11.5A]

A less awkward representation of the cyclic form of glucose than the Fischer projection is the Haworth projection formula. [Sec. 11.5B]

Haworth projections of six-membered ring sugars are known as pyranoses, and Haworth projections of five-membered ring sugars are called furanoses. [Secs. 11.5B, 11.6]

Both aldoses and ketoses contain the carbonyl function and therefore can form cyanohydrins, oximes, hydrazones, and phenylhydrazones. [Sec. 11.7A]

Pyranose and furanose structures can react further with alcohols or phenols to form acetals or ketals. The general term glycoside is used to denote any carbohydrate with an acetal or ketal group. [Sec. 11.7B]

The carbon–oxygen–carbon linkage that bonds the two components of the acetal is called the glycosidic linkage. [Sec. 11.7B]

The $C=O$ group of aldoses and ketoses can be reduced to the corresponding alcohols. The resulting polyhydric alcohols or polyols are given the class name of glycitols. [Sec. 11.7C]

The OH groups of aldoses and ketoses can react with acids to form esters. [Sec. 11.7C]

When aldoses and ketoses are treated with an excess of phenylhydrazine, they are converted into highly crystalline compounds called osazones. [Sec. 11.8A]

Osazone-producing reactions occur at C-1 and C-2 only. Therefore, sugars that have identical configurations at the other carbons give the same osazone. [Sec. 11.8A]

Sugars that react with Tollens' and Benedict's or Fehling's reagents are called reducing sugars. Those that do not react are classified as nonreducing sugars. [Sec. 11.8B]

Lactose and sucrose are two naturally occurring disaccharides of nutritional value. Maltose, isomaltose, and cellobiose are disaccharides obtained upon partial hydrolysis of starch and cellulose. All disaccharides contain a glycosidic linkage. A disaccharide may be a reducing sugar or a nonreducing sugar depending on whether a hemiacetal group is present or not. [Sec. 11.10]

Cellulose, starch, and glycogen are examples of polysaccharides consisting entirely of D-glucose units joined by glycosidic linkages. [Sec. 11.11]

Key Terms

photosynthesis	triose	hemiacetals
polyhydroxy aldehydes	tetrose	hemiketals
polyhydroxy ketones	pentose	anomeric carbon
aldoses	hexose	pyranose
ribose	mutarotation	osazones
glucose	monosaccharides	furanose
galactose	disaccharides	glycoside
ketoses	oligosaccharides	glycosidic linkage
fructose	polysaccharides	glycitols

Benedict's reagent	lactose	cellobiose
Fehling's reagent	sucrose	cellulose
reducing sugars	maltose	starch
nonreducing sugars	isomaltose	glycogen

Exercises

Definitions [Secs. 11.1–11.8]
11.1 Define or illustrate **(a)** aldohexose; **(b)** photosynthesis; **(c)** ketopentose; **(d)** disaccharide; **(e)** α-furanose; **(f)** β-pyranose; **(g)** a reducing sugar; **(h)** D family; **(i)** aglycone; **(j)** glycosidic linkage.

Structure of Monosaccharides; Mutarotation [Secs 11.2–11.6]
11.2 Draw the structures of the following.
    (a) L(−)-Glucose (open-chain form)   **(b)** L(+)-Fructose (open-chain form)
    (c) α-D-Ribofuranose   **(d)** β-D-Galactopyranose
11.3 2-Deoxy-D-glucose is being tested as a drug against certain venereal diseases. Draw the structure of the compound (open-chain form).
11.4 Explain why there are 16 open-chain aldohexoses and 32 cyclic structures.
11.5 **(a)** Define mutarotation.
    (b) Write equations that illustrate how D-ribose undergoes mutarotation

Reactions of Monosaccharides [Secs. 11.5–11.8]
11.6 Draw the structures (using Fischer projections) for the products expected from the reaction of D-galactose with these reagents.
    (a) Excess phenylhydrazine   **(b)** $NaBH_4$
    (c) HCN (two stereoisomers)   **(d)** H_2NOH
11.7 D-Glucose is used as a source of energy by the body. The first step in the metabolic pathway of glucose (glycolysis) is the formation of glucose 6-phosphate. Draw its structure starting from
    (a) The open-chain form of D-glucose   **(b)** α-D-Glucopyranose
11.8 Refer to Figure 11.1 to answer these questions.
    (a) Which two D-aldopentoses would be reduced to meso alcohols by $NaBH_4$?
    (b) Which two would be reduced to optically active alcohols?
11.9 **(a)** Write an equation for the formation of ethyl α-glucoside (pyranose form).
    (b) Are glycosides more stable under acidic or alkaline conditions?
11.10 **(a)** Refer to Figure 11.1 and draw the structures of the osazones that would be produced by reaction of each of the four D-aldopentoses with excess phenylhydrazine.
    (b) How many *different* osazones are formed altogether from the four sugars?
11.11 Which aldohexose and which ketohexose give the same osazone as D-glucose?

Structures and Reactions of Disaccharides and Polysaccharides [Secs. 11.10, 11.11]
11.12 Starting from the open-chain form of D-arabinose, draw the following.
    (a) The α-furanose form (Haworth projection)
    (b) The β-furanose form (Haworth projection)
    (c) The disaccharide formed by condensation between the α form (acting as a hemiacetal) and the C-5 OH group of the β form (acting as an alcohol). Is the resulting disaccharide a reducing or a nonreducing sugar?
    (d) The nonreducing disaccharide formed by the condensation between the α and β forms.

11.13 Gentiobiose is a disaccharide that may also be called β-D-glucose-(1,6)-β-D-glucose (pyranose form).
    (a) Draw its structure.
    (b) Determine whether it is a reducing or a nonreducing sugar.
11.14 Name the monosaccharides obtained by the hydrolysis of
    (a) Lactose   **(b)** Maltose   **(c)** Sucrose   **(d)** Cellobiose
    (e) Starch   **(f)** Cellulose   **(g)** Glycogen
11.15 Draw the structure of the disaccharide obtained when two α-D-glucopyranose molecules are joined by a 1,4-glycosidic linkage. On the basis of your structure, answer these questions about the disaccharide.
    (a) Is it a reducing or a nonreducing sugar? Explain your answer.
    (b) Can it mutarotate? Why or why not?

12

Carboxylic Acids and Their Derivatives

The most important class of organic acids is the carboxylic acids. All carboxylic acids contain the **carboxyl group,** $-\overset{\overset{\text{O}}{\|}}{\text{C}}-\textbf{OH,}$ which can be condensed to **—COOH** or **—CO$_2$H.** (The name is a combination of the two components of the group, —C=O, *carbo*nyl, and —OH, hydro*xyl*). As suggested by the name, carboxylic acids exhibit acidic properties in aqueous solutions. They turn blue litmus red and neutralize bases to form water and a salt.

Carboxylic acids are classified as aliphatic or aromatic depending on whether an R or an Ar residue is attached to the carboxyl group.

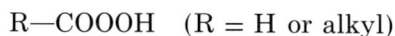

R—COOOH   (R = H or alkyl)   Ar—COOH   (Ar = C$_6$H$_5$—)

Aliphatic acid   Aromatic acid

Typical examples are acetic acid and benzoic acid.

CH$_3$—COOH

Acetic acid
(aliphatic)

Benzoic acid
(aromatic)

Because of their wide distribution and abundance in natural products, many carboxylic acids were known long before the advent of the IUPAC system. The common names of carboxylic acids all end in *-ic acid.* The names are derived from Latin or Greek and relate to their natural sources. For example, formic acid is the acid that gives the characteristic sting to an ant bite (from the Latin *formica,* ant); acetic acid is vinegar (from the Latin *acetum,* vinegar); and butyric acid is the compound that gives rancid butter its putrid smell (from the Latin *butyrum,* butter).

Long straight-chain carboxylic acids with even numbers of carbons, which were first isolated from fats and waxes, are called fatty acids. They are discussed in Chapter 13.

In the IUPAC system the acids are named in the usual way. The ending *-e* of the corresponding alkane is replaced by *-oic acid.* Thus, HCOOH, a one-carbon acid, is called *methanoic acid,* and CH_3COOH is called *ethanoic acid.* Table 12.1 gives the IUPAC and common names of the first five straight-chain carboxylic acids.

If substituents are present on the acid chain, their positions are located by Greek letters in the common nomenclature. The carbon adjacent to the carboxyl carbon is assigned the letter α (alpha), the next carbon β (beta), and so on. In the IUPAC system numbers are used to assign positions. The carboxyl carbon is numbered 1, the next carbon 2, and so on.

Common nomenclature:

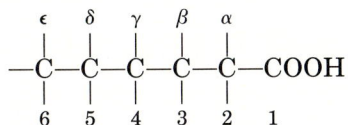

$$\overset{\epsilon}{-C}-\overset{\delta}{C}-\overset{\gamma}{C}-\overset{\beta}{C}-\overset{\alpha}{C}-COOH$$

IUPAC nomenclature: 6   5   4   3   2   1

For example,

$$CH_3-\underset{\underset{CH_3}{|}}{CH}-\underset{\underset{CH_3}{|}}{CH}-COOH$$

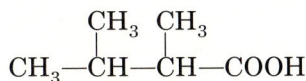

Common name: α,β-Dimethylbutyric acid
IUPAC name: 2,3-Dimethylbutanoic acid

Table 12.1 IUPAC and Common Names of Some Normal Carboxylic Acids

No. of carbon atoms	Formula	IUPAC name	Common name
1	HCOOH	Methanoic acid	Formic acid
2	CH_3COOH	Ethanoic acid	Acetic acid
3	CH_3CH_2COOH	Propanoic acid	Propionic acid
4	$CH_3(CH_2)_2COOH$	Butanoic acid	Butyric acid
5	$CH_3(CH_2)_3COOH$	Pentanoic acid	Valeric acid

When the carboxyl group is attached to a saturated ring, the acid is named as a cycloalkane carboxylic acid.

Cyclopropanecarboxylic
acid

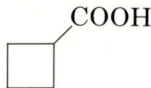

Cyclobutanecarboxylic acid

Cyclopentanecarboxylic
acid

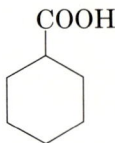

Cyclohexanecarboxylic
acid

Aromatic carboxylic acids are generally called by their common names.

Common name: Benzoic acid
IUPAC name: Benzenecarboxylic acid
(used as germicide and
as food preservative)

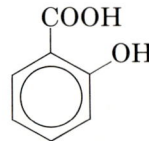

Salicylic acid
2-Hydroxybenzenecarboxylic acid
(raw material for aspirin)

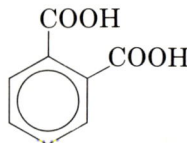

Common name: Phthalic acid
IUPAC name: Benzene-1,2-dicarboxylic acid
(medicinal uses; synthetic perfumes)

Terephthalic acid
Benzene-1,4-dicarboxylic acid
(constituent of Dacron)

The four simplest of the dicarboxylic acids—acids that contain two carboxyl groups—are known almost exclusively by their common names.

HOOC—COOH

Oxalic acid

HOOC—CH$_2$—COOH

Malonic acid

HOOC—CH$_2$CH$_2$—COOH

Succinic acid

HOOC—CH$_2$CH$_2$CH$_2$—COOH

Glutaric acid

Problem 12.1 Give both the IUPAC and common names for the following compounds

(a) $CH_3\overset{\overset{\displaystyle Br}{|}}{C}HCOOH$　　　　**(b)** $CH_3\overset{\overset{\displaystyle CH_3}{|}}{\underset{\underset{\displaystyle CH_3}{|}}{C}}CH_2COOH$　　　**(c)** [cyclohexane ring]—CH_2COOH

(d) [benzene ring]—CH_2CH_2COOH

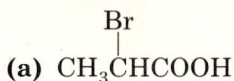

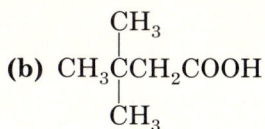

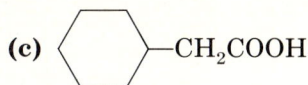

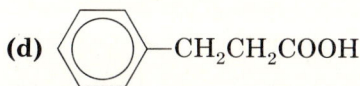

Problem 12.2　Write the structures for these acids.
(a) α,β-Dihydroxypropionic acid　　　　**(b)** 3,4-Dimethyl-2-ethylpentanoic acid
(c) α,β-Dimethylvaleric acid　　　　　　**(d)** γ-Phenylbutyric acid

---

Problem 12.3　There is something wrong with each of the names given. Write the structure and give the correct common and IUPAC names for each compound.
(a) α-Methylpentanoic acid　　　　**(b)** 3-Isopropylvaleric acid
(c) 2-Methyl-2-ethylacetic acid

Physical Properties of Carboxylic Acids　**12.2**

　　　The carboxyl group may be regarded as a water molecule in which one hydrogen atom is replaced by a C=O group. Therefore, like water, carboxylic acids are capable of hydrogen bonding (Fig. 12.1). Thus, the first four aliphatic acids (formic through butyric) are completely miscible in water. Higher members of the series are less soluble because the long alkyl chain gives them alkane-like characteristics. Aromatic acids are insoluble in water.

　　　The boiling points of carboxylic acids indicate a greater degree of association than for alcohols of comparable molecular weights. For example, acetic acid (mol wt = 60) boils at 118°C, whereas n-propyl alcohol (mol wt = 60) boils at only 97°C. In fact, simple carboxylic acids exist as hydrogen-bonded dimers (Fig. 12.2) and therefore behave as if they were of much higher molecular weight.

　　　Table 12.2 compares some physical properties of a few carboxylic acids and the alcohols of corresponding molecular weights.

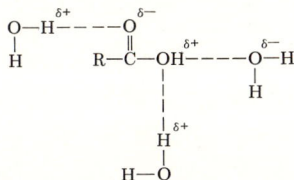

Figure 12.1　Association of a carboxylic acid with water molecules through hydrogen bonding.

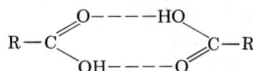

Figure 12.2 Hydrogen-bonded dimer of a carboxylic acid.

The first nine aliphatic acids are colorless liquids that have sharp, acrid odors. Vinegar, a 5% solution of acetic acid in water, has its characteristic sharp odor because of the acetic acid in the solution. Pure acetic acid is called *glacial acetic acid* because it solidifies into ice-like crystals at temperatures slightly below normal room temperature (about 17°C). Butyric acid smells like rancid butter and strong cheese. Acids of five to ten carbons have goat-like smells because they are present in the skin secretion of goats. Higher acids are wax-like solids and are practically odorless. Aromatic acids are also high-melting odorless solids.

Table 12.2 Some Physical Properties of Acids and Alcohols of Corresponding Molecular Weights

Structure	Name	Mol wt	Bp (°C)	Solubility in H_2O at 25°C
HCOOH	Formic acid	46	100	Very soluble
CH_3CH_2OH	Ethyl alcohol	46	78	Very soluble
CH_3COOH	Acetic acid	60	118	Very soluble
$CH_3CH_2CH_2OH$	n-Propyl alcohol	60	97	Very soluble
$CH_3(CH_2)_3COOH$	Valeric acid	102	187	4.0 g/100 g H_2O
$CH_3(CH_2)_4CH_2OH$	n-Hexyl alcohol	102	156	0.6 g/100 g H_2O
⬡—COOH	Benzoic acid	122	250	Insoluble
⬡—CH_2CH_2OH	2-Phenylethanol	122	220	Insoluble

12.3 Acid Strength

The strength of an acid depends on the extent it ionizes. Common mineral acids, such as HCl or HNO_3, ionize completely and are considered therefore to be strong acids.

$$HCl \longrightarrow H^+ + Cl^-$$

Carboxylic acids are weak acids because they are incompletely ionized and exist in equilibrium with a solution of their ions.

$$RCOOH \rightleftharpoons H^+ + RCOO^-$$

The strength of an acid can be measured quantitatively by determining the value of its **ionization constant, K_a**. An acid with a higher K_a value is stronger than one with a lower K_a. Strong acids have K_a values higher than 10^{-2}, and weak acids have K_as of 10^{-2} or less. The K_as of most carboxylic acids fall in the

acids, formic acid (higher K_a) being 10 times stronger than acetic acid
(lower K_a).

    HCOOH $K_a = 1.8 \times 10^{-4}$ CH$_3$COOH $K_a = 1.8 \times 10^{-5}$

    Formic acid   Acetic acid

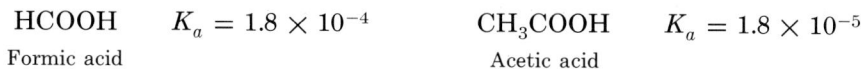

    A more convenient way of expressing acid strength quantitatively is in term
of their **pK_as**. The pK_a is defined as

    $\text{p}K_a = -\log K_a$ (1)

From (1) it can be shown that acid strength is inversely related to the pK_a; the
lower the pK_a, the stronger the acid, and vice versa. Formic acid, the stronger
acid, has a pK_a of 3.74, whereas acetic acid, the weaker acid, has a pK_a of 4.74.

Problem 12.4   Arrange the acids listed in order of decreasing acid strength:
formic acid (p$K_a = 3.74$); benzoic acid (p$K_a = 4.20$); butyric acid (p$K_a = 4.82$)

Acid Strength and Acid Structure   *12.4*

    Carboxylic acids are much more acidic than are alcohols even though both
classes of compounds contain an OH group. Why is this so? The answer lies in
the structures of their conjugate bases, the **carboxylate anion** and the alkoxide
anion, respectively.

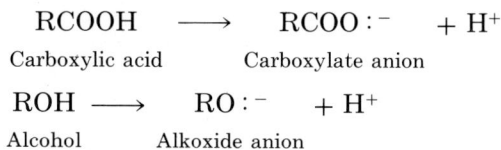

    RCOOH $\longrightarrow$ RCOO:$^-$ $+$ H$^+$

    Carboxylic acid   Carboxylate anion

    ROH $\longrightarrow$ RO:$^-$ $+$ H$^+$

    Alcohol   Alkoxide anion

Carboxylic acids are stronger acid than alcohols because carboxylate anions are
stabilized by resonance, whereas alkoxide anions are not, so that less energy is
needed to remove an H$^+$ from acids than is required to remove a proton from
alcohols (Fig. 12.3).

    Having explained why carboxylic acids are stronger acids than alcohols, can
we predict the relative acid strengths among carboxylic acids? The answer is yes.
Any factor that *stabilizes* the carboxylate anion of an acid will give it *greater*
acid strength than an acid lacking that factor. Conversely, any factor that
destabilizes the carboxylate anion of an acid will make that acid *less* strong.

    If we compare the acid strength of an unsubstituted carboxylic acid with
one that bears an electron-withdrawing substituent on the R (or Ar) portion of
the molecule, invariably we find that the stronger acid is the one with the electron-withdrawing group. By dispersing the negative charge, the electron-withdrawing group gives the carboxylate anion greater stability than the carboxylate
anion of the unsubstituted acid has. Examples of common electron-withdrawing,
and acid-strengthening, groups are

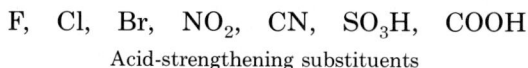

    F, Cl, Br, NO$_2$, CN, SO$_3$H, COOH

              Acid-strengthening substituents

No resonance: the
negative charge is
localized on one
oxygen atom.

Resonance hybrid: the
negative charge is spread
over two oxygen atoms

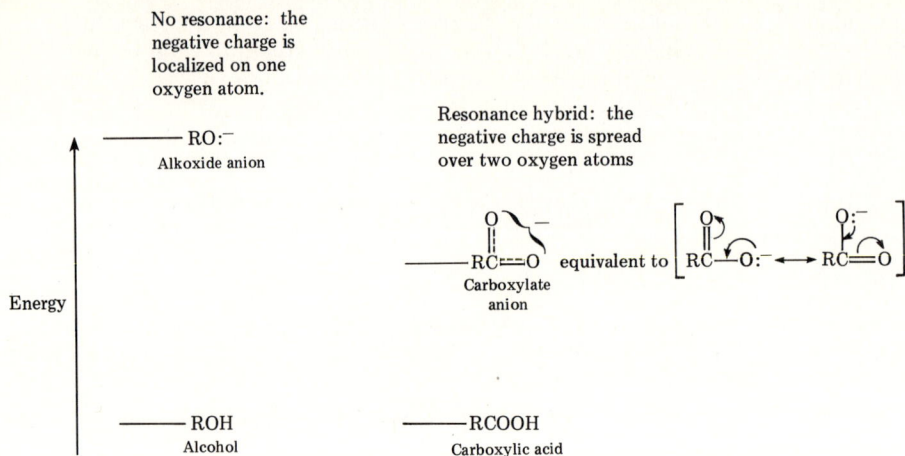

Figure 12.3 Comparison of acid strengths between carboxylic acids and alcohols.

Table 12.3 illustrates the acid-strengthening effect of chloro substituents on acetic acid.

Conversely, addition of any electron-donating substituent on the R (or Ar) portion of an acid decreases its acid strength relative to the parent unsubstituted acid. The electron-donating group destabilizes the carboxylate anion by intensifying the negative charge. The most common electron-donating, acid-weakening, substituent is the alkyl group (methyl, ethyl, propyl, and so on). A comparison of the acidities of formic acid and acetic acid illustrates the effect of alkyl substitution on acid strength. Formic acid (no alkyl group) is a stronger acid than acetic acid (one alkyl group).

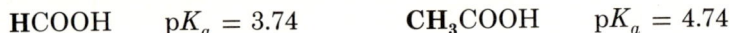

$$\text{HCOOH} \qquad \text{p}K_a = 3.74 \qquad\qquad \text{CH}_3\text{COOH} \qquad \text{p}K_a = 4.74$$

Table 12.3 Comparison of Acid Strengths of Acetic Acid and Chlorinated Acetic Acids

Name	Structure	$\text{p}K_a$	Relative acid strength
Acetic acid	CH_3COOH	4.7	1
Chloroacetic acid	ClCH_2COOH	2.8	80
Dichloroacetic acid	Cl_2CHCOOH	1.3	2,800
Trichloroacetic acid	Cl_3CCOOH	0.7	11,000

Table 12.4 Comparison of Acid Strengths of Butyric Acid and the Monochlorinated Butyric Acids

Name	Structure	$\text{p}K_a$	Relative acid strength
Butyric acid	$\text{CH}_3\text{CH}_2\text{CH}_2\text{COOH}$	4.82	1
α-Chlorobutyric acid	$\text{CH}_3\text{CH}_2\text{CHClCOOH}$	2.85	92
β-Chlorobutyric acid	$\text{CH}_3\text{CHClCH}_2\text{COOH}$	4.05	6
γ-Chlorobutyric acid	$\text{ClCH}_2\text{CH}_2\text{CH}_2\text{COOH}$	4.52	2

The electronic effect of a substituent transmitted through single bonds, be it electron-withdrawing or electron-donating, is called the **inductive effect.** The inductive effect becomes less pronounced with distance from the COOH group. This fact is illustrated in Table 12.4, which compares the acidities of the monochlorinated butyric acids. As can be seen, α-chlorobutyric acid is about 15 times more acidic than β-chlorobutyric acid.

Problem 12.5 Arrange the acids in each set in order of increasing acid strength.

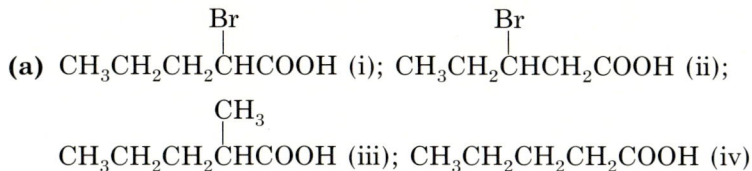

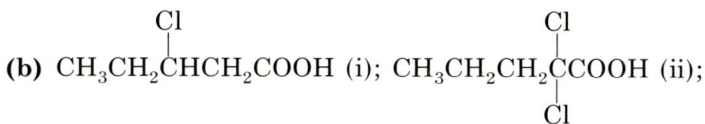

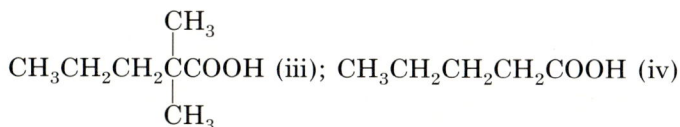

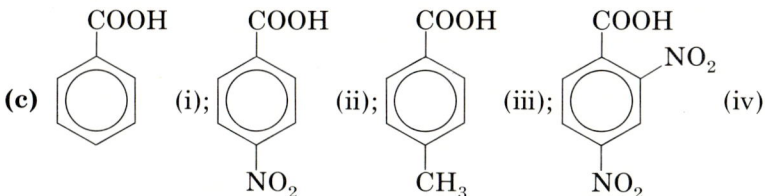

(a) CH$_3$CH$_2$CH$_2$CHCOOH (i); CH$_3$CH$_2$CHCH$_2$COOH (ii);
 (Br substituents)

CH$_3$CH$_2$CH$_2$CHCOOH (iii); CH$_3$CH$_2$CH$_2$CH$_2$COOH (iv)
 (CH$_3$ substituent)

(b) CH$_3$CH$_2$CHCH$_2$COOH (i); CH$_3$CH$_2$CH$_2$CCOOH (ii);
 (Cl substituents)

CH$_3$CH$_2$CH$_2$CCOOH (iii); CH$_3$CH$_2$CH$_2$CH$_2$COOH (iv)
 (CH$_3$ substituents)

(c) (structures with COOH) (i); (ii); (iii); (iv)
 (with NO$_2$, CH$_3$, NO$_2$ substituents)

Preparation of Carboxylic Acids *12.5*

We will consider three methods of preparation. They are presented here and discussed in the subsequent sections.

1. Oxidation
 a. Of primary alcohols or aldehydes

$$RCH_2OH \xrightarrow{[O]} RCHO \xrightarrow{[O]} RCOOH \quad ([O] = \text{oxidizing agent})$$

[Sec. 12.6A]

b. Of alkylbenzenes

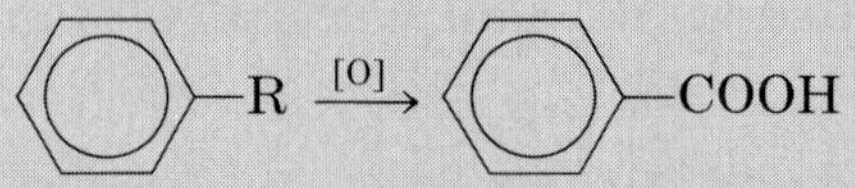

$$\bigcirc\!\!-R \xrightarrow{[O]} \bigcirc\!\!-COOH \qquad \text{[Sec. 12.6B; also 5.10B]}$$

2. Carbonation of Grignard reagents

$$RMgX + CO_2 \longrightarrow RCO_2MgX \xrightarrow{H^+} RCOOH \qquad \text{[Sec. 12.7]}$$

3. Hydrolysis of nitriles

$$RC\!\equiv\!N \xrightarrow{H_2O,\ H^+} RCOOH + NH_4^+ \qquad \text{[Sec. 12.8]}$$

12.6 Preparation of Acids by Oxidation

A Oxidation of Primary Alcohols or Aldehydes

Primary alcohols are oxidized quickly to carboxylic acids by potassium permanganate, $KMnO_4$, or by a mixture of potassium dichromate, $K_2Cr_2O_7$, and sulfuric acid. The alcohol is oxidized first to an aldehyde and then to a carboxylic acid. For convenience we represent the oxidizing agent by [O] instead of using its actual formula. We encountered this reaction earlier (Sec. 7.8).

$$CH_3CH_2OH + [O] \longrightarrow CH_3CHO + [O] \longrightarrow CH_3\textbf{COOH}$$

Ethyl alcohol $\qquad\qquad$ Acetaldehyde $\qquad\qquad$ Acetic acid

$$\bigcirc\!\!-CH_2OH + [O] \longrightarrow \bigcirc\!\!-CHO + [O] \longrightarrow \bigcirc\!\!-\textbf{COOH}$$

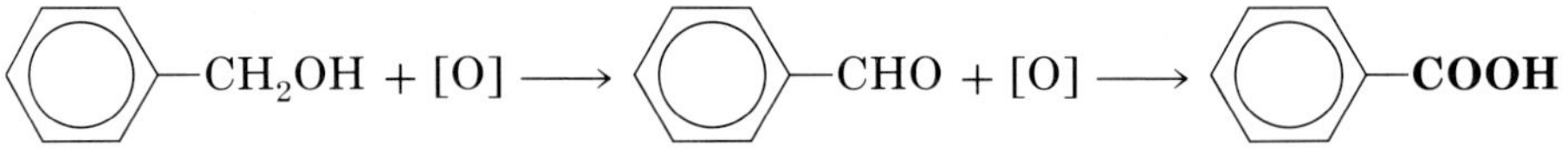

Benzyl alcohol $\qquad\qquad$ Benzaldehyde $\qquad\qquad$ Benzoic acid

B Oxidation of Alkylbenzenes

Vigorous oxidation of alkylbenzenes yields benzoic acid. The oxidation proceeds all the way back to the carbon directly attached to the benzene ring (see also Sec. 5.10B).

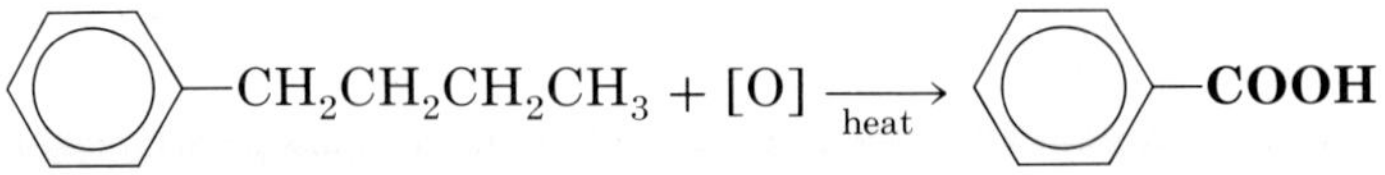

$$\bigcirc\!\!-CH_2CH_2CH_2CH_3 + [O] \xrightarrow{\text{heat}} \bigcirc\!\!-\textbf{COOH}$$

n-Butylbenzene $\qquad\qquad$ Benzoic acid

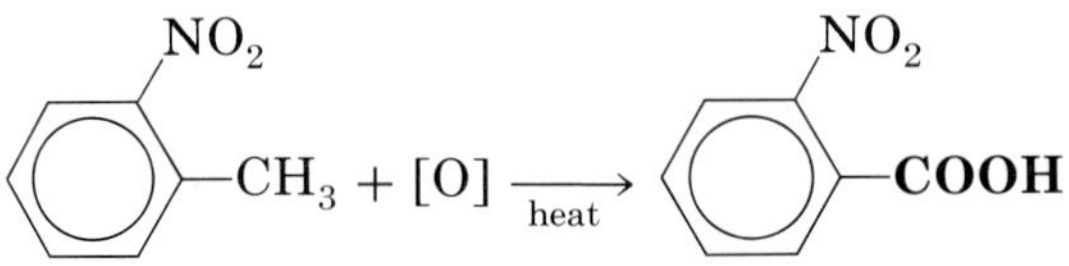

$$\overset{NO_2}{\bigcirc}\!\!-CH_3 + [O] \xrightarrow{\text{heat}} \overset{NO_2}{\bigcirc}\!\!-\textbf{COOH}$$

o-Nitrotoluene $\qquad\qquad$ o-Nitrobenzoic acid

Problem 12.6 Write the structure of the acid produced on oxidation of each compound.

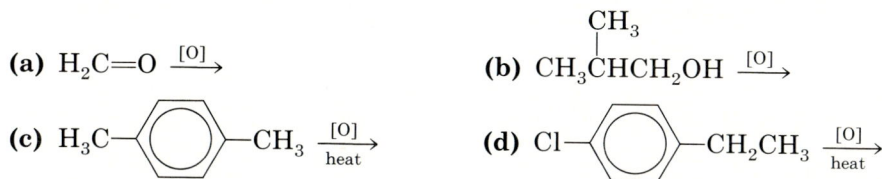

(a) $H_2C{=}O \xrightarrow{[O]}$

(b) $CH_3\overset{\overset{\displaystyle CH_3}{|}}{C}HCH_2OH \xrightarrow{[O]}$

(c) $H_3C{-}\langle\bigcirc\rangle{-}CH_3 \xrightarrow[\text{heat}]{[O]}$

(d) $Cl{-}\langle\bigcirc\rangle{-}CH_2CH_3 \xrightarrow[\text{heat}]{[O]}$

Carbonation of Grignard Reagents *12.7*

The addition of Grignard reagents to carbonyl compounds is discussed in Section 10.14. The addition of Grignard reagents to CO_2 in the form of dry ice proceeds in a similar fashion and yields the halomagnesium salt of a carboxylic acid. Hydrolysis of the salt gives an acid with one carbon more than the original Grignard reagent.

General equation

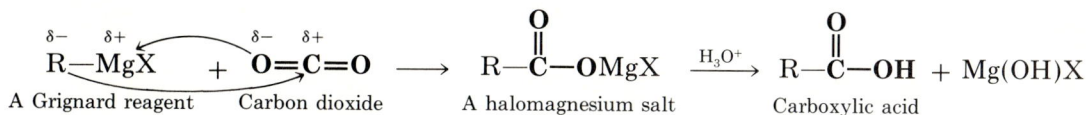

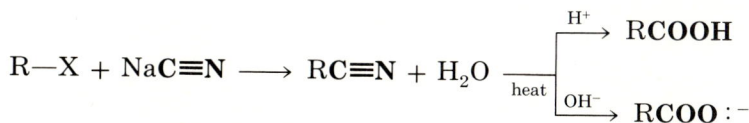

$$\overset{\delta-\ \ \delta+}{R{-}MgX} \ + \ \overset{\delta-\ \delta+}{O{=}C{=}O} \ \longrightarrow \ R{-}\overset{\overset{\displaystyle O}{\|}}{C}{-}OMgX \ \xrightarrow{H_3O^+} \ R{-}\overset{\overset{\displaystyle O}{\|}}{C}{-}OH \ + \ Mg(OH)X$$

A Grignard reagent Carbon dioxide A halomagnesium salt Carboxylic acid

Specific example

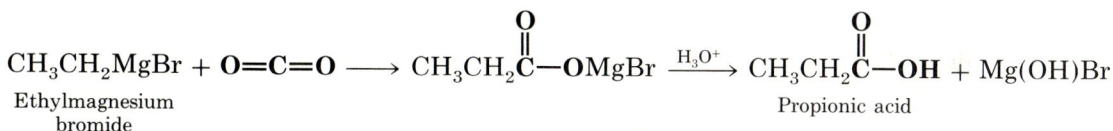

$$CH_3CH_2MgBr + O{=}C{=}O \longrightarrow CH_3CH_2\overset{\overset{\displaystyle O}{\|}}{C}{-}OMgBr \xrightarrow{H_3O^+} CH_3CH_2\overset{\overset{\displaystyle O}{\|}}{C}{-}OH + Mg(OH)Br$$

Ethylmagnesium bromide Propionic acid

Hydrolysis of Nitriles *12.8*

Nitriles are compounds with the general formula $RC{\equiv}N$ or $ArC{\equiv}N$. They are prepared by reacting a 1° or 2° alkyl halide with a cyanide salt. Acid hydrolysis of a nitrile yields a carboxylic acid. Alkaline hydrolysis yields a carboxylate salt. The resulting acid or salt both contain one carbon more than the starting alkyl halide.

General equation

$$R{-}X + NaC{\equiv}N \longrightarrow RC{\equiv}N + H_2O \ \xrightarrow[\text{heat}]{} \begin{cases} \xrightarrow{H^+} RCOOH \\ \xrightarrow{OH^-} RCOO{:}^- \end{cases}$$

Specific examples

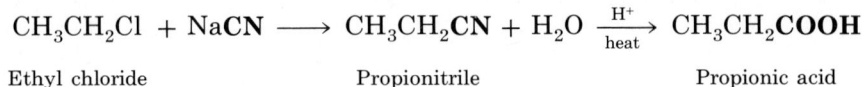

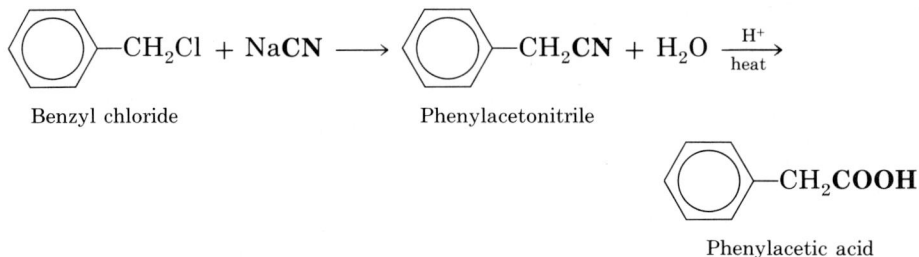

$$CH_3CH_2Cl + NaCN \longrightarrow CH_3CH_2CN + H_2O \xrightarrow[\text{heat}]{H^+} CH_3CH_2COOH$$

Ethyl chloride   Propionitrile   Propionic acid

$$\text{(C}_6\text{H}_5)-CH_2Cl + NaCN \longrightarrow \text{(C}_6\text{H}_5)-CH_2CN + H_2O \xrightarrow[\text{heat}]{H^+}$$

Benzyl chloride   Phenylacetonitrile

$$\text{(C}_6\text{H}_5)-CH_2COOH$$

Phenylacetic acid

Problem 12.7 Complete these reactions.

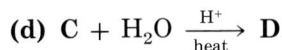

(a) $\text{(C}_6\text{H}_5)-CH_2CH_2Br + Mg \xrightarrow{\text{dry ether}} A$

(b) $A \xrightarrow[\text{(2) H}_3\text{O}^+]{\text{(1) CO}_2} B$

(c) $CH_3Cl + NaCN \longrightarrow C$

(d) $C + H_2O \xrightarrow[\text{heat}]{H^+} D$

(e) $Cl-CH_2CH_2-Cl + 2\,NaCN \longrightarrow E$

(f) $E + H_2O \xrightarrow[\text{heat}]{H^+} F$

Problem 12.8 Write equations that illustrate a good method for synthesizing the following acids. Use reagents as needed.
(a) Propanoic acid from 1-propanol
(b) Propanoic acid from ethyl chloride (two ways)
(c) Cyclopentanecarboxylic acid from chlorocyclopentane (two ways)
(d) Benzoic acid from chlorobenzene (Grignard method only)

12.9 Reactions of Carboxylic Acids

The reactions of acids are basically of two types.

1. Reaction with bases to form salts

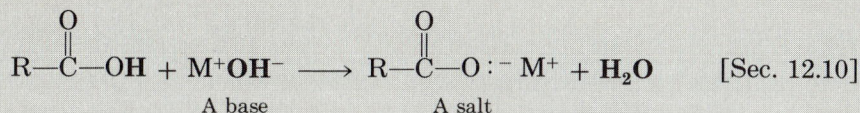

$$R-\overset{\overset{\displaystyle O}{\|}}{C}-OH + M^+OH^- \longrightarrow R-\overset{\overset{\displaystyle O}{\|}}{C}-O:^- M^+ + H_2O \qquad [Sec.\ 12.10]$$

A base   A salt

The carboxyl hydrogen is replaced by a metal ion, M^+.

2. Reaction with nucleophiles to form acid derivatives

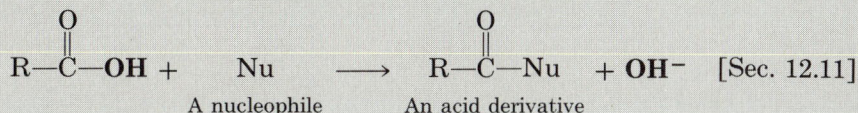

$$R-\overset{\overset{\displaystyle O}{\|}}{C}-OH + \quad Nu \quad \longrightarrow \quad R-\overset{\overset{\displaystyle O}{\|}}{C}-Nu \quad + OH^- \qquad [Sec.\ 12.11]$$

A nucleophile   An acid derivative

The OH of the carboxyl group is replaced by a nucleophile, a process known as nucleophilic substitution.

Reaction with Bases: Salt Formation   *12.10*

  Carboxylic acids react quantitatively with bases to form water-soluble salts. The water-soluble crystalline salts may be obtained by reacting equivalent amounts of acid and base and evaporating the solution to dryness.

  Salts of organic acids are named in much the same way as inorganic salts. The metal cation is named first, followed by the name of the carboxylate anion. The latter is named by dropping the *-ic acid* ending from the name of the parent acid and replacing it with *-ate*.

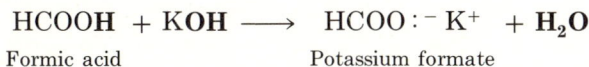

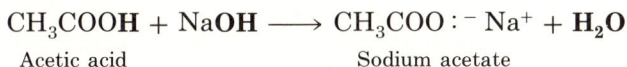

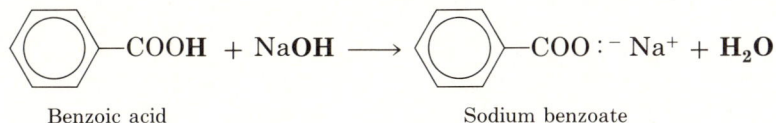

$$HCOOH + KOH \longrightarrow HCOO:^- K^+ + H_2O$$
Formic acid   Potassium formate

$$CH_3COOH + NaOH \longrightarrow CH_3COO:^- Na^+ + H_2O$$
Acetic acid   Sodium acetate

$$\bigcirc\!\!-COOH + NaOH \longrightarrow \bigcirc\!\!-COO:^- Na^+ + H_2O$$
Benzoic acid   Sodium benzoate

  Carboxylate salts are used commercially in a variety of applications. Sodium acetate is used in dyeing. Sodium propionate, $CH_3CH_2COO:^- Na^+$, and calcium propionate, $(CH_3CH_2COO:^-)_2Ca^{2+}$, are used in bread to prevent molding. Sodium benzoate is a food preservative.

  Carboxylic acids will also react with a weak base like sodium bicarbonate, $NaHCO_3$, to form water-soluble salts. Carbon dioxide is liberated as a by-product of the reaction. Weaker acids like phenols react only with strong bases ($NaOH$ or KOH) and will not react with $NaHCO_3$. (Picric acid and 2,4-dinitrophenol are exceptions in that they do react with sodium bicarbonate).

Table 12.5 Solubility Tests[a] for Various Classes of Compounds

	Soluble salt formed with	
	NaOH	**NaHCO₃**
Neutral compounds[b]	No	No
Phenols	Yes	No
Carboxylic acids	Yes	Yes

[a] The usual test is whether about 50 mg of the compound does or does not dissolve in 1 or 2 ml of a 5% solution of each base (NaOH or $NaHCO_3$).

[b] Neutral compounds include alkanes, alkenes, alkynes, aromatic hydrocarbons, alkyl halides, alcohols, ethers, aldehydes, ketones, and esters.

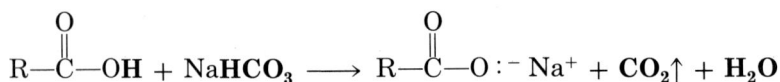

$$R-\overset{\overset{\displaystyle O}{\|}}{C}-OH + NaHCO_3 \longrightarrow R-\overset{\overset{\displaystyle O}{\|}}{C}-O:^- Na^+ + CO_2\uparrow + H_2O$$

$$\bigcirc\!\!\!\!\bigcirc-OH + NaHCO_3 \longrightarrow \text{No reaction}$$

The reaction with sodium bicarbonate can therefore be used as a simple qualitative test to distinguish carboxylic acids from phenols or other compounds. Table 12.5 recapitulates the discussion.

Problem 12.9 Describe a simple test that will distinguish between each pair of compounds.
(a) Phenol and benzyl alcohol   **(b)** Phenol and benzoic acid

---

12.11 Carboxylic Acid Derivatives

When the OH of a carboxylic acid is replaced by a nucleophile, :Nu, a **carboxylic acid derivative** is produced.

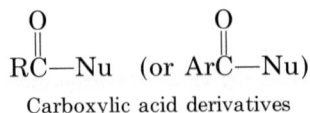

$$R\overset{\overset{\displaystyle O}{\|}}{C}-Nu \quad (or \ Ar\overset{\overset{\displaystyle O}{\|}}{C}-Nu)$$

Carboxylic acid derivatives

The $R\overset{\overset{\displaystyle O}{\|}}{C}-$ portion of acid derivatives is called the **acyl group.** (Aromatic carboxylic acid derivatives contain the aroyl group, $Ar\overset{\overset{\displaystyle O}{\|}}{C}-$).

The carboxylic acid derivatives presented in this chapter are

$$\underset{\substack{\text{Acyl chloride}\\\text{Acid chloride}}}{RC\!-\!Cl} \qquad \underset{\text{Ester}}{RC\!-\!OR'} \qquad \underset{\text{Amide}}{RC\!-\!N} \qquad \underset{\text{Acid anhydride}}{RC\!-\!O\!-\!CR}$$

In the next section we discuss the nomenclature of carboxylic acid derivatives. The sections that follow treat the methods of their preparation and their reactions.

Nomenclature of Acid Derivatives   *12.12*

Acyl chlorides, or **acid chlorides,** are named by replacing the *-ic acid* ending of the parent acid by *-yl chloride*. For example,

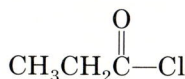

$$CH_3CH_2\overset{O}{\overset{\|}{C}}\!-\!Cl \qquad\qquad \underset{\text{Benzoyl chloride}}{\bigcirc\!-\!\overset{O}{\overset{\|}{C}}\!-\!Cl}$$

IUPAC name: Propanoyl chloride
Common name: Propionyl chloride

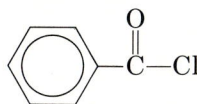

Esters are named as if they were formed from replacement of the carboxyl hydrogen by an alkyl group. (They are not actually formed this way; see Sec. 12.14.) The alkyl group is named first followed by the name of the parent acid with the ending *-ate* in place of *-ic acid*.

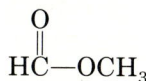

$$H\overset{O}{\overset{\|}{C}}\!-\!OCH_3 \qquad\qquad CH_3\overset{O}{\overset{\|}{C}}\!-\!OCH_2CH_3$$

IUPAC name: Methyl methanoate   Ethyl ethanoate
Common name: Methyl formate   Ethyl acetate

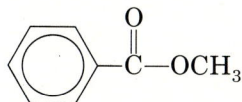

$$\bigcirc\!-\!\overset{O}{\overset{\|}{C}}\!-\!OCH_3 \qquad\qquad CH_3\overset{O}{\overset{\|}{C}}\!-\!OCH_2\!-\!\bigcirc$$

IUPAC name: Methyl benzoate   Benzyl ethanoate
Common name: Benzyl acetate

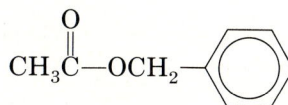

Amides are named by replacing the *-oic acid* or *-ic acid* of the parent acid's name by *-amide*. One or two substituents on the nitrogen atom give an N-substituted or an N,N-disubstituted amide. In such a case the substituents are named first.

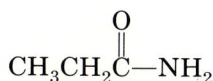

$$CH_3CH_2\overset{O}{\overset{\|}{C}}\!-\!NH_2 \qquad\qquad \underset{\text{Benzamide}}{\bigcirc\!-\!\overset{O}{\overset{\|}{C}}\!-\!NH_2}$$

IUPAC name: Propanamide
Common name: Propionamide

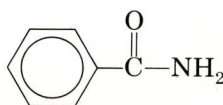

$$CH_3\overset{\displaystyle O}{\overset{\|}{C}}-N\overset{\displaystyle CH_3}{\underset{\displaystyle CH_3}{}}$$

$$H\overset{\displaystyle O}{\overset{\|}{C}}-N\overset{\displaystyle CH_3}{\underset{\displaystyle CH_2CH_3}{}}$$

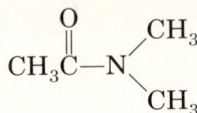

IUPAC name:	*N,N*-Dimethylethanamide	*N*-Ethyl-*N*-methylmethanamide
Common name:	*N,N*-Dimethylacetamide	*N*-Ethyl-*N*-methylformamide

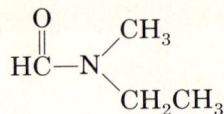

An **anhydride** is named by replacing the word *acid* with *anhydride* in the name of the acid from which it was formed. This rule applies also to cyclic anhydrides formed from dicarboxylic acids.

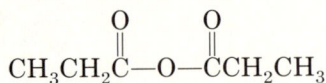

$$CH_3CH_2\overset{\displaystyle O}{\overset{\|}{C}}-O-\overset{\displaystyle O}{\overset{\|}{C}}CH_2CH_3$$

Propanoic anhydride
Propionic anhydride

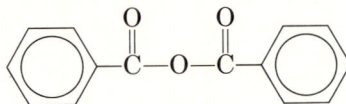

Benzoic anhydride

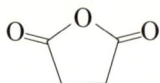

Succinic anhydride

Problem 12.10 Write the structure for each of these compounds.

(a) Formyl chloride
(b) *p*-Nitrophenylacetyl chloride
(c) Isopropyl butyrate
(d) Phenyl benzoate
(e) *N*-Methylformamide
(f) α-Chlorobutyramide
(g) Formic anhydride
(h) Glutaric anhydride

Problem 12.11 Name these compounds.

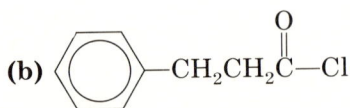

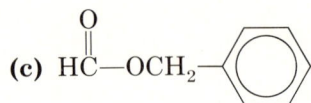

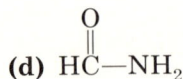

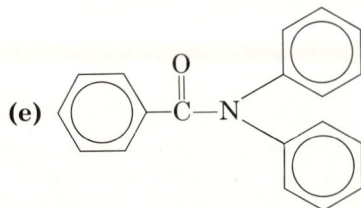

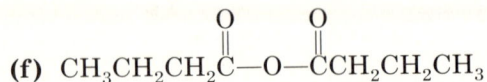

(a) $CH_3CH_2CH_2\overset{\displaystyle O}{\overset{\|}{C}}-Cl$

(b) $-CH_2CH_2\overset{\displaystyle O}{\overset{\|}{C}}-Cl$

(c) $H\overset{\displaystyle O}{\overset{\|}{C}}-OCH_2-$

(d) $H\overset{\displaystyle O}{\overset{\|}{C}}-NH_2$

(e) $-\overset{\displaystyle O}{\overset{\|}{C}}-N$

(f) $CH_3CH_2CH_2\overset{\displaystyle O}{\overset{\|}{C}}-O-\overset{\displaystyle O}{\overset{\|}{C}}CH_2CH_2CH_3$

Nucleophilic Substitution of Acids and Acid Derivatives **12.13**

Most reactions of acids and acid derivatives proceed by a common mechanism: **nucleophilic substitution.**

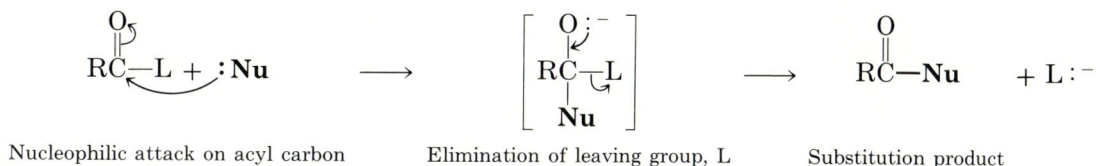

$$RC{-}L + :Nu \longrightarrow \left[\begin{array}{c} O:^- \\ RC{-}L \\ Nu \end{array} \right] \longrightarrow RC{-}Nu \quad + L:^-$$

Nucleophilic attack on acyl carbon　　Elimination of leaving group, L　　Substitution product

The leaving group, L, may be OH, OR, Cl, OCOR or NH_2.

Having presented the general mechanism of nucleophilic substitution of acid derivatives, we now look into specific preparations and reactions of esters, acid chlorides, acid anhydrides, and amides.

Esters from Carboxylic Acids: Esterification **12.14**

Esters are by far the most important acid derivatives, and this is why we discuss them first. One method of preparing esters is to treat a carboxylic acid with a primary or secondary alcohol. The reaction is called **esterification.** A small amount of mineral acid catalyst (H^+) is required to speed up esterification. Esterification is a reversible reaction.

General equation

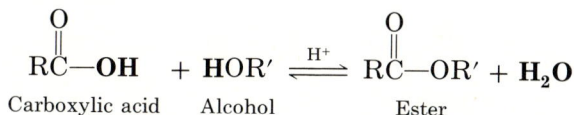

$$RC{-}OH + HOR' \underset{}{\overset{H^+}{\rightleftharpoons}} RC{-}OR' + H_2O$$

Carboxylic acid　　Alcohol　　　Ester

Specific example

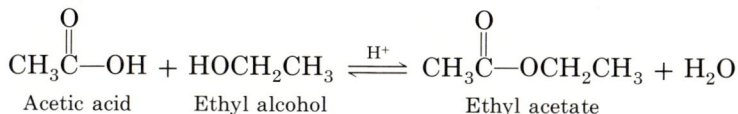

$$CH_3C{-}OH + HOCH_2CH_3 \underset{}{\overset{H^+}{\rightleftharpoons}} CH_3C{-}OCH_2CH_3 + H_2O$$

Acetic acid　　Ethyl alcohol　　　Ethyl acetate

If the acid and alcohol functions are part of the same molecule, *intramolecular* esterification is possible. The product, in such cases, is a *cyclic ester* or **lactone.** When the OH group is located on the fourth or fifth carbon of the acid chain, five- or six-membered lactones are easily formed.

$$
\begin{array}{c}
\underset{\text{H}_2\text{C}}{\overset{\displaystyle\overset{\text{O}}{\|}}{\text{C}-\text{OH}}}\ \overset{\text{OH}}{|} \\
\underset{\text{H}_2\text{C}-\text{CH}_2}{}
\end{array}
\ \underset{}{\overset{\text{H}^+}{\rightleftharpoons}}\
\begin{array}{c}
\overset{\displaystyle\overset{\text{O}}{\|}}{\text{C}} \\
\text{H}_2\text{C}\qquad\text{O} \\
\text{H}_2\text{C}-\text{CH}_2
\end{array}
\ +\ \textbf{H}_2\textbf{O}
$$

γ-Hydroxybutyric acid γ-Butyrolactone
(OH on fourth carbon) (five-membered lactone)

$$
\begin{array}{c}
\underset{\text{H}_2\text{C}}{\overset{\displaystyle\overset{\text{O}}{\|}}{\text{C}-\text{OH}}}\ \overset{\text{OH}}{|} \\
\text{H}_2\text{C}\qquad\text{CH}_2 \\
\text{CH}_2
\end{array}
\ \overset{\text{H}^+}{\rightleftharpoons}\
\begin{array}{c}
\overset{\displaystyle\overset{\text{O}}{\|}}{\text{C}} \\
\text{H}_2\text{C}\qquad\text{O} \\
\text{H}_2\text{C}\qquad\text{CH}_2 \\
\text{CH}_2
\end{array}
\ +\ \textbf{H}_2\textbf{O}
$$

δ-Hydroxyvaleric acid δ-Valerolactone
(six-membered lactone)

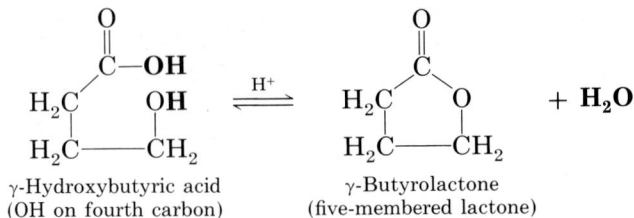

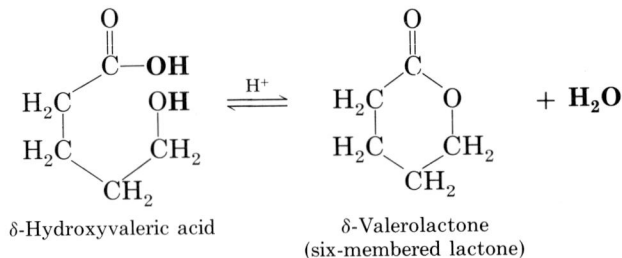

Some lactones are useful as intermediates in organic synthesis, and some occur naturally. Ascorbic acid (vitamin C) and the lactone of mevalonic acid (a biochemical precursor of cholesterol) are two important naturally occurring lactones.

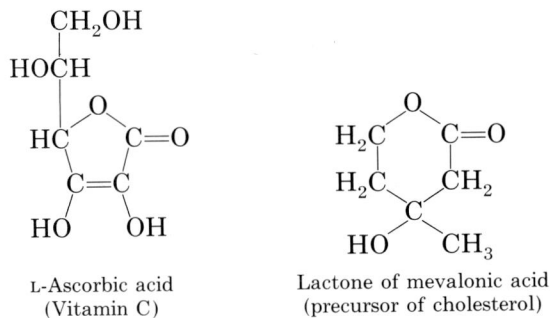

$$
\begin{array}{c}
\text{CH}_2\text{OH} \\
| \\
\text{HOCH} \\
\overset{\text{O}}{} \\
\text{HC}\qquad\text{C}=\text{O} \\
\text{C}=\text{C} \\
\text{HO}\qquad\text{OH}
\end{array}
\qquad\qquad
\begin{array}{c}
\overset{\text{O}}{} \\
\text{H}_2\text{C}\qquad\text{C}=\text{O} \\
\text{H}_2\text{C}\qquad\text{CH}_2 \\
\text{C} \\
\text{HO}\qquad\text{CH}_3
\end{array}
$$

L-Ascorbic acid Lactone of mevalonic acid
(Vitamin C) (precursor of cholesterol)

Although esterification is a reversible reaction, good yields of ester can be produced if the alcohol is used as the solvent. Another way of driving the reaction to completion is to remove the ester from the reaction mixture as soon as it is formed.

A nonreversible method for preparing an ester is to treat an acid chloride with an alcohol (see Sec. 12.18).

Problem 12.12 Write the equation for the esterification of 1-propanol with benzoic acid. Name the ester formed.

Problem 12.13 Draw the structures and name the alcohols and the acids needed to make these esters.

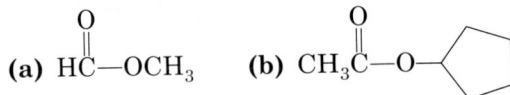

(a) $\text{H}\overset{\displaystyle\overset{\text{O}}{\|}}{\text{C}}-\text{OCH}_3$ (b) $\text{CH}_3\overset{\displaystyle\overset{\text{O}}{\|}}{\text{C}}-\text{O}-\!\!\bigpentagon$

Esterification is a nucleophilic substitution reaction. The steps in the mechanism are

Step 1. Protonation of the acyl group oxygen.

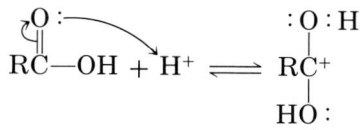

$$\underset{RC-OH}{\overset{O:}{\|}} + H^+ \rightleftharpoons \underset{\underset{\overset{|}{HO:}}{\overset{|}{RC^+}}}{\overset{:O:H}{}}$$

Step 2. Attack by alcohol nucleophile on positively charged carbon, followed by proton transfer.

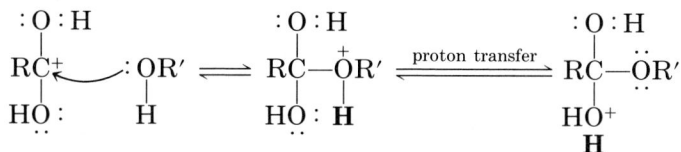

$$\underset{\overset{|}{HO:}}{\overset{:O:H}{RC^+}} \quad :\underset{H}{OR'} \rightleftharpoons \underset{\overset{|}{HO:}\ \mathbf{H}}{\overset{:O:H}{RC-OR'}} \xrightarrow{\text{proton transfer}} \underset{\underset{\mathbf{H}}{\overset{|}{HO^+}}}{\overset{:O:H}{RC-OR'}}$$

Step 3. Elimination of H_2O.

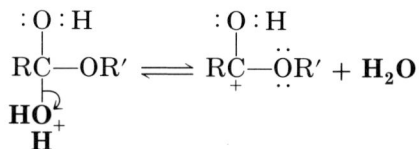

$$\underset{\underset{\mathbf{H}}{\overset{|}{HO^+}}}{\overset{:O:H}{RC-OR'}} \rightleftharpoons \underset{+}{\overset{:O:H}{RC-OR'}} + \mathbf{H_2O}$$

Step 4. Regeneration of proton catalyst and formation of ester.

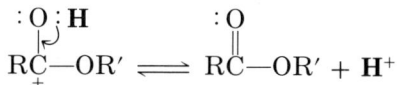

$$\underset{+}{\overset{:O:\mathbf{H}}{RC-OR'}} \rightleftharpoons \underset{RC-OR'}{\overset{:O}{\|}} + \mathbf{H^+}$$

Problem 12.14 Based on the mechanism just shown, complete the esterification reaction here and indicate whether the ^{18}O of the acid will appear in the product ester or in the water.

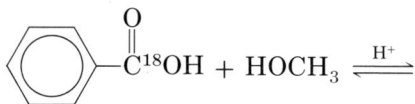

$$\bigcirc\!\!-\!\!\overset{O}{\overset{\|}{C}}{}^{18}OH + HOCH_3 \xrightarrow{H^+} \rightleftharpoons$$

Esters are found in virtually all living things and figure prominently in biochemical processes. Fats, oils, and waxes are naturally occurring esters of

14 334

3**Carboxylic Acids and
Their Derivatives**

Table 12.6 Characteristic Flavors of Some Esters

Flavor	Name	Structure
Apricot	n-Pentyl butyrate	$CH_3CH_2CH_2\overset{\displaystyle O}{\overset{\|}{C}}-O(CH_2)_4CH_3$
Banana	n-Pentyl acetate	$CH_3\overset{\displaystyle O}{\overset{\|}{C}}-O(CH_2)_4CH_3$
Orange	n-Octyl acetate	$CH_3\overset{\displaystyle O}{\overset{\|}{C}}-O(CH_2)_7CH_3$
Pineapple	Ethyl butyrate	$CH_3CH_2CH_2\overset{\displaystyle O}{\overset{\|}{C}}-OCH_2CH_3$
Rum	Ethyl formate	$H\overset{\displaystyle O}{\overset{\|}{C}}-OCH_2CH_3$
Wintergreen	Methyl salicylate	(benzene ring with $-\overset{\displaystyle O}{\overset{\|}{C}}-OCH_3$ and OH)

high molecular weight. Their properties are discussed in Chapter 13. Low-molecular-weight esters are pleasant-smelling substances that are responsible for the flavors and fragrances of fruits and flowers. Very careful blending of several or even dozens of esters is necessary to imitate the fragrance of a natural product, although the main aroma may be due to a single ester. Table 12.6 gives examples of the characteristic flavors of some esters.

Lower-molecular-weight esters are excellent solvents for many organic compounds, and their main use is for this purpose.

A very common ester of medicinal importance is acetylsalicylic acid (aspirin).

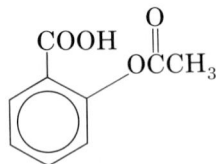

(benzene ring with $COOH$ and $O\overset{\displaystyle O}{\overset{\|}{C}}CH_3$)

Acetylsalicylic acid
(Aspirin)

Besides its well-known analgesic and antipyretic properties, aspirin has been shown recently to reduce the risks of recurrence of a stroke or heart attack.

Other esters are suitable textile fibers. The best-known polyester, Dacron, is polyethylene terephthalate, formed from the polymerization of ethylene glycol and terephthalic acid.

$$n \ HOOC-\langle\bigcirc\rangle-COOH + n \ HOCH_2CH_2OH \xrightarrow{\text{heat}}$$

Terephthalic acid   Ethylene glycol

$$\left[\!-CH_2CH_2OOC-\langle\bigcirc\rangle-COOCH_2CH_2-\!\right]_n$$

Polyethylene terephthalate
(Terylene; Dacron)

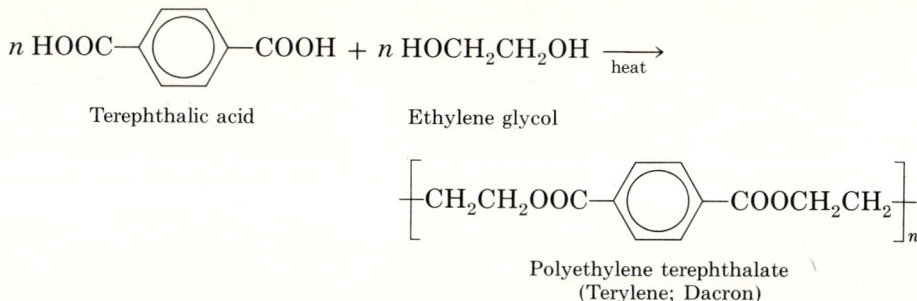

This polyester is also marketed successfully under other names, such as Teryl-ene. It has great strength, considerable stiffness, and remarkable resistance to creasing. For this reason, garments made of Dacron polyester are called perma-nent press. The stiffness of Dacron fibers, which often irritate the skin, is allevi-ated by blending them with wool or cotton.

Reactions of Esters 12.17

The reactions of esters go by the mechanism shown, in Section 12.13: nucle-ophilic attack on the acyl carbon followed by elimination of the OR group.

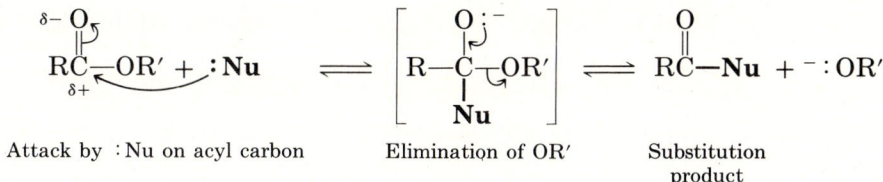

$$\overset{\delta-}{O}\overset{\parallel}{\underset{\delta+}{RC}}-OR' + :Nu \rightleftharpoons \left[R-\overset{O:^-}{\underset{\underset{Nu}{|}}{C}}-OR' \right] \rightleftharpoons \overset{O}{\overset{\parallel}{RC}}-Nu + {}^-:OR'$$

Attack by :Nu on acyl carbon   Elimination of OR'   Substitution
                                                       product

A   Acid-Catalyzed Hydrolysis of Esters

When a compound is broken down by the action of water, the reaction is called **hydrolysis.** The hydrolysis of an ester gives a carboxylic acid and an alcohol. The reaction is catalyzed by strong mineral acids (H^+). Hydrolysis of an ester is the reverse of the acid-catalyzed esterification reaction discussed earlier (Sec. 12.14). Therefore acid-catalyzed hydrolysis, like acid-catalyzed esterifica-tion, is an equilibrium reaction that does not go to completion.

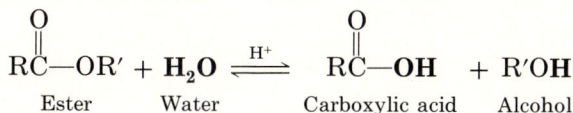

$$\overset{O}{\overset{\parallel}{RC}}-OR' + H_2O \overset{H^+}{\rightleftharpoons} \overset{O}{\overset{\parallel}{RC}}-OH + R'OH$$

Ester   Water   Carboxylic acid   Alcohol

Excess water can be used to drive the equilibrium to the right. For example,

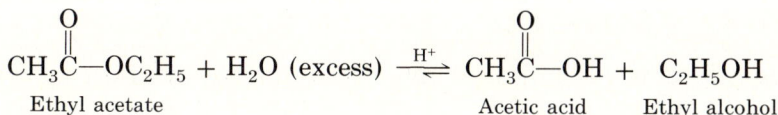

$$CH_3\overset{O}{\overset{\parallel}{C}}-OC_2H_5 + H_2O \ (\text{excess}) \overset{H^+}{\rightleftharpoons} CH_3\overset{O}{\overset{\parallel}{C}}-OH + C_2H_5OH$$

Ethyl acetate   Acetic acid   Ethyl alcohol

Problem 12.15 Consider the mechanism of esterification in Section 12.15 and other mechanisms illustrated previously in this chapter, and formulate a reasonable mechanism for the acid-catalyzed hydrolysis ethyl acetate.

B Alkaline Hydrolysis of Esters: Saponification

An ester can be hydrolyzed irreversibly if hydrolysis is carried out in an alkaline solution. *Alkaline hydrolysis* of an ester is called **saponification** because soap is the product of alkaline hydrolysis of esters of glycerol and long-chain fatty acids (see Sec. 13.6).

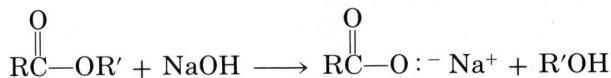

$$\overset{O}{\overset{\|}{R C}} - O R' + NaOH \longrightarrow \overset{O}{\overset{\|}{R C}} - O:^- Na^+ + R'OH$$

Treatment of the salt with mineral acid regenerates the organic acid.

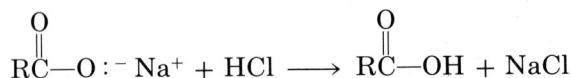

$$\overset{O}{\overset{\|}{R C}} - O:^- Na^+ + HCl \longrightarrow \overset{O}{\overset{\|}{R C}} - OH + NaCl$$

C Alcoholysis: Transesterification

Alcoholysis is the acid-catalyzed reaction between an ester and an alcohol to give an equilibrium mixture with another ester and another alcohol. This reaction is sometimes referred to as ester interchange, or **transesterification.**

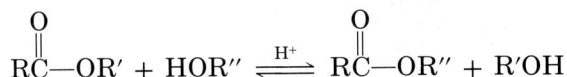

$$\overset{O}{\overset{\|}{R C}} - O R' + HOR'' \overset{H^+}{\rightleftharpoons} \overset{O}{\overset{\|}{R C}} - O R'' + R'OH$$

D Ammonolysis of Esters

Ammonolysis is the reaction of esters with ammonia to form an amide and an alcohol.

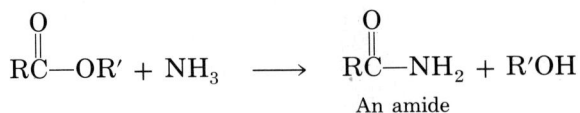

$$\overset{O}{\overset{\|}{R C}} - O R' + NH_3 \longrightarrow \overset{O}{\overset{\|}{R C}} - NH_2 + R'OH$$

$$\text{An amide}$$

The amide group, $-\overset{O}{\overset{\|}{C}}-N$, is widely distributed in nature, especially in protein molecules.

E Reduction of Esters

The reduction of esters with lithium aluminum hydride, $LiAlH_4$ produces two moles of alcohol: one from the acid part of the ester, the other from the alcohol part.

$$\underset{\text{Methyl acetate}}{CH_3\overset{\overset{\displaystyle O}{\|}}{C}\text{—}OCH_3} + LiAlH_4 \longrightarrow \underset{\text{Ethanol}}{CH_3CH_2OH} + \underset{\text{Methanol}}{CH_3OH}$$

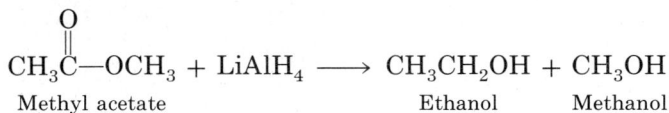

F   Esters and Grignard Reagents

The carbonyl group of an ester can react with a Grignard reagent to form a *tertiary alcohol*. The overall reaction is

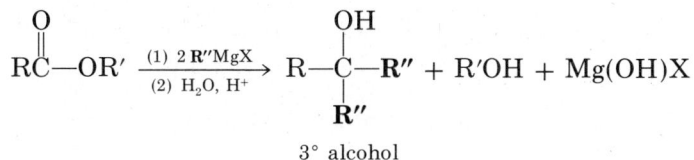

$$R\overset{\overset{\displaystyle O}{\|}}{C}\text{—}OR' \xrightarrow[\text{(2) } H_2O,\ H^+]{\text{(1) } 2\ \mathbf{R''MgX}} \underset{\underset{\displaystyle \mathbf{R''}}{\overset{\displaystyle OH}{\underset{\displaystyle |}{R\text{—}\overset{|}{C}\text{—}\mathbf{R''}}}}{} + R'OH + Mg(OH)X$$

3° alcohol

Two of the three alkyl groups of the 3° alcohol have their origins in the Grignard reagent, and the third alkyl group comes from the acyl portion of the ester.

Problem 12.16   Complete these reactions and name the organic products.

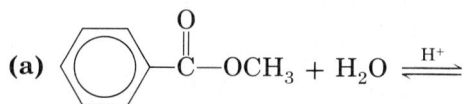

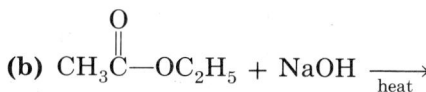

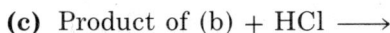

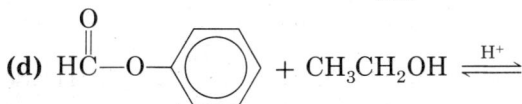

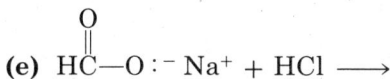

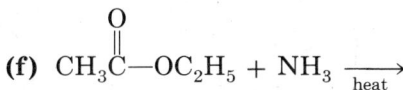

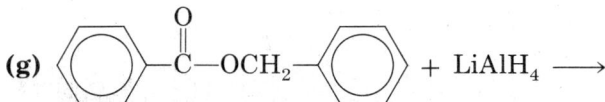

(a) $\langle\bigcirc\rangle\text{—}\overset{\overset{\displaystyle O}{\|}}{C}\text{—}OCH_3 + H_2O \underset{}{\overset{H^+}{\rightleftharpoons}}$
(b) $CH_3\overset{\overset{\displaystyle O}{\|}}{C}\text{—}OC_2H_5 + NaOH \xrightarrow[\text{heat}]{}$

(c) Product of (b) + HCl $\longrightarrow$
(d) $H\overset{\overset{\displaystyle O}{\|}}{C}\text{—}O\text{—}\langle\bigcirc\rangle + CH_3CH_2OH \overset{H^+}{\rightleftharpoons}$

(e) $H\overset{\overset{\displaystyle O}{\|}}{C}\text{—}O:^- Na^+ + HCl \longrightarrow$
(f) $CH_3\overset{\overset{\displaystyle O}{\|}}{C}\text{—}OC_2H_5 + NH_3 \xrightarrow[\text{heat}]{}$

(g) $\langle\bigcirc\rangle\text{—}\overset{\overset{\displaystyle O}{\|}}{C}\text{—}OCH_2\text{—}\langle\bigcirc\rangle + LiAlH_4 \longrightarrow$

(h) $H\overset{\overset{\displaystyle O}{\|}}{C}\text{—}OCH_3 + 2\ C_2H_5MgCl \longrightarrow$
(i) Product of (h) + $H_2O \xrightarrow{H^+}$

Acid Chlorides: Preparation   *12.18*

Acid halides other than acid chlorides can be made, but acid chlorides are commonly used because they can be prepared more easily and economically. Acid chlorides are reactive compounds and are important intermediates for the synthesis of other acid derivatives (Sec. 12.19). They are prepared by reaction of a carboxylic acid with phosphorus chlorides (PCl_5 or PCl_3) or, more frequently, with thionyl chloride ($SOCl_2$). Thionyl chloride is a better reagent because all the by-products are gases and are easily removed.

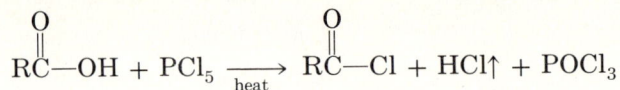

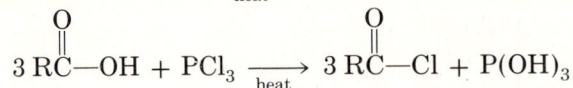

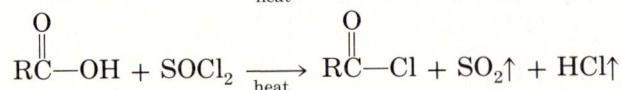

$$\underset{\overset{\|}{O}}{RC}-OH + PCl_5 \xrightarrow[heat]{} \underset{\overset{\|}{O}}{RC}-Cl + HCl\uparrow + POCl_3$$

$$3\,\underset{\overset{\|}{O}}{RC}-OH + PCl_3 \xrightarrow[heat]{} 3\,\underset{\overset{\|}{O}}{RC}-Cl + P(OH)_3$$

$$\underset{\overset{\|}{O}}{RC}-OH + SOCl_2 \xrightarrow[heat]{} \underset{\overset{\|}{O}}{RC}-Cl + SO_2\uparrow + HCl\uparrow$$

Acid chlorides are low-boiling liquids of irritating odors. Many are lachrymators (tear formers), and some are actually used in tear gas. This is why reactions with acid chlorides are usually carried out under a hood or in a well-ventilated laboratory.

12.19 Acid Chlorides: Reactions

Acid chlorides are very reactive compounds because the inductive effect of the chlorine atom further diminishes the electron density on the acyl carbon. Attack by a nucleophile on the acyl carbon is therefore enhanced. The mechanism of nucleophilic substitution of acid chlorides is similar to the one illustrated in Section 12.13 for acid derivatives.

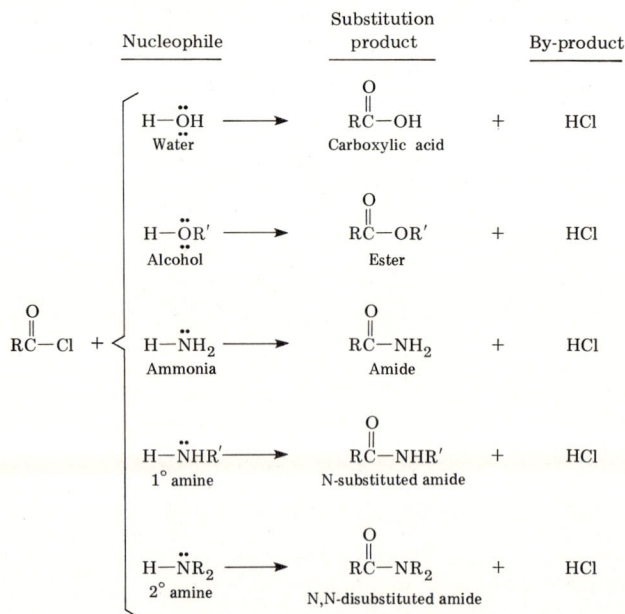

Figure 12.4 Nucleophilic substitution reactions of acid chlorides.

$$\overset{\delta-}{\underset{\delta+}{R\overset{O}{\overset{\|}{C}}}}Cl + :Nu \longrightarrow \left[R-\overset{\overset{O:^-}{\|}}{\underset{Nu}{C}}Cl \right] \longrightarrow R\overset{O}{\overset{\|}{C}}-Nu + {}^-:Cl$$

| Nucleophilic attack on acyl carbon | Elimination of Cl | Substitution product |

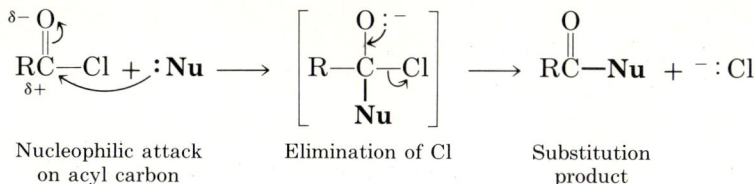

The attacking nucleophiles may be water, alcohol, ammonia, or amines. The substitution products are carboxylic acids, esters, amides, or substituted amides, respectively (Fig. 12.4).

Acid chlorides also form aromatic ketones via the Friedel–Crafts acylation, a reaction encountered previously (Sec. 10.9).

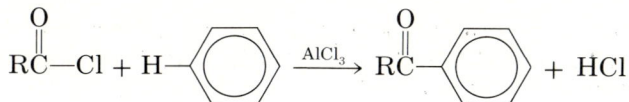

$$R\overset{O}{\overset{\|}{C}}-Cl + H-\bigcirc \xrightarrow{AlCl_3} R\overset{O}{\overset{\|}{C}}-\bigcirc + HCl$$

Treatment of acid halides with salts of carboxylic acids yields acid anhydrides, a preparation that is discussed in the next section.

Problem 12.17 Complete these reactions and name the organic products.

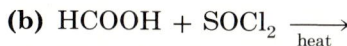

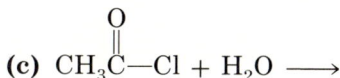

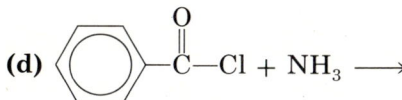

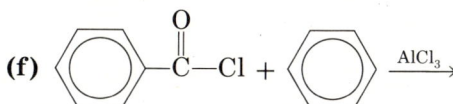

(a) $\bigcirc$—COOH + NaOH $\longrightarrow$ **(b)** HCOOH + SOCl$_2$ $\xrightarrow{heat}$

(c) $CH_3\overset{O}{\overset{\|}{C}}-Cl + H_2O \longrightarrow$ **(d)** $\bigcirc-\overset{O}{\overset{\|}{C}}-Cl + NH_3 \longrightarrow$

(e) $CH_3\overset{O}{\overset{\|}{C}}-Cl + CH_3OH \underset{}{\overset{H^+}{\rightleftharpoons}}$ **(f)** $\bigcirc-\overset{O}{\overset{\|}{C}}-Cl + \bigcirc \xrightarrow{AlCl_3}$

Acid Anhydrides: Preparation *12.20*

Anhydrides are compounds that may be thought of as being formed by loss of water between two molecules of an acid.

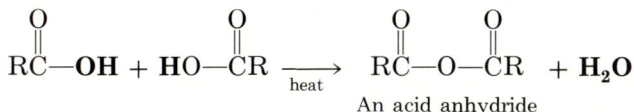

$$R\overset{O}{\overset{\|}{C}}-OH + HO-\overset{O}{\overset{\|}{C}}R \xrightarrow{heat} R\overset{O}{\overset{\|}{C}}-O-\overset{O}{\overset{\|}{C}}R + H_2O$$

An acid anhydride

Except for cyclic anhydrides, which are prepared this way, direct dehydration is seldom practiced. Most anhydrides are prepared by reaction between the sodium salt of the acid and an acid chloride.

$$\underset{\substack{\| \\ O}}{R\overset{\displaystyle O}{C}}-O:^-\,Na^+ + Cl-\underset{\substack{\| \\ O}}{\overset{\displaystyle O}{C}}R \longrightarrow \underset{\substack{\| \\ O}}{R\overset{\displaystyle O}{C}}-O-\underset{\substack{\| \\ O}}{\overset{\displaystyle O}{C}}R + NaCl$$

If the R groups of the acid salt and acid chloride are identical, we get a *simple
anhydride.*

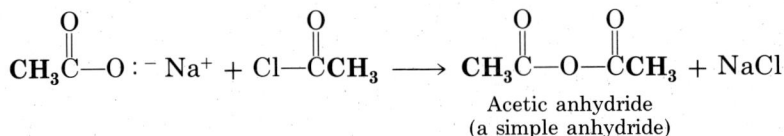

$$CH_3\overset{\displaystyle O}{\underset{\|}{C}}-O:^-\,Na^+ + Cl-\overset{\displaystyle O}{\underset{\|}{C}}CH_3 \longrightarrow CH_3\overset{\displaystyle O}{\underset{\|}{C}}-O-\overset{\displaystyle O}{\underset{\|}{C}}CH_3 + NaCl$$

Acetic anhydride
(a simple anhydride)

If the R groups are not the same, we get a *mixed anhydride.*

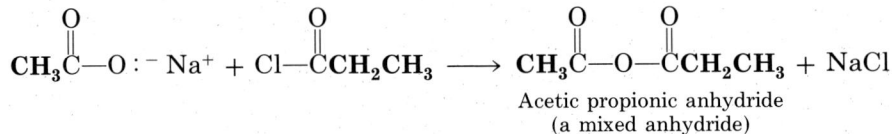

$$CH_3\overset{\displaystyle O}{\underset{\|}{C}}-O:^-\,Na^+ + Cl-\overset{\displaystyle O}{\underset{\|}{C}}CH_2CH_3 \longrightarrow CH_3\overset{\displaystyle O}{\underset{\|}{C}}-O-\overset{\displaystyle O}{\underset{\|}{C}}CH_2CH_3 + NaCl$$

Acetic propionic anhydride
(a mixed anhydride)

Acetic anhydride is by far the most important acid anhydride. It is used in
the production of synthetic rubber, fibers, lacquers, photographic films, cigarette
filters, magnetic tape, and thermoplastic moldings.

Cyclic anhydrides are prepared by intramolecular dehydration.

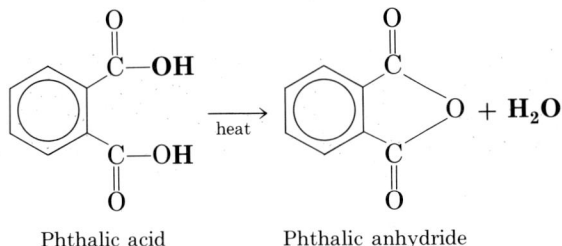

Phthalic acid   Phthalic anhydride

12.21 Acid Anhydrides: Reactions

The reactions of acid anhydrides with water, alcohols, ammonia, or amines
parallel those already shown for the acid chlorides (Sec. 12.19).

The by-product in all reactions of acid anhydrides is always a carboxylic
acid. Figure 12.5 summarizes the reactions of acid anhydrides with several nucleophiles.

Problem 12.18 Complete these reactions and name the organic products.

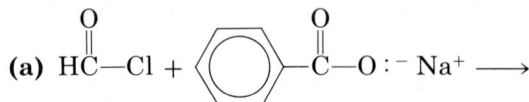

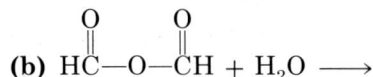

(a) $H\overset{\displaystyle O}{\underset{\|}{C}}-Cl + \langle\!\!\bigcirc\!\!\rangle-\overset{\displaystyle O}{\underset{\|}{C}}-O:^-\,Na^+ \longrightarrow$

(b) $H\overset{\displaystyle O}{\underset{\|}{C}}-O-\overset{\displaystyle O}{\underset{\|}{C}}H + H_2O \longrightarrow$

(c) 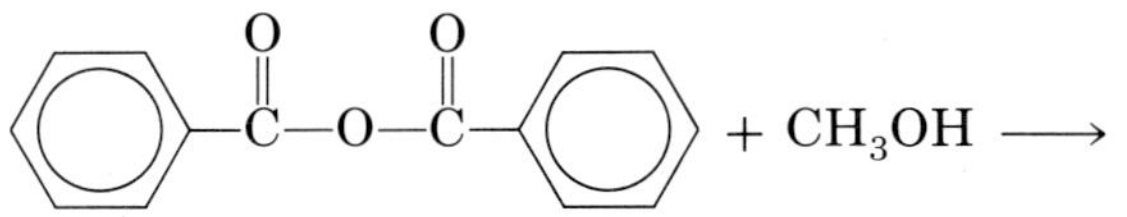 $\displaystyle C_6H_5-\overset{\overset{\displaystyle O}{\|}}{C}-O-\overset{\overset{\displaystyle O}{\|}}{C}-C_6H_5$ + CH_3OH $\longrightarrow$

(d) 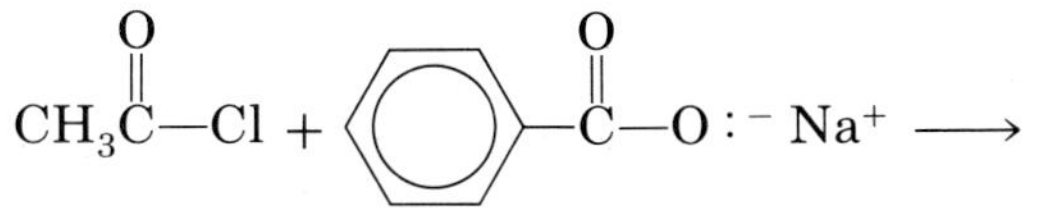 $\displaystyle CH_3\overset{\overset{\displaystyle O}{\|}}{C}-Cl$ + $\displaystyle C_6H_5-\overset{\overset{\displaystyle O}{\|}}{C}-O:^-\ Na^+$ $\longrightarrow$

(e) 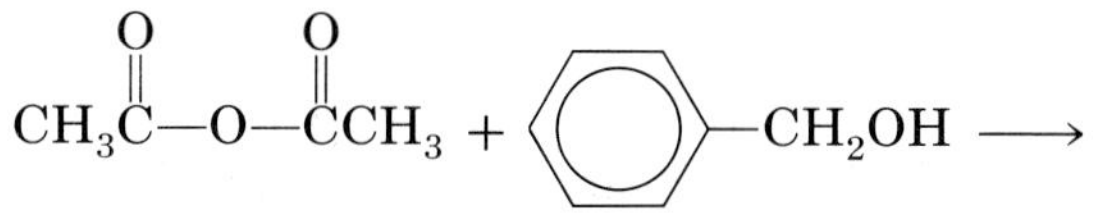 $\displaystyle CH_3\overset{\overset{\displaystyle O}{\|}}{C}-O-\overset{\overset{\displaystyle O}{\|}}{C}CH_3$ + $C_6H_5-CH_2OH$ $\longrightarrow$

(f) $HOOC-CH_2CH_2CH_2-COOH$ $\xrightarrow[\text{heat}]{}$

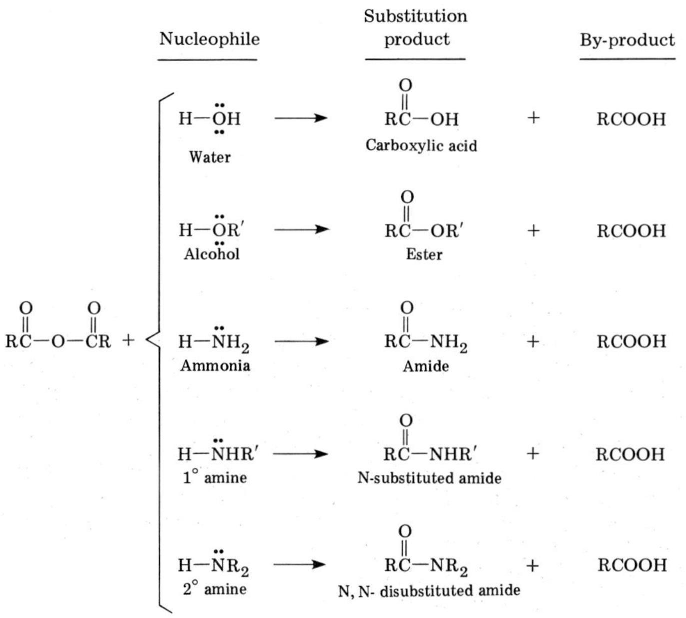

Nucleophile	Substitution product		By-product
$H-\overset{..}{\underset{..}{O}}H$ Water	$\displaystyle R\overset{\overset{\displaystyle O}{\|}}{C}-OH$ Carboxylic acid	$+$	$RCOOH$
$H-\overset{..}{\underset{..}{O}}R'$ Alcohol	$\displaystyle R\overset{\overset{\displaystyle O}{\|}}{C}-OR'$ Ester	$+$	$RCOOH$
$H-\overset{..}{N}H_2$ Ammonia	$\displaystyle R\overset{\overset{\displaystyle O}{\|}}{C}-NH_2$ Amide	$+$	$RCOOH$
$H-\overset{..}{N}HR'$ 1° amine	$\displaystyle R\overset{\overset{\displaystyle O}{\|}}{C}-NHR'$ N-substituted amide	$+$	$RCOOH$
$H-\overset{..}{N}R_2$ 2° amine	$\displaystyle R\overset{\overset{\displaystyle O}{\|}}{C}-NR_2$ N, N- disubstituted amide	$+$	$RCOOH$

with reactant $\displaystyle R\overset{\overset{\displaystyle O}{\|}}{C}-O-\overset{\overset{\displaystyle O}{\|}}{C}R$ +

Figure 12.5 Nucleophilic substitution reactions of acid anhydrides.

Amides 12.22

  Amides are commonly prepared in the laboratory by the ammonolysis of acid chlorides (Sec. 12.19) or acid anhydrides (Sec. 12.21). Amides of biological importance occur in nature in some plants and animals, and many are synthesized in the laboratory. For example, nicotinamide, the amide of nicotinic acid (niacin), is essential in the diet to prevent pellagra. Acetanilide and a derivative,

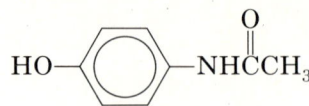

p-hydroxyacetanilide, are used as pain killers. Lidocaine is a widely used local anesthetic.

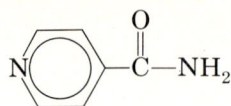

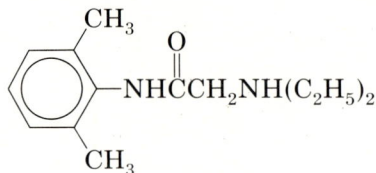

$$
\underset{\substack{\text{Nicotinamide}\\ \text{(prevents pellagra)}}}{\text{N}\langle\bigcirc\rangle\overset{\displaystyle O}{\overset{\|}{C}}-NH_2}
\qquad
\underset{\substack{\text{Acetanilide}\\ \text{(a pain killer)}}}{\langle\bigcirc\rangle-NH\overset{\displaystyle O}{\overset{\|}{C}}CH_3}
\qquad
\underset{\substack{p\text{-Hydroxyacetanilide}\\ \text{(a pain killer)}}}{HO-\langle\bigcirc\rangle-NH\overset{\displaystyle O}{\overset{\|}{C}}CH_3}
$$

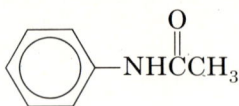

$$
\underset{\substack{\text{Lidocaine}\\ \text{(a local anesthetic)}}}{\overset{\displaystyle CH_3}{\underset{\displaystyle CH_3}{\langle\bigcirc\rangle}}-NH\overset{\displaystyle O}{\overset{\|}{C}}CH_2NH(C_2H_5)_2}
$$

Proteins are polyamides containing hundreds of amide linkages.

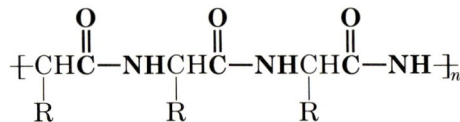

$$
\underset{\text{Polyamide polymer}}{\Big[CH\overset{\displaystyle O}{\overset{\|}{C}}-NHCH\overset{\displaystyle O}{\overset{\|}{C}}-NHCH\overset{\displaystyle O}{\overset{\|}{C}}-NH\Big]_n}
\quad
\begin{matrix} | \\ R \end{matrix}
\begin{matrix} | \\ R \end{matrix}
\begin{matrix} | \\ R \end{matrix}
$$

Simpler polyamides make up the industrially important Nylon 66, used in the production of stockings and other textiles, and in the manufacture of brushes and plastic toys. Nylon 66 is made from the reaction of hexamethylene-diamine and adipic acid.

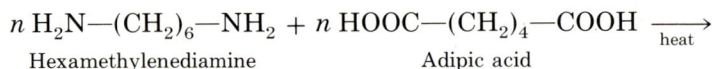

$$
n\,\underset{\text{Hexamethylenediamine}}{H_2N-(CH_2)_6-NH_2} + n\,\underset{\text{Adipic acid}}{HOOC-(CH_2)_4-COOH} \xrightarrow{\text{heat}}
$$

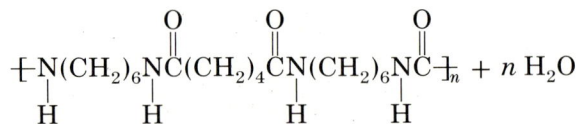

$$
\underset{\substack{\text{Nylon 66}\\ \text{(a polyamide)}}}{\Big[N(CH_2)_6N\overset{\displaystyle O}{\overset{\|}{C}}(CH_2)_4\overset{\displaystyle O}{\overset{\|}{C}}N(CH_2)_6N\overset{\displaystyle O}{\overset{\|}{C}}\Big]_n} + n\,H_2O
$$

Amides can be hydrolyzed in acid or in alkaline solution. Acid-catalyzed hydrolysis produces the free organic acid and an ammonium salt.

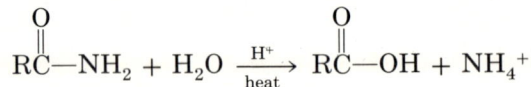

$$
R\overset{\displaystyle O}{\overset{\|}{C}}-NH_2 + H_2O \xrightarrow[\text{heat}]{H^+} R\overset{\displaystyle O}{\overset{\|}{C}}-OH + NH_4^+
$$

Base-catalyzed hydrolysis produces a carboxylate salt and free ammonia.

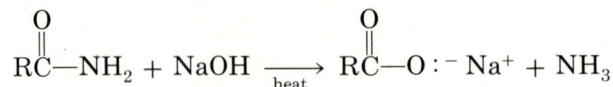

$$
R\overset{\displaystyle O}{\overset{\|}{C}}-NH_2 + NaOH \xrightarrow{\text{heat}} R\overset{\displaystyle O}{\overset{\|}{C}}-O:^-Na^+ + NH_3
$$

Amides, when treated with lithium aluminum hydride, are reduced to amines. The net reaction is the conversion of the C=O in amides to CH_2.

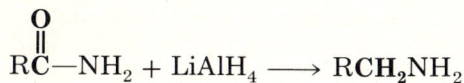

$$RC\overset{\overset{\displaystyle O}{\|}}{}-NH_2 + LiAlH_4 \longrightarrow RCH_2NH_2$$

$$RC\overset{\overset{\displaystyle O}{\|}}{}-NHR' + LiAlH_4 \longrightarrow RCH_2NHR'$$

$$RC\overset{\overset{\displaystyle O}{\|}}{}-NR_2 + LiAlH_4 \longrightarrow RCH_2NR_2$$

Simple amides can be reduced to amines containing *one less carbon atom* by reaction with alkaline hypohalite solution.

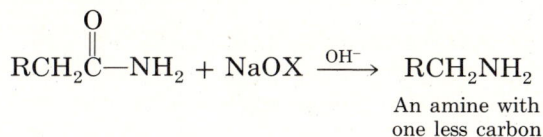

$$RCH_2C\overset{\overset{\displaystyle O}{\|}}{}-NH_2 + NaOX \xrightarrow{OH^-} RCH_2NH_2$$

An amine with one less carbon

Problem 12.19 Complete these reactions.

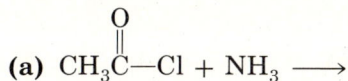

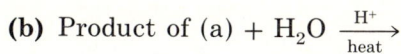

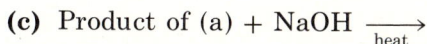

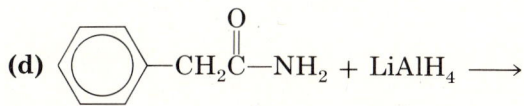

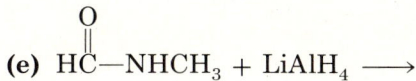

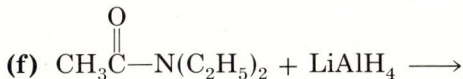

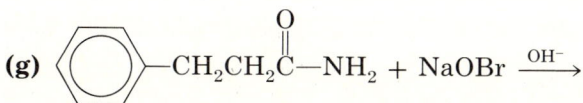

(a) $CH_3C\overset{\overset{\displaystyle O}{\|}}{}-Cl + NH_3 \longrightarrow$

(b) Product of (a) + $H_2O \xrightarrow[\text{heat}]{H^+}$

(c) Product of (a) + NaOH $\xrightarrow[\text{heat}]{}$

(d) $\langle\bigcirc\rangle-CH_2C\overset{\overset{\displaystyle O}{\|}}{}-NH_2 + LiAlH_4 \longrightarrow$

(e) $HC\overset{\overset{\displaystyle O}{\|}}{}-NHCH_3 + LiAlH_4 \longrightarrow$

(f) $CH_3C\overset{\overset{\displaystyle O}{\|}}{}-N(C_2H_5)_2 + LiAlH_4 \longrightarrow$

(g) $\langle\bigcirc\rangle-CH_2CH_2C\overset{\overset{\displaystyle O}{\|}}{}-NH_2 + NaOBr \xrightarrow{OH^-}$

Summary of Concepts and Reactions

The most important class of organic acids is the carboxylic acids. All carboxylic acids contain the carboxyl group, —COOH.   [Sec. 12.1]

The common names of carboxylic acids are often derived from the Latin or Greek and relate to their natural sources; -ic acid is the ending for all common names. The ending -e of the corresponding alkanes is replaced by -oic acid in the IUPAC nomenclature.   [Sec. 12.1]

Dicarboxylic acids are acids that contain two carboxyl groups.   [Sec. 12.1]

The first four aliphatic acids (formic through butyric) are completely miscible in water.   [Sec. 12.2]

The boiling points of simple carboxylic acids are higher than the boiling points of alcohols of comparable molecular weight. This shows that acids have a higher degree of hydrogen bonding than do alcohols.   [Sec. 12.2]

Compared with the common mineral acids, carboxylic acids are weak acids with a K_a (at 25°C) around 10^{-4} to 10^{-5}.   [Sec. 12.3]

Carboxylic acids are stronger acids than alcohols because the carboxylate anion is stabilized by resonance whereas the alkoxide anion is not.   [Sec. 12.4]

Electron-withdrawing substituents on the acid chain increase the acidity of an acid.   [Sec. 12.4]

Electron-donating substituents on the acid chain decrease the acidity of an acid.   [Sec. 12.4]

The electronic effect of a substituent transmitted through single bonds is called the inductive effect.   [Sec. 12.4]

Basically, carboxylic acids are prepared by the following three methods.

Oxidation of 1° alcohols or aldehydes and alkylbenzenes

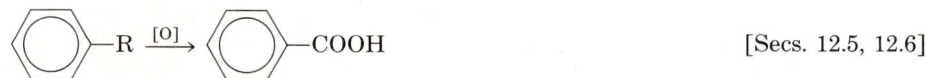

$$RCH_2OH \xrightarrow{[O]} RCHO \xrightarrow{[O]} RCOOH \qquad ([O] = \text{oxidizing agent})$$

$$\langle\bigcirc\rangle\!-\!R \xrightarrow{[O]} \langle\bigcirc\rangle\!-\!COOH \qquad\qquad\qquad [\text{Secs. 12.5, 12.6}]$$

Carbonation of Grignard reagents

$$RMgX + CO_2 \longrightarrow RCO_2MgX \xrightarrow{H^+} RCOOH \qquad [\text{Secs. 12.5, 12.7}]$$

Hydrolysis of nitriles

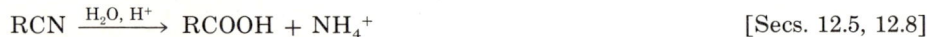

$$RCN \xrightarrow{H_2O,\ H^+} RCOOH + NH_4^+ \qquad [\text{Secs. 12.5, 12.8}]$$

The reactions of acids are basically of two types.

Reactions with bases to form salts

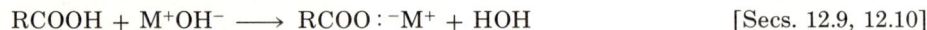

$$RCOOH + M^+OH^- \longrightarrow RCOO\!:^-M^+ + HOH \qquad [\text{Secs. 12.9, 12.10}]$$

Reactions with nucleophiles to form acid derivatives

$$RCOOH + :Nu \longrightarrow \overset{\displaystyle O}{\underset{\displaystyle \|}{R}C\!-\!Nu} \qquad [\text{Secs. 12.9, 12.11}]$$

The acid derivatives are formed from acids by nucleophilic attack on the acyl carbon followed by elimination of OH.

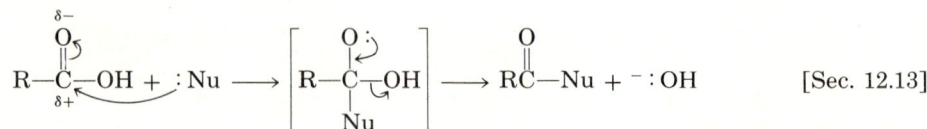

$$R\!-\!\overset{O}{\underset{\delta+}{\overset{\|}{C}}}\!-\!OH + :Nu \longrightarrow \left[R\!-\!\overset{O:}{\underset{Nu}{\overset{\|}{C}}}\!-\!OH \right] \longrightarrow RC\!-\!Nu + {}^-\!:OH \qquad [\text{Sec. 12.13}]$$

The acid derivatives discussed in this chapter are

Esters, RC—OR′ (with O double-bonded to C)   [Secs. 12.14–12.17]

Acid chlorides, RC—Cl (with O double-bonded to C)   [Secs. 12.18, 12.19]

Acid anhydrides, RC—O—CR (with O double-bonded to each C)   [Secs. 12.20, 12.21]

Amides, RC—N< (with O double-bonded to C)   [Sec. 12.22]

Key Terms

carboxylic acids

carboxyl group, —C—OH (with O double-bonded to C)
acid strength
ionization constant, K_a
pK_a
carboxylate anion
inductive effect

carboxylic acid derivative

acyl group, R—C— (with O double-bonded to C)
acyl chlorides, acid chlorides
esters
amides
anhydrides

nucleophilic substitution
esterification
lactone
hydrolysis
saponification
alcoholysis
transesterification
ammonolysis

Exercises

Structure and Nomenclature of Carboxylic Acids and Derivatives [Secs. 12.1, 12.12]

12.1 Write structural formulas for the following acids.
(a) Isobutyric acid
(b) 3-Chlorohexanoic acid
(c) *m*-Bromobenzoic acid
(d) 2,4-Dihydroxybenzoic acid
(e) 3-Bromo-4-phenylpentanoic acid
(f) β-Methylvaleric acid
(g) Phthalic acid
(h) α-Bromosuccinic acid
(i) α-Chloropropionic acid
(j) *trans*-3-Methylcyclohexanecarboxylic acid

12.2 Write the structure of each of the following compounds.
(a) Potassium butyrate
(b) Calcium acetate
(c) Ethyl butyrate
(d) *n*-Propyl acetate
(e) *t*-Butyl benzoate
(f) Methyl *p*-nitrobenzoate
(g) Dimethyl malonate
(h) Butanoyl chloride
(i) *m*-Nitrobenzoyl chloride
(j) α-Chloropropionyl chloride
(k) Acetic anhydride
(l) Phthalic anhydride
(m) Acetamide
(n) *m*-Nitrobenzamide
(o) *N*-Ethylformamide
(p) *N*,*N*-Dimethylpropionamide

12.3 There is something wrong with each of the names given. Write the structure and give the correct IUPAC name for each compound.
(a) α-Bromohexanoic acid
(b) 3-Methylvaleric acid
(c) 2-Ethylpropionic acid
(d) α-Bromo-3-chlorobutyric acid

12.4 Describe the most convenient method for naming these acids.

(a) $CH_3(CH_2)_4COOH$

(b)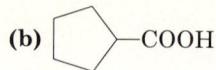

(c) $CH_3(CH_2)_2CHBrCOOH$

(d)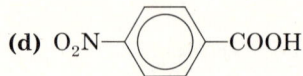

(e) CF_3COOH

(f)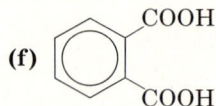

(g) $HOOC—COOH$

(h) $HOOCCH_2CHClCOOH$

12.5 Name each of these compounds.

(a) $CH_3(CH_2)_2COO^- Na^+$

(b) $(HCOO^-)_2 Ba^{2+}$

(c) $CH_3CH_2COOCH_3$

(d) $HCOOCH_3$

(e) $C_6H_5COOCH_2CH_3$

(f) CH_3CH_2COBr

(g) $(CH_3CO)_2O$

(h) $HCONH_2$

(i)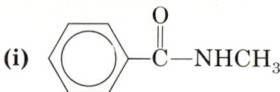

(j)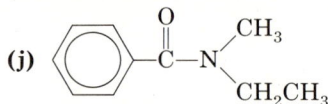

Physical Properties [Sec. 12.2]

12.6 Arrange the compounds in each set in order of increasing solubility in water. Give reasons for the orders you select.

(a) $CH_3(CH_2)_3CH_3$ (i); CH_3CH_2COOH (ii); $CH_3(CH_2)_2CH_2OH$ (iii).

(b) $CH_3(CH_2)_4CH_3$ (i); $CH_3CH_2COO^-Na^+$ (ii); $CH_3(CH_2)_3COOH$ (iii).

12.7 Arrange each set of compounds in Exercise 12.6 in order of increasing boiling points. Give reasons for the orders you select.

12.8 Explain why the boiling point of ethyl acetate is 87°C lower than that of butyric acid even though the molecular weights of the two compounds are the same.

Acid Strength and Acid Structure [Secs. 12.3, 12.4]

12.9 Arrange the acids listed in order of increasing acid strength.

(a) Lactic acid, K_a 1.38×10^{-4} (b) Pyruvic acid, K_a 3.16×10^{-3}

(c) Fluoroacetic acid, K_a 2.19×10^{-3} (d) Acetoacetic acid, K_a 2.63×10^{-4}

12.10 Arrange each set of compounds in order of increasing acid strength.

(a) CH_3COOH (i); CH_3CH_2OH (ii); $HCOOH$ (iii).

(b) CH_3CH_2COOH (i); $O_2NCH_2CH_2COOH$ (ii); $CH_3CHCOOH$ (iii);

NO_2

$(CH_3)_2CHCOOH$ (iv)

(c) CH_3CH_2COOH (i); FCH_2CH_2COOH (ii); CH_3CF_2COOH (iii); $CH_3CHFCOOH$ (iv)

(d)

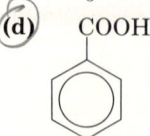   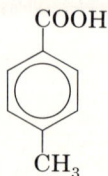   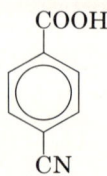   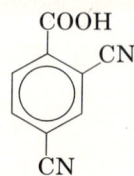

(i)   (ii)   (iii)   (iv)

12.11 Outline a procedure for the removal of acetic acid (bp 118°C) from a mixture that also contains *n*-butyl alcohol (bp 117°C).

12.12 What simple chemical tests can be used to distinguish the following pairs of compounds?

(a) Benzoic acid and methyl benzoate   **(b)** Pentanoic acid and 1-hexanol

(c) Benzoic acid and benzaldehyde   **(d)** Butyric acid and 1-hexene

Preparation and Reactions of Carboxylic Acids and Derivatives [Secs. 12.5–12.8, 12.10–12.22]

12.13 By means of equations, show how you would synthesize each acid.

    (a) Butanoic acid from 1-butanol

    (b) Butanoic acid from 1-chloropropane

    (c) Butanoic acid from 1-bromopropane (by a method other than **(b)**)

    (d) *p*-Bromobenzoic acid from *p*-bromotoluene

    (e) Malonic acid from chloroacetic acid

    (f) Succinic acid from ethylene

12.14 Name the organic products expected for the following reactions and write structures for the reactants and products.

    (a) Butyric acid + aqueous sodium hydroxide $\longrightarrow$

    (b) Propionic acid + thionyl chloride $\longrightarrow$

    (c) Benzoyl chloride + sodium benzoate $\longrightarrow$

    (d) Propionyl chloride + methanol $\longrightarrow$

    (e) Succinic anhydride + water $\longrightarrow$

    (f) Acetyl chloride + ammonia $\longrightarrow$

12.15 Show how each of the following compounds can be prepared from the appropriate acid.

    (a) Calcium propionate   **(b)** Methyl benzoate

    (c) Trifluoroacetic anhydride   **(d)** Formamide

    (e) *m*-Nitrobenzoyl chloride   **(f)** *N,N*-Dimethylphenylacetamide

    (g) Succinic anhydride

12.16 Write equations to show the reactions of butyryl chloride with

    (a) Water

    (b) Methyl alcohol

    (c) Phenol

    (d) Sodium acetate

    (e) Ammonia

    (f) Methylamine, CH_3NH_2

$$\overset{\displaystyle H}{\underset{\displaystyle |}{}}$$

    (g) Diethylamine, $CH_3CH_2-N-CH_2CH_3$

    (h) Benzene + $AlCl_3$

12.17 Write equations to show the reactions of propionic anhydride with

    (a) Water   **(b)** Methyl alcohol

    (c) Ammonia   **(d)** Methylamine, CH_3NH_2

    (e) Diethylamine, $(CH_3CH_2)_2NH$

12.18 Write equations to show the reactions of ethyl propanoate with

    (a) H_2O (excess), H^+, heat

    (b) Aqueous KOH, heat

    (c) $CH_3CH_2CH_2OH$ (excess), H^+, heat

    (d) Ammonia

    (e) $LiAlH_4$

    (f) Ethylmagnesium chloride followed by H_2O, H^+

12.19 Write equations to show the reactions of propionamide with

    (a) H_2O, H^+, heat   **(b)** $LiAlH_4$ **(c)** NaOCl, OH^-

12.20 Give structures of the starting materials and of compounds **A** through **J**.

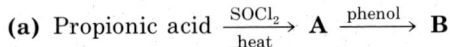

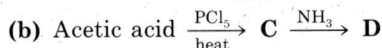

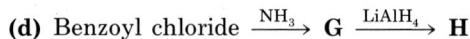

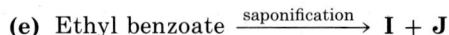

(a) Propionic acid $\xrightarrow[\text{heat}]{\text{SOCl}_2}$ **A** $\xrightarrow{\text{phenol}}$ **B**

(b) Acetic acid $\xrightarrow[\text{heat}]{\text{PCl}_5}$ **C** $\xrightarrow{\text{NH}_3}$ **D**

(c) Succinic anhydride $\xrightarrow{\text{CH}_3\text{OH (1 mole)}}$ **E** $\xrightleftharpoons{\text{CH}_3\text{CH}_2\text{OH, H}^+}$ **F**

(d) Benzoyl chloride $\xrightarrow{\text{NH}_3}$ **G** $\xrightarrow{\text{LiAlH}_4}$ **H**

(e) Ethyl benzoate $\xrightarrow{\text{saponification}}$ **I** + **J**

12.21 By means of a series of equations, show how you would accomplish the synthesis of each of the following compounds from the indicated starting material. You may use any other reagent needed.
(a) Propionyl chloride from *n*-propyl alcohol
(b) *n*-Butyl butanoate from *n*-butyl alcohol
(c) Benzamide from benzene
(d) Benzoic anhydride from toluene
(e) Isopropyl propanoate from propene

12.22 Write a mechanism for the reaction of acetyl chloride with water.

12.23 Write a mechanism for the reaction of ethyl acetate with ammonia.

12.24 A liquid, $C_6H_{12}O_2$, was hydrolyzed with water and acid to give an acid (**A**) and an alcohol (**B**). Oxidation of **B** with chromic acid produced **A**. What was the structure of the original compound? Write equations for the two reactions.

12.25 An unknown ester, $C_5H_{10}O_2$, was hydrolyzed with water and acid to produce an acid (**A**) and an alcohol (**B**). Treatment of **B** with PBr_3 gave an alkyl bromide (**C**). When **C** was treated with KCN, a product (**D**) was formed that, on hydrolysis with water and acid, gave acid **A**. Give the structure and the name of the original ester. Identify **A** though **D** and write equations for the reactions described.

13

Lipids

The lipids include a large number of substances of different types. For this reason no concise definition of lipids based on a distinguishing functional group can be given, as has been done with other classes of organic compounds discussed so far. For our purposes, a lipid will be any substance that meets these two criteria.

1. A lipid is insoluble in water but is soluble in organic solvents of low polarity such as chloroform, ether, and benzene.
2. A lipid is a constituent of the cell.

This definition, although somewhat vague, focuses first on one common characteristic of lipids: their *nonpolar nature*. We shall soon see that for a large number of lipids this nonpolarity is due to the presence of one or more fatty acid residues containing long aliphatic hydrocarbon chains. The second part of our definition excludes any man-made compounds not found within the cell. Also excluded are carbohydrates, proteins, and nucleic acids, for these components of the cell are insoluble in nonpolar organic solvents.

Lipids have a great variety of biological functions. The fats in our diet are lipids. Fats serve as a source of energy. Because they are poor conductors of heat, fats help to maintain optimum body temperature by providing thermal insulation. In addition, they serve as a covering of vital organs, protecting them from mechanical shock. Moreover, the nonpolar character of fats makes it possible for them to transport nonpolar materials through biological fluids. Examples of materials carried by fats, and also classified as lipids, are vitamins A, D, E, and K.

Lipids also are found as structural components of the cell wall. Often these lipids contain phosphorus (*phospholipids*) or are associated with a carbohydrate (*glycolipids*). As components of the cell wall, these lipids play a role in the permeability of cell membranes. A number of *hormones,* the regulators of enzymatic reactions, also are classified as lipids. It is clear that lipids constitute a biologically important group of compounds.

13.1 Classification of Lipids

Lipids are classified into three broad categories on the basis of their molecular structure and their hydrolysis products. Each category is subdivided into several groups.

1. Simple lipids
    a. Triglycerides (fats and oils)
    b. Waxes
2. Compound lipids
    a. Phospholipids
        (1) Phosphoglycerides
        (2) Sphingolipids
    b. Glycolipids
3. Derived lipids
    a. Steroids
    b. Fat-soluble vitamins
    c. Prostaglandins

13.2 Simple Lipids

The simple lipids are esters. They are subdivided into two groups, depending on the nature of the alcohol component. The *triglycerides* are esters of a trihydroxy alcohol, glycerol, with three fatty acid molecules. The *waxes* are esters of fatty acids with long-chain monohydroxy alcohols that have from 26 to 34 carbon atoms. Because fatty acids are constituents of both triglycerides and waxes, it is appropriate to begin our study of simple lipids by examining the fatty acids as a group in itself, after which we will consider the triglycerides and the waxes, in that order.

13.3 Fatty Acids in Simple Lipids

The naturally occurring fatty acids found in simple lipids have the following structural characteristics.

1. Most are linear, long-chain monocarboxylic acids that range from C_{12} to C_{26}, the C_{16} and C_{18} acids being the most abundant.

2. Almost all of them have an even number of carbon atoms (C_{12}, C_{14}, C_{16}, etc.).

3. They may be saturated, if all carbon atoms are single bonded, or they may be unsaturated with one or more double bonds between carbon atoms.

4. Almost all of the unsaturated fatty acids are of the *cis* configuration.

Examples are

$$CH_3-CH_2-CH_2-CH_2-CH_2-CH_2-CH_2-CH_2-CH_2-CH_2-CH_2-CH_2-CH_2-CH_2-CH_2-CH_2-CH_2-COOH$$

Stearic acid (18:0)*
(a saturated fatty acid)

$$CH_3-CH_2-CH_2-CH_2-CH_2-CH_2-CH_2-CH_2-CH=CH-CH_2-CH_2-CH_2-CH_2-CH_2-CH_2-CH_2-COOH$$

cis

Oleic acid (18:1)
(an unsaturated fatty acid)

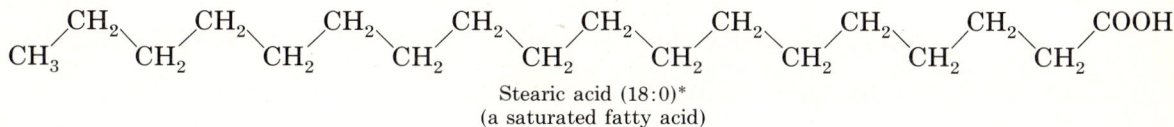

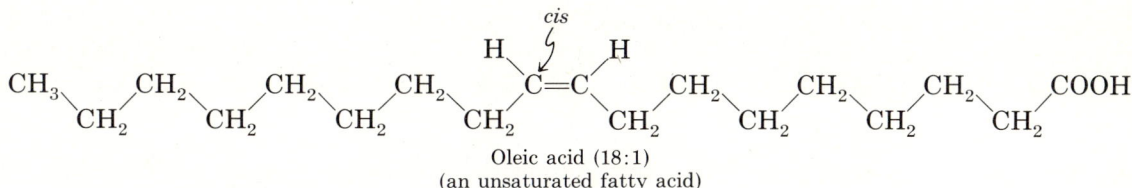

The names, structural formulas, melting points, and relative abundances of common fatty acids are given in Table 13.1. As shown, the melting points of the saturated fatty acids increase with the number of carbon atoms. Lauric acid (C_{12}, saturated) has the lowest melting point in the series; stearic acid (C_{18}, saturated) has the highest. This order is understandable if you consider that one member of the series differs from the next by two CH_2 groups. For example, lauric acid has ten $-CH_2-$ groups, whereas myristic has twelve. Because of the difference in the lengths of the alkyl chains, the van der Waals forces of attraction between lauric acid molecules are weaker than those between myristic acid molecules. The same reasoning applies for the other members of the series, palmitic acid and stearic acid. Because the attractive forces between lauric acid molecules are weaker than those between myristic acid molecules, less energy (in the form of heat) is needed to disrupt the orderly crystalline structure of lauric acid than is required to disrupt the structure of myristic acid. For the same reasons, the melting point of myristic acid is lower than that of palmitic acid, which in turn is lower than the melting point of stearic acid.

Table 13.1 Common Fatty Acids Found in Simple Lipids

Common name of fatty acid	Formula	Notation*	Mp (°C)	Relative abundance
Lauric acid	$CH_3(CH_2)_{10}COOH$	12:0	44	Small
Myristic acid	$CH_3(CH_2)_{12}COOH$	14:0	54	Intermediate
Palmitic acid	$CH_3(CH_2)_{14}COOH$	16:0	63	Great
Stearic acid	$CH_3(CH_2)_{16}COOH$	18:0	70	Great
Oleic acid	$CH_3(CH_2)_7CH=CH(CH_2)_7COOH$	18:1	13	Great
Linoleic acid	$CH_3(CH_2)_4(CH=CHCH_2)_2(CH_2)_6COOH$	18:2	−5	Great
Linolenic acid	$CH_3(CH_2CH=CH)_3(CH_2)_7COOH$	18:3	−11	Small
Arachidonic acid	$CH_3(CH_2CH=CHCH_2)_4(CH_2)_2COOH$	20:4	−50	Very small

*The first number indicates the number of carbon atoms; the second, the number of double bonds.

If we compare the C_{18} fatty acids, we find that their melting points *decrease as the degree of unsaturation increases.* For example, stearic acid (C_{18}, no double bond) has a higher melting point than oleic acid (C_{18}, one double bond). Linolenic acid (C_{18}, three double bonds) has the lowest melting point of all the acids in the series. We can account for this melting point sequence by referring back to the structures of stearic acid and oleic acid. Note that stearic acid has a saturated alkyl chain, which gives it an orderly zigzag arrangement. This regular arrangement enables one molecule of stearic acid to fit closely with the next one, thus forming a tightly packed crystalline structure. In oleic acid, on the other hand, the *cis* configuration about the carbon–carbon double bond produces a bend in the molecule. This bend prevents molecules of oleic acid from approaching one another as closely in the solid state. As a result, less heat energy is needed to convert oleic acid from the solid state to the liquid form than is required for stearic acid. A diagrammatic summary of the relationship between melting points and structures is shown in Figure 13.1.

Linoleic acid with two double bonds is more severely bent than oleic acid, and linolenic acid with three double bonds is even more severely bent. Applying the same reasoning used previously, it should not surprise us that the melting points of these acids decrease in the order given.

If we were asked to predict the solubility characteristics of the acids in Table 13.1, all we would have to do is apply the familiar "like dissolves like" rule. A glance at the structure of either stearic acid or oleic acid (page 351) shows that the nonpolar portion (the long aliphatic hydrocarbon chain) completely overshadows the polar segment (the COOH group). The same is true for any of the fatty acids listed in Table 13.1. We would therefore conclude that fatty acids are soluble in nonpolar organic solvents but insoluble in a polar solvent such as water. And we would be right.

As the next sections will demonstrate, the physical properties and most of the chemical properties of triglycerides are due primarily to the characteristics of their constituent fatty acids.

Figure 13.1 **(a)** Close packing among saturated fatty acids. Lauric acid has a lower melting point than stearic acid because there are fewer forces (van der Waals forces) of attraction between the shorter alkyl chains in the C_{12} acid than in the C_{18} acid. **(b)** The *cis*-9,10 double bond in oleic acid prevents close approach between molecules in the crystal, thus lowering the melting point.

The most abundant lipids are the triacylglycerols, commonly known as **triglycerides.** Triglycerides are esters of three molecules of fatty acid and **glycerol.**

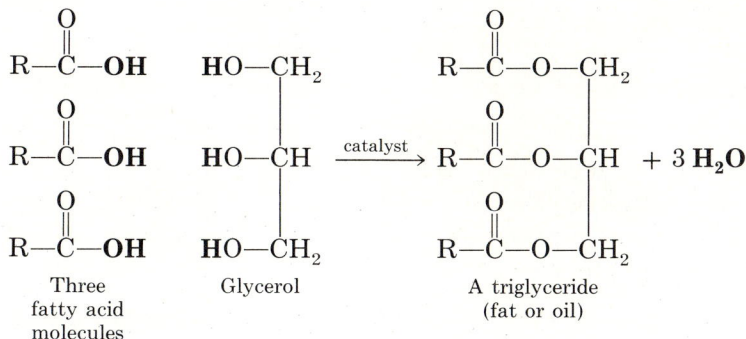

$$
\begin{array}{ccc}
\begin{array}{c} O \\ \parallel \\ R-C-OH \\ \\ O \\ \parallel \\ R-C-OH \\ \\ O \\ \parallel \\ R-C-OH \end{array}
&
\begin{array}{c} HO-CH_2 \\ \\ \\ HO-CH \\ \\ \\ HO-CH_2 \end{array}
\xrightarrow{\text{catalyst}}
&
\begin{array}{c} O \\ \parallel \\ R-C-O-CH_2 \\ \\ O \\ \parallel \\ R-C-O-CH \\ \\ O \\ \parallel \\ R-C-O-CH_2 \end{array}
+ 3\,H_2O
\end{array}
$$

| Three fatty acid molecules | Glycerol | A triglyceride (fat or oil) |

When the fatty acid residues (R) are the same, the triglyceride is known as a **simple triglyceride.** An example of a simple triglyceride is glyceryl tripalmitate, or tripalmitin.

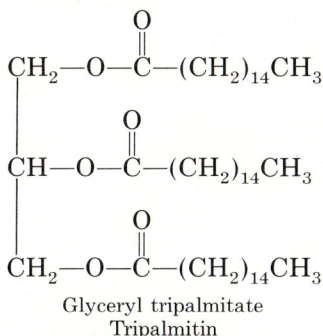

$$
\begin{array}{l}
O \\
\parallel \\
CH_2-O-C-(CH_2)_{14}CH_3 \\
\\
\qquad\quad O \\
\qquad\quad \parallel \\
CH-O-C-(CH_2)_{14}CH_3 \\
\\
\qquad\quad O \\
\qquad\quad \parallel \\
CH_2-O-C-(CH_2)_{14}CH_3
\end{array}
$$

Glyceryl tripalmitate
Tripalmitin

When the triglyceride contains different R groups, it is called a **mixed triglyceride. Fats** are complex mixtures of simple and mixed triglycerides that exist in the solid form at room temperature. Fats are mostly of animal origin. **Oils** are also complex mixtures of simple and mixed triglycerides. They occur as liquids at room temperature and are mostly of vegetable origin. The difference between fats and oils is, therefore, one of melting point.

Because all fats and oils contain glycerol, it is obvious that the difference in melting point is due to variations in the structures of the fatty acid residues, R. Other differences in physical and chemical properties between fats and oils must also be caused by variations in the structures of their fatty acid components. The fatty acid composition of some fats and oils is shown in Table 13.2. The data in Table 13.2 reveal that fats contain a higher percentage of saturated fatty acids and that oils contain a greater proportion of unsaturated fatty acids. In view of the previous discussion (Sec. 13.3), these findings are entirely expected.

The data in the table also show that there is a wide range of fatty acid compositions in fats and oils from a given source. This composition range is related to the diet of the animal and to the climatic conditions under which plants grow.

Table 13.2 Fatty Acid Composition of Some Fats and Oils

Fat or oil	Average composition of fatty acids (%)						
	Lauric acid	Myristic acid	Palmitic acid	Stearic acid	Oleic acid	Linoleic acid	Linolenic acid
Lard		1–2	25–30	12–16	40–50	5–10	1
Butterfat^a	2–5	8–14	25–30	9–12	25–35	2–5	
Beef tallow		3–5	25–30	20–30	40–50	1–5	
Coconut fat^b	45–48	16–18	8–10	2–4	5–8	1–2	
Olive oil			8–16	2–3	70–85	5–15	
Soybean oil			10	3	25–30	50–55	4–8
Cottonseed oil		1	20–25	1–2	20–30	45–50	
Safflower oil			6	3	13–15	75–78	
Linseed oil					20–35	15–25	40–60

^{a} Also 3–4% butyric acid and 1–3% each of C_6, C_8, and C_{10} acids
^{b} Also 5–9% each of C_8 and C_{10} acids

13.5 Hydrolysis in the Presence of Acids or Enzymes

Most reactions of fats and oils are completely predictable. As esters, they undergo hydrolysis. When hydrolysis is brought about in the presence of an acid catalyst, the products are glycerol and three molecules of fatty acid. Digestion of fats or oils is also an instance of hydrolysis. In digestion, the catalysts are enzymes called **lipases.** The general equation for the hydrolysis of a fat or an oil is

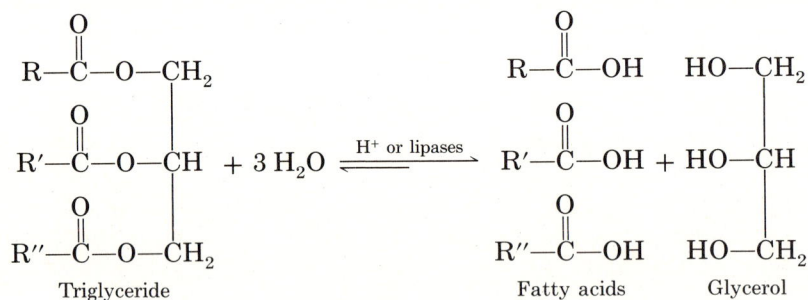

Triglyceride + 3 H_2O $\xrightleftharpoons{\text{H}^+ \text{ or lipases}}$ Fatty acids + Glycerol

13.6 Alkaline Hydrolysis: Saponification

Hydrolysis under acidic conditions is an equilibrium reaction. When hydrolysis is brought about in an alkaline medium (NaOH or KOH), the reaction is called **saponification.** Saponification, which is an irreversible process, produces glycerol and the salts of the fatty acids, or **soaps.**

$$
\begin{array}{l}
\text{R}-\overset{\overset{\displaystyle O}{\|}}{\text{C}}-\text{OCH}_2 \\[4pt]
\text{R}-\overset{\overset{\displaystyle O}{\|}}{\text{C}}-\text{OCH} \quad + \ 3\,\text{NaOH} \ \xrightarrow[\text{heat}]{\text{H}_2\text{O}} \ 3\,\text{R}-\overset{\overset{\displaystyle O}{\|}}{\text{C}}-\text{O}:^{-}\text{Na}^{+} \ + \ \text{HO}-\text{CH} \\[4pt]
\text{R}-\overset{\overset{\displaystyle O}{\|}}{\text{C}}-\text{OCH}_2
\end{array}
$$

HO—CH₂ and HO—CH₂ appear with glycerol

A triglyceride   Soap   Glycerol

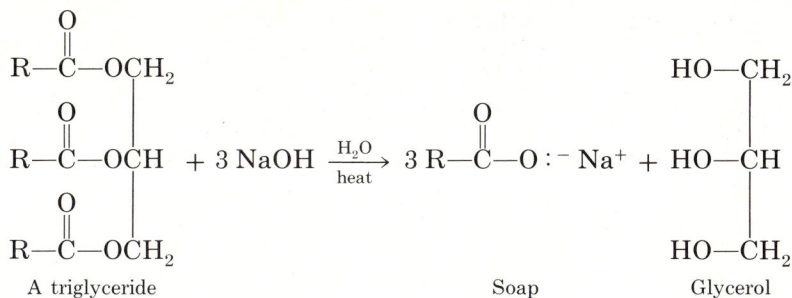

Sodium stearate, the sodium salt of stearic acid, is the principal active agent in common household soap. It is prepared from the saponification of glyceryl tristearate, a constituent of lard and beef tallow.

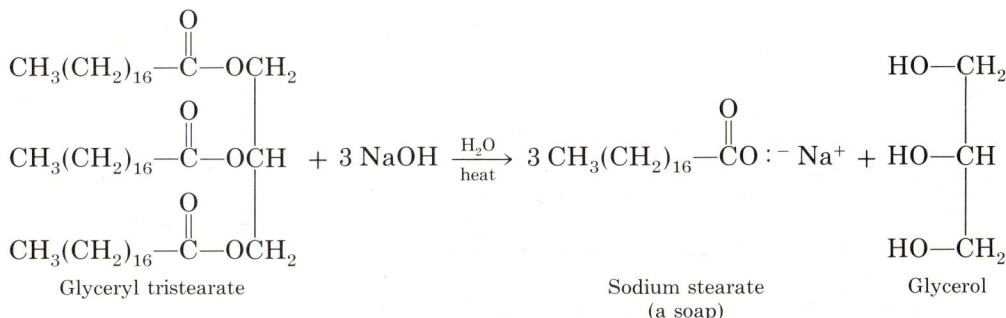

$$
\begin{array}{l}
\text{CH}_3(\text{CH}_2)_{16}-\overset{\overset{\displaystyle O}{\|}}{\text{C}}-\text{OCH}_2 \\[4pt]
\text{CH}_3(\text{CH}_2)_{16}-\overset{\overset{\displaystyle O}{\|}}{\text{C}}-\text{OCH} \quad + \ 3\,\text{NaOH} \ \xrightarrow[\text{heat}]{\text{H}_2\text{O}} \ 3\,\text{CH}_3(\text{CH}_2)_{16}-\overset{\overset{\displaystyle O}{\|}}{\text{C}}\text{O}:^{-}\text{Na}^{+} \ + \ \text{HO}-\text{CH} \\[4pt]
\text{CH}_3(\text{CH}_2)_{16}-\overset{\overset{\displaystyle O}{\|}}{\text{C}}-\text{OCH}_2
\end{array}
$$

Glyceryl tristearate   Sodium stearate   Glycerol
                                (a soap)

The potassium salts of fatty acids are softer and more soluble in water than sodium soaps. For this reason, they are frequently used in liquid soaps or shaving creams.

Problem 13.1   Given the structure

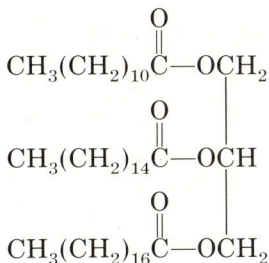

$$
\begin{array}{l}
\text{CH}_3(\text{CH}_2)_{10}\overset{\overset{\displaystyle O}{\|}}{\text{C}}-\text{OCH}_2 \\[4pt]
\text{CH}_3(\text{CH}_2)_{14}\overset{\overset{\displaystyle O}{\|}}{\text{C}}-\text{OCH} \\[4pt]
\text{CH}_3(\text{CH}_2)_{16}\overset{\overset{\displaystyle O}{\|}}{\text{C}}-\text{OCH}_2
\end{array}
$$

(a) Is it a simple or a mixed triglyceride? Explain.
(b) Name the products of hydrolyzing it in the presence of acid.
(c) Name the products of saponification of this triglyceride.

13.7 How Do Soaps Clean?

Soap consists of two parts: a highly polar COO$\colon^-$ Na$^+$ head and a long, nonpolar hydrocarbon tail. The polar segment is soluble in water (hydrophilic) but insoluble in grease (lipophobic). The nonpolar segment is soluble in grease (lipophilic) but insoluble in water (hydrophobic).

$$CH_3CH_2CH_2CH_2CH_2CH_2CH_2CH_2CH_2CH_2CH_2CH_2CH_2CH_2CH_2CH_2CH_2\overset{\displaystyle O}{\overset{\|}{-C}}-O^- \text{ Na}^+$$

Hydrophobic or lipophilic hydrocarbon chain
(nonpolar part of soap)

Hydrophilic
or lipophobic
(polar part
of soap)

This duality in properties is responsible for the cleansing action of soap. The lipophilic portion dissolves water-insoluble grease spots. The hydrophilic portion dissolves in the water wash and keeps the entire combination in the aqueous solution. A film of suds forms on the surface of the water, lowering the surface tension and permitting the emulsification of the grease particles. Stirring in a washing machine or scrubbing causes the grease spots to break down into tiny droplets. A cluster of soap molecules surrounding a grease droplet is called a **micelle.** The organic part of soap is dissolved in the grease, and the inorganic

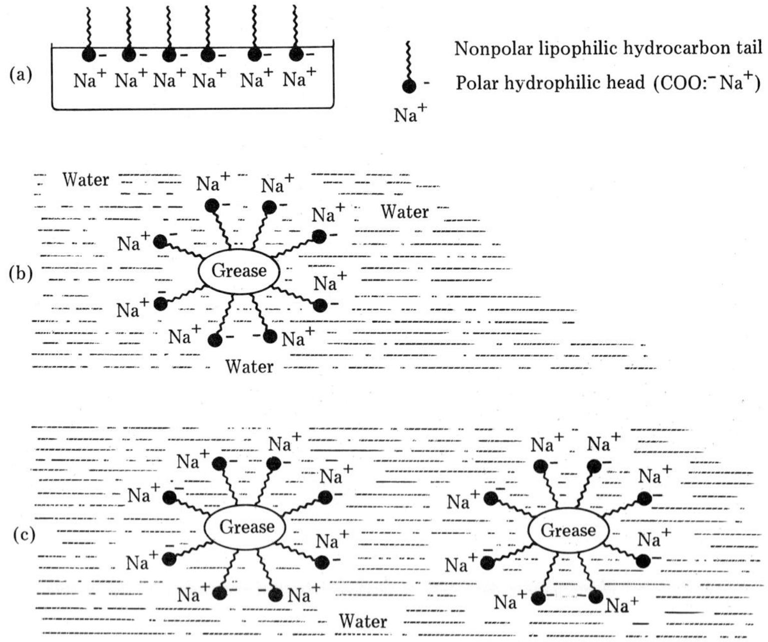

Figure 13.2 **(a)** Soap molecules in water showing orientation of the hydrophilic polar heads toward the surface of the water. **(b)** A soap micelle is a spherical cluster of soap molecules with the hydrophobic chains in the center, toward the grease spot, and the hydrophilic part projecting outward in water. **(c)** Soap micelles repel one another, thus keeping the grease droplets in a stable emulsion.

part projects outward in the water. The surfaces of the micelles, which are charged, repel one another and do not coalesce, thus forming an emulsion (Fig. 13.2). The cleansing action of soap is therefore due to its ability to lower the surface tension of water and to emulsify the grease particles.

As a cleansing agent, common household soap is not always effective. The calcium and magnesium ions present in hard water react with soap to form insoluble calcium and magnesium soaps that precipitate as an undesirable scum. As a result, laundry, dishes, or other items remain unclean.

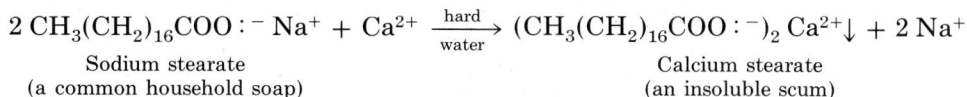

$$2\ CH_3(CH_2)_{16}COO:^-\ Na^+ + Ca^{2+} \xrightarrow[\text{water}]{\text{hard}} (CH_3(CH_2)_{16}COO:^-)_2\ Ca^{2+}\downarrow + 2\ Na^+$$

Sodium stearate
(a common household soap)

Calcium stearate
(an insoluble scum)

The scales found inside boilers and water kettles and the dirty rings in bathtubs come from the precipitation of a scum. Addition of water softeners such as Zeolite or Calgon is necessary to prevent the formation of a scum.

Synthetic Detergents (Syndets) *13.8*

To overcome the problems associated with laundering in hard water, the chemical industry has developed a number of **synthetic detergents (syndets).** Structurally, syndets resemble soaps: they contain a long hydrophobic portion and a short hydrophilic portion. Like soaps, synthetic detergents can stabilize water–grease emulsions. The main advantage of syndets over soaps is that their calcium or magnesium salts do not precipitate. Therefore, syndets can be used even in hard water. Synthetic detergents are classified as anionic, cationic, and nonionic, depending on their structures.

Anionic detergents resemble soap in that their water-soluble head is negatively charged. Typical anionic detergents are the linear alkyl sulfates (LAS) and alkyl benzene sulfonates (ABS).

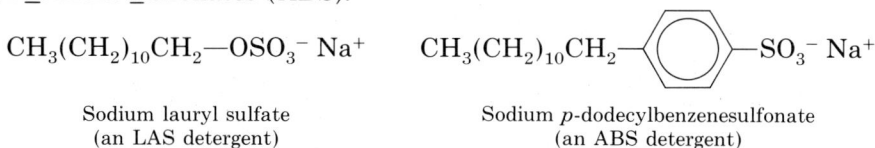

$$CH_3(CH_2)_{10}CH_2\!-\!OSO_3^-\ Na^+ \qquad CH_3(CH_2)_{10}CH_2\!-\!\!\!\bigcirc\!\!\!-\!SO_3^-\ Na^+$$

Sodium lauryl sulfate
(an LAS detergent)

Sodium p-dodecylbenzenesulfonate
(an ABS detergent)

Until recently, LAS and ABS detergents were fortified with enzymes, phosphate, and sulfate "builders." These "builders" were eventually banned because they polluted streams and rivers. LAS detergents are **biodegradable;** that is, they are broken down by microorganisms and will not pollute our waters. Because of this, they are preferable to ABS detergents, which resist decomposition.

Cationic detergents have a positively charged water-soluble portion, so they are also known as **invert soaps.**

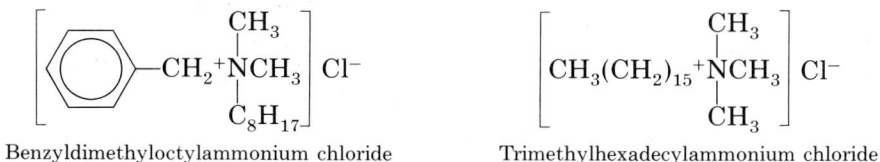

$$\left[\bigcirc\!\!-\!CH_2\overset{+}{N}\!\!\begin{array}{c}CH_3\\|\\CH_3\\|\\C_8H_{17}\end{array}\right]Cl^- \qquad\qquad \left[CH_3(CH_2)_{15}\overset{+}{N}\!\!\begin{array}{c}CH_3\\|\\CH_3\\|\\CH_3\end{array}\right]Cl^-$$

Benzyldimethyloctylammonium chloride

Trimethylhexadecylammonium chloride

Invert soaps are good cleansers and many possess germicidal properties. For this reason, people with skin problems frequently use them.

Nonionic detergents are neutral, and their hydrophilic portion functions by a hydrogen-bonding mechanism. They are used primarily in dishwashing liquids.

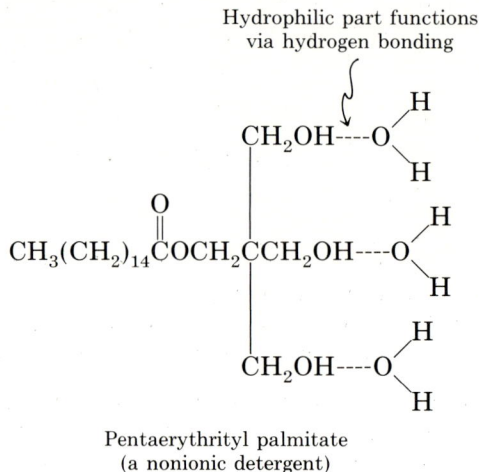

Hydrophilic part functions
via hydrogen bonding

$$CH_2OH\text{----}O\begin{matrix}H\\H\end{matrix}$$

$$CH_3(CH_2)_{14}\overset{\overset{\displaystyle O}{\|}}{C}OCH_2\overset{\displaystyle |}{C}CH_2OH\text{----}O\begin{matrix}H\\H\end{matrix}$$

$$CH_2OH\text{----}O\begin{matrix}H\\H\end{matrix}$$

Pentaerythrityl palmitate
(a nonionic detergent)

Problem 13.2 Define or illustrate **(a)** soap; **(b)** detergent; **(c)** micelle; **(d)** biodegradable.

13.9 Hydrogenation of Triglycerides

Hydrogenation of triglycerides is an important industrial reaction. The product of hydrogenation depends on the conditions under which the reaction is carried out. When unsaturated triglycerides are treated with hydrogen in the presence of metal catalysts (usually nickel) at low pressures, the product is, as expected, a saturated triglyceride. The reaction involves simply the addition of hydrogen to an alkene.

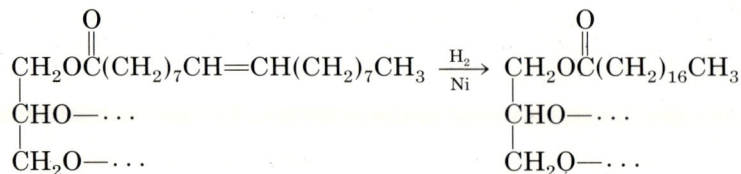

$$\begin{matrix} CH_2O\overset{\overset{\displaystyle O}{\|}}{C}(CH_2)_7CH{=}CH(CH_2)_7CH_3 \\ CHO\text{---}\cdots \\ CH_2O\text{---}\cdots \end{matrix} \xrightarrow[\text{Ni}]{H_2} \begin{matrix} CH_2O\overset{\overset{\displaystyle O}{\|}}{C}(CH_2)_{16}CH_3 \\ CHO\text{---}\cdots \\ CH_2O\text{---}\cdots \end{matrix}$$

Hardening of vegetable oils is brought about by the controlled hydrogenation of unsaturated triglycerides. Crisco, Spry, and Snodrift are examples of partially hydrogenated vegetable oils, as are margarine and butter substitutes.

Hydrogenolysis of Triglycerides 13.10

When hydrogenation is carried out at high pressure and temperature and in the presence of copper chromite catalyst, $CuCr_2O_4$, triglycerides are cleaved. The process is called **hydrogenolysis.** The products of hydrogenolysis are glycerol and long-chain alcohols. A typical example of this reaction is the reduction of coconut oil, which yields a large amount of lauryl alcohol.

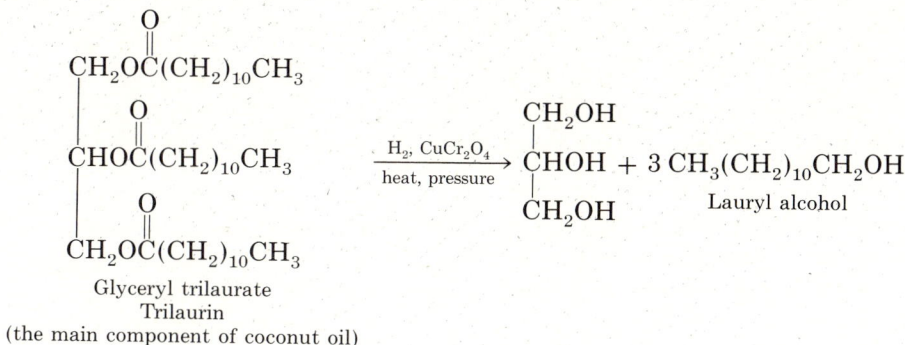

$$
\begin{array}{l}
\quad\quad\quad\overset{\displaystyle O}{\overset{\|}{}} \\
CH_2OC(CH_2)_{10}CH_3 \\
\quad\quad\quad\overset{\displaystyle O}{\overset{\|}{}} \\
CHOC(CH_2)_{10}CH_3 \\
\quad\quad\quad\overset{\displaystyle O}{\overset{\|}{}} \\
CH_2OC(CH_2)_{10}CH_3
\end{array}
\xrightarrow[\text{heat, pressure}]{H_2,\ CuCr_2O_4}
\begin{array}{l}
CH_2OH \\
CHOH \\
CH_2OH
\end{array}
+ 3\ CH_3(CH_2)_{10}CH_2OH
$$

Glyceryl trilaurate
Trilaurin
(the main component of coconut oil)

Lauryl alcohol

The lauryl alcohol can be esterified with sulfuric acid to produce lauryl hydrogen sulfate, whose sodium salt is a detergent.

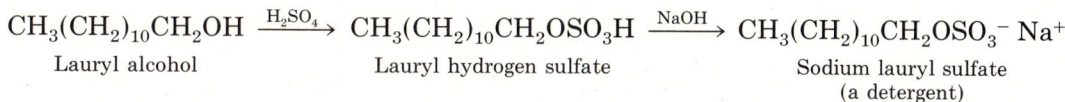

$$
CH_3(CH_2)_{10}CH_2OH \xrightarrow{H_2SO_4} CH_3(CH_2)_{10}CH_2OSO_3H \xrightarrow{NaOH} CH_3(CH_2)_{10}CH_2OSO_3^-\ Na^+
$$

Lauryl alcohol   Lauryl hydrogen sulfate   Sodium lauryl sulfate (a detergent)

Oxidation of Triglycerides 13.11

The presence of one or more double bonds in the fatty acid residues of triglycerides makes them susceptible to oxidation. Atmospheric oxygen slowly oxidizes unsaturated triglycerides into various products. The oxidation process, which is somewhat complicated, results in the eventual breaking of the carbon–carbon double bond. The products of oxidation contain shorter-chain keto acids and hydroxyacids. These oxidative products possess an unpleasant odor and flavor. *Rancid* oils are those in which this oxidation process has occurred.

When polyunsaturated oils are exposed to atmospheric oxygen, the oxidation products polymerize to form a tough, resinous film that is impermeable to water. For this reason, polyunsaturated oils such as linseed oil are called *drying oils*. They are used in protective coatings, such as oil-based paints and varnishes, in which the drying oils are mixed with pigments or coloring matter.

13.12 Analysis of Fats and Oils

One of the oldest qualitative methods for detecting fats and oils is the **acrolein test.** This test indicates the presence of glycerol, a component of all fats and oils. The test is conducted by heating a sample of the unknown substance in the presence of potassium bisulfate, $KHSO_4$. An acrid-smelling gas, acrolein, is released. The reaction for the acrolein test is

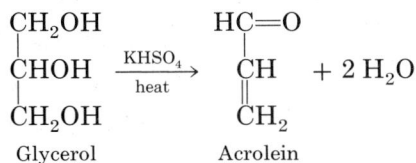

$$
\begin{array}{ccc}
CH_2OH & & HC\!=\!O \\
| & \xrightarrow[\text{heat}]{KHSO_4} & | \\
CHOH & & CH \quad + 2\,H_2O \\
| & & \| \\
CH_2OH & & CH_2 \\
\text{Glycerol} & & \text{Acrolein}
\end{array}
$$

Quantitative methods for identifying fats and oils are based on the *iodine number* and the *saponification number.*

A quantitative measure of the degree of unsaturation of a fat or an oil is expressed by its iodine value. The iodine value is based on the reaction between iodine chloride, ICl, or iodine bromide, IBr, with any carbon–carbon double bond in the fatty acid residue.

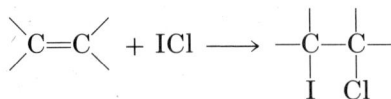

$$
\begin{array}{ccc}
\diagdown & & | \quad | \\
C\!=\!C \quad + ICl \longrightarrow & & -C-C- \\
\diagup \quad \diagdown & & | \quad | \\
& & I \quad Cl
\end{array}
$$

The reaction is quantitative: every $C\!=\!C$ bond adds one molecule of ICl. In practice, the reaction is conducted by using an excess of ICl. The leftover ICl is then measured to determine the amount of ICl that reacted with the fat or oil. The iodine value is generally given in terms of the **iodine number,** which is the number of grams of iodine, I_2, that would be absorbed by 100 grams of fat or oil.*

Saturated fatty acids, because they have no $C\!=\!C$ bonds, do not add ICl, and their iodine number is zero. Oleic acid, which has one $C\!=\!C$, has an iodine number of 90, whereas linoleic acid, which has two double bonds, has an iodine number of 181. The higher the iodine number, the higher the degree of unsaturation of a fat or oil. Natural fats that have a preponderance of saturated fatty acid residues have low iodine number values (10–70). Vegetable oils, especially the so-called *polyunsaturated* vegetable oils, have significantly higher iodine numbers (see Table 13.3).

A second quantitative method for characterizing a fat or oil is based on its saponification number. The **saponification number** is defined as the number of milligrams of KOH needed to saponify one gram of fat or oil. The saponification number of a fat or an oil varies inversely with its average molecular weight. Fats and oils that contain a preponderance of high-molecular-weight fatty acid residues (C_{18}) have lower saponification number values than do those that contain a larger proportion of low-molecular-weight fatty acid residues (C_{14} or lower). The saponification numbers for some fats and oils are shown in Table 13.3.

---

* Iodine, I_2, does not add to $C\!=\!C$ bonds. This is why the more reactive ICl is used. In the calculations, however, we figure the amount of I_2 that would have reacted by the same pathway (1 molecule of I_2 per $C\!=\!C$).

**Table 13.3 Iodine and Saponification Numbers of
Some Fats and Oils**

Fat or oil	Iodine number	Saponification number
Lard	46–70	195–203
Butter	26–28	210–230
Beef tallow	30–48	190–200
Olive oil	79–90	187–196
Soybean oil	127–138	189–195
Cottonseed oil	105–114	190–198
Safflower oil	140–156	188–194
Linseed oil[a]	170–185	187–195

[a] Not edible.

Problem 13.3 A triglyceride has a molecular weight of 890 and contains four double bonds. Calculate **(a)** the saponification number and **(b)** the iodine number.

For many years the determination of the fatty acid composition of fats and oils was a tedious, time-consuming task. With the advent of **gas-liquid chromatography (glc),** the analysis has become a matter of routine. The procedure is as follows. A fat is hydrolyzed, yielding a mixture of fatty acids and glycerol. The nonvolatile fatty acids are converted to volatile esters, usually the methyl esters. The latter are subjected to glc analysis. Each component as it is separated in the apparatus is recorded as a peak on a chart. The series of peaks recorded on the chart constitutes the *chromatogram.* Identification of each component is made by noting the position of the peak in the chromatogram, and the relative amount of each component is determined by measuring the area under each peak. A chromatogram of the methyl esters of fatty acids separated by glc is shown in Figure 13.3.

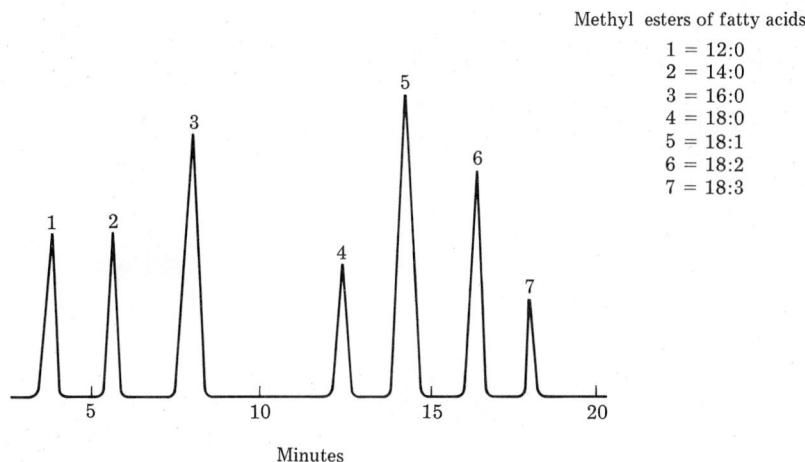

Methyl esters of fatty acids

1 = 12:0
2 = 14:0
3 = 16:0
4 = 18:0
5 = 18:1
6 = 18:2
7 = 18:3

Figure 13.3 A chromatogram of the methyl esters of fatty acids.

13.13 Waxes

Waxes are esters with the general formula $R\overset{\displaystyle O}{\overset{\|}{C}}$—$OR'$, where both R and R' are long hydrocarbon chains. Waxes are widespread in nature and occur usually as mixtures. Because of their long hydrocarbon chains, waxes are insoluble in water. They serve primarily as protective coatings on the surfaces of animals and plants. The waxy coating on certain fruits and leaves, for example, acts as a barrier to minimize loss of water. This coating on plants also serves to protect them from abrasive damage and possibly from infection. Insects also secrete waxes; the main constituent of beeswax obtained from the honeycomb of bees is

myricyl palmitate, $CH_3(CH_2)_{14}\overset{\displaystyle O}{\overset{\|}{C}}O(CH_2)_{29}CH_3$. In addition, waxes play an important role in waterproofing the surfaces of birds and animals, such as sheep. The damage done to birds as a result of accidental oil spills is dramatically illustrated in Figure 13.4. The oil, together with the detergents used to disperse the oil, dissolves the waxy layers that cover the feathers, causing the birds to lose their buoyancy.

The waxes just discussed should not be confused with household paraffin waxes. Chemically, these paraffin waxes are a mixture of straight-chain alkanes that have from 26 to 30 carbon atoms.

Figure 13.4 Pictures of oil-smeared birds, such as this pathetic duck, have focused national attention on one of the consequences of offshore oil seepage. [United Press International Photo.]

Hydrolysis of simple lipids yields an alcohol and fatty acids. **Compound lipids** are so named because upon complete hydrolysis they yield "other substances" in addition to an alcohol and fatty acids. Compound lipids may be subdivided into two main categories, the phospholipids and the glycolipids, which are distinguished by the nature of some of these "other substances."

A   Phospholipids

Phospholipids are lipids that contain a phosphate group. They are found in the outer membranes of most cells and are especially prevalent in the brain and nerve tissues. Phospholipids are classified into one of two groups, depending on the alcohol that is esterified with phosphoric acid.

Phosphoglycerides.   The most common type of phospholipids are the phosphoglycerides. **Phosphoglycerides** are derivatives of phosphatidic acid.

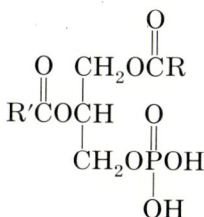

$$
\begin{array}{l}
\quad\quad\quad\quad O \\
\quad\quad\quad\quad \| \\
O\quad CH_2OCR \\
\| \quad\; | \\
R'COCH\quad O \\
\quad\quad | \quad\;\; \| \\
\quad\quad CH_2OPOH \\
\quad\quad\quad\;\; | \\
\quad\quad\quad\;\; OH
\end{array}
$$

Phosphatidic acid

Most phosphoglycerides found in living organisms are *esters* of phosphatidic acid. These esters are named as phosphatidyl derivatives or by common names. The most common types of phosphatidyl derivatives are those of choline, which are known as **lecithins,** and of ethanolamine and serine, which are called **cephalins** (see Fig. 13.5). A lecithin or cephalin molecule has polar and nonpolar sections, the same structural features that are present in soaps and detergents. Lecithins and cephalins, like soaps and detergents, are good emulsifying agents.

Sphingolipids.   The other main group of phospholipids is the sphingolipids. **Sphingolipids** contain a long-chain unsaturated alcohol, *sphingosine,* but no glycerol. The most abundant sphingolipid is *sphingomyelin.*

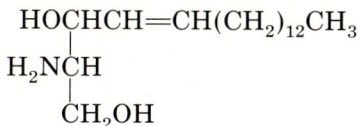

$$
\begin{array}{l}
HOCHCH{=}CH(CH_2)_{12}CH_3 \\
\; | \\
H_2NCH \\
\quad | \\
\quad CH_2OH
\end{array}
\qquad
\begin{array}{l}
\quad\quad O\quad HOCHCH{=}CH(CH_2)_{12}CH_3 \\
\quad\quad \| \quad\quad\;\; | \\
RC{-}NHCH\quad O \\
\quad\quad\quad\;\; | \quad\;\; \| \\
\quad\quad\quad\;\; CH_2OPOCH_2CH_2\overset{+}{N}(CH_3)_3 \\
\quad\quad\quad\quad\quad\; | \\
\quad\quad\quad\quad\quad\; O^-
\end{array}
$$

Sphingosine   Sphingomyelin

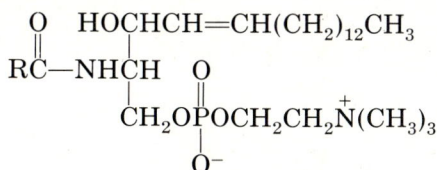

Sphingomyelin is similar in general size, shape, and polar characteristics to the lecithins because the hydrocarbon chain of sphingosine is equivalent to one of the fatty acids of lecithins. Sphingomyelin is a component of the sheath-like structure that surrounds the nerve fibers. This sheath-like structure, like the insulation around an electrical wire, prevents short circuits in the transmission of impulses by the nerves.

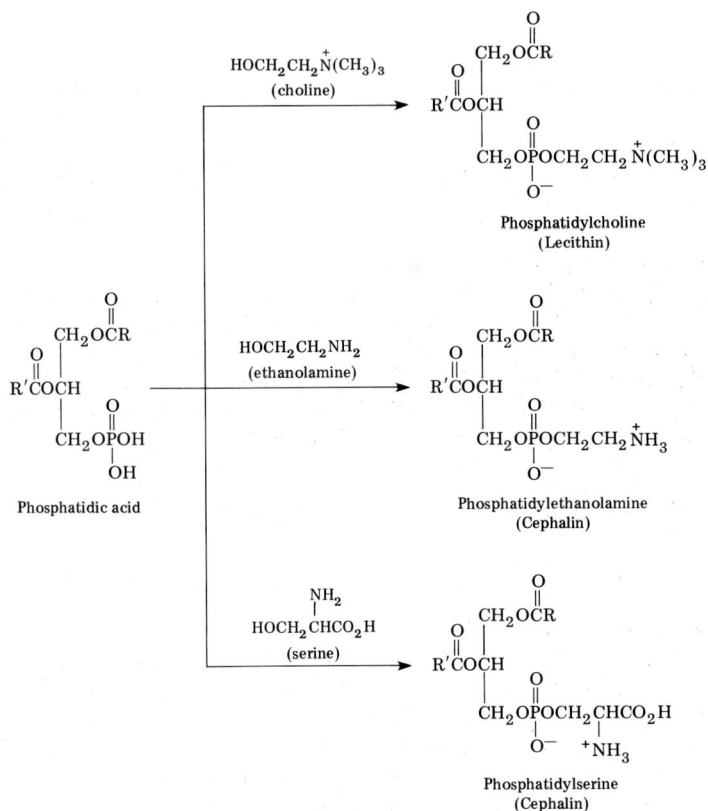

Figure 13.5 Structures of lecithin, the phosphatidyl ester of choline, and of cephalins, the phosphatidyl esters of ethanolamine and serine.

B Glycolipids

Glycolipids contain a sugar molecule, a fatty acid, and sphingosine. No phosphorus is present. The sugar molecule is usually galactose. Because glycolipids occur in abundance in the white matter of the brain, they are frequently referred to as *cerebrosides*.

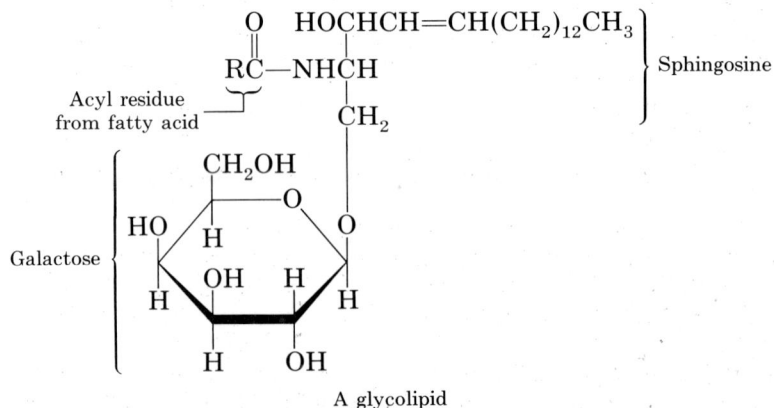

A glycolipid

Derived Lipids *13.15*

The derived lipids are truly a hodgepodge group of compounds in both structure and function. Their only common physical characteristic is that they are insoluble in water. We will briefly consider a few examples of this group of compounds.

A Steroids

Biochemically, the most important class of derived lipids is a group called the **steroids.** Although steroids exist in a great variety of structural forms, all of them show the same general 17-carbon structure of three six-membered rings fused to a five-membered ring. This arrangement is called the steroid nucleus and has the impressive name perhydrocyclopentanophenanthrene (Fig. 13.6b). The four rings are identified by the letters A through D, and the 17 carbon atoms are numbered as shown.

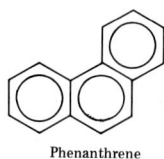

Phenanthrene

(a)

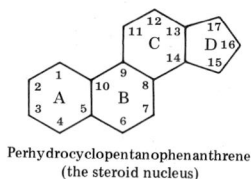

Perhydrocyclopentanophenanthrene
(the steroid nucleus)

(b)

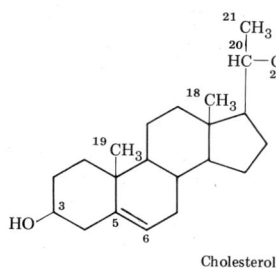

Cholesterol

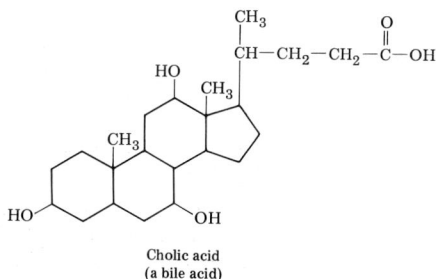

Cholic acid
(a bile acid)

The bile salts are good emulsifying agents, and are important in the digestion and assimilation of fats in the intestines.

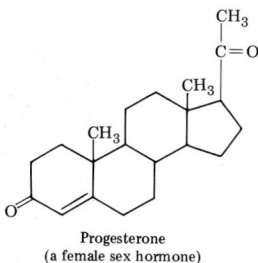

Progesterone
(a female sex hormone)

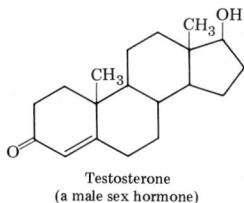

Testosterone
(a male sex hormone)

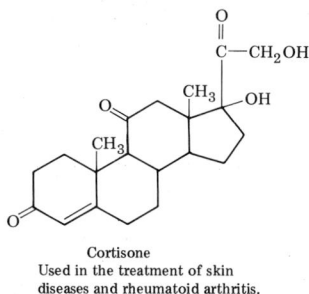

Cortisone
Used in the treatment of skin diseases and rheumatoid arthritis.

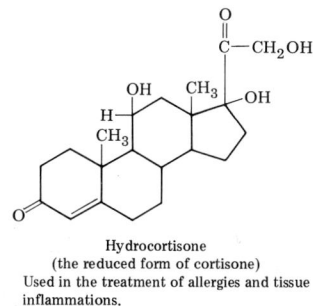

Hydrocortisone
(the reduced form of cortisone)
Used in the treatment of allergies and tissue inflammations.

Figure 13.6 Phenanthrene **(a)** is a common aromatic hydrocarbon that contains three six-membered rings. Perhydrocyclopentanophenanthrene **(b)** is the parent compound from which all steroids are derived. Below these are shown the formulas of six steroids.

365

Cholesterol (Fig. 13.6) is by far the most widely distributed and abundant steroid in animal and human tissues. In humans, cholesterol forms about 0.2% of the total body weight. It is concentrated in the brain and in the spinal chord, but small amounts are present in all cells. In addition to the steroid nucleus, cholesterol has an alcohol function at C-3, a double bond between C-5 and C-6, an eight-carbon alkyl chain at C-17, and two methyl groups. One methyl group is attached to C-10, at the junction of the A and B rings, and is designated as the C-19 angular methyl. The second methyl group is attached to C-13, at the junction of the C and D rings; it is designated as the C-18 angular methyl. Many steroids contain these angular methyl groups as well as a substituent at C-17.

The blood serum of normal individuals contains about 200 mg of total cholesterol (that is, free cholesterol and esters of cholesterol) per 100 ml. Cholesterol is also found in the bile liquid and is an elimination product. If the dietary intake of cholesterol is too high, it may precipitate out from solution. Precipitation from the blood serum leads to the constriction of blood vessels, which in turn reduces the blood flow, resulting in high blood pressure. Precipitation of cholesterol in the bile liquid forms gallstones. Gallstones are very painful and may block the normal bile liquid flow, causing a jaundice condition.

Apart from cholesterol, which is abundant, a number of often-important steroids occur in only small amounts. Among the important naturally occurring steroids are the bile acids, the male and female sex hormones, and the adrenocortical hormones. Some typical steroids are shown in Figure 13.6.

B   Fat-Soluble Vitamins

Another group of important derived lipids is the fat-soluble vitamins. These include vitamins D, A, E, and K, which must be supplied in the diet.

There are several D vitamins. All are colorless, odorless, crystalline alcohols with steroid-like structures. When present in the diet, the D vitamins prevent rickets. Vitamin D_2, or calciferol, is produced when ergosterol is irradiated with ultraviolet light. Ergosterol is a sterol of plant origin.

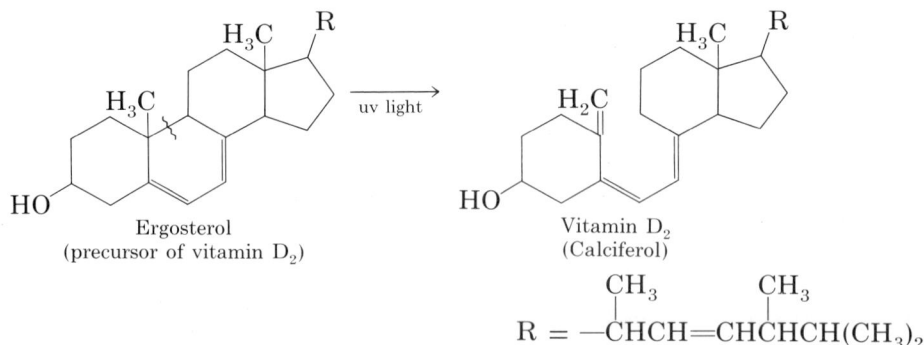

Ergosterol
(precursor of vitamin D_2)

uv light

Vitamin D_2
(Calciferol)

$$R = -\underset{\underset{CH_3}{|}}{C}H CH = CH \underset{\underset{CH_3}{|}}{C}H CH(CH_3)_2$$

Vitamin D_3 is obtained upon irradiation of 7-dehydrocholesterol, a steroid of animal origin. In both cases the vitamins are formed by rupture of the B ring of the steroid precursors.

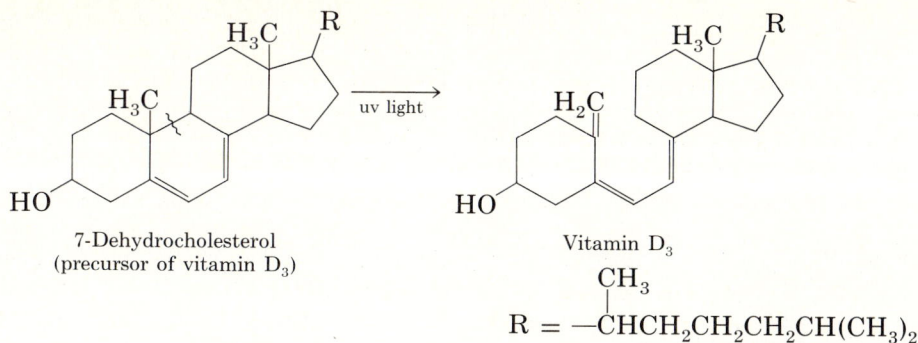

7-Dehydrocholesterol
(precursor of vitamin D$_3$)

uv light →

Vitamin D$_3$

$$R = -\overset{\underset{\displaystyle CH_3}{|}}{C}HCH_2CH_2CH_2CH(CH_3)_2$$

Vitamins A, E, and K (Fig. 13.7) are complex molecules made up in part of a simple repeating unit, the **isoprene unit,** a sequence of five carbon atoms that resembles isoprene, 2-methyl-1,3-butadiene. Although isoprene itself does not occur in nature, its biologically active counterpart, isopentenyl pyrophosphate, does. Isopentenyl pyrophosphate is formed from mevalonic acid by a series of enzyme-catalyzed reactions.

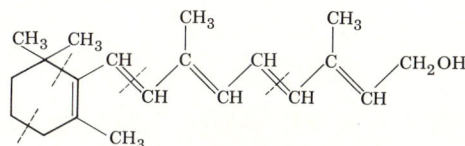

Vitamin A$_1$
(Retinol)
Deficiency causes poor growth, weight loss, xerophthalmia (eye disease), night blindness.

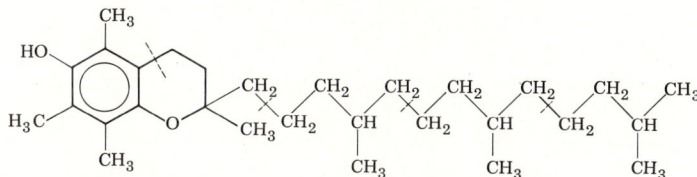

Vitamin E
(α-Tocopherol)
Deficiency causes sterility, muscular dystrophy, and liver diseases in experimental animals; effects on human beings are not clear.

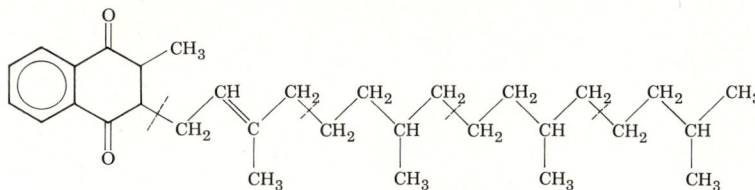

Vitamin K$_1$
Deficiency causes hemorrhages (because the blood lacks a coagulating factor).

Figure 13.7 Structures and deficiency effects of vitamins A, E, and K. Dashed lines show the positions of the five-carbon isoprene units.

Isoprene

Isoprene unit

Mevalonic acid

Isopentenyl pyrophosphate

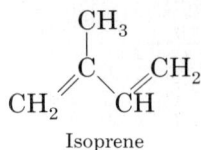

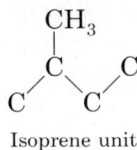

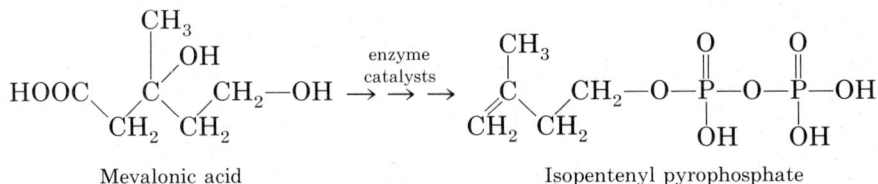

C Prostaglandins

Chemically, the **prostaglandins** are a family of 20-carbon unsaturated fatty acids. They are characterized by a five-membered ring and two side chains: a seven-carbon chain that ends with a carboxyl group and an eight-carbon chain with a hydroxyl attached to C-15.

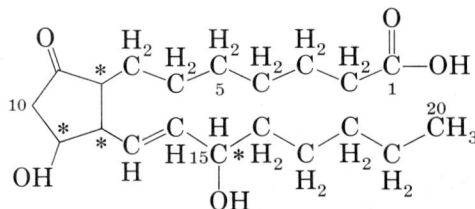

Prostaglandin

Originally thought to be produced by the prostate gland, from which they derive their name, prostaglandins have been found in minute quantities in nearly all mammalian tissues and fluids examined. Recently, prostaglandins have assumed increasing biological significance. Some have called prostaglandins the panacea drug. They act as regulators of a wide variety of bodily functions. For example, they have the ability to contract or relax smooth muscles, reduce inflammation of tissues, relieve arthritic pain, induce labor, control fertility, alleviate asthmatic attacks—a truly remarkable series of effects. Although the mechanism of their action is not well understood, the stereochemistry of the molecule is clearly of primary importance in determining the biological function of these compounds. The prostaglandin shown here has four chiral centers (indicated by asterisks). There are therefore 16 (4^2) possible stereoisomers.

Summary of Concepts and Reactions

A lipid is any substance that is (1) insoluble in water but soluble in nonpolar organic solvents and (2) a constituent of the cell.   [Sec. 13.1]

Lipids are classified into three broad categories: (1) simple lipids, (2) compound lipids, and (3) derived lipids.   [Sec. 13.1]

Simple lipids are esters, and they are subdivided into two groups: (1) triglycerides and (2) waxes. [Sec. 13.2]

Naturally occurring fatty acids have these structural characteristics. (1) Most are linear, long-chain monocarboxylic acids that range from C_{12} to C_{26}, with the C_{16} and C_{18} acids being the most abundant. (2) Almost all of them have an even number of carbon atoms. (3) They may be saturated or unsaturated. (4) Almost all of the unsaturated fatty acids are of the *cis* configuration. [Sec. 13.3]

The most abundant lipids are the triglycerides. [Sec. 13.4]

Triglycerides are esters of three molecules of fatty acid and glycerol. [Sec. 13.4]

When the fatty acid residues are the same, a triglyceride is called a simple triglyceride; when they are not the same, it is called a mixed triglyceride. [Sec. 13.4]

Fats are complex mixtures of simple and mixed triglycerides that exist in the solid form at room temperature. [Sec. 13.4]

Oils are complex mixtures of simple and mixed triglycerides that exist in the liquid form at room temperature. [Sec. 13.4]

Hydrolysis of a fat or an oil in the presence of an acid catalyst or by the action of enzymes (lipases) produces fatty acids and glycerol. [Sec. 13.5]

Hydrolysis of a fat or oil under alkaline conditions is called saponification. [Sec. 13.6]

Saponification produces glycerol and the salts of the fatty acids, or soaps. [Sec. 13.6]

Soaps consist of two parts: a highly polar head (hydrophilic) and a long, nonpolar hydrocarbon tail (lipophilic). [Sec. 13.7]

Synthetic detergents, or syndets, structurally resemble soaps, but unlike soaps their calcium or magnesium salts do not precipitate. [Sec. 13.8]

Synthetic detergents are classified into three types: (1) anionic, (2) cationic, and (3) nonionic. [Sec. 13.8]

Hardening of vegetable oils is brought about by the controlled hydrogenation of unsaturated triglycerides. [Sec. 13.9]

When hydrogenation is carried out at high pressure and temperature in the presence of a $CuCr_2O_4$ catalyst, triglycerides are cleaved. This process is called hydrogenolysis. [Sec. 13.10]

The presence of one or more double bonds in the fatty acid residues of the triglycerides makes them susceptible to oxidation. [Sec. 13.11]

The acrolein test is a simple chemical test for detecting fats and oils. A quantitative measure of the degree of unsaturation of a fat or oil is based on its iodine number. Another quantitative method for characterizing a fat or oil is based on its saponification number. The fatty acid composition of fats and oils can be determined by gas–liquid chromatography. [Sec. 13.12]

Waxes are esters with the general formula RCOOR′, where both R and R′ are long hydrocarbon chains. [Sec. 13.13]

Compound lipids may be subdivided into phospholipids and glycolipids. [Secs. 13.14A, B]

Steroids, fat-soluble vitamins, and prostaglandins are examples of derived lipids. [Sec. 13.15A–C]

Key Terms

fatty acids	fats	micelles
triglycerides	oils	synthetic detergents
glycerol	lipases	(syndets)
simple triglyceride	saponification	anionic detergent
mixed triglyceride	soaps	biodegradable

cationic detergent
invert soap
nonionic detergent
hydrogenolysis
acrolein test
iodine number
saponification number

gas-liquid chromatography (glc)
waxes
compound lipids
phospholipids
phosphoglycerides
lecithins
cephalins

sphingolipids
glycolipids
steroids
cholesterol
fat-soluble vitamins
isoprene unit
prostaglandins

Exercises

Nomenclature of Fatty Acids [Sec. 13.3]

13.1   Name the fatty acids described by these notations.

    (a) 12:0   **(b)** 16:0   **(c)** 18:1   **(d)** 18:2

Physical Properties and Structures of Fatty Acids and Triglycerides
[Secs. 13.2–13.4]

13.2   Explain why fatty acids are insoluble in water.

13.3   Define or illustrate the following.

    (a) Triglyceride   **(b)** Simple triglyceride   **(c)** Mixed triglyceride

    (d) Fat   **(e)** Oil   **(f)** Glycerol

13.4   Arrange the following triglycerides in order of increasing melting point: tristearin, triolein, trilinolein.

Reactions of Fats and Oils [Secs. 13.5, 13.6, 13.9–13.12]

13.5   Deduce the structure of an unknown fat from these data.

    (1) The fat is optically active

    (2) Hydrolysis under acidic conditions yields glycerol, two equivalents of lauric acid, and one equivalent of stearic acid.

13.6   Write an equation for the conversion of trilinolein to a solid fat.

13.7   What are the products of hydrogenolysis of tristearin?

13.8   Explain what is meant by **(a)** a rancid oil and **(b)** a drying oil.

13.9   **(a)** Write the structure of a mixed triglyceride formed of glycerol and palmitic, oleic, and linolenic acids.

    (b) Write an equation for the saponification of this triglyceride.

    (c) Write an equation for its digestion.

    (d) Write an equation for a reaction that would reduce its iodine number to zero.

    (e) Calculate its iodine number.

    (f) Is it likely to be a liquid or a solid at room temperature?

Soaps and Detergents [Secs. 13.6–13.8]

13.10 Identify each compound as a soap, a detergent, or neither.

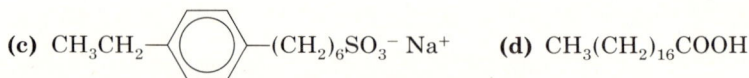

    (a) $CH_3(CH_2)_{12}COO^- Na^+$ **(b)** $CH_3(CH_2)_4 \overset{\displaystyle |}{\underset{\displaystyle CH_3}{CH}}(CH_2)_6 OSO_3^- K^+$

    (c) CH_3CH_2—⟨◯⟩—$(CH_2)_6 SO_3^- Na^+$ **(d)** $CH_3(CH_2)_{16}COOH$

13.11 Define **(a)** anionic detergent and **(b)** cationic detergent.

13.12 What type of detergent most resembles a soap?

13.13 Explain the significance of **(a)** the acrolein test, **(b)** iodine number, and **(c)** saponification number.

13.14 Is a triglyceride with an iodine number of 135 more likely to be a solid or a liquid at room temperature?

13.15 Calculate **(a)** the iodine number and **(b)** the saponification number of triolein (mol wt = 884).

Waxes [Sec. 13.13]

13.16 Starting with triplamitin as the only organic substance, show by means of equations how you would prepare a wax.

Compound Lipids and Derived Lipids [Secs. 13.14, 13.15]

13.17 List the products of complete hydrolysis of **(a)** phosphoglycerides, **(b)** sphingolipids, and **(c)** glycolipids.

13.18 Would you expect a lecithin to be more or less soluble in water than a fat? Explain your answer.

13.19 Draw the structures of **(a)** a lecithin and **(b)** a cephalin that contain palmitic and oleic acids.

13.20 Draw the structure of androsterone, a steroid that has the following substituents in the steroid nucleus: OH at position 3, =O at position 17, and methyl groups at positions 18 and 19.

13.21 Draw the structure of cholesterol and indicate the positions of the chiral centers with asterisks. How many possible stereoisomers of cholesterol are there?

13.22 The structure of β-carotene, the compound largely responsible for the color of carrots, is

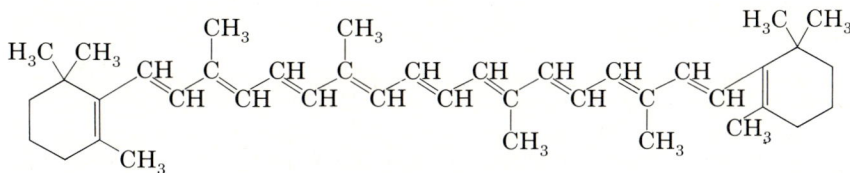

β-Carotene

(a) How many isoprene units are there in β-carotene?

(b) β-Carotene is a precursor of vitamin A (Fig. 13.7). Show by equation how vitamin A can be derived from β-carotene.

13.23 **(a)** What are prostaglandins? How many carbons do they contain?

(b) What fatty acid listed in Table 13.1 is the most likely precursor of prostaglandins?

14

Amines and Other Nitrogen Compounds

Amines and their derivatives are important nitrogen compounds. They are used as intermediates in organic reactions, and they are found in many biological systems. The amino acids that make up the proteins, the purine and pyrimidine bases that make up DNA and RNA, the molecules of the genetic code, the mood-controlling (psychomimetic) drugs used in the treatment of mental illness, and the alkaloid drugs used in chemotherapy and psychopharmacology are all amine derivatives. Some are discussed in this chapter; others are discussed in subsequent chapters.

14.1 Structure and Classification of Amines

Amines are compounds that may be thought of as having been derived from ammonia by replacement of one, two, or three hydrogens by alkyl or aryl groups. Aliphatic amines contain *only alkyl* groups bonded directly to the nitrogen atom. Aromatic amines are those in which one or more aryl groups are bonded directly to nitrogen.

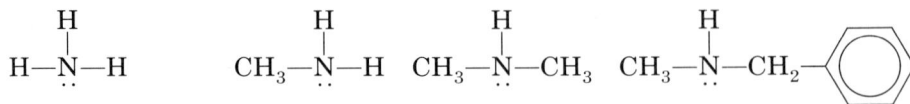

$$H-\overset{\overset{\displaystyle H}{|}}{\underset{\cdot\cdot}{N}}-H \qquad CH_3-\overset{\overset{\displaystyle H}{|}}{\underset{\cdot\cdot}{N}}-H \quad CH_3-\overset{\overset{\displaystyle H}{|}}{\underset{\cdot\cdot}{N}}-CH_3 \quad CH_3-\overset{\overset{\displaystyle H}{|}}{\underset{\cdot\cdot}{N}}-CH_2-\bigcirc$$

Ammonia   Aliphatic amines

Aromatic amines

Amines are classified as **primary (1°), secondary (2°),** or **tertiary (3°)** according to the number of R or Ar groups *attached to the nitrogen atom.*

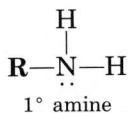

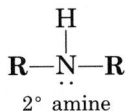

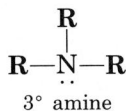

1° amine   2° amine   3° amine

When a fourth hydrocarbon group is attached to nitrogen, a **quaternary ammonium salt results.**

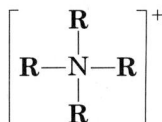

Quaternary ammonium salt

Note that amines are not classified in the same way as alcohols. Alcohols are classified according to the *kind of carbon atom to which the OH group is attached,* not according to the number of hydrocarbon groups attached to the oxygen atom. Thus, whereas *t*-butyl alcohol is a tertiary alcohol (because three carbons are attached to the carbinol carbon), *t*-butylamine is a primary amine (because only one carbon is attached directly to the nitrogen atom).

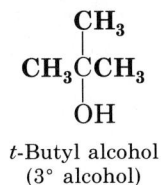

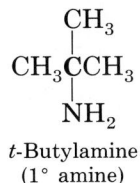

t-Butyl alcohol   *t*-Butylamine
(3° alcohol)   (1° amine)

Problem 14.1 Classify the following compounds as 1°, 2°, or 3°.

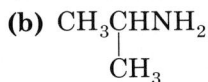

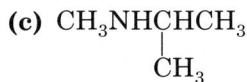

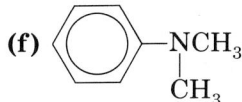

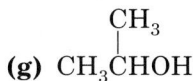

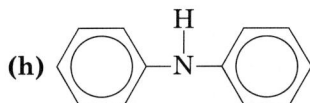

(a) CH_3CH_2OH (b) CH_3CHNH_2 (c) $CH_3NHCHCH_3$
                          | |
                         CH_3 CH_3

(d) NH   (e) —NH_2 (f) —NCH_3
                                                                              |
                                                                             CH_3

CH_3
|
(g) CH_3CHOH (h)

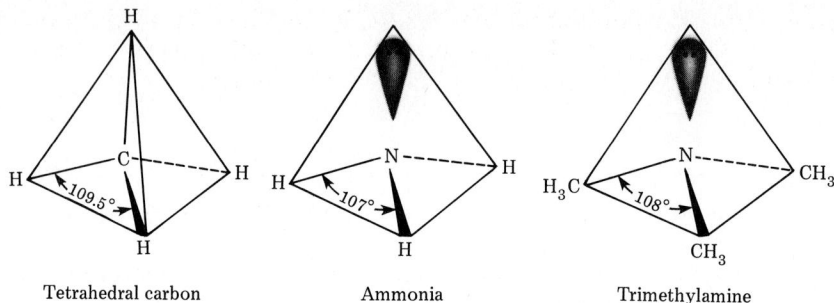

Figure 14.1 The tetrahedral geometry of methane, ammonia, and trimethylamine. In ammonia and in amines one corner of the tetrahedron is occupied by an electron pair.

Problem 14.2 Indicate which amines in Problem 14.1 are aliphatic and which are aromatic.

Problem 14.3 Draw structures and classify each as 1°, 2°, or 3°.
(a) The eight isomeric amines of formula $C_4H_{11}N$
(b) The five isomeric amines of formula C_7H_9N, each of which contains a benzene ring

The nitrogen of ammonia and amines is sp^3 hybridized and therefore tetrahedral (Fig. 14.1). Three corners of the tetrahedron are occupied by H or C. The fourth corner is occupied by the nonbonded electron pair of the nitrogen.

As we will soon learn, these nonbonded electrons are central to the understanding of the properties of amines.

14.2 Nomenclature of Amines

Simple aliphatic amines are named by listing, in alphabetical order, the alkyl groups attached to the nitrogen atom and adding the suffix -*amine,* all as one word. If two or three identical alkyl groups are attached to the nitrogen, the prefixe *di-* or *tri-* is added to the name of the amine. For example,

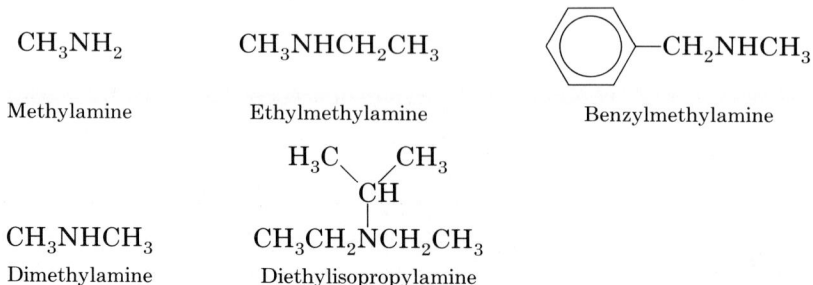

CH_3NH_2

Methylamine

$CH_3NHCH_2CH_3$

Ethylmethylamine

—CH_2NHCH_3

Benzylmethylamine

CH_3NHCH_3

Dimethylamine

H_3C CH_3 CH $CH_3CH_2NCH_2CH_3$

Diethylisopropylamine

If the amine is complicated, the IUPAC system is used. In this system the amino group ($—NH_2$) is considered the substituent, and its position on the chain is indicated by the lowest possible number. For example,

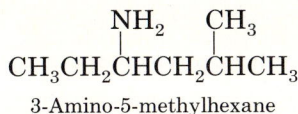

3-Amino-5-methylhexane

$H_2NCH_2CH_2CH_2CH_2CH_2CH_2NH_2$

1,6-Diaminohexane
(Hexamethylenediamine)

The amino group is also considered a substituent if it is part of a molecule that contains another functional group.

$H_2NCH_2CH_2OH$

2-Aminoethanol

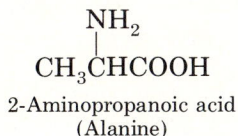

2-Aminopropanoic acid
(Alanine)

$H_2NCH_2CH_2CH_2OH$

3-Amino-1-propanol

Amine salts are named by replacing the suffix *-amine* by *ammonium,* followed by the name of the anion, which is written as a second word.

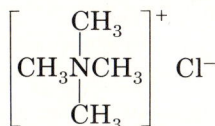

Tetramethylammonium chloride

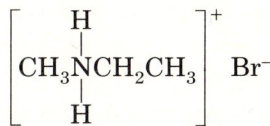

Ethylmethylammonium bromide

Aromatic amines are usually named as derivatives of **aniline.** The prefixes ortho (*o-*), meta (*m-*), and para (*p-*) are used to locate the position of a substituent. If the substituent is a methyl group, the compound is called *toluidine.* If hydrocarbon groups are attached on the nitrogen atom, the letter *N* is prefixed to the alkyl or aryl group name.

Aniline

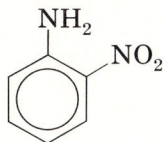

p-Nitroaniline

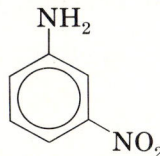

o-Nitroaniline

m-Nitroaniline

p-Hydroxyaniline
(p-Aminophenol)

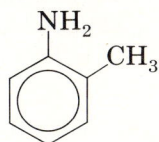

o-Toluidine

p-Toluidine

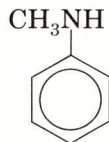

N-Methylaniline

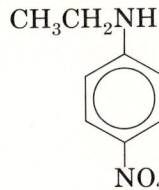

p-Nitro-N-ethylaniline

Problem 14.4 Draw the structure for each of the following compounds.
(a) Isopropylamine   **(b)** Cyclohexylamine

(c) Diphenylamine
(e) *p*-Chloroaniline

(d) Isopropylmethylammonium bromide
(f) *N*-Isopropyl-*N*-methylaniline

Problem 14.5 Give a correct name for each of the following compounds.

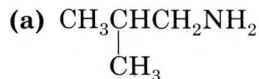

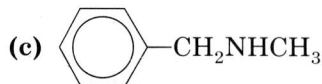

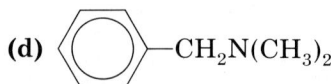

(a) CH$_3$CHCH$_2$NH$_2$
    |
    CH$_3$

(b) O$_2$N—⬡—NH$_2$

(c) ⬡—CH$_2$NHCH$_3$

(d) ⬡—CH$_2$N(CH$_3$)$_2$

(e) [(CH$_3$)$_2$N(C$_2$H$_5$)$_2$]$^+$ Cl$^-$

(f) ⬠—NH$_2$

14.3 Physical Properties of Amines

Low-molecular-weight aliphatic amines (methyl-, dimethyl-, and trimethyl-amines) are colorless gases that are soluble in water. Like ammonia, they form basic solutions. They have characteristically unpleasant odors that resemble the odors of ammonia and dead fish.

Amines containing four to eleven carbon atoms are liquids. They are also characterized by unpleasant odors. Putrescine and cadaverine, two diamines found in decaying flesh, are examples.

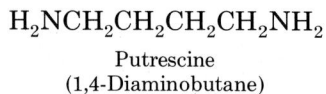

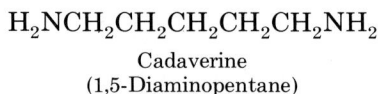

H$_2$NCH$_2$CH$_2$CH$_2$CH$_2$NH$_2$
Putrescine
(1,4-Diaminobutane)

H$_2$NCH$_2$CH$_2$CH$_2$CH$_2$CH$_2$NH$_2$
Cadaverine
(1,5-Diaminopentane)

Higher-molecular-weight amines are solids.

Because they possess a polar $^{\delta-}$N—H$^{\delta+}$ bond, primary and secondary amines are capable of intermolecular hydrogen bonding. Therefore their boiling points are higher than those of alkanes of comparable molecular weight but lower than those of alcohols of similar molecular weight (Table 14.1)

Problem 14.6 The boiling point of methyl alcohol (mol wt = 32) is 72°C higher than the boiling point of methylamine (mol wt = 31). Suggest a good explanation for this difference.

Tertiary amines are also polar compounds, but because hydrogen is not bonded to nitrogen, these amines are incapable of intermolecular hydrogen bonding. Consequently, they have lower boiling points than primary and second-

Table 14.1 Differences in Physical Properties Between Amines and Alkanes and Alcohols of Comparable Molecular Weight

Structure	Name	Mol wt	Bp (°C)	Solubility in H_2O (25°C)
CH_3CH_3	ethane	30	−89	insoluble
CH_3NH_2	methylamine	31	−7.5	very soluble
CH_3OH	methyl alcohol	32	64.5	very soluble
$CH_3CH_2CH_3$	propane	44	−42	insoluble
$CH_3CH_2NH_2$	ethylamine	45	17	very soluble
CH_3NHCH_3	dimethylamine	45	7.5	very soluble
CH_3CH_2OH	ethyl alcohol	46	78	very soluble
$CH_3(CH_2)_2CH_3$	n-butane	58	−0.5	insoluble
$CH_3(CH_2)_2NH_2$	n-propylamine	59	49	very soluble
$CH_3CH_2NHCH_3$	ethylmethylamine	59	35	very soluble
$(CH_3)_3N$	trimethylamine	59	3	very soluble
$CH_3CH_2CH_2OH$	n-propyl alcohol	60	97	very soluble
$CH_3CHOHCH_3$	isopropyl alcohol	60	82.5	very soluble

ary amines of identical molecular weights. However, their boiling points are higher than those of alkanes of similar molecular weight.

All amines are capable of forming hydrogen bonds with water, and amines with up to six carbons show appreciable solubility in water.

Basicity of Amines *14.4*

Amines are **bases** because the nitrogen atom has a nonbonded pair of electrons. This nonbonded electron pair can be donated to an acid's proton to form an ammonium salt.

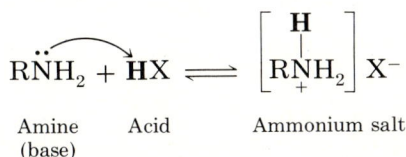

$$R\ddot{N}H_2 + HX \rightleftharpoons \left[R\overset{H}{\underset{+}{N}H_2} \right] X^-$$

Amine   Acid   Ammonium salt
(base)

The **base strength** depends on the degree of availability of the nonbonded electron pair; the more available the electron pair on N is to an acid, the stronger the base, and vice versa.

Quantitatively, the strength of a base is expressed by its **basicity constant, K_b.** A large value of K_b signifies a strong base; a small value of K_b indicates a weak base. Any base with a K_b lower than 10^{-2} is considered a weak base. Examples of weak bases are

$$\ddot{N}H_3 \quad (K_b = 1.8 \times 10^{-5}) \qquad CH_3\ddot{N}H_2 \quad (K_b = 4.4 \times 10^{-4})$$

Ammonia   Methylamine

although methylamine (larger K_b) is a stronger base than ammonia (lower K_b).

By analogy with acids (Sec. 12.3), a more convenient method of expressing base strength is through the **pK_b**. The pK_b is given by the equation

$$pK_b = -\log K_b \qquad (1)$$

From (1), it can be shown that the larger the pK_b, the weaker the base, and vice versa. Also, a one-unit difference in pK_b is equivalent to a 10^1 or tenfold difference in K_b, and a two-unit difference in pK_b is equivalent to a 10^2 or hundredfold difference in K_b, and so on.

Table 14.2 lists the K_b and pK_b values of some amines. From the table we see that the pK_bs of methylamine and dimethylamine are 3.36 and 3.29, respectively, making them stronger bases than ammonia ($pK_b = 4.74$). By examining the availability of their nonbonded electrons, we can explain their greater basicity. You will recall (Sec. 12.4) that methyl groups are electron donors. As such, they increase the electron density about the nitrogen atom to which they are attached, and therefore the nonbonded electron pairs are made more available for reaction with an acid.

$$CH_3-NH_2 \qquad\qquad CH_3-NH-CH_3$$

One electron-donating group   Two electron-donating groups

Examination of Table 14.2 reveals also that aliphatic amines are considerably more basic than aromatic amines. For example, the basicity of aniline turns out to be almost a million times weaker than that of methylamine, a characteristic attributed to resonance interactions. The unshared pair of electrons in the resonance hybrid is not localized on the nitrogen atom as it is in ammonia and aliphatic amines. Rather, it is distributed over the aromatic ring, thus making it less available for sharing in reaction with a Lewis acid.

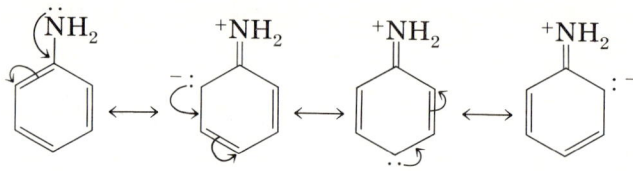

Table 14.2 Basic Strength of Some Amines

Name	Structure	K_b	pK_b
Ammonia	NH_3	1.8×10^{-5}	4.74
Methylamine	CH_3NH_2	4.4×10^{-4}	3.36
Dimethylamine	$(CH_3)_2NH$	5.1×10^{-4}	3.29
Aniline	C₆H₅—NH₂	4.2×10^{-10}	9.38
o-Nitroaniline	C₆H₄(NO₂)—NH₂	5×10^{-15}	14.3

Nitroaniline is even more weakly basic than aniline because the electron-withdrawing effect of the nitro group on the ring lowers the electron density on the nitrogen atom, making the electrons even less available for sharing with a Lewis acid (see also Sec. 5.12).

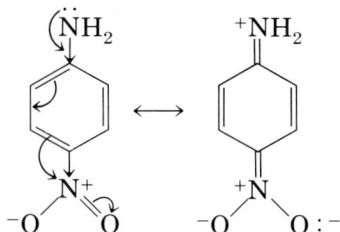

To summarize,

1. Electron-releasing groups on the nitrogen atom of amines increase the basicity of amines, and electron-withdrawing groups decrease the basicity.
2. Resonance effects in aromatic amines lower their basicity.
3. Electron-withdrawing groups on the aromatic ring lower the basicity even more.

Problem 14.7 Arrange each set of compounds in order of increasing basicity.
(a) CH_3NH_2 (i); $ClCH_2NH_2$ (ii); Cl_2CHNH_2 (iii); CH_3NHCH_3 (iv)

(b) $(CH_3)_2NH$ (i); CH_3—⟨◯⟩—NH_2 (ii); NC—⟨◯⟩—NH_2 (iii); ⟨◯⟩—NH_2 (iv)

Preparation of Amines **14.5**

Three methods of preparing amines are summarized here. Each method is discussed in subsequent sections.

1. Reduction of nitro compounds, nitriles, and amides

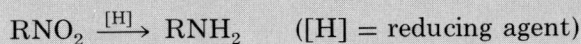

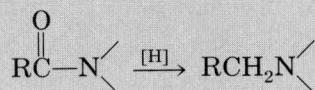

$$RNO_2 \xrightarrow{[H]} RNH_2 \quad ([H] = \text{reducing agent})$$

$$RCN \xrightarrow{[H]} RCH_2NH_2$$

$$RC\!\!\overset{O}{\underset{}{\|}}\!\!-N\big< \xrightarrow{[H]} RCH_2N\big<$$

[Sec. 14.6]

2. Alkylation of ammonia

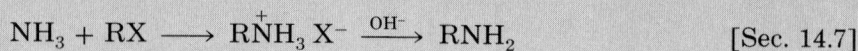

$$NH_3 + RX \longrightarrow R\overset{+}{N}H_3 X^- \xrightarrow{OH^-} RNH_2 \qquad \text{[Sec. 14.7]}$$

3. The Hofmann degradation of amides

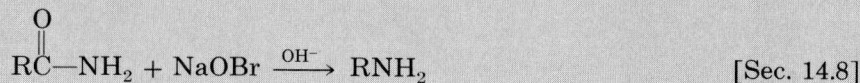

$$\underset{\substack{\|\\O}}{RC}-NH_2 + NaOBr \xrightarrow{OH^-} RNH_2 \qquad \text{[Sec. 14.8]}$$

14.6 Reduction of Nitro Compounds, Nitriles, and Amides

Several classes of organic compounds that already contain nitrogen may be reduced to amines. Most commonly used starting materials are nitro compounds, nitriles, and amides. The reduction may be accomplished either catalytically with hydrogen or directly which chemical reducing agents. Catalytic hydrogenation works well with nitro compounds and nitriles to give *primary amines.*

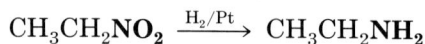

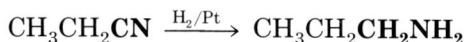

$$CH_3CH_2NO_2 \xrightarrow{H_2/Pt} CH_3CH_2NH_2$$

$$CH_3CH_2CN \xrightarrow{H_2/Pt} CH_3CH_2CH_2NH_2$$

The carbonyl function of amides is not easily reduced catalytically with hydrogen. A particularly useful reducing agent is **lithium aluminum hydride, LiAlH₄.** Lithium aluminum hydride is capable of reducing all three classes of nitrogen compounds cited to amines. Reduction of amides by LiAlH₄ is an especially versatile method since one can prepare primary amines, secondary amines, or tertiary amines by using the proper amide as starting material.

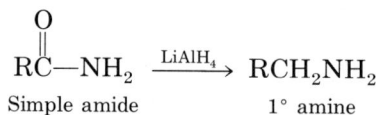

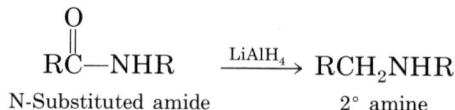

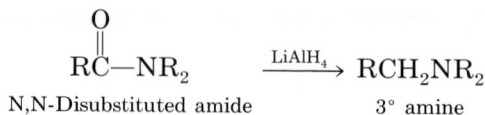

$$\underset{\substack{\|\\O}}{RC}-NH_2 \xrightarrow{LiAlH_4} RCH_2NH_2$$

Simple amide ⟶ 1° amine

$$\underset{\substack{\|\\O}}{RC}-NHR \xrightarrow{LiAlH_4} RCH_2NHR$$

N-Substituted amide ⟶ 2° amine

$$\underset{\substack{\|\\O}}{RC}-NR_2 \xrightarrow{LiAlH_4} RCH_2NR_2$$

N,N-Disubstituted amide ⟶ 3° amine

Aniline, by far the most important aromatic amine, is prepared by reduction of nitrobenzene. The reducing agent most frequently used is tin and hydrochloric acid, which gives anilinium chloride. Treatment of the latter with NaOH liberates the free amine.

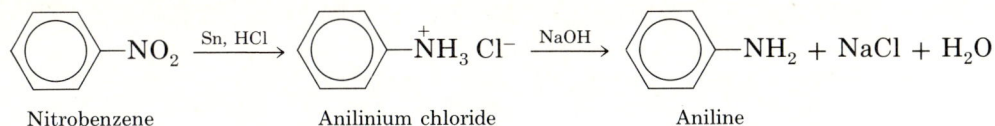

$$\text{Nitrobenzene} \xrightarrow{\text{Sn, HCl}} \text{Anilinium chloride} \xrightarrow{\text{NaOH}} \text{Aniline} + NaCl + H_2O$$

Problem 14.8 Complete each of the following reactions.

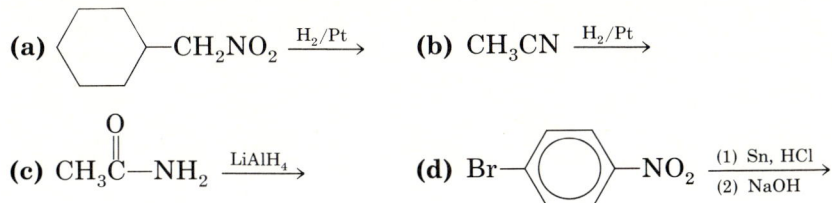

(a) $\text{(cyclohexyl)}-CH_2NO_2 \xrightarrow{H_2/Pt}$ (b) $CH_3CN \xrightarrow{H_2/Pt}$

(c) $CH_3\overset{O}{\overset{\|}{C}}-NH_2 \xrightarrow{LiAlH_4}$ (d) $Br-\text{(benzene)}-NO_2 \xrightarrow[\text{(2) NaOH}]{\text{(1) Sn, HCl}}$

Alkylation of Ammonia   *14.7*

The nonbonded electron pair on the nitrogen makes ammonia an excellent nucleophile. Ammonia is capable of attacking primary or secondary alkyl halides in an S_N2 reaction (see also Sec. 8.6) to give an alkylammonium salt.

$$\overset{..}{N}H_3 \quad + \quad R-X \quad \longrightarrow \quad R\overset{+}{N}H_3 \ X^-$$

$$\text{Ammonia} \quad \text{1° or 2° halide} \quad \text{Alkylammonium salt}$$

Subsequent treatment of the alkylammonium salt with a strong base (NaOH) liberates the free amine.

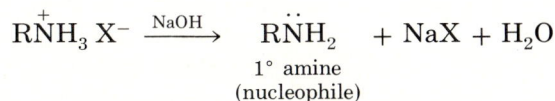

$$R\overset{+}{N}H_3 \ X^- \xrightarrow{\text{NaOH}} R\overset{..}{N}H_2 \quad + NaX + H_2O$$

$$\underset{\substack{\text{1° amine} \\ \text{(nucleophile)}}}{}$$

Since the net result is the replacement of a hydrogen of ammonia by an alkyl group, the reaction is called **alkylation of ammonia.**

A problem with this reaction is the difficulty in stopping it at the primary amine stage. The reason is that the primary amine produced is itself a good nucleophile and continues to react with the alkyl halide by replacement of the remaining hydrogens to give secondary amines and a tertiary amine as well. Furthermore, the tertiary amine, although it has no hydrogen, is also a nucleophile and will attack a fourth molecule of alkyl halide to give a quaternary ammonium salt. The sequence of reactions is illustrated here.

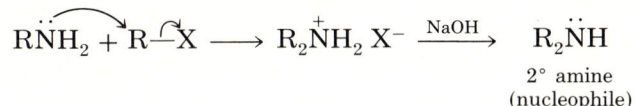

$$R\overset{..}{N}H_2 + R-X \longrightarrow R_2\overset{+}{N}H_2 \ X^- \xrightarrow{\text{NaOH}} R_2\overset{..}{N}H$$

$$\underset{\substack{\text{2° amine} \\ \text{(nucleophile)}}}{}$$

$$R_2\overset{..}{N}H + R\overset{\frown}{-}X \longrightarrow R_3\overset{+}{N}H\ X^- \xrightarrow{\text{NaOH}} R_3\overset{..}{N}$$

3° amine
(nucleophile)

$$R_3\overset{..}{N} + R\overset{\frown}{-}X \longrightarrow R_4\overset{+}{N}\ X^-$$

Quaternary ammonium salt

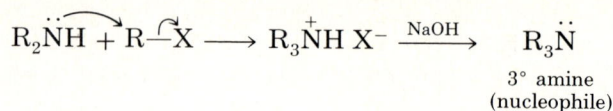

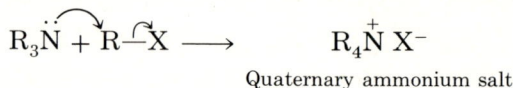

Alkylation is a useful reaction when both the nitrogen nucleophile and the halogen atom are part of the same molecule. In such cases, intramolecular S_N2 displacement takes place and good yield of a single product is formed. The synthesis of nicotine is an example of such reaction.

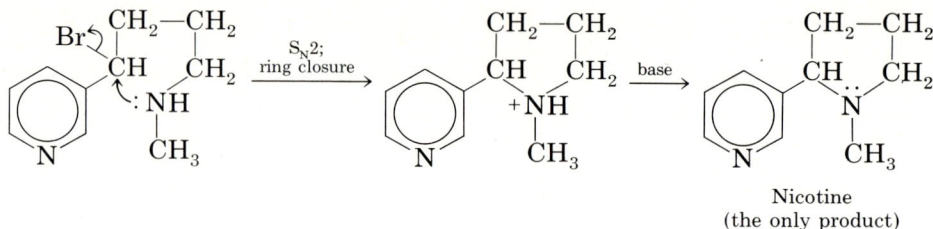

Nicotine
(the only product)

Problem 14.9 Write the structures of the organic products **A, B,** and **C** in these reactions.

(a) $NH_3 + CH_3CH_2Br \longrightarrow$ **A**

(b) **A** + NaOH $\longrightarrow$ **B**

(c) $(CH_3)_3N + CH_3CH_2Cl \longrightarrow$ **C**

14.8 Hofmann Degradation of Amides

An unusual but reliable method of preparing primary amines is the **Hofmann degradation of amides.** The reaction involves the conversion of a simple amide to a primary amine by the action of sodium hypobromite, NaOBr. We call the reaction a degradation because the primary amine contains one less carbon atom than the parent amide. Examples of the reaction are

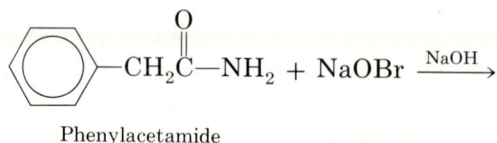

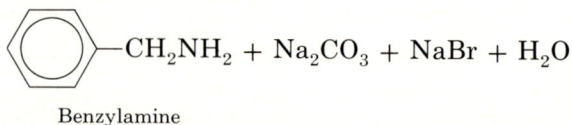

$$\underset{\text{Acetamide}}{CH_3\overset{\overset{\displaystyle O}{\|}}{C}-NH_2} + NaOBr \xrightarrow{\text{NaOH}} \underset{\text{Methylamine}}{CH_3NH_2} + Na_2CO_3 + NaBr + H_2O$$

Phenylacetamide

Benzylamine

Familiar Reactions of Amines **14.9**

In this section we discuss reactions of amines that should be familiar to you.

A   Salt Formation

Since amines are bases (Sec. 14.4), they react with acids to form ammonium salts. Conversion of amines to salts is a useful reaction because the ammonium salts are soluble in water but insoluble in organic solvents such as ethyl ether. Thus, it is possible to separate an amine from nonbasic organic compounds by converting the amine to an ammonium salt with aqueous HCl and extracting the salt in water. Treatment of the aqueous phase with a strong base (NaOH) releases the free amine. For example, the separation of a mixture of an amine and a water-insoluble but ether-soluble ketone can be accomplished by the scheme shown.

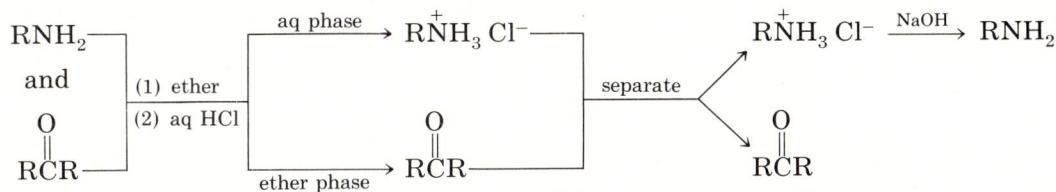

$$
\begin{array}{c}
RNH_2 \\
\text{and} \\
\underset{\displaystyle RCR}{\overset{\displaystyle O}{\|}}
\end{array}
\quad
\begin{array}{c}
\text{(1) ether} \\
\text{(2) aq HCl}
\end{array}
\quad
\begin{array}{c}
\xrightarrow{\text{aq phase}} R\overset{+}{N}H_3\,Cl^- \\
\\
\xrightarrow{\text{ether phase}} \underset{\displaystyle RCR}{\overset{\displaystyle O}{\|}}
\end{array}
\quad
\text{separate}
\quad
\begin{array}{c}
R\overset{+}{N}H_3\,Cl^- \xrightarrow{\text{NaOH}} RNH_2 \\
\\
\underset{\displaystyle RCR}{\overset{\displaystyle O}{\|}}
\end{array}
$$

Problem 14.10   A solution consists of a mixture of *n*-butylbenzene (bp = 183°C), aniline (bp = 184°C), and valeric acid (bp = 189°C). The mixture cannot be separated into its pure components by fractional distillation because of the closeness of their boiling points. Moreover, all three compounds are not very soluble in water but are soluble in ether. Suggest a simple, practical method that will quickly separate the three compounds.

B   Alkylation

Amines, like ammonia, react with primary and secondary alkyl halides to give *alkylated* amines. This is another reaction which depends on the electron pair of nitrogen. As has already been pointed out (Sec. 14.7), complete alkylation eventually leads to a quaternary ammonium salt. Two quaternary ammonium salts of biological significance are **choline** and **acetylcholine.**

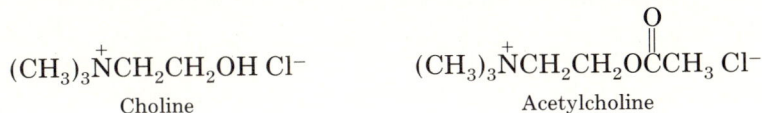

$$(CH_3)_3\overset{+}{N}CH_2CH_2OH\ Cl^-$$
Choline

$$(CH_3)_3\overset{+}{N}CH_2CH_2O\overset{\displaystyle O}{\overset{\displaystyle \|}{C}}CH_3\ Cl^-$$
Acetylcholine

Choline, which can be found in egg yolks, meats, and fish, is essential to growth. It is also involved in carbohydrate and protein metabolism as well as in fat transport. Choline is also the precursor of acetylcholine, a substance in the brain that transmits nerve impulses. There is much evidence today linking cells that use acetylcholine to the formation of memory. Thus, adding choline to the diet may help the elderly conquer the pervasive problem of memory loss.

C   Amide Formation

Conversions of amines to amides with acid chlorides and acid anhydrides are illustrated in Sections 12.19 and 12.21. Amide formation by these methods requires the amine nitrogen to have at least one replaceable hydrogen atom. Primary amines yield N-substituted amides and secondary amines N,N-disubstituted amides. Tertiary amines, which have no H atom on the nitrogen, cannot be converted to amides. The treatment of an acid chloride with the three classes of amines is shown here.

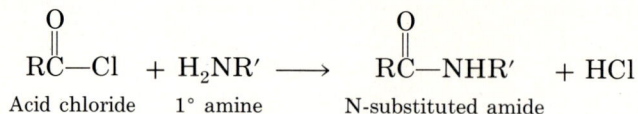

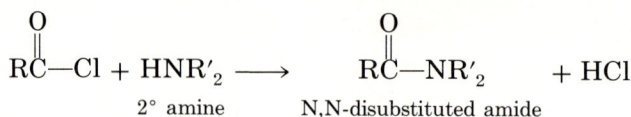

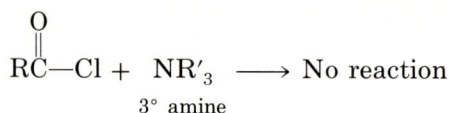

$$\underset{\text{Acid chloride}}{\text{RC}\overset{\text{O}}{\overset{\|}{}}\text{—Cl}} \;+\; \underset{\text{1° amine}}{\text{H}_2\text{NR}'} \;\longrightarrow\; \underset{\text{N-substituted amide}}{\text{RC}\overset{\text{O}}{\overset{\|}{}}\text{—NHR}'} \;+\; \text{HCl}$$

$$\text{RC}\overset{\text{O}}{\overset{\|}{}}\text{—Cl} + \underset{\text{2° amine}}{\text{HNR}'_2} \;\longrightarrow\; \underset{\text{N,N-disubstituted amide}}{\text{RC}\overset{\text{O}}{\overset{\|}{}}\text{—NR}'_2} \;+\; \text{HCl}$$

$$\text{RC}\overset{\text{O}}{\overset{\|}{}}\text{—Cl} + \underset{\text{3° amine}}{\text{NR}'_3} \;\longrightarrow\; \text{No reaction}$$

Problem 14.11   Write the structure of the organic product in each of these reactions. If no reaction takes place state so.

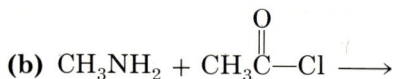

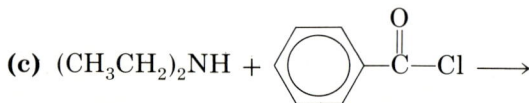

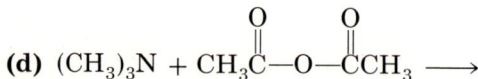

(a) $(\text{CH}_3\text{CH}_2)_3\text{N} + \text{CH}_3\text{Br} \longrightarrow$

(b) $\text{CH}_3\text{NH}_2 + \text{CH}_3\overset{\text{O}}{\overset{\|}{\text{C}}}\text{—Cl} \longrightarrow$

(c) $(\text{CH}_3\text{CH}_2)_2\text{NH} +$ ⟨C₆H₅⟩$\text{—}\overset{\text{O}}{\overset{\|}{\text{C}}}\text{—Cl} \longrightarrow$

(d) $(\text{CH}_3)_3\text{N} + \text{CH}_3\overset{\text{O}}{\overset{\|}{\text{C}}}\text{—O—}\overset{\text{O}}{\overset{\|}{\text{C}}}\text{CH}_3 \longrightarrow$

An interesting class of amides are the *sulfonamides*. Sulfonamides are prepared by treating primary or secondary amines with sulfonyl chlorides, the acid chlorides of sulfonic acids.

General equation

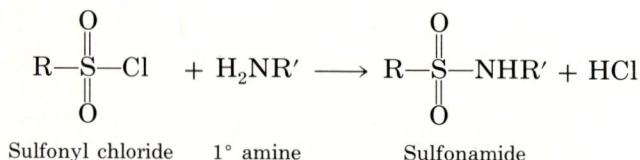

$$\underset{\text{Sulfonyl chloride}}{\text{R}\overset{\text{O}}{\underset{\text{O}}{\overset{\|}{\underset{\|}{\text{—S—}}}}}\text{Cl}} \;+\; \underset{\text{1° amine}}{\text{H}_2\text{NR}'} \longrightarrow \underset{\text{Sulfonamide}}{\text{R}\overset{\text{O}}{\underset{\text{O}}{\overset{\|}{\underset{\|}{\text{—S—}}}}}\text{NHR}'} + \text{HCl}$$

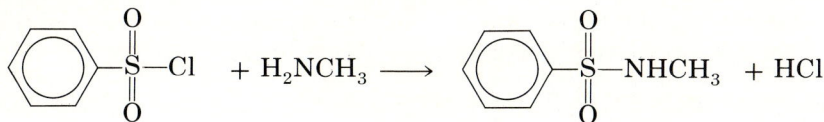

Benzenesulfonyl chloride $+$ H$_2$NCH$_3$ $\longrightarrow$ *N*-Methylbenzenesulfonamide $+$ HCl

A biologically important sulfonamide used to prevent bacterial infection is **sulfanilamide,** the first sulfa drug. Over 10,000 compounds containing sulfanilamide units have to date been tested and prepared for antibacterial activity. A few examples are

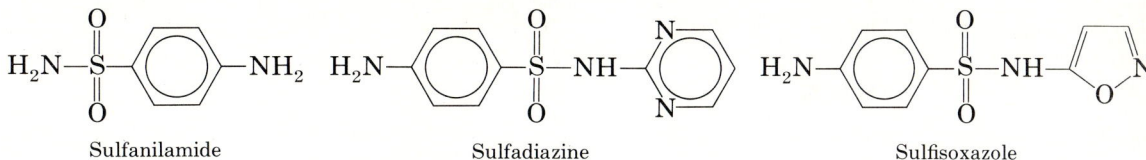

Sulfanilamide   Sulfadiazine   Sulfisoxazole

Today, antibiotics have largely replaced the sulfa drugs, but some are still used as powerful chemotherapeutic agents for certain organisms, particularly in urinary tract infections.

Visual Test for Amines: 14.10
Reaction with Nitrous Acid

We can differentiate among the three classes of amines (1°, 2°, and 3°) by examining the different products each forms when treated with nitrous acid, HNO$_2$. Nitrous acid is generated in the laboratory by treating a cold solution of sodium nitrite, NaNO$_2$, with aqueous HCl.

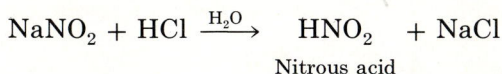

$$\text{NaNO}_2 + \text{HCl} \xrightarrow{\text{H}_2\text{O}} \text{HNO}_2 + \text{NaCl}$$
$$\text{Nitrous acid}$$

The **nitrous acid test** works as follows: when treated with cold nitrous acid (0–5°C), primary aliphatic amines yield nitrogen gas (seen as bubbles) and a mixture of other products.

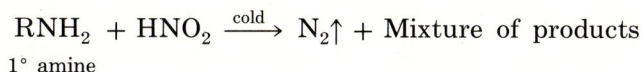

$$\text{RNH}_2 + \text{HNO}_2 \xrightarrow{\text{cold}} \text{N}_2\uparrow + \text{Mixture of products}$$
1° amine

Secondary aliphatic amines react more slowly under the same conditions and form water-insoluble, oily yellow derivatives called **nitrosamines** (R$_2$N—N=O, a nitrosamine).

The formation of nitrosamines with secondary amines not only is relevant to the analysis of amines but also is of concern to our health. It is common knowledge that most processed meats contain sodium nitrite, $NaNO_2$, as an additive. Processed meats also contain various secondary amines, sometimes at elevated levels. The functions of the sodium nitrite are to maintain the red color of the meat and to prevent botulism. These benefits are offset by the fact that, on ingestion, nitrous acid is liberated because of the low pH in the stomach. The nitrous acid in turn rapidly reacts with the secondary amines also ingested to form nitrosamines. Since there is ample evidence that nitrosamines are potent *carcinogens* (cancer-causing substances), there is danger that excessive consumption of processed meats could lead to cases of cancer that would not otherwise occur. Although this reaction has not been observed in humans, it is well documented that in rodents the administration of nitosamines, or of a secondary amine together with sodium nitrite, causes cancer (Fig. 14.2).

Tertiary aliphatic amines form water-soluble ammonium salts when treated with cold nitrous acid.

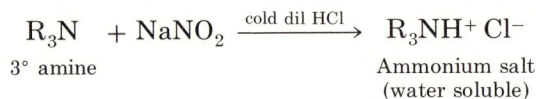

$$R_3N \ + \ NaNO_2 \ \xrightarrow{\text{cold dil HCl}} \ R_3NH^+ \ Cl^-$$
$$\text{3° amine} \qquad\qquad\qquad \text{Ammonium salt}$$
$$\text{(water soluble)}$$

Because of the solubility of the ammonium salt in aqueous solution, there is no visible sign of a reaction.

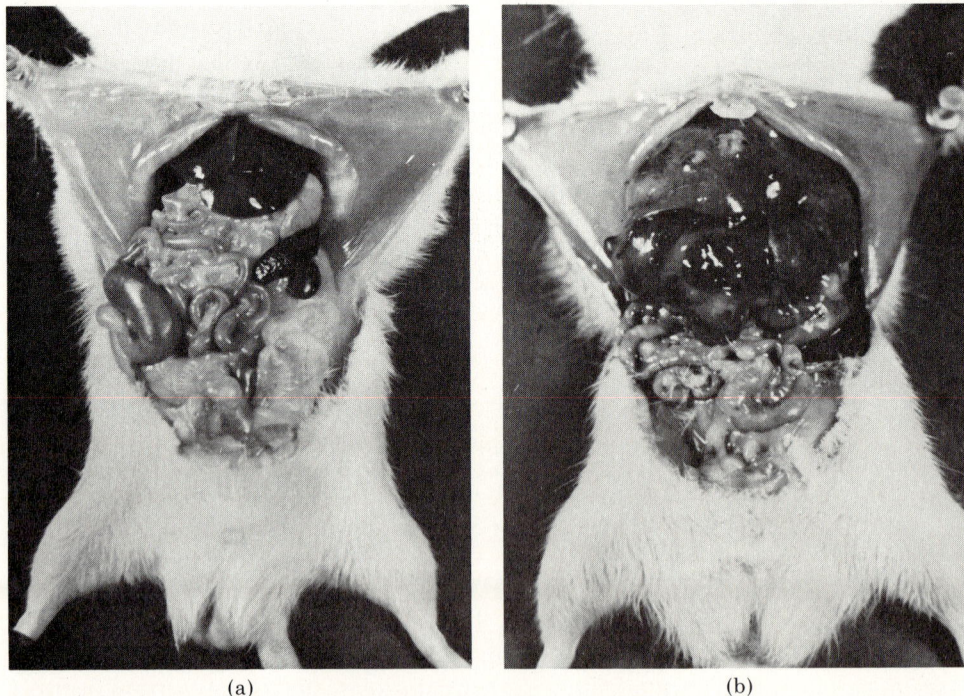

(a)   (b)

Figure 14.2 (a) Abdominal organs of a normal rat. **(b)** Abdominal organs of a rat that was administered dimethylnitrosamine, a powerful carcinogen. Large liver-tumor masses are visible. [From J. C. Arcos, "Cancer: chemical factors in environment," part 1. *American Laboratory, 10*(6), 65 (1978). Reprinted with permission.]

Aromatic primary amines can be distinguished from primary aliphatic amines because they react with cold nitrous acid to form highly reactive compounds called **diazonium salts** *without evolution of nitrogen gas*. Only when the diazonium salt is warmed to room temperature will bubbles of nitrogen gas appear.

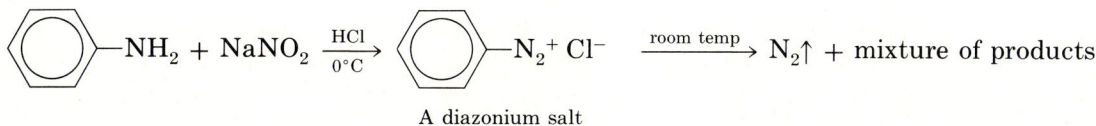

A diazonium salt

In summary, when an amine is treated with nitrous acid,

1. If nitrogen gas is produced at 0°C, the amine is 1° aliphatic.
2. If an oily yellow layer separates from the aqueous layer, the amine is 2° aliphatic.
3. If there is no visible reaction, the amine is 3° and aliphatic.
4. If nitrogen gas bubbles are not formed at 0°C but do appear at room temperature, the amine is a 1° aromatic amine.

Problem 14.12 An amine **(A)** has the formula C_2H_7N. When **A** was treated with nitrous acid at 0°C, a reaction took place, and an oily yellow layer separated from the reaction mixture. Write a structure for **A** that is consistent with the data given, and give an equation for the reaction.

Conversion of Diazonium Salts *14.11*

Aromatic diazonium salts are very useful for preparing a host of substituted benzene derivatives (Fig. 14.3). The substitution reactions of diazonium salts are particularly useful as routes to products that otherwise cannot be prepared by direct electrophilic aromatic substitution (see Sec. 5.9). The following example will illustrate this point.

Example 14.1 Starting from benzene, synthesize 2,4-dinitrofluorobenzene (also known as Sanger's reagent), an important compound used in protein chemistry.

Solution First draw the structure of the starting material and that of the product.

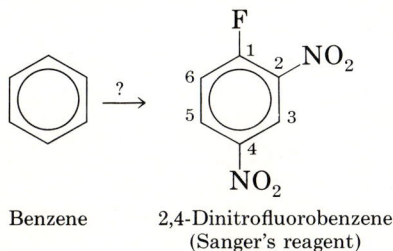

Benzene   2,4-Dinitrofluorobenzene
(Sanger's reagent)

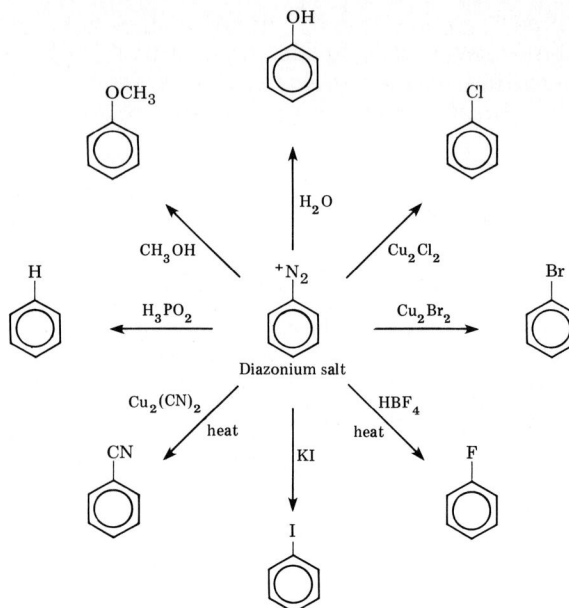

Figure 14.3 Important substitution reactions of diazonium salts.

The problem here is to introduce three substituents, one —F and two —NO$_2$, in the proper sequence. You know that —F is an ortho,para director, whereas the nitro groups have a meta-directing effect (Sec. 5.11). Obviously, the —F group must be introduced first, followed by nitration of fluorobenzene.

Desired sequence

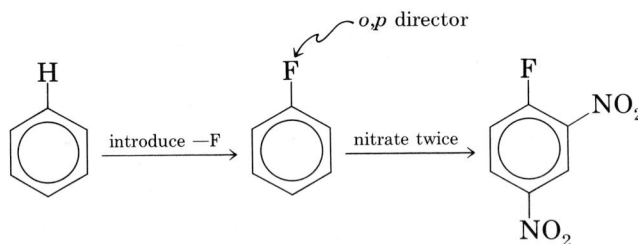

(1) *Synthesis of fluorobenzene.* The fluoro group can be introduced by nitration of benzene, reduction of the nitro group to the amine, formation of the diazonium compound, and treatment with fluoboric acid, HBF$_4$.

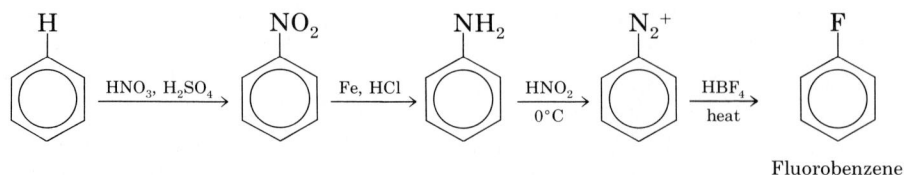

(2) *Nitration of fluorobenzene.* Once formed, fluorobenzene is nitrated, giving a mixture of *o-* and *p*-nitrofluorobenzene.

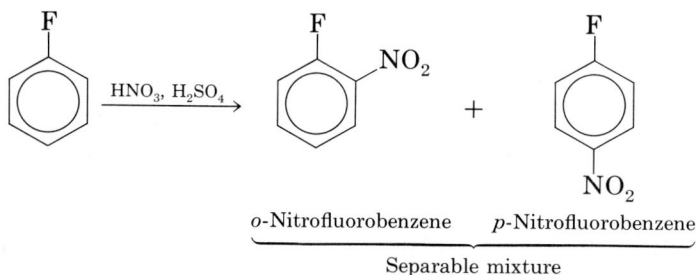

o-Nitrofluorobenzene   p-Nitrofluorobenzene
Separable mixture

(3) *Nitration of p-nitrofluorobenzene.* The mixture of o- and p-nitrofluorobenzene may be separated. Nitration of p-nitrofluorobenzene gives exclusively the desired product because each substituent directs the incoming nitro group to position 2.

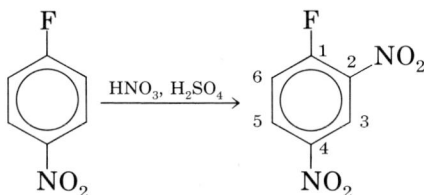

Problem 14.13 Show by means of equations how you would prepare
(a) Benzonitrile, C_6H_5CN, from benzene
(b) *m*-Bromophenol from nitrobenzene
(c) *m*-Dichlorobenzene from benzene
(d) Benzene from nitrobenzene

Azo Compounds *14.12*

When aromatic diazonium salts are treated with phenols or aromatic amines, a reaction takes place *without the loss of nitrogen.* The result is an **azo compound,** formed by the coupling of the two aromatic rings through the *azo group,* —N=N—. Coupling takes place preferentially at the position para to the activating group. If the para position is occupied, coupling then occurs at one of the ortho positions.

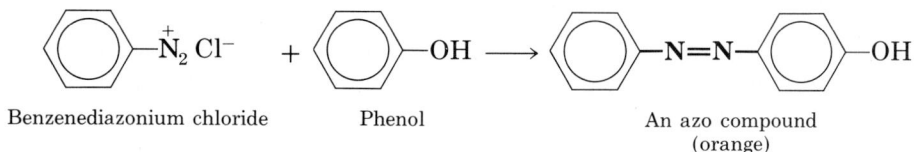

Benzenediazonium chloride   Phenol   An azo compound
(orange)

N-Methylaniline

An azo compound
(yellow)

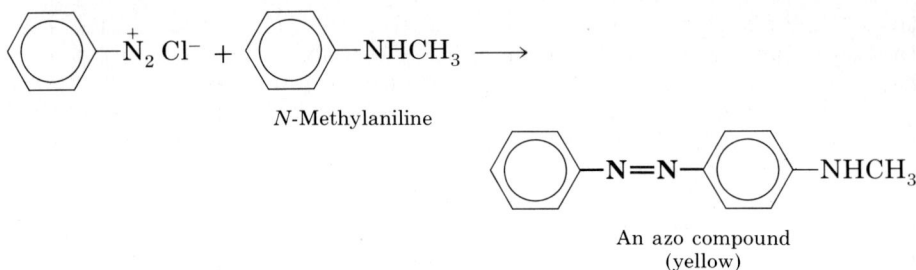

Azo compounds are colored and represent one important class of dyes known as azo dyes. By varying the design of the structures taking part in the coupling reaction, it is possible to control the colors of the azo dyes.

The azo coupling reaction is also used by forensic laboratories in the detection of marijuana. The three major constituents of marijuana, tetrahydrocannabinol (THC), cannabinol (CBN), and cannabidiol (CBD) are phenolic compounds that can react with *p*-nitrobenzenediazonium chloride to give a reddish brown color

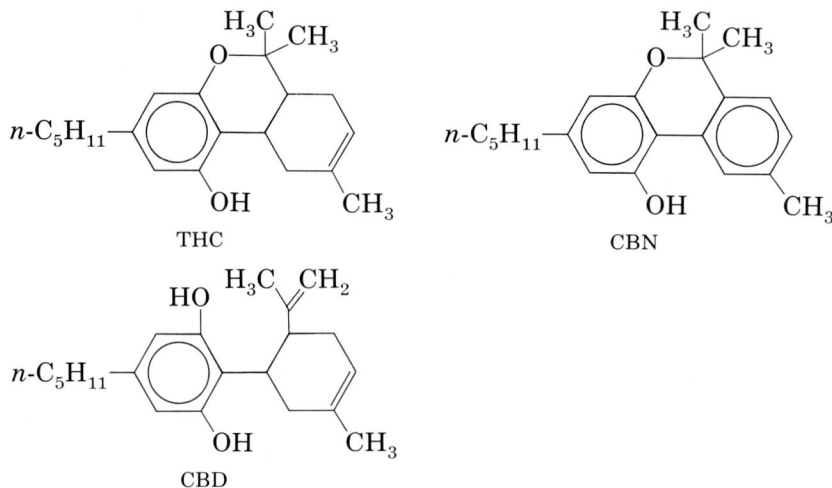

THC

CBN

CBD

Problem 14.14 Write the structures of the azo compounds formed by the coupling of *p*-nitrobenzenediazonium chloride and cannabinol (CBN).

14.13 Biogenic Amines and Alkaloids

A Biogenic Amines

Many of the drugs used by so many Americans for pleasure more often than for health, such as benzedrine, methedrine (speed), marijuana, LSD, barbiturates, heroin, morphine, and other addictive drugs, are amines or amine derivatives. They affect the nervous system and alter perception and natural responses to the environment. Drugs that have the effect of changing a person's mental or emotional state are referred to as **psychomimetic drugs.**

It is believed that these drugs interact with the nervous system of the body by substituting for, or mimicking (hence, the term *psychomimetic*), naturally occurring agents called **neurotransmitters.** Neurotransmitters are amine-like substances known as **biogenic amines** that transmit messages from one nerve cell to another. Through their action, psychomimetic drugs inhibit normal neural functions or cause the natural neural process to continue unabated rather than stop when it normally would.

Amphetamine, called "uppers," are examples of psychomimetic drugs that structurally resemble the biogenic amines adrenaline (epinephrine) and noradrenaline (norepinephrine).

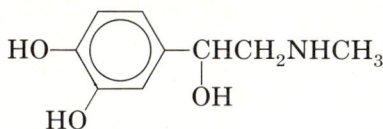

Adrenaline

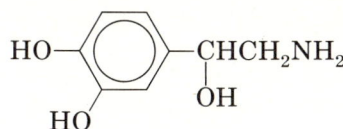

Noradrenaline

Adrenaline and noradrenaline are also known as catecholamines because they are structurally related to the phenolic compound catechol. Table 14.3 gives the structures and lists the psychological effects of some amphetamines.

Barbiturates, called "downers" because they depress the activities of many systems of the body, are derivatives of barbituric acid.

Table 14.3 Structures and Physiological Effects of Some Amphetamines

Name	Structural formula	Physiological effect
Amphetamine or benzedrine (1-phenyl-2-aminopropane)		increases alertness, lessens fatigue, counteracts depression, and decreases appetite
Methedrine or "speed" (*N*-methylamphetamine)		causes visual and auditory hallucinations
STP (2,5-dimethoxy-4-methylamphetamine)		causes hallucinations
Phenylephrine hydrochloride		clears sinuses (used in nose drops)

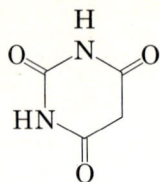

Barbituric acid

Barbiturates are addictive, and their effects decrease with constant use, thus requiring the user to ingest ever larger doses. The lethal dosage remains the same, however, and the individual runs the danger of a fatal overdose. Barbiturates should *never* be taken together with alcoholic beverages because both are depressants and their combined effect may equal that of a fatal dose of barbiturate taken singly.

Some barbiturates are lised in Table 14.4. Note that the effects differ with different alkyl substituents on the barbituric acid nucleus.

Table 14.4   Structures and Physiological Effects of Some Barbiturates

Name	R	R'	Physiological effects
Barbituric acid	$-H$	$-H$	toxic and habit forming; depresses central nervous system, slows breathing and heart rate
Veronal	$-CH_2CH_3$	$-CH_2CH_3$	sedative; prescribed for anesthetic as well as sedative purposes; produces depression of central nervous system and consequent sedation
Phenobarbital	$-CH_2CH_3$	$-C_6H_5$	sedative; sometimes prescribed to relieve teething pain in babies
Amobarbital (Amytal)	$-CH_2CH_3$	$-CH_2CH_2\overset{\displaystyle CH_3}{\underset{\displaystyle }{C}}HCH_3$	sedative; habit forming; causes drowsiness and loss of appetite
Secobarbital (Seconal)	$-\overset{\displaystyle CH_3}{\underset{\displaystyle }{C}}HCH_2CH_2CH_3$	$-CH_2CH=CH_2$	same as amobarbital

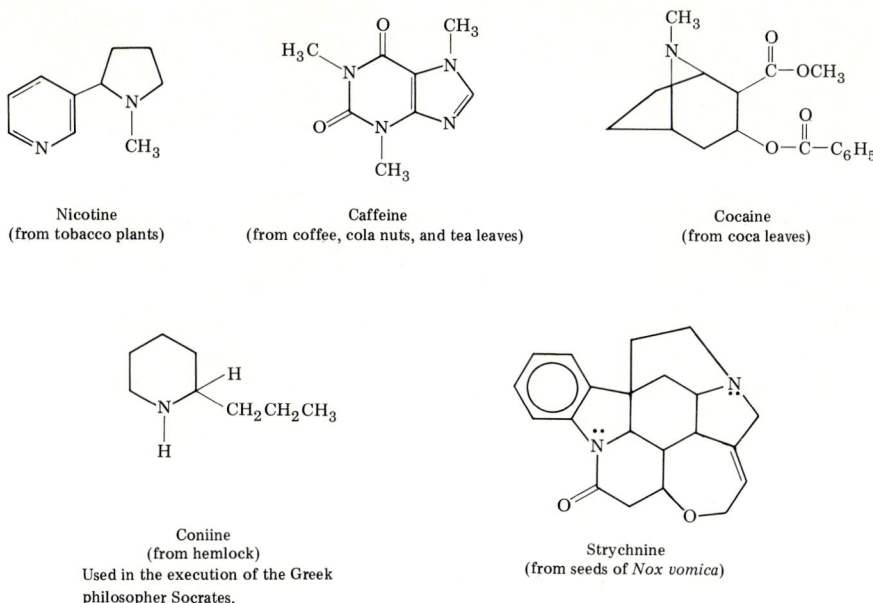

Nicotine
(from tobacco plants)

Caffeine
(from coffee, cola nuts, and tea leaves)

Cocaine
(from coca leaves)

Coniine
(from hemlock)
Used in the execution of the Greek
philosopher Socrates.

Strychnine
(from seeds of *Nox vomica*)

Figure 14.4 Structures of several alkaloids illustrating their heterocyclic nature.

B Alkaloids

Some of the most interesting nitrogen compounds found in plants are the **alkaloids.** The name, derived from *alkali,* denotes the basic properties of these compounds, which are due to the presence of an amine function or a nitrogen present with carbon atoms in a ring. Alkaloids therefore belong to a general group of compounds known as **heterocyclic compounds,** which contain one or more atoms other than carbon in a ring. The structures of some alkaloids are shown in Figure 14.4.

Alkaloids have profound effects on the activities of the central nervous system of higher animals, a characteristic attributed to the compound's ability to either stimulate or disrupt the activities of biogenic amines. With the exception of marijuana from the cannabis plant and myristicine from the nutmeg plant, almost all known major hallucinogens are alkaloids. Table 14.5 lists some of the major alkaloid hallucinogens.

The Nitrogens in Our Food **14.14**

Some nitrogen-containing compounds are added to our foods by manufacturers to give them pleasurable tastes or to act as preservatives. For example, the calorie-free sugar substitute *saccharin* is about 300 times sweeter than the

Table 14.5 Structures, Sources, and Physiological Effects of Some Major Alkaloid Hallucinogens

Name	Structural formula	Source	Physiological effects
Mescaline (3,4,5-trimethoxy-phenetylamine)		Peyote cactus of southwestern U.S.[a]	causes hallucinations in which the user sees extremely rich colors
Serotonin		beef serum	similar to those of mescaline
Scopolamine (truth serum)		belladonna plant	supposedly induces a person to reveal what he or she wants to hide, a questionable property in view of the hallucinogenic effects of the drug
Lysergic acid diethylamide (LSD) (ergot alkaloid)		Ergot, a fungus growing on rye[b]	produces strong hallucinogenic effects with as little as 50–100 μg (10^{-6} g) of the drug; can cause emotional trauma, fright, and terror
Morphine		*Papaver som-niferum* (raw opium poppy)	produces sleep and drowsiness; can alleviate pain effectively, although it normally is not used as a pain killer because of its habit-forming properties
Heroin		diacetyl derivative of morphine	relieves pain; produces state of euphoria and false sense of well-being; extremely habit forming; overdoses often fatal

[a] Used by the Mescaleros Apaches in their religious rites.
[b] First synthesized in 1938 by the Swiss chemist Arnold Hofmann.

household sugar sucrose. Despite its sweetness, saccharin leaves a slightly bitter aftertaste. For this reason research on commercial sweeteners has been directed toward developing artificial sweeteners that do not have this disadvantage.

One such group of sweeteners is the cyclamates. Because of their low caloric content, these compounds were used extensively in diet foods. In 1969 the Food and Drug Administration banned their use when tests indicated that massive doses of cyclamate induced cancer and chromosomal damage in rats.

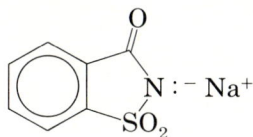

Saccharin
(sodium salt)

Cyclamate
(sodium salt)

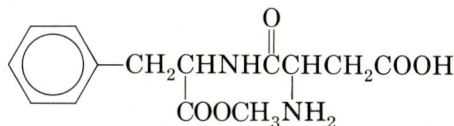

Aspartame

A few years ago a new artificial sweetener, aspartame, which is much sweeter than saccharin and has none of its aftertaste, was developed by the G. D. Searle Co. After early controversy regarding the safety of this sugar substitute, the FDA approved use of aspartame in 1981.

A natural, sweet-tasting protein called *monellin,* whose structure is still unknown, has recently been isolated from a species of African wild berries. It is several thousand times sweeter than sucrose and so far has not shown any adverse effects on test animals. Until its structure is known, however, it cannot be synthesized in the laboratory.

Monosodium glutamate (MSG), the sodium salt of the amino acid glutamic acid, has long been used as a flavor enhancer in a variety of food products. Until recently, MSG has been added to baby food, despite its lack of nutritional value, to make the food more palatable to the mothers, many of whom tasted the food before feeding it to their children. The use of monosodium glutamate in baby food was discontinued as a result of the controversy surrounding the biological effects of this compound on growing infants.

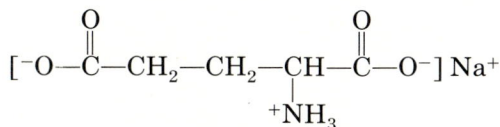

Monosodium glutamate

Summary of Concepts and Reactions

Amines are compounds that may be thought of as having been derived from ammonia by replacement of one, two, or three hydrogens by alkyl or aryl groups.   [Sec. 14.1]
When a fourth group is attached to nitrogen, we have a quaternary ammonium salt.
[Sec. 14.1]
Amines have higher boiling points than alkanes of comparable molecular weight but lower boiling points than alcohols of comparable molecular weight.   [Sec. 14.3]
All amines are capable of forming hydrogen bonds with water, and amines of up to six carbons are soluble in water.   [Sec. 14.3]

Amines are bases because the nitrogen atom possesses a nonbonded pair of electrons that can be shared by an acid proton.

$$R\ddot{N}H_2 + H^+ \longrightarrow R\overset{+}{N}H_3$$

[Sec. 14.4]

Aliphatic amines are more basic than are aromatic amines. [Sec. 14.4]
The three basic preparations of amines are
1. Reduction of nitro compounds, nitriles, and amides.

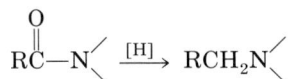

$$RNO_2 \xrightarrow{[H]} RNH_2$$

$$RCN \xrightarrow{[H]} RCH_2NH_2$$

$$R\overset{O}{\overset{\|}{C}}-N{\big\langle} \xrightarrow{[H]} RCH_2N{\big\langle}$$

[Sec. 14.6]

2. Alkylation of ammonia.

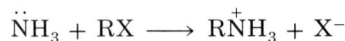

$$\ddot{N}H_3 + RX \longrightarrow R\overset{+}{N}H_3 + X^-$$

$$R\overset{+}{N}H_3 \xrightarrow{base} RNH_2$$

[Sec. 14.7]

3. The Hofmann degradation of amides.

$$R\overset{O}{\overset{\|}{C}}-NH_2 + NaOBr \xrightarrow{OH^-} RNH_2$$

[Sec. 14.8]

Amines react with strong mineral acid to form salts. [Sec. 14.9A]
Reactions of amines with primary or secondary alkyl halides yield alkylated amines or quaternary ammonium salts. [Sec. 14.9B]
Primary and secondary amines react with acid chlorides or acid anhydrides to form substituted amides. Tertiary amines cannot be converted to amides. [Sec. 14.9C]
A biologically important sulfonamide used to prevent bacterial infection is sulfanilamide, the first sulfa drug. [Sec. 14.9C]
The three classes of amines can be differentiated visually by means of the nitrous acid test. [Sec. 14.10]
The nitrogen of aromatic diazonium salts can be replaced by other groups to give many substitution products. [Sec. 14.11]
Aromatic diazonium salts react with phenols and aromatic amines to yield derivatives called azo compounds. [Sec. 14.12]
All azo compounds are colored, and many are used as synthetic dyes. [Sec. 14.12]
Neurotransmitters are amine-like substances, known as biogenic amines, that transmit messages from one nerve cell to another. [Sec. 14.13A]
Alkaloids are nitrogen compounds found in plants that have profound effects on the activities of the central nervous system. [Sec. 14.13B]
Alkaloids belong to a general group of compounds called heterocyclic compounds because they contain one or more atoms other than carbon in a ring. [Sec. 14.13B]
Some nitrogen-containing compounds are added to our foods by manufacturers to give them pleasurable tastes or to act as preservatives. [Sec. 14.14]

amines
primary (1°) amines
secondary (2°) amines
tertiary (3°) amines
quaternary ammonium
   salt
aniline
bases
base strength

basicity constant, K_b
pK_b
lithium aluminum hydride
alkylation of ammonia
Hofmann degradation of
   amides
choline
acetylcholine
sulfanilamide

nitrous acid test
nitrosamines
diazonium salts
azo compounds
psychomimetic drugs
neurotransmitters
biogenic amines
alkaloids
heterocyclic compounds

Exercises

Structure, Nomenclature, and Classification of Amines [Secs. 14.1, 14.2]

14.1 Write structures for the following compounds

(a) Benzylamine
(b) 2-Phenyl-2-aminopropane
(c) *p*-Nitroaniline
(d) Tri-*n*-propylamine
(e) 1,4-Diaminobutane
(f) Methylcyclohexylamine
(g) Tetraethylammonium chloride
(h) *N*-Ethylaniline
(i) *p*-Aminobenzoic acid
(j) Diethylisopropylamine
(k) α-Naphthylamine
(l) 2,4,6-Tribromoaniline

14.2 Name the following compounds

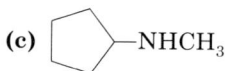

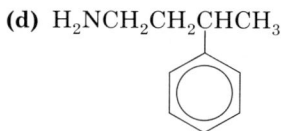

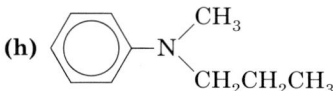

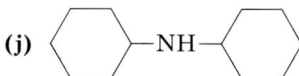

(a) $CH_3CH_2CH_2NH_2$

(b) Br—⟨⟩—NH_2

(c) ⟨⟩—$NHCH_3$

(d) $H_2NCH_2CH_2CHCH_3$ (with phenyl substituent)

(e) $CH_3CHCH_2CH_2CH_3$ (with NH_2 substituent)

(f) ⟨⟩—$NHC(CH_3)_3$

(g) CH_3NHCH_2—⟨⟩

(h) ⟨⟩—N⟨ CH_3 / $CH_2CH_2CH_3$

(i) $[(CH_3)_2N(CH_2CH_3)_2]^+ OH^-$

(j) ⟨⟩—NH—⟨⟩

14.3 Classify the amines in Exercises 14.1 and 14.2 as (a) aliphatic or (b) aromatic.

14.4 Which compounds in Exercises 14.1 and 14.2 can be classified as (a) primary amine, (b) secondary amine, (c) tertiary amine, (d) quaternary ammonium salt?

Structures of Amine Derivatives [Secs. 14.6–14.12]

14.5 Draw structures for these compounds.

(a) Benzamide   (b) Azobenzene

(c) *p*-Chlorobenzenesulfonamide   (d) *N*-Methyl-*N*-nitrosoaniline

(e) *N,N*-Dimethylformamide   (f) Benzenediazonium chloride

Physical Properties, Basicity [Secs. 14.3, 14.4]

14.6 Arrange the following compounds in order of increasing boiling point. Give reasons for the order you select.

$$CH_3(CH_2)_4CH_3 \qquad CH_3(CH_2)_4NH_2 \qquad CH_3(CH_2)_2N(CH_3)_2 \qquad H_2N(CH_2)_4NH_2$$
$$\text{(i)} \qquad\qquad\qquad \text{(ii)} \qquad\qquad\qquad\qquad \text{(iii)} \qquad\qquad\qquad\qquad \text{(iv)}$$

14.7 Arrange the compounds in each group in order of increasing base strength. Outline reasons for the orders you select.

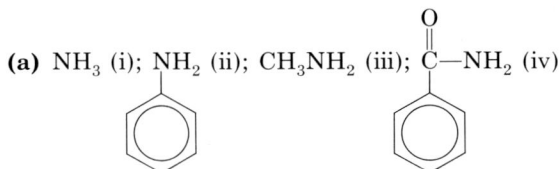

(a) NH_3 (i); NH_2 (ii); CH_3NH_2 (iii); $\overset{\displaystyle O}{\overset{\|}{C}}-NH_2$ (iv)

(b) $CH_3\overset{\displaystyle O}{\overset{\|}{C}}NH_2$ (i); $CF_3\overset{\displaystyle O}{\overset{\|}{C}}NH_2$ (ii); $CH_3CH_2NH_2$ (iii); $CH_3CH_2NHCH_2CH_3$ (iv)

14.8 Taking advantage of acidic and basic properties, outline a procedure for separating a mixture of benzoic acid, aniline, and chlorobenzene.

Preparation of Amines [Secs. 14.5–14.8]

14.9 Identify each lettered product in the following reaction sequences.

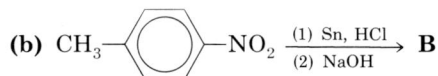

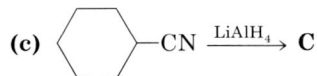

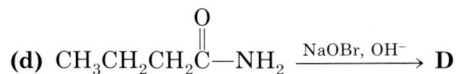

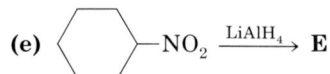

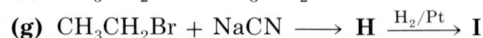

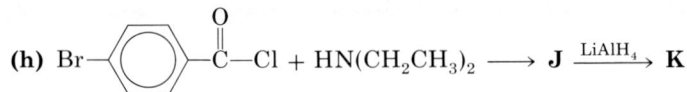

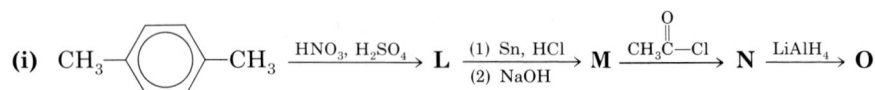

(a) $CH_3CH_2NH_3{}^+\,Cl^- + NaOH \longrightarrow$ **A**

(b) $CH_3-\!\!\!\!\bigcirc\!\!\!\!-NO_2 \xrightarrow[\text{(2) NaOH}]{\text{(1) Sn, HCl}}$ **B**

(c) $\bigcirc\!\!\!\!-CN \xrightarrow{\text{LiAlH}_4}$ **C**

(d) $CH_3CH_2CH_2\overset{\displaystyle O}{\overset{\|}{C}}-NH_2 \xrightarrow{\text{NaOBr, OH}^-}$ **D**

(e) $\bigcirc\!\!\!\!-NO_2 \xrightarrow{\text{LiAlH}_4}$ **E**

(f) $CH_3CH_2Br + CH_3NH_2 \longrightarrow$ **F** $\xrightarrow{\text{NaOH}}$ **G**

(g) $CH_3CH_2Br + NaCN \longrightarrow$ **H** $\xrightarrow{\text{H}_2/\text{Pt}}$ **I**

(h) $Br-\!\!\!\!\bigcirc\!\!\!\!-\overset{\displaystyle O}{\overset{\|}{C}}-Cl + HN(CH_2CH_3)_2 \longrightarrow$ **J** $\xrightarrow{\text{LiAlH}_4}$ **K**

(i) $CH_3-\!\!\!\!\bigcirc\!\!\!\!-CH_3 \xrightarrow{\text{HNO}_3,\ \text{H}_2\text{SO}_4}$ **L** $\xrightarrow[\text{(2) NaOH}]{\text{(1) Sn, HCl}}$ **M** $\xrightarrow{\overset{\displaystyle O}{\overset{\|}{CH_3C}}-Cl}$ **N** $\xrightarrow{\text{LiAlH}_4}$ **O**

14.10 Outline a suitable method to carry out each conversion. The products indicated should be obtained in a high state of purity (that is, uncontaminated by traces of secondary or tertiary amines).

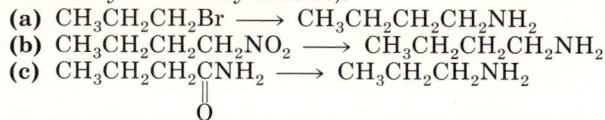

(a) $CH_3CH_2CH_2Br \longrightarrow CH_3CH_2CH_2CH_2NH_2$

(b) $CH_3CH_2CH_2CH_2NO_2 \longrightarrow CH_3CH_2CH_2CH_2NH_2$

(c) $CH_3CH_2CH_2\underset{\underset{O}{\|}}{C}NH_2 \longrightarrow CH_3CH_2CH_2NH_2$

Preparations and Reactions of Amines and Amine Derivatives [Secs 14.5–14.12]

14.11 Identify each lettered product in the following reaction sequences.

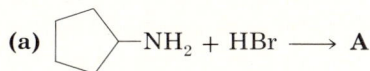

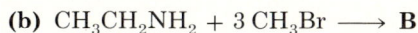

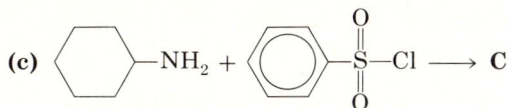

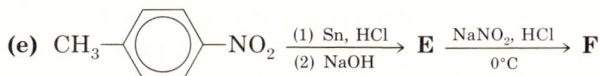

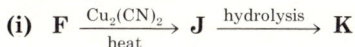

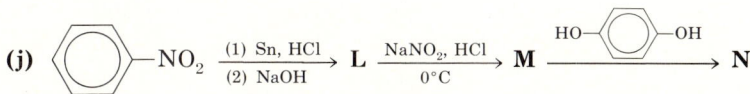

(a) ⬠—NH_2 + HBr $\longrightarrow$ **A**

(b) $CH_3CH_2NH_2$ + 3 $CH_3Br \longrightarrow$ **B**

(c) ⬡—NH_2 + ⬡—$\overset{\overset{O}{\|}}{\underset{\underset{O}{\|}}{S}}$—Cl $\longrightarrow$ **C**

(d) $CH_3NHCH_2CH_3$ $\xrightarrow[0°C]{NaNO_2, HCl}$ **D**

(e) CH_3—⬡—NO_2 $\xrightarrow[(2)\ NaOH]{(1)\ Sn,\ HCl}$ **E** $\xrightarrow[0°C]{NaNO_2,\ HCl}$ **F**

(f) **F** $\xrightarrow{H_2O}$ **G**

(g) **F** $\xrightarrow[heat]{HBF_4}$ **H**

(h) **F** $\xrightarrow{KI}$ **I**

(i) **F** $\xrightarrow[heat]{Cu_2(CN)_2}$ **J** $\xrightarrow{hydrolysis}$ **K**

(j) ⬡—NO_2 $\xrightarrow[(2)\ NaOH]{(1)\ Sn,\ HCl}$ **L** $\xrightarrow[0°C]{NaNO_2,\ HCl}$ **M** $\xrightarrow{HO—⬡—OH}$ **N**

14.12 Nylon 66 can be made by condensation between 1,6-diaminohexane and adipic acid, $HOOC(CH_2)_4COOH$, as starting materials. Show how the two raw materials can be synthesized from 1,4-dichlorobutane.

14.13 Write the structural formulas for the dyes obtained when these compounds interact.

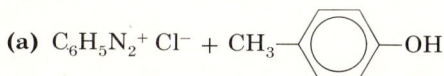

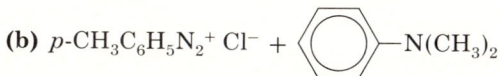

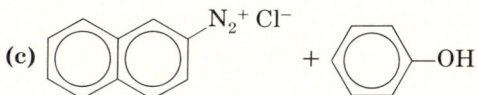

(a) $C_6H_5N_2^+$ Cl^- + CH_3—⬡—OH

(b) p-$CH_3C_6H_5N_2^+$ Cl^- + ⬡—$N(CH_3)_2$

(c) ⬡⬡—N_2^+ Cl^- + ⬡—OH

14.14 Write equations for the following conversions (some may involve more than one step).
  (a) Toluene to *p*-toluidine
  (b) Acetamide to methylamine
  (c) Ethyl bromide to *n*-propylamine
  (d) Benzene to azobenzene
  (e) Toluene to *p*-methylphenol (*p*-cresol)
  (f) Nitrobenzene to fluorobenzene

Visual Tests and Structure Identification [Secs. 14.10–14.12]

14.15 Describe a simple chemical test that would distinguish between each pair of compounds. Tell exactly what you would do and see.
  (a) 1-Aminobutane and diethylamine
  (b) Diethylamine and triethylamine
  (c) 1-Aminobutane and aniline
14.16 An amine **(A)** has the formula C_3H_9N. When **A** was treated with nitrous acid at $0°C$, a reaction took place and an oily yellow layer separated from the reaction mixture. Write a structure for **A** and give an equation for the reaction.
14.17 An amide **(A)**, $C_{13}H_{11}NO$, was heated with aqueous sodium hydroxide. When cooled, the mixture consisted of two separate layers. The top oily layer contained compound **B,** which was soluble in acid. The bottom layer on acidification gave a precipitate, **C.** Treatment of **B** with cold nitrous acid produced a compound, **D,** that formed a dye with phenol. Treatment of **C** with $SOCl_2$ followed by NH_3 produced compound **E,** C_7H_7NO. When **E** was treated with alkaline NaOBr, compound **B** was formed. Write structures for **A** through **E** and give equations for all reactions.

Biogenic Amines [Sec. 14.13]

14.18 Define **(a)** biogenic amine, **(b)** psychomimetic drug, **(c)** amphetamines, **(d)** barbiturates, and **(e)** heterocyclic compounds.

Amino Acids, Peptides, and Proteins

The name protein comes from the Greek *proteios,* meaning "of prime importance." The name is appropriate, for proteins represent one of the most important groups of compounds. Proteins are a major constituent in every living cell. They play an important role in the cell wall, in the liquid portion of the cell, and in the various cellular particles and other structures. The structural material of muscle, skin, hair, and other body tissues is made up of proteins. Proteins also have a variety of other functions. As enzymes, they catalyze biochemical reactions; as hormones, they regulate the activity of various organs in the body; as antigens, they invade the organism to bring it disease or even death; and as antibodies, they counteract the adverse effects of antigens.

Chemically, proteins are complex, high-molecular-weight compounds that contain carbon, hydrogen, oxygen, and nitrogen. Many proteins also contain sulfur, phosphorus, and traces of other elements. The nitrogen content of most proteins is about 16%, which makes proteins the main source of nitrogen intake for our bodies, without which we could not survive. The chemical makeup of proteins is complex, and it is best to study this class of compounds by considering first the amino acids, which are the building blocks of proteins.

Structure and Classification of Amino Acids 15.1

Because amino acids are the building blocks of proteins, a study of the structures and properties of amino acids is essential to the understanding of proteins. There are more than 200 naturally occurring amino acids; only about

20 are found in proteins. All amino acids that are found in proteins have an amino group (—NH$_2$) adjacent to (α-) the carboxyl group, hence the name **α-amino acids.** With the exception of proline and hydroxyproline, which have a secondary amino group, all other amino acids have a primary amino function. The general formula of α-amino acids is

Common to
most amino ⟶   COOH
acids | ⟵ α carbon
           H$_2$N—C—H
                      |
                      R ⟵ Side chain
                   (differs from one amino
                       acid to another)

The structures of α-amino acids differ from one another in the nature of the R group. The R group is called the **side chain.** The side chain varies considerably, from a simple hydrogen atom in glycine to a more complex structure such as

the guanidine derivative, $\begin{matrix} H_2N \\ \diagdown \\ CNHCH_2CH_2CH_2- \\ \diagup \\ HN \end{matrix}$, in arginine. The differences

in structure of the side chain serve as a convenient way for classifying amino acids into various categories, as illustrated in Table 15.1, which gives the full and abbreviated names and the structures of the 21 amino acids that are usually found in proteins.

Although amino acids could be named by the IUPAC system, the usual practice is to use the common names listed in Table 1.1. The trivial names reflect the source or property of the substance. For example, asparagine is so named because it was first isolated from asparagus juice, and glycine comes from *glykis,* the Greek word for sweet. Standard three-letter abbreviations, derived from the common names, are used to designate the amino acids that make up peptides and proteins.

To facilitate remembering the names and structures of the amino acids listed in Table 15.1, we have classified the compounds into four general categories, depending on the kind of side chains present.

In the first category we have grouped amino acids with *nonpolar,* or *hydrophobic,* side chains (entries 1–10). The nonpolar side chains may be aliphatic (entries 1–6), aromatic (entries 7 and 8), or thiol derivatives (entries 9 and 10). The solubility of the amino acids in water decreases rapidly as the size of the side chain increases: glycine is quite soluble, whereas tryptophan and phenylalanine have limited solubilities. Proline is the only amino acid with the amino group incorporated into a five-membered ring. Because proline has a ring that includes the α carbon, the presence of this amino acid in a protein affects profoundly its three-dimensional structure.

Amino acids in the second category have *polar but uncharged* side chains at the pH of the human body. These amino acids (entries 11–16) are much more soluble in water than the first group. The hydroxyl and sulfhydryl (SH) groups often play a key role in the proteins of which they are a part. Asparagine and glutamine (entries 15 and 16) are amide derivatives of aspartic and glutamic acids, respectively. Aspartic acid and glutamic acid (entries 17 and 18) are classified as *acidic amino acids* (the third category) because they contain two acidic groups and only one basic amino group. The side chains of these amino acids are negatively charged at the physiological pH (that is, the carboxyl group does not

Table 15.1 Amino Acids Commonly Found in Proteins

Name	Abbre-viation	Structural formula	pI[a]

A. Neutral Amino Acids with Nonpolar Side Chains

1. Glycine — Gly

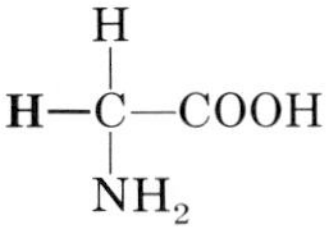

$$H-\underset{\underset{NH_2}{|}}{\overset{\overset{H}{|}}{C}}-COOH \qquad 6.1$$

2. Alanine — Ala

$$CH_3-\underset{\underset{NH_2}{|}}{\overset{\overset{H}{|}}{C}}-COOH \qquad 6.1$$

3. Valine — Val

$$\underset{CH_3}{\overset{CH_3}{>}}CH-\underset{\underset{NH_2}{|}}{\overset{\overset{H}{|}}{C}}-COOH \qquad 6.0$$

4. Leucine — Leu

$$\underset{CH_3}{\overset{CH_3}{>}}CH-CH_2-\underset{\underset{NH_2}{|}}{\overset{\overset{H}{|}}{C}}-COOH \qquad 6.0$$

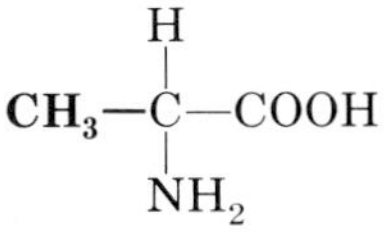

5. Isoleucine — Ile

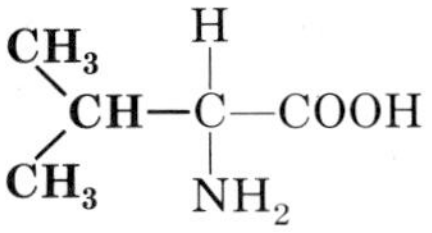

$$CH_3-CH_2-\underset{\underset{CH_3}{|}}{CH}-\underset{\underset{NH_2}{}}{\overset{\overset{H}{|}}{C}}-COOH \qquad 6.0$$

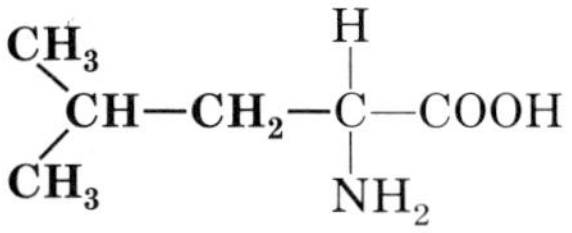

6. Proline — Pro

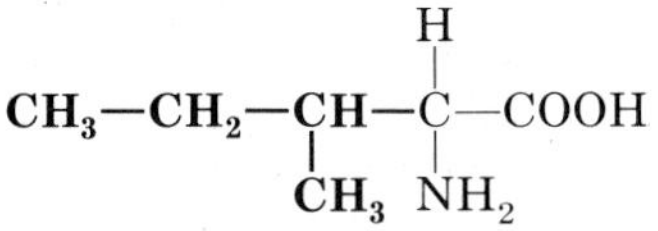

$$\underset{H_2C-NH}{\overset{CH_2}{H_2C}}CH-COOH \qquad 6.4$$

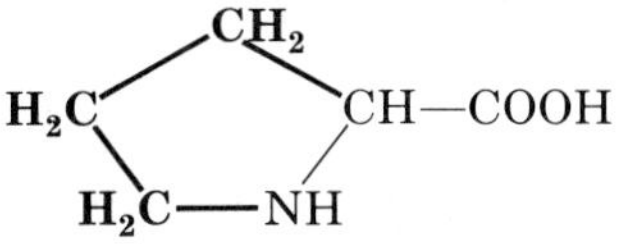

7. Phenylalanine — Phe

$$\bigcirc\!\!-CH_2-\underset{\underset{NH_2}{|}}{\overset{\overset{H}{|}}{C}}-COOH \qquad 5.9$$

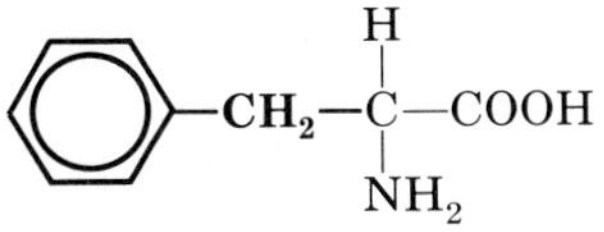

8. Tryptophan — Trp

$$\overset{C-CH_2-\underset{\underset{NH_2}{}}{\overset{\overset{H}{|}}{C}}-COOH}{\underset{NH}{CH}} \qquad 5.9$$

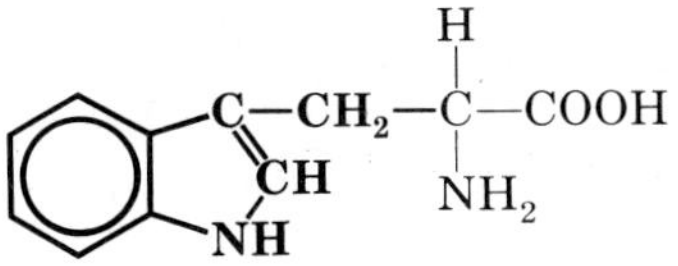

9. Methionine — Met

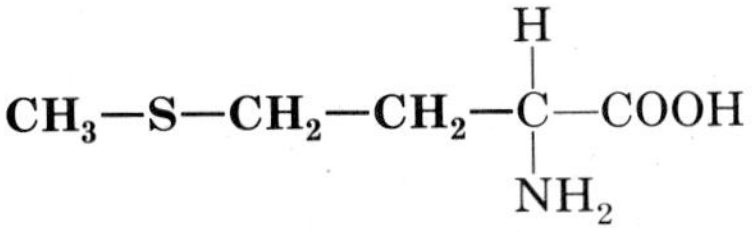

$$CH_3-S-CH_2-CH_2-\underset{\underset{NH_2}{|}}{\overset{\overset{H}{|}}{C}}-COOH \qquad 5.8$$

10. Cystine — (Cys)$_2$

$$\left._2\right(S-CH_2-\underset{\underset{NH_2}{|}}{CH}-COOH) \qquad 5.1$$

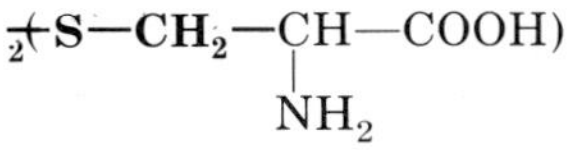

[a]pI stands for isoelectric point, a term that is defined in Section 15.5.

Table 15.1   (continued)

Name	Abbre-viation	Structural formula	pI^a

B. Neutral Amino Acids with Polar but Uncharged Side Chains

11. Serine   Ser

$$HO-CH_2-\overset{\overset{\displaystyle H}{|}}{\underset{\underset{\displaystyle NH_2}{|}}{C}}-COOH$$ 5.7

12. Threonine   Thr

$$CH_3-\overset{\overset{\displaystyle H}{|}}{\underset{\underset{\displaystyle OH}{|}}{CH}}-\overset{\overset{\displaystyle H}{|}}{\underset{\underset{\displaystyle NH_2}{|}}{C}}-COOH$$ 6.5

13. Cysteine   Cys

$$HS-CH_2-\overset{\overset{\displaystyle H}{|}}{\underset{\underset{\displaystyle NH_2}{|}}{C}}-COOH$$ 5.1

14. Tyrosine   Tyr

$$HO-\bigcirc-CH_2-\overset{\overset{\displaystyle H}{|}}{\underset{\underset{\displaystyle NH_2}{|}}{C}}-COOH$$ 5.7

15. Asparagine   Asn

$$\underset{NH_2}{\overset{O}{\diagdown}}C-CH_2-\overset{\overset{\displaystyle H}{|}}{\underset{\underset{\displaystyle NH_2}{|}}{C}}-COOH$$ 5.4

16. Glutamine   Gln

$$\underset{NH_2}{\overset{O}{\diagdown}}C-CH_2-CH_2-\overset{\overset{\displaystyle H}{|}}{\underset{\underset{\displaystyle NH_2}{|}}{C}}-COOH$$ 5.7

C. Acidic Amino Acids with Negatively Charged Side Chains

17. Aspartic acid   Asp

$$\underset{HO}{\overset{O}{\diagdown}}C-CH_2-\overset{\overset{\displaystyle H}{|}}{\underset{\underset{\displaystyle NH_2}{|}}{C}}-COOH$$ 3.0

18. Glutamic acid   Glu

$$\underset{HO}{\overset{O}{\diagdown}}C-CH_2-CH_2-\overset{\overset{\displaystyle H}{|}}{\underset{\underset{\displaystyle NH_2}{|}}{C}}-COOH$$ 3.2

D. Basic Amino Acids with Positively Charged Side Chains

19. Lysine   Lys

$$\underset{NH_2}{\overset{}{}}CH_2-CH_2-CH_2-CH_2-\overset{\overset{\displaystyle H}{|}}{\underset{\underset{\displaystyle NH_2}{|}}{C}}-COOH$$ 9.7

Table 15.1 (continued)

405

15.2 Essential
Amino Acids

Name	Abbre-viation	Structural formula	pI[a]
20. Histidine	His	(Imidazole group) $C-CH_2-\overset{H}{\underset{NH_2}{C}}-COOH$	7.6
21. Arginine	Arg	(Guanidine group) $C-N-CH_2-CH_2-CH_2-\overset{H}{\underset{NH_2}{C}}-COOH$	10.8

exist as —COOH but rather as the ion —COO⁻). Amino acids 15–18 are very soluble in water.

The amino acids in the fourth category (entries 19–21) have side chains that are positively charged at the physiological pH range. They are called *basic amino acids* because they contain two basic functions but only one acidic —COOH group.

In addition to the 21 amino acids listed in Table 15.1, which are widely distributed in all proteins, others occur, often at high concentration, in a few proteins. For example, hydroxyproline and hydroxylysine represent more than 12% of *collagen,* the structural protein of bones and cartilage.

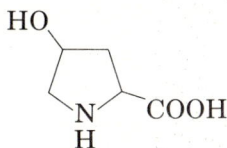

Hydroxyproline

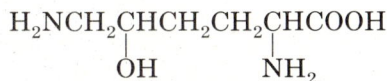

$H_2NCH_2CHCH_2CH_2CHCOOH$ with OH and NH_2

Hydroxylysine

Essential Amino Acids *15.2*

Ten amino acids are shaded in Table 15.1. These are the **essential amino acids,** those that the body cannot manufacture at all or cannot produce in sufficient amounts to maintain good health. The requirement of each essential amino acid is in the range of 1–2 g per day. They must be supplied to us from an outside source: the proteins we eat. Not all proteins contain all the essential amino acids; those that do are called **complete proteins.** *Casein* (the protein from milk), for example, is a complete protein: it contains 19 common amino acids and all the essential ones. *Zein* (corn protein), on the other hand, is not a complete protein because it is deficient in two essential amino acids, lysine and tryptophan. Nutritionally, it is best to eat proteins from a variety of sources to ensure an adequate supply of the essential amino acids in the diet.

15.3 Stereochemistry: D and L Families

With the exception of glycine (where R = H), all the amino acids listed in Table 15.1 have four different groups attached to the α carbon. This carbon is therefore a *chiral center*. As a result, there are two possible structures, one the mirror image of the other. By analogy with the convention adopted in carbohydrate chemistry, we divide amino acids into D and L families (Sec. 6.12A).

D-Amino acids are those in which the NH_2 group points to the right of the α carbon in the Fischer representation of an amino acid (the —COOH is at the top and the carbon chain is at the bottom). Conversely, L-amino acids have the NH_2 pointing to the left.

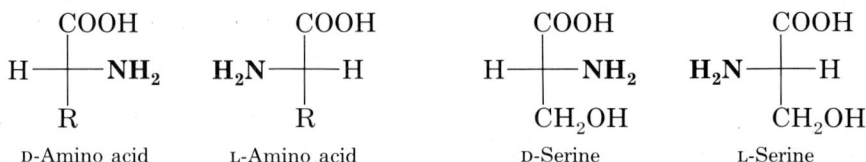

D-Amino acid   L-Amino acid   D-Serine   L-Serine

Although amino acids with the D configuration are found as part of certain peptides with antibiotic properties, *only L-amino acids are incorporated in proteins.* As with sugars, the D and L designations refer to configurations only; they have nothing to do with the sign of optical rotation, (+) or (−), which must be determined experimentally.

15.4 Ionic Properties of Amino Acids

For simplicity, the α-amino and the α-carboxyl groups of the amino acids are shown as nonionized forms, $RCHNH_2COOH$, in Table 15.1. In fact, the two groups rarely exist as such. The reason is simple. The acidic group (—COOH) and the basic function (—NH_2) are attached to the same carbon, and we know that acids react with bases to form salts, which are *ionic compounds*. We could expect the proton from the —COOH group to be transferred to the basic —NH_2 group in an internal acid-base reaction. This indeed does happen in the physiological environment (at pH 7).

Internal acid-base reaction in amino acids

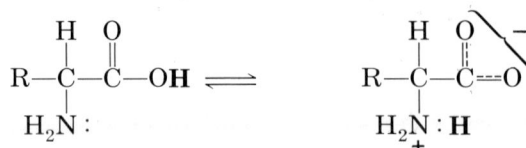

Dipolar structure of an α-amino acid
(a zwitterion)

The resulting structure, which has a positive and a negative charge within the same molecule, is called a *dipolar ion* or **zwitterion** (from the German *zwitter,* double). Like the nonionized structure, the dipolar ionic structure of the amino acids that carry no charge on the R group (categories 1 and 2 in Table 15.1) have a net charge of zero and are therefore electrically neutral (the net charge refers to the algebraic sum of the positive and negative charges in the molecule). The dipolar ionic structure, which is salt-like, nevertheless describes the properties of amino acids better than the nonionized structure. For example, all amino acids are high-melting-point solids that often decompose on heating before reaching their melting points. Also, all amino acids dissolve more easily in water than in organic solvents.

The zwitterionic structure is in agreement also with the **amphoteric properties** of amino acids (that is, their ability to react with acids as well as with bases). When the dipolar ion is treated with a strong acid (HCl, for example), the carboxylate anion ($-COO^-$) picks up a proton and is converted to a carboxyl group ($-COOH$). As a result, the dipolar ion is transformed to a *cationic form,* which has a *net positive charge.*

In acid

$$H_3\overset{+}{N}-\underset{\underset{R}{|}}{\overset{\overset{COO:^-}{|}}{C}}-H \quad + \overset{\downarrow}{H}^+ \rightleftharpoons H_3\overset{+}{N}-\underset{\underset{R}{|}}{\overset{\overset{COOH}{|}}{C}}-H$$

Dipolar ion   Cationic form
(net charge of zero)   (net positive charge at low pH)

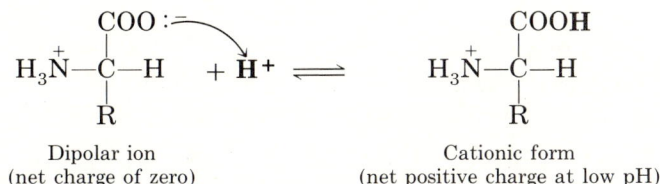

In strongly basic solution amino acids exist as *anionic structures,* which have a *net negative charge.*

In base

$$H_2\overset{+}{N}-\underset{\underset{\mathbf{H}\ R}{|}}{\overset{\overset{COO:^-}{|}}{C}}-H \quad + :OH^- \rightleftharpoons H_2N-\underset{\underset{R}{|}}{\overset{\overset{COO:^-}{|}}{C}}-H \quad + H_2O$$

Anionic form
(net negative charge
at high pH)

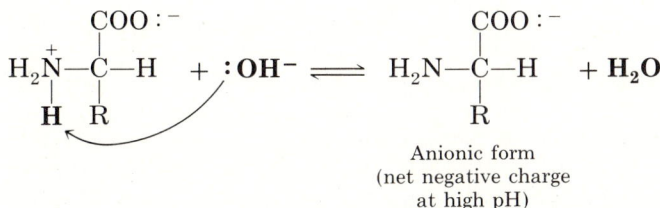

Because of their amphoteric properties, amino acids (and proteins) can function as *biological buffers* and can thus maintain the pH of the body.

Problem 15.1 Draw (1) the dipolar ionic form, (2) the cationic form, and (3) the anionic form of **(a)** glycine and **(b)** alanine.

Isoelectric Points 15.5

From the preceding discussion it is obvious that the exact structure of an amino acid depends on the pH. The pH at which an amino acid exists as a zwitterion is known as the **isolectric point** or the **pI.** At the isoelectric point, amino acids (and proteins) have a zero net charge and are electrically neutral. The isoelectric points of the 21 amino acids commonly found in proteins are given in Table 15.1. Note that each amino acid has a particular pI value, which does not usually coincide with pH 7.0. When the pH of a solution is greater than the pI of the compound, the amino acid will carry a net negative charge (anionic form). Conversely, when the pH of the solution is below the pI value, the amino acid exists predominantly in the cationic form (net positive charge).

The net charges associated with the structures of amino acids (and proteins) are the basis of two important separation techniques: electrophoresis and ion-exchange chromatography. These techniques are explained in the next section.

Problem 15.2 Indicate what structure (zwitterionic, cationic, or anionic) will predominate at pH 6.0 for **(a)** leucine, **(b)** aspartic acid, and **(c)** lysine. (*Hint:* Consult Table 15.1 for the pI values.)

15.6 Separation of Amino Acids

One important part of the analysis of a peptide or a protein is the determination of its amino acid composition. Usually this is done by splitting the peptide or protein into its constituent amino acids. The next step involves the separation and identification of the amino acids in the mixture.

A Electrophoresis

Electrophoresis is one technique used to separate and identify a mixture of amino acids. The separation is based on the difference in net charges of the components in the mixture. Essentially, the procedure consists of applying a mixture of amino acids onto an *inert support,* which may be paper, cellulose

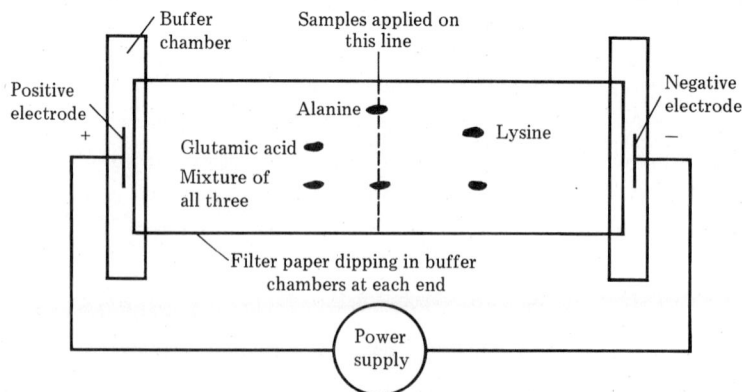

Figure 15.1 Paper electrophoresis (schematic). The spots on the paper show how three amino acids can be separated and identified by this method at pH 6.0.

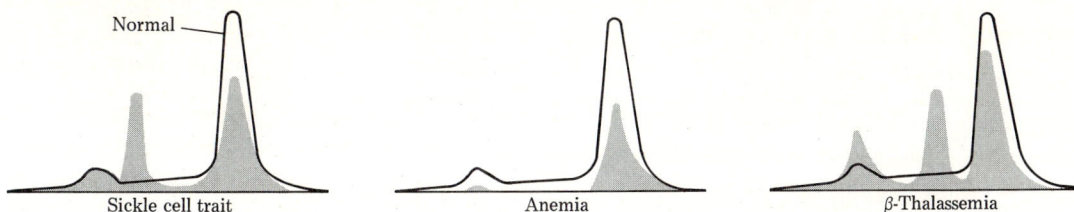

Figure 15.2 Electrophoretic patterns of normal and abnormal hemoglobin. [Courtesy of Gelman Instrument Company.]

acetate, or some polymeric gel. The ends of the support are immersed into a chamber containing a buffer solution of known pH. A direct current is then passed through the solution. Amino acids with a *net positive charge migrate toward the cathode (the negative pole),* whereas those that are *negatively charged move toward the anode (the positive pole).* Amino acids that exist as *zwitterions remain immobile.* Figure 15.1 shows an electrophoresis apparatus.

After the direct current is turned off, the positions of the various amino acids can be detected visually. This is done by spraying the support with a developing agent, such as *ninhydrin,* which reacts with amino acids to give a purple color. (If some ninhydrin were to spill on your skin, it would turn purple because your skin is made of proteins.) Treatment of the support with ninhydrin produces a series of colored bands, known as an *electrophoretic pattern.* The electrophoretic pattern of an unknown mixture may be compared with that of a standard for identification purposes. Electrophoresis is an invaluable technique not only for identification and separation purposes but also as a medical diagnostic tool (see Fig. 15.2).

B   Ion-Exchange Chromatography

Ion-exchange chromatography is another technique used for the selective separation of amino acid mixtures (and proteins) based on the difference in their net charges. Chemically, ion-exchange resins are insoluble polymeric materials that contain ionizable groups on their surfaces. If the ionizable groups are positively charged (H^+, Na^+, etc.), the resin is called a *cation-exchange resin.* Conversely, if the ionizable groups are negatively charged, the resin is referred to as an *anion-exchange resin.* Physically, the resins appear as spherical particles, represented schematically in Figure 15.3.

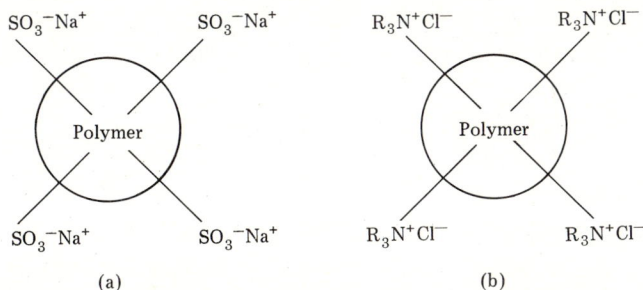

Figure 15.3 **(a)** A cation exchange resin; Na^+ is the ionizable group. **(b)** An anion-exchange resin; Cl^- is the ionizable group.

The exchange process between a cation-exchange resin and amino acids in their cationic form (AA^+) is illustrated in Figure 15.4a. The ionizable positive ions from the resin (Na^+ ions in the illustration) are replaced by positively charged amino acids. The amino acids are held onto the resin by electrostatic attraction between opposite charges (SO_3^- from the resin and AA^+). The same kind of attractive forces exist between negatively charged amino acids (AA^-) and anion-exchange resins (Fig. 15.4b).

Experimentally, ion-exchange chromatography is carried out as follows. A slurry of an ion-exchange resin is poured into a cylindrical column to a height of several centimeters. A small volume (about 1 ml) of a solution containing a mixture of amino acids is placed on top of the column. Positively charged amino acids interact with the resin in the column and become bound to it. The degree of binding depends on the magnitude of the net charges on the different amino acids: the greater the net positive charge, the greater the forces of attraction between AA^+ and the resin particles and the slower the movement of these amino acids along the column. Amino acids are made to emerge selectively from the column by running solutions of gradually increasing pH through it. The effluents emerging from the column are collected systematically as fractions. These fractions can then be analyzed by a suitable technique, such as one that involves the use of ninhydrin as a visualizing reagent. A typical ion-exchange column is shown in Figure 15.5.

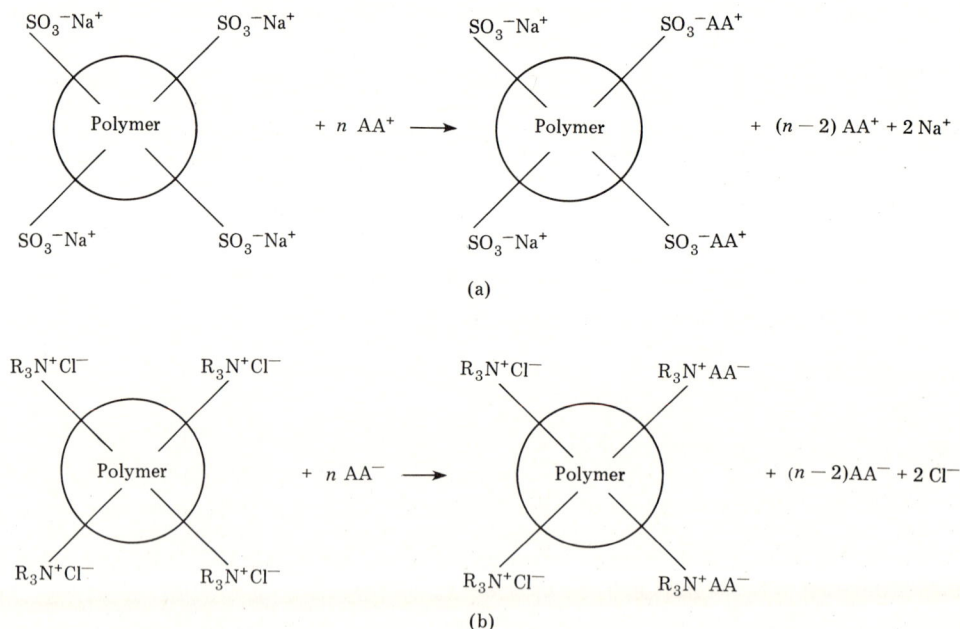

(a)

(b)

Figure 15.4 **(a)** Exchange process between positively charged amino acids and a cation-exchange resin, showing the replacement of two ionizable Na^+ ions by two positively charged amino acids, AA^+. **(b)** Exchange process between negatively charged amino acids and an anion-exchange resin, showing the replacement of two ionizable Cl^- ions by two negatively charged amino acids, AA^-.

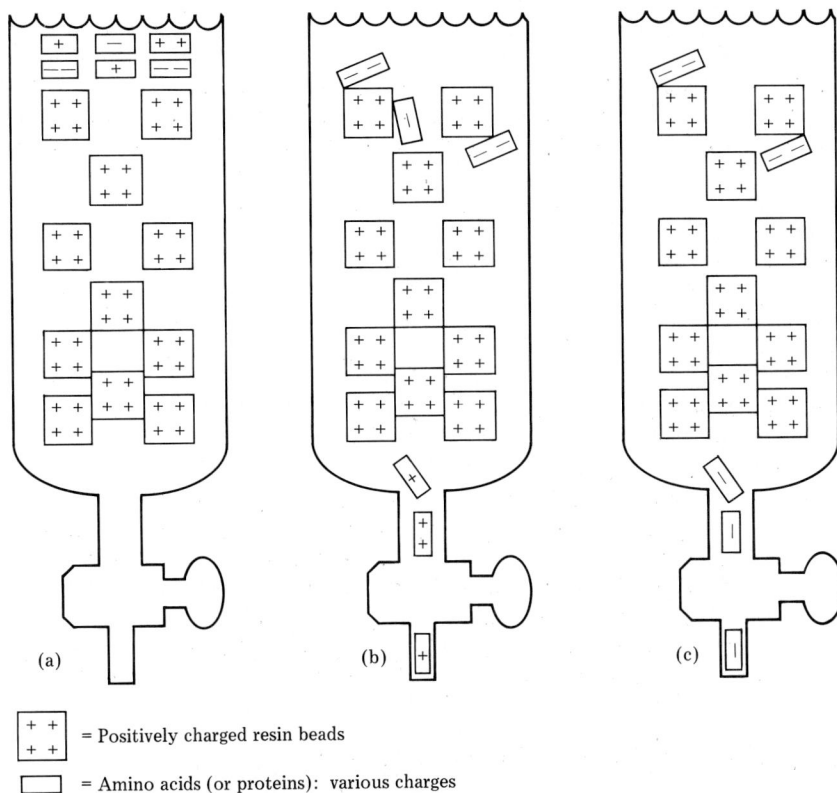

= Positively charged resin beads

= Amino acids (or proteins): various charges

Figure 15.5 **(a)** Ion exchange column with a mixture of amino acids (or proteins) of various charges. **(b)** Amino acids (or proteins) with the same net charge pass through; those with the opposite charge stick. **(c)** Increasing the pH of the solution gradually achieves separation of weakly held amino acids (or proteins).

There are now commercially available "automated amino acid analyzers." As the name implies, these machines are capable of separating and analyzing mixtures of amino acids automatically. The results of such an analysis are recorded as chromatograms (Fig. 15.6).

Reactions of Amino Acids **15.7**

Amino acids undergo many of the reactions characteristic of the functional groups.

A Reactions of the Amino Group

Three reactions of the amino group are especially useful in connection with the analysis of amino acids, peptides, and proteins.

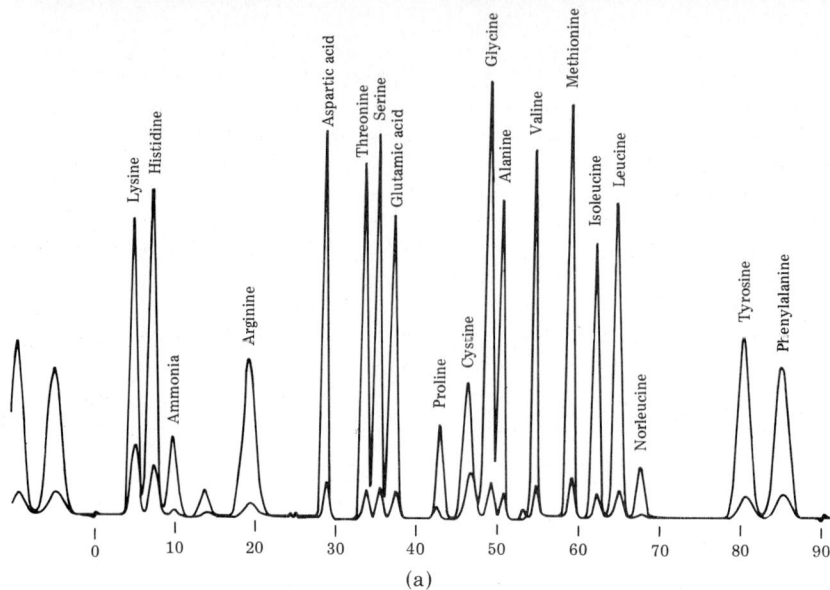

(a)

(b)

Figure 15.6 **(a)** Chromatogram of an amino acid mixture separated by an amino acid analyzer such as shown in **(b).** The area under each peak is proportional to the concentration of acid present. [(a) courtesy of Perkin-Elmer Corporation; (b) 121 MB Amino Acid Analyzer, Beckman Instruments, Inc., Spinco Division.]

Reaction with nitrous acid. You will recall that primary aliphatic amines react with nitrous acid to liberate N_2 and produce a mixture of organic compounds. Amino acids with primary α-amino groups also react with nitrous acid. The products are an α-hydroxy acid, water, and nitrogen gas.

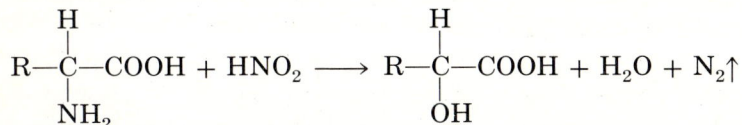

$$\underset{\underset{NH_2}{|}}{\overset{\overset{H}{|}}{R-C-COOH}} + HNO_2 \longrightarrow \underset{\underset{OH}{|}}{\overset{\overset{H}{|}}{R-C-COOH}} + H_2O + N_2\uparrow$$

The reaction is quantitative, and it is used in the *Van Slyke method* to estimate the number of primary α-amino groups in amino acids, peptides, and proteins. The Van Slyke determination of amino nitrogen consists of trapping the evolved N_2 gas and measuring its volume. Proline and hydroxyproline, which contain no primary α-amino groups, do not give off N_2.

Reaction with ninhydrin. **Ninhydrin** reacts with ammonia, NH_3, to give a purple-colored dye. Ninhydrin also produces a purple complex with primary amines that have a hydrogen attached to the carbon holding the amino group, $\diagup\!\!\!\!\diagdown\!\!CH-NH_2$. Most amino acids fall in this category and will therefore give a positive test with ninhydrin.

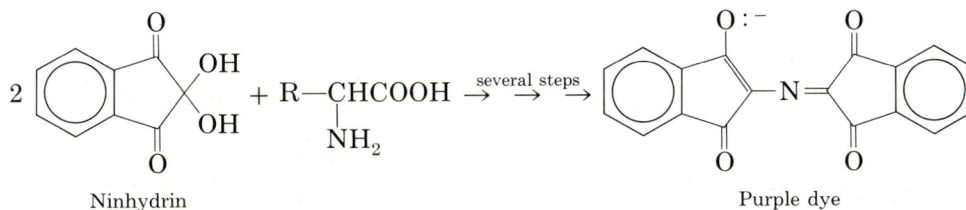

Ninhydrin   Purple dye

Problem 15.3 Which amino acid(s) listed in Table 15.1 will give a negative ninhydrin test?

Ninhydrin has been used extensively in conjunction with ion-exchange chromatography to estimate quantitatively the concentration of amino acids. The intensity of the color produced with the reagent can be related directly to the concentration of amino acid. Ninhydrin is also used qualitatively to detect the position of amino acids in electrophoresis, in paper chromatography, and in thin-layer chromatography.

Reaction with 2,4-dinitrofluorobenzene (Sanger's reagent). The 1° amino group of amino acids reacts with 2,4-dinitrofluorobenzene (DNFB) to yield a yellow product, the 2,4-dinitrophenyl derivative, or DNP-amino acid.

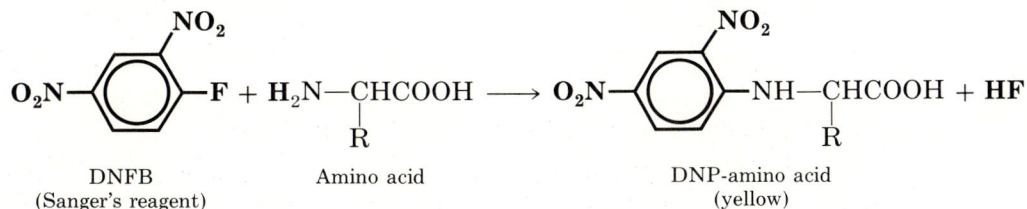

DNFB
(Sanger's reagent)   Amino acid   DNP-amino acid
                                                                       (yellow)

Peptides and proteins with free NH_2 groups also react with DNFB to yield yellow DNP derivatives (see Sec. 15.10). DNFB is also known as Sanger's reagent, after its discoverer, who first used it to establish the structure of insulin.

B   Reactions of the R groups

Among the reactions of the side chains, the most significant is the formation of a disulfide bond. The reaction involves the thiol (—SH) groups of two cysteine molecules to form cystine.

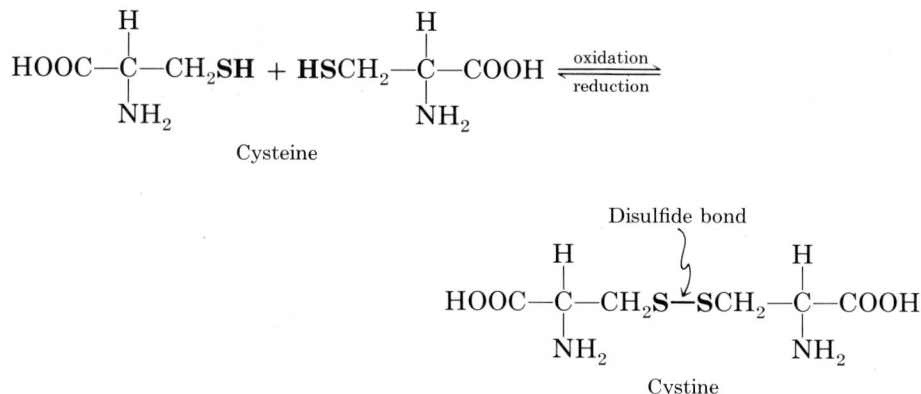

$$
\underset{\text{Cysteine}}{HOOC-\underset{\underset{NH_2}{|}}{\overset{\overset{H}{|}}{C}}-CH_2\mathbf{SH}} \;+\; \mathbf{HS}CH_2-\underset{\underset{NH_2}{|}}{\overset{\overset{H}{|}}{C}}-COOH \quad \underset{\text{reduction}}{\overset{\text{oxidation}}{\rightleftharpoons}}
$$

Disulfide bond

$$
\underset{\text{Cystine}}{HOOC-\underset{\underset{NH_2}{|}}{\overset{\overset{H}{|}}{C}}-CH_2S\!-\!SCH_2-\underset{\underset{NH_2}{|}}{\overset{\overset{H}{|}}{C}}-COOH}
$$

The disulfide bond in cystine can easily be broken with a reducing agent to regenerate the —SH groups.

Disulfide bonds between cysteine residues occur frequently in proteins, where they serve as *bridges* connecting different portions of the molecule. An example of such disulfide bridges is shown in Figure 15.7 for a hypothetical molecule containing 20 amino acids.

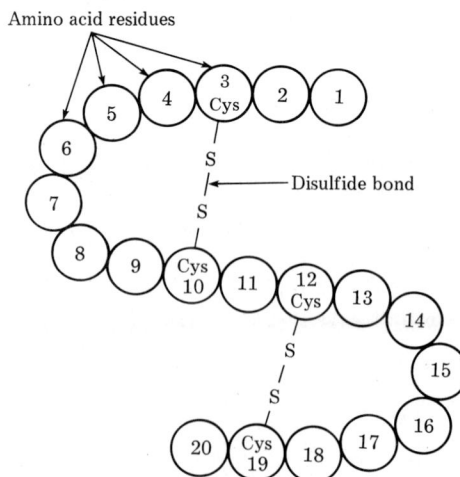

Figure 15.7 Two disulfide bonds between cysteine residues.

The presence of disulfide bonds has a direct effect on the shape of the molecule, which in turn determines whether or not the molecule will function biologically. For example, the enzyme ribonuclease has four disulfide bridges (Fig. 15.14). If any is broken, the shape of the molecule is altered and the catalytic function is lost. The hormone insulin is another example of a protein in which disulfide bonds play a crucial role. The insulin molecule is made up of two chains of amino acids connected via —S—S— bridges. Rupture of the disulfide bonds connecting the two chains causes loss of activity in the molecule (Fig. 15.13).

C Reactions of the Carboxyl Group

The carboxyl group of amino acids may be converted to any of the derivatives of carboxylic acids discussed in Chapter 12. Of these, amides, produced when the α-carboxyl group of one amino acid reacts with the α-amino group of another amino acid are, without a doubt, the most important.

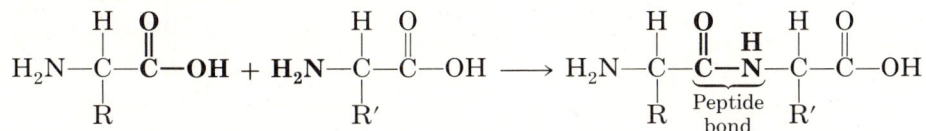

$$H_2N-\underset{R}{\underset{|}{C}}\overset{H}{\underset{|}{}}-\overset{O}{\underset{||}{C}}-OH \;+\; H_2N-\underset{R'}{\underset{|}{C}}\overset{H}{\underset{|}{}}-\overset{O}{\underset{||}{C}}-OH \longrightarrow H_2N-\underset{R}{\underset{|}{C}}\overset{H}{\underset{|}{}}-\underset{\text{Peptide bond}}{\overset{O}{\underset{||}{C}}}-\underset{}{\overset{H}{\underset{|}{N}}}-\underset{R'}{\underset{|}{C}}\overset{H}{\underset{|}{}}-\overset{O}{\underset{||}{C}}-OH$$

The amide link formed in the manner shown is called a **peptide bond.** Since all amino acids are incorporated into proteins through peptide bonds, it is appropriate that we discuss peptides and peptide bonds in greater detail.

Problem 15.4 Draw the structures of the following compounds and indicate the significance of each in analysis of amino acids (and proteins).
(a) Sanger's reagent   **(b)** Ninhydrin

The Peptide Bond; Classification of Peptides *15.8*

Early in this century, Emil Fischer showed that all proteins, despite the diversity of their biological functions, have one structural feature in common: *their amino acids are connected to one another via peptide bonds.* In the process of peptide bond formation, water is given off and the product is called a **peptide.** A peptide made by combining two amino acids is called a *dipeptide.* A dipeptide is composed of *two* amino acid residues, linked by *one* peptide bond. The amino acid residues may be the same or different.

If a third amino acid is joined to a dipeptide in the same manner, the product is a *tripeptide.* A tripeptide contains *three* amino acid residues linked by *two* peptide bonds. Similar combinations of four, five, and six amino acids give a tetrapeptide, a pentapeptide, and a hexapeptide, respectively, and so on. The term *oligopeptide* is used to denote any peptide having up to *ten* amino acid residues.

Peptides containing more than *ten* amino acid units are called **polypeptides. Proteins** are polypeptides; most proteins have 100 or more amino acid residues. There is no clear line of demarcation that distinguishes polypeptides from proteins. Thus, insulin, although it contains only 51 amino acid units, is generally considered a small protein.

Although cyclic peptides do exist, most peptides are chain-like molecules with an $-NH_2$ group at one end of the chain and a $-COOH$ group at the other end, for example,

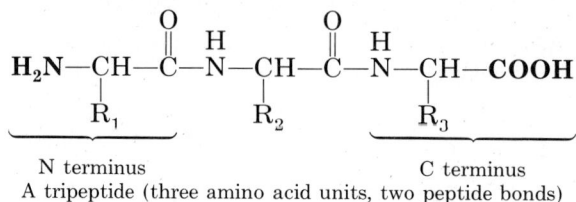

$$H_2N-CH-\overset{\overset{O}{\|}}{C}-\overset{\overset{H}{|}}{N}-CH-\overset{\overset{O}{\|}}{C}-\overset{\overset{H}{|}}{N}-CH-COOH$$

$$\underbrace{ R_1 } \qquad R_2 \qquad \underbrace{R_3 }$$

N terminus   C terminus
A tripeptide (three amino acid units, two peptide bonds)

The $-\overset{\overset{O}{\|}}{C}-\overset{\overset{H}{|}}{N}-C-$ sequence of bonds in a peptide is referred to as its **backbone,** and the R groups are called the **side chains.** The amino acid unit with the free amino group is known as the **N-terminal amino acid** or the **N terminus.** The one with the free carboxyl group is called the **C-terminal amino acid** or the **C terminus.**

By convention, the structure of a peptide is written with the N terminus on the left and the C terminus on the right. The amino acid units are numbered consecutively, beginning from the N terminus as residue number 1 and ending with the C terminus.

In naming a peptide, we also start from the N terminus and proceed in order until the C terminus is reached. Each amino acid residue, with the exception of the C terminus, is named by replacing the suffix *-ine* in the parent amino acid by *-yl.* The C terminus retains the name of the amino acid. For example, a tripeptide made of glycine, alanine, and serine in the order given is called glycylalanylserine. Its structure is

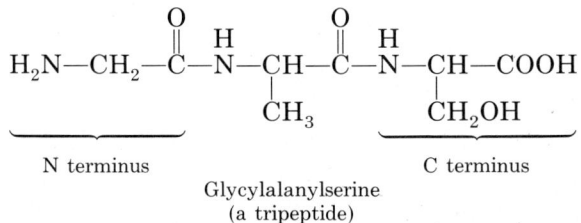

$$H_2N-CH_2-\overset{\overset{O}{\|}}{C}-\overset{\overset{H}{|}}{N}-\underset{\underset{CH_3}{|}}{CH}-\overset{\overset{O}{\|}}{C}-\overset{\overset{H}{|}}{N}-\underset{\underset{CH_2OH}{|}}{CH}-COOH$$

N terminus   C terminus
Glycylalanylserine
(a tripeptide)

Because the structural formulas of even the simpler peptides are cumbersome to draw, chemists and biochemists use abbreviated structures for most purposes. The amino acid units in a peptide or a protein are given the three-letter symbols shown in Table 15.1. The peptide bond connecting one amino acid residue to another may be spelled out as in Figure 15.8a; more commonly the acid abbreviations are connected by hyphens (Fig. 15.8b).

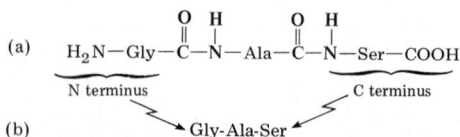

(a) $\quad H_2N-Gly-\overset{\overset{O}{\|}}{C}-\overset{\overset{H}{|}}{N}-Ala-\overset{\overset{O}{\|}}{C}-\overset{\overset{H}{|}}{N}-Ser-COOH$

N terminus   C terminus

(b)   Gly-Ala-Ser

Figure 15.8 Abbreviated structures for peptides (the N-terminal amino acid unit is at the left, the C-terminal at the right).

A number of small peptides occur in the free state in nature. These small peptides play a variety of roles in biological systems: some function as catalysts in biochemical reactions, some have antibiotic properties, some are hormones, and so on. Often their structures are characterized by unusual features, such as the presence of an uncommon amino acid, or of amino acids with the D configuration rather than the usual L form (in proteins, the amino acids always belong to the L family). In some cases, the unusual feature may be a cyclic structure or a unique type of peptide bond.

Some of the most exciting discoveries made in recent years are the **enkephalins** and the **endorphins.** These substances are part of the brain's natural pain-control system. Physiologically, enkephalins and endorphins behave much like morphine, the narcotic drug derived from the opium poppy. Enkephalins and endorphins, like morphine, have a strong affinity for certain regions of the brain that bind opiate drugs. Recent experiments have suggested that several methods used to treat chronic pain—acupuncture, direct electrical stimulation of the brain, and even hypnosis—may act by causing the brain to release enkephalins or endorphins. This hypothesis is based on the finding that the effectiveness of all these methods of treatment can be blocked almost entirely by the administration of naxolone, a drug that prevents the binding of morphine to the opiate receptor of the brain. Chemically, both enkephalins and endorphins are peptides. Enkephalins are pentapeptides; the two that have been discovered to date differ only in their C-terminal amino acids.

Tyr-Gly-Gly-Phe-Met   Tyr-Gly-Gly-Phe-Leu

Met-enkephalin   Leu-enkephalin

Endorphins, "the morphine within," are polypeptides. In general, they incorporate as part of their structure one of the enkephalins. The structure of one endorphin, β-endorphin, is

Tyr-Gly-Gly-Phe-Met-Thr-Ser-Glu-Lys-Ser-Gln-Thr-Pro-Leu-Val

Gln-Gly-Lys-Lys-His-Ala-Asn-Lys-Val-Ile-Ala-Asn-Lys-Phe-Leu-Thr

β-Endorphin
(Note the Met-enkephalin component.)

Additional examples of naturally occurring peptides are shown in Figure 15.9. *Bacitracin A* and *Gramicidin S* are antibiotics. They contain D- and L-amino acids and the uncommon amino acid orinithine (Orn). Gramicidin S also has a completely cyclic structure.

Oxytocin and *vasopressin* are hormones secreted by the pituitary gland. These two compounds are structurally similar: both are partially cyclic peptides with a disulfide bond bridging residues 1 and 6, and both possess an amide function at the C terminus instead of the usual carboxyl group. Oxytocin and vasopressin also have the same sequence of amino acids, with the exception of residues 3 and 8. Although small, these structural differences result in dramatic differences in the physiological functions of the two compounds. Oxytocin brings on uterine contraction to induce labor and enhances the ejection of milk from the mammary glands. Vasopressin, on the other hand, controls the excretion of water from the kidneys (anti-diuretic) and causes the constriction of blood vessels, thus raising the blood pressure.

$$H_2N-Ile-Cys-Leu-D-Glu-Ile-Lys \overset{\overset{\text{D-Orn—Ile}}{\diagup}}{\diagdown} \; _{Asp—His}^{\text{D-Phe}}$$

Bacitracin A

$$
\begin{array}{c}
Leu \\
Orn \quad D\text{-Phe} \\
Val \qquad Pro \\
Pro \qquad Val \\
D\text{-Phe} \quad Orn \\
Leu
\end{array}
$$

Gramicidin S

$$H_2N-CH_2-CH_2-CH_2-\overset{\overset{\textstyle H}{|}}{\underset{\underset{\textstyle NH_2}{|}}{C}}-COOH$$

Ornithine (Orn)

Human oxytocin

Human vasopressin

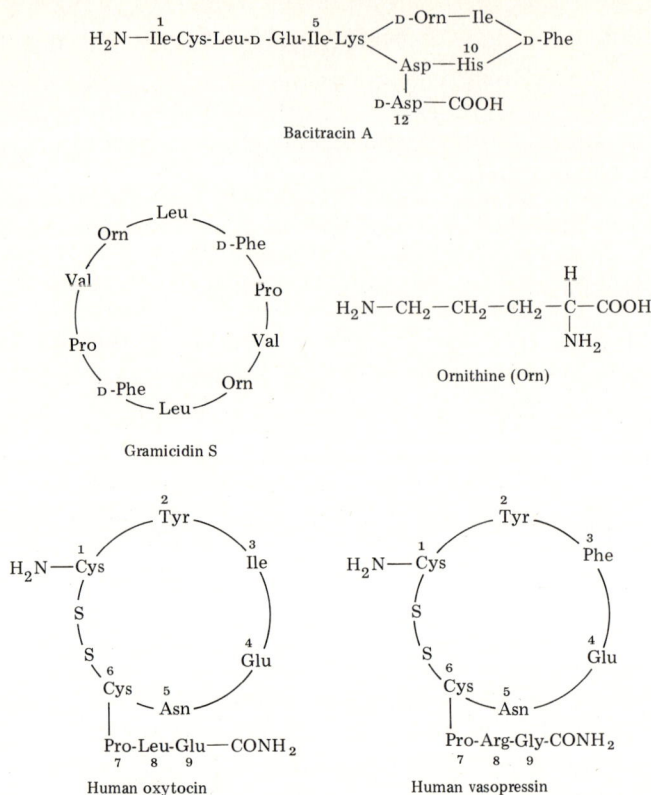

Figure 15.9 Examples of naturally occurring peptides with unusual structures.

15.10 Structure Determination of Peptides; Primary Structure of Proteins

A major objective of peptide chemists is to determine the structure of a peptide. The reason for doing this is simple: there is a strong correlation between the structure of a peptide and its biological function. Once the structure of a peptide is known, a chemist can then synthesize it, whether it be a hormone, an antibiotic, or any other peptide that may be needed in larger quantity than is available from natural sources. Or a chemist may want to modify a natural peptide, to enhance or to reduce its normal biological function.

To understand the problem of determining the structure of a peptide, let us use an analogy. Peptides may be compared to words. Just as words are made up from an alphabet of letters, peptides are made up from an "alphabet" of amino acids. And just as the arrangement of letters gives a particular meaning to a word, the order of amino acids determines the function of a peptide or a protein. For example, if we replace the *d* in foo*d* by an *l*, the meaning of the word is changed altogether. Similarly, if we substitute, for example, isoleucine for leucine at the eighth position in oxytocin, we find that the resulting peptide has lost

(a)   Val-His-Leu-Thr-Pro-Glu^6-Glu- · · ·

(b)   Val-His-Leu-Thr-Pro-Val^6-Glu- · · ·

Figure 15.10 **(a)** Partial composition of the beta chain of normal hemoglobin (Hb-A). **(b)** Partial composition of the beta chain of sickle cell hemoglobin (Hb-S).

much of its ability to stimulate lactation and to induce labor. Various genetic disorders are due to minor alterations in the structure of peptides. A classical example is sickle cell anemia, in which the ability of hemoglobin to transport oxygen to the tissues is considerably reduced. The result is often death at a premature age for those afflicted with the disease. One form of sickle cell anemia arises when a single amino acid, valine, replaces glutamic acid at the sixth position of a beta chain of hemoglobin, one of the four chains, each consisting of approximately 140 amino acids, that make up the hemoglobin molecule (Fig. 15.10).

The cases just discussed are good examples of the effect of structure on function. Now that you understand this principle, the next step is to learn how to arrive at the structure of a peptide. This section is limited to chemical methods of structure determination, which give us the sequence of amino acids in a peptide. When the peptide is a protein, this sequence is referred to as its **primary structure.** To establish the structure of a peptide we must determine.

1. The *kind* of amino acids that make up the peptide.
2. The *number* of amino acids of each type.
3. The *order* in which these amino acids are arranged in the peptide chain.

The procedure, too, has three parts. In the first step, the peptide chain is broken into its amino acid components through chemical *hydrolysis*. In the second step, the solution, or hydrolyzate, that contains the amino acid mixture from the first step is *analyzed* to determine its amino acid composition. Finally, in the third step, the *pieces are fitted together* to obtain the correct sequence. Let us see the details of how each of the three steps is carried out.

1. Complete chemical hydrolysis.   To identify all the amino acids present in a peptide or protein one resorts to *chemical hydrolysis*. Usually, chemical hydrolysis is carried out by heating the peptide with 6 M HCl in a sealed tube at 110°C for about 24 hours. Under these conditions *all* peptide bonds are broken, giving a mixture of all amino acids present.

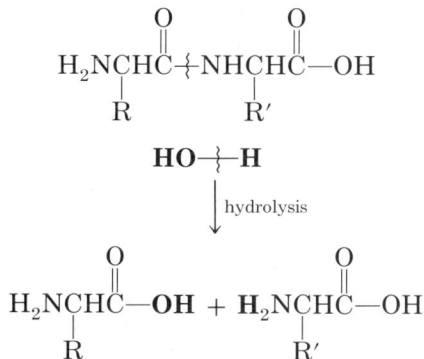

$$
\underset{\displaystyle R}{H_2NCHC} \overset{\displaystyle O}{\overset{\|}{C}} \!\!+\!\! NH \underset{\displaystyle R'}{CHC} \overset{\displaystyle O}{\overset{\|}{C}} \!\!-\!\! OH
$$

$$
HO \!\!+\!\! H
$$

$$\downarrow \text{hydrolysis}$$

$$
\underset{\displaystyle R}{H_2NCHC} \overset{\displaystyle O}{\overset{\|}{C}} \!-\! OH \;+\; \underset{\displaystyle R'}{H_2NCHC} \overset{\displaystyle O}{\overset{\|}{C}} \!-\! OH
$$

2. Analysis of hydrolyzate. When hydrolysis is completed, the relative abundance of each amino acid in the hydrolyzate is identified and measured. This may be done by any suitable method mentioned earlier: electrophoresis, ion-exchange chromatography, or automated amino acid analysis.

At this point, all that is known is the amino acid composition of the peptide or protein—that is, the kinds of amino acids and the number of each kind that is present. The correct sequence of the amino acid residues is still not known. An example will show us why.

Example 15.1 Suppose you are given the tripeptide

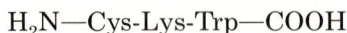

H$_2$N—Cys-Lys-Trp—COOH

(I)

as an unknown and are asked to determine its structure.

Solution If you subject I to chemical hydrolysis and analyze the hydrolyzate, you will find equal amounts of cysteine, lysine, and tryptophan. What is not known, however, is the order in which these three amino acids are joined together. Any one of these six combinations is possible.

(Ia) H$_2$N—Cys-Lys-Trp—COOH (the actual structure)
(Ib) H$_2$N—Cys-Trp-Lys—COOH
(Ic) H$_2$N—Lys-Cys-Trp—COOH
(Id) H$_2$N—Lys-Trp-Cys—COOH
(Ie) H$_2$N—Trp-Cys-Lys—COOH
(If) H$_2$N—Trp-Lys-Cys—COOH

---

3. Amino acid sequence determination. The third step in the determination of peptide or protein structure is more complex than the preceding two. The example just discussed tells us why: we are still faced with six possibilities in the attempt to elucidate the structure of a simple tripeptide. Imagine the number of possible permutations had we decided to find out the structure of insulin, a peptide hormone that contains 16 of the 20 common amino acids. Insulin has a molecular weight of about 6500 and is known to contain a total of 51 amino acid residues (see Fig. 15.13). The number of possible combinations after steps 1 and 2 is truly astronomical! Nevertheless, Frederick Sanger established the structure of insulin and received a Nobel prize for this feat. The sequence of amino acids in a number of larger proteins has also been decoded. For example, the structures of ribonuclease, an enzyme composed of 124 amino acid residues, and of human growth hormone, a protein consisting of 188 amino acid units, have been established.

The methods used to determine the amino acid sequence in peptides may be subdivided into two parts. The first phase consists of hydrolyzing the initial peptide chain into smaller segments by means of certain enzymes. Each segment is, in turn, analyzed for its amino acid content. The second phase involves determination of the amino acids that are present at the C terminus and at the N terminus in the original peptide as well as in fragments obtained through enzymatic hydrolysis. The pieces are then fitted together to arrive at the complete sequence. The details for carrying out the third step are outlined in the following paragraphs.

Partial hydrolysis (*use of enzymes*).   The partial hydrolysis of peptides is undertaken by means of enzymes known generally as **proteases.** Proteases readily hydrolyze peptide bonds only between *specific* amino acid residues. Thus, the three proteases chymotrypsin, pepsin, and trypsin all hydrolyze peptide linkages, but each has a preference as to which peptide bond is broken.

Chymotrypsin selectively splits the peptide bond at the *carboxyl* position of *phenylalanine, tyrosine,* and *tryptophan* (the amino acids with the aromatic R groups).

Pepsin preferentially cleaves the peptide bond at the *amino* position of phenylalanine, tyrosine, and tryptophan.

Trypsin preferentially hydrolyzes the peptide bond at the *carboxyl* position of *basic* amino acids, such as arginine and lysine.

The specific action of the three proteases on a heptapeptide is illustrated in Figure 15.11.

Now that you know how these proteases work, you are ready to learn how they can help in determining the sequence of amino acid residues in peptides and proteins. For instance, treatment of cysteyllysyltryptophan (the unknown tripeptide in Example 15.1) with pepsin yields the free amino acid tryptophan and a dipeptide. The only structures that are compatible with this result are

Ia   H₂N—Cys-Lys-Trp—COOH   (the correct structure)

Ic   H₂N—Lys-Cys-Trp—COOH   (possible, but incorrect)

We have narrowed the six possible combinations obtained through chemical hydrolysis down to two by simply treating the tripeptide with pepsin. To arrive at the correct ultimate choice, we must proceed to the second part of amino acid sequence determination, called **end-group analysis.**

End-group analysis.   By end-group analysis we mean the identification of the C- and N-terminal amino acids in a peptide chain using specific tests. The C terminus is determined enzymatically by treating the peptide with **carboxypeptidase.** Carboxypeptidase is an enzyme that hydrolyzes peptides from the C terminus only.

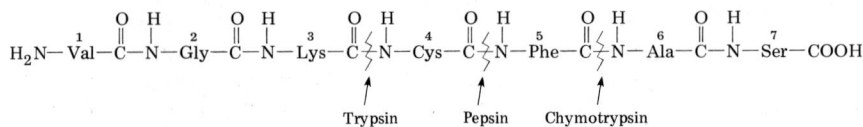

Figure 15.11 Products of hydrolysis of a heptapeptide.
Trypsin: H₂N—Val-Gly-Lys—COOH + H₂N—Cys-Phe-Ala-Ser—COOH
Pepsin: H₂N—Val-Gly-Lys-Cys—COOH + H₂N—Phe-Ala-Ser—COOH
Chymotrypsin: H₂N—Val-Gly-Lys-Cys-Phe—COOH + H₂N—Ala-Ser—COOH

The N-terminal residue may also be determined enzymatically by treating the peptide chain with **aminopeptidase.** Aminopeptidase is an enzyme that cleaves peptides from the N terminus only.

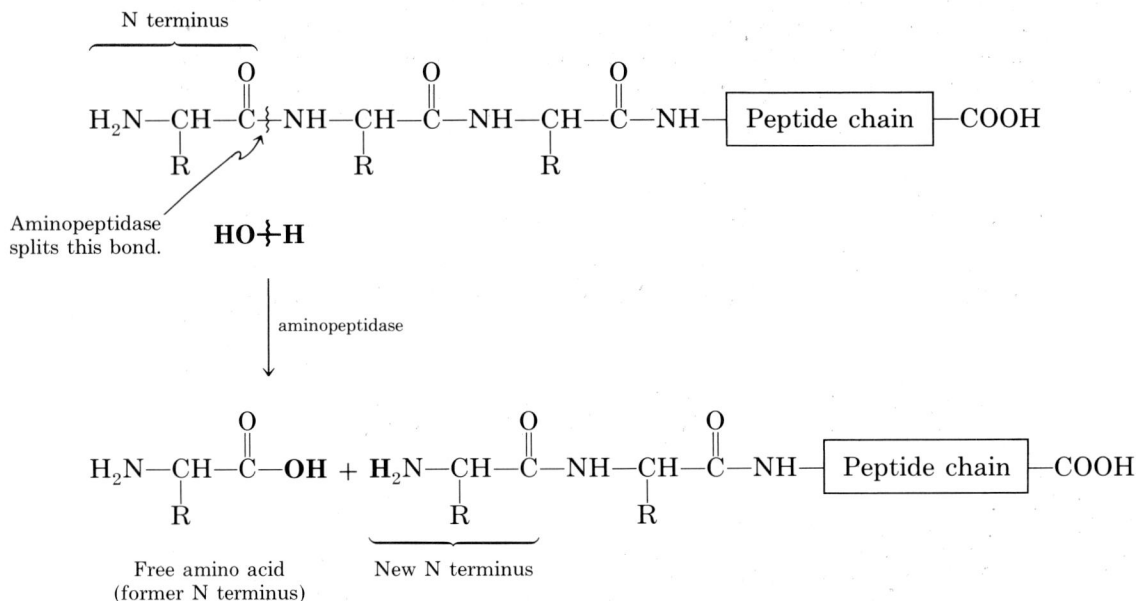

By following the increase in amino acids liberated by these enzymes at regular time intervals, it is possible to determine the sequence of amino acid residues in peptides. This method is also called *time-course analysis*. For example, a time-course analysis of the tripeptide (I)

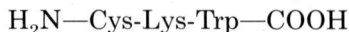

H₂N—Cys-Lys-Trp—COOH

(I)
Tripeptide

with carboxypeptidase gives the results shown in Figure 15.12a. The carboxypeptidase treatment liberates first tryptophan, indicating that it is the C-terminal amino acid. Simultaneously, the dipeptide H₂N—Cys-Lys—COOH is generated. As time goes on, the concentration of cysteyllysine increases. This dipeptide is in turn hydrolyzed by carboxypeptidase, releasing equal amounts of cysteine and lysine. The results of carboxypeptidase treatment therefore tell us that tryptophan is the C-terminal amino acid, but do not inform us as to the sequence of the other two residues.

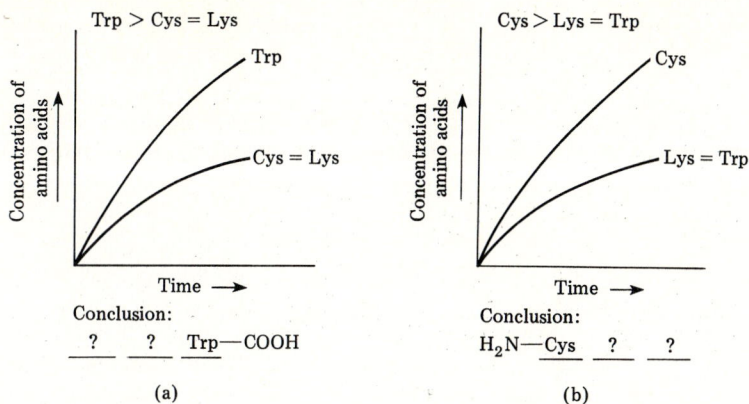

Figure 15.12 Time-course analyses of tripeptide I: **(a)** with carboxypeptidase and **(b)** with aminopeptidase.

Similarly, a time-course analysis of tripeptide I with aminopeptidase gives the results shown in Figure 15.12b. The aminopeptidase treatment reveals that cysteine is at the N terminus, but by itself it does not specify the position of the other two residues.

However, if the information of the two time-course analyses is combined, we arrive at the complete sequence.

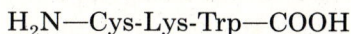

H_2N—Cys-Lys-Trp—COOH

Problem 15.5 What results would you expect from the time-course analysis of the tripeptide H_2N—Lys-Cys-Trp—COOH with **(a)** carboxypeptidase and **(b)** aminopeptidase?

The N-terminal amino acid may also be identified chemically with Sanger's reagent, 2,4-dinitrofluorobenzene. Recall that DNFB reacts with primary amino groups to give dinitrophenyl (DNP) derivatives. For a peptide, the formation of a DNP derivative with the N-terminal amino acid is

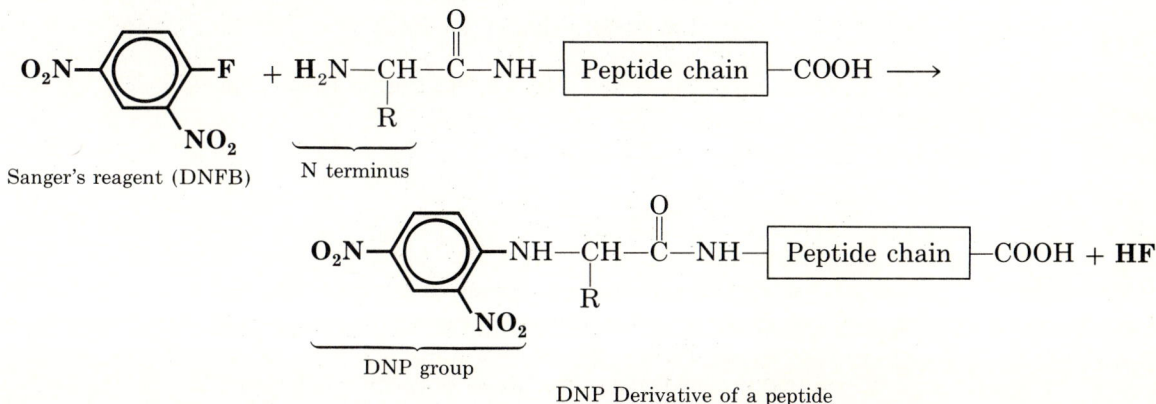

When the product is hydrolyzed in acid, the peptide linkages are split, but the DNP bond remains attached to the N-terminal amino acid. The hydrolyzate

thus contains a mixture of amino acids and the DNP of the terminal amino acid. The latter, which is yellow, is easily isolated and characterized.

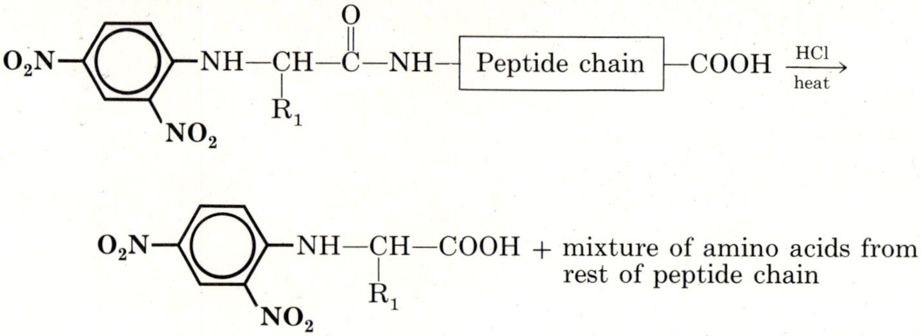

$$O_2N-\underset{NO_2}{\bigcirc}-NH-CH-\underset{R_1}{\overset{O}{\overset{\|}{C}}}-NH-\boxed{\text{Peptide chain}}-COOH \xrightarrow[\text{heat}]{\text{HCl}}$$

$$O_2N-\underset{NO_2}{\bigcirc}-NH-\underset{R_1}{CH}-COOH + \text{mixture of amino acids from rest of peptide chain}$$

DNP–N-terminal amino acid
(yellow)

Now that we have outlined the methods used for determining the structure of peptides, let us review them by working out an example.

Example 15.2 Determine the sequence of amino acids in a heptapeptide (I) from the following data.

After complete hydrolysis and amino acid analysis the heptapeptide was found to contain the following amino acids in the amounts shown.

Ala (2 moles)   Leu (1 mole)   Ser (1 mole)

Asp (1 mole)   Phe (1 mole)   Val (1 mole)

The reactions performed and the results obtained are summarized in the following scheme.

I
- (A) (1) DNFB / (2) hydrolysis → DNP-Asp
- (B) carboxypeptidase → Ala > Val
- (C) aminopeptidase → Asp > Ala
- (D) chymotrypsin → Dipeptide + Pentapeptide
    (II)   (III)

II
- (E) (1) DNFB / (2) hydrolysis → DNP-Val

III
- (F) carboxypeptidase → Phe > Ser
- (G) aminopeptidase → Asp > Ala

Solution Interpretation of the data.
Reaction A: Aspartic acid is the N-terminal amino acid.

(I) H₂N—Asp ___ ___ ___ ___ ___ ___

Reaction B: Alanine is the C-terminal amino acid, preceded by valine.

(I) H_2N—Asp ___ ___ ___ ___ ___ Val-Ala—COOH

Reaction C: Aspartic acid, the N-terminal amino acid, is followed by alanine.

(I) H_2N—Asp-Ala ___ ___ ___ Val-Ala—COOH

Reaction D: Chymotrypsin splits a peptide bond in which an aromatic amino acid furnishes the —COOH group; the only aromatic amino acid is phenylalanine, which must be the C-terminal of either the dipeptide (II) or the pentapeptide (III).

Reaction E: In the dipeptide, valine is the N-terminal amino acid; the other acid must be alanine (see reaction B).

(II) H_2N—Val-Ala—COOH

Reaction F: The pentapeptide contains phenylalanine as the C-terminal amino acid, preceded by serine.

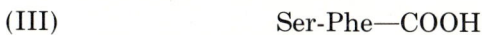

(III)   ___ ___ ___ Ser-Phe—COOH

Reaction G: The pentapeptide contains aspartic acid as the N-terminal amino acid, followed by alanine.

(III) H_2N—Asp-Ala ___ ___ ___

By combining the information from reactions F and G we arrive at

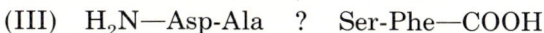

(III) H_2N—Asp-Ala ? Ser-Phe—COOH

If we correlate all the results we find

Reactions		1	2	3	4	5	6	7
A, B, C	(I)	Asp	Ala				Val	Ala
D, E	(II)						Val	Ala
F, G	(III)	Asp	Ala		Ser	Phe		

One amino acid is missing: leucine, which can only be at position 3. The complete sequence is therefore

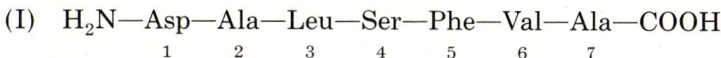

(I) H_2N—Asp—Ala—Leu—Ser—Phe—Val—Ala—COOH
      1   2   3   4   5   6   7

Now try to solve this problem.

Problem 15.6 Determine the sequence of amino acids in an octapeptide (I) from the following data.

After complete hydrolysis and amino acid analysis, the octapeptide was found to contain the following amino acids.

Ala (1 mole)   Gly (2 moles)   Ser (1 mole)   Val (1 mole)
Asp (1 mole)   Leu (1 mole)   Tyr (1 mole)

The reactions performed and the results obtained are summarized in this scheme.

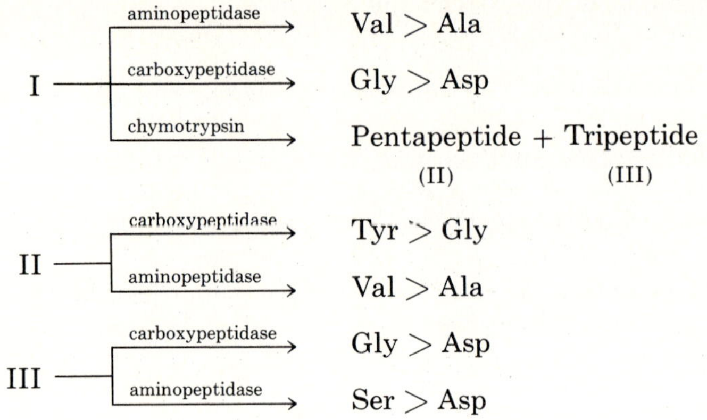

$$
\begin{array}{ll}
\text{I} & \xrightarrow{\text{aminopeptidase}} \quad \text{Val} > \text{Ala} \\
& \xrightarrow{\text{carboxypeptidase}} \quad \text{Gly} > \text{Asp} \\
& \xrightarrow{\text{chymotrypsin}} \quad \text{Pentapeptide} + \text{Tripeptide} \\
& \qquad\qquad\qquad\qquad\quad \text{(II)} \qquad\qquad \text{(III)}
\end{array}
$$

$$
\begin{array}{ll}
\text{II} & \xrightarrow{\text{carboxypeptidase}} \quad \text{Tyr} > \text{Gly} \\
& \xrightarrow{\text{aminopeptidase}} \quad \text{Val} > \text{Ala}
\end{array}
$$

$$
\begin{array}{ll}
\text{III} & \xrightarrow{\text{carboxypeptidase}} \quad \text{Gly} > \text{Asp} \\
& \xrightarrow{\text{aminopeptidase}} \quad \text{Ser} > \text{Asp}
\end{array}
$$

15.11 Specific Properties of Proteins

Much of what we learned about amino acids and peptides in the previous sections applies equally to proteins. Like the component amino acids, proteins are optically active molecules. Proteins, like amino acids and peptides, carry electrical charges and have characteristic isoelectric points. The techniques used to determine the amino acid sequence in a peptide apply to proteins to give their *primary structures*. The primary stuctures of insulin, ribonuclease, and myoglobin shown in Figures 15.13, 15.14, and 15.17 were determined by chemical and enzymatic methods similar to those discussed in the previous section.

In several respects, however, proteins differ from smaller peptides. For example, in a great many proteins the polypeptide chain is associated with other groups, and such proteins are classified as **conjugated proteins.** The nonprotein portion is known as the **prosthetic group.** The prosthetic groups may be carbohydrates, lipids, nucleic acids, or phosphates. In hemoglobin, the prosthetic groups are heme molecules (Fig. 15.19).

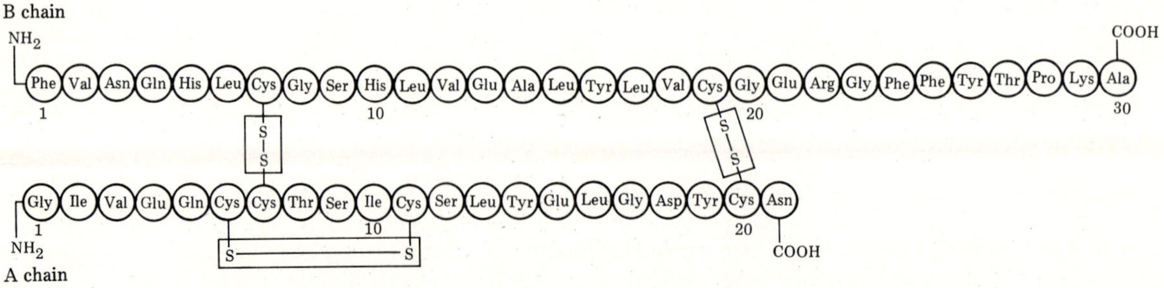

Figure 15.13 Primary structure of human insulin, a hormone essential for the regulation of carbohydrate metabolism.

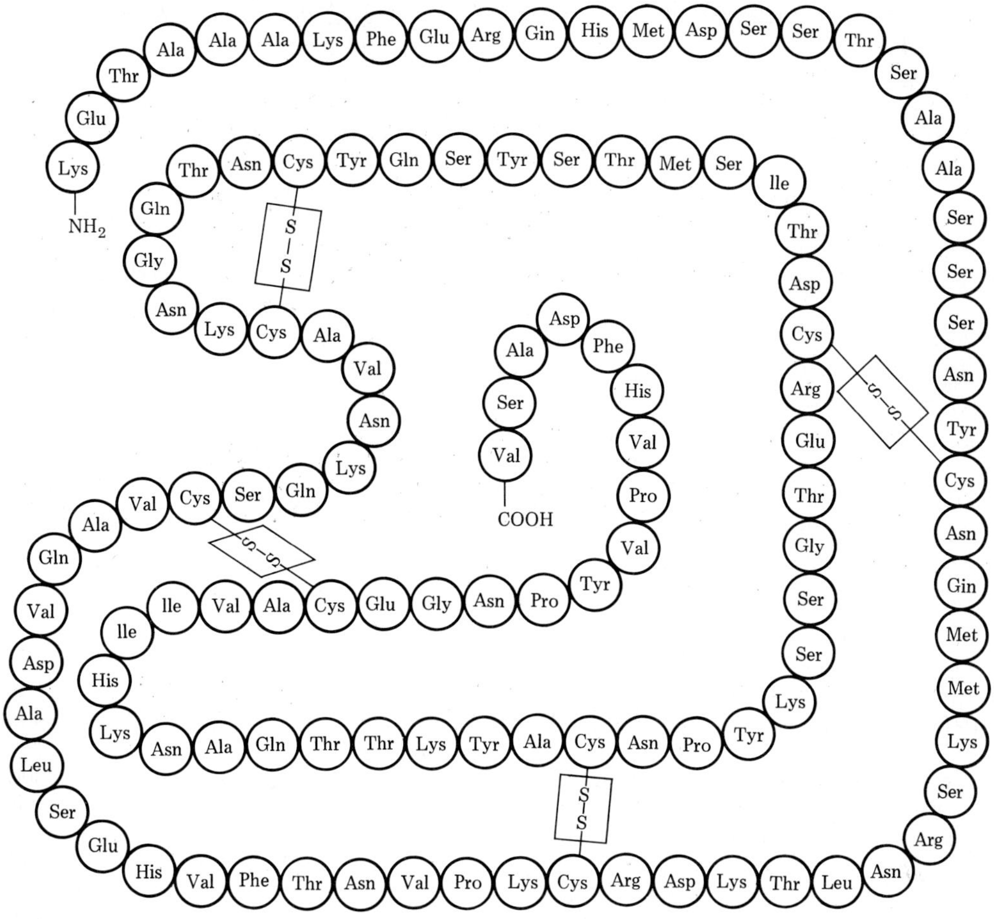

Figure 15.14 Primary structure of pancreatic ribonuclease. Ribonuclease is an enzyme that catalyzes the hydrolysis of ribonucleic acid.

In contrast to small peptides, which are crystalline compounds, many proteins are amorphous solids. The solubility of proteins also varies considerably. Some proteins are water soluble; others are insoluble in water. Some require small concentrations of salts to bring them into solution; others are soluble only in dilute bases or acids. Certain proteins are insoluble in pure water or absolute alcohol, but become soluble in aqueous alcohol.

All proteins, when they dissolve, form *colloidal solutions*. The particles in a colloidal solution do not go through certain membranes (cellophane, for example), in contrast to the particles in a true solution, which diffuse through the pores of the membrane.

Proteins undergo **denaturation.** Denaturation is a process whereby the protein molecule undergoes drastic changes in its physical and biological properties, *even though its chemical composition remains the same.* The primary structure of the protein is not changed. Denaturation is brought about by a variety of conditions and reagents such as heat, ultraviolet light, changes in pH, organic solvents, heavy metals, and so on. In some cases denaturation is a reversible phenomenon: The original properties of the protein are restored when the agent causing the denaturation is removed. Often, however, denaturation is an irreversible process. For example, a 70% alcohol solution is an effective sterilizing agent because it destroys microorganisms, which are protein-like in character, by penetrating them and causing irreversible denaturation. The testing for foreign protein in urine is done by heating the sample. The presence of proteins, a symptom of kidney malfunction or possible diabetes, is indicated by the formation of a precipitate brought about by irreversible denaturation.

How do we account for the differences in properties between proteins and smaller peptides? One factor is, undoubtedly, the great size of proteins. For example, the fact that protein molecules form colloidal solutions may be ascribed to the large size of the molecules. To understand the denaturation process we must take into account the three-dimensional shapes of proteins.

15.12 Three-Dimensional Shapes of Proteins

The primary structure of a protein gives us the sequence of amino acid units but reveals nothing about the three-dimensional shape of the molecule, which is also very important for its function. To determine the three-dimensional structure it is necessary to have crystals of the pure protein. The molecule is photographed by means of x rays and the picture is then analyzed. The pioneering work of Linus Pauling contributed a great deal to the unraveling of the factors that influence the complex shapes of proteins and to our understanding of some of their properties. A system of structural organization for proteins has been devised in which the structures are designated as primary (1°), secondary (2°), tertiary (3°), and quaternary (4°).

The primary structure is the sequence of amino acids in the protein, a subject with which you should be thoroughly familiar by now.

15.13 Secondary Structure of Proteins

The **secondary structure** of proteins is due to hydrogen bonding between peptide bonds. Hydrogen bonds occur between the $C{=}O$ of one peptide bond and the $N{-}H$ of another.

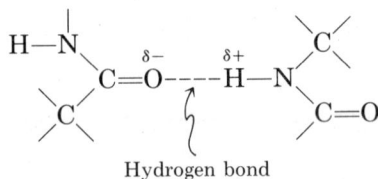

Hydrogen bond

There are two types of secondary structures found in proteins: the helix and the pleated sheet. The **α helix** is the result of hydrogen bonding between peptide groups within the *same molecule*. A schematic diagram of such a helical arrangement of the backbone in a polypeptide is shown in Figure 15.15.

Soluble proteins are known to have the α-helix structure in various portions

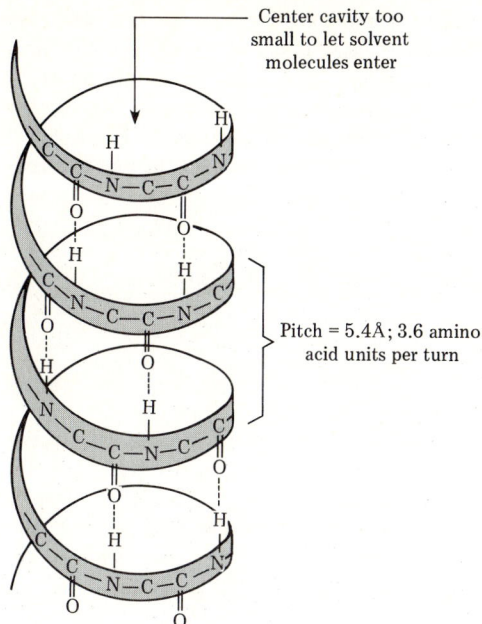

Figure 15.15 The α-helix structure (intramolecular hydrogen bonding).

of the molecule. The helix has several interesting characteristics: there are 3.6 amino acid units per turn; the distance between turns (the pitch) is 5.4 Å. The diameter of the helix is 10–11 Å, which is too small to let solvent molecules enter the space.

The **pleated-sheet** arrangement is the second type of secondary structure found in proteins. It is due to hydrogen bonding between peptide groups in *dif-*

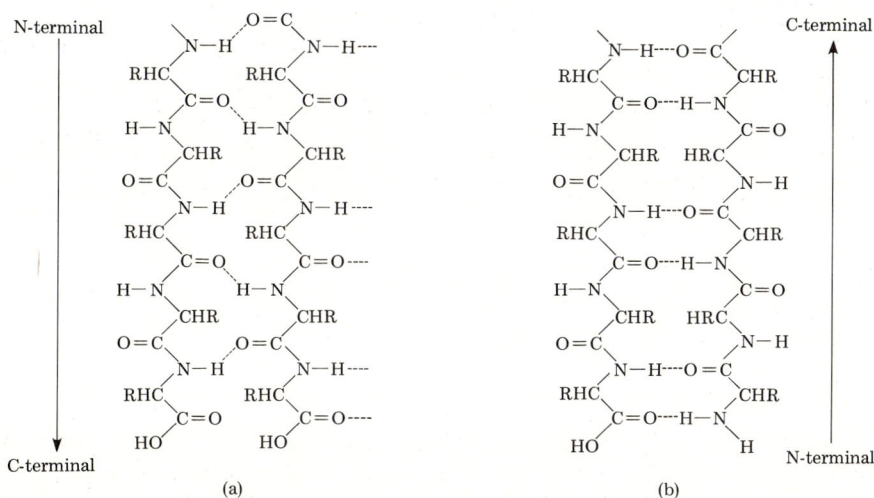

Figure 15.16 The pleated-sheet structure (intermolecular hydrogen bonding): **(a)** parallel arrangement and **(b)** antiparallel arrangement.

ferent polypeptide chains. Proteins with this structure are quite insoluble in aqueous solvents. There are two possible arrangements of the protein chains that form the pleated sheet. One is the *parallel* arrangement, in which the chains all run in the same direction. The second is the *antiparallel* arrangement, in which the —COOH end of one chain is next to the —NH$_2$ end of another in an alternate manner. Figure 15.16 shows the two kinds of pleated-sheet structures.

15.14 Tertiary Structure of Proteins

Most soluble proteins do not occur as randomly coiled molecules. Usually, they exist as tight, compact structures with definite shapes, as illustrated in Figure 15.17 for myoglobin.

The forces responsible for the three-dimensional shapes of proteins are due to interactions between the side chains of the amino acid units. The **tertiary structure** refers to the shapes resulting from the interactions among the R groups. These interactions include *covalent (disulfide) bonds, ionic bonds, hydrogen bonds,* and *hydrophobic bonds.* The four kinds of interactions associated with the side chains are illustrated in Figure 15.18.

The disulfide bond is the only covalent bond involved in the 3° structure of proteins. About 50–100 kcal/mole of energy is required to break it. The other bonds are much weaker; they can be affected by changes in temperature, solvent, pH, salt concentrations, and so on. We may explain the denaturation of proteins as being the result of disruptions of their secondary and tertiary structures.

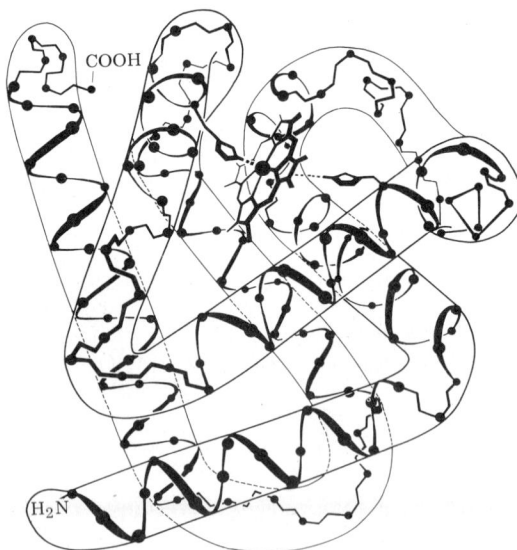

Figure 15.17 Three-dimensional structure of myoglobin, a respiratory pigment. The primary structure is the sequence of dots, which represent the amino acids; the secondary structure is the clearly visible helical arrangement down the length of the amino acid chain; and the tertiary structure is the folding and wrapping of the chain into its actual conformation. [Adapted from "The hemoglobin molecule" by M. F. Perutz. Copyright © 1964 by Scientific American, Inc. All rights reserved.]

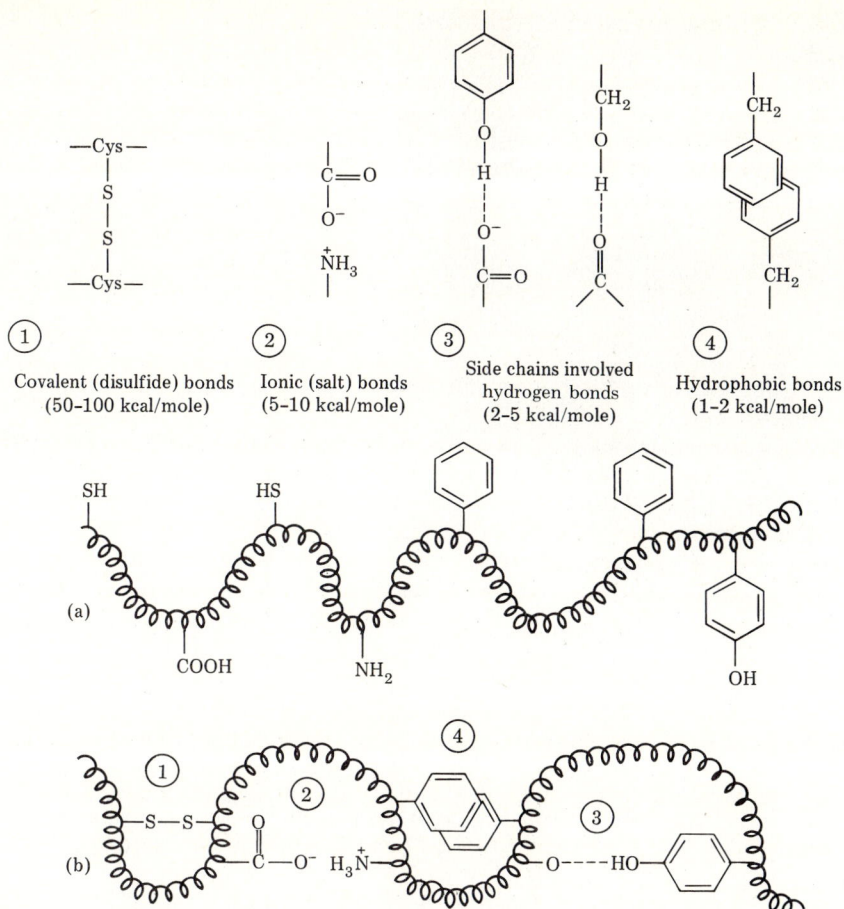

Figure 15.18 Interactions associated with the tertiary structure of protein: **(a)** random protein, no tertiary structure, and **(b)** protein with a tertiary structure.

The 3° structure plays an extremely important role when the protein functions as an enzyme. This is because enzymes, to be active, must bind with the substrate, the molecule on which they act. The binding between an enzyme and a substrate occurs only if the shape of the enzyme fits with that of the substrate, much like a lock and a key.

Quaternary Structure of Proteins *15.15*

In some proteins, the molecule as a whole contains more than one polypeptide chain. Each polypeptide chain is considered as a subunit. The **quaternary structure** refers to the state of aggregation of the subunits in the total molecule and to the forces that keep the subunits together. These forces are usually electrostatic in nature and do not involve covalent bonding. Hemoglobin, for exam-

(a)

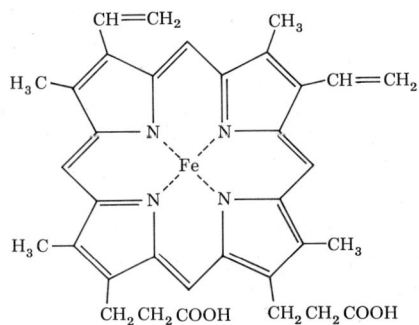

(b)

Figure 15.19 (a) A three-dimensional model of hemoglobin (the alpha chains are white and the beta chains are shaded). **(b)** The structure of heme, the prosthetic group responsible for carrying oxygen to tissues. The heme shown here is found in hemoglobin, myoglobin, and most of the cytochromes. [(a) reprinted by permission from M. F. Perutz et al., *Nature, 219*, 131. Copyright © 1968 Macmillan Journals Limited.]

ple, consists of four subunits: a pair of polypeptide chains containing 141 amino acids each (α chains) and a pair of β chains made up of 146 amino acid residues each. Each chain is associated with a heme molecule, the prosthetic group responsible for carrying oxygen to the tissues. The four chains in the complete protein are closely associated to form a compact multiunit structure, as shown in Figure 15.19.

Many other proteins have more than one polypeptide chain. *Collagen* is made up of three strands wound together like a rope. The *polio virus* contains 130 subunits, each with an average molecular weight of 27,000. The record belongs to the *tobacco mosaic virus,* which contains a total of 2130 subunits clustered around a nucleic acid molecule.

Disruption of quaternary structure may or may not affect biological function. Enzymes often become inactive when they are split into smaller units.

Summary of Concepts and Reactions

All amino acids found in proteins have an amino group (—NH$_2$) on the alpha carbon, the carbon adjacent to the carboxyl group (—COOH); hence the name α-amino acids.

[Sec. 15.1]

The structures of α-amino acids differ from one another in the nature of the R group. The R group is called the side chain. [Sec. 15.1]

Out of the 21 naturally occurring amino acids found in proteins, 10 cannot be manufactured or produced in sufficient quantities by the body to sustain good health. These constitute the essential amino acids. [Sec. 15.2]

With the exception of glycine, which is achiral, all other amino acids found in proteins are optically active and belong to the L family. [Sec. 15.3]

Amino acids are amphoteric substances. Depending on the pH, they can exist as zwitterions, cations, or anions. [Sec. 15.4]

The pH at which an amino acid exists as a zwitterion is called the isoelectric point or pI. [Sec. 15.5]

The two basic techniques used to separate and identify a mixture of amino acids are (1) electrophoresis and (2) ion-exchange chromatography. [Sec. 15.6]

Reactions of amino acids can be subdivided into those that involve the NH_2, R, and the COOH group, respectively. Reactions with nitrous acid, ninhydrin, and Sanger's reagent involve the NH_2 group exclusively. They are important for analytical purposes. [Sec. 15.7A]

A common reaction of the side chain is formation of the disulfide bond. [Sec. 15.7B]

The most important reaction of all is the formation of the peptide bond. [Secs. 15.7C, 15.8]

A peptide bond is an amide bond formed when the α-carboxyl group of one amino acid molecule reacts with the α-amino function of another. The product of reaction is a peptide. [Sec. 15.7C, 15.8]

Peptides containing more than ten amino acid units are called polypeptides. Proteins are polypeptides; most proteins have 100 or more amino acid residues. [Sec. 15.8]

The sequence of amino acids in a protein is referred to as its primary structure and is determined by chemical and enzymatic methods. [Sec. 15.10]

The polypeptide chain in a protein can associate with other groups that are nonprotein-like. The nonprotein portion is called the prosthetic group, and the protein is a conjugated protein. [Sec. 15.11]

Proteins undergo denaturation. Denaturation is a process whereby a protein molecule undergoes drastic changes in its physical and biological properties, even though its chemical composition remains the same. [Sec. 15.11]

The three-dimensional shape of proteins depends not only on the amino acid sequence (primary structure) but also on the secondary (2°), tertiary (3°) and quaternary (4°) levels of organization. [Sec. 15.12]

A 2° structure involves hydrogen bonding between peptide bonds (α-helix or pleated-sheet structures). [Sec. 15.13]

A 3° structure involves interaction of the side chains in the protein. [Sec. 15.14]

A 4° structure refers to the state of aggregation of the peptide chains in the protein. [Sec. 15.15]

Key Terms

α-amino acids
side chain
essential amino acids
complete proteins
zwitterion
amphoteric properties
isoelectric point, pI

electrophoresis
ion-exchange
   chromatography
ninhydrin
Sanger's reagent
peptide bond
peptide

proteins
backbone
N-terminal amino acid
N terminus
C-terminal amino acid
C terminus
enkephalins

endorphins	end-group analysis	secondary structure
primary structure	carboxypeptidase	α helix
proteases	aminopeptidase	pleated sheet
chymotrypsin	conjugated protein	tertiary structure
pepsin	prosthetic group	quaternary structure
trypsin	denaturation	

Exercises

Nomenclature, Structure, and Classification of Amino Acids [Secs. 15.1–15.3]

15.1 Name and write the formula of one amino acid for each of the following classes.
  (a) A neutral amino acid with a nonpolar side chain
  (b) A neutral amino acid with a polar but uncharged side chain
  (c) An acidic amino acid
  (d) A basic amino acid

15.2 **(a)** What is meant by an essential amino acid? Are any of the amino acids you named in Exercise 15.1 essential amino acids?
  (b) What is a complete protein?

15.3 **(a)** Name and draw the structure of an optically inactive amino acid found in proteins.
  (b) Optically active amino acids found in proteins belong to the L family. Explain what is meant by L.

Ionic Properties; pI [Secs. 15.4–15.6]

15.4 **(a)** Amino acids and proteins have amphoteric properties. Explain.
  (b) Draw the (1) zwitterionic, (2) cationic, and (3) anionic forms of serine.

15.5 **(a)** Define pI.
  (b) What amino acid (or protein) structure is present at the pI?

15.6 Consult Table 15.1 and complete the table here by indicating which form (anionic, cationic, zwitterionic) of the given amino acids will predominate at the pH values listed.

	Valine	Aspartic acid	Arginine
pH 6.0		anionic	
pH 3.0	cationic		
pH 10.8			zwitterionic

15.7 Predict the direction of migration, using (0) = stationary, $(-)$ = toward cathode, and $(+)$ = toward anode, of the following amino acids during electrophoresis at pH 1.0, pH 3.2, pH 6.0, and pH 11.1
  (a) Valine   **(b)** Glutamic acid   **(c)** Lysine

Reactions of Amino Acids [Sec. 15.6]

15.8 Complete the following equations.

  (a) $CH_3\underset{\underset{\displaystyle NH_2}{|}}{C}HCOOH \xrightarrow{\ HNO_2\ }$

  (b) $O_2N-\underset{}{\bigcirc}\!\!-F + H_2NCH_2COOH \longrightarrow$ (with NO_2 group on ring)

15.9 (a) Define peptide bond

(b) How many peptide bonds are there in a dipeptide? in a pentapeptide?

15.10 Label each of the following as a dipeptide, tripeptide, and so on.

(a) Ala-Gly

(b) Phe-Glu-Glu

(c) $H_2NCH_2\overset{O}{\overset{\|}{C}}NHCHCH\overset{O}{\overset{\|}{C}}NHCHCOOH$
                    CH_3 CH_3

(d) $H_2NCH_2\overset{O}{\overset{\|}{C}}(NHCH_2\overset{O}{\overset{\|}{C}})_4NHCH_2COOH$

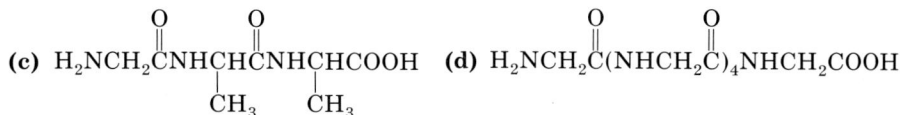

15.11 Draw in abbreviated form and name all tripeptides that could possibly be synthesized from glycine, alanine, and tyrosine.

Identification of Primary Structure of Peptides [Sec. 15.10]

15.12 Complete hydrolysis of a tetrapeptide (I) gave one molecule each of Ala, Cys, Phe, and Val. Treatment of I with Sanger's reagent followed by hydrolysis yielded DNP-Val. Treatment of I with carboxypeptidase gave Phe > Cys. What is the structure of the tetrapeptide

15.13 The complete hydrolysis of a peptide (I) followed by analysis of its amino acid content gave Ala (2 moles), Phe (1 mole), and Ser (1 mole).

(a) Using abbreviated structures, write all possible sequences for this peptide. For each structure show the position of the peptide bonds as well as the location of the N- and C-terminal amino acids.

(b) Treatment of I with carboxypeptidase gave Ala > Phe. What structures remain possible for I?

(c) Treatment of I with aminopeptidase gave Ala > Ser. What is the structure of I?

15.14 List the fragments found upon hydrolysis of the peptide

Ala-Glu-Lys-Pro-Leu-Phe-Cys-Tyr-Gly

with (a) pepsin, (b) chymotrypsin, and (c) trypsin.

15.15 Determine the sequence of amino acids in the decapeptide (I) from the following data.

I $\xrightarrow{\text{complete hydrolysis}}$ 1 mole each of
Ala, Asp, Glu, Gly, Leu, Lys, Phe, Ser, Tyr, and Val

I
- $\xrightarrow{\text{carboxypeptidase}}$ Asp > Leu
- $\xrightarrow{\text{aminopeptidase}}$ Ala > Val
- $\xrightarrow{\text{pepsin}}$ dipeptide + tripeptide + pentapeptide

	Dipeptide	Tripeptide	Pentapeptide
Aminopeptidase			Tyr > Lys
Carboxypeptidase	Glu > Phe	Ser > Val	Asp > Leu

Shapes and Structures of Proteins [Secs. 15.11–15.15]

15.16 What is meant by a conjugated protein? Give an example of a prosthetic group.

15.17 Differentiate among 1°, 2°, 3°, and 4° structures of proteins.

15.18 What is denaturation? Illustrate diagrammatically the denaturation of a protein.

15.19 Why is the shape of a protein important?

16

Nucleosides, Nucleotides, and Nucleic Acids

The material that we now call nucleic acid was first isolated in 1869 by a Swiss physiologist, Friedrich Miescher. **Nucleic acids** were so named because Miescher's new substance turned out to have acidic properties and was found in the nuclei of cells. We know today that nucleic acids are distributed in other parts of the cell as well as in the nucleus, either in the free state or combined with proteins as nucleoproteins. We also know that, like polysaccharides and proteins, nucleic acids are biological polymers. The repeating units in nucleic acids are the **nucleotides,** and for this reason nucleic acids are also called **polynucleotides.** *Nucleotides,* in turn, are substances that on complete hydrolysis yield three components: nitrogen-containing bases, pentose sugars, and phosphoric acid. The number of nucleotide units varies, depending on the kind of nucleic acid, from fewer than one hundred units up to several million. There are two families of nucleic acids: **deoxyribonucleic acid (DNA)** and **ribonucleic acid (RNA).** DNA is found in cell nuclei and contains deoxyribose as the only pentose sugar. RNA is present primarily in the cytoplasm. On hydrolysis, RNA yields ribose as the only sugar component.

Nucleic acids perform two major biological functions. Each is of vital importance. DNA is directly responsible for storing and passing on hereditary traits, for it is the genetic material itself. Both DNA and RNA are associated with the synthesis of proteins in living organisms.

To explain the biological roles of DNA and RNA we must understand how these compounds function on a molecular level. In fact, two of the most important scientific breakthroughs in history were the discovery of the double-helix structure of DNA and the "cracking" of the genetic code. In both cases, the main achievement consisted of interpreting the properties of nucleic acids in terms of their molecular structures.

Let us therefore consider first the structural components of nucleic acids, after which we shall examine the way in which these parts are assembled to give the polynucleotides. Finally, we shall describe how the structures of nucleic acids offer an explanation of their biological roles.

Hydrolytic Products of DNA and RNA *16.1*

Controlled hydrolysis of nucleic acids with aqueous acid, base, or certain enzymes gives successively smaller fragments.

Polynucleotides (DNA and RNA)

$\downarrow$ H_2O, catalyst

Oligonucleotides

$\downarrow$ H_2O, catalyst

Mononucleotides (base-sugar-phosphate)

$\downarrow$ H_2O, catalyst

Nucleosides (base-sugar) + Phosphoric acid

$\downarrow$ H_2O, catalyst

Base + Sugar

The ultimate hydrolytic products of DNA and RNA are phosphoric acid, a sugar, and four nitrogenous bases (Table 16.1).

Table 16.1 Products of Complete Hydrolysis of DNA and RNA

Product	From DNA	From RNA
Acid	Phosphoric	Phosphoric
Sugar	D-2-Deoxyribose	D-Ribose
Nitrogenous bases		
Purines	{ Adenine { Guanine	Adenine Guanine
Pyrimidines	{ Cytosine { Thymine	Cytosine Uracil

16.2 Sugar Components of Nucleotides

The only sugar found in DNA is **2-deoxyribose,** and the only sugar present in RNA is **ribose.** Both are D-aldopentoses and exist in the β-furanose form (see Sec. 11.6).

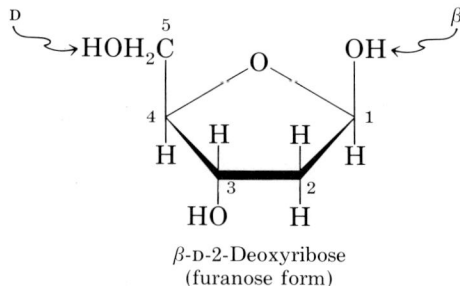

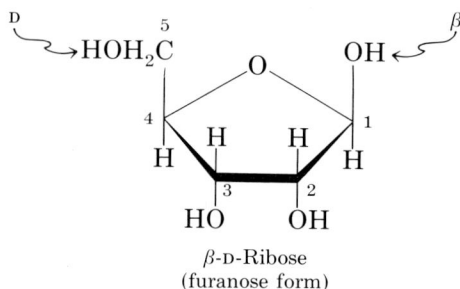

β-D-2-Deoxyribose
(furanose form)

β-D-Ribose
(furanose form)

The difference in the structures of deoxyribose and ribose is the basis of one analytical method for distinguishing DNA from RNA.

16.3 Nitrogenous Bases

There are two classes of nitrogen-containing bases in nucleic acids. These are derivatives of **purine** and **pyrimidine.** The structural formula of the parent compound of each series and the system used for numbering the ring positions are

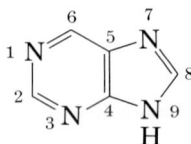

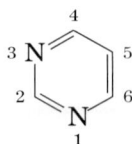

Purine   Pyrimidine

As you can see, both purine and pyrimidine are *heterocyclic* compounds because they contain more than one kind of atom in their respective rings. Purine is essentially a pyrimidine with a fused five-membered imidazole ring. We encountered the imidazole ring as part of the structure of histidine (Sec. 15.1).

The two most common purine bases found in nucleic acids are **adenine** and **guanine.** Adenine and guanine are distributed in DNA and in RNA.

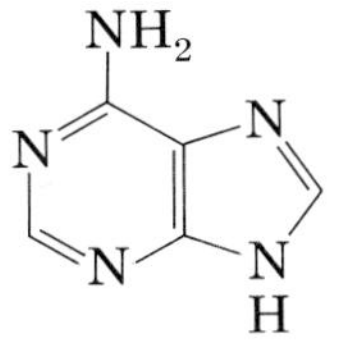

Adenine
(6-Aminopurine)

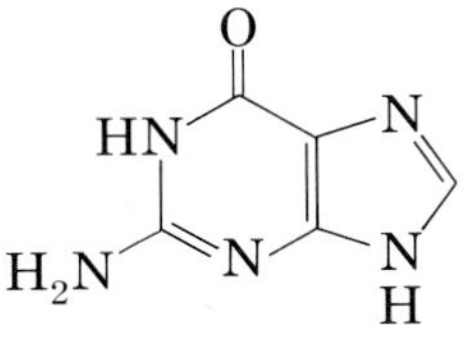

Guanine
(2-Amino-6-oxypurine)

The three most common pyrimidine bases are **cytosine, uracil,** and **thymine.**

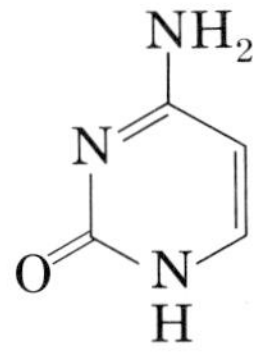

Cytosine
(2-Oxy-4-aminopyrimidine)

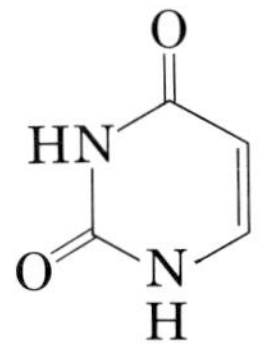

Uracil
(2,4-Dioxypyrimidine)

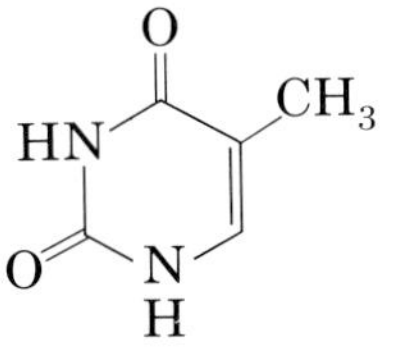

Thymine
(2,4-Dioxy-5-methylpyrimidine)

Cytosine is present in all nucleic acids. Uracil is found in RNA but is absent from DNA. The reverse is true for thymine; it is present in DNA only. The presence or absence of thymine can be used as another analytical method for distinguishing DNA from RNA.

Problem 16.1 Draw structural formulas for the following purines and pyrimidines, which are found in small amounts in some nucleic acids.
(a) 2-Methyladenine   **(b)** 8-Methylguanine
(c) 2,6-Dihydroxypurine (xanthine)   **(d)** 5-Methylcytosine

---

Nucleosides *16.4*

A **nucleoside** is a nitrogenous base joined to one of the two pentoses, D-ribose or D-2-deoxyribose. The bond between the sugar and the base is always between C-1 of the sugar and the nitrogen in position 9 of purine bases, or the nitrogen in position 1 of pyrimidine bases. To distinguish between the positions of the atoms in the bases from those in the sugars, the atoms in the sugar are given *primed numbers.* Two examples of nucleosides are shown in Figure 16.1.

Nucleosides, as you can see, are β-*N*-glycosides (Sec. 11.7B). Like all glycosides, nucleosides are stable to alkali, but are readily hydrolyzed in aqueous acid to form a base and a pentose.

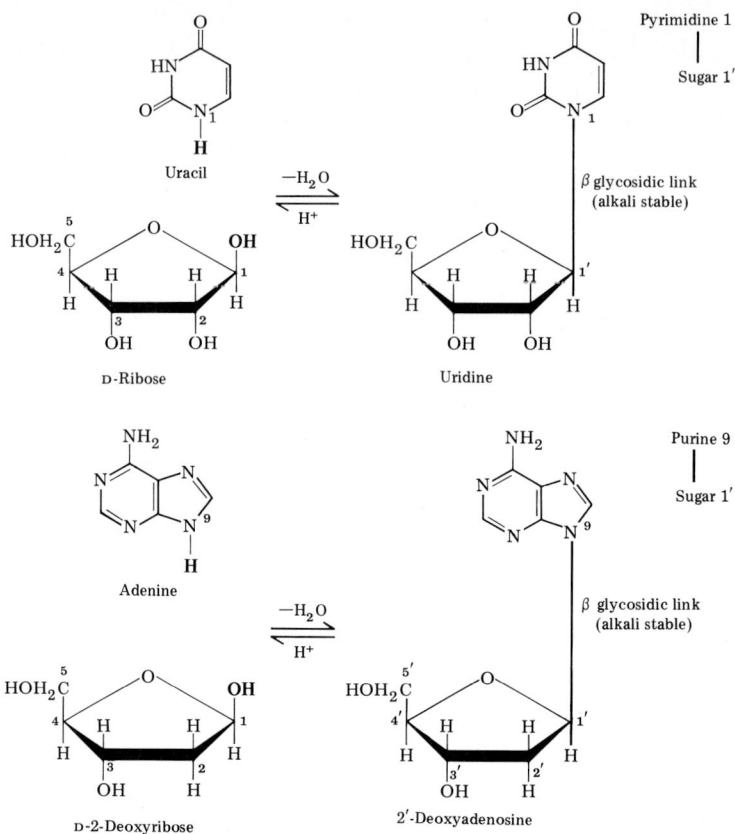

Figure 16.1 Uridine and deoxyadenosine, two typical nucleosides.

The nomenclature of nucleosides is given in Table 16.2. The suffix *-osine* is used to denote nucleosides containing a purine base. The suffix *-idine* implies that the nitrogenous base in the nucleoside is a pyrimidine. The sugar is not mentioned if it is ribose, but if it is deoxyribose we denote its presence by the use of the prefix *deoxy-*.

Table 16.2   Nomenclature of Nucleosides

Base	Ribose nucleoside	Deoxyribose nucleoside
Purines		
Adenine	Adenosine	Deoxyadenosine
Guanine	Guanosine	Deoxyguanosine
Pyrimidines		
Uracil	Uridine	Deoxyuridine[a]
Cytosine	Cytidine	Deoxycytidine
Thymine	Thymidine[b]	Deoxythymidine

[a] Not usually found in DNA.
[b] Not usually found in RNA.

Nucleotides are simply phosphate esters of the nucleosides just discussed.

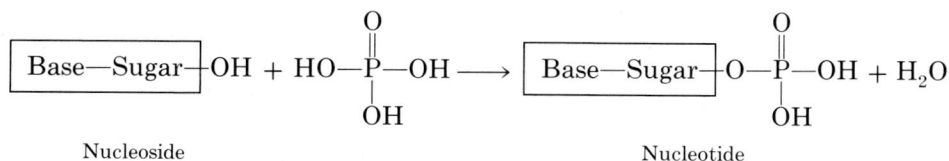

$$\boxed{\text{Base—Sugar}}\text{—OH} + \text{HO—}\overset{\overset{\displaystyle O}{\|}}{\underset{\underset{\displaystyle OH}{|}}{P}}\text{—OH} \longrightarrow \boxed{\text{Base—Sugar}}\text{—O—}\overset{\overset{\displaystyle O}{\|}}{\underset{\underset{\displaystyle OH}{|}}{P}}\text{—OH} + H_2O$$

Nucleoside   Nucleotide

The phosphate ester bond is formed with one of the hydroxyl groups of the sugar, acting as an alcohol. With a ribonucleoside, the phosphate ester bond can be formed at three possible sites: the —OH group at C-2′, C-3′, and C-5′. Thus, in naming a nucleotide we must specify the point of attachment of the phosphate, as illustrated with the 5′-phosphate esters of uridine and of deoxyadenosine.

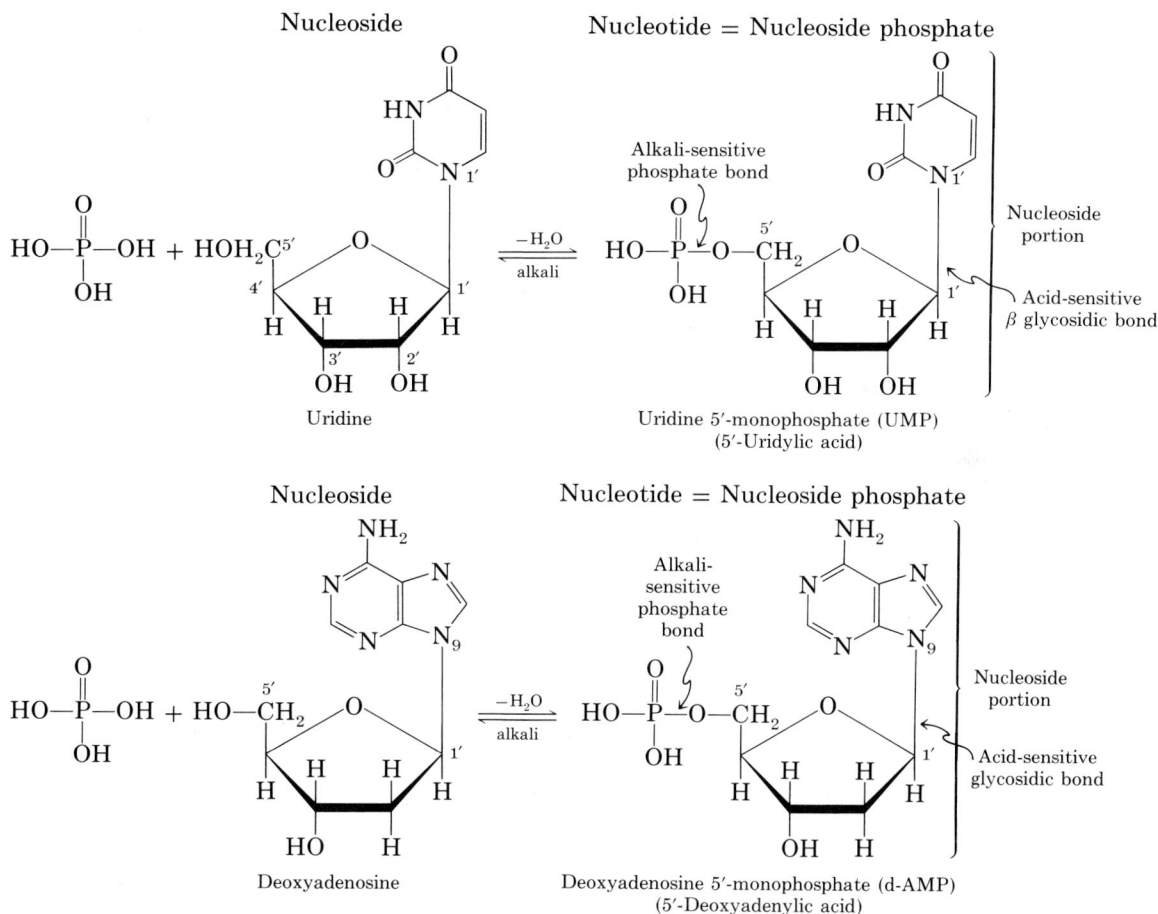

Nucleoside   Nucleotide = Nucleoside phosphate

Alkali-sensitive phosphate bond

Nucleoside portion

Acid-sensitive β glycosidic bond

Uridine

Uridine 5′-monophosphate (UMP)
(5′-Uridylic acid)

Nucleoside   Nucleotide = Nucleoside phosphate

Alkali-sensitive phosphate bond

Nucleoside portion

Acid-sensitive glycosidic bond

Deoxyadenosine

Deoxyadenosine 5′-monophosphate (d-AMP)
(5′-Deoxyadenylic acid)

Note that we have two ways to name nucleotides. For *nucleoside phosphates,* we simply name the nucleoside and give a primed number to the phosphate group to indicate its position. For *acids,* we change the suffix to *-ylic acid,* which denotes a nucleotide, and again give a primed number to the phosphate group to indicate its position, as in 5'-uridylic acid. The names of common 5'-mononucleotides derived from ribonucleic acids and deoxyribonucleic acids are listed in Table 16.3. Other isomers in addition to the 5' shown are possible.

Problem 16.2 Draw the structure of, and name as an acid, **(a)** deoxyadenosine 3'-monophosphate and **(b)** cytidine 3'-monophosphate.

In addition to being the building blocks of nucleic acids, certain nucleotides play important biological roles of their own. A few of these will be mentioned now.

Cyclic adenosine 3',5'-monophosphate (cyclic AMP) is a mononucleotide in which the phosphate group is joined to carbons 3' and 5' of ribose. This substance is important as a hormone.

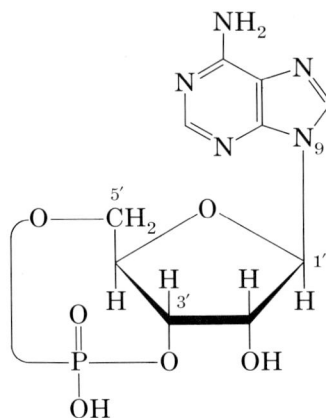

Cyclic AMP
(a hormone)

Table 16.3   Nomenclature of 5'-Mononucleotides Derived from RNA and DNA

RNA	DNA
adenosine 5'-monophosphate (AMP) or 5'-adenylic acid	deoxyadenosine 5'-monophosphate (d-AMP) or 5'-deoxyadenylic acid
guanosine 5'-monophosphate (GMP) or 5'-guanylic acid	deoxyguanosine 5'-monophosphate (d-GMP) or 5'-deoxyguanylic acid
cytidine 5'-monophosphate (CMP) or 5'-cytidylic acid	deoxycytidine 5'-monophosphate (d-CMP) or 5'-deoxycytidylic acid
uridine 5'-monophosphate (UMP) or 5'-uridylic acid	deoxythymidine 5'-monophosphate (d-TMP) or 5'-deoxythymidylic acid

One or two phosphate groups can be added to the first phosphate group of a mononucleoside by means of the *pyrophosphate bond*.

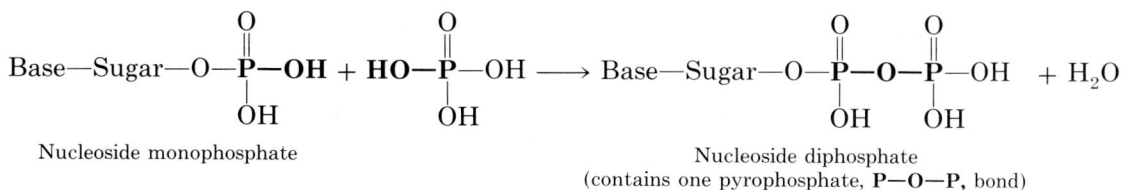

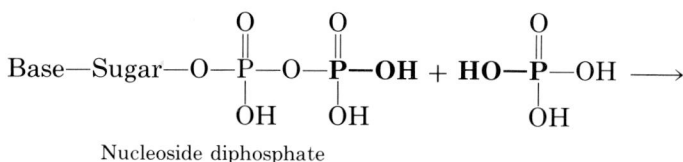

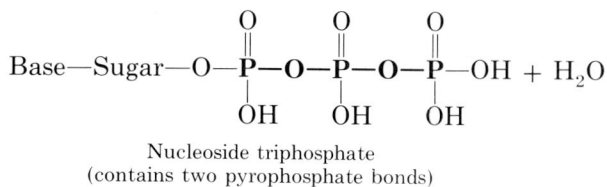

$$\text{Base—Sugar—O—} \overset{\overset{\textstyle O}{\|}}{\underset{\underset{\textstyle OH}{}}{P}} \text{—OH} + \text{HO—} \overset{\overset{\textstyle O}{\|}}{\underset{\underset{\textstyle OH}{}}{P}} \text{—OH} \longrightarrow \text{Base—Sugar—O—} \overset{\overset{\textstyle O}{\|}}{\underset{\underset{\textstyle OH}{}}{P}} \text{—O—} \overset{\overset{\textstyle O}{\|}}{\underset{\underset{\textstyle OH}{}}{P}} \text{—OH} + H_2O$$

Nucleoside monophosphate

Nucleoside diphosphate
(contains one pyrophosphate, P—O—P, bond)

$$\text{Base—Sugar—O—} \overset{\overset{\textstyle O}{\|}}{\underset{\underset{\textstyle OH}{}}{P}} \text{—O—} \overset{\overset{\textstyle O}{\|}}{\underset{\underset{\textstyle OH}{}}{P}} \text{—OH} + \text{HO—} \overset{\overset{\textstyle O}{\|}}{\underset{\underset{\textstyle OH}{}}{P}} \text{—OH} \longrightarrow$$

Nucleoside diphosphate

$$\text{Base—Sugar—O—} \overset{\overset{\textstyle O}{\|}}{\underset{\underset{\textstyle OH}{}}{P}} \text{—O—} \overset{\overset{\textstyle O}{\|}}{\underset{\underset{\textstyle OH}{}}{P}} \text{—O—} \overset{\overset{\textstyle O}{\|}}{\underset{\underset{\textstyle OH}{}}{P}} \text{—OH} + H_2O$$

Nucleoside triphosphate
(contains two pyrophosphate bonds)

The molecules formed in this way are called nucleoside diphosphates and nucleoside triphosphates. Three such compounds are ADP, ATP, and GTP.

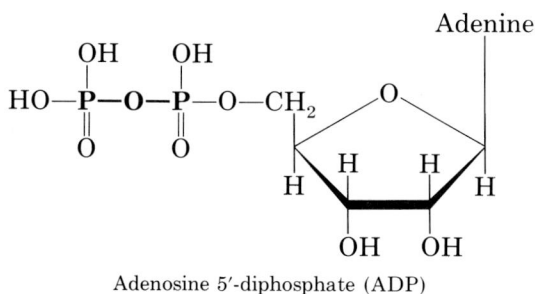

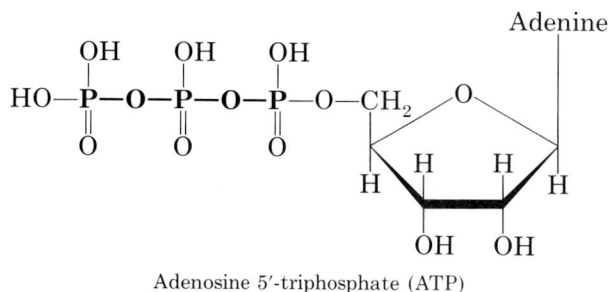

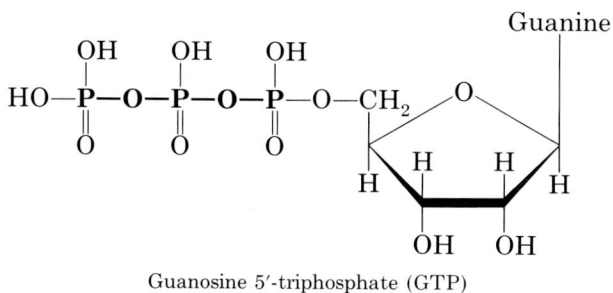

Adenosine 5'-diphosphate (ADP)

Adenosine 5'-triphosphate (ATP)

Guanosine 5'-triphosphate (GTP)

ADP, ATP, and GTP fulfill vital roles in cell metabolism as **high-energy compounds.** High-energy compounds serve as energy stores for the cell. The energy is trapped in the pyrophosphate bonds, two of which are present in ATP and GTP and one of which is found in ADP. When pyrophosphate bonds are broken through hydrolysis, the stored energy is released and is passed on to energy-requiring reactions. The most common high-energy compound found in the body is ATP.

Some nucleotides are *coenzymes,* the prosthetic groups of enzyme proteins. **Coenzyme A,** derived from adenosine diphosphate, plays a central role in carbohydrate, fat, and amino acid metabolisms as a carrier of acetyl groups and fatty acid acyl groups. Another coenzyme, **nicotinamide adenine dinucleotide (NAD$^+$),** is essential in many biological oxidation-reduction reactions.

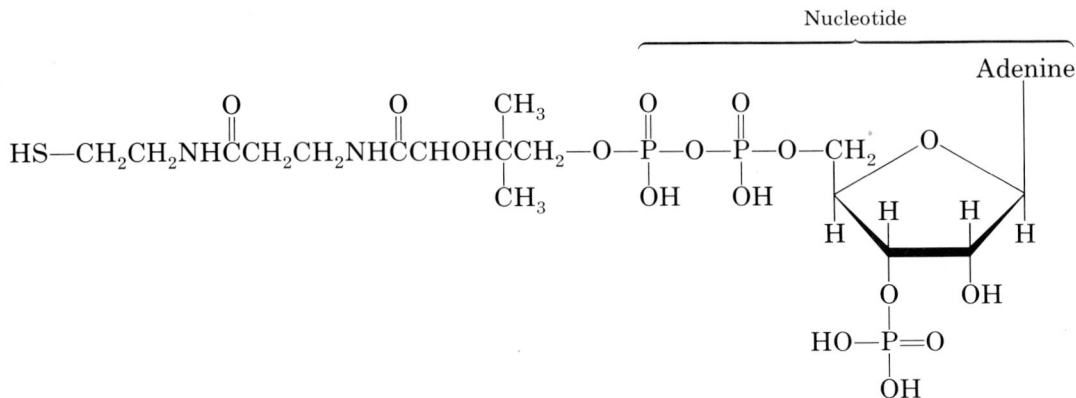

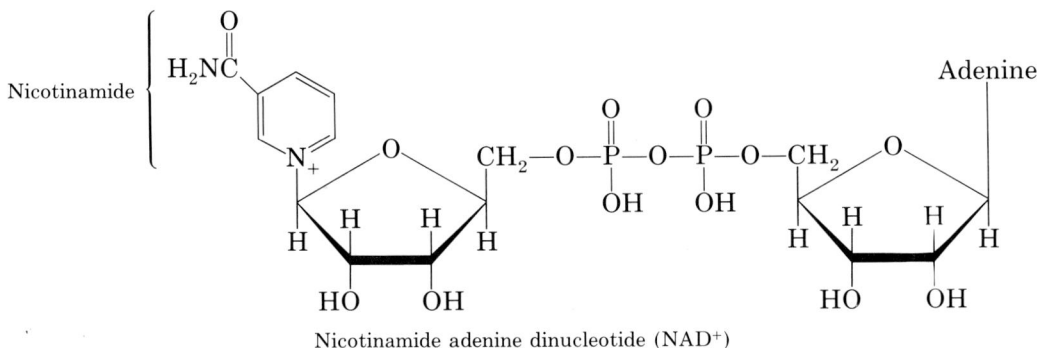

Coenzyme A (CoA)

Nicotinamide adenine dinucleotide (NAD$^+$)

16.6 Primary Structure of Nucleic Acids

Now that you are familiar with the structures of nucleotides, you are ready to learn how these building blocks of nucleic acids are assembled to give a polynucleotide. In both DNA and RNA, the nucleotides are joined together by means of a phosphate ester bond between the 3′ carbon —OH group of the pentose of one nucleotide and the 5′ carbon —OH group of the pentose of another nucleotide unit. Thus there is a phosphate *diester* bond between one nucleotide and the next. Figures 16.2 and 16.3 show the structural formulas for segments of DNA and RNA, respectively, each containing four nucleotides. The sugar–phosphate sequence constitutes the backbone of each chain, with the nitrogenous bases projecting upward. The bases shown in these figures are those

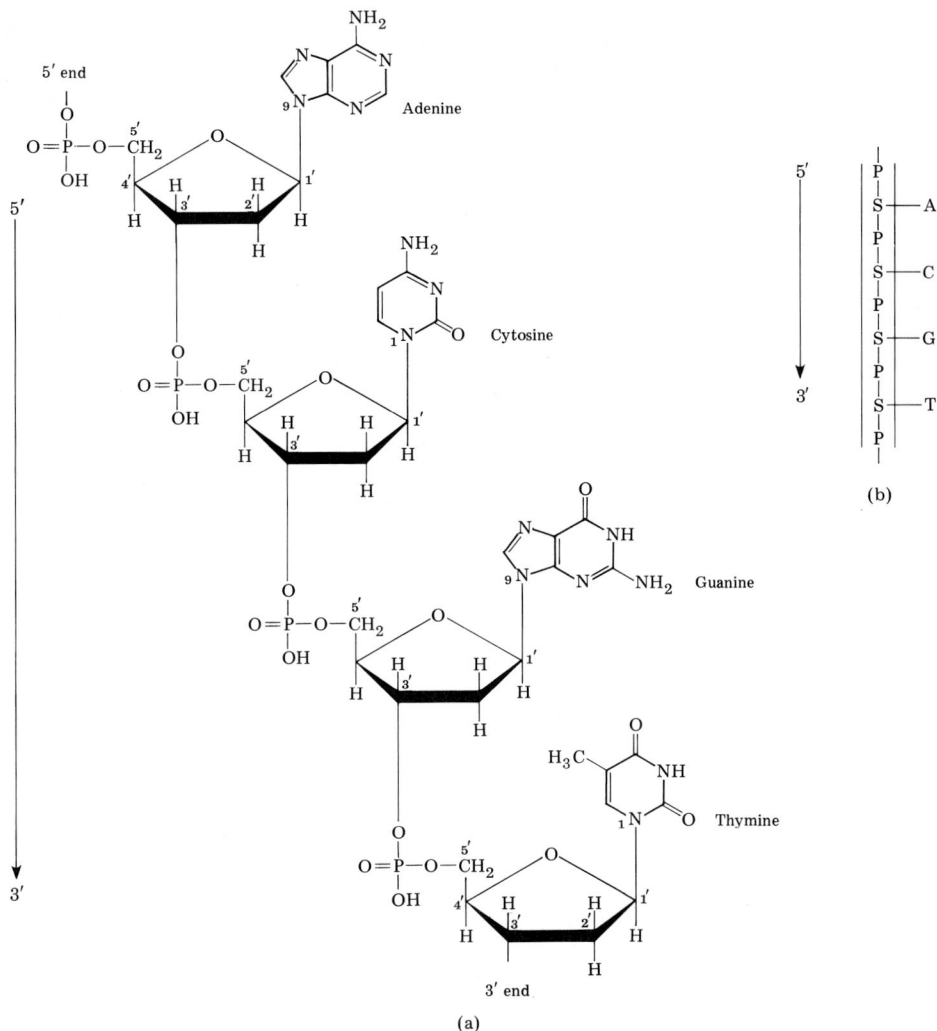

Figure 16.2 (a) A segment of DNA, showing the typical bases joined to 2-deoxyribose.
(b) The corresponding shorthand notation.

that are commonly found in DNA and in RNA. Note the phosphate diester
bridges running in the $5'{\rightarrow}3'$ direction (when reading from top to bottom).

Because it is cumbersome to write the structural formula of a polynucleotide, we shall use a shorthand method to indicate its structure (see Figs. 16.2b
and 16.3b).

1. The sugar–phosphate backbone is indicated by the letters S and P enclosed
   between two parallel lines.
2. The bases jutting out of the backbone are indicated by their initials.
3. The β glycosidic linkage is represented by a line connecting the sugar to a
   base.
4. The direction of the phosphate diester bridges is shown by an arrow numbered at each end.

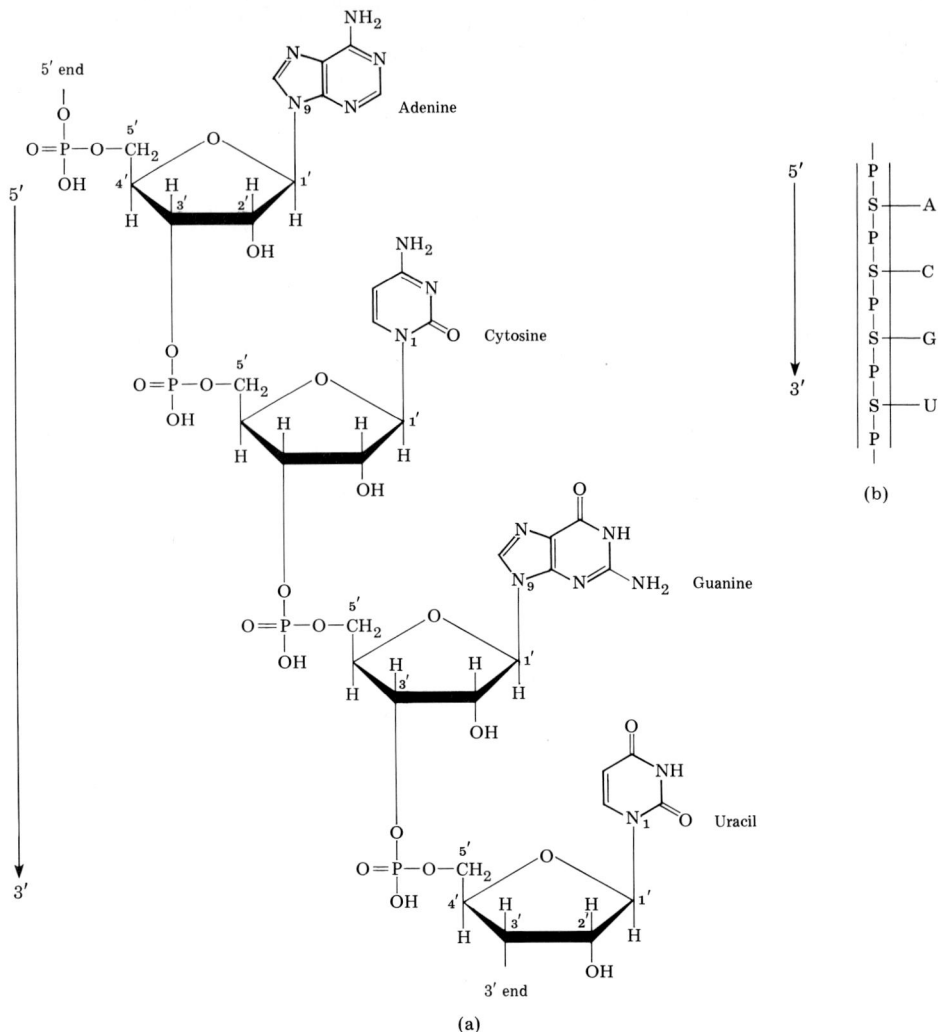

Figure 16.3 **(a)** A segment of RNA, showing the typical bases joined to ribose. **(b)** The corresponding shorthand notation.

It took approximately 70 years from the time of Miescher's discovery of DNA in 1869 to arrive at the structure of polynucleotides. There were several reasons for this slow progress, the primary one being that only a handful of scientists were interested in nucleic acids. Until the early 1940s, this lack of interest was caused by one major misconception: most scientists thought that the carriers of genetic traits were the proteins. The false assumption arose as follows. It was generally known that chromosomes contain almost all the DNA, *but they also contain almost as much protein.* It seemed reasonable to assume that if genetic traits have a molecular basis, then the most likely candidate for this role would be the protein component. After all, the most significant attribute of the gene is that it is able to carry a great deal of information and to transmit it from one generation to another. The 20-odd amino acid units that make up proteins can be arranged in an almost infinite number of combinations. DNA, on the other hand, has only four such variables, its four bases: adenine (A), guanine (G), cytosine (C), and thymine (T). The amount of information that could be encoded by four variables seemed too limited to assign the genetic role to DNA. It appeared logical, therefore, to those who were looking for the secret of the gene to devote their attention to the protein part of chromosomes. It was logical, but it proved to be erroneous.

DNA as the Genetic Material *16.7*

In 1944 a team of biochemists headed by Ostwald T. Avery, working at Rockefeller University, published a paper that revolutionized the thinking of the scientific community. In their publication, the Rockefeller University scientists announced that *the material responsible for transmitting hereditary characteristics from one generation to another is DNA and not the protein part of chromosomes.* Avery and co-workers based their conclusion on the following experimental data. They had taken one strain (type III) of a pneumonia-causing bacterium and painstakingly isolated its DNA, *freed from any protein.* They then added the DNA extract of the type III bacterium to another strain (type II) of live pneumonia-causing bacterium. When the type II strain was allowed to multiply, it was found that the new generation of bacteria contained type III strain. The obvious conclusion is that the active material that caused the genetic transformation of type II to type III must have been carried by the type III DNA extract.

Avery's publication created a surge of interest in nucleic acid research. Soon other examples of transformations from strain to strain were reported among bacteria; in every case, the transforming agent was DNA. By the late 1940s biochemists all agreed: the genetic material was DNA.

The Three-Dimensional Structure of DNA: *16.8*
The Double Helix

After the nature of the genetic material had been identified, there remained a crucial question to be answered: How does the structure of DNA account for its genetic rule? More specifically, how does the structure of DNA explain the process of **self-replication,** whereby each cell in a living organism is the exact duplicate of all other cells? The answer to this question had to await the elucidation of the three-dimensional shape of the DNA molecule by Watson and Crick.

Meanwhile, Erwin Chargaff and co-workers undertook a meticulous analysis of the base composition of the DNAs from various sources. They found that regardless of the source of the DNA, the number of adenines is always equal to the number of thymines. Similarly, the number of guanines is always equal to the number of cytosines. Chargaff and co-workers also found that although the *ratio* of A/T or G/C equals one, the *amount* of these pairs of bases is *not* the same. Table 16.4 compares the distribution of bases in DNAs from various

Table 16.4 Comparison of Base Composition of DNAs from Various Sources

Source	Percentage				Ratios		
	A	G	C	T	A/T	G/C	Purines/ pyrimidines
Animals							
Marine crab	47.3	2.7	2.7	47.3	1.00	1.00	1.00
Frog	26.3	23.5	23.8	26.4	1.00	0.99	0.99
Chicken (liver)	30.3	22.0	19.7	28.0	1.08	1.12	1.10
Rat (liver)	28.6	21.4	21.5	28.4	1.01	1.00	1.00
Human (liver)	30.3	19.5	19.9	30.3	1.00	0.98	0.99
Plants							
Carrot	26.7	23.1	23.2	26.9	0.99	1.00	0.99
Peanut	32.1	17.6	18.0	32.2	1.00	0.98	0.99
Bacteria							
E. coli	23.8	26.0	26.4	23.8	1.00	0.98	0.99

sources. Chargaff's findings were of great help in working out the three-dimensional structure of DNA. Until then it had been thought that a DNA molecule was made up of the polynucleotide segments shown in Figure 16.2a, repeated uniformly a number of times so as to give the same amount of each base. Chargaff's analyses disproved that notion.

Another important advance in the investigation of DNA structure came when M. H. F. Wilkins, in England, obtained clear x-ray diffraction patterns of DNAs from various sources. The x-ray patterns displayed remarkable similarities, despite their different origins. Careful analysis of the data showed that all DNA molecules had identical thicknesses. Furthermore, the x-ray data also indicated that the same pattern recurred along the length of the molecule at a distance of every 34 Å.

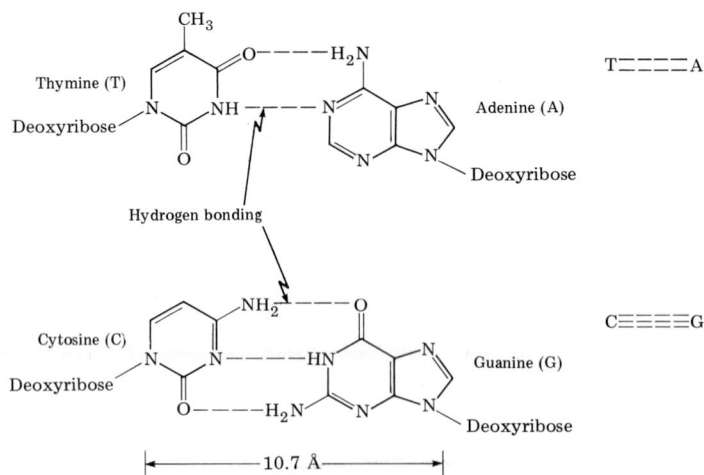

Figure 16.4 Hydrogen bonding between thymine and adenine (T====A) and between cytosine and guanine (C≡≡≡≡G). The dimensions of the T====A pair are the same as those of C≡≡≡≡G pair.

Combining the observations of Chargraff and the x-ray diffraction data of Wilkins, and inspired by Pauling's announcement of the α-helix structure for proteins (Sec. 15.13), Watson and Crick proposed their now-famous double helix model for DNA.

The three-dimensional structure of DNA that emerged was that of two polynucleotide chains wrapped around each other in a spiral fashion. Both chains are right-handed helices. The outer portion of each strand is composed of deoxyribose units joined by phosphate diester bridges; together these form the backbone of the molecule. The two chains are said to be *antiparallel* because they run in opposite directions (that is, one chain in a 3′→5′ direction, the other in a 5′→3′ direction). Projecting inward from the backbone are the bases. The bases are stacked on top of one another, 3.4 Å apart, in a parallel fashion much like the rungs of a ladder. The bases are paired in a very specific manner: the adenines of one strand are paired to the thymines of the other strand. Similarly, the guanines of one chain are paired to the cytosines of the other. The base pairs, which are termed **complementary bases,** are held together by hydrogen bonding (see Fig. 16.4).

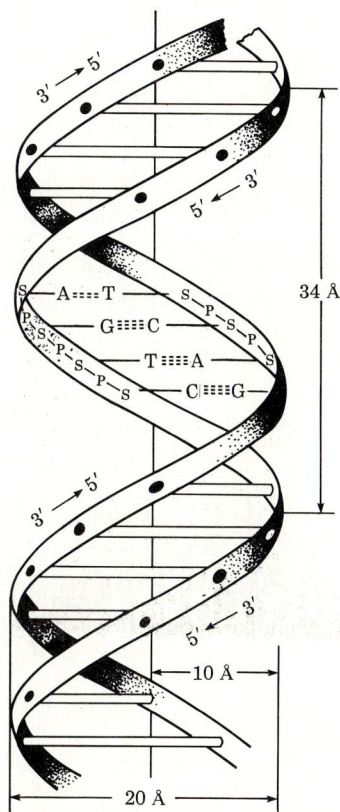

Figure 16.5 A schematic representation of the double helix structure of DNA proposed by Watson and Crick. The backbone contains deoxyribose (S) and phosphate diester bonds (P). The two strands run in opposite directions. A====T is the adenine–thymine pairing, and G≡≡≡≡C is the guanine–cytosine pairing. The backbones of the two strands of the double helix are separated by a distance of 10 Å. A complete turn of the helix is repeated every 34 Å.

The hydrogen bonds between the bases hold the two strands of the double helix together and give it its three-dimensional stability. The bonds also serve to maintain a constant distance between the two chains, thus accounting for the identical thickness of DNA molecules observed in Wilkins' x-ray data.

Finally, to account for the recurring x-ray patterns along the length of the molecule, Watson and Crick postulated that every 34 Å distance corresponded to a complete turn of the helix. A schematic diagram of the DNA double helix is shown in Figure 16.5.

16.9 DNA Replication

The double helix model of DNA not only was consistent with the physical and chemical properties of the molecule, but it also provided a beautifully simple explanation for the way in which a gene duplicates itself. The propagation of a

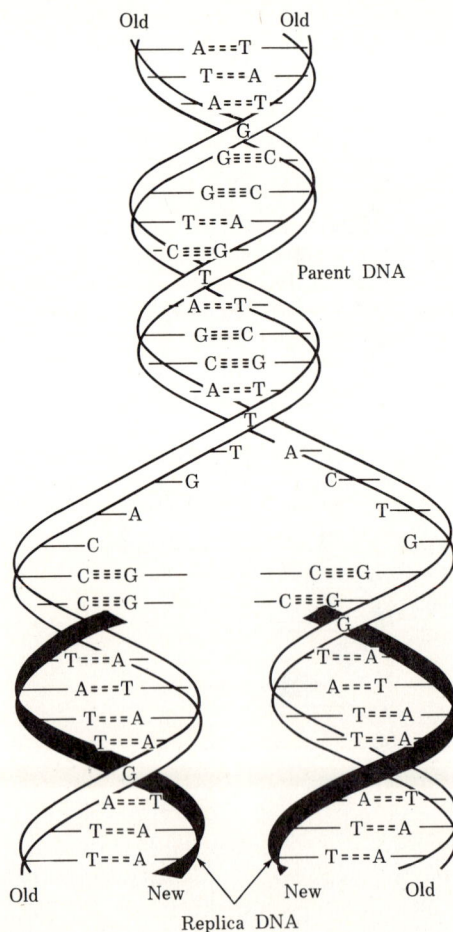

Figure 16.6 The overall scheme of the replication of DNA, showing the simultaneous duplication of the two strands.

species requires a faithful reproduction of its genes at cell division. A gene, as you recall, is the carrier of hereditary characteristics; on a molecular level, a gene is simply a small segment of a DNA molecule.

A probable mechanism by which DNA replication takes place is illustrated in Figure 16.6. The process starts with the unwinding of the two chains in the parent DNA. As the two strands separate, each can serve as a master copy, or **template,** for the construction of a new partner. This is done by bringing the appropriate nucleotides in place and linking them together. Because the bases must be paired in a specific manner (adenine with thymine and guanine with cytosine, each newly built strand is not identical but is complementary to the old one. Thus when replication is completed, we have two DNA molecules, *each identical to the original.* Each of the new molecules is a double helix that has one old strand and one new strand to be transmitted to daughter cells.

The mechanism of replication just described was proposed originally by Watson and Crick. It has been tested and confirmed experimentally by others. In 1962, Watson, Crick, and Wilkins were awarded the Nobel prize for Medicine and Physiology for "their discoveries concerning the molecular structure of nucleic acids, and its significance for information transfer in living material." This statement brings us to the second important function of nucleic acids: protein synthesis.

Nucleic Acids and Protein Synthesis *16.10*

If, as we have stated, a gene is a small section of DNA, then the information carried by that gene must also be contained in that small segment of the DNA molecule. Because the gene for "blue eyes" is different from the gene for "nose," it is reasonable to infer that the portion of DNA that corresponds to "blue eyes" has a different structure from the segment of DNA that translates into "nose." The only structural variables in DNA are its four bases. We must therefore conclude that *the information carried by a gene is encoded in its sequence of bases.*

How does a gene express the information it carries? Usually in the form of proteins, particularly enzymes, which are essential for the growth and development of any living organism.

This raises, however, a serious question: How can a molecule consisting of four variable bases (adenine, cytosine, guanine, and thymine) supply the information needed to direct the synthesis of proteins, which usually contain 20 odd amino acids? Obviously, one base in DNA is not enough to specify one amino acid. But if a one-base code is not enough, what about a two-base code? A pair of bases provides 16 (4^2) possible combinations, which is still inadequate to specify 20 or so different amino acids. A three-base code, on the other hand, allows 4^3 or 64 different possibilities, which is more than enough to code for all the amino acids in a protein. A triplet of bases in DNA that designates a specific amino acid is called a **codon.** The validity of the codon hypothesis has been proved experimentally by the deciphering of the genetic code. But before we consider the genetic code (Sec. 16.11), let us address ourselves to the broader question of how nucleic acids direct the synthesis of proteins.

A From DNA to RNA

The process of translating the information contained in DNA into protein requires three kinds of ribonucleic acid. These three types of RNA have essentially the composition described previously (Sec. 16.1); they differ in their molecular sizes and in their functions.

Messenger RNA, or mRNA, is a polyribonucleotide of molecular weight between 200,000 and 50 million. Messenger RNA is synthesized in the nucleus by a mechanism much like that of DNA replication. A section of the DNA double helix untwists, and one of its strands acts as a template from which mRNA is derived. The mRNA when formed is a single-stranded molecule whose order of bases is governed by the sequence of bases in the DNA template. Thus the mRNA molecule has the same sequence of bases as the DNA strand that complements the DNA template strand, with *one exception:* uracil rather than thymine appears in mRNA as the base complementary to adenine. The synthesis of mRNA as governed by the order of bases in DNA is called **transcription.** The net result of transcription is shown in the diagram in Figure 16.7.

Once formed, mRNA leaves the nucleus and migrates into the cytoplasm, where protein synthesis occurs. The function of mRNA, as its name implies, is to act as a messenger to bring the information from DNA, which is in the nucleus, to the cytoplasm, specifically to the ribosomes.

Ribosomes are granular particles of very high molecular weight that are found in the cytoplasm. Ribosomes are composed of about 40% protein and 60% **ribosomal RNA, or rRNA.** Each ribosome consists of two subunits, one approximately twice the size of the other. When the two units are brought together, the ribosome is complete and active. We can think of an active ribosome as a protein factory. In this factory, the groove below the smaller subunit binds the mRNA, and the larger subunit attaches the third type of RNA involved in protein synthesis, transfer RNA, as illustrated in Figures 16.10 and 16.11.

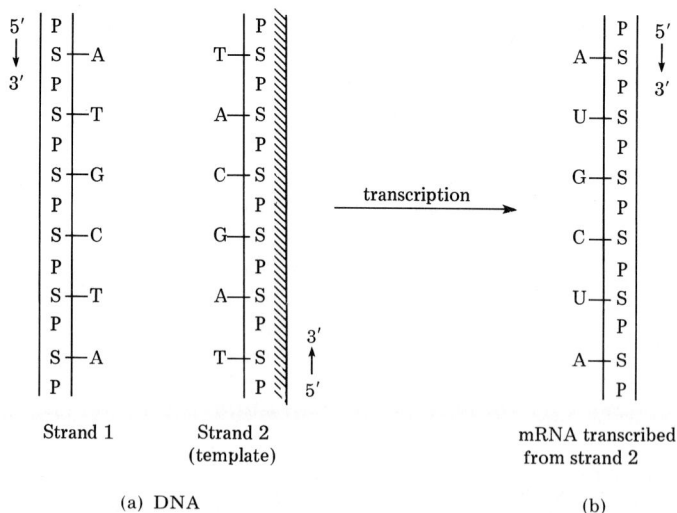

Figure 16.7 **(a)** A schematic diagram of a section of double-stranded DNA; strand 2 acts as the template for the transcription process. **(b)** mRNA transcribed from strand 2 of DNA; note that its bases are the same as those of strand 1, except that U replaces T.

Transfer RNA, or **tRNA,** is the smallest of the RNAs and is found in the cytoplasm. There are a large number of tRNAs: at least one for each of the amino acids found in protein. Despite their variety, all tRNAs have several features in common. They are soluble in water and have a molecular weight of about 25,000. Structurally, most tRNAs are single-stranded molecules made up of 75–80 nucleotides linked by phosphate diester bridges. Certain regions of the molecule are helical in shape because of hydrogen bonding between internal base pairs. Although each tRNA is specific for a particular amino acid, all tRNAs have the same sequence of three bases at one end of the chain. The order of these bases is cytosine, cytosine, and adenine (C, C, A). The function of tRNA is to transport a given amino acid to the ribosome, the site of amino acid assembly into protein. An amino acid is transported to the ribosome by forming an ester bond between its α —COOH group and the 3′ carbon —OH group of the ribose attached to the terminal adenine. The formation of this bond is catalyzed by enzymes.

Because there is a different tRNA for every amino acid, there must be specific segments in tRNA to recognize the proper amino acid. One such segment includes a series of three nucleotides that we call an **anticodon.** The anticodon contains a triplet of bases on tRNA that is complementary to a triplet of bases, or codon, on mRNA.

A schematic representation of a tRNA molecule is shown in Figure 16.8.

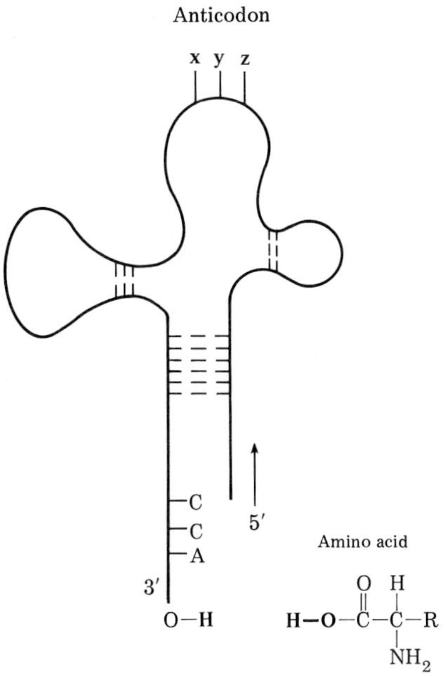

Figure 16.8 A schematic diagram of a tRNA molecule. The solid line represents the backbone and the dashed lines represent hydrogen bonds between internal base pairs. The anticodon is indicated by the letters xyz at the end of one loop. The potential formation of the ester bond between the α COOH group of an amino acid and the C-3′ OH group of ribose attached to the terminal adenine is also illustrated.

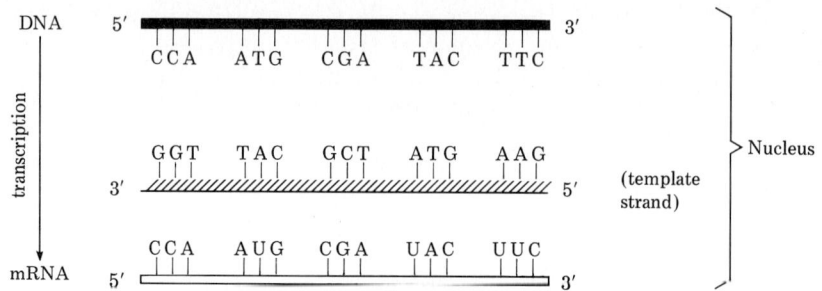

Figure 16.9 The imprinting of a portion of DNA code on a strip of messenger RNA.

B Protein Synthesis

The process of building a protein begins when the DNA imprints a portion of its code on a strip of messenger RNA (transcription). In Figure 16.9, the bases are grouped in triplets (or codons) for clarity, although in actuality they are the same distance apart. The messenger RNA strip migrates into the cytoplasm and is picked up by a ribosome. It is then inserted in the groove between the two subunits of the ribosome. The ribosome proceeds to read off the message encoded in the mRNA word by word (codon by codon), like a tape recorder (Fig. 16.10).

Each codon on mRNA designates a specific amino acid. As a ribosome (or a series of ribosomes, called a polyribosome, or **polysome**) moves along the mRNA and reads off a codon, the designated amino acid is picked up by its tRNA and brought into position on the ribosome. The tRNA, with its amino acid, is held in place on the ribosome by pairing (via hydrogen bonds) its anticodon with the codon on the mRNA. In this way, one by one, following the exact order dictated by mRNA (and ultimately by DNA), the proper amino acids are brought into line and assembled to give a polypeptide. As a tRNA transfers its amino acid on the growing polypeptide chain, it leaves the ribosome and is free to pick up another amino acid from the amino acid pool in the cell fluid.

As the ribosome continues to move along the mRNA strip, it eventually reaches a "nonsense" codon for which no amino acid is called for. This "nonsense" codon orders it to stop. At this point the ribosome releases its two subunits and leaves the mRNA, thus freeing the completed protein. The process by which a protein is manufactured from the message dictated by mRNA is

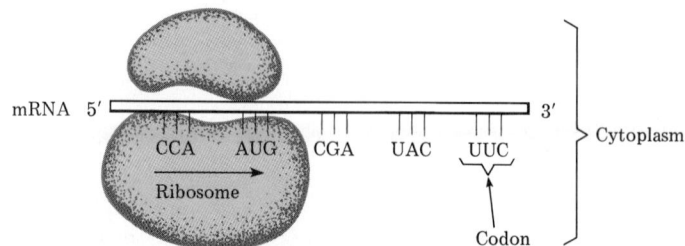

Figure 16.10 As the mRNA passes between the grooves in the ribosome, the message encoded in the mRNA is read codon by codon.

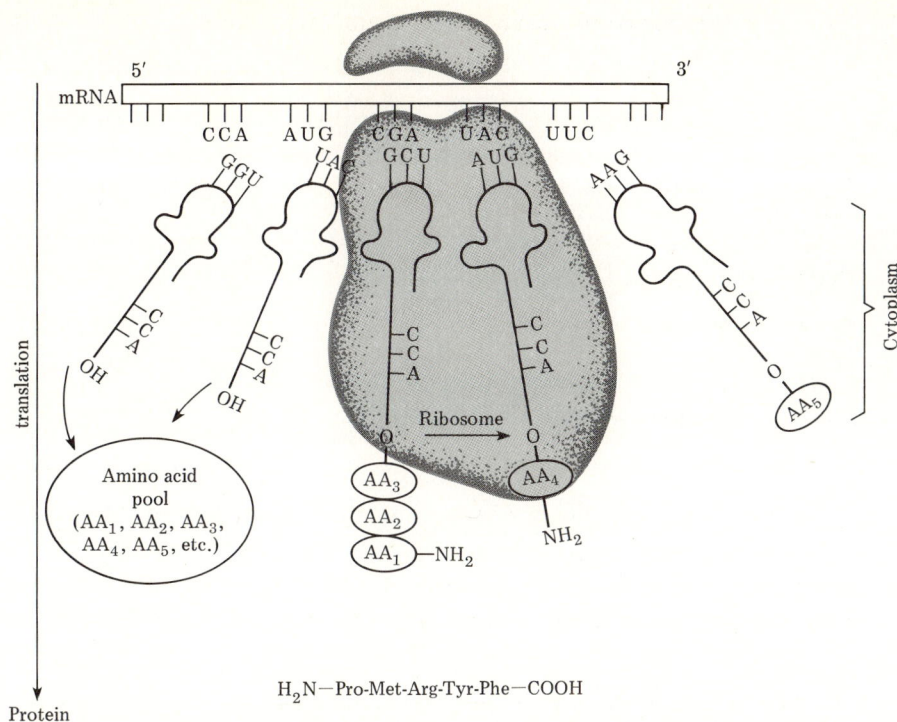

Figure 16.11 The translation of five codons in a segment of mRNA.

called **translation.** The translation of the five codons on the mRNA segment in our example is shown in Figure 16.11.

The actual mechanism of translation is rather complicated. It requires several enzymes, high-energy compounds, and other factors. For this reason, we will not discuss it here.

The Genetic Code **16.11**

We learned in the preceding section how the sequence of codons in DNA ultimately dictates the sequence of amino acids in proteins. Once the nature of the triplet code was recognized, the next step was to decipher it: that is, to find out which codon specifies which amino acid.

One obvious approach to the key of the code is: to isolate a gene, determine its sequence of bases, and compare that sequence of bases with the order of the amino acids in the protein built under the directives of that gene. Unfortunately, until very recently, this approach proved to be impossible. Therefore, another method was used to crack the genetic code.

The breakthrough came in a series of clever experiments initiated by Marshall Nirenberg at the National Institutes of Health (NIH). What the NIH biochemist did may be summarized as follows. In a control experiment Nirenberg made a broth containing ribosomes, tRNAs, amino acids, and enzymes, and

Table 16.5   The Genetic Code: RNA Codons and the Amino Acids for Which They Code

		Gln	CAG			CUU		UCC
Ala	GCA		CAA			UUA		UCU
	GCC					UUG		
	GCG	Glu	GAA				Thr	ACA
	GCU		GAG	Lys	AAA			ACG
					AAG			ACC
Arg	AGA	Gly	GGA					ACU
	AGG		GGC	Met	AUG[a]			
	CGA		GGG				Trp	UGG
	CGG		GGU	Phe	UUU			
	CGC				UUC		Tyr	UAC
	CGU	His	CAC					UAU
			CAU	Pro	CCA			
Asn	AAC				CCC		Val	GUA
	AAU	Ile	AUA		CCG			GUG
			AUC		CCU			GUC
Asp	GAC		AUU					GUU
	GAU			Ser	AGC			
		Leu	CUA		AGU			
Cys	UGC		CUC		UCA		UAA } Stop
	UGU		CUG		UCG		UAG } signals

[a]AUG, which codes for methionine, is also a chain-initiating codon (a start signal).

then incubated the mixture. With only these components present, no protein was synthesized. Nirenberg then repeated the experiment, but this time he added a synthetic mRNA, *polyuridylic acid* (poly-U), to the broth. A very dramatic result was obtained. In the broth was a high-molecular-weight polypeptide that contained only one kind of amino acid: *phenylalanine.* The first word of the **genetic code** had been deciphered: *the UUU codon stands for phenylalanine.* In the same manner it was found that the base sequence UGU stands for cysteine, and that the codon GUG represents valine. By using and extending the same strategy, the complete genetic dictionary was deciphered (see Table 16.5). Nirenberg's monumental contribution in determining the genetic code earned him the Nobel prize for Medicine and Physiology in 1968.

Several aspects of the genetic code are worthy of comment.

1. The code appears to be **universal**—that is, it applies to all living organisms.
2. The code is said to be **degenerate,** meaning that *more than one codon may designate one amino acid.* For example, glycine is specified by four codons: GGA, GGC, GGG, and GGU.
3. The **degeneracy** is not random but rather is **systematic,** meaning that *in most cases the first two bases remain the same and only the third varies* (as in glycine).
4. The code contains several **"nonsense" codons** that do not call for any amino acid. These codons are thought to be the *start* and *stop signals* for polypeptide synthesis.

A number of diseases are of genetic origin. Recall, for example, that one form of sickle cell anemia occurs when a single amino acid, valine, is substituted for glutamic acid at position 6 in the β chain of hemoglobin (Fig. 15.10). Because proteins are manufactured under the direction of nucleic acids, the disease may ultimately be ascribed to a deviation from the normal order of bases in DNA. The change in the normal order of bases is called a **mutation.** If we consult the genetic code dictionary, we see that the GAG codon calls for glutamic acid. If a mistake is made and the codon GUG is instead transcribed in mRNA, then valine will be incorporated in the protein. How this mistake is made is not exactly clear. But we do know that a great number of agents can interact with the bases in nucleic acids in such a way as to cause mutations. These agents include x rays, ultraviolet and radioactive radiations, and a large variety of chemical reagents. The mutagenic effects of nitrous acid (HNO_2) will serve as an example.

Potential sources of HNO_2 include $NaNO_2$ (used as a food preservative in products such as frankfurters and sausages) and the nitrogen dioxide present in smog and in automobile exhausts.

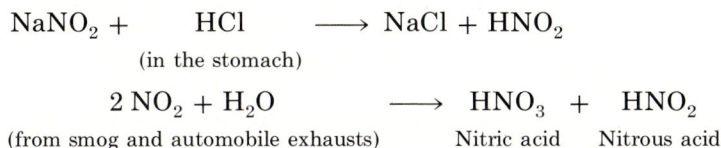

$$NaNO_2 + \quad HCl \quad \longrightarrow \quad NaCl + HNO_2$$
(in the stomach)

$$2\,NO_2 + H_2O \quad \longrightarrow \quad HNO_3 \quad + \quad HNO_2$$
(from smog and automobile exhausts) \quad Nitric acid \quad Nitrous acid

Nitrous acid causes chemical transformation of adenine and cytosine into hypoxanthine and uracil. In each case the $C-NH_2$ grouping has been replaced by a $C=O$ grouping. Because each modified base has a complementary base different from the original, the wrong information will be perpetuated during replication and protein synthesis (Fig. 16.12). This, in effect, represents a mutation.

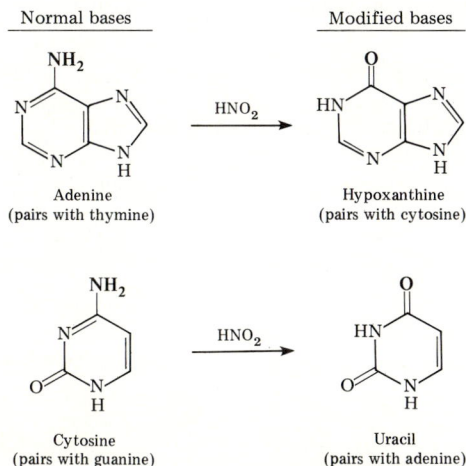

Figure 16.12 Mutagenic effect of nitrous acid.

457

Mutations are not always harmful, however. In agriculture, for example, the genetic makeup of certain varieties of wheat and of rice have been modified to produce mutant plants that are richer in protein. In industry, mutant organisms have been used beneficially to produce higher yields of products and to develop more powerful antibiotics.

16.13 Gene Splicing

Dramatic advances have been made in recent years in the field of genetic manipulation. This research, namely **recombinant DNA,** involves the splicing of a gene from one organism into the chromosomes of another organism.

Gene splicing has already proved extremely beneficial. For example, the gene coding for insulin was grafted onto certain bacteria that became factories for producing the hormone. Human growth hormone produced in bacteria that have been manipulated by gene splicing techniques has been tested in humans (early in 1981) with promising results. In June 1981, scientists from the U.S. department of Agriculture announced the production of an effective vaccine, through the use of gene splicing, against foot-and-mouth disease. The discovery could have enormous potential benefit for the world's supply of food, especially in Africa and Latin America where the disease is widespread and devastating to the livestock.

Although such genetic manipulations could ultimately lead to cure for a wide variety of diseases, concern has been voiced about the possibility of producing new strains of pathogenic organisms against which there is no known defense. Until now, however, the direst perils of gene-splicing research have been avoided while some of its best promises are being realized at a fast pace.

Summary of Concepts and Reactions

Nucleic acids are biological polymers that consist of repeating units called nucleotides. Nucleotides are substances that on complete hydrolysis yield three components: nitrogen-containing bases, pentose sugars, and phosphoric acid.   [Sec. 16.1]

There are two families of nucleic acids: deoxyribonucleic acid (DNA) and ribonucleic acid (RNA).   [Sec. 16.1]

DNA is found in cell nuclei and contains D-2-deoxyribose as the only pentose sugar; RNA is present primarily in the cytoplasm and on hydrolysis yields D-ribose as the only sugar component.   [Sec. 16.2]

There are two classes of nitrogen-containing bases in nucleic acids. These are derivatives of purine and pyrimidine, both of which are heterocyclic compounds.   [Sec. 16.3]

The two most common purine bases found in nucleic acids are adenine and guanine. The three most common pyrimidine bases are cytosine, uracil, and thymine.
                                                                                                            [Sec. 16.3]

A nucleoside is a nitrogenous base joined to one of the two pentoses, D-ribose or D-2-deoxyribose. The bond between the sugar and the base is always between C-1 of the sugar and the nitrogen in position 9 of purine bases or the nitrogen in position 1 of pyrimidine bases.   [Sec. 16.4]

Nucleotides are phosphate esters of nucleosides.

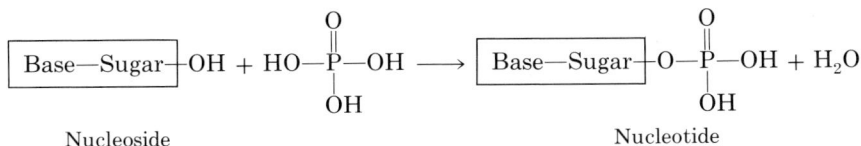

$$\boxed{\text{Base—Sugar}}\text{—OH} + \text{HO—}\overset{\displaystyle O}{\underset{\displaystyle OH}{\overset{\|}{P}}}\text{—OH} \longrightarrow \boxed{\text{Base—Sugar}}\text{—O—}\overset{\displaystyle O}{\underset{\displaystyle OH}{\overset{\|}{P}}}\text{—OH} + H_2O$$

Nucleoside   Nucleotide

[Sec. 16.5]

In addition to being the building blocks of nucleic acids, some nucleotides play important biological roles of their own. Examples of such nucleotides are cyclic AMP, ADP, ATP, GTP, Coenzyme A, and nicotinamide adenine dinucleotide (NAD$^+$).   [Sec. 16.5]

In both DNA and RNA, the nucleotides are joined together by means of a phosphate ester bond between the 3' carbon —OH group of the pentose of one nucleotide and the 5' carbon —OH group of the pentose of another nucleotide unit.   [Sec. 16.6]

The material responsible for transmitting hereditary characteristics from one generation to another is DNA and not the protein part of chromosomes.   [Sec. 16.7]

The three-dimensional structure of DNA consists of two polynucleotide chains wrapped around each other in a spiral fashion (a double helix). The base pairs or complementary bases of the double helix are held together by hydrogen bonds.   [Sec. 16.8]

During replication, the two helical strands of DNA separate and each serves as a master copy or template for the construction of a new partner.   [Sec. 16.9]

The information carried by a gene is encoded in its sequence of bases. A triplet of bases in DNA that designates a specific amino acid is called a codon.   [Sec. 16.10A]

The process of translating information contained in DNA into protein requires three kinds of ribonucleic acid: (1) messenger RNA or mRNA, (2) ribosomal RNA or rRNA, and (3) transfer RNA or tRNA.   [Sec. 16.10A]

The synthesis of mRNA as governed by the order of bases in DNA is called transcription.   [Sec. 16.10A]

The process by which a protein is manufactured from the message dictated by mRNA is called translation.   [Sec. 16.10B]

The genetic code tells us which triplet of bases in mRNA corresponds to which amino acid in a protein.   [Sec. 16.11]

The change in the normal order of bases in DNA is called mutation.   [Sec. 16.12]

Recombinant DNA involves the splicing of genes from one organism into the chromosomes of another.   [Sec. 16.13]

Key Terms

nucleic acids
nucleotides
polynucleotides
deoxyribonucleic acid
    (DNA)
ribonucleic acid (RNA)
2-deoxyribose
ribose
purine
pyrimidine

adenine
guanine
cytosine
uracil
thymine
nucleoside
high-energy compounds
coenzyme A
nicotinamide adenine
    dinucleotide (NAD$^+$)

self-replication
complementary bases
template
codon
messenger RNA (mRNA)
transcription
ribosomes
ribosomal RNA (rRNA)
transfer RNA (tRNA)
anticodon

polysome	universal code	nonsense codons
translation	degenerate code	mutation
genetic code	systematic degeneracy	recombinant DNA

Exercises

Names and Structures of RNA and DNA Components [Secs. 16.1–16.3]

16.1 Name the products of complete hydrolysis of **(a)** RNA and **(b)** DNA.

16.2 Draw the structures of the components **(a)** found in DNA but absent in RNA and **(b)** found in RNA but absent in DNA.

Nitrogenous Bases; Nucleosides [Secs. 16.3, 16.4]

16.3 5-Fluorouracil and 6-mercaptopurine are two compounds that have been used in the treatment of leukemia. Write their structures.

16.4 True or false (if false, explain why)—
  (a) Nucleosides result from the combination of a purine or pyrimidine base with a phosphate group.
  (b) Nucleosides are stable to alkali but are readily hydrolyzed in acid solution.
  (c) Urosine is the name of the nucleoside made up of uracil and ribose.
  (d) Hydrolysis of 2-deoxyguanosine yields 2-deoxyribose and guanosine.
  (e) A nucleoside is a nucleotide from which a phosphate group has been removed.

Nucleotides [Sec. 16.5]

16.5 Give another name for each of the following nucleotides.
  (a) Adenosine 5′-pyrophosphate   **(b)** 3′-Guanidylic acid
  (c) Deoxycytidine 5′-monophosphate

16.6 Write the structures of the compounds in Exercise 16.5.

16.7 What is meant by high-energy compounds? Give an example.

Structure and Functions of Nucleic Acids [Secs. 16.6–16.11]

16.8 What type of bond links the pentose with the bases in nucleic acids?

16.9 **(a)** What is meant by complementary bases?
  (b) What type of bond holds complementary bases together?

16.10 Describe the structural features of double helix DNA.

16.11 What are the two main functions of DNA?

16.12 Give the main functions of each of the following.
  (a) mRNA   **(b)** tRNA   **(c)** rRNA

16.13 In this description of a portion of a DNA molecule, the bases of one strand (A) are given.

5′ ――――――――――――――――――――――――――― 3′   DNA strand A
     CCT TCA AGT ACG TTT AAA AAC TAC CCA

  (a) Complete the diagram by showing the sequence of bases in strand B.
  (b) What would be the sequence of bases in the mRNA strand that is complementary to strand A?
  (c) What would be the order of amino acids in the peptide translated from mRNA?

16.14 Angiotesin II is a peptide that produces a marked increase in blood pressure. Its amino acid sequence is

Asp-Arg-Val-Tyr-Val-His-Pro-Phe

(a) List the base sequence in the mRNA strand that corresponds to the amino acid sequence in angiotesin II.

(b) List the base sequence in the corresponding DNA segment.

Mutations [Sec. 16.12]

16.15 Write the structure of the modified base obtained when guanine is treated with nitrous acid.

Definitions

16.16 Define or describe **(a)** DNA; **(b)** RNA; **(c)** replication; **(d)** translation; **(e)** transcription; **(f)** complementary bases; **(g)** codon; **(h)** anticodon; **(i)** genetic code; **(j)** code degeneracy; **(k)** mutation.

17

Spectral Methods of Structure Determination

Throughout our studies we have relied on chemical tests and solubility tests to classify and identify organic compounds. Such classical methods of structure determination have the advantage that they can be carried out inexpensively in test tubes and other readily available equipment. Chemical techniques, however, are generally time consuming, especially if the molecule is one with some degree of complexity. A more serious disadvantage is that fairly large amounts of samples are needed for each test, and often the unknown is destroyed during the testing procedure.

Over the past decades, the development of instrumentation that allows the chemist to determine various spectral properties of organic molecules has supplemented, and in some cases even supplanted, the chemical methods. Spectral measurements can yield, in a few minutes, more information about the structure of a compound than the classical chemical techniques could sometimes even after months of work. An added advantage is that minute amounts of the same sample can be used to perform several spectral analyses, since most spectroscopic methods are nondestructive. The high cost of instrumentation, however, has hindered the wider utilization of spectroscopy in laboratories with limited financial resources. In this chapter we introduce the three most commonly used **spectral methods: ultraviolet (uv), infrared (ir),** and **nuclear magnetic resonance (nmr)** spectroscopy.

Principles of Absorption Spectroscopy 17.1

Spectroscopy is the study of the interactions of molecules with light (electromagnetic radiation) of specific energy. No matter what kind of electromagnetic radiation falls on a molecule, certain general principles always apply.

A Quantization of Molecular Energy Levels

Normally, molecules exist in the most stable (lowest-energy-level) state, called the *ground state*. When a molecule absorbs radiation, its energy is raised from the ground state to a less stable (higher-energy-level) state, or *excited state*. However, the energy of a molecule cannot assume continuous spectrum values; it is restricted to certain discrete values of energy levels. Since only certain energy values are permitted, we say that the energy is *quantized*.

B The Nature of Radiant Energy

The properties of radiant energy reveal a duality of nature. In some respect, its properties are those of a wave; in others, it can be thought of as consisting of small bundles of energy, called *photons*. The energy, E, of a photon is related to the frequency and the wavelength of the radiation by the equation

$$E_{\text{photon}} = h\nu = hc/\lambda$$

where h = Planck's constant
ν = frequency
c = velocity of light
λ = wavelength.

The equation shows that the energy of a photon is directly related to the frequency of the beam of radiation; thus, the greater the frequency, the greater the energy. The equation also shows that frequency and wavelength are inversely related; thus, the higher the frequency, the shorter the wavelength; and the greater the energy of the photon, the shorter the wavelength of the radiation.

Figure 17.1 shows a schematic diagram of parts of the electromagnetic spectrum of interest to organic chemists.

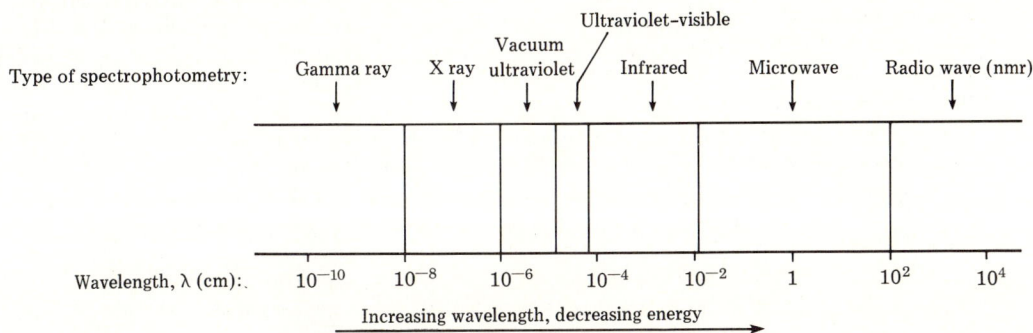

Figure 17.1 The electromagnetic spectrum.

C   Interaction of Photons with Molecules: Absorption Spectrum

When radiation falls on a molecule, the photons may or may not be absorbed, depending on the structure of the molecule and the energy of the photons (or the wavelength of the light). To be absorbed, the energy of a photon must correspond exactly to the energy difference between two energy levels in the molecule. Otherwise, no light is absorbed. Since only certain wavelengths of radiation are absorbed by a molecule, it is possible to obtain an **absorption spectrum** by measuring the amount of light absorbed as a function of the wavelength of the radiation. Absorption spectra can be recorded as graphs by means of instruments, called spectrometers.

Let us now look into what kind of structural information can be obtained from each of the three commonly used types of spectroscopy: ultraviolet–visible, infrared, and nuclear magnetic resonance.

17.2 Ultraviolet–Visible Spectrometry

A   Qualitative Analysis

The photons of the uv–visible portion of the electromagnetic spectrum possess much more energy than radiation in the ir or nmr regions. The energy of the radiation in the uv–visible region is high enough to cause the valence electrons of *certain molecules* to be raised from the ground state to the excited state. This kind of phenomenon is called **electronic excitation,** and ultraviolet and visible spectra are associated with electronic excitation.

Molecules, depending on their structure, may have the following kinds of valence electrons.

σ electrons as in C—C
π electrons as in C=C, C=O
n (nonbonding) electrons as in C—Ö, C=Ö

In the 200–750 nm (1 nanometer $= 10^{-9}$ meter) region, which is the spectral range most accessible to ordinary uv–visible spectrometers, absorption does *not* take place with molecules containing only σ electrons or *isolated* C=C bonds. Thus, alkanes and alkenes, such as ethylene, do not absorb (are transparent) above 200 nm because too much energy is needed to excite such electrons. Liquids such as hexane, which contain only σ electrons, or ethanol and water, which contain only σ electrons and nonbonding electrons, are essentially transparent to uv–visible light. For this reason, such liquids are commonly used as solvents for measuring the uv spectra of other absorbing compounds.

Conjugated alkenes *do* absorb in the uv–visible region. This is because the π electrons in such systems are held less tightly than are those of isolated alkenes. Compare, for example, the two isomeric dienes.

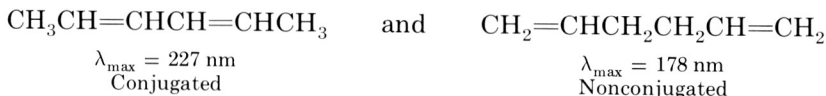

$CH_3CH=CHCH=CHCH_3$ and $CH_2=CHCH_2CH_2CH=CH_2$
$\lambda_{max} = 227$ nm $\lambda_{max} = 178$ nm
Conjugated   Nonconjugated

The wavelength of maximum absorption (λ_{max}) of the conjugated diene is at 227 nm, which is within the spectral range of ordinary uv spectrometry. The nonconjugated diene, with its two isolated C=C units, does not absorb in ordinary uv spectrometry.

As the number of conjugated double bonds increases, less and less energy is required to excite one of the π electrons, and absorption takes place at longer wavelengths. This shift to longer wavelengths is called *red shift*. In addition, the intensity of the absorption, denoted by ε values, generally increases (see Table 17.1). Systems with enough conjugated double bonds eventually absorb above 400 nm and appear colored to the eye. As an example, β-carotene a compound with eleven conjugated C=C bonds has a strong band at 450 nm (λ_{max} 450 nm, ε 140,000) and appears yellow to the eye. Table 17.1 summarizes the absorption characteristics of some conjugated alkenes.

Aromatic compounds also strongly absorb in the uv–visible region. The absorption spectra of aromatic compounds are more complex than are those of aliphatic compounds but follow the expected trends. Thus, in going from benzene to naphthalene to compounds with larger number of fused rings, we find that less energy is required to excite the π electrons. Therefore, λ_{max} increases, as does the intensity of the absorption band (ε). For example,

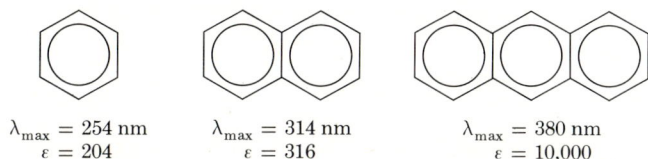

$\lambda_{max} = 254$ nm $\lambda_{max} = 314$ nm $\lambda_{max} = 380$ nm
$\varepsilon = 204$ $\varepsilon = 316$ $\varepsilon = 10,000$

Ultraviolet spectrometry is useful also in the identification of aldehydes and ketones. We find that even nonconjugated aldehydes and ketones have a very weak ($\varepsilon = 10\text{--}30$) absorption band in the 200–300 nm region. This weak band arises from excitation of one of the nonbonding electrons of the C=Ö group to the next higher energy level. For example,

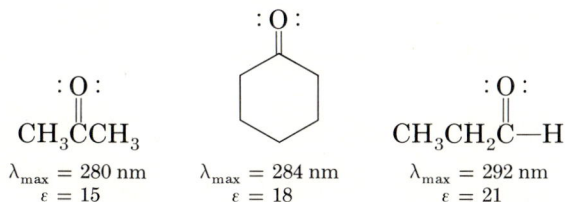

$\lambda_{max} = 280$ nm $\lambda_{max} = 284$ nm $\lambda_{max} = 292$ nm
$\varepsilon = 15$ $\varepsilon = 18$ $\varepsilon = 21$

Compounds having a carbonyl group in conjugation with a carbon–carbon double bond show two absorption bands, one due to C=C and one due to C=O.

Table 17.1 Absorption Data for Some Conjugated Alkenes

Compound	λ_{max} (nm)	ε
CH_2=CHCH=CH_2	217	21,000
CH_2=CHCH=CHCH=CH_2	258	35,000
(CH_2=CHCH=CH_2)$_2$	287	52,000

As expected, both bands are shifted to longer wavelengths by conjugation. Compare, for example,

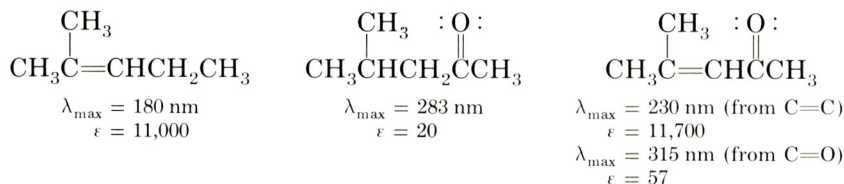

$$CH_3C{=}CHCH_2CH_3$$

with a CH_3 substituent

$\lambda_{max} = 180$ nm
$\varepsilon = 11,000$

$$CH_3\overset{CH_3}{\underset{|}{C}}HCH_2\overset{:O:}{\overset{\|}{C}}CH_3$$

$\lambda_{max} = 283$ nm
$\varepsilon = 20$

$$CH_3\overset{CH_3}{\underset{|}{C}}{=}CH\overset{:O:}{\overset{\|}{C}}CH_3$$

$\lambda_{max} = 230$ nm (from C=C)
$\varepsilon = 11,700$
$\lambda_{max} = 315$ nm (from C=O)
$\varepsilon = 57$

In summary, uv–visible spectrometry provides a powerful tool for distinguishing between certain types of compounds, especially between conjugated and nonconjugated systems. Thus, aromatic compounds and conjugated systems absorb strongly in the 200–750 nm region. Alkanes and isolated alkenes are transparent in this spectral range. An absorption band of weak intensity in the 200–300 nm region is typical for nonconjugated aldehydes and ketones.

Problem 17.1 Which of these pairs of isomeric compounds could easily be distinguished by uv–visible spectrometry? Explain your answers briefly.

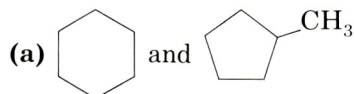

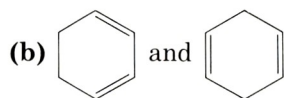

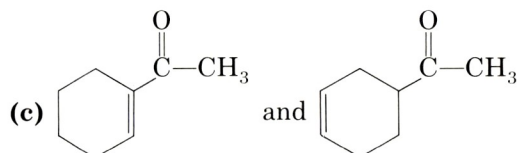

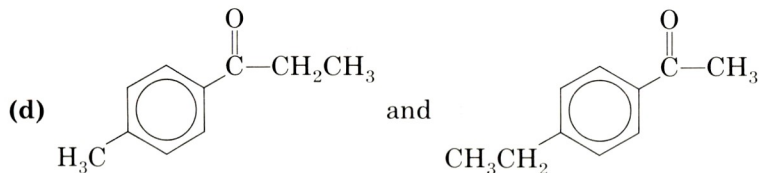

(a) and

(b) and

(c) and

(d) and

B   Quantitative Analysis

The greatest use of uv–visible spectroscopy is in quantitative analysis, especially in biochemical and clinical work. The concentration of dilute solutions or the amounts of impurities in a sample can be determined with great accuracy by uv–visible spectra. This is the basis of the routine use of automated clinical analyzers in hospitals and medical laboratories. In practice, the procedure is as follows.

1. Place a dilute solution of the compound in a nonabsorbing solvent, such as water, ethanol, or cyclohexane, in a cell of accurately known length. Quartz cells, which are transparent to uv radiation, must be used for samples that absorb in the uv range (200–400 nm).
2. Record the spectrum of the sample. Figure 17.2 shows the uv spectrum of a dilute solution of 1,3-cyclohexadiene.

The wavelength of absorption is usually reported as λ_{max}, the wavelength of the maximum absorption. The λ_{max} for 1,3-cyclohexadiene is 256 nm. The vertical axis is a record of the **absorbance, A,** of the sample. Absorbance is defined by the equation

$$A = \log I_0/I \tag{1}$$

where I_0 = intensity of the light beam that strikes the sample
I = intensity of the light beam after it has passed through the sample

Obviously, when no light is absorbed by the sample, I and I_0 are identical and $\log I_0/I = 0 = A$.

For samples that *do* absorb light, the value of A, in dilute solution, is directly proportional to the number of molecules in the path of the light. Thus, the absorbance is directly proportional to the concentration of the sample and the

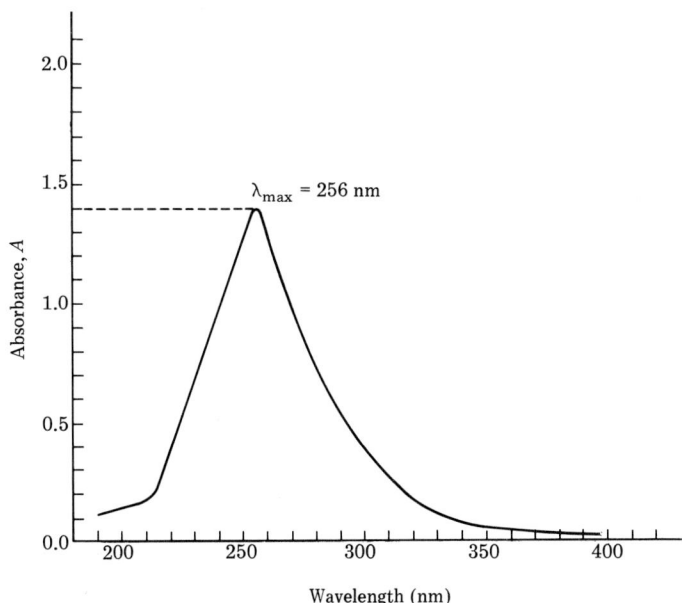

Figure 17.2 Ultraviolet spectrum of 1,3-cyclohexadiene, 1.75×10^{-4} M in n-hexane, 1.0 cm cell.

sample path length (the internal cell length). The exact relationship is

$$A = \varepsilon \cdot c \cdot l \qquad (2)$$

where A = absorbance

ε = molar absorptivity, a constant for a particular compound at a given wavelength*

c = concentration of sample, in moles per liter (M)

l = cell length, in centimeters

Equation (2), known as the **Beer–Lambert law,** sometimes simply called *Beer's law,* can be rearranged to determine the concentration of a sample. The mathematical statement is

$$c = \frac{A}{\varepsilon \cdot l} \qquad (3)$$

Thus, if the value of ε for a pure sample is known, it is a simple matter to take a uv–visible spectrum and use the experimental values of A and l to determine c according to equation (3).

Let us work out an example.

Example 17.1 A bottle of cyclohexane is known to be contaminated with benzene. At 254 nm, benzene has a molar absorptivity of 204 and cyclohexane has a molar absorptivity of zero. A uv spectrum of the contaminated cyclohexane taken in a 2.0 cm cell shows an absorbance of 0.070. What is the molar concentration of benzene?

Solution Using equation (3);

$$c = \frac{A}{\varepsilon \cdot l}$$

and substituting;

$$c = \frac{0.070}{204 \times 2.0} = 0.00017 \, M$$

Now solve the following problems.

Problem 17.2 A solution of camphor in cyclohexane has an absorbance of 1.48 at 295 nm in a 5.0 cm cell. The molar absorptivity of camphor at this wavelength is 14. What is the concentration of camphor?

Problem 17.3 Cyclopentanone (0.043 M in a 1.0 cm cell) has λ_{max} at 288 nm with an absorbance of 0.85. What is the molar absorptivity of cyclopentanone at 288 nm?

---

* Although the units of ε are $M^{-1} \, cm^{-1}$, it is usually shown as a unitless quantity. Values of ε vary over a wide range from 2 or 3 to over 200,000.

Infrared (ir) radiation has less energy than do rays in the uv–visible region. Absorption of energy in the ir region is associated with changes in the **frequency of vibrations** of atoms or groups of atoms in a molecule. The vibrational excitation may be due to **stretching** or **bending** of a bond. Figure 17.3 shows the various types of molecular vibrations due to stretching and bending of bonds.

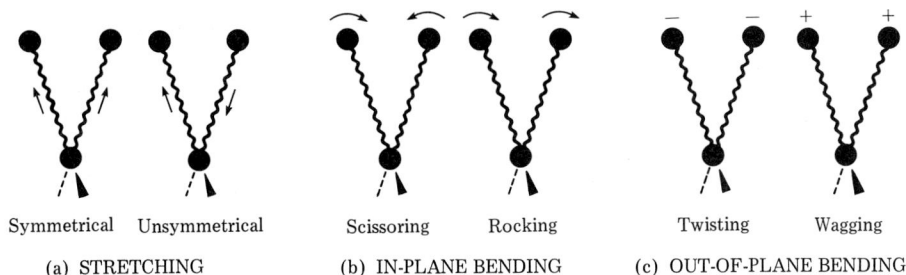

Symmetrical	Unsymmetrical	Scissoring	Rocking	Twisting	Wagging
(a) STRETCHING		(b) IN-PLANE BENDING		(c) OUT-OF-PLANE BENDING	

Figure 17.3 Kinds of vibrational excitations. Atoms are represented by balls and bonds by springs. **(a)** Stretching vibrations. **(b)** In-plane bending. **(c)** Out-of-plane bending (+ denotes above the plane, − below the plane).

Since even the simplest organic molecule contains several covalent bonds, absorption of ir radiation produces many vibrational patterns. For this reason, ir absorption spectra are usually complex. However, because the position of an ir absorption band depends on the strengths of the bonds and on the relative masses of the atoms involved in the vibration, we find that certain functional groups absorb at unique frequencies. Thus, the stretching frequencies of C—H, O—H, C=C, C≡C, C—C, C=O, and many other bonds are observed in different regions of the ir spectrum. Often, the absence of these characteristic absorption bands in a particular region of a spectrum is as important in structure determination as their presence.

Infrared spectra are usually recorded as plots of percent transmittance of light versus **wavenumbers,** which are frequencies in reciprocals of wavelengths in centimeters (cm^{-1}). Occasionally, especially in the older literature, we find ir spectra recorded as plots of percent transmittance versus wavelengths in microns (μ; 1 μ = 10^{-4} cm).

The region accessible to most ir spectrometers ranges from 4000 to 660 cm^{-1} (2.5–16 μ). Table 17.2 lists the characteristic absorption frequencies from some common types of bonds. You should consult this table whenever you are asked to solve problems dealing with ir spectra. The ir spectra of representative organic compounds are illustrated in Figure 17.4. In these spectra, the positions of characteristic ir absorption bands have been labeled to help you identify the functional group present in each compound.

To summarize, ir spectra are extremely valuable for detecting the presence or absence of functional groups.

Table 17.2 Characteristic Infrared Absorption Frequencies for Some Common Groups

Frequency range (cm^{-1})	Group	Class of compounds
	Stretching vibrations	
3700–3200	O—H[a]	alcohols, phenols
3500–3100	N—H[b]	1° and 2° amines, amides
3320–3000	≡C—H	terminal alkynes
3100–3000	=C—H	alkenes
	C—H	aromatics
3000–2800	—C—H	alkanes
3000–2500	O—H----O	carboxylic acids (H bonded)
2260–2240	C≡N	nitriles[c]
2260–2100	C≡C	alkynes
1820–1600	C=O[c]	aldehydes, ketones, carboxylic acids, and derivatives
1680–1500	C=C	alkenes, aromatics
1560–1490; 1360–1320	—NO$_2$	nitro compounds
1200–1000	C—O	alcohols, ethers, carboxylic acids, esters
	Bending vibrations	
1470–1430	—CH$_2$—; C—H	
1470–1375	CH$_3$	
960–900	C=C—H	
1600–1500	aromatic ring (often weak)	
900–700	Ar—H	

[a] In dilute solutions appears as a sharp peak of variable intensity between 3700 and 3500 cm^{-1}. For the pure compound or in concentrated solutions a strong broad band appears in the 3400–3200 cm^{-1} region.
[b] Concentration effects are similar to those for O—H: two bands for 1° amines, one for 2° amines.
[c] Nonconjugated. Conjugation of a multiple bond lowers the stretching frequency by about 30 cm^{-1} for every C=C.

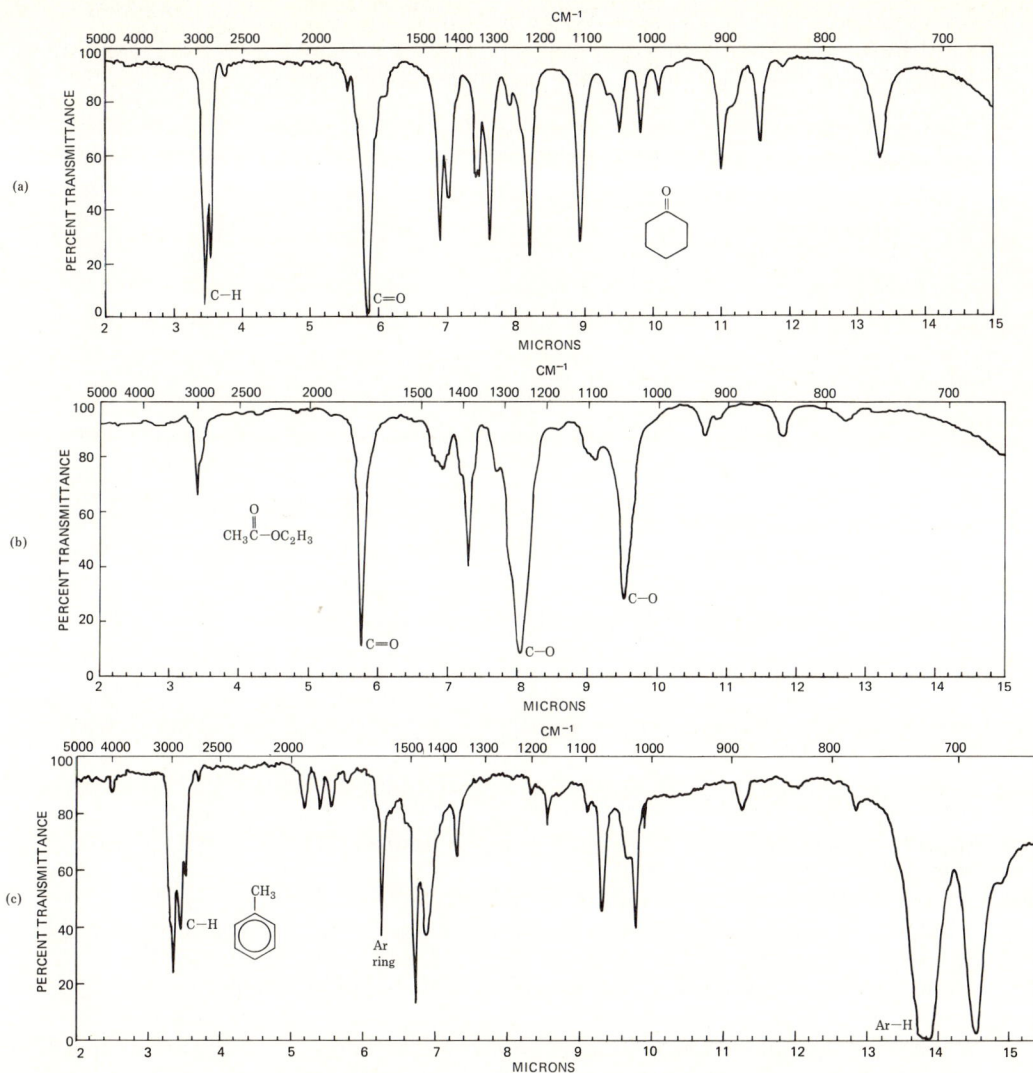

Figure 17.4   Infrared spectra of **(a)** cyclohexanone, **(b)** ethyl acetate, and **(c)** toluene.

Problem 17.4   Consult Table 17.2 and determine which of these pairs of compounds could be distinguished easily by ir spectrometry. Cite characteristic absorption bands you would expect to be present or absent for each.

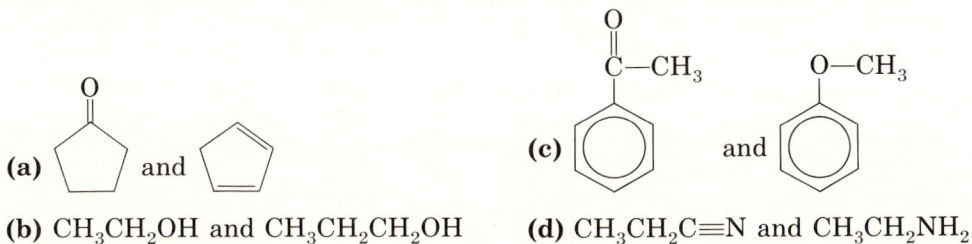

(a) and

(c) and

(b) CH_3CH_2OH and $CH_3CH_2CH_2OH$

(d) $CH_3CH_2C{\equiv}N$ and $CH_3CH_2NH_2$

Problem 17.5 Identify the unknown, C_2H_6O, from the ir spectrum shown here. (Assign as many of the absorption bands as you can.)

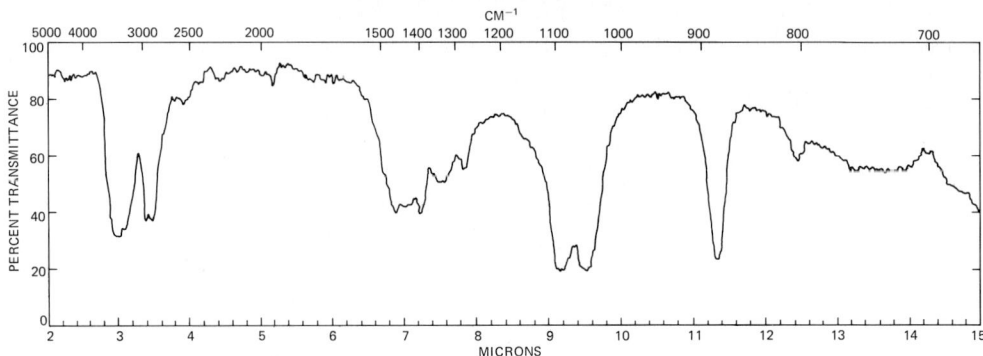

17.4 Nuclear Magnetic Resonance Spectra

Certain nuclei possess a *magnetic dipole*. Such nuclei, when held in a magnetic field, tend to align themselves parallel to the applied field. Their behavior is similar to that of the magnetic needle of a compass that aligns itself parallel to the earth's magnetic field. Among nuclei that have a magnetic dipole are the isotopes of certain elements such as ^1H, ^{13}C, ^{15}N, ^{19}F, and so on. To the organic chemist, the most important nucleus is the **proton, ^1H,** because it is present in almost all organic molecules.

The basis of nmr technique is as follows. An organic compound is placed in a strong, homogeneous magnetic field, which causes most of the protons to align themselves parallel to the applied field. The sample is then irradiated with a radio frequency signal. At a particular combination of magnetic field strength and radio frequency, energy is absorbed. Absorption of energy causes the protons to be flipped from the parallel alignment (a low-energy state) to an antiparallel alignment (an excited state). This absorption of energy is recorded as a band, called an **nmr signal.**

The great value of nmr spectrometry lies in the fact that *not* all protons absorb at the same magnetic field–radio frequency conditions. Experimentally, we find that protons in different structural environments absorb the energy of a particular radio frequency at slightly different values of applied magnetic field strengths.* Some protons require a stronger applied field to absorb the constant radio frequency signal; others require less. Protons that require a stronger applied magnetic field to produce an nmr signal are said to be **shielded;** those that require a smaller value are said to be **deshielded.** The shielding or deshielding effect depends on the electronic environment of the proton. As a general rule, a high electron density around a proton has a shielding effect.

---

* The reverse procedure of applying a constant magnetic field and a variable radio frequency gives the same result theoretically but is more difficult to attain instrumentally.

An nmr spectrum thus consists of a series of signals, each corresponding to a particular type of proton in a molecule. Let us now examine what types of information are obtainable from an nmr spectrum.

A   How Many Nmr Signals to Expect in the Nmr Spectrum

A set of chemically equivalent protons give one nmr signal. For example, the following compounds all give one nmr signal since the protons in each are in chemically equivalent environments.

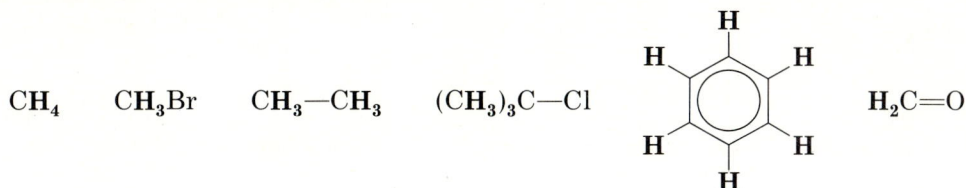

$$CH_4 \qquad CH_3Br \qquad CH_3{-}CH_3 \qquad (CH_3)_3C{-}Cl \qquad \qquad H_2C{=}O$$

A molecule with two sets of protons in different chemical environments will give an nmr spectrum with two signals, a molecule with three different kinds of protons will give three nmr signals, and so on. For example, toluene has two sets of protons, labeled (a) and (b), that are chemically distinct, and the nmr spectrum of toluene shows two separate signals (see Fig. 17.5). Examples of other molecules with distinct sets of protons are shown in Figure 17.6.

Problem 17.6   How many signals would the nmr spectrum of each of these compounds show? Label each type of proton with a letter.
(a) Cyclopentane   **(b)** acetone   **(c)** *p*-Xylene
(d) Ethanol   **(e)** Ethyl methyl ether   **(f)** Bromocyclopropane

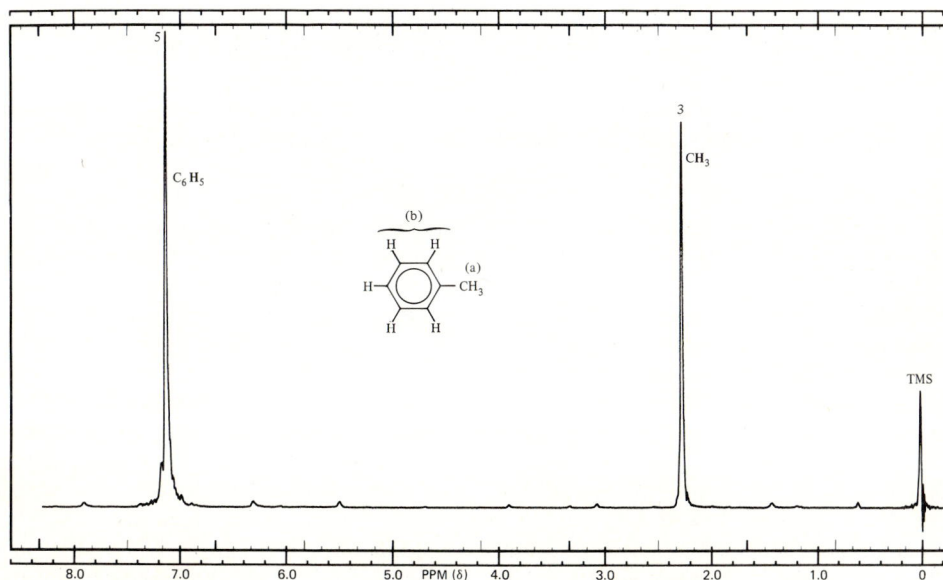

Figure 17.5   Nuclear magnetic resonance spectrum of toluene.

(a) (b)
CH_3-CH_2-Cl

(a) O (b)
$CH_3-\overset{\parallel}{C}-O-CH_3$

(a) (b) (a)
$CH_3-CH-CH_3$
|
Br

(a)

H H

H—⟨ring⟩—$CH_2-\overset{O}{\overset{\parallel}{C}}-CH_3$ (b) (c)

H H

(a) (b) (c)
$CH_3-CH_2-CH_2-Br$

(a) O (d)
H C—O—CH_3
\ /
C=C
/ \
(b) H H (c)

Molecules with two nmr signals   Molecules with three nmr signals   Molecule with four nmr signals

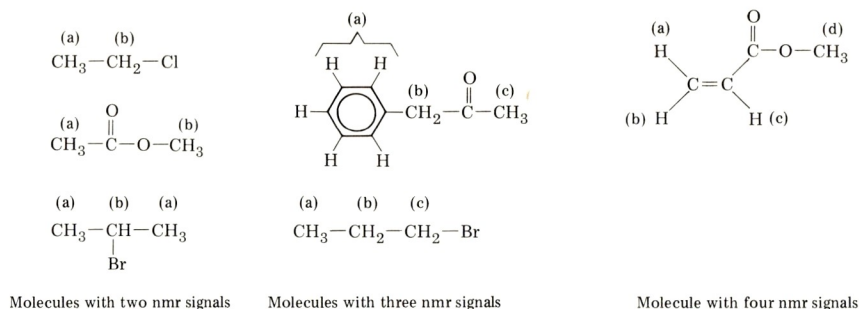

Figure 17.6 Some representative molecules with two, three, and four nmr signals.

B The Intensity of the Nmr Signal: Number of Equivalent Protons

The intensity of an nmr signal is directly proportional to the number of protons causing that signal. An nmr spectrum therefore tells us not only how many different kinds of protons are present in a molecule but also *how many* protons of each type there are. The *intensity* of a signal is obtained by measuring the *relative areas* under each peak, a process called integration. Integration of the signal is done automatically by the nmr spectrometer and is shown either as a step graph or as a printout. The nmr spectrum of toluene (Fig. 17.5), for example, shows two signals, one due to the five equivalent aromatic protons and one due to the methyl protons. Integration of the two signals reveals them to be in the ratio of 5 to 3, the same as the ratio of the number of each type of protons.

C The Position of the Nmr Signal: The Chemical Shift

From what we have learned so far, an nmr spectrum gives us two important pieces of information about a compound: the different kinds of protons present and the relative numbers of each type of proton. The *position* of an nmr signal is a third source of valuable information regarding the structure of a compound. The position of a signal, called the **chemical shift (δ),** gives us an indication of the *type of proton* present. For example, the chemical shift due to protons on an aromatic ring is different from that of protons on an alkyl group or of the proton on an aldehyde group. Table 17.3 shows the chemical shifts for representative protons. The chemical shifts in Table 17.3 are expressed in parts per million (ppm) relative to a reference standard. The reference compound is **tetramethylsilane (TMS),** $(CH_3)_4Si$, whose signal is taken as 0.0 ppm.

When we want the nmr spectrum of a compound, we add a drop or two of TMS to a solution of the sample and measure the position of the signals relative to that of TMS. Since the protons of organic compounds, with few exceptions, are deshielded relative to those of TMS, the signals will appear **downfield** (to the left) of TMS. The magnitude of δ is indicative of the amount of shielding and deshielding. A high δ value means that the protons are effectively deshielded. Conversely, a chemical shift close to 0.0 ppm signifies that the proton(s) are well shielded **(upfield)** and need a greater applied magnetic field to produce the nmr signal. Figure 17.7 shows the nmr spectrum of *t*-butyl acetate. Note that the chemical shifts of the methyl protons in *t*-butyl acetate are different from the chemical shift of the CH_3 protons in toluene (Fig. 17.5).

Table 17.3 Typical Chemical Shifts of Protons

Type of proton	δ (ppm)	Type of proton	δ (ppm)
CH$_3$ group protons		CH$_2$ group protons	
CH$_3$—R	0.8–1.2	RCH$_2$R	1.1–1.5
CH$_3$—CR=C	1.6–1.9	RCH$_2$Ar	2.5–2.9
CH$_3$—Ar	2.2–2.5	$\overset{\text{O}}{\overset{\|}{\text{RCH}_2\text{CR}}}$	2.5–2.9
$\overset{\text{O}}{\overset{\|}{\text{CH}_3-\text{C}-\text{R}}}$	2.1–2.4	RCH$_2$OH	3.2–3.5
		RCH$_2$OAr	3.9–4.3
$\overset{\text{O}}{\overset{\|}{\text{CH}_3-\text{C}-\text{Ar}}}$	2.4–2.6	$\overset{\text{O}}{\overset{\|}{\text{RCH}_2\text{OCR}}}$	3.7–4.1
$\overset{\text{O}}{\overset{\|}{\text{CH}_3-\text{C}-\text{OR}}}$	1.9–2.2	RCH$_2$Cl	3.5–3.7
CH$_3$—N$\diagdown$	2.2–2.6	R$_2$CH— group protons	
		R$_3$CH	1.4–1.6
CH$_3$—OR	3.2–3.5	R$_2$CHOH	3.5–3.8
$\overset{\text{O}}{\overset{\|}{\text{CH}_3-\text{OCR}}}$	3.6–3.9	Ar$_2$CHOH	5.7–5.8
Proton on sp^2 carbon		Proton on N or O	
R$_2$C=CHR	5.0–5.7	ROH	3–6
		ArOH	6–8
Ar—H	6.0–7.5	RCOOH	10–12
RCHO	9.4–10.4	R$_2$NH	2–4

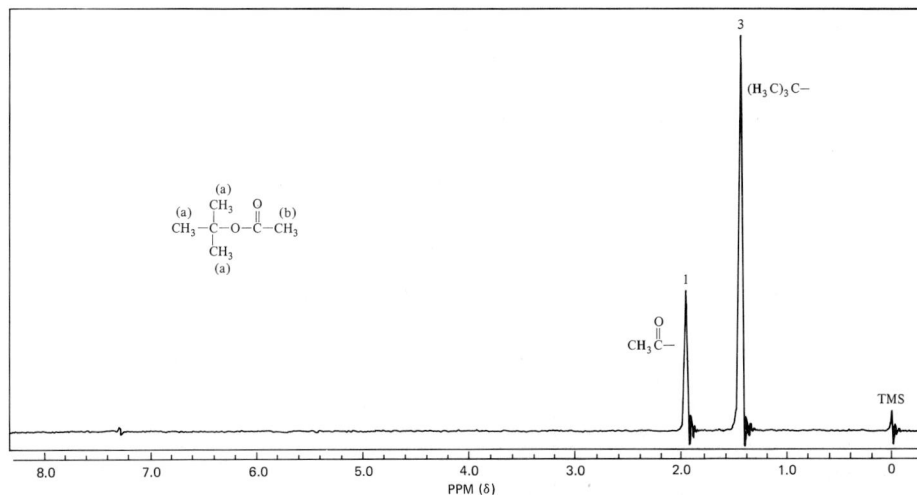

Figure 17.7 Nmr spectrum of *t*-butyl acetate.

D   Spin-Spin Coupling: The ($n + 1$) Rule

The nmr spectra we have considered up to now contained a series of single peak signals, each representing protons in different chemical environments—the intensity of each peak is proportional to the number of protons it represents. The nmr spectrum of ethyl chloride (Fig. 17.8) shows another important feature of nmr spectra.

As expected, the nmr spectrum of ethyl chloride consists of two signals: one due to the CH_3 protons and one due to the CH_2 protons with relative area ratio of 3:2. *But Figure 17.8 shows that the two signals are split.* The splitting of the signals is called **spin-spin coupling.** In general, spin-spin coupling occurs between *nonequivalent protons on adjacent carbons.* When nonequivalent protons are more than one carbon away from each other (as in the spectra of Figs. 17.5 and 17.7), they do not generally show splitting; each peak is a *singlet.* When splitting does take place, it is possible to predict the multiplicity of the splitting pattern by the **($n + 1$) rule.** This rule states that a signal is split into ($n + 1$) peaks, where n represents the number of adjacent equivalent protons. Thus, the CH_3 signal in ethyl chloride is split into three parts, *a triplet* ($n + 1$ adjacent CH_2 protons), and the CH_2 signal is split into four parts, *a quartet* ($n + 1$ adjacent CH_3 protons).

Problem 17.7   Describe the nmr spectrum of each compound, indicating (1) the number of signals; (2) the splitting pattern of each signal, (3) the relative area of each signal, and (4) the relative position of each signal.
(a) Cl_2CHCH_2Cl **(b)** $CH_3CH_2OCH_2CH_3$

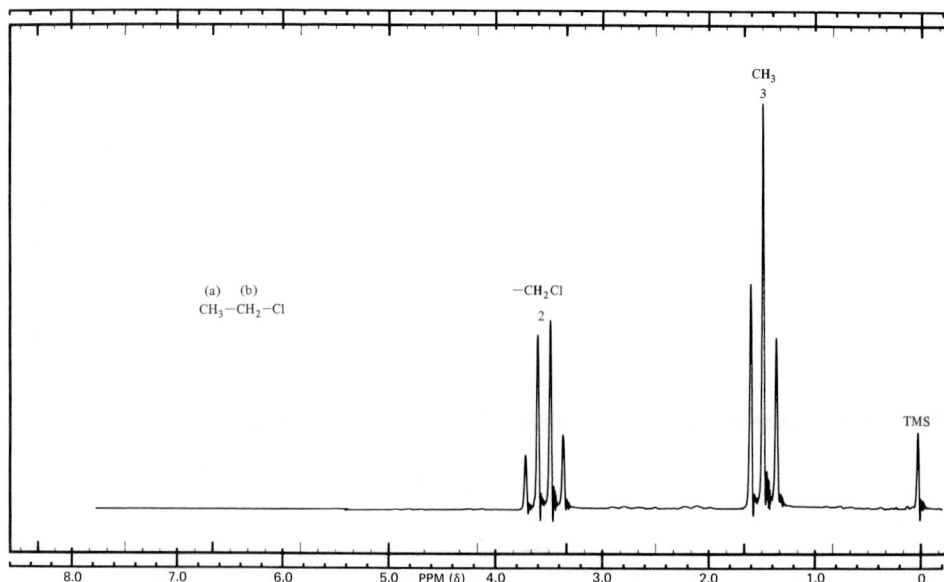

Figure 17.8   Nmr spectrum of ethyl chloride.

Problem 17.8 Deduce the structural formula of the compound $C_2H_3Br_3$ from its nmr spectrum.

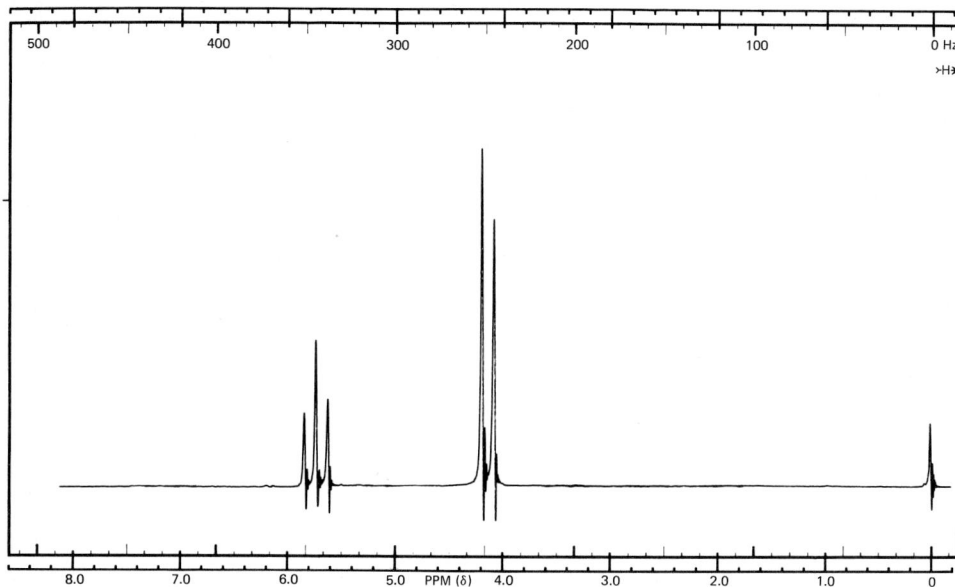

[From *Introduction to Organic Chemistry*, 2nd ed., by Andrew Streitwieser, Jr., and Clayton H. Heathcock. Copyright © 1981 by Macmillan Publishing Co., Inc.]

In summary, the nmr spectrum gives us four types of information.

1. The number of signals indicates the different types of protons.
2. The relative areas of the signals tell us the relative numbers of protons represented by each signal.
3. The positions of the signals (chemical shifts) indicate what type of proton is present.
4. The splitting of the signal (spin-spin coupling) into two or more peaks indicates the interaction of protons of one type with protons of another type on adjacent carbons. The number of these peaks is given by the $(n + 1)$ rule.

Identifying a Compound from Its Spectra: **17.5** An Illustrative Example

Now that you are familiar with the three spectroscopic methods discussed in the chapter, let us apply the principles you learned by determining the structure of a compound from its spectra.

Let us work out the structure of a compound **(X)**, $C_6H_{12}O_2$, whose uv spectrum shows no absorption above 200 nm, using the ir and nmr spectra in Figure 17.9.

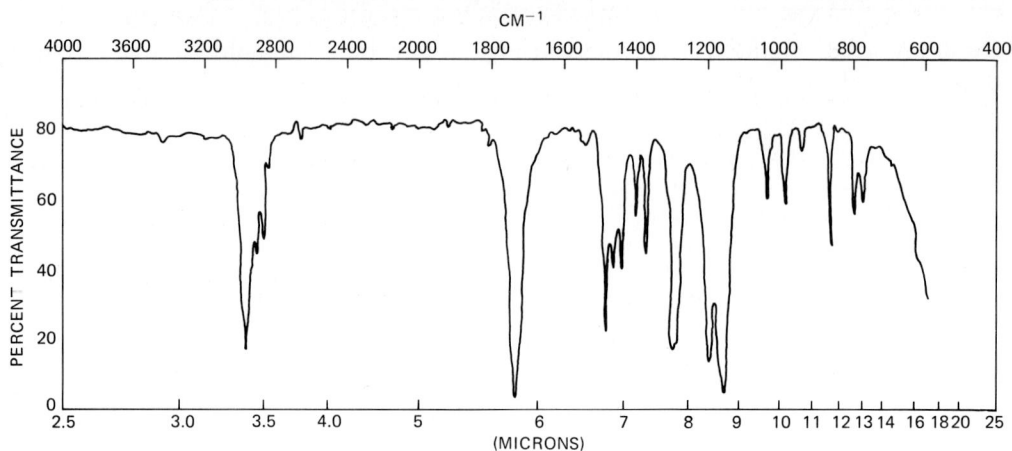

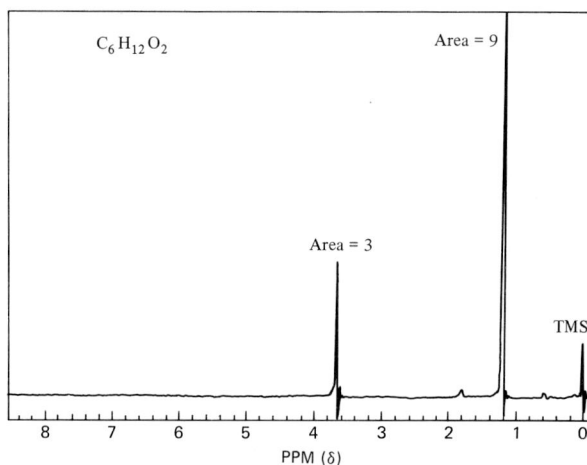

Figure 17.9 Infrared and nmr spectra of unknown **(X)**, $C_6H_{12}O_2$.

Solution. The best way to approach problems of this kind is to remember first what type of information one can obtain from each spectroscopic method. Namely,

1. The uv spectrum tells us something about the unsaturation of the carbon skeleton.
2. The ir spectrum provides an idea of the functional groups that are present (or absent).
3. The nmr spectrum indicates the different kinds of hydrogens present and how many hydrogens of each type there are.

Coming back to our example, we learn, from the lack of absorption in the uv spectrum, that compound **X** contains neither a conjugated system nor an aromatic ring. Furthermore, the lack of absorption in the uv spectrum above 200 nm also eliminates a ketone or an aldehyde as a possibility.

A look at the ir spectrum shows several strong bands. Consulting Table 17.2, we deduce the following.

1. The band around 3000–2800 cm^{-1} is due to C—H stretching on a saturated carbon. (The lack of absorption to the left of 3000 cm^{-1} eliminates a C—H stretch on a vinylic, aromatic, or acetylenic hydrogen.)
2. The strongest band in the spectrum, around 1730–1740 cm^{-1}, is due to C=O. Since the uv spectrum eliminates an aldehyde or a ketone as a possibility, we must consider **X** to be either a carboxylic acid or an ester. We can discount a carboxylic acid as a possibility, since the ir spectrum shows no broad absorption band in the 3000–2500 cm^{-1} region due to —OH. The absence of a COOH group is further confirmed by the nmr spectrum. The latter reveals no signal in the region ascribed to the COOH proton (near 10–12 ppm).
3. The bands in the 1200–1100 cm^{-1} region of the ir spectrum are due to C—O, confirming that compound **X** is most likely an ester.

Let us now turn to the nmr spectrum. Since there are only two signals, in the ratio 3:9, there must be two different types of protons: one set of 3 equivalent protons and a second set of 9 equivalent protons. Since each signal is a singlet, this means that each set of equivalent protons is on *nonadjacent* carbons.

Starting from the left, the position of the signal at δ 3.7 ppm suggests a CH$_3$ group next to an electronegative oxygen, CH$_3$—O—. The second signal, δ 1.2 ppm, suggests a (CH$_3$)$_3$C— group.

Putting all the pieces together, we conclude that **X** is an ester containing a CH$_3$ and a (CH$_3$)$_3$C— group. The structure of **X** we deduce is

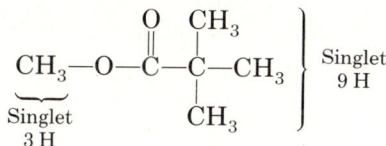

$$\underbrace{CH_3}_{\substack{\text{Singlet} \\ \text{3 H}}} \!\!\! -O-\overset{\displaystyle O}{\overset{\displaystyle \|}{C}}-\overset{\displaystyle CH_3}{\underset{\displaystyle CH_3}{\overset{\displaystyle |}{\underset{\displaystyle |}{C}}}}-CH_3 \left.\right\} \begin{array}{c} \text{Singlet} \\ \text{9 H} \end{array}$$

The alternate possibility

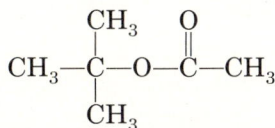

$$CH_3-\overset{\displaystyle CH_3}{\underset{\displaystyle CH_3}{\overset{\displaystyle |}{\underset{\displaystyle |}{C}}}}-O-\overset{\displaystyle O}{\overset{\displaystyle \|}{C}}-CH_3$$

would give similar uv and ir spectra, but the position of the nmr signals would be quite different (see Fig. 17.7).

Summary of Concepts and Reactions

The three most commonly used spectral methods are ultraviolet (uv), infrared (ir), and nuclear magnetic resonance (nmr).   [Sec. 17.1]

Spectroscopy is the study of the interactions of molecules with light (electromagnetic radiation) of specific energy.   [Sec. 17.1]

Molecules absorb only certain specific wavelengths of radiation, and it is therefore possible to obtain an absorption spectrum.   [Sec. 17.1C]

Ultraviolet-visible spectrometry enables us to distinguish between conjugated and nonconjugated systems. Thus aromatic compounds and conjugated compounds absorb strongly in the 200–750 nm range, which is the spectral range most accessible to ordinary uv–visible spectrometers. [Sec. 17.2A]

The greatest use of uv–visible spectroscopy is in quantitative analysis, especially in biochemical and clinical work. The concentration of dilute solutions or the amounts of impurities in a sample can be determined with great accuracy by uv–visible spectra. [Sec. 17.2B]

Infrared (ir) radiation has less energy than do rays in the uv–visible region. Absorption of energy in the ir region is associated with changes in the frequency of vibrations of atoms or groups of atoms in a molecule. The vibrational excitation may be due to stretching or bending of a bond. [Sec. 17.3]

Infrared spectra are usually recorded as plots of percentage transmittance of light versus wavenumbers, which are frequencies in reciprocal of wavelengths in centimeters (cm^{-1}). [Sec. 17.3]

The stretching frequencies of C—H, O—H, C=C, C≡C, C—C, C=O, and many other bonds are observed in different regions of the ir spectrum (4000 to 660 cm^{-1}). The presence or absence of these characteristic absorption bands in a particular region of a spectrum is important in the structure determination of a compound. [Sec. 17.3]

An nmr spectrum consists of a series of signals, each corresponding to a particular type of proton in a molecule. The scale following shows the approximate position of the nmr signal (δ, ppm) of a particular proton in a compound. [Sec. 17.4]

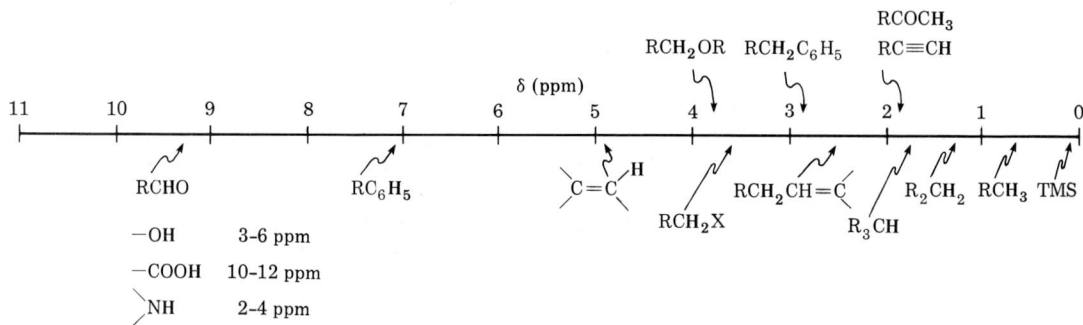

—OH	3-6 ppm			
—COOH	10-12 ppm			
NH	2-4 ppm			

Key Terms

ultraviolet (uv)
  spectral method
infrared (ir) spectral
  method
nuclear magnetic
  resonance (nmr)
  spectroscopy
absorption spectrum
electronic excitation

absorbance, A
Beer–Lambert law
infrared (ir) radiation
frequency of vibrations
stretching of a bond
bending of a bond
wavenumbers
proton, ^1H
nmr signal

shielded
deshielded
chemical shift, δ
tetramethylsilane (TMS)
downfield
upfield
spin-spin coupling
$(n + 1)$ rule

Exercises

Definitions of Terms [Secs. 17.1–17.4]

17.1 Identify, illustrate, or define **(a)** nanometer; **(b)** ν; **(c)** λ_{max}; **(d)** conjugated system; **(e)** chemical shift; **(f)** δ scale; **(g)** TMS; **(h)** downfield shift; **(i)** spin-spin coupling; **(j)** $(n + 1)$ rule

17.2 A chemist prepared the two isomeric ketones **A** and **B,** placed them in separate flasks, but forgot to label them. How could you differentiate the two by uv spectroscopy?

$$CH_3CH_2CH=CHCCH_3 \qquad\qquad CH_3CH=CHCH_2CCH_3$$
$$\text{A} \qquad\qquad\qquad\qquad\qquad \text{B}$$

17.3 A solution of phenol in water was placed in a 1.0 cm cell. The uv spectrum showed an absorbance of 0.73 at 270 nm. The molar absorptivity of phenol at this wavelength is 1450. What is the concentration of phenol?

Choosing the Correct Spectral Method of Analysis [Secs. 17.2–17.4]

17.4 Which method, uv–visible, ir, or nmr spectrometry, would be best suited for solving each problem?

(a) Distinguish between ⬡—C—H and ⬡—C—CH₃.

(b) Distinguish between ⬠ and ⬠.

(c) Distinguish between $CH_3CH_2CHCH_2CH_2CH_2CH=CH$—⬡ and

$$CH_3CH_2CCH_2CH_2CH_2CH=CH-\bigcirc.$$

(d) Determine whether a sample of an aldehyde contains a carboxylic acid as an impurity.

Nmr Spectroscopy [Sec. 17.4]

17.5 How many signals (ignoring splitting patterns) would you see in the nmr spectra of these compounds?
    (a) Butanone   **(b)** Cyclobutane   **(c)** *p*-Xylene   **(d)** 2-Propanol

17.6 The nmr signal of the boldfaced proton is split into two peaks. Which atom in this molecule causes the splitting of the signal?

$$\begin{matrix} \mathbf{H} & Cl \\ | & | \\ Br-C-C-H \\ | & | \\ Br & Cl \end{matrix}$$

17.7 Predict the signal pattern of the CH_3 protons in the nmr spectra of these compounds.

    (a) CH_3CHBr_2 **(b)** CH_3C-OH **(c)** CH_3CH_2Br **(d)** $(CH_3)_2CHOH$

Identifying a Compound from Its Spectra [Secs. 17.2–17.5]

17.8 The uv spectrum for a compound with formula C_3H_6O shows a weak absorption band at 280 nm. The nmr spectrum shows only one signal, a singlet. What is the structure of the compound?

17.9 Deduce the structures of the isomeric compounds with molecular formula $C_2H_4Br_2$ from the following nmr data.
  (1) The nmr spectrum of isomer A contains one signal, a singlet.
  (2) The nmr spectrum of isomer B contains two signals in the area ratio of $3:1$. One signal is a doublet ($\delta\ 2.47$) and the second signal is a quartet ($\delta\ 5.9$).

17.10 A compound with molecular formula C_2H_4O has the following spectral characteristics.
  (1) Uv–visible spectrum: transparent above 200 nm.
  (2) Ir spectrum: strong absorption in 1200–1000 cm^{-1} region; no absorption in 1820–1600 and 3700–3200 cm^{-1} regions.
  (3) Nmr spectrum: one signal, a singlet.
  What isomeric structures are ruled out by the spectra? What is a possible structure for the compound?

17.11 Refer to Tables 17.2 and 17.3 and deduce the structure of a compound, $C_9H_{10}O_2$, that is consistent with the ir and nmr spectra given here. The compound absorbs strongly in the uv region.

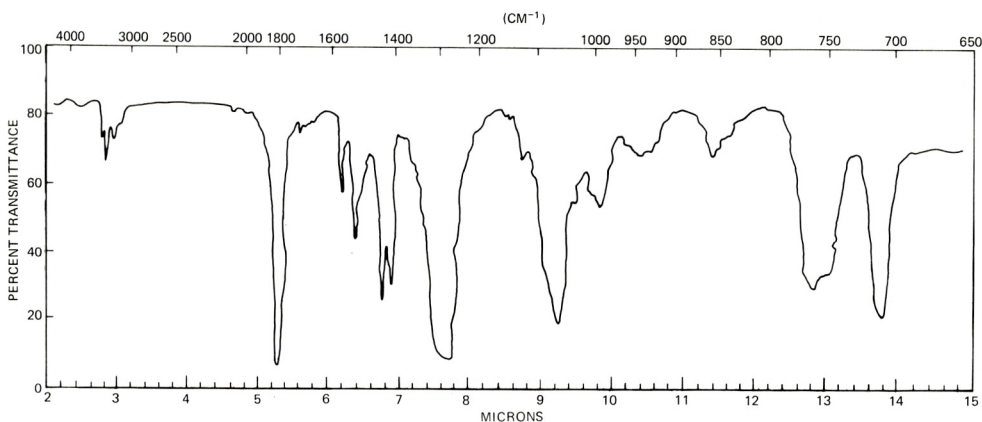

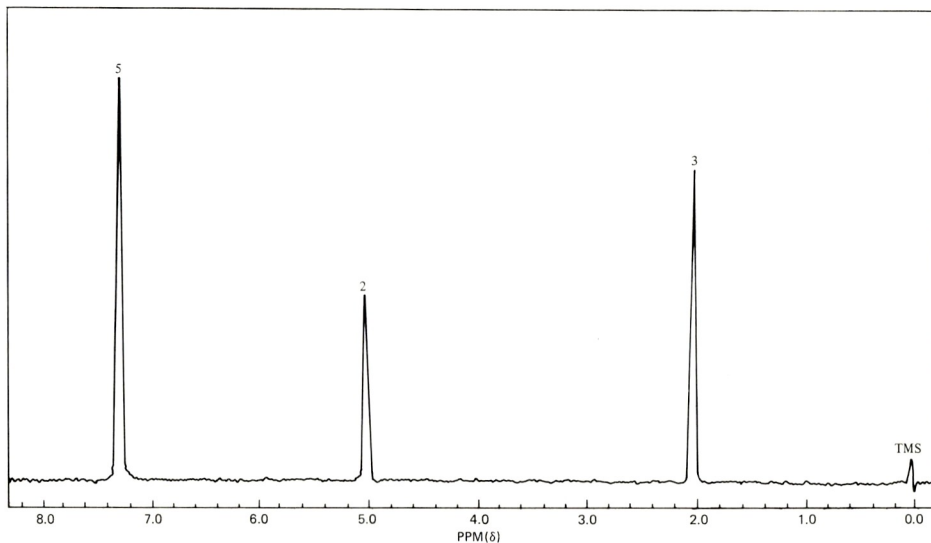

Exercises 1–9 deal with classes of compounds covered in Chapters 1–8.

1. The nmr spectrum of compound **A**, C_5H_{12}, reveals only one signal, a singlet. What is the structure of **A**?

2. Compound **B**, C_8H_{12}, is thought to be one of these isomers.

C≡CH   CH=CH₂   CH₂CH₃

    (i)   (ii)   (iii)   (iv)

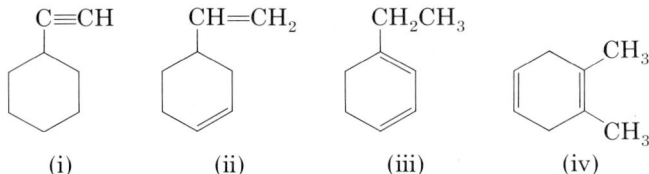

   Deduce which structure is consistent with the following data.
   (1) **B** + silver nitrate ⟶ No reaction
   (2) The uv spectrum of **B** shows an absorption maximum above 200 nm.

3. Compound **C**, C_6H_{12}, is transparent to uv light above 200 nm. The test with Br_2 in CCl_4 is positive. The nmr spectrum of **C** shows one signal only. Suggest a reasonable structure for compound **C**.

4. Compound **D**, C_8H_{10}, is soluble in concentrated sulfuric acid and absorbs strongly in the uv region. The nmr spectrum of **D** shows three signals: singlet (δ 7.3 ppm, intensity 5), quartet (δ centered at 2.7 ppm, intensity 2), and triplet (δ centered at 1.3 ppm, intensity 3). Draw the structure of **D** that is consistent with the data.

5. Compound **E**, C_7H_8O, reacts with metallic sodium with liberation of a gas. The uv spectrum shows λ_{max} around 260 nm. The ir spectrum contains a broad absorption band around 3300 cm^{-1}. The nmr spectrum reveals three signals: singlet (δ 7.3 ppm, intensity 5), singlet (δ 4.6 ppm, intensity 2), and singlet (δ 2.4 ppm, intensity 1). The position of the third nmr signal varies with concentration. Suggest a reasonable structure for compound **E**.

6. Compound **F**, C_3H_8O, is transparent to uv light above 200 nm. The ir spectrum shows a broad absorption band around 3300 cm^{-1} and a strong absorption band around 1100 cm^{-1}. The nmr spectrum reveals three signals: singlet (δ 4.8 ppm, intensity 1), septet (δ centered at 4.0 ppm, intensity 1), and doublet (δ 1.2 ppm, intensity 6). Suggest a structure for compound **F** that is in agreement with the data.

7. Compound **G**, $C_5H_{11}Cl$, does not absorb in the uv region above 200 nm. Its nmr spectrum shows two signals: singlet (δ 3.2 ppm, intensity 2) and singlet (δ 1.0 ppm, intensity 9). Propose a structure for compound **G** that is consistent with the data given.

8. Compound **H**, C_7H_7Br, when treated with silver nitrate, yields a precipitate immediately. The uv spectrum shows an absorption maximum at 260 nm. The nmr spectrum reveals two signals: singlet (δ 7.2 ppm, intensity 5) and singlet (δ 4.2 ppm, intensity 2). Suggest a reasonable structure for compound **H**.

9. The nmr spectrum of compound **I**, $C_3H_6Cl_2$, contains two signals: triplet (δ 3.7 ppm, intensity 4) and quintet (δ 2.2 ppm, intensity 2). Suggest a reasonable structure for compound **I**.

Exercises 10–18 deal with classes of compounds covered in Chapters 9–17.

10. The nmr spectrum of compound **J**, C_2H_6O, shows one signal only, a singlet. Deduce the structure of **J**.

11. Compound **K**, an ether of formula $C_4H_8O_2$, does not absorb uv light. The nmr spectrum shows one signal, a singlet (δ 3.6 ppm). What is the structure of compound **K**?

12. Compound **L**, $C_6H_{12}O$, gives a yellow precipitate with 2,4-dinitrophenylhydrazine and a yellow precipitate with sodium hydroxide + iodine. It has the following spectral characteristics.

(1) Uv–visible spectrum: λ_{max} 280 nm ($\varepsilon = 23$).

(2) Ir spectrum: strong absorption band at 1715 cm^{-1}.

(3) Nmr spectrum, two signals: singlet (δ 2.0 ppm, intensity 3) and singlet (δ 1.0 ppm, intensity 9). Suggest a structure for compound **L** that is consistent with the data.

13. Compound **M**, $C_9H_{10}O$, gave a positive 2,4-dinitrophenylhydrazine test but a negative Tollens' test. The nmr spectrum showed three signals: singlet (δ 7.2 ppm, intensity 5), singlet (δ 3.5 ppm, intensity 2), and singlet (δ 2.0 ppm, intensity 3). Suggest a reasonable structure for compound **M**.

14. Compound **N**, C_4H_8O gave a positive Tollens' test. The ir spectrum showed a strong absorption band at about 1725 cm^{-1}. The nmr spectrum had four signals: triplet (δ 9.7 ppm, intensity 1), quartet (δ 2.4 ppm, intensity 2), sextet (δ 1.7 ppm, intensity 2), and triplet (δ 1.0 ppm, intensity 3). Suggest a reasonable structure for compound **N**.

15. Compound **O**, $C_3H_6O_2$, has the following spectral characteristics.
    (1) Ir spectrum: one broad absorption band extending from 2500 to 3000 cm^{-1} and another absorption band around 1720 cm^{-1}.
    (2) Nmr spectrum, three signals: singlet (δ 11.2 ppm, intensity 1), quartet (δ 2.4 ppm, intensity 2), and triplet (δ 1.2 ppm, intensity 3). Deduce a reasonable structure for compound **O**.

16. Compound **P**, $C_3H_6O_2$, has the following spectral characteristics.
    (1) Ir spectrum: a strong absorption band at 1735 cm^{-1} and absorptions in the 1000–1200 cm^{-1} region.
    (2) Nmr spectrum, two signals: singlet (δ 3.7 ppm, intensity 3) and singlet (δ 2.1 ppm, intensity 3). Deduce a reasonable structure for compound **P**.

17. Compound **Q**, $C_4H_7BrO_2$, is insoluble in water but soluble in NaOH and in NaHCO$_3$. The ir spectrum contains a strong absorption band extending from 3000 to 2500 cm^{-1} as well as an absorption around 1730 cm^{-1}. The nmr spectrum contains four signals: singlet (δ 11.0 ppm, intensity 1), triplet (δ 4.2 ppm, intensity 1), quintet (δ 2.1 ppm, intensity 2), and triplet (δ 1.1 ppm, intensity 3). Suggest a reasonable structure for compound **Q**.

18. Compound **R**, $C_4H_8O_2$, showed the following spectral characteristics.
    (1) Ir spectrum: strong absorption bands at about 1735 cm^{-1} and about 1200 cm^{-1}; no broad absorption band in the 2500–3000 cm^{-1} region.
    (2) Nmr spectrum, three signals: quartet (δ 4.1 ppm, intensity 2), singlet (δ 2.0 ppm, intensity 3), and triplet (δ 1.3 ppm, intensity 3). Suggest a reasonable structure for compound **R**.

Selected Answers
to Problems

Chapter 1

1.1 $\text{Mg}\overset{\times}{\times} + \cdot\overset{\cdot}{\text{O}}: \longrightarrow \text{Mg}^{2+}\left[\overset{\times\cdot}{\underset{\cdot\cdot}{\times\text{O}}}:\right]^{2-}$

1.2 **(a)** $^{\delta+}\text{H}-\text{Cl}^{\delta-}$ **(b)** $^{\delta-}\text{O}-\text{H}^{\delta+}$ **(c)** $^{\delta+}\text{C}-\text{Cl}^{\delta-}$

1.3 **(a)** $\text{H}-\overset{\cdot\cdot}{\underset{\cdot\cdot}{\text{O}}}-\text{H} + \text{H}^+ \longrightarrow \left[\text{H}-\overset{\overset{\displaystyle\text{H}}{|}}{\underset{\cdot\cdot}{\text{O}}}-\text{H}\right]^+$

             Lewis   Lewis
             base   acid

1.4 **(a)** Two   **(c)** Zero

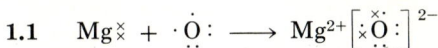

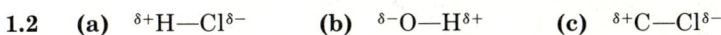

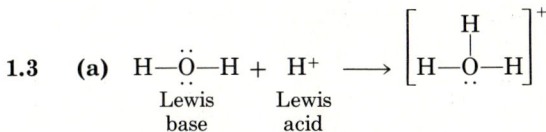

1.5 Structural formula is given first, then molecular formula.

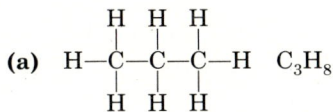

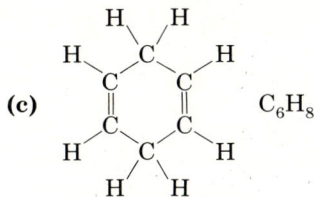

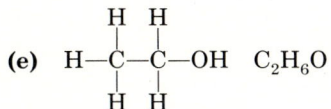

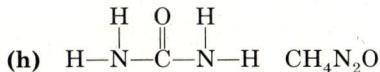

(a) $\begin{array}{ccc} \text{H} & \text{H} & \text{H} \\ | & | & | \\ \text{H}-\text{C}-\text{C}-\text{C}-\text{H} \\ | & | & | \\ \text{H} & \text{H} & \text{H} \end{array}$ C_3H_8

(c) (cyclohexadiene structure) C_6H_8

(e) $\begin{array}{cc} \text{H} & \text{H} \\ | & | \\ \text{H}-\text{C}-\text{C}-\text{OH} \\ | & | \\ \text{H} & \text{H} \end{array}$ $\text{C}_2\text{H}_6\text{O}$

(h) $\begin{array}{ccc} \text{H} & \text{O} & \text{H} \\ | & \| & | \\ \text{H}-\text{N}-\text{C}-\text{N}-\text{H} \end{array}$ $\text{CH}_4\text{N}_2\text{O}$

1.6 **(a)** Partially condensed formula: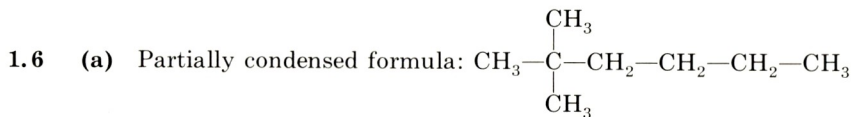

fully condensed formula: $(CH_3)_3CCH_2CH_2CH_2CH_3$ or $(CH_3)_3C(CH_2)_3CH_3$

1.9 **(a)** Greater bond dissociation energy: $O{=}C{=}O$; a $C{=}O$ bond is stronger than a $C{-}O$ bond. Greater bond length: $H_3C{-}O{-}CH_3$; a $C{-}O$ bond is longer than a $C{=}O$ bond.

1.10 **(a)** Alkene   **(b)** Ether   **(c)** Alcohol

Chapter 2

2.2 **(a)** Same   **(c)** Isomers   **(d)** Unrelated

2.6 **(a)** 2,2-Dimethylpentane   **(e)** 4-Ethyl-6-isopropyl-2-methylnonane

2.7 **(a)** 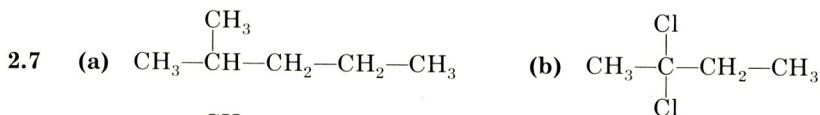   **(b)**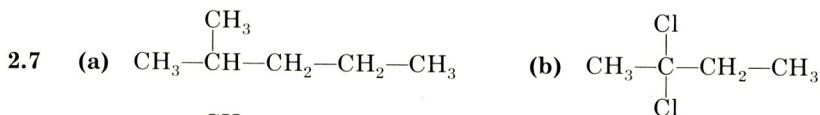

2.8 **(a)** 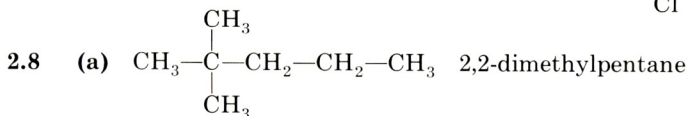 2,2-dimethylpentane

2.10 **(a)**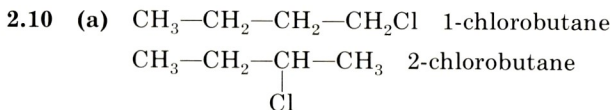

$CH_3{-}CH_2{-}CH_2{-}CH_2Cl$ 1-chlorobutane

$CH_3{-}CH_2{-}\underset{\underset{Cl}{|}}{CH}{-}CH_3$ 2-chlorobutane

2.11 **(a)**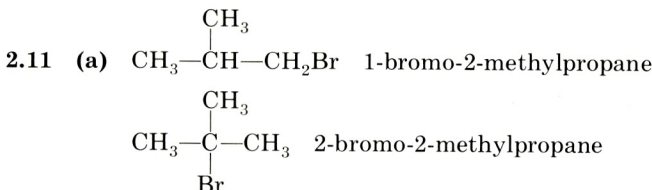

$CH_3{-}\underset{\underset{CH_3}{|}}{\overset{\overset{CH_3}{|}}{C}}{-}CH_3$ 2-bromo-2-methylpropane

2.14 **(a)** 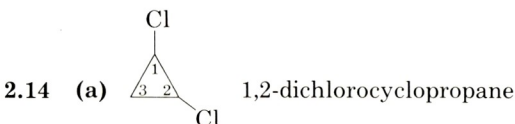 1,2-dichlorocyclopropane

(b) 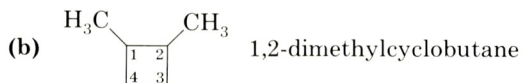 1,2-dimethylcyclobutane

2.15 **(a)** 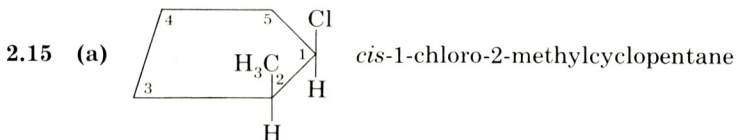 *cis*-1-chloro-2-methylcyclopentane

(b) 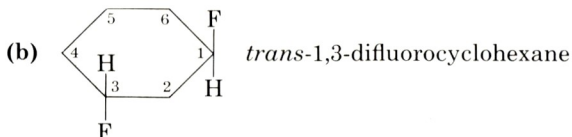 *trans*-1,3-difluorocyclohexane

Chapter 3

3.1 **(a)**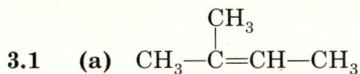

3.2 **(a)** 6-Ethyl-3,3,6-trimethyl-4-nonene   **(e)** 3-Bromo-4-methylcyclopentene

3.3 **(a)** 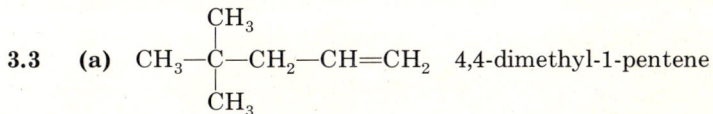 4,4-dimethyl-1-pentene

(d) 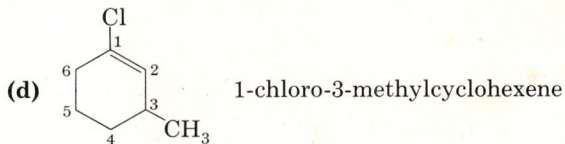 1-chloro-3-methylcyclohexene

3.5 **(a)**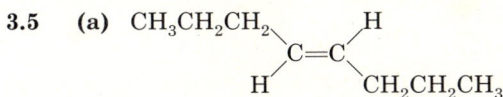

3.6 **(a)** *Z*-1-Bromo-1-chloropropene

3.7 **(a)**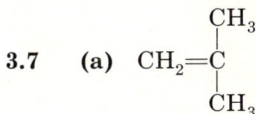

3.8 **(a)** Major product: $CH_3CH_2CH=CHCH_3$
        minor product: $CH_3CH_2CH_2CH=CH_2$

3.12 **(a)**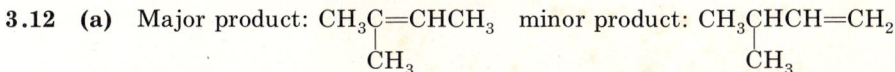

3.13 **(a)** $CH_3CHBrCHBrCH_3$

3.15 **(a)**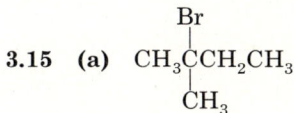

3.17 **(c)**

3.18 **(c)**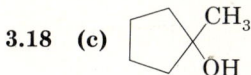

3.20 **(b)**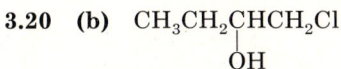

3.21 **A** = cyclohexane

3.22 **(b)**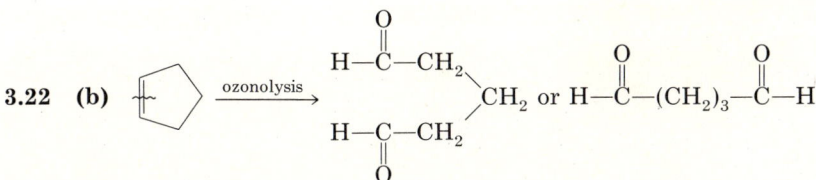

3.23 **(a)** $CH_3\overset{\overset{\displaystyle CH_3}{|}}{C}\!\!=\!\!\overset{\overset{\displaystyle CH_3}{|}}{C}CH_3$

3.25 **(a)** $CH_3-CH\!=\!CH-CH_2-Cl$ allylic chlorination

3.26 **(a)** $CH_3CH_2OH \xrightarrow[\text{heat}]{H^+} CH_2\!=\!CH_2 \xrightarrow{H_2/Pt} CH_3CH_3$

Chapter 4

4.1 **(b)**

(c) $CH_3-\overset{\overset{\displaystyle CH_3}{|}}{C}\!=\!CH-CH_2-CH\!=\!CH-CH_3$

4.3 **(a)** 1,3-Pentadiene   **(d)** 1,3-Cyclopentadiene

4.4 **(a)** $CH_3CH_2\overset{\overset{\displaystyle Br}{|}}{C}HCH\!=\!CHCH_3 + CH_3\overset{\overset{\displaystyle Br}{|}}{C}HCH_2CH\!=\!CHCH_3$
(1,2-addition products)

$CH_3\overset{\overset{\displaystyle Br}{|}}{C}HCH\!=\!CHCH_2CH_3$ (1,4-addition product)

4.6 **(a)** Lycopene: 8 isoprene units

4.7 **(a)** $HC\!\equiv\!CCH_2CH_2CH_3$ **(d)**

4.8 **(a)** $CH_3CH_2CH_2C\!\equiv\!CH$ 1-pentyne

4.10 **(a)** $HC\!\equiv\!CH + Na \xrightarrow{\text{liq } NH_3} HC\!\equiv\!C:^- Na^+ + CH_3Br \longrightarrow HC\!\equiv\!CCH_3$

$CH_3C\!\equiv\!CH + Na \xrightarrow{\text{liq } NH_3} CH_3C\!\equiv\!C:^- Na^+ + CH_3Br \longrightarrow CH_3C\!\equiv\!CCH_3$
2-Butyne

4.11 **(a)** $HC\!\equiv\!CH \xrightarrow{HBr} H_2C\!=\!CHBr \xrightarrow{HBr} H_3CCHBr_2$

4.12 **(a)** Enol product: $CH_3CH_2CH_2\overset{\overset{\displaystyle OH}{|}}{C}\!=\!CH_2$ keto product: $CH_3CH_2CH_2\overset{\overset{\displaystyle O}{||}}{C}CH_3$

Chapter 5

5.1 61.0 kcal/mole

5.2 Structures with aromatic character: **(b), (d), (e), (f)**

5.3 **(a)** o-Difluorobenzene   **(b)** p-Chlorostyrene
(f) 2-Chloro-2-phenylbutane

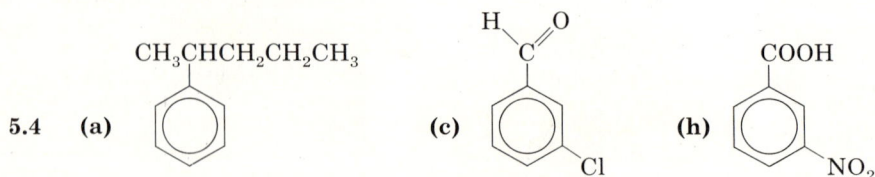

5.4 **(a)** $CH_3CHCH_2CH_2CH_3$

(c)

(h)

5.5 (a) 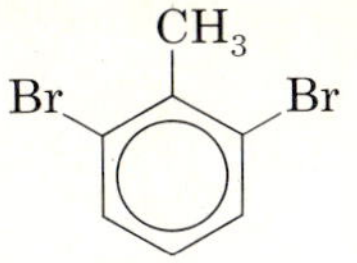 (d) 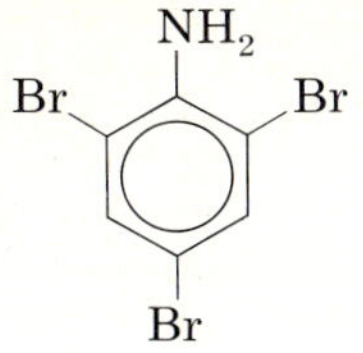 (g) 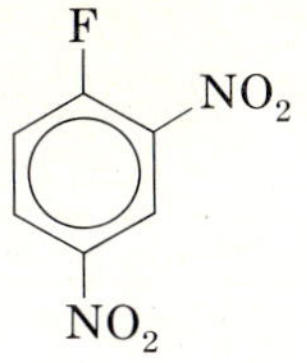

5.6 (b) 2,4,5-Trinitroethylbenzene (c) 2-Benzyl-4-bromophenol
(f) 1,4-Dibromotetraphenylbenzene

5.7 (a) 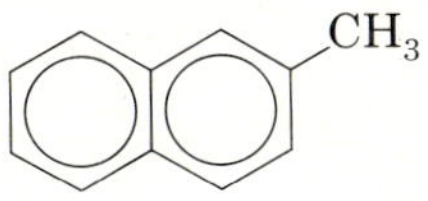 (b) 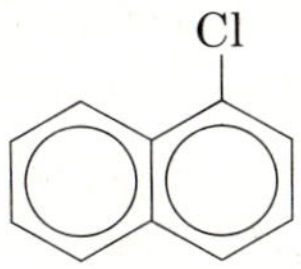 (h) 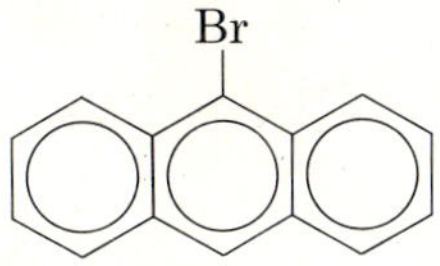

(k) 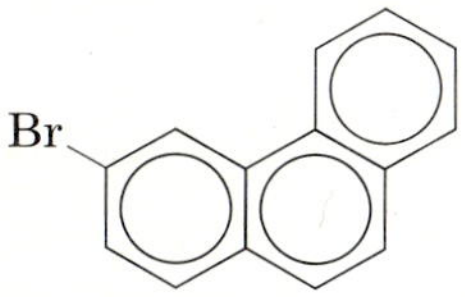

5.8 (a) 2-Nitronaphthalene or β-nitronaphthalene (c) 9-Methylanthracene

5.9 (a) 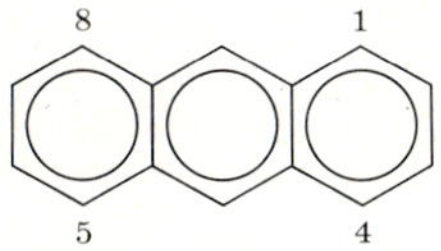 positions 1, 4, 5, and 8 are equivalent.

5.10 (a) 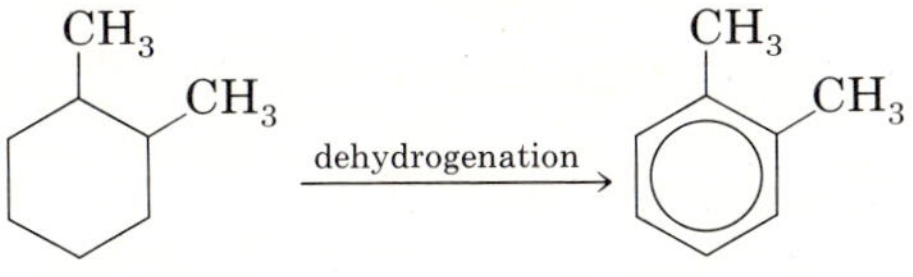

o-Xylene

5.12 (a) 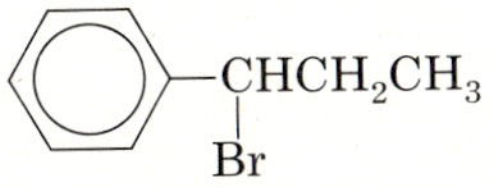 1-bromo-1-phenylpropane

5.13 (a) 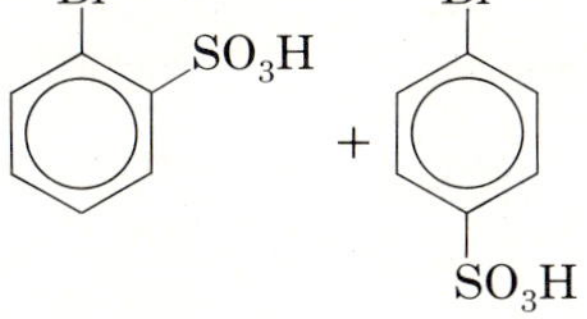 (c) 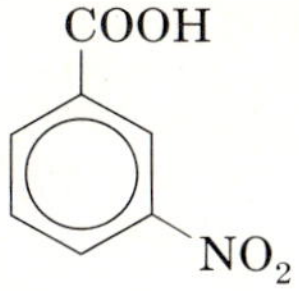

(e) 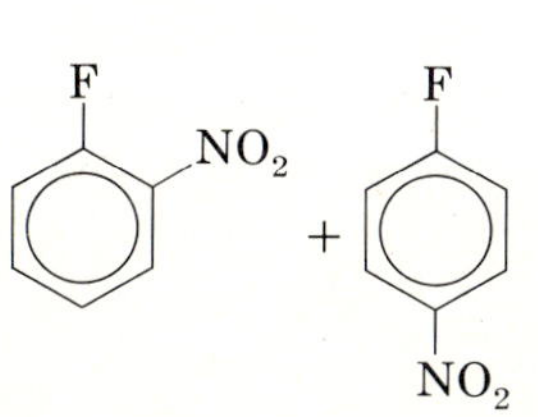 (g) 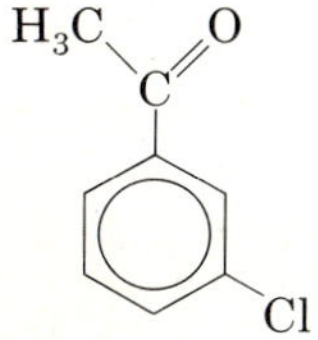

5.14 **(c)**

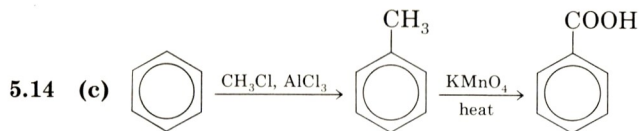

Chapter 6

6.1 $[\alpha] = -96.2°$

6.2 Chiral objects: **(c), (d), (e), (f)**

6.3 **(a)**

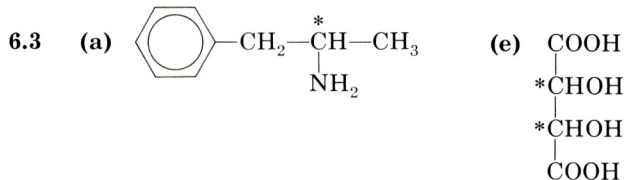

6.4 **(a)** Diastereomers: I and III

6.7 **(b)** Enantiomers: (iii) and (iv)

6.8 **(a)** Enantiomers: **(a)** and **(i)**; diastereomers: **(d)** and **(i)**

6.9 Identical: **(a)** and **(b)**; diastereomers: **(a)** and **(c)**

6.10 D configuration: **(a), (c),** and **(d)**; L configuration: **(b)**

6.11 **(a)** *R*   **(c)** *S*

6.12 **(a)**   **(c)**

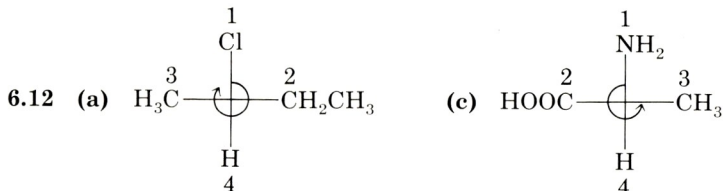

Chapter 7

7.1 **(a)** *n*-Propyl alcohol, 1°   **(c)** Benzyl alcohol, 1°
      (d) Cyclopropyl alcohol, 2°

7.2 **(a)** Phenylcarbinol

7.3 **(c)** 2-Propanol   **(f)** Cyclohexanol   **(h)** 2-Propen-1-ol

7.4 **(a)** 3-Hexanol   **(f)** 1-Methylcyclopentanol   **(h)** 2-Buten-1-ol

7.5 1° alcohols: **(c), (h), (i),** and **(j)**; 2° alcohols: **(a), (b), (d), (e),** and **(g)**; 3° alcohol: **(f)**

7.6 **(a)**

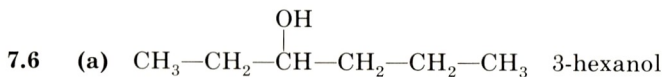

3-hexanol

(e)

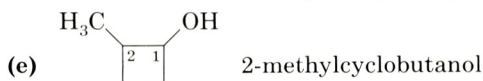

2-methylcyclobutanol

(i)

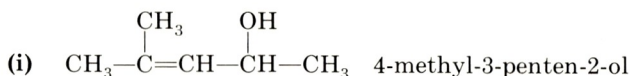

4-methyl-3-penten-2-ol

7.7 **(a)** 2,3-Butanediol   **(d)** 1,4-Cyclohexanediol

7.8 **(a)** 2-Methylphenol or *o*-methylphenol
    (f) 4-Hydroxyphenol or *p*-hydroxyphenol
    (j) 4-Methyl-1-naphthol

7.9 **(a)** (lowest boiling point) 2-Methylpropane < butane < 2-propanol < 1-propanol (highest)

7.10 **(a)** (least soluble) Pentane < 1-hexanol < 1-pentanol < 1-butanol (most soluble)

7.12 (most acidic) (ii) < (i) < (iv) < (iii) (least acidic)

7.14 **(a)** $2 CH_3CH_2OH + 2 Na \longrightarrow H_2\uparrow + 2 CH_3CH_2O^- Na^+$

7.16 **(a)** Reduction   **(f)** Oxidation

7.18 **(b)** $\overset{\displaystyle O}{\overset{\|}{HC}}{-}OH$ **(c)** [benzene ring]$-\overset{\displaystyle O}{\overset{\|}{C}}H$

7.19 **(a)** CH_3ONO_2 **(d)** $CH_3O\overset{\displaystyle O}{\overset{\|}{P}}(OH)_2$

7.20 **(b)** $CH_3CH_2CH_2CH_2ONO$ *n*-butyl nitrite

    (f) $CH_3CH_2O\overset{\displaystyle O}{\overset{\|}{P}}OCH_2CH_3$ triethyl phosphate
$\qquad\qquad\;\, \underset{OCH_2CH_3}{}$

7.21 (least reactive) (ii) < (i) < (iii) (most reactive); (ii) a 1° alcohol, (i) a 2° alcohol, and (iii) a 3° alcohol

7.22 **(a)** CH_3CH_2Br **(c)** No reaction

7.23 **(b)** $CH_3CH{=}CHCH_3$ (Saytzeff product)

7.24 **(a)** $CH_3CH_2\underset{\underset{OH}{|}}{C}HCH_3$ (Markovnikov product)

    (d) [cyclohexane ring with CH_3 and OH substituents] (anti-Markovnikov product)

7.25 **(a)** $CH_3CH_2CH_2{-}SH$ **(c)** $CH_3CH_2{-}S{-}S{-}CH_2CH_3$

Chapter 8

8.1 **(a)** Alkyl, 1°   **(c)** Alkyl, 2°   **(d)** Vinyl   **(j)** Allyl

8.2 **(a)** Dichloromethane   **(f)** 2-Bromopropane
    (h) 3-Chloro-1-propene

8.4 **(a)** (lowest boiling point) Fluorobenzene < chlorobenzene < bromobenzene < iodobenzene (highest)

8.5 **(b)** $O_2N{-}$[benzene ring with Br substituent]$-CH_3$ (electrophilic substitution)

(d) O_2N—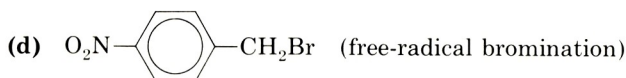—CH_2Br (free-radical bromination)

(e) 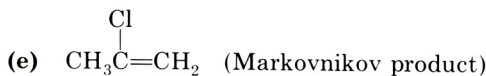   (Markovnikov product)

(f) $ClCH_2CH\text{=}CH_2$ (allylic, free-radical chlorination)

8.6   **(a)** $CH_3CH_2CH_2OH$ alcohol   **(b)** $CH_3\text{—}O\text{—}CH_2CH_3$ ether

8.8   (fastest)   (i) > (ii) > (iii) > (iv)   (slowest)

8.9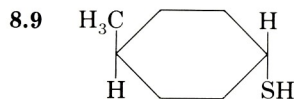

8.12   (fastest)   (ii) > (iv) > (iii) > (i)   (slowest)

8.13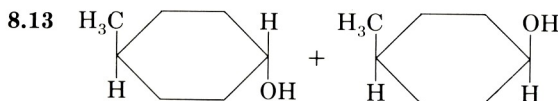

8.15   **A** = 2-bromobutane, $CH_3\underset{\overset{\displaystyle |}{Br}}{C}HCH_2CH_3$

8.18   **(a)** $CH_3CH_2CH_2Br \xrightarrow{\text{Mg, dry ether}} CH_3CH_2CH_2MgBr$ (= **A**)

Chapter 9

9.1   **(a)**   Methyl-*n*-propyl ether or 1-methoxypropane   **(d)**   Allyl ether

9.2   **(b)** $CH_3CH_2\underset{\overset{\displaystyle |}{OCH_2CH_3}}{C}HCH_2CH_2CH_2CH_3$ **(c)**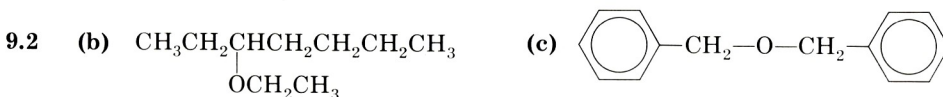

(e) $CH_3CH\text{=}CH\underset{\overset{\displaystyle |}{OCH_3}}{C}HCH_2CH_3$

9.5   **(a)** $2\,CH_3CH_2CH_2OH \xrightarrow[\text{heat}]{H_2SO_4} CH_3CH_2CH_2\text{—}O\text{—}CH_2CH_2CH_3 + H_2O$

9.6   The only suitable combination is (i) + (iii).

9.8   **A** = 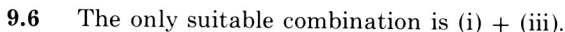$\text{—}OCH_3$ **B** = $\text{—}OH$

Chapter 10

10.1   **(a)** $CH_3CH_2CH_2CH_2CH_2\text{—}\overset{\overset{\displaystyle O}{\|}}{C}H$ **(f)** $CH_3\underset{\overset{\displaystyle |}{Cl}}{C}\text{=}CHCH_2\text{—}\overset{\overset{\displaystyle O}{\|}}{C}H$

10.2   **(a)**   Formaldehyde or methanal   **(b)**   *p*-Chlorobenzaldehyde

10.3   **(b)** $CH_3\text{—}\overset{\overset{\displaystyle O}{\|}}{C}\text{—}CH_2CH_2CH_2CH_3$ 2-hexanone

(e) $\underset{\underset{O}{\overset{O}{\parallel}}}{CH_3-C-CH_2CH=CHCH_3}$ 4-hexen-2-one

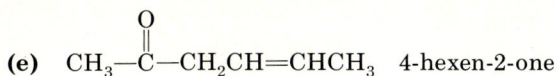

10.4 **(a)** Cyclobutanone   **(b)** Benzyl ethyl ketone or 1-phenyl-2-butanone
(d) 3-Chlorocyclopentanone

10.5 (lowest boiling point) (i) < (iii) < (ii) < (iv) (highest)

10.6 (least soluble in water) Hexanal < pentanal < methyl ethyl ketone < acetone (most soluble)

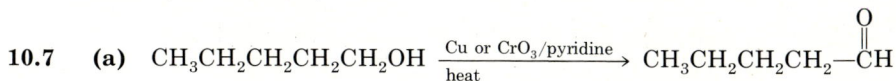

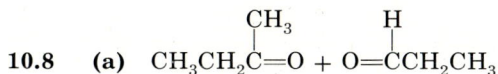

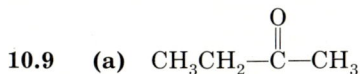

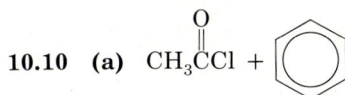

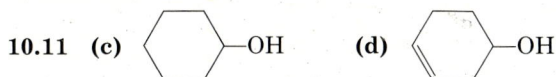

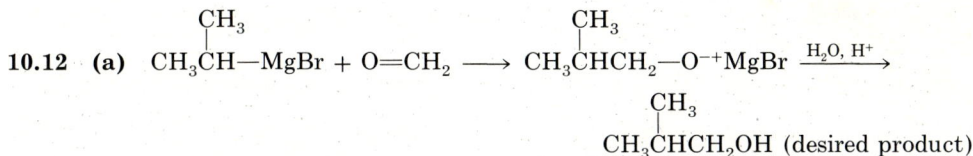

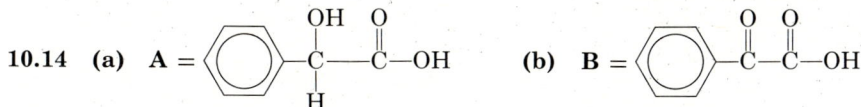

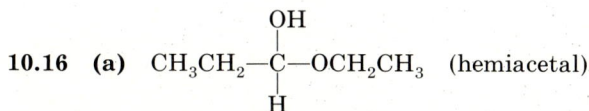

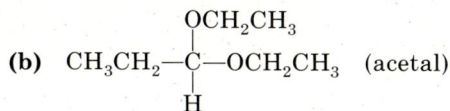

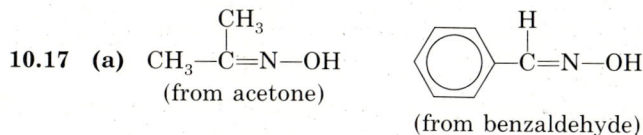

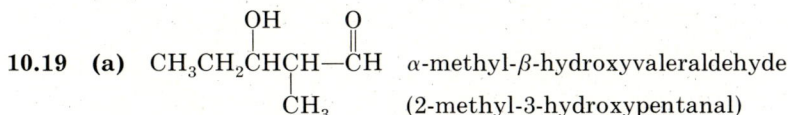

10.7 **(a)** $CH_3CH_2CH_2CH_2CH_2OH \xrightarrow[\text{heat}]{\text{Cu or CrO}_3/\text{pyridine}} CH_3CH_2CH_2CH_2-\overset{\overset{O}{\parallel}}{C}H$

10.8 **(a)** $CH_3CH_2\overset{\overset{CH_3}{|}}{C}{=}O + O{=}\overset{\overset{H}{|}}{C}CH_2CH_3$

10.9 **(a)** $CH_3CH_2-\overset{\overset{O}{\parallel}}{C}-CH_3$

10.10 **(a)** $CH_3\overset{\overset{O}{\parallel}}{C}Cl +$ ⬡

10.11 **(c)** ⬡—OH   **(d)** ⬡—OH

10.12 **(a)** $CH_3\overset{\overset{CH_3}{|}}{C}H-MgBr + O{=}CH_2 \longrightarrow CH_3\overset{\overset{CH_3}{|}}{C}HCH_2-O^{-+}MgBr \xrightarrow{H_2O, H^+}$

$CH_3\overset{\overset{CH_3}{|}}{C}HCH_2OH$ (desired product)

10.14 **(a)** $A =$ ⬡$-\overset{\overset{OH}{|}}{\underset{\underset{H}{|}}{C}}-\overset{\overset{O}{\parallel}}{C}-OH$ **(b)** $B =$ ⬡$-\overset{\overset{O}{\parallel}}{C}-\overset{\overset{O}{\parallel}}{C}-OH$

10.16 **(a)** $CH_3CH_2-\overset{\overset{OH}{|}}{\underset{\underset{H}{|}}{C}}-OCH_2CH_3$ (hemiacetal)

(b) $CH_3CH_2-\overset{\overset{OCH_2CH_3}{|}}{\underset{\underset{H}{|}}{C}}-OCH_2CH_3$ (acetal)

10.17 **(a)** $CH_3-\overset{\overset{CH_3}{|}}{C}{=}N-OH$ ⬡$-\overset{\overset{H}{|}}{C}{=}N-OH$
(from acetone)   (from benzaldehyde)

10.19 **(a)** $CH_3CH_2\overset{\overset{OH}{|}}{C}H\underset{\underset{CH_3}{|}}{C}H-\overset{\overset{O}{\parallel}}{C}H$ α-methyl-β-hydroxyvaleraldehyde
(2-methyl-3-hydroxypentanal)

10.20 **(a)** $CH_3CH_2CH{=}\overset{\displaystyle O}{\overset{\|}{C}}{-}\underset{\displaystyle CH_3}{CH}$

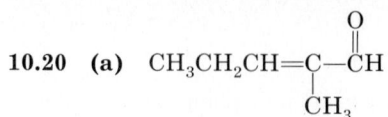

10.23 Compounds that give positive iodoform test: **(a), (c), (d), (g),** and **(h)**

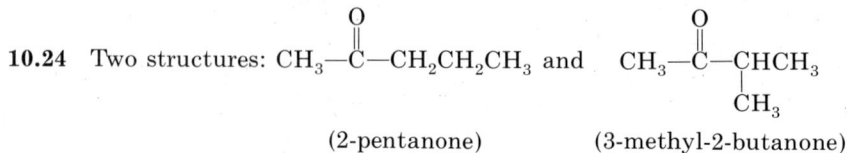

10.24 Two structures: $CH_3{-}\overset{\displaystyle O}{\overset{\|}{C}}{-}CH_2CH_2CH_3$ and $CH_3{-}\overset{\displaystyle O}{\overset{\|}{C}}{-}\underset{\displaystyle CH_3}{CHCH_3}$

                (2-pentanone)   (3-methyl-2-butanone)

Chapter 11

11.1 **(a)**

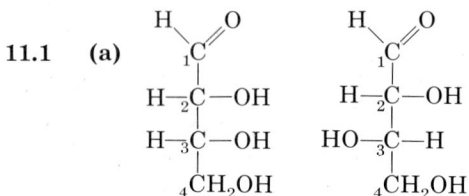

11.3 **(a)** $[\alpha] = +65.5°$ **(b)** The β form predominates in equilibrium mixture

11.5

Hemiacetal OH at	C-2	C-3	C-4	C-5	C-6
Resulting ring size	3	4	5	6 (most common)	7

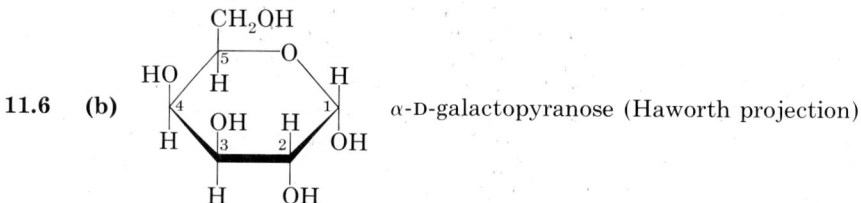

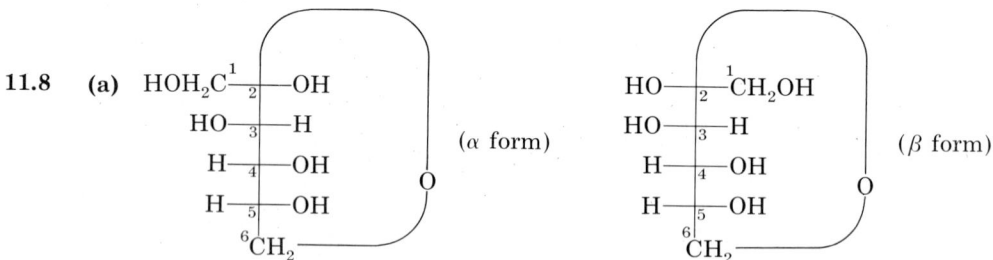

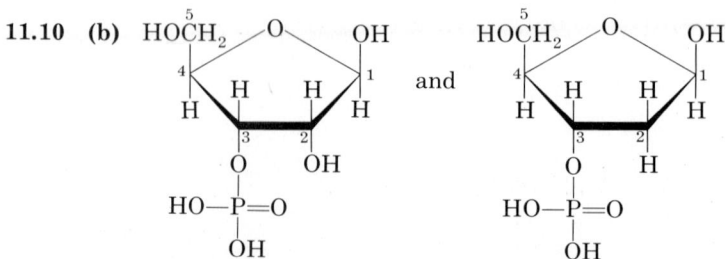

11.6 **(b)**

α-D-galactopyranose (Haworth projection)

11.8 **(a)**

(α form)   (β form)

11.10 **(b)**

and

11.11 **(b)**

$$^{1}\text{CH}_2\text{OH}$$
$$^{2}\text{C}{=}\text{O}$$
$$\text{H}{-}^{3}{-}\text{OH}$$
$$\text{H}{-}^{4}{-}\text{OH}$$
$$^{5}\text{CH}_2\text{OH}$$

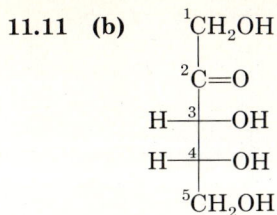

11.12 Yes; because there is a hemiacetal function in its structure

11.15 Maltose

Chapter 12

12.1 Common name is given first, then IUPAC name.
    (a) α-Bromopropionic acid; 2-bromopropanoic acid
    (c) Cyclohexylacetic acid; cyclohexylethanoic acid

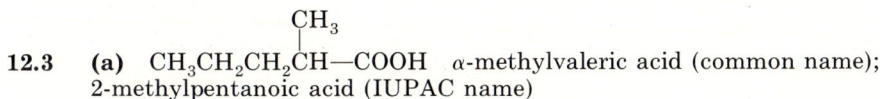

12.2 **(a)** HO—CH$_2$CH—COOH   **(d)** ⬡—CH$_2$CH$_2$CH$_2$—COOH
               |
               OH

12.3 **(a)** CH$_3$CH$_2$CH$_2$CH—COOH α-methylvaleric acid (common name);
                    |CH$_3$
2-methylpentanoic acid (IUPAC name)

12.4 (strongest acid) Formic acid > benzoic acid > butyric acid (weakest)

12.5 **(a)** (weakest acid) (iii) < (iv) < (ii) < (i) (strongest)

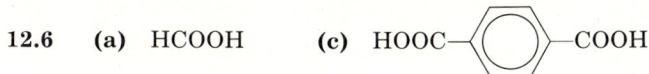

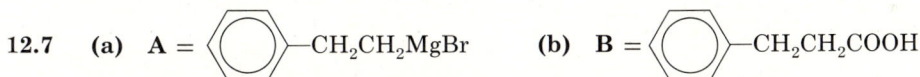

12.6 **(a)** HCOOH   **(c)** HOOC—⬡—COOH

12.7 **(a)** **A** = ⬡—CH$_2$CH$_2$MgBr   **(b)** **B** = ⬡—CH$_2$CH$_2$COOH

    (f) **F** = HOOC—CH$_2$CH$_2$—COOH

12.8 **(b)** CH$_3$CH$_2$Cl $\xrightarrow{\text{Mg, dry ether}}$ CH$_3$CH$_2$MgCl $\xrightarrow[\text{(2) H}_3\text{O}^+]{\text{(1) CO}_2}$ CH$_3$CH$_2$COOH

    or CH$_3$CH$_2$Cl $\xrightarrow{\text{NaCN}}$ CH$_3$CH$_2$CN $\xrightarrow[\text{heat}]{\text{H}_2\text{O, H}^+}$ CH$_3$CH$_2$COOH

    (d) ⬡—Cl $\xrightarrow{\text{Mg, dry ether}}$ ⬡—MgCl $\xrightarrow[\text{(2) H}_3\text{O}^+]{\text{(1) CO}_2}$ ⬡—COOH

12.10 **(a)** H—C(=O)—Cl   **(d)** ⬡—C(=O)—O—⬡   **(e)** H—C(=O)—N(H)(CH$_3$)

    (g) H—C(=O)—O—C(=O)—H

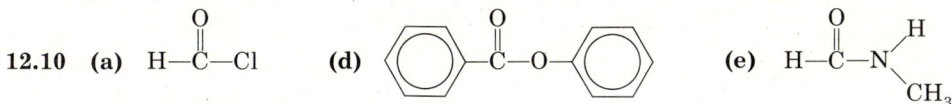

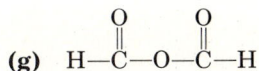

12.11 The name given first is the common name.
    (a) Butyryl chloride; butanoyl chloride
    (c) Benzyl formate; benzyl methanoate   **(e)** *N*,*N*-Diphenylbenzamide

12.13 **(a)** HCOOH formic acid (IUPAC: methanoic acid) + HOCH$_3$ methyl alcohol (IUPAC: methanol)

12.14 The ^{18}O will be part of the water molecule in the product, H$_2$^{18}O.

12.16 **(a)** ⬡—COOH benzoic acid + CH$_3$OH methyl alcohol (methanol)

(d) $\overset{O}{\overset{\|}{HC}}$—OCH$_2CH_3$ ethyl formate (ethyl methanoate) + ⬡—OH phenol

(e) $\overset{O}{\overset{\|}{HC}}$—OH formic acid (methanoic acid)

(g) 2 ⬡—CH$_2$OH benzyl alcohol

12.17 **(a)** ⬡—COO$^-$ Na$^+$ sodium benzoate

(b) $\overset{O}{\overset{\|}{HC}}$—Cl formyl chloride (methanoyl chloride)

(e) CH$_3$$\overset{O}{\overset{\|}{C}}$—OCH$_3$ methyl acetate (methyl ethanoate)

12.18 **(a)** ⬡—$\overset{O}{\overset{\|}{C}}$—O—$\overset{O}{\overset{\|}{C}}$—H benzoic formic anhydride (benzoic methanoic anhydride)

(b) 2 H—$\overset{O}{\overset{\|}{C}}$—OH formic acid (methanoic acid)

(f) glutaric anhydride

12.19 **(a)** CH$_3$$\overset{O}{\overset{\|}{C}}$—NH$_2$ + HCl **(c)** CH$_3$$\overset{O}{\overset{\|}{C}}$—O$^-$ Na$^+$ + NH$_3$

(d) ⬡—CH$_2$CH$_2$NH$_2$ **(g)** ⬡—CH$_2$CH$_2$NH$_2$

Chapter 13

13.1 **(a)** A mixed triglyceride because it contains different fatty acid residues.
(b) Glycerol, sodium laurate, sodium palmitate, sodium stearate

13.3 **(a)** Saponification number = 189 **(b)** Iodine number = 114

14.1 Primary (alcohol or amine): **(a)**, **(b)**, and **(e)**; secondary (alcohol or amine): **(c)**, **(d)**, **(g)**, and **(h)**; tertiary: **(f)**

14.2 Aliphatic amines: **(b)**, **(c)**, and **(d)**; aromatic amines: **(e)**, **(f)**, and **(h)**

14.3 Primary:

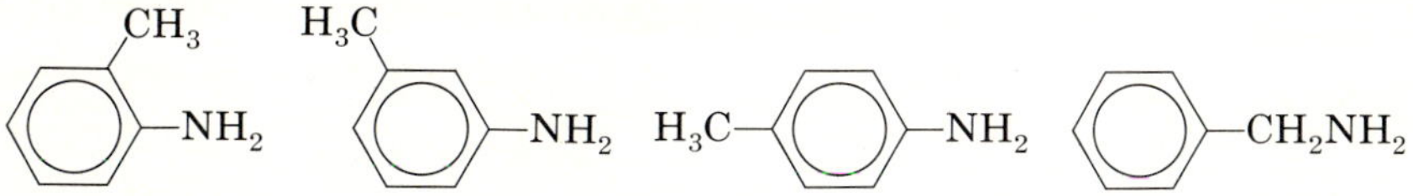

Secondary:

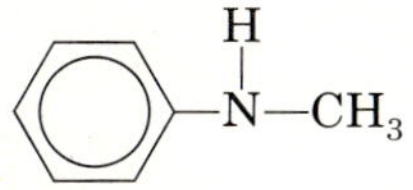

14.4 **(a)** $CH_3\overset{CH_3}{\underset{}{CH}}-NH_2$ **(b)**

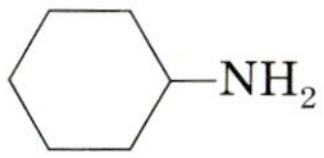

(d) $\left[CH_3-\overset{H}{\underset{H}{N^+}}-\overset{CH_3}{CHCH_3} \right] Br^-$ **(f)**

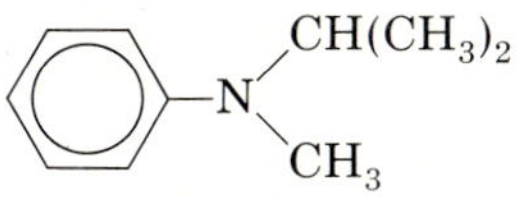

14.5 **(b)** *p*-Nitroaniline   **(d)** Benzyldimethylamine
  (e) Diethyldimethylammonium chloride

14.7 **(a)** (least basic)   (iii) < (ii) < (i) < (iv)   (most basic)

14.9 $A = CH_3CH_2\overset{+}{N}H_3\ Br^-$ $B = CH_3CH_2NH_2$ $C = (CH_3)_3\overset{+}{N}CH_2CH_3\ Cl^-$

14.11 **(a)** $(CH_3CH_2)_3\overset{+}{N}CH_3\ Br^-$ **(d)** No reaction

14.12 $A = CH_3NHCH_3$

14.13 **(b)**

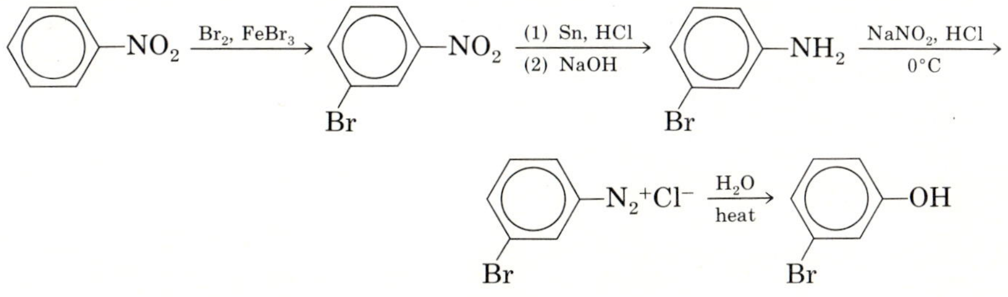

(d)

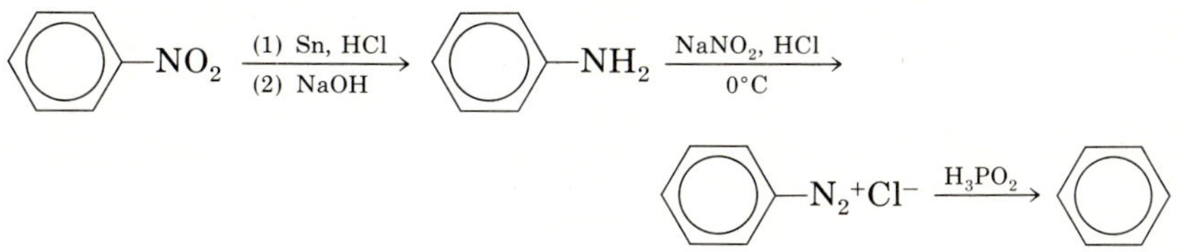

Chapter 15

15.1 **(a)** (1) $H_3\overset{+}{N}-CH_2COO^-$; (2) $H_3\overset{+}{N}-CH_2COOH$; (3) $H_2N-CH_2COO^-$

15.2 **(a)** Zwitterion   **(c)** Cationic form

15.3 Proline

15.4 **(a)** 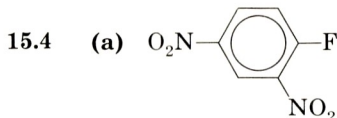 **(b)**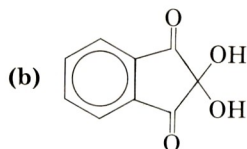

15.5 **(a)** Trp > Cys = Lys **(b)** Lys > Cys = Trp

15.6 I = Val-Ala-Leu-Gly-Tyr-Ser-Asp-Gly

Chapter 16

16.1 **(a)** 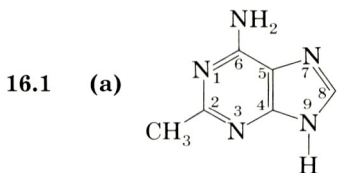 **(d)**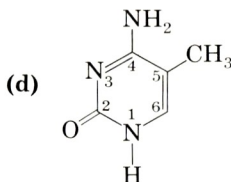

16.2 **(a)** 3′-Deoxyadenylic acid **(b)** 3′-Cytidylic acid

Chapter 17

17.1 Pairs **(b)** and **(c)** because in each pair one structure has a conjugated system and the other an isolated system

17.2 0.021 M

17.3 19.7 $M^{-1}cm^{-1}$ or 19.7

17.4 Pairs **(a), (c),** and **(d)** because there are different functional groups in each pair.

17.5 **A** = ethanol

17.6 **(a)** One signal only **(c)** Two signals **(e)** Three signals

17.8 1,1,2-Tribromoethane

Index

References to figures are indicated by (f), to text problems by (p), and to tables by (t) following the page number.

Functional Groups and Classes of Organic Compounds

Class	General formula	Functional group	Specific examples
Alkane	RH	C—C (single bond)	H_3C—CH_3
Alkene	R—CH=CH_2	C=C (double bond)	H_2C=CH_2
Alkyne	R—C≡CH	C≡C (triple bond)	HC≡CH
Alkyl halide	RX	—X (X = F, Cl, Br, I)	H_3C—Cl
Alcohol	R—OH	—OH	H_3C—OH
Ether	R—O—R′	$-\overset{\mid}{\underset{\mid}{C}}-O-\overset{\mid}{\underset{\mid}{C}}-$	H_3C—O—CH_3
Aldehyde	$R-\overset{O}{\overset{\|}{C}}H$	$-\overset{O}{\overset{\|}{C}}-H$	$H-\overset{O}{\overset{\|}{C}}-H$, $H_3C-\overset{O}{\overset{\|}{C}}-H$
Ketone	$R-\overset{O}{\overset{\|}{C}}-R′$	$-\overset{\mid}{\underset{\mid}{C}}-\overset{O}{\overset{\|}{C}}-\overset{\mid}{\underset{\mid}{C}}-$	$H_3C-\overset{O}{\overset{\|}{C}}-CH_3$
Carboxylic acid	$R-\overset{O}{\overset{\|}{C}}-OH$	$-\overset{O}{\overset{\|}{C}}-OH$	$H-\overset{O}{\overset{\|}{C}}-OH$, $H_3C-\overset{O}{\overset{\|}{C}}-OH$
Ester	$R-\overset{O}{\overset{\|}{C}}-OR$	$-\overset{O}{\overset{\|}{C}}-OR$	$H-\overset{O}{\overset{\|}{C}}-OCH_3$, $H_3C-\overset{O}{\overset{\|}{C}}-OCH_3$
Amine	R—NH_2	$-\overset{\mid}{\underset{\mid}{C}}-NH_2$	H_3C—NH_2